BIOLOGICAL SCIENCE

An Ecological Approach

SEVENTH EDITION

BIOLOGICAL SCIENCE
An Ecological Approach

SEVENTH EDITION
BSCS Green Version

BSCS
The Colorado College
Colorado Springs, Colorado 80903

Revision Team

Jean P. Milani, BSCS, Revision Coordinator
William H. Leonard, Clemson University, Clemson, South Carolina
Thomas R. Manney, Kansas State University, Manhattan, Kansas
Kenneth G. Rainis, Ward's Natural Science Establishment, Inc.
Gordon E. Uno, University of Oklahoma, Norman, Oklahoma
Katherine A. Winternitz, BSCS

KENDALL/HUNT PUBLISHING COMPANY
2460 Kerper Boulevard P.O. Box 539 Dubuque, Iowa 52004-0539

Contributors

Barbara Andrews, Colorado Springs, Colorado
Charles R. Barman, Indiana University at Kokomo
Don Chmielowiec, Ward's Natural Science Establishment, Inc.
Edward Drexler, Pius XI High School, Milwaukee, Wisconsin
Vernon L. Gilliland, Liberal Senior High School, Liberal, Kansas
Lynn Griffin, Cherry Creek High School, Englewood, Colorado
Donald P. Kelley, South Burlington, Vermont
Joseph D. McInerney, BSCS
George Nassis, Ward's Natural Science Establishment, Inc.
Larry N. Norton, Quantum Technology, Inc.
Chester Penk, Quantum Technology, Inc.
Barry Schwartz, Chatfield High School, Littleton, Colorado
John Settlage, Boston, Massachusetts
Carol Leth Stone, Alameda, California
Richard R. Tolman, Brigham Young University, Provo, Utah
David A. Zegers, Millersville University, Millersville, Pennsylvania

Safety Consultant

Kenneth G. Rainis, Ward's Natural Science Establishment, Inc.

BSCS Production Staff

Judy L. Rasmussen
Christine A. Schemmel
Katherine A. Winternitz

Artists

Carl J. Bandy
Bill Border
Kathy M. Foraker
Randall E. Kochevar
Marjorie C. Leggitt

BSCS Administrative Staff

Roger G. Olstad, Chair, Board of Directors
Joseph D. McInerney, Director
Rodger W. Bybee, Associate Director
Larry Satkowiak, Business Manager

Grateful acknowledgement is made to the following individuals for permission to reprint from their work:

D. Nelkin and L. Tancredi, *Dangerous Diagnostics: The Social Power of Biological Information* (New York: Basic Books, 1989).

N. A. Holtzman, *Proceed with Caution: Predicting Genetic Risks in the Recombinant DNA Era* (Baltimore, MD: Johns Hopkins University Press, 1989).

Acknowledgements continue on page 798.

Editorial, design, and production services provided by The Book Department, Inc., Boston, Massachusetts

Cover photograph by Marshall Noice

Content Reviewers

Kenneth J. Andrews, Colorado College, Colorado Springs, Colorado
John Alcock, Arizona State University, Tempe, Arizona
John E. Cadle, The Academy of Natural Sciences, Philadelphia, Pennsylvania
James J. Ebersole, Colorado College, Colorado Springs, Colorado
John Lynn, Louisiana State University, Baton Rouge, Louisiana
Thomas R. Manney, Kansas State University, Manhattan, Kansas
Jeff Murray, M.D., University of Iowa Hospital and Clinics, Iowa City, Iowa
Richard D. Storey, Colorado College, Colorado Springs, Colorado
Gordon E. Uno, University of Oklahoma, Norman, Oklahoma
Kent M. Van De Graaff, Brigham Young University, Provo, Utah
Barbara L. Winternitz, Colorado College, Colorado Springs, Colorado

Teacher Reviewers

Richard Benz, Wickliffe High School, Chardon, Ohio
Richard Bergholz, Montesano High School, Montesano, Washington
Richard E. Carman, Wadena Senior High School, Wadena, Minnesota
Stuart Caudill, Ashbrook High School, Gastonia, North Carolina
Ned Hatfield, Bonny Eagle High School, West Boxton, Maine
Gregory S. Hennemuth, Lake Region Union High School, Orleans, Vermont
Peggy Keeling, St. Paul Academy and Summit School, St. Paul, Minnesota
Rex Miller, Roy J. Wasson High School, Colorado Springs, Colorado
Patricia D. Schmitt, The Bishop's School, Encinitas, California
Gary Smith, Katella High School, Anaheim, California
Paula J. Thorpe, Woodland Park High School, Woodland Park, Colorado
Holly Williams, Storm King School, Cornwall-on-Hudson, New York
Sherry Yarema, Naperville Central High School, Naperville, Illinois

Acknowledgements

The following teachers contributed comments and suggestions for this edition:

Amy Adams, Washington Middle School, Green Bay, Wisconsin; Bruce Alexander, Chapel Hill High School, Pittsboro, North Carolina; Mary Ashton, Orange High School, Pepper Pike, Ohio; Shirley Ballack, Joliet Central High School, Joliet, Illinois; Bob Beaudoin, Franklin Middle School, Green Bay, Wisconsin; Barbara R. Beitch, Hamden Hall Country Day School, Hamden, Connecticut; William Benner, Joliet Township High School, Peotone, Illinois; Richard Benz, Wickliffe High School, Wickliffe, Ohio; Rick Berken, Preble High School, Green Bay, Wisconsin; Dave Birschbach, East High School, Central Office, Green Bay, Wisconsin; Judy Bond, Lakeland High School, Lakeland, Florida; Luann R. Bridle, Forsyth Country Day School, Lewisville, North Carolina; Kirk Brill, Southeast Polk High School, Runnells, Iowa; Don Brown, Joliet West High School, Joliet, Illinois; Don Buntman, West High School, Green Bay, Wisconsin; Doug Cameron, St. Andrews, Tennessee; Stuart Caudill, Ashbrook High School, Gastonia, North Carolina; Lyle Chee, Mid Pacific Institute, Hawaii.

James Colman, Bow Basin Junior and Senior High School, Bow Basin, Wyoming; Gale Cook, Northeast High School, Lincoln, Nebraska; Kathleen L. Davis, Roosevelt High School, Wyandotte, Michigan; Ken Davis, Woodway High School, Spokane, Washington; W. C. Dicks, Northville High School, Novi, Michigan; Mary Diedrich, West High School, Depere, Wisconsin; Lucille M. Dostie, Wiscasset High School, Wiscasset, Maine; Clarice Douoguih, Montclair Kimberley Academy, Montclair, New Jersey; Mark Eberhard, St. Clair High School, New Baltimore, Michigan; John L.

Eckert, Hutchinson High School, Hutchinson, Kansas; Rod Epp, Hastings Senior High, Hastings, Nebraska; Jon Forney, Grant High School, Grant, Nebraska; Claudia R. Fowler, Louisiana State University Laboratory School, Baton Rouge, Louisiana; Patricia Foy, Charter Oke High School, Corina, California; Conradt Fredell, Clear Creek High School, Idaho Springs, Colorado; Martha Friedlander, Dexter, Michigan; Francis H. Gately Jr., Algonquin Regional High School, Northboro, Massachusetts; Vernon Gilliland, Liberal, Kansas.

Mary Ann Gillis, Franklin Middle School, Green Bay, Wisconsin; James Glock, West High School, Green Bay, Wisconsin; Dwane Grace, Celina Senior High School, Celina, Ohio; Mark E. Greene, Algonquin Regional High School, Northboro, Massachusetts; Howard Grimm, Upper Arlington High School, Arlington, Ohio; Eileen T. Grosso, Minooka High School, Minooka, Illinois; Warren Hagestuen, Park Center Senior High, Brooklyn Park, Minnesota; Thomas Hagewood, Schuyler Central High School, Schuyler, Nebraska; Russell Hanseter, Seymour High School, Seymour, Wisconsin; Charles Hawkins, Franklin Middle School, Green Bay, Wisconsin; Philip Heck, Middle Township High School, Cape May Court House, New Jersey; Dale Hertel, Fargo South High School, Fargo, North Dakota; Holly Heverly, Lake Region Union High School, Orleans, Vermont; George Hrab, Montclair Kimberley Academy, Montclair, New Jersey; Louise Huey, Clemson, South Carolina; Stephanie Hunter, Richmond, Virginia; John E. Hutchins, Hanover High School, Hanover, New Hampshire; Debora Hutchison, Washington Middle School, Green Bay, Wisconsin; Robert Hutter,

Southwest High School, Green Bay, Wisconsin; Leon Jacques, Edison Middle School, Green Bay, Wisconsin; P. R. Janda, Shelton High School, Shelton, Washington; Marion V. Jaskot, Scotch Plains–Fanwood High School, Scotch Plains, New Jersey; Ron Johnson, Sycamore High School, Sycamore, Illinois; Connie Jones, Enka High School, Enka, North Carolina; Judy Jones, Chapel Hill High School, Chapel Hill, North Carolina; Bob Keel, Sierra High School, Colorado Springs, Colorado; Tim Kennedy, Celina Senior High School, Celina, Ohio; Ed Kidder, Richmond, Virginia; Kenneth Kiesner, Edison Middle School, Green Bay, Wisconsin; Bruce Kilmer, Edison Middle School, Green Bay, Wisconsin; Michael L. Kimmel, Conneaut High School, Conneaut, Ohio; Dan Kirsch, Bay Port High School, Green Bay, Wisconsin; Lanny R. Kizer, Loup City High School, Loup City, Nebraska; Robert Klaper, Wheaton Central High School, Wheaton, Illinois; Michael Kobe, School City of Hammond Indiana, Hammond, Indiana; Evelyn F. Kolojychrek, Cranberry Area School District, Seneca, California; Robin Krause, Knob Noster High School, Knob Noster, Missouri; Steve Krings, Southwest High School, Green Bay, Wisconsin.

Laurel Krol, School City of Hammond Indiana, Hammond, Indiana; Beverly Lauring, St. Petersburg, Florida; Sandy Leotta, Casper, Wyoming; Stanley A. Likes, Pawnee High School, Pawnee, Illinois; Bill Lipp, Fargo South High School, Fargo, North Dakota; Jerry Loynachan, Wheaton Central High School, Wheaton, Illinois; Barbara Mack, Conway, South Carolina; Susan W. Majors, Weston High School, Weston, Massachusetts; Gary Manfready, Mulberry, Florida; James Martin, Wheaton North High School, Wheaton, Illinois; Nancy Martinez, Emery High School, Huntington, Texas; Larry M. Maurer, Concord High School, Wilmington, Delaware; Paul D. McIver, Englewood High School, Englewood, Colorado; Philip P. Merman, Fairfield High School, Fairfield, Connecticut; James H. Meyer, Cedar Falls High School, Cedar Falls, Iowa; Robert D. Miller, Parkland High School, Orofield, Pennsylvania; John Monsma, N. Michigan Christian High School, McBain, Michigan; Ted Munnezke, The Principia Upper School, St. Louis, Missouri; Ken Nicholson, Arlington Memorial High School, Arlington, Vermont; Susan Nuemann, Chapel Hill High School, Chapel Hill, North Carolina; Wayne Olm, Preble High School, Green Bay, Wisconsin; Jeff Olofson, St. Francis Community High School, St. Francis, Kansas; Albert A. Perez, Leilehua High School, Wahiawa, Hawaii; John Petrie, Milan High School, Milan, Michigan; Bob Pfister, Kaukauna High School, Kaukauna, Wisconsin; Randy Phillips, East High School, Green Bay, Wisconsin; Frank Pirman, Preble High School, Green Bay, Wisconsin; Herbert T. Potter, Rio Mesa High School, Oxnard, California; Leon Raether, West High School, Green Bay, Wisconsin; Cathy Rager, Bretwood Junior and Senior High School, Brentville, Pennsylvania; Kathleen Ranwez, Moore Junior High, Arvada, Colorado; Fran Reinke, George Rogers Clark Junior and Senior High School, Hammond, Indiana.

Raybun Reynolds, Kathleen Senior High School, Winter Haven, Florida; Joan Richardson, Greenfield High School, Greenfield, Massachusetts; Sheila Richardson, Lakeview Academy, Atlanta, Georgia; Doug Rosendahl, Winona Senior High School, Winona, Minnesota; Robb Ross, Liberal High School, Liberal, Kansas; Pamela Roth, Caseville Public Schools, Caseville, Michigan; W. Bruce Roup, Holly Junior and Senior High School, Holly, Colorado; Roger Santille, Upper Arlington High School, Arlington, Ohio; Glenn Schlender, Southwest High School, Green Bay, Wisconsin; Gene Schmidt, Monroe High School, Fairbanks, Alaska; Robert Schoob, Bolingbrook High School, Bolingbrook, Illinois; Lisa Schweiner, Edison Middle School, Green Bay, Wisconsin; Calvin Scott, Illiara Christian High School, Lansing, Illinois; Sylvia Scyphers, Ariton, Alabama; Terri Seeley, George Walton Academy, Lawrenceville, Georgia; Tina Servais, Edison Middle School, Green Bay, Wisconsin; Jeanne Shaw, Liberty High School, Brentwood, California; Albert J. Shelley, Chesapeake High School, Baltimore, Maryland.

Forest H. Shoemaker, Stordley Lake High School, Broomfield, Colorado; Chris Smith, Bethel High School, Hampton, Virginia; Frances M. Smith, Easley, South Carolina; William G. Smith, Moorestown Friends School, Moorestown, New Jersey; Glenn M. Snyder, Wheat Ridge High School, Wheat Ridge, Colorado; Kris Spiegler, Hempstead High School, Iowa; Margot Steinberg, United Nations International School, New York, New York; Grace Stephen, Estes Park High School, Estes Park, Colorado; Sarah Stewart, Quincy Senior High, Quincy, Illinois; Lois C. Stoner, Chestnut Ridge Senior High School, New Paris, Pennsylvania; Susan Straten, Montclair Kimberley Academy, Montclair, New Jersey; Edgar R. Stuhr, Lexington High School, Lexington, Massachusetts; Richard Summers, Willard High School, Willard, Missouri; Jean Paul Thibault, Cape Elizabeth High School, Cape Elizabeth, Maine; Brother Robert Thomas, De La Salle High School, Minneapolis, Minnesota; Tony Toston, Summit High School, Colorado; Judith A. Treharne, Ocean Township High School, Oakhurst, New Jersey; Frank Turner, McKinley High School, North Canton, Ohio; Lloyd A. Turner, San Fernando High School, San Fernando, California; Howie Usher, Mingus Union High School, Cottonwood, Arizona; Gene Vander Velden, East High School, Green Bay, Wisconsin.

Grandon Voorhis, Glen Rock High School, Glen Rock, New Jersey; Ruth B. Vredeveld, Girls Preparatory School, Chattanooga, Tennessee; Larry Wakeford, Chapel Hill High School, Carrboro, North Carolina; Randyll Warehime, Farrington High School, Honolulu, Hawaii; Mary Watson, Bartlesville Mid-High School, Bartlesville, Oklahoma; David Wehner, Roosevelt High School, Lansing, Michigan; Maryanne Weins, Marine City High School, Marine City, Michigan; Linden Welle, Fort Morgan High School, Fort Morgan, Colorado; Steve Wester, Odell High School, Odell, Nebraska; Joyce Whitney, Edison Middle School, Green Bay, Wisconsin; Michelle Will, Shoshone High School, Shoshone, Idaho; Jim Willard, Schenck High School, East Millinocket, Maine; Robert Williams, Canton High School, Plymouth, Michigan; Tom Younk, Preble High School, Central Office, Green Bay, Wisconsin.

Contents

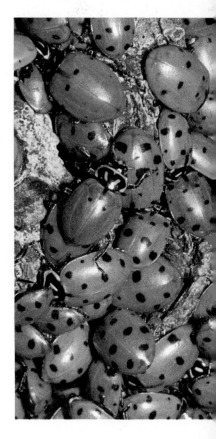

Section Two
Continuity in the Biosphere

Section Four

Functioning Organisms in the Biosphere

Section Five

Patterns in the Biosphere

Biology Today Features

Foreword

In this the closing decade of the twentieth century, an ecological approach to biology seems particularly appropriate. More than ever before, we have become aware of the human potential for permanently altering our environment—often to the detriment of future generations. An altered environment affects not only humans but also all other creatures, and there is no way to measure the cost of lost biodiversity. At the same time, we are beginning to develop the technology needed for sustainable development of the earth—development that allows economic growth but also preserves environmental choices for our children and our children's children.

The task of managing our environment will not be easy. Science can illuminate the problems, and technology can provide some possible solutions, but the choices depend on individual values and require international cooperation. Ecological problems transcend national boundaries—they are global in extent. Acid rain, for example, does not fall only on the country where the causative pollutants are emitted; destruction of rain forests and burning of fossil fuels produce greenhouse gases that contribute to global warming and the destruction of the ozone layer; and winds and air currents carry radioactive fallout far beyond the site of a nuclear disaster such as Chernobyl.

If we are to succeed in managing our environment, we must realize that the human species is a part of nature—a part whose existence is sustained by a finite planet and whose continuance depends on maintenance of that planet's resources, both geological and biological. We hope the information you encounter in this book will help you understand the interrelationships among all living things, particularly the role of humans in the web of life. Information means little, however, if we cannot apply it to the solution of personal, societal, or global problems and to the improvement and protection of the earth and its web of life. We hope this book will provide you with some of the knowledge and skills required for this important task.

BSCS Green Version Revision Team
Jean P. Milani, Coordinator

March 1991

Section One

The World of Life: The Biosphere

Except for the most desolate and forbidding regions of the polar ice caps, all of the earth teems with life. The study of life is called biology, and it includes not only a study of the structure and function of individual organisms, but also a study of the relationships between organisms themselves and between organisms and their environment.

Sea horses, mushrooms, lilacs, mosquitoes, and parakeets are just a small fraction of the types of life found on earth. Of all the diverse forms of life, however, humans have the greatest effect on the earth and on other organisms. Section One examines the relationships between some familiar living things and begins to explain just how profound the effects of human activities can be.

What do these lights represent? Give as many causes for these lights as you can. Show this photo to someone who is not in your class and ask them what they think the photo will look like in 1999.

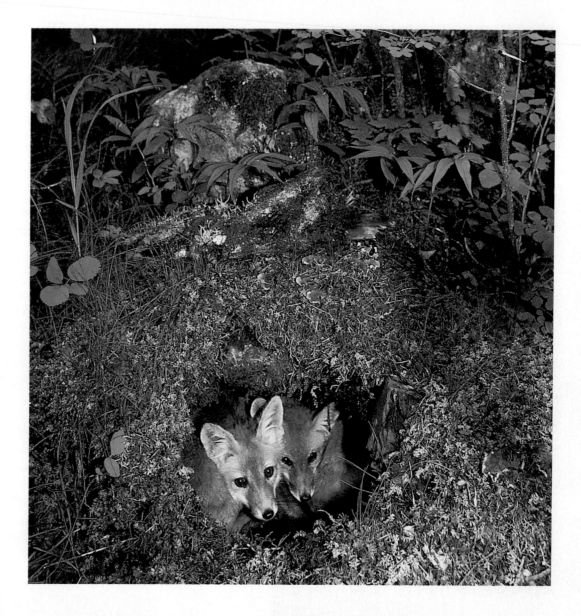

As secure as they may appear to be, these red fox kits (*Vulpes vulpes*) are dependent on many things in their environment. What three examples of such dependence are visible in this photo?

The Web of Life 1

All living things are part of a complex network of interactions called the web of life. Much as the strands of silk in a spider's web bind one segment to another, feeding and other interactions bind all organisms to one another. Like other animals, humans form a part of the web of life. Frequently we forget our connections to the other organisms that inhabit the earth. We are reminded of them, however, whenever our actions have visible effects on our surroundings.

Because of our numbers and our use of natural resources, we have the greatest impact of any living thing on the web of life. We often affect it unknowingly because many of the interrelationships are difficult to see or to understand. We may not notice that something has gone wrong until the problem is so large that we no longer can ignore it. Even then, our tendency to consider ourselves separate from the other living things that surround us affects our ability to see how the problem concerns us.

If the human population is to continue to thrive on earth, we must begin to understand the interrelationships among all living things. We must learn to modify our actions in order to decrease our impact on earth and the life it supports. Only when we have done this can we begin to solve such problems as global warming and acid rain. This chapter introduces the basic interrelationships among living things and provides a foundation on which to build solutions to some of the problems facing humankind.

Interactions Among Living Things

Guidepost: Why is it difficult to study an organism in isolation from its surroundings?

INVESTIGATION 1.1 The Powers of Observation

You can explore the complex interrelationships among living things in many ways. In this investigation, you will make careful observations and then report and verify those observations.

Materials (per class)
10X hand lens or stereomicroscope   labeled specimens of organisms
millimeter ruler

Procedure
1. Located around the room are sets of organisms—living things—or parts of organisms. Each set contains four specimens. (A specimen is a sample individual or, in

the case of large plants, a characteristic part, such as a leaf.) Each set is labeled with a group name and a number. Each specimen is labeled with a letter.

2. Work in teams of two to four. Each team will begin with a different set of specimens. You will have approximately 10 minutes to observe and to describe the set.

3. Select one person on your team to take notes. Observe the four specimens and make notes on differences you see among them. If appropriate, take measurements. Remember, the differences must be in the organisms, not in their containers.

4. When your team has decided on the differences between the four specimens, choose one. On a separate sheet of paper, write as complete a description as possible of the specimen. Do not use the specimen's letter on your description sheet. Write the group number and letter of the specimen described on another slip of paper and give it to your teacher.

5. When your teacher signals that time is up, place the description your group has written next to the set of specimens.

6. When your teacher tells you, move to the specimen set having the next highest number. (The team with the highest number will go to Set 1.) You will have several minutes to observe the set. Select one team member to take notes. Then, read the description next to the set and decide as a team which of the specimens it describes.

7. In your data book, prepare a table with these headings:

Set Number	Specimen Fitting Description

Record the letter of the specimen in your table.

8. Repeat Steps 6 and 7 until you return to your starting point.

9. Your teacher will list the correct answers on the chalkboard. Check your table against this list. If your conclusions do not agree with the list, recheck the group of specimens. Did you miss anything? Was the description complete?

10. Wash your hands thoroughly before leaving the laboratory.

Discussion

1. Which was easier, writing a clear description or selecting the specimen another team had described? Why?

2. For each description, what information could be added to make it clearer? Does everyone in the class agree on what could be added? Why or why not?

3. For each description, what information could be removed and still leave the description clear? Does everyone agree on what information to remove? Why or why not?

4. Was there a set of specimens that you would have liked to describe? Why?

5. In what ways did reporting and verifying observations in this investigation increase your knowledge of these organisms?

1.1   Organisms Interact with Organisms They Eat

Living things may affect one another in many ways. They may crowd each other, protect each other, poison each other, or give each other shelter. At the most basic and important level, however, living things consume one another. Feeding relationships are the major strands in the web of life.

Consider a brightly colored grasshopper sitting on a young plant and chewing a leaf. The grasshopper (see Figure 1.1a) is red, yellow, and black

a

b

Figure 1.1 What activities are illustrated here?

and very small. When the leaf is completely eaten, the grasshopper jumps toward another plant but lands on a sticky thread. The thread is just one of many that are carefully woven together into a large, shiny trap—a web. As the grasshopper struggles to free itself from the thread, more threads stick to it. With each movement, the grasshopper sends a vibration from thread to thread and finally to the maker of the trap. In the blink of an eye, a large brown and yellow spider (see Figure 1.1b) seizes the grasshopper and kills it with poison injected from the spider's fangs. Digestive fluid from the spider's mouth liquefies the grasshopper's body, and the spider sucks up the resulting broth. Later, the spider drops the empty body of the grasshopper on the ground and waits for another insect.

This story is repeated every day all around the world. Animals eat plants or other animals or both. Organisms are connected to each other in the web of life by their need for energy to live and to reproduce. Green plants, such as the poppies in Figure 1.2, get energy directly from sunlight and use it to make their food. Animals that eat plants or other animals get energy indirectly. No matter where the energy comes from, without it, an organism dies. If the spider web is broken by a falling twig, it must be repaired, or the trap can no longer catch food for the spider. The spider's link to the supply of energy cannot be broken if the spider is to live. If grasshoppers eat all the leaves of the young plant, the plant can no longer make its own food, and it dies.

Many things can change the relationships between the plant, the grasshopper, and the spider. Caterpillars, such as the one shown in Figure 1.3 on page 8, may eat the plant, which means less food for the grasshoppers.

Figure 1.2 What relationship is shown between the poppies and the sun?

Figure 1.3 A gypsy moth caterpillar eating a leaf. How is the food supply of other organisms affected?

Without enough rain, the plant may die before it can reproduce or provide food for an animal. The spider's web may trap flies and moths as well as grasshoppers. A bird may eat the grasshopper or even the spider or caterpillar. Just as the spider's web has many connected threads, in the web of life many organisms are connected to each other. The more closely we view the world, the more complex this web becomes and the more we become aware of the impact of humans on the web. The smallest change in the web can have a major effect on the connected organisms.

You have seen how plants and animals interact within the web of life. Nonliving things, such as temperature, sunlight, and rainfall, also affect the growth of plants and animals. The study of the living and nonliving parts of the environment and how they affect organisms is called **ecology**. Scientists who study ecology are called ecologists. Ecology is only one aspect of **biology**, the study of living things. In this textbook, you will study many areas of biology, but you will focus on ecology.

1.2   Plants, Animals, and Other Organisms Make Up a Food Chain

Not far from the spider's web discussed in Section 1.1 is a raspberry bush. Underneath the bush, a rabbit (see Figure 1.4) finds shelter and a place to hide from animals that may kill it. The bush is an ideal place to hide because its thorns can tear clothes or dig into the flesh of larger animals, and its low-hanging red fruit provides the rabbit with food. A small bird feasts on the fruit near the top of the bush. Because rabbits usually do not eat raspberry leaves, the rabbit ventures out to look for grasses to eat once the berries are gone. Its movement is spotted by a hungry fox, which slinks forward and suddenly makes a leap for the rabbit. The rabbit looks up just

Figure 1.4 Describe all the relationships you see among the plants and animals and their environment.

Figure 1.5 What role do decomposers play in a food web?

in time, and a wild chase begins. This time, the rabbit reaches safety in another raspberry bush.

Not far from the rabbit's bush, however, is the fox's den. Near the den seen in Figure 1.5 are scraps of fur and flesh, the remains of last week's rabbit. The fox had carried this rabbit to the den and eaten most of it. What he did not eat, he left to rot. **Microorganisms**—organisms too small to be seen with the unaided eye, such as bacteria—began to break down the uneaten portions, causing them to decay. Flies laid eggs on the remains, and the eggs hatched into wormlike maggots that fed on the flesh.

Rabbits, raspberries, and the other nearby plants and animals play various roles in the web of life. The green plants use light energy to make food. Because they make their own food, they are called **producers**. Animals cannot make their own food, so they eat plants or other animals or both. Organisms that are unable to make their own food are called **consumers**. During the process of decay, consumers that break down the bodies of dead plants and animals are called **decomposers**. Bacteria and mushrooms, such as those shown in Figure 1.6, are examples of decomposers. The producers, consumers, and decomposers that live and interact in one area form a **community**.

Figure 1.6 What role do mushrooms play in a food web?

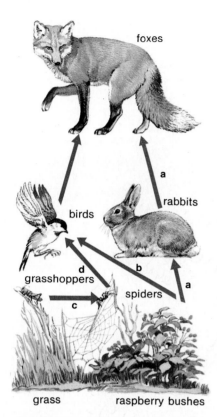

Figure 1.7 There is more than one food chain here. How many can you find?

The raspberry bush, the rabbit, and the fox can be connected in a **food chain**, a pathway that tells what eats what. Several food chains are shown in Figure 1.7. In Path a, the rabbit eats raspberries from the bush, and the fox eats the rabbit. Path b illustrates a shorter food chain—the bird eats the raspberries. These two food chains are connected by the raspberry bush. Section 1.1 described a food chain in which a spider ate a grasshopper that ate a plant (Path c). If the bird that eats raspberries also eats the spider, two more food chains are connected (Path d).

When all the food chains in a community are connected to each other, a **food web** is formed. Figure 1.7 does not include all the plants a rabbit might eat, all the animals that might eat a rabbit, or any decomposers. What would the web look like if there were two spiders instead of just one? Figure 1.8 shows a larger food web that includes the organisms discussed in this chapter. You can see that a food web can be large and complex.

Food webs and food chains tend to keep living organisms in balance. The rabbits live off the green plants, and many other animals, including humans, live off the rabbits. This might appear to be hard on the rabbits, but rabbits produce many offspring in a short time. Imagine how many rabbits there would be if they reproduced without control. They soon would be so numerous that they would eat all the plants. Without the plants, the rabbits would starve. Foxes and other animals that eat rabbits may help to keep the rabbit population under control. Disease or lack of food also may help to keep the rabbit population from growing too large. These controls, or checks, apply to all living organisms, including humans, and are just one part of the balance of nature. Investigation 1.2 may help you understand your place in the web of life.

Concept Review

1. What parts of the environment does an ecologist study?
2. How do producers differ from consumers?
3. How is a food chain related to a food web?
4. Explain how reproduction and death are part of the balance of nature.
5. In what way do decomposers differ from other consumers?

INVESTIGATION 1.2   You and the Web of Life

How do you fit into the web of life? By relating the food you eat for one day to the plants and animals from which it came and to the other organisms with which those plants and animals interact, you can begin to form a picture of your role in the biosphere.

Materials (per person)
data book
pen

Procedure
1. In your data book, list all the foods you ate yesterday.
2. Separate these items into foods that came from animals and foods that came from plants. Remember, many foods are combinations of different foods. List the ingre-

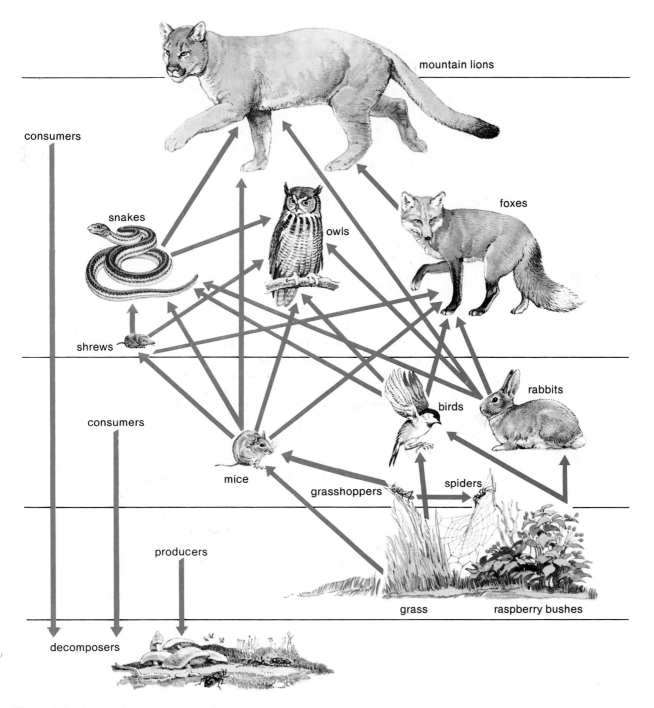

consumers

snakes

owls

foxes

mountain lions

shrews

consumers

mice

grasshoppers

spiders

birds

rabbits

producers

grass

raspberry bushes

decomposers

Figure 1.8 Can you find any more relationships?

dients of each food separately. Then indicate whether each was from a plant or animal and the type of plant or animal. For example, if you had cake for lunch you should list: flour—plant, wheat; sugar—plant, sugar beets; eggs—animal, chicken. For every animal you have listed, list several foods that it eats. For example, if you had milk with your cake, list a cow and then list grass and corn as food the cow eats.

3. Across the bottom of a blank sheet of paper, write the names of all the plants you have mentioned. In a row above this row, list all plant-eating animals (herbivores) that eat any of the plants in the first row. From each plant, draw a line to every animal that eats it.

4. Above the herbivore line, enter the names of all animals (carnivores) from your list that eat other animals. From each herbivore, draw a line to the carnivores that eat that herbivore. The food web you have drawn shows some of the interrelationships between the plants and animals that provided your food for one day.

Discussion

1. Which, if any, of the items you listed as foods for animals are also foods that you could eat?

2. Did you include organisms that might compete with you for your food? Add as many as you can to your food web.

3. What about the role of decomposers in your food web? Indicate with lines how they might be involved.

Going Further

Research the origins of several of the plants and animals that appear in your food web. What were the original ranges of those species? Where were they domesticated? Where are they raised now? What are the requirements of those plants and animals? How might they be affected by global climatic change? What environmental alterations are necessary for the production of those foods? In what other ways could these plants and animals be produced?

Guidepost: Where do organisms obtain the matter and energy they require?

Matter and Energy—The Foundations of Life

1.3   All Biological Activity Requires Energy

Food chains and food webs are based on the flow of energy and matter from one organism to another organism. The details of this flow are developed throughout this course. Here, you will look at just the broad outline.

All of an organism's activities require energy. Imagine the marathon runner shown in Figure 1.9a trying to run a race without having eaten high-energy foods. The activity does not have to be very great to require energy; even the movement of a tiny one-celled organism across a drop of water requires energy (see Figure 1.9b). Whenever you see any type of biological activity, you need to ask, "Where does the energy come from to support this activity?"

Where do you get your energy? It may take some imagination to see energy in a hamburger and a pile of french fries. There is energy in this food, however. It is **chemical energy**. Chemical energy is found in the structure of the molecules that make up the meat and the potatoes. Other forms of energy include electrical, mechanical, heat, light, and nuclear energy. The most important form of energy for you is the chemical energy stored in the food you eat. You begin to release this energy as you digest your food. Most of the energy from your food is released within your cells in a complex series of chemical reactions. You use this energy to grow and to develop.

Figure 1.9 Where do these organisms get their energy?

1.4 Photosynthesis Supplies Food Energy

Remember that you are part of many food chains and one giant food web, as you saw in Investigation 1.2. The hamburger that contains chemical energy came from a cow; cows eat only grasses and grains. A grass plant and a potato do not eat other organisms, so where do they get their energy? All green plants grow in light. In the process of **photosynthesis**, plants absorb light energy and convert it to the chemical energy found in sugars. The sugars formed by photosynthesis provide food for the plant. The plant then can use the energy in the sugars to grow and reproduce. Energy that is not used may be stored in the form of starch to be used at a later time. The potato plant, for example, stores energy as starch. When you eat the potato, you benefit from the chemical energy stored in it.

Photosynthesis is the basis for almost all of the food energy in the world. After grass makes its own food in photosynthesis, it uses some of that food to grow. Thus, some of the energy that is captured from sunlight is used before it reaches the cow, even if the cow eats the whole plant. Only certain bacteria are able to make their own food in other ways.

Because no animal can make its own food, it must get its energy from plants or other animals. The cow eats the grasses and uses the chemical energy in them to grow, to produce milk, and to move across the pasture. The energy not used by the cow remains in the waste products dropped in the pasture. This energy is not completely lost to all organisms. Decomposers break down the cow dung and use the energy from it for their own growth and reproduction. Decomposers also get energy from the bodies of the cows and plants that die. This series of steps is referred to as the flow of energy through a system.

The conversion of chemical energy to the energy used by living organisms is not efficient. Whenever an organism breaks down its food, some of the energy escapes as heat. If you touch your arm, for example, it feels warm. This is because some of the chemical energy stored in the fries and

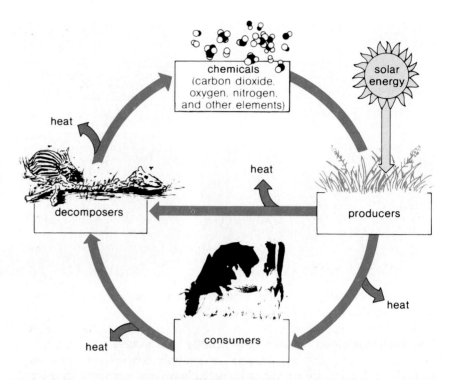

Figure 1.10   Why is it impossible for organisms to recycle energy?

hamburger you ate is converted to heat energy. Your body uses the rest of the chemical energy to keep you alive and growing. Although some heat energy keeps the body warm, which allows biological activities to continue, most of it is lost to the air. No matter where energy comes from, all the energy that enters a food web eventually is lost from the community in the form of heat (see Figure 1.10). No organism can use heat energy for growth. Therefore, energy must enter a community continually, beginning with photosynthesis, or the community will die. The source of chemical energy for almost all communities in the world is light energy from the sun. If the sun were to burn out, life on earth would cease.

1.5   Matter Is Used to Build Living Things

Unlike energy, which flows one way through a food web, matter does not leave the food web. Instead, it changes from one form to another within the web. This cycling of matter is important in the web of life. Most living things get their energy directly or indirectly from the sun, but all living things get their matter, or substance, from the earth and the air around them. You probably already know that all matter is made of minute particles called **atoms**. Matter that is made of a single type of atom forms an **element**. Most elements, however, occur as **molecules** (MOL uh kyoolz), which are usually combinations of two or more atoms. A **compound** is matter that is made up of more than one type of atom chemically combined. Water, for example, is a compound made of atoms of two different elements—hydrogen and oxygen (see Figure 1.11).

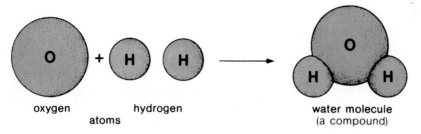

Figure 1.11 Water is a compound made up of two elements—hydrogen and oxygen.

Because living things are so different from nonliving things, scientists once believed that living matter contained unique elements. We now know this is not the case. Of the more than 100 different elements found on the earth, only about 30 are used in the makeup of organisms. Most of these elements, such as the hydrogen found in a water molecule and the carbon found in the carbon dioxide of the air, are very common. Figure 1.12 shows the proportions of elements in humans.

Plants and animals are made up of many different compounds, but the atoms used to make up these compounds occur all around you in the non-living world. For example, plants use the simple compounds of carbon dioxide and water in photosynthesis, but only small amounts of chemical energy are present in these compounds. During photosynthesis, the plants build complex compounds. Using light energy, they link together the atoms from carbon dioxide and water to make sugars. Both energy and matter are stored in sugars. A plant can use these sugars as a source of energy, as discussed in Section 1.4, or the plant can use the sugar molecules to make other molecules that it needs to build its body. To do this, a plant rearranges the atoms in the sugar molecules and adds new atoms. Sugars, therefore, are a food—a substance that an organism can break down to get energy for growth, body maintenance, and repair. Food also is matter that can be used to build the structure of the body.

When an animal eats a plant as food, both energy and matter are passed from one organism to another in a food web. Unlike the flow of energy in a food web, however, the flow of matter is not one way. Matter cycles within

Figure 1.12 Elements present in the human body.

energy
(as light)

B i o s p h e r e

matter
as simple
compounds

matter
as
complex
compounds

energy
(mostly as heat)

Figure 1.13 Explain the relationships energy and matter have with a community.

a community. Plants use carbon dioxide, water, and other substances during photosynthesis. These same substances are given off by organisms after they have used a plant as food. Other plants then can use these materials again to continue the process of photosynthesis.

In this manner, the same matter is used over and over again in a community. Matter travels in cycles from the nonliving environment into food webs and back to the nonliving environment. It then enters the food webs once again in photosynthesis. Figure 1.13 summarizes the relationship between matter and energy in the biosphere.

Pollution affects the cycling of matter. Pollutants can tie up materials that would ordinarily cycle between the living and nonliving parts of the world. If the pollutants can be changed to other materials, they can be made less harmful, and perhaps some of the matter can return to normal cycling.

Concept Review

1. What is the source of energy for almost all living things?
2. How are light and chemical energy related in photosynthesis?
3. How does matter get from the soil and air to animals?
4. How is the flow of energy through a community different from that of matter?
5. In what way do simple compounds differ from complex compounds?
6. Explain why energy flows and chemicals cycle through food webs.

INVESTIGATION 1.3   Field Observation

Life is everywhere, although at times it may be difficult for you to see it. Some organisms are widespread, others found only in certain places. All organisms, however, need food, water, and protection to survive. In this investigation, you will sharpen your observation skills and try to explain the presence or absence of organisms in a study area.

Materials (per person)
data book
pen

Procedure

1. Look around you. What type of environment are you in? Is it next to a city street, in a park, or in the country? Being as specific as possible, record your answer in your data book.
2. What time of day is it? What are the weather conditions? Record your answers.
3. What type of life do you see? The type of life you observe may be flowering plants in a planter, people in cars, cows and grain in a field, or squirrels in trees. Record as many types of life as you can and be sure to note where you see them.

   CAUTION: Do not handle any organism you observe, even if you are certain it is harmless. Unnecessary handling may be harmful to the organism and yourself.
4. What type of birds do you see? Are they pigeons or some other bird? Where do you see them? Record your answers.
5. Find a patch of ground away from the other members of your class. Look at the area closely. Are there any plants growing there? Do you see any other organisms? How many? Record what you see.

6. What might explain the presence of these organisms?

7. What types of organisms common to your area, such as ducks or lizards, do you *not* see? Write down a few examples.

8. What might explain why some organisms are *not* present?

9. How would you determine if the time of day or the weather influences the types of organisms you see?

Discussion

1. Share your observations with the other members of the class. What organisms did they see? Did everyone see the same ones?

2. Where were the organisms found? Why do you think they were found there?

3. Do you think the same organisms could live in the downtown area of a large city? In the country? Why or why not?

4. What organisms were not found? Share your explanations of their absence.

Going Further

Use plant and animal keys to identify some of the organisms you observed.

Studying the Living World

1.6 The Biosphere Is Home to All Living Things

After studying many different food webs, biologists have concluded that the spider, the rabbit, and you are all part of one worldwide web of life. All organisms are part of this web, and every organism is tied to another by the energy and the matter it needs for life.

The living world forms only a thin layer around the nonliving world. This layer is called the **biosphere** (BY oh sfir). It includes all the organisms and the air, soil, and water surrounding them. The biosphere extends from the bottom of the oceans to the air above the earth.

Because you are a living organism, you are part of the biosphere. You have interrelationships with your family, friends, and teachers. Because you are part of the web of life, you also have relationships with all other organisms. Some relationships are obvious, such as those with the plants and animals you eat. However, you may not recognize the relationship that exists between you and animals that eat the same foods you eat. The relationship between you and a grasshopper that eats and damages lettuce is indirect, but any damage to your food plants can affect your food supply.

Humans have far-reaching effects on the biosphere. Data being collected today show that many human activities are straining and destroying the delicate balance of nature in the biosphere. In the spider's web, pulling a single thread affects the entire web. So, too, do your actions affect the world. What you, your friends, and your family do affects the rest of the biosphere in many subtle ways.

The 1989 *Exxon Valdez* disaster is a dramatic example of the impact of humans on the biosphere. The tanker disgorged more than 10 million gallons of crude oil into Alaska's Prince William Sound (see Figure 1.14), and it will be many years before the wildlife in this area recovers. Yet, when compared with the far-reaching effects of excessive energy consumption

Guidepost: How do humans fit into the biosphere?

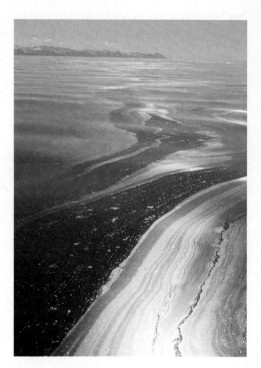

Figure 1.14 Why is an oil spill bad?

Ecologist

Tropical forests supply most of the oxygen that air-breathing organisms need for life. The destruction of these forests by burning and cutting releases a tremendous amount of carbon dioxide into the atmosphere. Increased carbon dioxide levels are one factor implicated in the predictions of global warming (see Chapter 4, Biology Today). Ariel Lugo is one of the many people who are studying the effects of tropical forest destruction. Dr. Lugo studies the role of tropical forest ecosystems for the United States Forest Service (a division of the Department of Agriculture) in Puerto Rico. Dr. Lugo was not always interested in ecology. In fact, he admits his passion originally was baseball. His family, however, encouraged him to study medicine. While at medical school, Dr. Lugo began to work with a prominent ecologist who was conducting radiation experiments in the rain forest. Dr. Lugo says, "After a summer of carrying cement blocks through the rain forest and after obtaining a D in embryology, I was sure that medicine was out and ecology was in—to the distress of my family."

Currently, Dr. Lugo is conducting research in three areas. First, he and a colleague, Dr. Sandra Brown, are analyzing the content of tropical forest soils that have been exposed to a variety of land-use practices. These analyses allow them to determine how organic matter in the soil changes over time under different types of use. These studies will help to determine the role and importance of tropical forests in the global carbon balance.

A second area of research involves the study of forested wetlands. Forested wetlands are lowland ecosystems, such as marshes or swamps, that are saturated with fresh or salt water. Saline (salty) forested wetlands consist mostly of mangrove trees, which grow on tropical coastlines and can tolerate seawater. Freshwater forested wetlands include high-elevation palm forests. Dr. Lugo's interests lie in the study of the structure and function of these ecosystems and how they change through time.

The third area of research involves the dynamics of tropical tree plantations. Dr. Lugo is comparing plantations with natural areas of similar age to determine what elements of the natural area remain in the artificial plantation and how the chemical cycling in each area differs. Understanding these patterns is important because plantation agriculture is one way of rehabilitating damaged lands and restoring natural forest ecosystems in areas where humans have devastated the tropical forest.

Dr. Lugo has never questioned his decision to quit medical school and pursue a career as an ecologist. Being an ecologist has allowed him to meet many talented people, travel the world, and visit some of the most exciting locations on earth.

and pollution around the world, the oil spill is a trivial occurrence. In 1989, for example, the population increased by about 87.5 million, further straining supplies of food and other resources. Deforestation and the burning of fossil fuels added at least 19 billion tons of carbon dioxide to the atmosphere, thus aggravating the global warming process. Another 11.3 million hectares (28 million acres) of tropical forest were destroyed along with the organisms living there. Many of these organisms will remain forever unknown. The hole in the ozone layer over the Antarctic remained alarmingly large, and scientists report that another hole is developing over the Arctic, increasing ultraviolet radiation to the earth and the subsequent risk of skin cancers.

These incidents show that our planet is in trouble. If we do not take action, our planet one day may become unfit as a human habitat. People are gradually beginning to recognize and deal with these problems. Among the international initiatives currently underway is a world climate conference. It is hoped that the conference can lay the groundwork for a global limitation on the emission of greenhouse gases and, thus, the stabilization of the world's climate. Two other goals for the conference are the drafting of a biodiversity treaty, which would provide financial incentives for the protection of tropical forests, and the development of an agreement to phase out the production of chlorofluorocarbons (CFCs)—the major culprits in the depletion of the ozone layer. There are other measures that should be taken under consideration. These include worldwide family planning to reduce population growth and new manufacturing techniques to reduce waste. At a more personal level, every individual can look into recycling discarded materials, insulating his or her home, and using mass transit and carpooling to cut energy use.

Because the United States has the technology to help other nations with environmental problems and because it consumes a large part of the world's resources, it has a major role to play in these and other areas.

1.7 A Hypothesis Is a Statement that Explains an Observation

Humans look at the biosphere from many different viewpoints, but each viewpoint reveals only a small part of the complex whole. In our working roles, we have limited points of view. An artist looks at living things for their beauty of color and form. A farmer looks for ways to manage the growth of living things. A biologist views the world scientifically.

As a science, biology does not deal with value judgments. Beliefs and value judgments differ from person to person. For example, some students looking at Figure 1.15 on page 20 might prefer the fawn, others the puppy. These are value judgments. They are based on how you feel about puppies and fawns.

Biology is concerned with data—observations that do not differ from one person to another. The Steller's jay in Figure 1.15 is blue. This is an observation that can be agreed on by everyone with normal color vision. Biologists collect and organize data about organisms. They use the data in various ways to gain more information. Science, however, is more than just the collection of data. It is a systematic way of looking at the world, of

Figure 1.15 "Which is nicer, the fawn or the puppy?" "Which bird is blue?" What is the difference between these two questions?

obtaining data, and of interpreting it. It is a continuous process of inquiry, the product of which is a body of knowledge. This body of knowledge is subject to change and revision as we acquire new information. Above all, science is a human endeavor because people are involved in the process of inquiry.

Science is one way of obtaining knowledge about the world around us. The methods of obtaining scientific knowledge involve a series of steps that often begin with observations of the living world. In the field with the spider and the rabbit, there may be many flowers with insects buzzing around them. A scientist may observe this field and ask the question, "What attracts bees to the bright red flowers in the field?" Library research may provide information about previous investigations of the question or suggest appropriate approaches to the problem. Reading and thinking about the question may produce thoughts such as, "Bees are attracted to this type of flower because of its red color." This thought is a **hypothesis** (hy POTH uh sis), a statement that explains an observation. A good hypothesis predicts a relationship between a cause and an effect. It can be tested by an experiment specifically designed to collect evidence that either will support or will not support the hypothesis. Evidence cannot *prove* a hypothesis, however, because it is always possible that new evidence may provide a better explanation. You can examine how hypotheses are tested in Investigation 1.4.

An understanding of science and of scientific methods is particularly important today as we face such problems as overpopulation, starvation, pollution, and the loss of forests. By using scientific methods to study our impact on the world around us, we can begin to understand how our actions affect the biosphere. Out of this understanding, we can begin to develop solutions that will minimize our impact and allow us to live in balance with the web of life.

Concept Review

1. What parts of the earth make up the biosphere?
2. How does a value judgment differ from data?
3. What is the relationship between hypotheses and observations?
4. How do humans fit in the biosphere? Why is understanding our role in the biosphere important?

INVESTIGATION 1.4   How Do Flowers Attract Bees? A Study of Experimental Methods

In this investigation, you will learn about the experimental process by examining in detail several possible experiments.

Materials (per team of 3)
3 data books
3 pens

Procedure
PART A   Controlling for Variables

1. Recall the hypothesis "Bees are attracted to this type of flower because of its red color." This hypothesis can be investigated if it is reworded as follows: "If bees are attracted to the flower by its red color, then an artificial flower cut out of red paper may attract the same bees to it when placed in the field." The "if" part of the hypothesis refers to the cause—what is being investigated. The "then" part of the hypothesis refers to the effect—the outcome of the experiment. How would you investigate this hypothesis? Record your answer in your data book.

2. Suppose you carried out an experiment in which you cut out paper flowers of several colors (for example, red, yellow, green, and blue) and placed these in a field. Suppose also that a bee landed on the red paper flower but not on the other paper flowers. Does this observation support your hypothesis? Explain.

3. Your observations provide information, or data, to help answer your question or to test your hypothesis. A scientist uses data to draw conclusions about an experiment. Do the above data prove your hypothesis was true? Why or why not?

4. How could you test the hypothesis that bees landed on the red paper flower because it was the most convenient, not because it was red?

5. You know that flowers differ in many ways. These differences are called **variables**. Some of the variables for flowers in a field are size, shape, odor, color, and the size, position, or shape of the petals. A good experiment tries to control all variables except the one being studied—the experimental variable. What is the experimental variable in the flower experiment described in Step 2?

6. The colored paper flowers that are not red are **controls**. Controls are the parts of an experiment that make it possible to eliminate variables other than the one being studied. In this case, you are testing the variable of color, making certain it is the red color that attracts the bees. What are some variables that should be controlled for in this experiment?

7. You want to improve this experiment by making it more valid. Other than differing in color, what must these flowers be like?

8. Assume for a moment that no bees land on the red paper flowers. Does that support or refute the hypothesis in Step 1?

9. Usually, once a hypothesis has been refuted, it does not have to be tested again. It is thereafter considered false, assuming the experiment is a valid one. Given your answer to Question 8, in this case would you need to continue the experiment? Why?

PART B   Analyzing an Experimental Design

10. Study the observation and hypothesis in Figure 1.16.

Flower 1   Flower 2

Figure 1.16   *Observation:* Bees are attracted to Flower 1 but not to Flower 2. *Hypothesis:* If bees are attracted to Flower 1 by scent, then adding scent to a flower that does not attract bees should cause the bees to be attracted to that flower.

11. Examine Experiment 1, represented in Figure 1.17. Here the scent of a flower that attracts bees is added to a flower that does not usually attract bees. What can you tentatively conclude from this experiment?

Flower 1   Flower 2
                         scent added

Figure 1.17   *Experiment 1:* Scent of Flower 1 is added to Flower 2.

12. What does Experiment 1 lack?
13. Is scent the only variable?
14. What other variables are there between these two types of flowers?

Outcome A

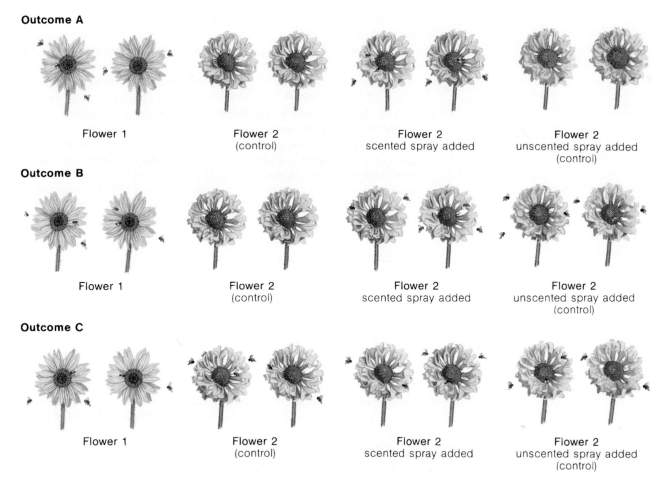

| Flower 1 | Flower 2
(control) | Flower 2
scented spray added | Flower 2
unscented spray added
(control) |

Outcome B

| Flower 1 | Flower 2
(control) | Flower 2
scented spray added | Flower 2
unscented spray added
(control) |

Outcome C

| Flower 1 | Flower 2
(control) | Flower 2
scented spray added | Flower 2
unscented spray added
(control) |

Figure 1.18 *Experiment 2:* This is a redesign of Experiment 1 to consider other variables that might be responsible for the original observation. Flower 2 in this case is a different type from Flower 1, although both appear similar.

15. Carefully study Experiment 2, represented in Figure 1.18. Be prepared to defend or criticize it. A, B, and C show different possible outcomes.

Discussion

1. Compare the results for Flower 2 in Experiments 1 and 2. What variables are being controlled in the design of Experiment 2?
2. In Experiment 2, the scent-producing substance from Flower 1 is dissolved in a liquid (water or another solvent harmless to flowers) to produce a spray. What is the purpose of the control flowers with nothing sprayed on them?
3. What is the purpose of the control flowers sprayed with unscented spray?
4. What hypothesis would account for Outcome A in Experiment 2?
5. What hypothesis would account for Outcome B?
6. What hypothesis would account for Outcome C?
7. How does Outcome B illustrate the difference between the two sets of control plants?
8. What other variables might it be helpful to control in Experiment 2?

9. Why would it be useful to try a similar experiment with more than one group of the same type of bees?

10. Why would it be useful to repeat the experiment with more of the same type of flowers?

11. How would you interpret the results if the *numbers* of bees visiting each type of flower differed noticeably?

12. How would results similar to Outcome A with different types of flowers strengthen your conclusions?

13. Experiment 2 is still an imperfect attempt to test the original hypothesis. Design an improved, but similar, experiment that might provide additional support for your conclusions. Consider the issues raised in Questions 9, 10, 11, and 12 above. Consider also what additional information you may need to accept or reject the hypothesis. Remember that you want to manipulate only one variable, so all other possible variables must be controlled. Describe your design and procedure in a short paragraph.

Summary

No organism is isolated from other living things. All organisms are part of a large, complex web of life. Plants are producers. They use the energy from the sun and simple compounds from the soil and air to produce molecules of food that allow them to grow and reproduce. Because animals are consumers and cannot do this, they depend, directly or indirectly, on plants as their source of energy and matter. Humans also rely on plants for food. Humans, however, have a greater effect on the biosphere than other organisms because their activities often result in permanent changes in the environment. Using the techniques of observation, hypothesis formation, and experimentation, we are becoming more aware of how those changes affect the lives of all organisms in the biosphere. In response to growing global problems, we are beginning to take responsibility for our actions and to develop plans that will allow us to manage our planet in a manner that will sustain it for generations.

Applications

1. What might happen to the balance in the food web of a pond if the number of one type of organism suddenly increased greatly? How long do you think such an increase would last?

2. Consider the pond in Question 1. Is your prediction affected by the type of organism involved? Consider first a producer, then a consumer.

3. Look at the fawn and the puppy in Figure 1.15. Draw a food web for each animal. Explain the role each animal plays in a food web.

4. After drawing a food web for the fawn and the puppy in Question 3, has your value judgment

about "which one is better" changed? Why or why not?

5. Suppose that the sun were to die out, leaving the earth in darkness. The earth would not freeze immediately, and not all living things would die at the same time. Which do you think would die first—plants, animals, or decomposers? Explain your answer.

6. Someone once said, "You are what you eat." If this is true, some people would say that your body is really rearranged plant fruits, roots, stems, and leaves. Would you agree or disagree? Explain.

7. Suppose you found a spider web and broke one of the strands at the outer edge. Do you

think the web would still function and insects still could be caught in it? What if you broke two, three, or more strands? Now, think about the web of life in your community. Suppose a natural disaster killed 10 percent of the birds in the community. Do you think the community would still function? What if 50 percent of the trees died?

8. Why do almost all food chains begin with photosynthetic organisms of some type?

9. What is the relationship between hypotheses and data?

Problems

1. Wildlife can be found everywhere, even in cities. Animals such as skunks, raccoons, and squirrels are common in cities. Birds, of course, are among the most common types of urban wildlife. Make a list of the plants and animals in your area and use it to build a food web.

2. Describe how building a road through the mountains might affect the organisms that live in the area. List as many organisms that might be affected as you can and try to show some of the relationships among the organisms. For example, cutting down trees affects the availability of nest sites for birds.

3. In making an extended journey into outer space, astronauts would have to take along a part of our biosphere. Design an efficient "package" of the biosphere for such a journey.

4. Many people have careers related to biology. How does the career of the nutritionist in your school cafeteria relate to energy flow and materials cycles in a food web?

5. Choose a current environmental issue in the United States or in your local community. After some research, present to your class the causes of the problem and the potential solutions.

6. Suppose that one morning your family car does not start. Which do you think might be the more likely reason: (a) the car is missing a hubcap, or (b) the battery is dead? Why does one of these reasons seem more likely than the other? How would you test your hypothesis?

7. Make a list of hypotheses you make in your daily life during a weekend. Explain why you made these hypotheses and not others.

8. What plants and animals lived in your area *before* any humans lived there? Construct a food web for that community.

Adelie penguins (*Pygoscelis adeliae*) in Antarctica. A population is a group of organisms of the same species that live and interact in the same place at the same time. What is the evidence of a population in this photo? What is the evidence that the penguins are the same species? Identify some of the variations among individuals. How can the penguins have these individual variations and still be a population?

Populations 2

A herd of mule deer moves slowly across a meadow at dusk. Nearby, a scrub jay lets out an angry squawk, and soon its voice is joined by several others. The deer suddenly stop, but the jays continue their chorus. A coyote quietly walks out of a clump of scrub oak. Because the coyote is not a threat to them, the deer do not run. Instead, they resume browsing on young twigs and leaves.

Although the participants may differ, scenes similar to this one are repeated everywhere. Each individual in these situations lives with other individuals—some like itself, some different. Many times, as with the coyote, only one individual can be seen. The coyote, however, lives with many other organisms in the meadow, and when it mates, it spends at least some time with another coyote. Individuals of the same type that interbreed with one another, like the deer or the coyotes, make up a population. In much the same way that the web of life connects individuals, it also connects populations. We, as humans, have the greatest effect on other populations because we have the greatest ability to change the environment to suit our needs. Suppose that an apartment complex were built in the meadow. How might the complex affect the deer, the scrub jays, and the coyotes? Our needs often may conflict with the needs of other organisms. This conflict has created many serious problems that all populations, including our own, must face. We are, however, the only population capable of solving the problems we have created. This chapter discusses the nature of populations, how they are interrelated, and the special characteristics of the human population.

Individuals, Populations, and Environment

Guidepost: How do populations change?

INVESTIGATION 2.1 Study of a Population

Because they reproduce rapidly and are small, yeast organisms are useful for studying populations. You will observe a population of yeast cells growing in a broth medium in a test tube.

This type of population is a closed population. In nature, open populations increase or decrease in size as organisms enter or leave them. In open populations, materials can cycle through the population. Conditions in a closed population are somewhat different, but you can estimate the rate of population growth more easily than in an open system. Using the LEAP-System™ (a computer interfacing system) with a turbidity probe, you can measure the turbidity, or cloudiness, of the medium to determine how the yeast population changes through time. By making cell counts, you can correlate changes in the number of yeast cells with the turbidity readings. Instructions for use of the LEAP-System are provided in the LEAP-System BSCS Biology Lab Pac 2.

 Read the investigation carefully and develop three hypotheses that might explain how a population changes during a period of time. Record these hypotheses in your data book and evaluate them with the data you collect during this investigation.

Materials (per team of 2)

2 pairs safety goggles	Apple IIe, Apple IIGS, Macintosh, or IBM
2 16-mm × 150-mm screw-cap test tubes, each with 10 mL sterile broth medium	computer
	LEAP-System™
	turbidity probe
2 18-mm × 150-mm test tubes	microscope
cover slip	test-tube rack
ruled microscope slide (2-mm × 2-mm squares)	metric ruler
	glass-marking pencil
1-mL graduated pipet	4 sheets of semilog graph paper
dropping pipet	

SAFETY   Put on your safety goggles.

Procedure

Day 0

1. In your data book, prepare a data table similar to Table 2.1.

2. Use the glass-marking pencil to label one screw-cap test tube *A* and the other *B*. Mark each test tube with your team symbol. Do not remove the caps.

3. Your teacher will transfer 0.1 mL of a yeast stock culture to test tube A. Gently invert the tube several times to distribute the cells evenly. Nothing will be added to test tube B. Slightly loosen the caps and store the two test tubes in the test-tube

---

TABLE 2.1 Number of Yeast Cells

Count	Day 0		Day 1		Day 2		Day 3		Day 4		Day 5		Day 6		Day 7	
Tube	A	B	A	B	A	B	A	B	A	B	A	B	A	B	A	B
First sample																
Second sample																
Total																
Average																
Dilution																
Average × dilution																
Turbidity reading																
Estimate of count																

rack and in the location specified by your teacher. What is the purpose of test tube B?

4. Test tube A contains the start of a new population. Review your hypotheses regarding what you think might happen during the next week.

5. To determine how rapidly the yeast population grows, you will take population counts during the course of the experiment. Refer to Figure 2.1 for the appearance of yeast cells. Tighten the cap and gently invert test tube A several times to distribute the yeast. Then use a dropping pipet to transfer 1 drop from test tube A to the grid of the slide. Carefully place a clean cover slip over the drop, trying not to trap any air bubbles. Examine the slide with the high-power lens of a microscope. **Note:** Yeast cells are difficult to see if the light is too bright.

6. To count the number of individual yeast cells in the center square, position the upper left corner of the square under the high-power lens and count the cells in that field. Move the slide to the upper right corner and count those cells. Continue clockwise to the lower right corner and then the lower left corner until you have counted all the yeast cells in the square. Make certain you are observing yeast cells and not other material. The cells often stick together, but count each cell in any clump separately. Buds also count as individuals.

7. Count at least 300 cells. If your count is less than 300 cells, count the cells in additional squares around the original square until you have counted 300 cells. Calculate the average number of cells per square by dividing the number of cells by the number of squares counted.

8. To obtain the number of cells per cubic centimeter (1 mL), multiply the number of cells by 2500. Each square is 2 mm x 2 mm; the cover slip is approximately 0.1 mm above the counting area; each square, therefore, has a volume of 2 mm x 2 mm x 0.1 mm = 0.4 mm³. There are 1000 mm³/cm³; therefore,

$$\frac{\text{cells}}{0.4\text{mm}} \times \frac{1000 \text{ mm}^3}{\text{cm}^3} = \frac{2500 \times \text{cells}}{\text{cm}^3}$$

To obtain the total number of cells in test tube A, multiply the final number by 10, because test tube A contains 10 mL of culture medium.

9. Have your partner count in the same way from a new sample. Record these counts in the data table and calculate the average of the two counts. Record this Day 0 population size in your data table and on the class master table.

10. Repeat Steps 5–9 for test tube B.

11. Use the turbidity probe (Figure 2.2) to take readings of both test tubes for about 10 seconds each. Remove the screw cap before placing the cap on the probe and replace after removing the test tube from the probe. Calculate the average of the readings (the computer takes one reading every second) and enter the averages under Day 0 in your data table and on the class master table. What do you predict will happen as the yeast organisms grow? Record your prediction in your data book.

12. Wash your hands thoroughly before leaving the laboratory.

Days 1–7

13. Invert test tube A several times to distribute the yeast organisms evenly. Using the turbidity probe, take readings for about 10 seconds. Record the average of the readings under Day 1 in your data table and on the class master table. Repeat the same procedure with test tube B.

14. Repeat the counting procedure in Steps 5–9, using one drop from test tube A and then one drop from test tube B. Record your data under Day 1 in your data table and on the class master table.

15. Based on the information you now have, you can start a graph that will help you estimate the yeast population size from the turbidity readings on the days you do

Figure 2.1 Yeast (×600). Note the buds still attached to the yeast cells. These will become new individuals.

Figure 2.2 The LEAP-System with turbidity probe.

not make counts. Draw *x* and *y* axis lines on a sheet of semilog graph paper. Label the *y* (vertical) axis *Number of Yeast Cells*. Label the *x* (horizontal) axis *Turbidity Readings*. The values on the *y* axis should be from 10^5 to 10^9. The values on the *x* axis should be from 0 to 100.

16. Make a mark on your graph at the point where the Day 0 turbidity reading and the yeast count for test tube A intersect. Label this point *Day 0*. Make a second mark on your graph at the point where the Day 1 turbidity reading and the yeast count intersect. Label this point *Day 1*. Use a ruler to draw a line connecting the two marks. Extend the line until it reaches both edges of the graph. This line now can be used to help you estimate your yeast counts on Days 2 and 3. Repeat for test tube B, using a dotted line.

17. Repeat Steps 13 and 16 each day (skip the weekend). Include Step 14 on Days 4 and 7. If your count is more than 300 or too many to count, make a dilution as follows. Use the 1-mL graduated pipet to transfer 0.9 mL of water to a clean test tube; label this test tube *D1*. Invert test tube A several times until the organisms are evenly distributed, and immediately transfer 0.1 mL to test tube D1. Invert test tube D1 several times to mix, and then use the dropping pipet to transfer one drop to the grid of the slide. Add a cover slip as in Step 5 and count the cells as in Step 6. If necessary, make another dilution by transferring 0.1 mL from test tube A to a fresh tube containing 1.9 mL of water; label this tube *D2*. Calculate the population size as in Step 8. What additional calculation must you make if you have made the first dilution? The second dilution? Record the calculated results under the appropriate day in the data table.

18. On Days 4 and 7, make an additional mark on your graph for each test tube at the points where your yeast counts and the turbidity readings intersect. If necessary, modify your lines after each mark is entered on the graph.

19. On Day 7, calculate the average of all the team readings from the class master table.

20. Each day, wash your hands thoroughly before leaving the laboratory.

Discussion

In this investigation, you estimated the population size in your test tube by counting the individuals in a sample. This method is called sampling. You also obtained an estimate based on the turbidity in the test tube. To increase the accuracy of your estimates and the computer readings, you took certain precautions. You inverted the test tube several times to distribute the organisms evenly. You used a counting procedure to estimate the population size on Days 0, 1, 4, and 7. The averaging procedure smoothed out chance differences between your counts. You also estimated the population size based on the turbidity readings obtained with the computer. On the class master table, you averaged the data obtained by all teams. That further smoothed out chance differences among the test tubes. The final count for each day's population is the average number of organisms predicted by the computer data.

1. On a fresh sheet of semilog graph paper, list the ages of the cultures (in days) on the horizontal axis. List the number of organisms per estimate on the vertical axis. Plot your own data for test tubes A and B, and then use a different color to plot the averages of all teams for both test tubes.

2. On the basis of your discussion of this investigation, explain similarities and differences among the graph lines representing data from different teams.

3. Is there any general trend in the graph line representing the average data of all teams? If so, describe it.

4. Review the three hypotheses you developed before starting this investigation. Are any of them supported by the data that you have accumulated? Are any of them not supported? Explain.

5. Why was it necessary to make the yeast counts on Days 0, 1, 4, and 7?

6. Were there changes in turbidity in test tube B? How can you account for those changes? How do those changes affect the accuracy of your readings for test tube A?

7. What limiting factors could have influenced the growth of the yeast population?

For Further Investigation

How does temperature affect the growth of a yeast population? Repeat the procedures, but incubate the cultures at a constant temperature 15° C above or below the average temperature at which the tubes were incubated before. Or, place one set of test tubes in a water bath at 30° C, and another set in a water bath at 50° C. On Day 2 of the experiment, place one set of test tubes in the freezer and another in the refrigerator. After one or two weeks, remove these test tubes, make counts (after the frozen tube thaws), and continue the experiment.

2.1   Populations Are Made Up of Individuals

In general, life processes occur in separate "packages." You are such a package—an individual. You carry on the activities of life within your body apart from the life processes that occur in the bodies of your parents, brothers, or sisters. Each cow in a herd, each corn plant in a field also is an individual.

A group of interbreeding individuals of the same type that live in a particular area is called a population. To define a population, you need to identify the type of individuals, the time, and the place. Thus, for example, you can refer to the population of pigeons in Cincinnati, Ohio, in 1989, the number of spring beauty flowers in an Oklahoma field in March, or the lady bird beetles shown in Figure 2.3. When biologists want to test hypotheses, they usually study populations. They do so because the individuals in a population usually vary. Looking at only a few individuals could

Figure 2.3   A population of ladybird beetles. How many differences can you see within this population? What is the range in number of spots present? What important characteristic of populations does this illustrate? What is responsible for these differences?

be misleading. For example, a biologist who studied only a few human individuals who happened to have brilliant blue eyes might jump to the conclusion that all humans have brilliant blue eyes. It is important to look at enough individuals to be sure the conclusions and observations hold true for the population.

By studying populations, scientists often can see what is happening throughout a community. As one population changes in size, it may affect many other populations. The more we know about one population, the more we can predict about others.

2.2   Four Rates Determine Population Size

The size of a population changes through time. Suppose a biologist counted 700 ponderosa pines on a hill in Colorado in 1980 (see Figure 2.4). In 1990, when the biologist counted the trees again, there were only 500. In other words, there were 200 fewer trees in 1990 than in 1980, a decrease in the population of ponderosa pines. This change in population may be expressed as a rate—the amount of change divided by the amount of time for the change to take place. The rate is an average. In this example, the rate of change in the number of trees divided by the change in time may be expressed as: -200 trees/10 years $= -20$ trees per year. To the biologist, this means each year there were 20 fewer trees in the population. Keep in mind, however, that this rate is an average. It is unlikely the trees disappeared on such a regular schedule. All of the trees may have been lost in one year due to a fire, or the decrease may have been caused by selective cutting during several years.

What does the decrease of 200 pine trees in 10 years represent? Because pine trees cannot wander away, they must have died or have been cut down. In this situation, then, the decrease represents the death rate, or **mortality** (mor TAL ih tee) rate, of the pine population. The number of deaths in the pine population per unit of time is the mortality rate. Mortality is not the only change that can affect a population, however. While some of the pines may have died, some young pine trees may have started to grow from seed. Death decreases a population; reproduction increases it. The rate at which reproduction increases the population is called the birthrate, or **natality** (nay TAL ih tee).

Organisms that can move have two other ways to bring about a change in population size. If you were studying the pigeon population in your city or town, you might discover that a certain number of pigeons flew into the city in one year and a certain number flew out. **Immigration** (im uh GRAY shun) occurs when one or more organisms move into an area where others of their type can be found. Immigration increases the population. **Emigration** (em uh GRAY shun) occurs when organisms leave the area. Emigration decreases the population. In any population that can move, then, natality and immigration increase the population, and mortality and emigration decrease the population. Thus, the size of any population is the result of the relationships among these rates.

Natality, mortality, immigration, and emigration rates apply to every population, including the human population. The sum of these rates makes up the growth rate of a population.

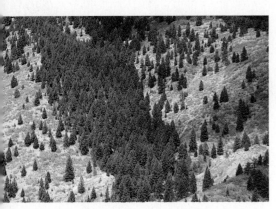

Figure 2.4 How did the biologist count the number of trees on this hillside?

2.3 The Environment Limits Population Size

In nature, a population varies between some upper and lower limit, depending upon its natality, mortality, immigration, and emigration rates. Each population is like a swing moving back and forth between two points, or between a high and a low number. The lower limit is, of course, zero. At that point, the population no longer exists. What is the upper limit on population size? It depends on the **environment** (en VY run ment), everything that surrounds and affects an organism. The environment may slow the individual's growth, may kill the individual, or may stimulate the individual's growth and reproduction. In any case, the environment affects individuals and, thus, the ultimate size of a population.

The environment is made up of two parts: the living part and the non-living part of an organism's surroundings. The living (or recently living) part is called the **biotic** (by OT ik) environment. For you, the biotic environment includes your neighbors, houseplants, the dog, the fleas on the dog, and all the organisms you eat. The nonliving part is called the **abiotic** (AY by OT ik) environment. The abiotic environment includes such things as living space, sunlight, soil, wind, and rain. Both biotic and abiotic factors affect the size of a population. Any biotic or abiotic factor that can slow the growth of a population, such as temperature, is called a **limiting factor**.

The climate is a group of abiotic limiting factors (weather factors) that affect all plants and animals. Weather factors include temperature, sunlight intensity, precipitation (rainfall, snowfall, fog), humidity (amount of moisture in the air), and wind. Although each factor may be measured alone, each factor affects the other, and together, they affect population size. The effect may be either direct or indirect.

In the northern part of the United States, for example, the first heavy frost kills almost all adult mosquitoes. In the autumn, when the population of insects in the north has dropped, swallows fly south. These population decreases are related to one weather factor—temperature. Temperature affects the mosquito population directly by killing most mosquitoes. By limiting the swallows' food supply, however, temperature affects the swallow population indirectly.

Water is another important abiotic factor. All living organisms need water. Almost all chemical reactions needed to keep an organism alive take place in water, and indeed, water molecules are a part of many chemical reactions. Although a few organisms can survive by becoming inactive when there is no water, most organisms die. Because water is essential for all living organisms, it often is a limiting factor. Seedlings, such as those shown in Figure 2.5, grow and gradually establish themselves when they receive enough water. Because of other climatic factors, however, adequate rain and snow alone may not ensure enough water. Wind, for example, speeds up the rate of water evaporation. Low humidity in the desert also results in an increased evaporation rate. Thus, high rates of evaporation can affect the survival of certain organisms, even when precipitation is adequate. In some cases, a high rate of evaporation together with low rainfall may permit only plants with special features to survive.

Another abiotic limiting factor for populations is space. Although every individual needs living space, some organisms need less space than others. For example, individual corn plants grow well when they are planted closely

Figure 2.5 A pine seedling. What environmental factors are important to the survival of this seedling?

Figure 2.6 (a) Plan of an orchard. Each dot represents one tree. What is the density of trees in the orchard? (b) Under natural conditions, organisms are rarely distributed evenly. Calculate the density of dusty clover in the field as a whole and then only in the northwest corner. Compare.

together. A mountain lion, on the other hand, usually requires many square kilometers for its range. In general, organisms that move about, such as animals, need more space than stationary organisms, such as plants.

The amount of space needed by a single organism is tied in part to a biotic factor—the availability of food energy. However, it also is affected by population **density**, the number of individuals in relation to the space the population occupies (see Figure 2.6). For example, in a classic experiment, illustrated in Figure 2.7 on page 35, mice in cages were given more than enough food each day. As the mice reproduced, the density of the population increased, and the cages became very crowded. Some female mice stopped taking care of their nests and young. Mice continued to be born, but many newborn mice died from neglect. Eventually, mortality of the young mice reached nearly 100 percent. The high mortality of the mouse pups kept the population density from increasing further.

2.4 Abiotic and Biotic Factors Work Together to Influence Population Size

Not only do abiotic and biotic factors in the environment influence one another, they also affect the size of a population. For example, a population is limited by the amount of matter and energy, or **resources**, available to it. In any particular space, even under the best conditions, there is a limit to matter and energy. In other words, resources are finite, or limited.

Figure 2.7 In this experiment, mice were provided with more than enough food (left). As a result (right), the population has grown dramatically. Predict what will happen if the mice experience a food shortage as well as overcrowding.

The organism's relationships with others around it can affect the availability of these resources. Predators, disease, and competition with other organisms for scarce resources may take their toll on a population. If an animal population is so large that it uses all the available water or food, it probably will decrease in size. Rabbits that exhaust the supply of grass and other foodstuffs in their area, for example, may die from starvation and disease. Usually, however, the size of the rabbit population is controlled by other populations in the environment. In this case, because of the coyotes that live in the same area, the rabbit population probably will not grow large enough to exhaust the food supply. Environmental stress brought about by abiotic factors, such as extreme cold or heat, also may limit population size.

The greatest number of individuals that a space can support indefinitely without degrading the environment is called its **carrying capacity**. Although a larger number of individuals may be supported temporarily in the space, the environment will suffer. Ultimately, the carrying capacity of the environment is the most important factor in determining population size.

A given valley in the mountains, for example, has a carrying capacity of only so many deer. Suppose the deer population there increases. Gradually, the types of plants the deer eat become scarce. A few deer may emigrate and look for food in other places. Those that stay in the valley become undernourished and weak. They are easily caught by mountain lions, killed by disease, or become too weak to search for food when the snow is deep and therefore starve. Because of this emigration and mortality, the deer population decreases, which means that fewer plants are eaten. As the plants that remain continue to grow, more food becomes available for the surviving deer. Deer stop emigrating, and a few hungry deer even move into the valley from the surrounding hills. All the deer now find plenty of food and become healthy. Mountain lions may still catch some of these healthy deer, but the other factors influencing deer mortality decrease. With less mortality and more immigration, the deer population again increases.

This deer population example illustrates **homeostasis** (hoh mee oh STAY sis), the tendency for a population to remain relatively stable in size.

Figure 2.8 Calculate the changes in the rat population between 1946 and 1947, 1947 and 1948, 1948 and 1949, and 1949 and 1950. What might account for these changes?

Through time, the deer population remains between the upper and lower limits that are defined by mortality, natality, and population movements.

2.5 Population Density May Fluctuate

Any population has a built-in, characteristic growth rate—the rate at which the group would grow if food and space were unlimited and individuals bred freely. Environmental factors do affect a population's growth rate, however. The interaction of the population's natural growth rate and the environment determines the density of the surviving population.

If you measure the density of a population at intervals during a given period of time, you seldom find any two consecutive measurements the same. Density increases or decreases continually. Most natural populations are open populations, or populations in which individuals are free to emigrate or immigrate and in which the birth and death rates fluctuate.

The graph in Figure 2.8 is based on data collected while doing studies of Norway rats in Baltimore, Maryland. In 1942, the city's health department conducted a poisoning campaign that seemed to wipe out the rat population in the city block from which the data were collected. Because it is difficult to count individuals in natural populations, it is possible that a few rats survived or that a few immigrated to the block from elsewhere. In either case, a new rat population was present and increasing by early 1945.

Look at the line for the later years of the rat study. The population decreased after a peak density was reached, yet the decrease did not continue. Open populations usually increase again, just as the rat population did. After an open population peaks, it again decreases, and then may stabilize. Natural populations normally show fluctuations, or ups and downs. Variables in the environment, such as climate, available food, or the activities of natural enemies, are the causes of the fluctuations.

Sometimes population fluctuations are fairly regular, and the peaks on a graph are at approximately equal distances. For example, populations of lemmings (see Figure 2.9) often peak every three or four years. Many of the animals that live in the northern parts of Europe, Asia, and North America show similar population cycles. Although the data show very regular cycles when they are plotted on a graph, the reasons for the seemingly regular cycles are not well understood. A combination of purely chance events also can produce apparently regular cycles.

Although populations may change cyclically, many population changes are permanent. If a population becomes extinct, for example, the change is permanent. Any permanent change in a population is a change in the community to which the population belongs. Permanent changes in one population also may affect other populations of organisms in the same community.

$\times$ 1/2

Figure 2.9 Lemmings are mouselike animals that live in the northern parts of North America, Europe, and Asia. What factors might affect the size of the lemming population?

INVESTIGATION 2.2 Population Growth

Four rates—natality, mortality, immigration, and emigration—determine a population's size. A given area can support only a certain number of individuals on a long-term basis. That number is known as the carrying capacity. Populations grow at different rates; the *doubling time* is the number of years required for a particular population to double its size. In this investigation, you will compare the growth of two populations and examine the roles played by carrying capacity, doubling time, and the four rates that determine a population's size.

Materials (per team of 3)

pocket calculator
graph paper

PART A Reindeer Population
Procedure

1. In 1911, 25 reindeer—4 males and 21 females—were introduced onto St. Paul Island, one of the Pribilof Islands in the Bering Sea near Alaska. St. Paul Island is approximately 106 km² in size (41 square miles), and is more than 323 km (200 miles) from the mainland. On St. Paul Island there were no predators of the reindeer, and no hunting of the reindeer was allowed. Study the graph in Figure 2.10 and answer the following questions (Steps 2 through 7) in your data book.

2. What was the size of the population at the beginning of the study? In 1920? What was the difference in the number of reindeer between 1911 and 1920? What was the average annual increase in the number of reindeer each year between 1911 and 1920?

3. What was the difference in population size between the years 1920 and 1930? What was the average annual increase in the number of reindeer in each of the years between 1920 and 1930?

4. What was the average annual increase in the number of reindeer in each of the years between 1930 and 1938?

5. During which of the three periods—1911–1920, 1920–1930, or 1930–1938—was the increase in the population of reindeer greatest?

6. What was the greatest number of reindeer found on St. Paul Island between 1910 and 1950? In what year did this occur?

7. In 1950, only eight reindeer were still alive. What is the average annual *decrease* in the number of reindeer in each of the years between 1938 and 1950?

Figure 2.10 Changes in the reindeer population on St. Paul Island between 1911 and 1950.

Discussion

1. Could emigration or immigration have played a major role in determining the size of the reindeer population? Explain your answer.

2. What might account for the tremendous increase in the population of reindeer between 1930 and 1938, as compared with the rate of growth during the first years the reindeer were on the island?

3. What effect might 2000 reindeer have on the island and its vegetation?

4. Consider all the factors an organism requires to live. What might have happened on the island to cause the change in population size between 1938 and 1950?

5. Beginning in 1911, in which time spans did the population double? How many years did it take each of those doublings to occur? What happened to the doubling time between 1911 and 1938?

6. If some of the eight reindeer that were still alive in 1950 were males and some females, what do you predict would happen to the population in the next few years? Why?

7. What evidence is there that the carrying capacity for reindeer on this island was exceeded?

8. What does this study tell you about unchecked population growth? What difference might hunters or predators have made?

PART B   Human Population
Procedure

8. On a piece of graph paper, plot the growth of the human population using the data in Table 2.2.

9. Use your graph to determine the doubling times for the human population between AD 1 and 1990. How much time elapsed before the human population of AD 1 doubled the first time? Is the amount of time needed for the human population to double increasing or decreasing? What does that indicate about how fast the human population is growing? Record your answers in your data book.

10. Extend your graph to the year 2000. What do you estimate the human population will be in that year?

TABLE 2.2   Human Population Growth Between AD 1 and 1990

Date AD	Human Population (millions)	Date AD	Human Population (millions)
1	250	1920	1800
1000	280	1930	2070
1200	384	1940	2300
1500	427	1950	2500
1650	470	1960	3000
	(Black Death)	1980	4450
1750	694	1985	4850
1850	1100	1990	5300
1900	1600		

Source: American Association for the Advancement of Science, 1951

11. Using the equations below, estimate the doubling time for the current population based on the rate of growth from 1980 to 1990. In what year will the present population double?

$$\text{Rate of growth (in percent)} = \frac{(\text{population in 1990} - \text{population in 1980}) \times 100}{\text{population in 1980} \times \text{number of years}}$$

$$\text{Doubling time} = \frac{70}{\text{rate of growth}}$$

Discussion

1. What similarities do you see between the graph of the reindeer population and your graph of the human population?
2. What are the three or four most important factors required to sustain a population?
3. In what ways is the earth as a whole similar to an island such as St. Paul? Does the earth have a carrying capacity? Explain your answer.
4. What might happen to the population of humans if the present growth rate continues?
5. What methods could be used to reduce the growth rate?
6. Cite a place in the world where population growth is a problem today. How is it a problem?
7. Cite a place in the world where population growth is *not* a problem today. Why is it not a problem?
8. Suggest several problems in your country that are related to the human population.
9. What are the most important three or four factors to think about with regard to the world population?

Figure 2.11 How are the environments of the polar bear and the penguin similar? How are they different?

2.6 Populations May Spread to Neighboring Areas

Environmental factors keep populations in a specific area between upper and lower limits, but what factors restrict the spread of populations into new areas? For example, although the climate and other abiotic conditions at the North Pole and the South Pole are similar, polar bears roam the ice floes only in the Arctic, and penguins are found only in the Antarctic (see Figure 2.11). Why are there no penguins near the North Pole and no polar bears near the South Pole?

Apparently an organism's ability to tolerate some environmental factors, such as cold climates, does not entirely explain its geographic range, or where it lives. Just because an organism *can* live in a particular place does not mean that it *does* live in that place. To understand the geographic ranges of penguins, polar bears, and other organisms, we need additional information.

All populations of living things have the tendency to spread from one place into others. This spreading of organisms is called **dispersal**. Have you ever noticed, for example, how a single dandelion in your yard can turn into 50 dandelions, seemingly overnight? Similarly, the European starling,

Figure 2.12 Dandelion seeds. Do dandelions emigrate or immigrate?

an exotic species, was introduced into this country on the East Coast, but it now is found everywhere.

In the case of organisms that can move, dispersal may occur through flying, swimming, walking, running, crawling, or burrowing. For organisms that cannot move, dispersal is passive. In other words, these organisms are spread from one place to another by other agents (see Figure 2.12). Seeds, spores, and eggs can remain alive for long periods. During this time, they may be carried great distances by air or water currents. They even may travel in the mud on a bird's foot. Sometimes organisms that can move may be transported by other agents much farther than they can travel themselves. Floating ice may carry a polar bear hundreds of kilometers, and a spider may be blown long distances by the wind. Of course, dispersal alone does not change an organism's geographic range. Unless the organisms are able to survive and reproduce in the new location, their range has not changed.

2.7 Barriers Can Prevent Dispersal

Because every type of organism has some means of dispersal, you might expect that all species would be found wherever environmental conditions were suitable. Actually, however, a broad geographic range is the exception, not the rule. What limits the dispersal of populations?

Physical barriers, such as mountains, prevent dispersal. For most terrestrial (land-dwelling) organisms, large bodies of water are effective barriers. On the other hand, for aquatic animals, land may be a barrier. Besides these physical barriers, there are ecological barriers. For organisms that live in the forest, grasslands and deserts may be barriers. Some barriers are behavioral. You might suppose that flying birds are found everywhere, but many birds that can fly great distances remain in very restricted regions. In these cases, distribution apparently is limited by the behavior of individuals in selecting their living area. For example, the tree pipit (*Anthus trivialis*) and the meadow pipit (*A. pratensis*) live in similar areas, except that the tree pipit breeds only in areas having at least one tall tree. Both pipits nest on the ground, and they feed on the same types of organisms. Both have similar songs when they fly, but at the end of its song, the tree pipit perches in a tree, and the meadow pipit lands on the ground (see Figure 2.13). This difference in behavior limits the dispersal of the two species.

An increase in a population of organisms that can move encourages emigration to less populated areas that may have resources. Sometimes dispersal takes place rapidly as the population grows. If the barriers are great, however, dispersal may be slow even though the population is increasing rapidly.

For organisms or their seeds that undergo passive dispersal, the means of transportation help to determine the rate of spread. Some types of trees with seeds that are carried away and buried by squirrels are estimated to spread only about 1.6 km in 1000 years. By contrast, organisms that are caught in a tornado may be carried 30 km in a few hours.

How does this discussion of dispersal and barriers explain the absence of polar bears from the Antarctic? First, assume that polar bears originated in the Arctic and that the population, like many populations, tended to disperse. Although polar bears could tolerate the cold temperatures at the

Figure 2.13 The distribution of tree and meadow pipits is limited by behavioral barriers.

South Pole, the temperate environment they would have to travel through to reach the South Pole would be too hot. Therefore, it would act as a barrier to their dispersal. This wide barrier has existed for at least as long as polar bears have existed. Thus far no part of the population has been able to move across it successfully. Similar reasoning may be applied to lack of dispersal by penguins.

Limiting factors not only shape the survival of populations but also the environments in which they can live. If a population has the characteristics needed for survival in an area, the individuals in that population will grow and reproduce. If not, the individuals will be unable to produce enough offspring to maintain the population. Eventually, the population will no longer exist in that area.

Concept Review

1. What components must be identified in order to define a population?
2. How do limiting factors affect population growth?
3. How does the availability of resources affect the growth and size of a population?
4. In what ways is water important to living things?
5. Explain the concept of carrying capacity and relate it to limiting factors.
6. In what ways is the density of a population important?
7. What barriers might affect the dispersal of a population?
8. What factors interact to determine the geographic range of a population?

Human Populations

2.8 Few Barriers Prevent Human Dispersal

You are one of more than 5.3 billion living humans on the earth today. In 1989, the human population grew by an estimated 87.5 million people—more people than live in California, New York, Texas, and Pennsylvania combined. Experts have estimated that at the current rate of growth, the human population on earth in the year 2025 will be double what it was in 1975. In other words, when you are about 50 years old, the world may be twice as crowded as it was when you were born. Think about what would happen if the student population in your school doubled, but the school's resources remained the same. You would have twice as many students, but the same number of teachers, classrooms, desks, and lunches. Could your school handle the increase? What would happen if the population doubled again?

Just like the student population in the crowded school, the human population keeps growing, but the available resources do not. Why is the human population growing so rapidly? Look at Figure 2.14 on the following page. Simply put, natality exceeds mortality. Our ability to control many diseases and provide a reliable food supply has resulted in a decrease in mortality for humans. In other words, we have temporarily disrupted homeostasis for the human population. More people produce more babies, and the human population continues to grow.

Guidepost: What are the characteristics of the human population, and how is this population affected by limiting factors?

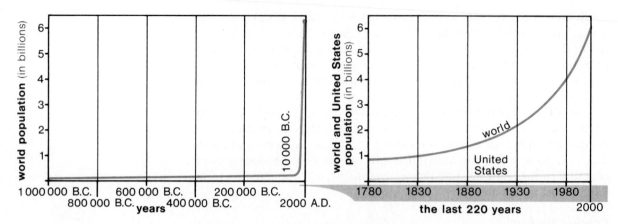

Figure 2.14   Human population growth from 1 000 000 BC to the year 2000. What might account for the abrupt increase in numbers?

Although the geographic range of a population is determined by its environment, the dispersal of the population within that range is affected by various barriers. Polar bears are not found in the Antarctic because they cannot tolerate heat, and they would have to disperse through a temperate climate. Think about all the places on the earth where humans are found. What barriers affect the dispersal of the human population? If we did not modify our environment by using central heating and air conditioning, where would humans be found? Would we be so widespread if we did not have cars, planes, trains, and ships?

Technology has allowed the human population to occupy every environment on earth and to move among these environments. Through technology, we have eliminated or reduced the incidence of some diseases that helped to control population growth in the past. Therefore, we now live longer on the average. The development of this technology has consumed many resources—resources that are limited. Given a growing population and a finite supply of resources, what may happen?

2.9   Earth's Carrying Capacity Is Limited

Food, water, and available space are some of the limiting factors that can affect the size and growth rate of the human population. Like all areas, the earth as a whole has a carrying capacity. Scientists do not agree about the exact carrying capacity of the earth for humans. If food were the only factor, some people claim the earth could support 7 to 8 billion people. Other people think that not even the current population can be maintained for long. Even with these disagreements however, all agree that the human population cannot grow forever. In areas where the carrying capacity for humans has been reached, population demands for resources—the necessities of life—have exceeded the supply. In these areas, there have been dramatic population crashes.

Humans need space for living. Although experiments on overcrowding, such as the one discussed in Section 2.3 involving mice, help to predict what may happen to other populations under similar conditions, it

Figure 2.15   As the population increases, is land more valuable for growing food, housing people, or as habitat for other organisms? What might be responsible for this warning?

usually is difficult to apply the results directly to humans. Part of the space needed by humans for living must be used to grow food (see Figure 2.15). Unfortunately, the amount of land on which food can be grown decreases each year because other demands for space increase. People want more space for their houses and other buildings, but each house occupies about one-eighth of a hectare. Eight houses occupy one hectare, which is enough land to provide food for three people.

Water is another limiting factor for the human population, just as it is for other populations. Without an adequate supply of water, humans die. Furthermore, the water must be sanitary. Because many diseases are spread by water, a source of pure drinking water is vital to the health of humans and other animals. Unclean conditions only increase the risk of exposure to disease. In some parts of the world, the only water available for drinking contains disease-causing organisms and traces of human wastes. Toxic chemicals that have seeped into wells and reservoirs of drinking water also can cause illnesses. In some parts of the United States, toxic chemicals have threatened the purity of water supplies.

INVESTIGATION 2.3   Water—A Necessity of Life

In this investigation, you will calculate how much water you use on a daily basis and then calculate the water needs for your family, your class, your school, your town, and your state. You also will discover some of the ways in which you use water indirectly.

Materials (per person)
pocket calculator (optional)
graph paper

Procedure
1. In your data book, record how many gallons of water you think you use in an average day. Later, you will compare this *estimated* daily water use with your *calculated* daily water use.

TABLE 2.3 Domestic Uses of Water

Activity	Amount Used (gallons)
Automatic dishwasher	15
Brushing teeth	2–10
Cooking a meal	5–7
Flushing toilet	3.5–8
Getting a drink	0.25
Housing cleaning	7
Leaking faucet	25–50/day
Shaving	20 (2/min)
Showering	20–25 (5/min)
Tub bathing	25–35
Washing dishes	30 (8–10/meal)
Washing machine	24–50
Washing hands	2
Watering lawn	10/min (102/1000 m²)

Faucet and toilet
leaks in New York City = 757 million/day

Source: From *Living in the Environment: An Introduction to Environmental Science,* 6th ed., by G. Tyler Miller, Jr. © 1990 by Wadsworth, Inc. Used by permission.

2. As a group, list all the ways in which the members of your class use water on a day-to-day basis.

3. Using the data in Table 2.3, determine your *individual* water use per day for each activity your class listed. Include your share of general family uses such as dishwasher and washing machine. Then determine your individual *total* water use per day.

4. Compare the individual water use you calculated with the water use you estimated. Is your calculated figure higher or lower than your estimated figure?

5. Find out how many people are in your school, including teachers and students. Find out how many people live in your town or city and your state.

6. Calculate the total amount of water that is used each day by your family, your class, your school, your town, and your state, for each of the activities the group listed in Step 2.

7. Calculate the total water use per day for all the activities combined for your family, your class, your school, your town, and your state.

8. Draw a bar graph to illustrate how much water is used by your class for each activity. Which activities require the most water?

Discussion

1. Many water uses are not obvious to most people. Consider, for example, how much water is necessary to raise one calf until it is fully grown (see Table 2.4). Why do you think so much water is needed to raise a calf?

2. Make a list of the ways you use water indirectly—for example, in the production of the food you eat or the materials you use.

TABLE 2.4 How to Make a Cow

Ingredients*

1	80-lb calf	2500	lbs corn
8	acres grazing land	350	lbs soybeans insecticides herbicides antibiotics hormones
12 000	lbs forage		
125	gals gasoline and various petroleum byproducts		
305	lbs fertilizer (170 lbs nitrogen, 45 lbs phosphorus, 90 lbs potassium)	1.2	million gals water, to be added regularly throughout
1.5	acres farmland		

*These are the ingredients used to produce choice beef. Lesser amounts are required for today's leaner beef.

Take one 80-pound calf, allow to nurse and eat grass for 6 months, then wean. Over next 10 months, feed 12 000 pounds of forage. Use about 25 gallons of petroleum to make fertilizer to add to the 1.5 acres of land. Set aside rest of gasoline to power machinery, produce electricity, and pump water. Plant corn and soybeans; apply insecticides and herbicides. At 24 months, feed cow small amounts of crop and transfer to feedlot. Add antibiotics to prevent disease and hormones to speed up fattening. During next 4 months, feed remaining crop mixed with roughage. Recipe yields about 440 usable pounds of meat—1000 7-ounce servings. Option: Bake the 2500 pounds of corn and 350 pounds of soybeans into bread and casseroles—18 000 8-ounce servings.

Source: Excerpted from *The Cousteau Almanac* by Jacques-Yves Cousteau. Copyright © 1980, 1981 by the Cousteau Society, Inc. Reprinted by permission of Bantam, Doubleday, Dell Publishing Group, Inc.

TABLE 2.5   Indirect Uses of Water

Agricultural		Industrial	
Item	*Gallons used*	*Item*	*Gallons used*
1 loaf bread	150	1 kg aluminum	2200
1 kg corn	374	1 car	100 000
1 kg cotton	4440	1 kilowatt of electricity	80
1 kg grain-fed beef	1760	1 gallon gasoline	10
1 kg rice	1232	1 kg paper	220
		1 kg steel	25
		1 kg synthetic rubber	660

Source: From Living in the Environment: An Introduction to Environmental Science, 6th ed., by G. Tyler Miller, Jr. © 1990 by Wadsworth, Inc. Used by permission.

3. Compare your list with Table 2.5. How many of the indirect uses of water shown in the table did you list?
4. How could you reduce your indirect use of water?
5. What could you do to reduce your direct use of water?
6. Is there any evidence that the water supply you use daily is decreasing in size or is being contaminated by pollutants? How could you go about obtaining this information?

For Further Investigation

1. Research actual water use for one of the items you listed in Question 2.
2. Examine your family utility bills for the past year and note the units of water used in each month. Each unit represents one cubic meter of water. If one gallon equals 0.004 cubic meters, calculate your family's average monthly water use in gallons. How does actual use compare with your calculations in this investigation?

2.10   Uneven Distribution of Food May Limit Population Size

In many parts of the world today, the food supply is a major limiting factor for humans. The overall production of food in the world may be enough to feed all the people, but food resources are unevenly distributed. Sometimes, getting enough of the right type of nutrition is not easy. Not all people have equal access to the same quality food sources. Living organisms need a wide variety of foods for maintenance and growth. If adequate amounts of food are available, a population may grow. Excess amounts can be harmful. Eating too much of a particular fat, for example, may lead to heart disease. Scarcity, on the other hand, may limit the growth of a population. Many places in the world experience periodic times of **famine** (FAM in), a great shortage of food. This lack of food may slow growth or even cause death. In such places as Asia, Africa, and Latin America, the number of people who must be fed is increasing far more quickly than the amount of food that can be grown to feed them.

Food energy is usually measured in units called calories. A calorie is the amount of energy required to raise the temperature of one gram of water one degree Celsius. The calorie you probably are familiar with, however, is the food calorie, which is really a kilocalorie or kcal—1000 calories. Teenage females need 1200 to 3000 kcals per day, depending on how active they are. Teenage males need 2100 to 3900 kcals, and a football player may consume more than 6000 kcals.

Although life-styles have changed drastically in industrialized countries in the last 100 years and many people have become less active, they have continued to eat about the same number of kcals. In other parts of the world, however, some people are not able to obtain enough kcals to maintain normal activity levels. Without a constant and sufficient supply of kcals, activity levels drop, and eventually muscles are broken down by the body to supply cells with energy.

The number of kcals is not all that is measured when judging the nutritional value of food. Humans need certain foods that are not abundant in plants. A meal of only corn bread or white rice is not nutritious. Animal products may provide enough nourishment and energy, but animals are costly to produce and to buy. To provide a human with 3000 kcals of food that meets all nutritional requirements, a farmer must raise about 30 000 kcals in plant substance. In the United States, almost 90 percent of those 30 000 kcals is used to feed animals raised for meat. In other parts of the world, many people cannot afford to buy meat or raise animals for meat, so they lack both the kcals and the proper foods needed for good health. This problem is not limited to any one part of the earth. Overall, the amount of land that can be farmed is limited; therefore as the population increases, the amount of land available for farming decreases. Although a small amount of land can be very productive, it is costly to maintain.

Most farmland in the United States, for example, is very productive, but this high productivity is maintained only by using huge amounts of fuel energy and fertilizers. Nonmechanized farming practices in tropical and warm climates may return 10 to 20 kcals in food per kcal of energy spent. In the United States, however, for each kcal of energy used to grow food, we get back only 0.1 kcal of energy. In other words, more energy is used to grow and transport food than is gotten back from the food itself. To continue to produce food in this way, we must use our supplies of oil and gas. Oil and gas, however, are nonrenewable resources—materials that never can be produced again. If we use all of these resources now, we will be unable to support our agricultural system in the future. In other words, we cannot keep growing food indefinitely with an energy-losing system.

2.11 Human Activities Change the Environment

Humans have modified the environment more, and in a shorter period of time, than any other organism has. For example, while the composition, temperature, and self-cleansing ability of the earth's atmosphere have varied since the formation of the planet, the pace of the changes in the past two centuries has quickened remarkably. These changes include "acid rain" and similar processes, urban smog, and a thinning of the protective ozone

Figure 2.16 How does the removal of these trees affect the environment?

layer that shields the earth from harmful radiation. Some scientists think the planet soon may begin to warm rapidly (see Chapter 4, Biology Today), causing drastic climatic changes. Certainly, some changes in the atmosphere can be attributed to natural causes, but the activities of humans account for the most rapid changes. These activities include the burning of fossil fuels (coal and petroleum) for energy, slash-and-burn agricultural methods, and deforestation (see Figure 2.16).

Human activities also affect the water supply. As the demand for water by the human population becomes greater than the supply, shortages will occur. Depletion of groundwater, for example, already is common in India, China, and the United States. Because deforestation removes the tree roots that hold soil in place and help retain water, deforestation leads not only to increased soil erosion but also to increased water runoff and flooding. Instead of entering the groundwater, rain runs into rivers.

2.12 Managing Earth Requires Global Cooperation

Our interactions with the environment are complicated by the sheer number of humans on earth. These numbers have increased the rate, scale, and complexity of those interactions. What once were local incidents of pollution now often involve several countries. For example, acid rain falls in most countries of Europe and North America. These days, preservation of the environment has many economic and political aspects because human development has led to environmental changes that affect the entire planet.

To move from causing these changes to managing them, we must ask three questions: What type of planet do we want? What type of planet can

Famine in Africa

In late 1984, the world learned of a severe food shortage in Africa. The shortage was greatest in the Sahel, a swath of land cutting across the northern desert. By early 1985, the famine had killed 300 000 people in Ethiopia and another 200 000 in Mozambique. Officially more than 30 countries in Africa are listed as famine areas. In the most desperate of these, whole populations may be wiped out. How did this happen?

This worsening situation is caused by a number of interacting factors:

■ Worldwide food distribution is uneven. If all the food produced in the world were divided equally, there would be enough to keep 6 billion people alive (2.5 billion on a diet typical of developed countries). Food is distributed unevenly because of differences in soil, climate, and income.

■ Africa has the fastest population growth of any continent. Every three weeks an additional one million people are added to the population.

■ A severe, 17-year drought has compounded the problem. Rainfall in the region can vary by as much as 40 percent from one year to the next. Prolonged droughts are not uncommon.

■ Much of Africa has poor soil. Three-tenths of the land is covered by desert or is too sandy to grow crops. Poor farming practices have degraded marginal soils, and much of the land has been overgrazed and deforested. These conditions lead to severe topsoil erosion and desertification (the desert spreads out to cover degraded areas).

■ Africa has poor food distribution systems. The lack of adequate, fast transportation hinders distribution. Often, perishable food rots before it reaches its destination.

■ Foreign aid is used to raise cash crops on irrigated land rather than to improve dry-land farming, grow food for domestic use, conserve soil and water resources, and replant trees.

In late 1984, a massive relief effort began, but it was stymied by corruption at the local level, mismanagement, and civil strife. Although the rains returned in 1985, hunger continues. The land has been so overworked and overgrazed that its carrying capacity has been permanently lowered.

Satellite images indicate that the land can recover some of its productivity. In good years of rain, the "green wave" of vegetation extends farther north into the Sahel than it does in dry years. Although high-quality vegetation will not return quickly to overgrazed areas, if controlled grazing programs are followed, much of the land may recover. New irrigation techniques help to conserve water, and planting several crops in the same field may increase the yield for all crops by reducing soil erosion, weed growth, and nutrient loss.

Experts agree that merely providing food for the hungry does not solve the problem. The only reasonable way to reduce hunger is to teach people how to produce food without degrading the environment. Organizations such as the Consultative Group on International Agricultural Research and the United Nations Environment Program are developing new crops and farming methods to increase self-sufficiency without harming the environment.

Vegetation is shown on the scale from little or none (tan) through heavy (purple). The year 1980 (upper map) had average rainfall. The year 1984 (lower map) was very dry.

a b

Figure 2.17 Old irrigation techniques (a) sprayed water over a large area. Some techniques (b) deliver water directly to the plants. How is this new technique beneficial?

we get? How do we get it? Although many individuals have begun to change their values, beliefs, and actions, it is as a global population that we are transforming the planet. Only by pooling our knowledge, coordinating our actions, and sharing the resources of the planet on a global scale can we hope to manage the earth in a sustainable fashion. Some steps already have been taken, such as the development of new irrigation techniques that save water by sending it directly to the roots of plants (see Figure 2.17). There is also a multinational agreement to cut in half the emission of some gases that may harm the atmosphere. Many other steps need to be taken, however, before the problems are under control.

Concept Review

1. What is the relationship between resources and the carrying capacity of a particular area?
2. Discuss the relationship between kcals and nutrition.
3. How have humans disrupted the homeostasis of our population?
4. Discuss the interaction of limiting factors for the human population. What is likely to happen in the future?
5. How have humans affected their environment?
6. If adequate amounts of food are produced overall, why do some countries experience famine?

Summary

We are all part of a population. Like all other populations, the human population is affected by both biotic and abiotic factors. Whereas some of these factors help to increase a population's size, other factors cause it to decrease. Populations have built-in growth rates, but their growth is limited by environmental factors that may act as barriers to dispersal and affect the density of a population. Any given area has a carrying capacity; that is, it has resources to support only so many individuals. Once the density of the population has approached the carrying capacity of the area, the necessities for life are in short supply, and the population size decreases. How long will it take before the carrying capacity of the earth for humans is reached? The human population is huge and growing. We affect other organisms because of the amount of resources we use and the amount of waste we produce in our daily lives. Through our use of resources and the environment, we change the world around us. When these changes occur, other populations are affected, not always for the better. These changes occur because all populations are linked together in the biosphere. Now that we recognize how our actions have affected the organisms around us, we are attempting to develop reasonable and workable management plans to reduce our impact on the environment and to extend the life of the human population and all other populations that share the earth.

Applications

1. Suppose you know the size of a population of ducks on your local lake in 1990, and you know how many ducks will be born and will die during the next 10 years. Can you determine the duck population size at the end of 10 years? Why or why not?

2. Is the carrying capacity of an area the same for all types of organisms that live there? Why or why not?

3. Humans can live almost anywhere and our population has steadily increased through time. Does this mean that limiting factors do not affect humans? Explain your answer.

4. Suppose there is a population of some type of animal in a huge space, such as rabbits on a large ranch or fruit flies in a big building. If you gave these animals all the food and water they needed, would their populations continue to increase in size forever? Explain your answer.

5. Why are the terms *mortality* and *natality*—as they are defined in this text—not applicable to individuals?

6. Discuss the present relationship between the human population and the carrying capacity of the earth.

7. Why is it difficult for people to see the effects they have on the biosphere?

8. A team of biologists studied a population of box turtles in an Ohio woodlot for a period of 10 years. They determined that the natality averaged 40 per year, the mortality 30 per year, immigration 3 per year, and emigration 8 per year. Was the population increasing or de-creasing? Was the area supplying box turtles to other places or vice versa?

9. In Question 8 what was the average annual change due to immigration and emigration? If the initial population was 15 turtles, what was the population at the end of 10 years?

10. Suppose a particular human population continues to reproduce and multiply, far outstripping the carrying capacity of the land. Food is in extremely short supply, water supplies are limited, and what water is available is polluted. Disease is everywhere. Discuss possible solutions to this problem and include the implications of each solution. For example, one solution is to supply food to the people, thus enabling them to live longer and raise their families. What are the implications of this solution? What solutions are ethically possible?

11. Name several ways that humans continue to change the biosphere and discuss the effects of these changes on the biosphere and other populations. What solutions can you identify?

12. It is seldom possible to count all the individuals in a natural population. How can a biologist study populations without such data?

Problems

1. Use a world almanac to determine how the population of the United States has changed through time. How much of the increase in population may be attributed to immigration? to natality? to mortality?

2. Suppose you had to collect all of the food you eat for yourself or your family. There are groups of people in the world—hunters and gatherers—who must do this on a daily basis.

Find out about a group of these people—what they eat, how much territory they need to locate their food, and how many people live in each group.

3. Individuals of certain species have been dispersed to new areas with dramatic consequences to both the population and the area in which the species is introduced. Find out what happened to the European starling after it was introduced to the United States, to the pig after it was introduced to the Hawaiian Islands, the African bee after it was introduced to South America, or the prickly pear cactus after it was introduced to Australia.

4. You have just been elected mayor of your city. Unfortunately, the population of your city will double within the next 10 years. How are you going to handle all of the changes? What plans do you have to make to handle the changes?

5. Obtain the census data for your state. Make a graph of these data, beginning with the first census after your state entered the union. How does the form of this graph compare with that for the population of the United States as a whole? Try to explain any differences.

6. Obtain some duckweed, an aquatic plant that grows on the surface of ponds. Tap the mass of plants with your finger to separate a single individual. Place the single duckweed plant in pond or aquarium water in a petri dish. Make counts of the numbers of individuals at intervals of two to four days. Keep a record of dates and numbers. Construct a graph to show the growth of the duckweed population.

7. Grains, such as wheat and rice, are a basic food throughout the world. The 1985 worldwide production of grains was approximately 1.7 billion tons. This was enough grain to provide 4.6 billion people with enough food energy for normal activity for a year. Still, hundreds of thousands of people in Asia, Africa, and South America were malnourished, and thousands of others starved to death. How can you explain this? *The World Almanac* and books such as L. Brown et al., *State of the World* (New York: W. W. Norton, 1990) may give you some insights.

8. Suppose two populations of a protozoan are grown in an aquarium but are separated by a barrier. Each population reaches 1000, the maximum that can be supported. Then the barrier is removed. If population A is a killer group that attacks population B, what will happen to the two populations in terms of dispersal and eventual population size?

9. Explanations of the present geographic ranges of organisms may depend not only on expansion from former ranges but also on reduction of former ranges. What are some factors that might decrease a group's range?

What does the ant appear to be doing with these aphids? Who is benefiting from this relationship? Is any organism harmed? Give two other examples of similar relationships in which neither party is harmed and one or both parties benefit.

Communities and Ecosystems

3

In the web of life, boundaries mean little. Individuals and populations interact with one another; they affect and are affected by their nonliving environments. Under favorable conditions, populations may grow larger and spill into new areas. Under unfavorable conditions, populations may decline and leave space for other organisms to move in. Boundaries constantly shift and change. This chapter discusses interacting populations within specific boundaries. Establishing boundaries, even if they are artificial or temporary, makes it possible to study one section of the web of life at a time.

Life in a Community

3.1 Many Interactions Are Indirect

All populations, including human populations, interact with one another in a complex web of relationships. The set of interacting populations present during one time in one place is called a community. In your community, there may be dogs, cats, trees, weeds, and humans that interact (see Figure 3.1). When you mow your lawn or your dog bites the mailman, for example, the interaction is very direct. Most of the time, however, the interactions are indirect, and people tend to ignore them. Suppose, for example, that you had vegetable soup for lunch and threw the can away. A garbage truck might haul the can to the dump, where it would fill with rainwater. A female mosquito might lay her eggs in the water, and her offspring, once mature, might fly from the dump and bite a person or another animal. Perhaps neither would have been bitten if you had not eaten the soup, but since you did, you have had an indirect relationship with the mosquitoes and their victims.

Every community is affected by the abiotic environment—sunlight, soil, wind, rain, and temperature. Together, a community and its physical environment make up an **ecosystem** (EE koh sis tum). Both the terms *community* and *ecosystem* can describe the same group of organisms, but the term *ecosystem* refers to the community and its abiotic environment. Although the biosphere is too big to study as a whole, it is possible to study a part of it—an ecosystem. In Investigation 3.1, you can study two of the abiotic factors in an ecosystem that affect organisms.

Guidepost: How do different populations affect each other?

Figure 3.1 How do these organisms interact?

53

INVESTIGATION 3.1   Abiotic Environment: A Comparative Study

Two important abiotic factors are temperature and relative humidity. Relative humidity is the percentage of water vapor *actually* in the air at any given temperature compared with the amount of water vapor that the air *could* hold at that temperature. Organisms usually lose water faster in an atmosphere with low relative humidity than in an atmosphere with high relative humidity.

▶ Read through the procedure and then develop hypotheses that describe expected differences in the three environments and at the four heights.

Materials (per team of 6)

3 watches	3 screw-top jars containing 30–50 mL
3 metersticks	distilled water
3 nonmercury thermometers	3 pieces of stiff cardboard
(−10°C to + 100°C)	3 umbrellas or other shade devices
3 nonmercury thermometers of same	table of relative humidities
range with cotton sleeves over the	(see Appendix 3)
bulbs	

Procedure

1. In your data book, prepare a table like Table 3.1, shown below.

2. Divide your team into three pairs of students. Each pair will measure the temperature in one of three types of environment: a dense cover of vegetation (a woods, a thicket, or a mass of shrubbery); a single layer of herbaceous vegetation (a meadow or a lawn, preferably not cut close to the ground); no vegetation (bare ground or a tennis court). The environments should be located as close to each other as possible. In each team pair, one person will read both thermometers and the other will fan the thermometers with the stiff cardboard and record the data in the data table. Before starting, the people on the team who will record the data should synchronize their watches and agree on the time at which each measurement will be made. Schedule at least 8 minutes between readings. Write these times on the first line of the data table.

3. Take four sets of measurements in each environment: at ground level, at 30 cm above the ground, at 90 cm above, and at 150 cm above.

4. Take readings on both thermometers at the same time. Position the thermometers at least 5 minutes before the readings are taken. For example, if the first reading is to be taken at 1:30 PM, both thermometers should be in the first position at 1:25 PM. Use the umbrellas to shield the thermometers from the direct rays of the sun. To take a wet-bulb reading, soak the sleeve of the thermometer in water and

TABLE 3.1   Temperature Readings

Location	Height			
	0 cm	*30 cm*	*90 cm*	*150 cm*
Time				
Dry-bulb temperature				
Wet-bulb temperature				
Relative humidity				

fan it vigorously for at least 2 minutes; then read the thermometer. Plan for about 8 minutes between readings to allow for positioning and the 5-minute wait.

5. Using both your dry-bulb and wet-bulb thermometer readings and the relative humidity table, determine the amount of water vapor in the air compared with the amount the air could hold at that temperature. (The necessary calculations were made when the table was constructed.)

Discussion

1. At ground level, which environment is coolest and most humid?

2. At ground level, which environment is warmest and least humid?

3. How do these two environments compare in temperature and humidity at higher levels above the ground?

4. At which level above the ground are all three environments most alike in temperature and humidity?

5. How does the greatest temperature difference *within* the same environment compare with the greatest temperature difference *among* the environments?

6. What differences among the three environments might account for differences in temperatures and relative humidities?

7. How do these differences show the interaction of biotic and abiotic factors in an ecosystem?

8. You have been examining the differences among environments. Now turn to differences within an environment. How does the temperature in each environment vary with respect to elevations? Is the variation the same for each environment? If not, in which is the variation greatest? How does the humidity in each environment vary with respect to elevation? Is the variation the same for each environment?

9. In weather forecasts, temperatures predicted for the center of a city often differ from those predicted for the suburbs. Relate this fact to the situations you have been observing.

10. What differences in temperature and humidity would be experienced by a beetle crawling on the ground in a meadow and a gnat hovering at 1.5 m above the meadow?

11. Whereas we may say the beetle and the gnat are in the same environment, small differences within that environment are often important to the existence of some organisms. We therefore can distinguish small environments within larger ones on the basis of measurements such as those you have made in this investigation. Would it be useful to measure such differences if you were studying the ecological relationships among cows in a meadow? Explain.

3.2   Many Populations Interact in the Florida River Community

An aquatic community lives in the rivers along the west coast of Florida. Among the largest individuals in this community are the river turtles shown in Figure 3.2 on page 56. The adult turtles in the community eat not only many of the plants that grow in the rivers but also the long, narrow blades of the tape grass, which is their favorite food.

Consumers that eat only plants are called **herbivores** (HER bih vorz). Unlike the adult turtles, young river turtles are **carnivores** (KAR nih vorz), consumers that eat other animals. The young turtles feast on snails, aquatic

Figure 3.2   What interactions do you see?

insects, and worms. Carnivores that eat both plants and animals are **omni-vores** (OM nih vorz). Because the adult turtles are herbivores and young turtles are carnivores in this community, they have different roles in the food web.

There are many other carnivorous animals in the community. These animals eat turtle eggs and young river turtles (see Figure 3.3). The eggs have the highest mortality rate. Female turtles lay their eggs in holes they

Figure 3.3   Four animals that eat young river turtles. Describe the direct and indirect relationships shown here.

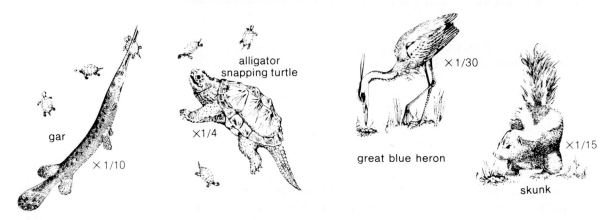

dig on land, but they do not guard their eggs. The eggs may be dug up and eaten by skunks, raccoons, or snakes. Unhatched turtles also are killed by molds that live in the soil and grow through the thin shells into the eggs. If the eggs survive, the young turtles that hatch still may be eaten by snakes or raccoons before they reach the river. Even if the young turtles reach the river, they are vulnerable to carnivores. Large fish, alligators, herons, and snapping turtles may eat them. For the young turtles that have survived the trip to the river, mats of floating plants, tangled tree roots, and sunken logs offer some protection from carnivores.

When the young turtles have grown larger, few organisms can kill them directly. Leeches may become attached to the turtles and suck their blood, but that usually does not kill the turtles. Turtles do die of disease, of accidents, and of old age, however. After they die, their bodies then become food for the decomposers that return all the substances in the turtle's body to the abiotic world.

Each of the relationships described so far is a direct one between river turtles and other types of organisms. Like all organisms, however, river turtles have many indirect relationships with other organisms. For example, plants provide not only food and hiding places for the turtles but also oxygen for the fish that might eat young turtles. Snails eat tape grass and therefore compete with the adult river turtles for the same food supply. Another type of turtle, the musk turtle, eats nothing but snails. By reducing the number of snails that eat tape grass, musk turtles increase the supply of tape grass and thus have an indirect effect on the river turtle population. Because many of the carnivores that eat young river turtles also eat musk turtles, the more musk turtles there are in the river, the less likely it is that young river turtles will be eaten.

3.3 A Niche Is the Role of an Organism

Adult river turtles and young river turtles have different functions in the community—different **niches** (NITCH ez). The niche of an organism is its role in the community: what it eats, what organisms eat it, and what indirect relationships it has with organisms. An organism's niche affects and is affected by its **habitat** (HAB ih tat)—the place where it lives. Thus, a niche includes both the biotic and the abiotic environments of the organism. No two types of organisms occupy exactly the same niche within a community. For example, there are many different types of turtles living in a Florida river. They live in the same habitat, but because each type lives in a manner distinct from the others and has different relationships with other members of the community, they are said to occupy different niches. Understanding a community requires understanding all the relationships among all the organisms and all their niches. That is not an easy task, even for a small community.

Consider the effect of adding just one more population, such as a human population, to the river community. Humans trap skunks, kill snakes, and catch fish that eat young river turtles. All of these activities indirectly reduce the mortality of young river turtles. Humans, however, also may dredge the rivers and thus kill the tape grass, which then will increase adult river turtle mortality.

INVESTIGATION 3.2 Competition

Organisms need certain resources to live. For survival and reproduction, they may need food, water, light, breeding space, mates, and other substances. If these resources are in short supply, the organisms may compete for them. Sometimes the competition between two groups of organisms is so intense that one group loses completely and no longer competes. The winner in such contest may be able to use the resource or may prevent the loser from using it. In these situations, the losing group of organisms usually moves to another place, where the competition is less intense or where they are better competitors, or the group dies out.

To study how competition works, you will compare it to something familiar. Here the groups are fast food restaurants (FFRs), and the resource in short supply is the consumer's money. Although the resource may be in short supply, obviously FFRs do coexist, often in close proximity to one another.

 Before beginning this investigation, develop several hypotheses to answer the question: "How might these FFRs avoid going out of business?" Another way of asking the same question is: "What might individual FFRs do to increase their own chances of survival?" Record your hypotheses in your data book.

Materials (per person)
none

Procedure
PART A Fast Food Restaurant Competition

1. Study the entrees from four nationally franchised FFRs shown in Table 3.2. The categories of entrees are general because each FFR has its own set of unique names for many entrees. Therefore, different types of hamburgers, for example, were grouped on the basis of size (all quarter-pound hamburgers were categorized as *large*, and anything smaller was labeled *regular*).

2. Figure 3.4 on page 59 shows three overlapping circles that contain different numbered areas. Draw three of these diagrams for use in comparing similar FFRs or use the diagrams provided by your teacher. Label the circles in the first diagram *Group A, Group B,* and *Group D*. Label the circles in the second diagram *Group A, Group C,* and *Group D*. Label the circles in the third diagram *Group B, Group C,* and *Group D*. Number the areas of each set of circles as shown in Figure 3.4. Each set of circles represents a neighborhood with three FFRs in it.

TABLE 3.2 Entrees of Four Nationally Franchised Fast Food Restaurants

Entree	A	B	C	D	Entree	A	B	C	D
Roast beef sandwich	X				Large cheeseburger		X	X	X
Turkey sandwich	X				Extra-large hamburger		X	X	X
Chicken pieces		X	X	X	Extra-large cheeseburger			X	X
Chicken sandwich	X		X	X	Salad bar	X		X	X
Fish sandwich	X	X	X		Baked potato				X
Breakfast	X	X	X		Chili				X
Children's meal		X	X	X	Hot dog	X			
Regular hamburger	X	X	X		Ham-and-cheese sandwich			X	
Regular cheeseburger	X		X		Bacon hamburger	X		X	X
Large hamburger		X	X	X					

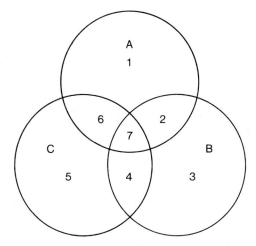

Figure 3.4 Diagram for comparison of FFR entrees.

3. Write the name of the entree that fits each category in the proper numbered area on each diagram. For example, one entree unique to Group A is a roast beef sandwich. Write *roast beef sandwich* in Area 1 of the Group A circle of the A-B-D set. Complete the list of entrees for all three sets of FFR species and use the information to answer Questions 4 and 5.

4. How is coexistence possible the for FFRs in each neighborhood?

5. Do your data support the following hypothesis, "Entree specialization explains coexistence"?

PART B Competition Among Birds

6. In the short-grass prairie of the Great Plains, yellow-headed and red-winged blackbirds live together in the same marshes. These two groups have basically the same niche. The only difference is that red-winged blackbirds are generalists. They can nest in a variety of habitats ranging from moist, brushy areas to moist meadows. They do, however, prefer to nest in extremely moist marshes at the edge of the open water of ponds and lakes. Yellow-headed blackbirds are specialists. They require the extremely wet part of the marsh, next to open water, to nest successfully. Because of their larger size and more aggressive nature, yellow-headed blackbirds will displace red-winged blackbirds in open wet marsh. Thus, the yellow-headed blackbird has a niche totally included within the niche of the red-winged blackbird, as shown in Figure 3.5 on page 60. Use this information to answer Questions 7 through 9.

7. How do the competitive abilities of red-winged blackbirds compare to those of yellow-headed blackbirds?

8. Why are all red-winged blackbirds not displaced by yellow-headed blackbirds?

9. Recall your information from Part A. Compare FFRs B and C. Using what you have learned about blackbirds, how is coexistence possible between these two FFRs?

Discussion

1. A new FFR chain, *E*, opens a restaurant one block from the established restaurant *B*. If *E* offers exactly the same menu and service as *B*, what do you think will happen, and why? What does this suggest to you about the chances of a particular group of fish that swims up a river into a lake and attempts to establish itself?

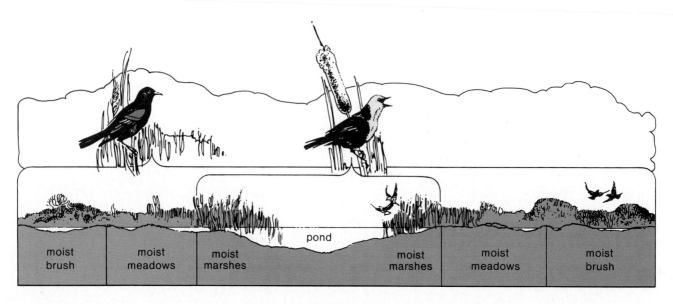

moist
brush

moist
meadows

moist
marshes

pond

moist
marshes

moist
meadows

moist
brush

Figure 3.5   Overlapping niches of red-winged and yellow-headed blackbirds.

2. Five similar groups of small, insectivorous birds live in the same spruce forest and apparently coexist. What variables might you investigate to explain this?

3. Competitive exclusion does not always occur among competing groups. How do the groups coexist?

4. How can two groups of organisms coexist when one group has a niche that overlaps the niche of the other?

5. How can the concepts of competition, specialization, and overlapping niches be applied to aspects of urban life?

3.4   Organisms May Benefit or Harm One Another

Many niches are easy to recognize. For example, algae, tape grass, and other green plants are producers. River turtles, snails, skunks, and fish are consumers. There are, however, other types of relationships that help to form the community's web of life. Each of these relationships involves at least two different organisms. In a **predator-prey** relationship, for example, one type of organism (the prey) is eaten by the other (the predator). In this relationship, one organism benefits from the relationship, but the other does not.

Another type of relationship, **competition**, benefits neither organism. Consider the Florida river community. Both snails and river turtles eat tape grass. The tape grass that is eaten by a turtle cannot be eaten by a snail. The grass eaten by a snail cannot be eaten by a turtle. Therefore, the presence of one of these herbivores is harmful to the other. Organisms may compete for such things as food, space, sunlight, nutrients, or water. The competi-

×1/3

Figure 3.6 Competition does not always involve food. What might the bluebird and the starling be competing for? Who do you think wins?

tion is always for a scarce resource (see Figure 3.6). In the Florida river community, if both tape grass and other plants are scarce, competition for tape grass by herbivores is intense. If plants are abundant, competition for the tape grass is reduced.

When different organisms live in direct, physical contact with one another, the relationship is called **symbiosis** (sim by OH sis). For example, lichens (LY kenz) are leaflike, hairlike, or crustlike structures that grow on rocks or the bark of trees, such as those shown in Figure 3.7. Lichens may be gray-green, black, yellow, or orange in color. Although a lichen appears to be a single organism, it is actually two different organisms that live in close association with one another. In the most common lichens, one partner is an alga, a microscopic producer that makes food by photosynthesis. The other partner is a fungus, a consumer that gets its food from the alga but in return provides moisture for the alga. Their relationship benefits both organisms.

In many symbiotic relationships, both partners benefit, or at least no one is harmed. Figure 3.8 shows a remora and a shark. Using the suction disk on the top of its head, the remora attaches itself to the shark. The shark is not harmed, and the remora expends little energy because the shark pulls it around. In addition, the remora eats small bits of the shark's prey as they float by.

A special, and common, form of symbiosis is **parasitism** (PAIR uh sih tiz um). In parasitism, one organism (the parasite) lives on or in another organism, using it as a food source. The food source (the host) usually remains alive during the interaction. A leech, for example, clings to a turtle's skin and sucks its blood. Parasitic microorganisms in the turtle absorb food directly from its blood. Like a leech, a tick on your skin sucks your blood. Plants also may have parasites. Some molds, microorganisms, and even other plants, such as mistletoe (see Figure 3.9 on page 62), are parasites. Large microorganisms may have smaller parasitic microorganisms in

Figure 3.7 How many different types of organisms do you see?

Figure 3.8 What relationship do you see?

Figure 3.9 Mistletoe (yellow-green) growing on the branch of a ponderosa pine. What relationship is involved here?

Guidepost: How are biotic and abiotic environmental factors related to each other in an ecosystem?

them. Unlike a predator, which kills its prey outright, a parasite may kill its host indirectly by weakening it. Once it is weakened, the host becomes more susceptible to disease or an easier prey for predators.

Organisms may play different roles within their community at different times. For example, recall the difference in roles between young river turtles and adults. Or, an organism may play more than one role in a community. Snapping turtles prey on young river turtles, which makes them predators, but they are also **scavengers** (SKAV en jerz), organisms that eat the flesh of dead animals they did not kill. You also play more than one role in your community. When you eat a salad, you are a herbivore. When you eat both salads and hamburgers, you are an omnivore. Every community includes many different types of organisms and an individual can have many types of relationships.

Concept Review

1. Give an example of a direct and an indirect interaction between two organisms.
2. Distinguish between a community and an ecosystem.
3. How does an organism's niche differ from its habitat?
4. How does symbiosis differ from a predator-prey relationship?
5. In what ways are predation and parasitism alike?

Ecosystem Structure

3.5 Boundaries of Ecosystems Overlap and Change

What are the boundaries of an ecosystem? You might think that the boundary in the Florida river example is easy to define—the edge of the water (see Figure 3.10). Turtles, however, crawl onto the river banks to lay their eggs. Herons get most of their food from the river, but they nest in tall trees near the water. Frogs are caught and eaten by raccoons on land. Although some organisms such as fish spend all of their lives in the river, other organisms in the river do not.

When abiotic factors are considered, the question of boundaries becomes complex. Energy for the entire ecosystem comes from the sun, which is a great distance from the river's shoreline. Rain falls from the sky and may carry soil from the riverbanks to the river. The wind may blow seeds into the river from communities that are many kilometers away. Thus, the boundaries of this one ecosystem can expand to include many ecosystems.

The Florida river is not a special case. All ecosystems are connected to others around them. A river ecosystem is linked to a forest; a forest ecosystem is linked to the grassland, and so on. In fact, all ecosystems on the earth are connected to one another to form the biosphere. Because they are connected, a change in one ecosystem may affect many others.

Because of the web of relationships in a community, when one population is affected by abiotic factors, many others in the same community are affected at the same time. If the temperature drops to abnormal lows in Florida, river turtle eggs and young river turtles may be killed while they

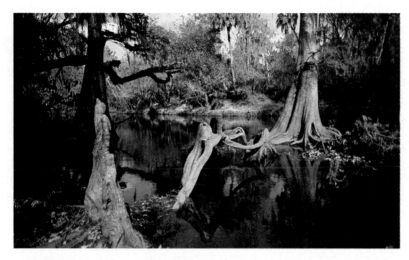

Figure 3.10   A Florida river. What is the boundary of this ecosystem?

are still on land. With fewer river turtles, more tape grass will grow, but there will be less food for snapping turtles. If a heavy rainfall washes mud into the river, light will not penetrate as deeply in the water, and fewer plants will be able to photosynthesize. That, in turn, will mean less oxygen and less food for herbivores. Thus, each environmental change has many effects. One change causes many others, as shown in Figure 3.11. In Investigation 3.3, you will examine the effects of one environmental change on seed germination.

INVESTIGATION 3.3   Acid Rain and Seed Germination

Acid rain is something of a misnomer because natural rainwater is acidic and has a pH of about 5.6. (pH is a measure of acidity; the lower the number, the greater the acidity. A pH of 7 is neutral.) Acid rain, therefore, refers to rain with a pH lower than 5.6. Acid rain is produced when sulfur and nitrogen compounds are released into the atmosphere where they combine with water to form sulfuric and nitric acids. Sulfur compounds may come from such natural sources as decomposing organic matter, volcanos, and geysers. The environmental problem known as acid rain, however, does not arise from natural sources. It is caused primarily by fossil fuel combustion. When coal, oil, and gas are burned, large amounts of sulfur and nitrogen are released as gases and combine with water to make the rain more acid than usual. Acid rain has many effects on an ecosystem. In this investigation, you will examine just one of them.

 Read through the procedure and develop a hypothesis about the ideal pH for bean seed growth.

Figure 3.11   What abiotic factors might produce the effect shown here?

Materials (per team of 2)

2 pairs of safety goggles	transparent metric ruler
2 lab aprons	graph paper
2 pairs of plastic gloves	colored chalk or magic markers
petri dish	water solutions ranging in pH from
glass-marking pencil	2 to 7
absorbent paper towels	rainwater
scissors	4 bean seeds

SAFETY   **Put on your safety goggles, lab apron, and gloves.**

Procedure

Day 0

1. In your data book, prepare a table in which to record seed lengths and averages for the duration of the investigation.

2. Cut four discs the size of the petri dish from the absorbent paper towel.

3. Dampen the discs with the water assigned to you. Record the pH of the water.

   WARNING: Solutions of pH 4 or less are irritants. Avoid skin/eye contact; do not ingest. Should a spill or splash occur, call your teacher immediately; flush the area with water for 15 minutes; rinse mouth with water.

4. Place two of the paper discs on the bottom of the petri dish.

5. Use the transparent ruler to measure the lengths of your four seeds in millimeters. Determine the average length and record under Day 0 in your data table.

6. Sketch the shapes of the seeds and note their color.

7. Arrange the seeds in the petri dish and cover with the two remaining discs. Make sure the discs are still moist. If not, add more of your assigned pH solution.

8. Place the lid on the petri dish and label with your team name.

9. Wash your hands thoroughly before leaving the laboratory.

Days 1–10

10. Remove the lid from the petri dish and lift off the discs covering the four seeds.

11. Measure the length of the seeds in millimeters, average the lengths, and record in your data table under the appropriate day.

12. Sketch the shapes of the seeds and note their color.

13. Cover the seeds with the paper discs. If necessary, moisten the paper with the assigned pH solution and replace the lid.

14. On a piece of graph paper, set up a graph labeled *Age in Days* on the horizontal axis and *Length of Seeds (mm)* on the vertical axis.

15. Plot the average length of your seeds for the two measurements (Day 0 and Day 1) you have made.

16. Repeat Steps 10–15 each day for the duration of the investigation. If the seeds begin to germinate during this time, include the length of any growth you observe in your measurements.

17. Combine your data with those of other students using the same pH solution and determine the average. Record the average on the class graph, using the color assigned to your particular pH.

18. Each day, wash your hands thoroughly before leaving the laboratory.

Discussion

1. Study the data on the completed class graph. What appears to be the optimal pH solution for successful bean seed germination?

2. What appears to be the worst pH solution for successful bean seed germination?

3. What pH do you think the rainwater has, based on the data gathered? Determine its pH.

4. What impact on local crops might an increased rain acidity have?

5. Do you think there is reason for concern about the rain in your area?

3.6 Most Communities Have More Producers Than Consumers

One of the most important abiotic factors that affects relationships in a community is energy. As discussed in Chapter 1, organisms in an ecosystem are tied together by the flow of energy and matter from one organism to another. The food chain that exists when a herbivore eats a plant and a carnivore eats a herbivore depends on the energy entering the community in the form of sunlight. Without the sun, there would be no green plants, no herbivores, and no carnivores. There are a few ecosystems that get their energy from another source. These unusual communities are described in Chapter 23.

The size of a community, therefore, is limited by the amount of energy entering it through its producers. The total amount of chemical energy stored by photosynthesis is the gross primary productivity of the community. Much of that energy is used by the producers to grow and to maintain themselves. The remaining energy, which is available to the consumers as food, is the net primary productivity of the community. One way to measure the net primary productivity of a community is to find the mass of all producers. Mass is a measure of the amount of matter in an object. Because much of the mass of living organisms is water, the producers first must be dried for a truer estimate of their mass. For example, the energy found in a fresh apricot and a dried apricot is the same, though the mass of the dried fruit is much less.

The greater the net primary productivity of a community, the greater the amount of food available to consumers. When net primary productivity is high, more consumers can live in the community. No matter how great the productivity, however, there is always some limit—the carrying capacity—to the number of organisms that can be found in any community.

Like a pyramid, which decreases in size from bottom to top, a community usually has many producers at its base, fewer herbivores in the middle, and even fewer carnivores at the top. Usually, the total mass of all the producers is much greater than that of the consumers.

Without more food producers than food consumers in a community, there will not be enough food to go around, and consumers will die. Because plants use for their own activities much of the energy they produce, a large number of plants is needed to support a few herbivores. Energy that has been used to keep a plant alive cannot be taken in by a herbivore. Because herbivores also use some of the energy to keep themselves alive, not all of the energy they take in is available to carnivores.

The loss of energy from a food chain may be represented by an **energy pyramid** (see Figure 3.12 on page 66), in which producers form the foundation. In general, the energy at each level of the pyramid is one-tenth that at the level below it. This concept often is referred to as the 10 percent rule because only 10 percent of the energy at any given level is available for use by organisms in the next higher level.

Because of this steplike decline of available energy in a food chain, the number of consumers supported at the top level is substantially smaller than the number of consumers and producers supported at the lower levels. Only about one one-thousandth of the chemical energy used by producers can flow all the way through a three-consumer food chain to a top predator.

Figure 3.12 An idealized energy pyramid. How much energy is lost at each level?

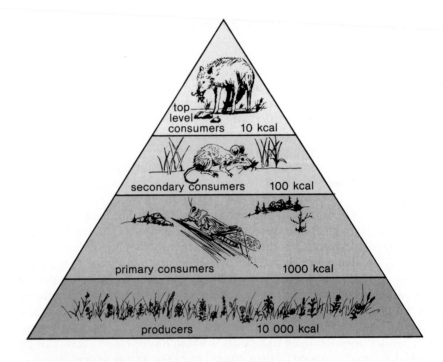

For example, in a small valley there may be millions of individual producers, such as grasses, but only a few hundred herbivores, such as rabbits. In this same valley, you may have to search carefully to find just one or two carnivores, such as foxes.

The concept of an energy pyramid has serious implications for the human population. Eating meat is an inefficient way to obtain the energy originally captured by plants. Recall the recipe for making a cow in Investigation 2.3. You can obtain far more kcals by eating a given amount of grain yourself than by eating the chicken or beef supported by the same amount of grain.

Concept Review

1. Why is it difficult to draw distinct boundaries between most ecosystems?
2. How are gross and net primary productivity of a community related?
3. Why is only part of the chemical energy produced by plants available for use by herbivores?
4. Explain the 10 percent rule. Why is it important in ecosystems?

Guidepost: How do the complex activities of humans affect ecosystems?

Ecosystem Stability and Human Influences

3.7   Humans Lower the Stability of Ecosystems

Ecosystems tend to be stable through time because of the great number of different types of organisms and interconnecting relationships found in the ecosystems. Usually, the greater the number of types of organisms and, thus, the greater the number of links in the food web, the more stable the community is thought to be. A large community is like a web with many threads. If only one of the threads is broken, the web as a whole may still function. If a disease kills many of the rabbits in a community, the foxes can eat mice and squirrels until rabbits reproduce or immigrate into the community. Even though many rabbits have died, the rest of the ecosystem remains intact. It may be changed, but it is not destroyed. This is the process of homeostasis, discussed in Chapter 2.

Human activities, however, can change ecosystems greatly. Humans disrupt homeostatic processes. In doing this, we can have positive effects on other members of the community, or we can have dramatic negative effects. Farmers, for example, eliminate many members of a food web when they raise crops to meet human needs (see Figure 3.13). To feed the human population, it is necessary to plant large fields of producers. To use machinery efficiently, a single crop is planted in a large field. One cornfield may be 16 hectares, contain 76 000 plants, and yield 7258 m³ of corn. Although a wild field of corn would provide energy for a great number of birds, insects, and animals that eat birds and insects, these consumers compete with humans for the crop. Frequently, they are eliminated by humans, as are other plants that compete with the corn for light, water, and nutrients. With such a large population of plants, however, plant diseases can spread easily. The disease might be a parasite of the corn. If the farmer is to grow crops economically, he or she must control the food web and reduce the pests and diseases. Often, such control involves widespread spraying of chemicals on crops, which may have unplanned side effects.

Figure 3.13 What members of the food web might be eliminated by crop dusting?

Sometimes humans change ecosystems in the quest for health. Because parasitic diseases, such as malaria and African sleeping sickness, which are carried by insects, cause illness or death, we try to eliminate the insects that carry them. To eliminate or to control these pests, we use poisons called insecticides, herbicides, and fungicides. All of these products may be included in the term **biocides**. The United States alone produces more than 636 million kg of biocides each year.

What effect do these biocides have on the relationships in a community? DDT serves as a typical example. At one time, marshes along the north shore of Long Island, New York, were sprayed with DDT to control mosquitoes. Later, microscopic organisms in the marsh waters were found to have about 0.04 parts per million(ppm) of DDT in their cells (see Figure 3.14 on page 68). While this is a low level of poison, no one had expected to find any poison in the organisms. The minnows, clams, and snails that ate these organisms had levels of DDT more than 10 times higher— between 0.5 and 0.9 ppm. The eels, flukes, and billfish that ate the snails and small fish had levels of DDT ranging from 1.3 to 2.0 ppm. The ospreys, herons, and gulls that ate the eels and minnows had levels of DDT between 10 and 25 ppm. Thus, the concentration of DDT in the tissues of the organisms in this food chain increased almost 10 million times from the amount in the seawater and nearly 625 times from the producer level at the bottom of the pyramid to the consumer level at the top.

Because each consumer eats a large number of producers, DDT and other chemicals become concentrated at upper levels in a food chain. Although the body of each producer can be broken down and used by the consumer, the DDT in the producer's body is not broken down. Each of

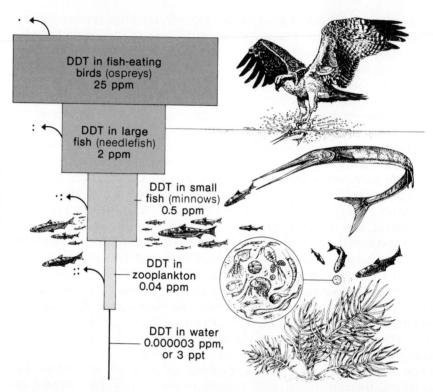

DDT in fish-eating
birds (ospreys)
25 ppm

DDT in large
fish (needlefish)
2 ppm

DDT in small
fish (minnows)
0.5 ppm

DDT in
zooplankton
0.04 ppm

DDT in water
0.000003 ppm,
or 3 ppt

Figure 3.14   What happens to the concentration of DDT as it travels up the food chain?

the producers, then, contains a little DDT, and these small amounts build up to a large amount within the consumer. The greater the concentration of DDT in an organism, the greater the damage. Ospreys and pelicans with high levels of DDT accumulated in their bodies, for example, cannot produce normal eggs. The egg shells are so thin that the eggs are crushed by the nesting parent. At one time, some of the osprey and pelican populations were nearly exterminated because of the effects of DDT. No one expected that the DDT used to control mosquitoes could affect so many other organisms far from the area where the poison was sprayed.

DDT also has been used to combat the mosquitoes that carry the microorganisms causing malaria. A person with malaria experiences successive bouts of chills, followed by high fever. In some cases, the person may die. At one time, malaria was a major health problem on the island of Borneo in Indonesia. To help the people of Borneo, workers from the World Health Organization sprayed remote villages and nearby areas with DDT. Most of the mosquitoes in the sprayed area died, but other organisms also were affected. Flies and cockroaches, the favorite food of the lizards found in Borneo, died from the DDT. The lizards gorged themselves on the DDT-poisoned insects, and they, too, began to die. Local cats ate the infected lizards and died. After the cats died, the rat population grew unchecked.

The rats in the villages carried a disease that affected humans. Now, the people on the island no longer had to worry about malaria, but they began to die from the disease carried by the rats. To restore the balance in the community, cats were brought into the country and parachuted into remote villages to eat the rats (Figure 3.15). This is another example of the unplanned and long-lasting effects that we, as humans, may have on other organisms and on ourselves because of interactions in communities and because of our own ignorance.

Some pest populations include individuals that are resistant to a biocide. These individuals vary in their ability to tolerate, detoxify, or avoid the poison. When most of the pests are killed by a biocide, these few individuals may survive and reproduce. They pass their biocide resistance to their offspring. Then, when people try to kill these offspring with the same biocide, the poison has little effect. To be effective, the strength or the amount of the spray must be increased, which then affects other organisms. In some cases, no amount of spray will control the pests. In these cases, a different pesticide may or may not be effective. Today, one can find many resistant disease-causing and disease-carrying organisms in places where biocides have been used often. DDT-resistant mosquitoes, herbicide-resistant weeds, and antibiotic-resistant bacteria are not uncommon. Although the use of DDT has been banned in the United States, U.S. chemical companies make DDT and export it to other nations that still use this biocide.

3.8 Humans Threaten the Diversity of Organisms

At the same time that human activities are raising the number of biocide-resistant organisms, many potentially important organisms are disappearing. Their disappearance also is due to human influence and results in a decrease in **biodiversity**—the number of different types of organisms that live on earth. Ecosystems with a large number of different types of organisms are usually quite stable. Ecosystems with a small number of different organisms are less stable. As our human population grows and we expand our activities, we occupy more land and therefore destroy the habitats of many organisms. Often this damage extends to beneficial and desirable organisms (see Figure 3.16 on page 70). The smog created by automobiles and industry is killing many types of trees over a wide area of southern California. The needles of ponderosa pines, for example, gradually turn brown, and the tops of palm trees have only small tufts of fronds. When this happens, photosynthesis is drastically reduced, and the plants die.

The Everglades National Park in southern Florida depends on a slowly moving sheet of water that flows from north to south. Drainage ditches built at the northern edge of the Everglades have decreased the flow of water over the entire area. As a result, many alligator holes (see Figure 3.17 on page 70) have dried up. These holes helped to contain fires in the Everglades. Now, destructive fires are more frequent in this national park.

Plants and animals in heavily populated areas are not the only ones threatened. Tropical rain forests, such as the one shown in Figure 3.18, are the most diverse ecosystems on the earth. They provide habitats for many different **species**, groups of organisms that only can reproduce successfully with others of the same type. Unfortunately, deforestation for lumber, grazing land, and other uses is destroying countless species every day. Two-thirds

Figure 3.15 Borneo rat patrol.

Figure 3.16   What might be killing the forest?

Figure 3.17   These holes are the aquatic homes of alligators. How could these holes help control fire?

of the world's species are located in the tropics and subtropics. If the current rate of habitat destruction in these areas continues, nearly half of the earth's species of plants, animals, and microorganisms will become **extinct**—gone forever—or severely threatened, during the next 25 years. To find a comparable rate of extinction (loss of species), you need to go back 65 million years, to the end of the Cretaceous period when dinosaurs and other organisms disappeared. Because there are more species today than there were 65 million years ago, however, the absolute number of species lost will be greater now. Hundreds of species of plants and animals, including the whooping crane and some rare pitcher plants (see Figure

Figure 3.18   Tropical rain forests are the most diverse ecosystems on the earth.

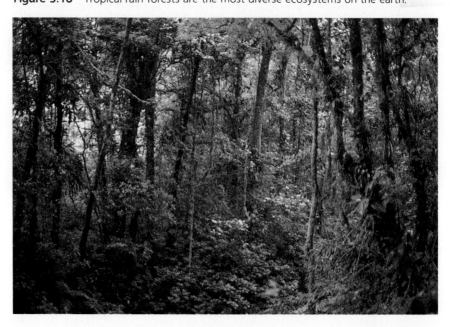

3.19), are threatened with extinction today. Although extinction is a natural process, the process has been greatly accelerated because humans have changed whole ecosystems.

Why does it matter if whooping cranes and pitcher plants become extinct? One argument against human-caused extinction comes from genetics. If all the crop plants in a field are genetically similar and one individual gets a disease, all the plants may die. As long as wild populations exist, a vast pool of genetic characteristics remains available. These characteristics may be used to prevent widespread death among the genetically similar individuals of our crops and domesticated animals (see Chapter 9, Biology Today). The extinction of each wild population erases genetic material that could mean healthy crops and animals. Once extinction occurs, the genetic material is gone forever.

A second argument against human-caused extinction is related to the instability of simplified ecosystems. Think of a field of corn as a simplified ecosystem. If something should happen to the corn, the whole ecosystem would collapse. To prevent such a collapse in a natural ecosystem, a community must have a wide diversity of species. The fewer the species in a community, the easier it is for homeostasis to be disrupted. Although new species are evolving all the time, the process is extremely slow compared to the rate at which humans are able to cause species to become extinct. Each time a species becomes extinct, the biosphere is simplified a little more, and it becomes more difficult to maintain the stable biosphere on which all life depends.

a

×1/20

b

Figure 3.19 What difference does it make if whooping cranes (a) and pitcher plants (b) become extinct?

Figure 3.20   Madagascar periwinkle.

A third argument comes from research on plants. The island of Madagascar, off the east coast of Africa, is the only known habitat of the Madagascar periwinkle (see Figure 3.20). This plant produces two chemicals not produced by other plants—vincristine (vin KRIS teen) and vinblastine (vin BLAS teen). Both of these chemicals are used to help combat Hodgkin's disease, a leukemialike disease that affects thousands of people each year. As the human population on Madagascar grew, the habitat for the periwinkle shrank almost to the point of making the periwinkle extinct. Fortunately, botanists collected and grew some of these plants before they were gone forever. The medicines made from the Madagascar periwinkle are worth millions of dollars each year and help many people with Hodgkin's disease to live longer. These medicines never would have been known if the plant had become extinct.

There are thousands of organisms in the world that we know very little about. More than 750 000 species of insects alone have been identified, but at least twice that many are thought to exist. If they become extinct, we will never know anything about them or the many other undiscovered species. The roles these species may play in ecosystems are not precisely known, although many bacteria and other microscopic organisms play a vital part in maintaining the homeostasis of ecosystems. What might happen if we lose species that are important in maintaining the atmosphere of the earth?

3.9   Humans Also Can Help Conserve Species

Figure 3.21   Three Sisters wilderness area in Oregon.

We can take many measures to decrease the loss of species diversity. One such action is the establishment of wilderness areas, such as the one shown in Figure 3.21. Only foot and horse travel is allowed in these areas. In some areas, camping may be restricted to designated areas. Some wilderness areas require a permit for use. What good are these places if more people cannot use them? Wilderness areas act as one type of habitat preservation. In these areas, the effect of human activities on local organisms is reduced. In this way, an "island" of unspoiled habitat may help to ensure the safety of some species. Many countries now have park or conservation systems similar to those in the United States.

The preservation of existing habitats may not be enough, however. Not all types of habitat are represented in preservation systems. The size of the areas set aside may not be adequate to stop the extinction of some species, particularly those requiring a large range or those sensitive to small climatic changes. Furthermore, many of the conservation areas do not have enough professional, well-trained, and well-equipped staff to manage them. What else can be done to slow the loss of species? Many countries have plans for replanting deforested areas, actively managing wildlife populations, and instituting nondestructive farming practices. Brazil, for example, recently developed a conservation system that encompasses almost 15 million hectares. These areas include good examples of many of Brazil's major ecosystems. China recently has replanted 30 million hectares with trees.

Many times, the loss of wild populations is due to habitat destruction, economic and cultural pressures, like those that have led to the overhunting of the African elephant, or perhaps to correctable environmental changes, such as introduced predators and competitors. Sometimes overall

Naturalist/Interpreter

Dave Marshall is a naturalist/interpreter for the Bear Creek Nature Center in Colorado. He grew up in Colorado Springs, close to the Rocky Mountains. His father was an outdoorsman who took Dave on many fishing trips to the nearby streams and lakes. This, along with his father's interest in birds, sparked Dave's curiosity about nature, and he decided to pursue a career related to natural history. Originally, he had planned to study forest management and ecology. After discovering a program in environmental interpretation in a college catalog, however, he decided that this new career "seemed to fit," and he went on to obtain degrees in environmental education and in resource and recreation management.

One of Dave's major duties is to conduct natural history programs for elementary school students at the Bear Creek Nature Center. He takes the children on short walks on the trails, shows them the exhibits at the center, introduces them to a display animal, such as a turtle or a garter snake, and gives a short talk on the specific program. Dave is in charge of the mammal natural history records, the insect collection, and the Bear Creek Nature Center's library. Dave also leads nature hikes to various types of natural areas.

Dave's favorite part of being a naturalist/interpreter is leading programs and meeting the public. He particularly enjoys working with children, and he feels a special satisfaction when he sees their curiosity and imagination stimulated by his program. Dave feels that being a good communicator is the most important quality for a naturalist/interpreter. He believes being an expert naturalist is less important than being able to communicate through words and writing what you know. By communicating what you know, you may stimulate people's curiosity about the natural world.

Other duties of a naturalist/interpreter include designing and constructing exhibits, preparing brochures and educational materials, providing information for visitors, doing natural history research, identifying plants and animals, caring for small animals, and training volunteers. Because naturalist/interpreters sometimes lead hikes into the back country, they must be in good physical health.

Many people interested in natural history become trained volunteers, or *docents*. Docents assist in all the duties of the naturalist/interpreter and often present programs and lead walks themselves. Volunteers also may become involved in programs with senior citizens and scouts. Docents with special skills or interests, such as photography and art, often are able to use these skills. No special education is required to become a volunteer—just a strong interest in natural history and a desire to share this knowledge with others.

Positions such as naturalist volunteers and naturalist/interpreters can be very rewarding because they provide a service. This service will be needed even more in the future as the relationship between humans and the environment becomes more critical.

population numbers can be increased by using such technologies as the establishment of breeding populations in captivity, the reintroduction of populations into the wild, and the relocation of populations from one "island" of habitat to another. For example, the Andean condor has been successfully reintroduced, many types of monkeys have been moved to new areas where their populations are less threatened, sandhill cranes have been used to incubate the eggs of whooping cranes, and the eggs of giant pandas have been artificially fertilized. In many cases, zoos and botanic gardens have established breeding populations of many different plants and animals. The need for these technologies may well increase in the years to come as the biosphere changes from a system of wild ecosystems to a system of managed nature preserves and zoos. These preserves may become the final refuge for many species.

Concept Review

1. How does the number of types of organisms in an ecosystem relate to the stability of that ecosystem?
2. What happens to the concentration of a biocide in the members of a food chain?
3. Explain how humans can influence ecosystems by changing the environment.
4. Why is maintaining biodiversity important?

Summary

Some of the relationships in a community are direct, but most are indirect. A single organism may have many types of relationships with other organisms, and these relationships define the organism's niche. Relationships include predation, competition, and types of symbiosis. The organism's habitat, on the other hand, is the abiotic environment in which it lives. The niches of organisms are distinct, but the boundaries between ecosystems are not clear because energy and matter pass from one ecosystem to another. In most ecosystems, there are more producers than consumers, and there is stability through time. Humans can affect the populations of producers or consumers by what they do. Sometimes, as when DDT is sprayed, the results are many, surprising, and undesirable. In some cases, human activity can lead to the extinction of another species, sometimes many species. This loss of biodiversity may have negative consequences for humans in the future. Humans can use different strategies, such as preserving wilderness areas and reintroducing populations, to help slow down the rate of species diversity loss.

Applications

1. How does the concept of ecosystems help us to understand the biosphere?
2. What would happen if all the turtles were killed within the Florida river community?
3. Predict what would happen to a community if the number of carnivores present suddenly were doubled.
4. A predator affects the population of its prey, but a parasite may not have the same effect on its host population. Why?

5. Explain the relationship between habitats and niches.

6. Why is it difficult to describe an organism's niche accurately?

7. Most organisms can survive only within a narrow temperature range. Does this mean that temperature is an environmental factor for which organisms compete? Explain your answer.

8. Suppose there are two different types of plants growing together in one pot. When the two plants are separated and planted in different pots, one plant dies and the other plant starts to grow much larger. What type of relationship may these plants have had?

9. Suppose you were trapped on a desert island with 12 chickens and a bushel of corn as your only food. You will not be rescued for a month, which is not a long enough time to grow more corn. Which would you do to survive: (a) eat the corn first, then the chickens; (b) eat the chickens first, then the corn; (c) feed the corn to the chickens, then eat the eggs laid by the chickens? Give reasons for your selection.

10. Although pesticides may be beneficial to farmers, they have serious side effects on ecosystems and nonpest organisms, including humans. Many people advocate using natural predators to control pest populations. What are some advantages to this solution? What are some drawbacks?

Problems

1. How might an ecologist find out whether the effect of one type of organism on another is beneficial, harmful, or neutral?

2. Study a community in your city or town. Gather information to answer these questions: (a) How does the community get its energy? (b) Which organisms would be present and which organisms would be absent if humans were not part of the community?

3. Investigate these two parasite-host relationships: (a) chestnut tree and chestnut blight; (b) whooping cough and humans. How are the hosts affected, and how have humans tried to control the parasites?

4. Describe your niche. Build lists of organisms that you affect and that affect you. Describe what each interaction is, and whether it is direct or indirect.

5. Do you think that biocides should be sprayed on your food or not? Write an essay describing the hazards of biocide use, and the potential benefits and problems associated with growing food entirely without biocides.

6. What does "endangered species" mean? You can obtain a list of the endangered plants and animals in your state from the state's natural heritage program, from *National Wildlife Magazine*, or from national wildlife organizations. Collect or create drawings, photos, or other illustrations of endangered species to display in your classroom. What can you do to keep these species from vanishing?

7. Two major wildlife problems are human activities, which can lead to or speed up the extinction of species, and the introduction of alien species into an ecosystem. How can the introduction of an alien species, such as the water hyacinth or the Norway rat, upset the balance in an ecosystem? Research an alien species and prepare a report for the class.

What happened to this area about three weeks before this photo was taken? Nearly all the energy stored in the organisms previously present was consumed. Give three specific examples of what remains in this environment that these seedlings are using.

Matter and Energy in the Web of Life

4

All living things are tied together by their need for matter and energy. Using energy from the sun and absorbing matter from the surrounding soil and air, producers, such as green plants, make their own food. Consumers, on the other hand, must obtain their matter and energy from other organisms. Humans, for instance, eat plants and animals to get their matter and energy. This matter and energy (food) is made up of biological molecules—the molecules that are found in all living things. We eat the molecules of plants and animals and then rearrange them to make our own molecules. This chapter examines these important molecules and investigates some of the characteristics of matter and energy.

Matter and Energy

Guidepost: How are matter and energy related in the biosphere?

4.1 Matter Is Made of Atoms

Biological molecules are **organic compounds**—compounds built of carbon combined with other elements. Like all other compounds, organic compounds can be broken down to the elements from which they were formed. For example, molecules of sugar, an organic compound, can be broken down to the elements carbon, hydrogen, and oxygen. Elements cannot be broken down to other types of substances. Carbon, iron, nitrogen, gold, silver, calcium, and chlorine are some elements with which you probably are familiar. Four elements—carbon, oxygen, hydrogen, and nitrogen—are common to all living systems.

Elements are composed of atoms, which are made of even smaller particles. Each atom has a core, or nucleus (NOO klee us), that contains positively charged particles called protons (PROH tahnz) and uncharged particles called neutrons (NOO trahnz). Rapidly orbiting the nucleus are one or more electrons (ee LEK trahnz), which are negatively charged particles. Because the number of electrons equals the number of protons, an atom is electrically neutral.

The numbers of particles in an atom determine what element is formed. For example, a hydrogen atom is made of one proton and one electron. (Hydrogen is the only element that does not have neutrons in its atoms.) An atom of carbon contains six protons, six neutrons, and six electrons.

77

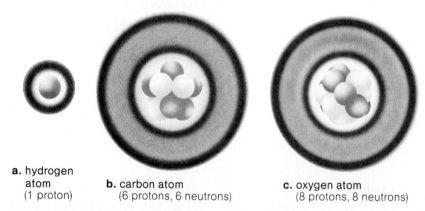

a. hydrogen atom (1 proton)

b. carbon atom (6 protons, 6 neutrons)

c. oxygen atom (8 protons, 8 neutrons)

Figure 4.1   Simplified models of (a) hydrogen, (b) carbon, and (c) oxygen. Red circles represent protons, yellow circles represent neutrons. The models show electrons as a cloud of negative charge around the nucleus. The number of electrons in the outer level determines chemical activity.

Oxygen atoms are composed of eight protons, eight neutrons, and eight electrons. Figure 4.1 shows models of these atoms.

Reactions between atoms depend on the number of electrons the atoms have. Sometimes a reaction involves an electron moving from one atom to another. For example, when atoms of sodium and atoms of chlorine react to form table salt, each sodium atom gives up an electron to a chlorine atom, as shown in Figure 4.2. As a result, the number of protons and the number of electrons in the sodium atom are no longer equal. Because the sodium atom has 11 protons but only 10 electrons, it has a positive charge. The chlorine atom, which captured the electron from the sodium atom, now has one more electron that it had originally. Consequently, it has a negative charge. A charged particle that has either a negative or positive

Figure 4.2   Sodium and chlorine can react to form the salt sodium chloride (NaCl). By losing one electron, sodium becomes a positive ion, and by gaining one electron, chlorine becomes a negative ion, chloride.

sodium atom   **a**   sodium ion (Na⁺)   **b**

11 electrons   (+)   10 electrons

1 electron

17 electrons   (−)   18 electrons

chlorine atom   chloride ion (Cl⁻)

sodium + chlorine ⟶ sodium chloride

charge is called an **ion** (EYE on). The chlorine atom has become an ion with a negative charge—a chloride ion. The sodium atom has become an ion with a positive charge. The positively charged sodium ions and the negatively charged chloride ions are attracted to each other and come together, forming sodium chloride, or table salt.

Often when atoms react, they do not gain or lose electrons. Instead, they share electrons. For example, in a molecule of water, one oxygen atom shares electrons with two hydrogen atoms (see Figure 4.3). Molecules of carbon dioxide, hydrogen gas, and oxygen gas also are formed by shared electrons. The attractions that hold atoms or ions together are called **chemical bonds**.

Figure 4.3   In a molecule of water, the oxygen atom forms an electron-sharing bond with each hydrogen atom.

4.2   Chemical Reactions Are Essential to Life

There are two basic types of chemical reactions in living cells. When sodium and chlorine combine to form table salt, or oxygen and hydrogen combine to form water, compounds are built up. Reactions that lead to the buildup of compounds are called **synthesis** (SIN thih sis) reactions. Compounds also may be broken down. When this happens, the reaction is known as a **decomposition** (de kom poh ZISH un) reaction. The digestion of foods involves decomposition reactions.

For chemical reactions to take place, the reacting substances must come in contact with each other. This happens most easily when the substances are in solution, that is, dissolved in water. When table salt dissolves in water, the sodium and chloride ions separate from each other, but they remain as ions in solution. Nonionic compounds, such as water, also can be converted into ions through a process called **ionization** (see Figure 4.4). Stated in simple terms, during ionization, water molecules separate into hydrogen ions (H^+) and hydroxide ions (OH^-). A hydrogen ion is a single proton—a hydrogen atom that has lost its only electron. The missing electron is held by the hydroxide ion, which consists of an oxygen atom, a hydrogen atom, and the electron the hydrogen atom has lost.

Although water is a common compound, it has unique properties, such as ionization, that make it essential for life. Only about one in 10 million molecules of water forms ions, but all life processes depend on this small

Figure 4.4   Water molecules ionize into hydrogen and hydroxide ions.

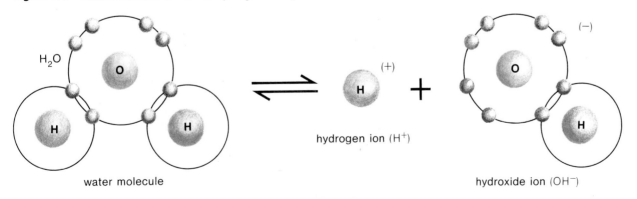

H_2O

water molecule

hydrogen ion (H^+)

hydroxide ion (OH^-)

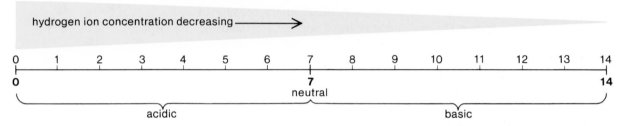

Figure 4.5   A pH scale. When is a substance acid? When is it basic?

Figure 4.6   The pH of several common substances.

percentage. Hydrogen and hydroxide ions are involved in most of the reactions that occur in organisms. If more hydrogen ions than hydroxide ions remain in solution after a reaction, the solution is said to be acidic. If more hydroxide ions than hydrogen ions remain, the solution is said to be basic, or alkaline. The relative levels of hydrogen and hydroxide ions are very important to organisms because of their effects on chemical reactions.

The hydrogen ion level of a solution is described by a range of numbers known as the **pH scale** (see Figure 4.5). The scale runs from 0 to 14. A solution that has the same number of hydrogen and hydroxide ions is neutral and has a pH of 7. Pure water has a pH of 7. As the hydrogen ion level rises, the solution becomes more acidic, and the pH drops. Thus, a solution with a pH of 2 is highly acidic. Solutions with a pH above 7 are basic. They have relatively low levels of hydrogen ions and high levels of hydroxide ions. Figure 4.6 shows the pH of several common substances.

Organisms have an internal pH that must remain fairly stable for chemical reactions to occur. Environmental factors may affect that stability in a variety of ways. In Investigation 4.1 you can discover how internal pH is regulated.

Even when they are dissolved in water, most atoms and molecules react extremely slowly, if at all. Not enough of them come into contact with each other. There are, however, certain substances that promote chemical reactions by increasing the chances of contact. These substances are called **catalysts** (KAT uh lists). Catalysts are present only in small amounts, and although they participate in the reactions, they themselves are not changed or used up in the reactions. Catalysts make it possible for reactions to occur at rates high enough to sustain life. The specialized and highly specific catalysts present in organisms are called **enzymes** (EN zymz). These are discussed in Section 4.11.

INVESTIGATION 4.1   Organisms and pH

Chapter 2 discussed the tendency for populations to maintain a relative stability, or homeostasis. Individual organisms and cells must maintain an internal homeostasis, but many factors can affect that stability—for example, the relative concentrations of hydrogen ions (H^+) and hydroxide ions (OH^-). The biochemical activities of living tissues frequently affect pH, yet life depends on maintaining a pH range that is normal for each tissue or system. Using a pH probe and the LEAP-System™, or a pH meter or wide-range pH paper if the LEAP-System is not available, you can compare the responses of several materials to the addition of an acid and a base. Instructions for use of the LEAP-System are provided in the LEAP-System BSCS Biology Lab Pac 2.

Before you begin, study the investigation and develop a hypothesis that answers the question "How do organisms survive and function despite metabolic activities that tend to shift pH toward either acidic or basic ends of the scale?"

Materials (per team of 4)

4 lab aprons
4 pairs of safety goggles
50-mL beaker or small jar
50-mL graduated cylinder
3 colored pencils
Apple IIe, Apple IIGs, or IBM computer
LEAP-System™
printer
pH probe
pH meter or wide-range pH paper
  (optional)

forceps (optional)
tap water
0.1M HCl in dropping bottle
0.1M NaOH in dropping bottle
sodium phosphate pH 7 buffer solution
liver homogenate
potato homogenate
egg white (diluted 1:5 with water)
warm gelatin suspension, 2% solution

SAFETY Put on your safety goggles and lab apron.

Procedure

1. In your data book, prepare a table similar to Table 4.1.
2. Pour 25 mL of tap water into a 50-mL beaker.
3. Record the initial pH by using a pH probe with the LEAP-System (Figure 4.7), a pH meter, or by dipping small strips of pH paper into the water and comparing the color change to a standard color chart.
4. Add 0.1M HCl a drop at a time. Gently swirl the mixture after each drop. Determine the pH after 5 drops have been added. Repeat this procedure until 30 drops have been used. Record the pH measurements in your table.

    CAUTION: 0.1M HCl is an irritant and may destroy clothing. Avoid skin/eye contact; do not ingest. Should a splash or spill occur, call your teacher immediately; flush the area with water for 15 minutes; rinse mouth with water.
5. Rinse the beaker thoroughly and pour into it another 25 mL of tap water. Record the initial pH of the water and add 0.1M NaOH drop by drop, recording the pH changes in exactly the same way as for the 0.1M HCl.

Figure 4.7 The LEAP-System with pH probe.

TABLE 4.1 Testing pH

Solution Tested	Tests with 0.1M HCl pH after addition of							Tests with 0.1M NaOH pH after addition of						
	0	5	10	15	20	25	30 drops	0	5	10	15	20	25	30 drops
Tap water														
Liver														
Potato														
Egg white														
Gelatin														
Buffer														

CAUTION: 0.1*M* NaOH is an irritant and may destroy clothing. Avoid skin/ eye contact; do not ingest. Should a spill or splash occur, call your teacher immediately; flush the area with water for 15 minutes; rinse mouth with water.

6. Using the biological material assigned by your teacher, repeat Steps 2–5. Record the data in your table.

7. Test the buffer solution (a nonliving chemical solution) using the same method outlined in Steps 2–5. Record the data in your table.

8. Wash your hands thoroughly before leaving the laboratory.

Discussion

1. Summarize the effects of HCl and NaOH on tap water.

2. What was the total pH change for the 30 drops of HCl added to the biological material? For the 30 drops of NaOH added? How do these data compare with the changes in tap water?

3. Run stripcharts to obtain individual graphs of the data. In your data book, prepare a simple graph of pH versus the number of drops of acid and base solutions added to tap water. Plot two lines—a solid line for changes with acid and a dashed line for changes with base. Using different colored solid and dashed lines, add the results for your biological material. Compare your graph to the graphs of teams who used a different biological material. What patterns do the graphs indicate for biological materials?

4. How do biological materials respond to changes in pH?

5. Use different colored solid and dashed lines to plot the reaction of the buffer solution on the same graph. How does the buffer system respond to the HCl and NaOH?

6. Is the pH response of the buffer system more like that of water or of the biological material?

7. How does the reaction of the buffer solution serve as a model for the response of biological materials to pH changes?

8. Would buffers aid or hinder the maintenance of homeostasis within a living cell in a changing environment?

9. What does the model suggest about a mechanism for regulating pH in an organism?

Figure 4.8 Is this person doing work?

4.3   Energy Makes Work and Order Possible

Chemical reactions usually involve energy. In general, synthesis reactions require an input of energy. Decomposition reactions usually release energy. What is energy, and how is it used in living systems?

Understanding matter is fairly easy. You can see it, you can touch it, and you can weigh it. Understanding energy is more difficult. In a general way, energy can be defined as the ability to do work or to cause change. It is work to move an arm, play tennis, jump the high jump, heat a house, or build a skyscraper (see Figure 4.8). Growing a leaf or a wing also can be considered work because energy is used in these processes. In a cell, energy is used to move substances and to build complex molecules. This, too, is work.

Energy also is required to make and maintain order. Living things are extremely complex. Their atoms and molecules are arranged in highly or-

From *Living in the Environment*, Third Edition, by G. Tyler Miller, Jr. © 1982 by Wadsworth, Inc. Used by permission.

Figure 4.9   The tendency toward increasing disorder of a system and its surroundings.

ganized systems. High levels of organization, however, can be unstable. If left to themselves, all systems tend to become simple and random, or disorganized. Buildings crumble, weeds grow in the garden, and dead organisms decay. Only by a continual input of energy can organization be maintained.

Think about your room. When it is clean and neat, it is an organized system. How long does it stay that way? The tendency is for your room to become disorganized with your books and clothes spread out in a random fashion (see Figure 4.9). To get your room organized again requires energy and work.

Like your clean room, a living organism such as a frog is an organized system. The frog maintains its internal organization by eating flies and other insects that contain energy. This energy keeps the frog alive and allows it to grow and reproduce. If the frog does not get enough food, it dies. As soil decomposers break down the frog's body, it becomes disorganized. The decomposers live by using the matter and energy remaining in the frog's body cells.

Many life processes tend to bring about an organized state with a minimum of randomness. That requires energy. Where does the energy come from?

Concept Review

1. How are the protons, neutrons, and electrons within an atom related to one another?
2. How are chemical bonds formed?
3. In terms of pH values, what is the difference between neutral, acidic, and alkaline solutions?
4. How is matter different from energy?
5. What is a catalyst?
6. Why do all organisms need energy?

Guidepost: What is the source of energy for living organisms and how is it used?

Energy For Life

4.4   Photosynthesis Is the Source of Your Energy

Biological activity, like all other types of activity, requires energy. Consumer organisms get their energy from the food they eat, but where do the producers get their energy? Usually, their energy comes from the sun. Because no organism can use light energy directly from the sun as a source of food energy, the energy must be converted to chemical energy.

In the process of photosynthesis, green plants (and some other organisms) convert light energy from the sun into chemical energy that is stored in complex food molecules. That chemical energy then can be used either by the plants themselves or by organisms that eat the plants. Because animals cannot make their own food, most animals depend on plants for their source of energy as well as matter. Directly or indirectly, therefore, photosynthesis is the source of energy for biological activity in almost all organisms on earth.

The first step in photosynthesis is the absorption of light energy by a green plant. The energy is absorbed primarily by **chlorophyll** (KLOR uh fil), a green pigment that gives plants their color. A plant also absorbs carbon dioxide molecules from the air and water molecules from the soil. Light energy, carbon dioxide, and water are the raw materials used to make sugars. The light energy absorbed by the plant is used to break down water molecules into hydrogen and oxygen. The hydrogen is combined with carbon dioxide to form sugar molecules. Thus, some of the light energy absorbed by the plant is stored in the sugar molecules as chemical energy. The oxygen is released into the air as oxygen gas. Figure 4.10 highlights these events.

Figure 4.10   The energy in sunlight is converted to chemical energy during photosynthesis.

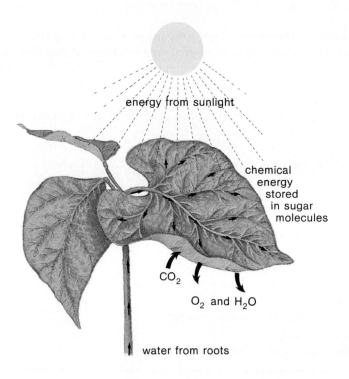

energy from sunlight

chemical energy stored in sugar molecules

CO_2

O_2 and H_2O

water from roots

When sugars are formed, several small molecules are linked together by chemical bonds. The energy used to form the sugars is stored in the structure of the molecules. When sugar molecules are broken down in a cell, the energy stored in them is released. That energy is used by the cell to do cellular work.

4.5   Energy Is Released as Food Is Broken Down

How is the energy stored in the structure of the molecules released? The major energy-releasing process is **cellular respiration** (SEL yoo ler   res pih RAY shun), a series of chemical reactions that occur in all living cells. During these reactions, the sugars made in photosynthesis are broken down, and energy is released.

The way in which different organisms release energy from food is remarkably similar. Foods are compounds of carbon, hydrogen, oxygen, and often other elements—building blocks for the cell. Chemically, foods are similar to substances such as wood, coal, and oil, which can be used as fuels because they contain chemical energy. During the chemical reactions that occur in burning, fuels are reduced to simpler compounds. The chemical energy from the fuel is released in the form of heat and light, as shown in Figure 4.11a.

The chemical energy in food is also released by chemical reactions. The chemical reactions in a cell, however, are quite different from those in a fire. When fuels burn, a large amount of energy is released in a short time. The sudden release of energy produces high temperatures—high enough to provide heat for cooking. Cells, however, would be destroyed by such high temperatures. Energy-releasing chemical reactions occur in cells at low temperatures.

In cells, the same total amount of energy is released as in burning, but it is released gradually, in many small steps that are controlled by enzymes (see Figure 4.11b). The cell uses the energy in food that is released during respiration to move substances and to carry out other cell work. Carbon dioxide and water molecules are the by-products that are formed as the food molecules are broken down. Notice that carbon dioxide and water are

Figure 4.11   A comparison of energy released during (a) burning and during (b) cellular respiration. How are they the same? How are they different?

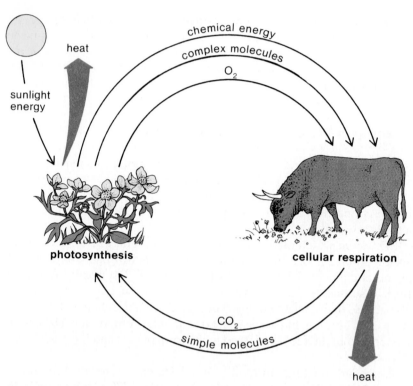

Figure 4.12   Photosynthesis and cellular respiration are two important links in an ecosystem.

the same molecules the plant uses to make sugars in photosynthesis. Figure 4.12 above summarizes the relationships between photosynthesis and cellular respiration.

4.6   Energy Is Used in Small Packets

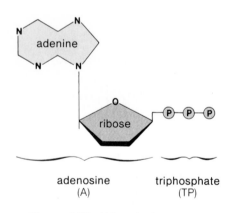

Figure 4.13   Where is the energy stored in ATP?

If you had one $100 bill, you might find it difficult to buy small things such as a hamburger, a pack of notebook paper, or a comb. It would be much easier if you had 100 $1 bills. So, too, with the energy in a cell. The large amounts of energy in food molecules are converted into "small change." This "small change" is the chemical energy stored in compounds such as **ATP**, adenosine triphosphate (uh DEN oh seen   try FOS fayt).

ATP is the most important of several energy-transfer compounds that are found in all organisms. These compounds are involved in all the energy processes of living cells. The energy released during respiration is temporarily transferred to molecules of ATP. As you can see in Figure 4.13, each ATP molecule is made up of a main section (A) to which are attached three identical groups of atoms called phosphates (TP).

As food molecules are broken down to simpler compounds, a great deal of energy is released. The energy is used to make ATP molecules. Energy is stored in ATP until it is released by reactions that remove the third phosphate group from ATP. This energy is used to help the cell do its work.

The work may be to move a muscle, to send a nerve impulse, to grow, or to form new compounds. Thus, ATP is a carrier of chemical energy in the cell.

Each ATP molecule releases a bit of energy whenever a phosphate group is broken off or transferred to another molecule. The molecule that remains, which has only two phosphate groups, is called **ADP**, adenosine diphosphate (uh DEN oh seen dy FOS fayt). Just as you cannot keep spending money from your pocket without eventually putting more money in, a cell cannot continue to "spend" its ATP without rebuilding it. As you can see in Figure 4.14, there is a continual ADP-ATP cycle. To make ATP molecules, an ADP molecule and a phosphate group are needed as well as chemical energy to combine the two. This energy comes from the breakdown of food molecules.

Figure 4.14   Explain the ATP-ADP cycle.

Concept Review

1.   In what way is photosynthesis important for all living organisms?
2.   What are the two products of photosynthesis?
3.   How are the reactions of photosynthesis and cellular respiration similar? How are they different?
4.   Is there more energy in one molecule of ATP or one molecule of sugar? How do you know?
5.   Explain the relationship between ATP and ADP.

Biological Molecules

INVESTIGATION 4.2   Compounds in Living Organisms

The compounds your body needs for energy and building materials are carbohydrates, proteins, fats, vitamins, and other nutrients. These compounds are present in the plants and animals you use as food. In this investigation, you will observe tests for specific compounds and then use those tests to determine which compounds are found in ordinary foods.

Guidepost: How are carbon-containing molecules important to living things?

Materials (per team of 4)

4 pairs of safety goggles
4 lab aprons
4 pairs of plastic gloves
250-mL beaker
10-mL graduated cylinder
6 18-mm × 150-mm test tubes
test-tube clamp
hot plate
Benedict's solution in dropping bottle

Biuret solution in dropping bottle
indophenol solution in dropping bottle
Lugol's iodine solution in dropping bottle
1% silver nitrate solution in dropping bottle
isopropyl alcohol (99%) in screw-top jar
brown wrapping paper
apple, egg white, liver, onion, orange, potato, or other foods of your choice

SAFETY   **Put on your safety goggles, lab apron, and gloves. Tie back long hair.**

TABLE 4.2   Reagent Tests of Known Food Substances

Food Substance	Reagent Test	Results
Gelatin	Biuret solution	
Glucose	Benedict's solution	
Starch	Lugol's iodine solution	
Vitamin C	Indophenol solution (0.1%)	
Sodium chloride	Silver nitrate solution (1%)	
Butter or vegetable oil	Brown paper	

Procedure

PART A   Test Demonstration

1. In your data book, prepare a table similar to Table 4.2.

2. Scientists use special chemical solutions, or reagents, to detect the presence of certain compounds. Observe the six reagent tests your teacher performs. In your table, describe the results of each test.

PART B   Compounds in Foods

3. In your data book, prepare a table similar to Table 4.3. Then, record the presence (+) or absence (−) of each substance in the foods you test.

WARNING: The reagents you will use in this procedure may be corrosive, poisonous, and/or irritants, and they may destroy clothing. Avoid skin and eye contact; do not ingest. Should a splash or spill occur, call your teacher immediately; flush the area with water for 15 minutes; rinse mouth with water.

4. Predict what substances you will find in each sample your teacher assigns to you. Then, test the samples as your teacher demonstrated or as described in Steps 5–10. Record the result of each test in your data book, using a + or −.

TABLE 4.3   Analysis of Compounds in Common Foods

Substance		Protein	Glucose	Starch	Vitamin C	Chloride	Lipid
Egg	Prediction						
	Test results						
Potato	Prediction						
	Test results						
Etc.	Prediction						
	Test results						

5. Protein test: Place 5 mL of the assigned food in a test tube. Add 10 drops of Biuret solution.

6. Glucose test: Add 3 mL of Benedict's solution to 5 mL of the assigned food. Place the test tube in a beaker of boiling water and heat for five minutes.

   WARNING: Use test-tube clamps to hold hot test tubes. Boiling water will scald, causing second-degree burns. Do not touch the beaker or allow boiling water to contact your skin. Avoid vigorous boiling. Should a burn occur, call your teacher immediately; place burned area under *cold* running water.

7. Starch test: Add 5 drops of Lugol's iodine solution to 5 mL of the assigned food.

8. Vitamin C test: Add 8 drops of indophenol to 5 mL of the assigned food.

9. Chloride test: Add 5 drops of silver nitrate solution to 5 mL of the assigned food.

10. Fat test: Rub the assigned food on a piece of brown wrapping paper. Hold the paper up to the light. When food contains only a small amount of fat, it may not be detected by this method. If no fat has been detected, place the assigned food in 10 mL of a fat solvent such as isopropyl alcohol (99%). Allow the food to dissolve in the solvent for about five minutes. Then, pour the solvent on brown paper. The spot should dry in about ten minutes. Check the paper.

    WARNING: Isopropyl alcohol is flammable; its vapors may explode. Keep away from heat and sparks. Extinguish any open flames in the area.

11. Wash your hands thoroughly before leaving the laboratory.

Discussion

1. How did your predictions compare with the test results?
2. Which of your predictions was totally correct?
3. Which foods contained all the compounds for which you tested?
4. On the basis of your tests, which food could be used as a source of protein? glucose? starch? vitamin C? fat?
5. How might the original colors of the test materials affect the results?

4.7 Carbon Is Found in All Living Things

Although organisms are composed of many different chemical elements, carbon is the central element for all living systems. Carbon atoms can join together to form chains or rings, as shown in Figure 4.15. Furthermore, carbon atoms can combine with hydrogen, oxygen, nitrogen, sulfur, and phosphorus to form a vast number of organic compounds. In fact, the atoms of the elements present in organic compounds can be arranged in so many ways that the variety of organic compounds is almost limitless. This variety ensures the uniqueness of each organism.

Organic compounds are the essential building blocks for organisms and are also their major source of chemical energy. Four basic types of carbon-containing molecules are found in all organisms. **Carbohydrates** (kar boh HY drayts) and **lipids** (LIP idz) are important energy-storing compounds. They also form part of the cell structure. The sugars produced in photosynthesis and used in respiration are carbohydrates. Starch is a large carbohydrate molecule made by joining many individual sugar molecules. Oils and fats are examples of lipids. **Proteins** (PROH teenz) function as enzymes and form part of the cell structure. Muscles ("meat") are composed largely of protein. **Nucleic** (noo KLEE ik) **acids** are the hereditary, or

straight chain

branched chain

ring

Figure 4.15 Carbon atoms can bond in several ways. The unconnected lines protruding from some of the carbon atoms show that any one of a number of elements can bond with the carbon in these positions.

genetic, material for all organisms. They also coordinate the activities of the cell.

Because all living organisms contain the same types of biological molecules, you can get the molecules you need for life from another organism. In fact, that is the only way you can get the molecules you need. To live, you must eat a plant or animal and rearrange its molecules and atoms into your own molecules and atoms.

4.8 Carbohydrates Are Used for Energy, Storage, and Building

Carbohydrates contain only the elements carbon, hydrogen, and oxygen. Hydrogen and oxygen are present in the same proportions as in water. The subunits, or building blocks, of carbohydrates are single sugars, such as glucose and fructose (see Figure 4.16a), which contain no more than seven carbon atoms in each molecule. Glucose molecules can be changed into other biological molecules within the cell. They are a major source of energy for most organisms. Two single sugars may bond together to form a double sugar, as shown in Figure 4.16b. The most familiar double sugar is sucrose, commonly called cane, beet, or table sugar. Sucrose is formed by a chemical reaction that combines a glucose and a fructose molecule.

In synthesis reactions, many glucose molecules may bond together to build complex carbohydrates such as starch and cellulose (SEL yoo lohs),

Figure 4.16 Single sugars (a) can combine to form double sugars (b). Starch, glycogen, and cellulose (c) are formed by linking together many glucose units.

glucose   fructose

a   (single sugars)

sucrose   water

b   (double sugar)

starch or glycogen

cellulose

c   (polysaccharides)

Figure 4.17 A fat molecule consists of fatty acids joined to a glycerol molecule. To form a fat, one molecule of glycerol combines with three molecules of fatty acids. The fatty acids in one fat may be alike or different. Examine the diagram. What is a by-product of this reaction?

represented in Figure 4.16c. Starch is an energy-storage compound in many plants and an important food source for humans. (See Chapter 15.) Cellulose, a major part of wood and cotton fibers, gives the cell walls surrounding plant cells their rigidity. In the human liver and muscles, carbohydrates are stored as glycogen, also called animal starch. Molecules of starch, cellulose, and glycogen consist of thousands of molecules of glucose units and have no fixed size.

4.9   Lipids Are Efficient Energy Storage Compounds

Like carbohydrates, lipids are composed only of carbon, hydrogen, and oxygen atoms, but lipids contain fewer oxygen atoms than do carbohydrates. Simple fats are the lipids most common in human diets and bodies. These fats are made of two types of lipid building blocks: glycerol, which is a single sugar, and fatty acids, which are chains of carbon and hydrogen with an acid grouping on one end (see Figure 4.17). Both building blocks can be formed in cells from glucose.

Both carbohydrates and lipids are important energy-storage compounds in organisms. A gram of fat, however, contains more than twice as much chemical energy as a gram of carbohydrate. Therefore, fats are better storage compounds. As animals prepare for winter when food is scarce, they eat large amounts of food. Much of this food energy is converted into fat, and the fat levels in their bodies increase dramatically. Lipids are also essential structural parts of all cells. Lipids include plant waxes and cholesterol (koh LES ter ol), in addition to simple fats. The chemical structure of cholesterol is illustrated in Figure 4.18. Although cholesterol is needed by the body, excess cholesterol in the blood has been linked to heart disease.

cholesterol

Figure 4.18 A diagram representing the structure of cholesterol.

Figure 4.19   (a) The formation of a dipeptide from two amino acids. Polypeptide chains (b) are formed from long strings of amino acids.

4.10   Muscles, Enzymes, and Many Cell Parts Are Made of Protein

Figure 4.20   The polypeptide chain in a myoglobin molecule is coiled and folded into a three-dimensional structure.

Protein molecules form part of the structure of every cell. Proteins also form enzymes and muscle tissue. Usually, protein molecules contain thousands of atoms—sometimes tens of thousands. The subunit of a protein molecule is an **amino** (uh MEEN oh) **acid**, shown on the left side of Figure 4.19a. Amino acids always contain at least four types of atoms: carbon, hydrogen, oxygen, and nitrogen. There are two amino acids that also contain sulfur. Twenty different types of amino acids can be found in a protein molecule. Green plants can synthesize all of these amino acids from simple materials. Animals cannot. They must get some amino acids ready-made in the food they eat. Unfortunately, not every type of food contains all the needed amino acids. Therefore, animals need a balanced diet of protein sources. Without it, protein-deficiency diseases may occur.

To synthesize a protein, amino acids must be linked together. First, two amino acids are linked together to form a dipeptide (DY PEP tyd), as shown in Figure 4.19a. A long chain of amino acids is a **polypeptide** (POL ee PEP tyd), illustrated in Figure 4.19b. Some proteins are made of only one polypeptide, but most are made of two or more polypeptides bonded together. Polypeptide chains are coiled and folded into complex three-dimensional structures to form proteins such as myoglobin, shown in Figure 4.20. The structure of the protein determines how it will function.

Thousands of different proteins can be made from various combinations of the 20 different types of amino acids. This variety is important in a living organism because each type of chemical reaction is controlled by a different type of enzyme, which is formed from proteins. The production of many different enzymes makes possible thousands of chemical reactions.

INVESTIGATION 4.3   Enzyme Activity

In this investigation, you will study several factors that affect the activity of enzymes. The enzyme you will use is catalase, which is present in most cells and found in high concentrations in liver and blood cells. You will use liver homogenate as the source of catalase. Catalase promotes the decomposition of hydrogen peroxide (H_2O_2) in the following reaction:

$$2H_2O_2 \xrightarrow{\text{catalase}} 2H_2O + O_2$$

Hydrogen peroxide is formed as a by-product of chemical reactions in cells. It is toxic and would kill cells if not immediately removed or broken down.

Materials (per team of 3)

3 pairs of safety goggles
3 lab aprons
50-mL beaker
2 250-mL beakers
10-mL and 50-mL graduated cylinders
reaction chamber
6 18-mm × 150-mm test tubes
forceps
square or rectangular pan

test-tube rack
nonmercury thermometer
filter-paper disks
buffer solutions: pH 5, pH 6, pH 7, pH 8
catalase solution
fresh 3% H_2O_2
ice
water bath at 37° C

SAFETY   **Put on your safety goggles and lab apron.
Tie back long hair.**

Procedure

In all experiments, make certain that your reaction chamber is scrupulously clean. Catalase is a potent enzyme, and if the chamber is not washed thoroughly, enough will adhere to the sides to make subsequent tests inaccurate. Measure all substances carefully. Results depend on comparisons between experiments, so the amounts measured must be equal or your comparisons will be valueless. Before you do the experiment, read through the instructions completely. Make sure that you have all the required materials on hand, that you understand the sequence of steps, and that each member of your team knows his or her assigned function.

PART A   The Time Course of Enzyme Activity

1. Prepare a table in your data book similar to Table 4.4 on page 94.
2. Obtain a small amount of stock catalase solution in a 50-mL beaker.
3. Obtain a reaction chamber and a number of filter-paper discs.
4. Place 4 catalase-soaked filter-paper discs high on *one* interior sidewall of the reaction chamber. (They will stick to the sidewall.) Prepare a disc for use in the reaction chamber by holding it by its edge with a pair of forceps and dipping it into

TABLE 4.4   Catalase Activity Under Various Conditions

Experiment	mL O$_2$ Evolved/30 sec																			
Reading	1	2	3	4	5	6	7	8	9	10	11	12	13	14	15	16	17	18	19	20
(Part A) Full concentration																				
(Part B) 3/4 concentration																				
Etc.																				

the stock catalase solution for a few seconds. Drain excess solution from the disc by holding it against the side of the beaker before you transfer it to the reaction chamber.

5. Stand the reaction chamber upright and carefully add 10 mL of 3% hydrogen per-oxide (H$_2$O$_2$) solution. *Do not allow the peroxide to touch the filter-paper discs.*

   Ⓡ   CAUTION: H$_2$O$_2$ is a reactive material. Avoid skin/eye contact; do not ingest. Should a splash or spill occur, call your teacher immediately; flush the area with water for 15 minutes; rinse mouth with water.

6. Tightly stopper the chamber.

7. Fill a pan almost full with water.

8. Lay the 50-mL graduated cylinder on its side in the pan so that it fills with water completely. If any air bubbles are present, carefully work these out by tilting the cylinder slightly. Turn the cylinder upside down into an upright position, keeping its mouth underwater at all times.

9. Making certain the side with the discs is at the top, carefully place the reaction chamber and its contents on its side in the pan of water.

10. Move the graduated cylinder into a position so that its mouth comes to lie directly over the tip of the dropping pipet extending from the reaction chamber, as shown in Figure 4.21. One member of the team should hold it in this position for the du-ration of the experiment.

11. Rotate the reaction chamber 180° on its side so that the hydrogen peroxide so-lution comes into contact with the soaked discs.

12. Measure the gas levels in the graduated cylinder at 30-second intervals for 10 min-utes. Record the levels in your data table.

Figure 4.21 Apparatus for measuring O$_2$ production in a reaction between catalase and hydrogen peroxide.

PART B   The Effect of Enzyme Concentration on Enzyme Activity

13. Test 3/4, 1/2, and 1/4 concentrations of enzyme solution, using the procedure for Part A with the following changes:
    a. 3/4 concentration: Use 3 catalase-soaked discs instead of 4.
    b. 1/2 concentration: Use 2 catalase-soaked discs as well as a 10-mL graduated cylinder.
    c. 1/4 concentration: Use 1 catalase-soaked disc and a 10-mL graduated cylinder.
14. Record all data in your data table.

PART C   The Effect of Temperature on Enzyme Activity

15. Add 10 mL of 3% H_2O_2 to each of two test tubes. Place one test tube in a beaker of ice water and the other in a beaker with water maintained at 37° C.
16. When the temperature of the chilled H_2O_2 reaches approximately 10° C, repeat Part A with the following changes:
    a. In Step 5, use 10 mL of chilled 3% H_2O_2.
    b. In Step 7, add ice to the pan to chill the water to approximately 10° C.
17. When the temperature of the warmed H_2O_2 reaches approximately 37° C, repeat Part A with the following changes:
    a. In Step 5, use 10 mL of warmed 3% H_2O_2.
    b. In Step 7, fill the pan with water warmed to approximately 37° C.
18. Record the data in your data table.

PART D   The Effect of pH on Enzyme Activity

19. Label 4 test tubes *pH 5*, *pH 6*, *pH 7*, and *pH 8*, respectively. Add to each of these 8 mL of 3% H_2O_2.
20. Add 4 mL of pH 5 buffer solution to the *pH 5* test tube, shaking well to ensure mixing. Do the same for the other buffer solutions.
21. Repeat Part A for each pH value, substituting the buffered 3% H_2O_2 solutions.
22. Record the results in your data table.
23. Wash your hands thoroughly before leaving the laboratory.

Discussion

1. In your data book, plot the data of Part A on a graph. Label the horizontal axis *Time (sec)*, and label the vertical axis *mL O_2 Evolved*. Does the action of catalase change through time?
2. Plot the data of Part B on the grid used for Part A and label the concentrations on the graph. Based on these data, how does enzyme activity vary with concentration?
3. Copy the graph for Part A and plot the data from Part C on it. Based on these data, how does temperature affect enzyme action?
4. Plot the results of all four runs of Part D on a third graph. How does pH affect enzyme action?
5. What is a buffer? Would Parts A, B, and C have been different if buffers had been used in them too? If so, how?
6. Summarize the general conditions necessary for effective enzyme action. Are these conditions the same for each enzyme? Why?
7. How would you design an experiment to show how much faster H_2O_2 decomposes in the presence of catalase than it does without the enzyme?
8. Inasmuch as the liver from which you obtained catalase was dead, how does it happen that the enzyme is still active?

4.11   Enzymes Catalyze Cell Reactions

Most enzymes are large, complex proteins made by the cell. Like other cat-alysts, enzymes promote reactions but are not used up in the reactions. They allow chemical reactions to take place at the temperature of the cell. Enzymes are needed in only small amounts, because one enzyme molecule can complete the same reaction thousands of times in a single minute.

As shown in Figure 4.22, the specific reaction catalyzed by an enzyme depends on the molecular structure and shape of a small area of the enzyme known as the **active site**, which can attract and hold only specific mole-cules. An enzyme must bind closely to the molecules—the **substrates** (SUB strayts)—on which it acts. Sometimes the enzyme changes shape slightly after the substrate binds, creating the necessary fit. Because only a few sub-strate molecules are enough alike in structure and shape to fit the active site, each enzyme can catalyze only specific chemical reactions.

To act as a catalyst, an enzyme must take part temporarily in a chemical reaction. The reacting molecules combine with the active site of an enzyme, forming an **enzyme-substrate complex**. The enzyme aligns the reacting molecules precisely and permits chemical changes to be completed rap-idly at low temperatures. Once the reactions are complete, the new mol-ecules break away, leaving the enzymes as they were before the reaction.

Enzymes catalyze both synthesis and decomposition reactions. Usually a different enzyme catalyzes each reaction. In a synthesis reaction, two or more small molecules combine with the enzyme. The enzyme provides the proper alignment, which enables these small molecules to join into one large molecule. In a decomposition reaction, the substrate combines with the enzyme and is split into two or more smaller molecules. Figure 4.23 on page 97 shows examples of both types of reactions.

Two aspects of enzyme activity are very important to cells. Enzyme re-actions are faster at high temperatures but only to a certain point. At higher temperatures, the enzymes may begin to lose their shape. Because fit is so

Figure 4.22   (a) A substrate binds to the enzyme at the active site. (b) Sometimes the active site changes shape after the substrate binds, bringing about the necessary fit. The binding produces an enzyme-substrate complex.

important for proper enzyme action, enzymes that lose their shape no longer function. Enzyme activity also varies with the pH of the solution. Thus, the temperature and the pH must be at the right levels for enzymes to act effectively, as you observed in Investigation 4.3.

4.12   Nucleic Acids Control the Activities of the Cell

Two types of nucleic acids are present in all cells and are vital to cell function: **RNA**, ribonucleic (ry boh noo KLEE ik) acid, and **DNA**, deoxyribonucleic (dee OK sih ry boh noo KLEE ik) acid. An organism may be made up of billions of cells, but each living cell has its own DNA and RNA. Information stored in DNA controls all cell activities and determines the genetic, or hereditary, characteristics of the cell and the organism. RNA is required for the synthesis of proteins, including enzymes.

Both DNA and RNA are made up of individual subunits called **nucleotides** (NOO klee oh tydz). Each nucleotide, in turn, is made up of three

Figure 4.23   In synthesis, two or more substrate molecules join at the active site, forming one larger molecule. In decomposition, the substrate combines with the enzyme and is split into two or more smaller molecules.

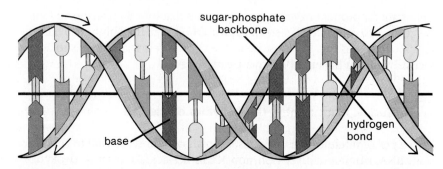

Figure 4.25   The structure of DNA. Alternating sugar and phosphate molecules connect each chain of nucleotides. Two chains are held together by specific pairing of the bases and coiled around a central axis to form a double helix. The axis is shown to help visualize the correct twisting. The two strands run in opposite directions, and there are 10 nucleotides per turn of the helix. Only a small portion of a DNA molecule is shown.

nucleotide

Figure 4.24   The parts of a nucleotide.

small molecules linked together: a phosphate group, a five-carbon sugar, and a nitrogen base (see Figure 4.24). The nitrogen base may be one of five different types. Each type is made up of carbon, hydrogen, oxygen, and nitrogen atoms. RNA nucleotides contain a single sugar called ribose (RY bohs). DNA, shown in Figure 4.25, contains a slightly different sugar called deoxyribose (dee OK sih RY bohs), which has one less oxygen atom than ribose. DNA molecules are made of two strands with thousands of linked nucleotides. The two strands are attached in a specific way and coiled to form a double helix. RNA molecules are usually smaller and may be single-stranded.

It is the information stored in DNA that determines the sequence of amino acids in proteins. This information also plays a major role in controlling when each protein is made. By controlling the synthesis of enzymes necessary for chemical reactions in the cell, DNA controls the activities of the cell. All the cells in an individual have the same unique genetic information in their DNA. That unique pattern of DNA results from the combination of genetic information provided by the individual's parents during sexual reproduction.

Concept Review

1. Why is the element carbon so important to living things?
2. What are the subunits of carbohydrates? of fats? of proteins? of nucleic acids?
3. Name one important function of each of the biological molecules named in Question 2.
4. Why is fat a better storage compound that starch?
5. How do enzymes work to catalyze a chemical reaction?
6. How is it possible for so many different types of proteins to be made with only 20 different types of amino acids?
7. What role do enzymes play in the release of energy from food?

The Carbon Cycle and Global Warming

Because carbon is present in many of the gases that compose the earth's atmosphere, it is a major determinant of climate. Although some carbon dioxide is released by the respiration of organisms, larger amounts are released when fossil fuels are burned and forests are cleared. Since the Industrial Revolution the atmospheric content of carbon dioxide has increased by about 25 percent.

As early as the nineteenth century, experts recognized that carbon dioxide in the atmosphere gives rise to a "greenhouse" effect. Think how a glass greenhouse allows sunlight to stream in but keeps heat from escaping. Similarly, carbon dioxide and other "greenhouse gases," such as methane, ozone, and nitrous oxide, allow sunlight to reach the earth but trap and reflect the heat. If there were no greenhouse gases, the earth would be cold and devoid of life. Too great a concentration of such gases, on the other hand, would cause the temperature to rise.

Human activity interrupts the carbon cycle in two ways. First, the burning of carbon-containing fuels adds carbon dioxide to the atmosphere. Second, the destruction of rain forests and other vegetation leaves less plant life to absorb carbon dioxide. When unharvested plant matter decays and when roots in the exposed soil react with oxygen, still more carbon dioxide enters the atmosphere.

Some scientists expect the increased levels of carbon dioxide, other gases, and water vapor to raise the earth's temperature from 1° C to 6° C in the next 100 years. Such an increase could affect global climate, food-producing patterns, and world population. Suppose, because of the rise in temperature, a population can no longer grow enough food to feed itself. If famine results, the population may migrate into a new area, compounding the problems of the population there and perhaps creating new ones.

Most experts agree on what the greenhouse effect is and how it works, but they disagree on the extent of the effects. Critics contend the computer models used to predict the greenhouse effect are so weak they do not account for the 0.5° C warming that has occurred in the last 100 years. Many experts feel that ignorance of the interactions between the oceans and the atmosphere limits their ability to predict effects. A new study, however, indicates that increased evaporation of the earth's oceans due to higher temperatures will amplify the greenhouse effect.

On the other hand, in February 1990, a federal government study indicated that the temperature in the southeastern United States actually has fallen 1° C during the past 30 years. Although the report does not disprove the occurrence of global warming, it should stimulate the debate.

1. Sunlight penetrating the atmosphere warms the earth's surface.

2. The earth's surface radiates heat to the atmosphere.

3. Greenhouse gases and water vapor absorb some and re-radiate part of them toward the earth.

4. When greenhouse gases build up in the atmosphere, more heat is trapped near the earth's surface.

Figure 4.26 Plants use the sugars made during photosynthesis in a variety of ways.

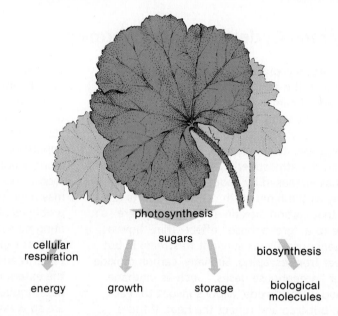

photosynthesis

sugars

cellular
respiration

biosynthesis

energy   growth   storage   biological
molecules

Guidepost: What is the source of matter for living things, and what happens to the matter?

Life Is Based on Carbon

4.13 Plants Make and Use Carbon-Containing Sugars

Recall from Section 4.4 that plants take up carbon atoms in the form of carbon dioxide from the air. During photosynthesis, they use the energy in sunlight to make sugars from carbon dioxide and water. In this way, the energy from sunlight and the carbon from carbon dioxide are stored in the sugars.

The sugars created during photosynthesis can be used in four ways by the plant, as shown in Figure 4.26. First, the plant may break down the sugar molecules immediately to release the stored energy. This happens during respiration. The energy that is released from the sugars during respiration may be used by plant cells to continue all the activities of life. Second, a plant may use the sugar molecules for growth. In this case, many sugar molecules are joined together to make the building materials necessary for more cells. Cellulose, for example, is one of these materials. Third, the plant may store sugars for future use. Starch is an important storage compound found in many parts of a plant. When the plant needs energy, starch is first broken down to individual sugar molecules. During respiration, the sugars are broken down to release energy. Fourth, sugar molecules may be converted into the other biological molecules needed for life.

4.14 Carbon Cycles Within an Ecosystem

As a plant grows, its body becomes larger. If the plant is eaten, the carbon in the plant is passed from producer to consumer. For the consumer to use the food, it must break down the plant body. As this happens, the carbon-containing molecules are broken apart, releasing both carbon atoms and energy. Much of the energy released is used for the activities of the consumer. Some of the carbon from the body of the plant is added to the body of the consumer. The rest of the carbon is exhaled into the air as carbon dioxide. For example, you take in carbon in all the foods you eat. You return carbon dioxide to the air every time you exhale. A plant also returns carbon dioxide to the air when it uses its own sugars as a source of energy. When another plant takes in this carbon dioxide during photosynthesis, the cycle of carbon through the community is complete.

Carbon dioxide also is returned to the air by decomposers. When producers or consumers die, decomposers begin their work. As its source of energy, a decomposer uses the energy locked in the bodies of dead organisms, and it uses the carbon from the bodies to build its own body. Carbon that is not used is returned to the air as carbon dioxide. Eventually, almost all the carbon that is taken in by plants during photosynthesis is returned to the air by the activity of decomposers.

Hundreds of millions of years ago, many energy-rich plant bodies were buried before decomposers could get to them. When that happened, the bodies slowly changed during long periods of time. They became a source of fuels—coal, oil, and natural gas. Today, when these fuels are burned, energy is released, and the carbon in the fuels is returned to the air as carbon dioxide. You can see that even the energy obtained from fuels is a result of photosynthesis. The process in which carbon is passed from one organism to another, then to the abiotic community, and finally back to the plants is called the **carbon cycle** (see Figure 4.27). Some of the other cycles found within ecosystems—the water cycle, the sulfur cycle, and the nitrogen cycle—are described in later chapters.

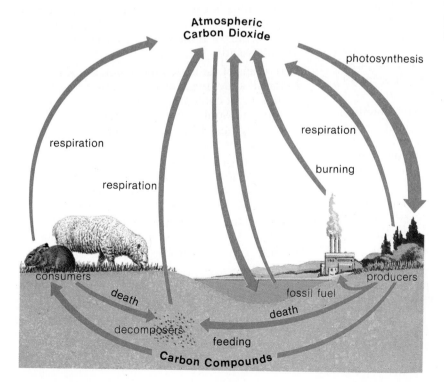

Figure 4.27 Where would you place humans in the carbon cycle?

Atmospheric
Carbon Dioxide

photosynthesis

respiration

respiration

respiration

burning

consumers

death

death

decomposers

feeding

fossil fuel

producers

Carbon Compounds

Concept Review

1. Of the four ways a plant uses the sugars it makes, which is the only way that does not add material to the plant?
2. How are producers, consumers, and decomposers involved in the carbon cycle?
3. What is the greenhouse effect, and how does it work?

Summary

Energy is needed to do work and to make and maintain order. Without energy, highly organized systems such as living things could not exist. The need for energy is continuous, and therefore, organisms must obtain and use energy throughout their lives. Organisms also must obtain matter to build up their bodies. Without matter, no organism could grow. Plants make sugars from simple molecules in the process of photosynthesis. They convert these sugars into other biological molecules, including proteins, lipids, and nucleic acids, all of which also contain carbon. These biological molecules are built up and broken down in many different chemical reactions. All chemical reactions are catalyzed by enzymes, and many of them produce ATP as a temporary energy-transfer molecule.

Applications

1. How would knowing the carbon cycle be useful to biologists in making a study of food chains or food webs?
2. The proteins in the cells of a wheat plant are different from the proteins in your cells. How can the differences be explained? What must happen when you use wheat as a nutrient for the formation of your proteins?
3. Many botanists believe that the concentration of carbon dioxide in the air was much greater during the Carboniferous period, when most of the large coal deposits were being formed, than it is at present. What might be the basis for their belief?
4. In what ways do you contribute to the carbon cycle?
5. Are animals needed for carbon to cycle within a community? Explain.
6. Do you think a sugar molecule has order and organization? How does this question relate to the use of sugar as a source of energy by living things?

7. Imagine you could count the number of structures in one of your cells. For each of the following pairs, choose one that would be more numerous in each cell: (a) atoms or protons; (b) nucleotides or DNA molecules; (c) water molecules or lipid molecules; (d) enzymes or amino acids.
8. In terms of calories, why do you think seeds (small structures containing young plants) are high in fat content? Examples are the peanut, corn, and sunflower seeds from which we get cooking oils. Why are such seeds good sources of food for humans?
9. Apply what you have learned about carbon dioxide and the carbon cycle to global warming. Suppose the temperature of the earth increased 5° C. What effects might this have? What if the temperature increased 15° C?

Problems

1. Make two lists: (1) all the ways you use energy in a day, and (2) all the different types of energy you use in a day (light, heat, mechanical, chemical).

2. One hundred years ago, the carbon dioxide in the atmosphere was measured at 0.0283 percent. Today the level is 0.0330 percent. What human activities during the last 100 years may have contributed to this increase? What might be some possible future consequences if this trend continues?

3. Many fats and cooking oils are called "polyunsaturated." Find out what this means in chemical terms.

4. Each day, you come in contact with many basic, acidic, and neutral solutions. You may have seen products that are labeled "pH-balanced" or "buffered." What does this mean? Make a list of such products or foods and explain how they affect you. Do you use these products or eat these foods because of their acidity or alkalinity, or doesn't it make any difference what their pH is?

5. *The Andromeda Strain*, by Michael Crichton, is a novel about a life form that comes to earth from another galaxy. Read it, or a similar science-fiction novel, and compare the extraterrestrial life form to life forms on earth. What is its chemical composition and structure? Is it based on carbon? Does it require energy?

6. Build three-dimensional models of DNA, RNA, and an enzyme, that can be hung from your classroom ceiling. After reading Chapter 5, create cell parts to hang from the ceiling and turn your classroom into a model of a large cell.

Section Two

Continuity in The Biosphere

Individuals within a population come and go, but the population itself can exist for a long time. Fossil evidence indicates the biosphere itself has endured for more than 3.5 billion years. Thus, there is a continuity in the biosphere, but there also is change. The fossil record reveals that change usually has occurred slowly and systematically—that homeostasis exists between organisms and their environment. How continuity is maintained in populations and how populations change over time are considered in Section Two.

A great gray owl (*Strix nebulosa*) and its young. What evidence is there for continuity in the biosphere? What factors and forces are responsible for this continuity?

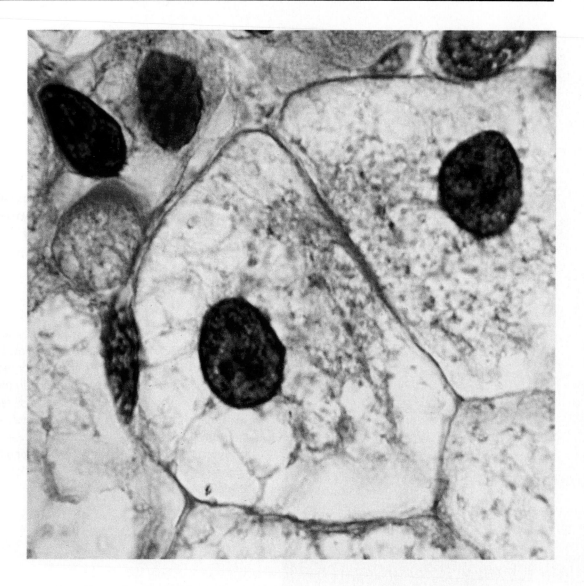

These liver cells from a salamander are enlarged approximately 250 times and are about 50 μm in diameter. What is the large pigmented spot near the center of each cell? Estimate the distance across this spot. How many darkly pigmented dots are inside these spots? What are these dots, and how large are they?

Continuity in Cells 5

In learning how the world of life is organized, you have looked at large living systems—populations, communities, ecosystems, and the entire biosphere. To learn more about these complex systems, you have read about some of the important biological molecules that make up organisms.

Biological molecules are arranged not randomly but in organized patterns. Amino acids, for example, are linked together to make proteins, and single sugars are linked together to form complex carbohydrates. Proteins, sugars, and various other compounds are arranged in patterns that make up the membranes and various other structures that are found within cells. In turn, cells are organized into more complex arrangements called tissues. Tissues are organized into organs, organs into body systems, and body systems into complete organisms.

Each of these levels of organization enables the organism to solve its problems of staying alive and of passing on its distinctive nature to its offspring. Even the simplest organisms have various parts that allow them to take in food, to move, to detect the environment, and to reproduce. Biologists study all these parts to learn how they function, how they interact, and how they permit life to continue. This chapter is about cells, the basic units of life—how they are constructed, how they work, and how they reproduce.

Cell Structure

Guidepost: What features do cells have in common?

5.1 Cells Are the Units of Life

All the organisms scientists have examined are made of tiny, highly active units called cells. Some organisms are unicellular. Each one consists of only a single cell that carries out all the activities of the organism. Others are multicellular and may be composed of trillions of cells. As illustrated in Figure 5.1, there are many different types of cells. In multicellular organisms, many of the cells are specialized for particular functions. Because cells are found in all organisms, scientists consider them to be the basic units of life.

The idea that cells are the basic units of life began to take shape in the 1700s and early 1800s. This idea, which now is known as the **cell theory**,

107

Figure 5.1 What similarities can you see among the groups of cells in these photos? (a) Unicellular green alga, (b) cross-section of a young maple stem, (c) cells from the lining of a cat stomach, (d) human nerve cells.

has undergone modifications through the years but can be stated generally as follows:

1. Cells, or products made by cells, are the units of structure and function in all organisms.
2. All cells come from preexisting cells.

Like other theories in science, the cell theory is a model that explains many observations and predicts the outcomes of future observations. Recall that a hypothesis is a testable statement that explains an observation. A theory, on the other hand, is a model that explains a great many observations and predicts the outcomes of future observations. Theories that serve as models in science, such as the cell theory, have been supported repeatedly by data resulting from testable hypotheses. In addition, almost no data opposing these theories have been found. Any theory, however, no matter how well established it has become, can be changed or discarded if new data do not support the model. All the theories that serve as models in science have been modified as new research has been performed and new information is found.

The Cell Theory

Much of the scientific progress that occurred in the sixteenth century depended on the development of glass lenses. By 1650, the art of grinding and polishing pieces of glass into lenses had so developed that it became possible to build good microscopes. Anton van Leeuwenhoek (1632–1723), a Dutch civil servant, was a master lens maker. Leeuwenhoek placed his lenses into simple microscopes. Using these microscopes, Leeuwenhoek discovered an amazing, invisible world. For 50 years, he described and made careful drawings of bacteria, detailed structures in small insects, and even sperm cells from humans, dogs, frogs, and insects.

In the late 1600s, an Englishman named Robert Hooke also made microscopes and used them to examine small objects. When Hooke examined the cork layer of bark from an oak tree, he observed rows of compartments. The compartments reminded him of the small cells in which monks lived in medieval monasteries. For that reason, he called the compartments *cells*.

Scientists gradually began to realize that cells are the fundamental units of living organisms. In 1838, two German biologists, Matthias Schleiden, a plant biologist (botanist), and Theodor Schwann, an animal biologist (zoologist), proposed the cell theory. According to the cell theory, all organisms consist of cells and cell products, and one can understand how living creatures are built and how they function if cells can be understood. Until the development of the cell theory, the emphasis had been on the cell walls. In looking at animal cells,

however, Schwann and others had been unable to find the boxlike cells seen in plants. Schwann interpreted his observations in a new way, emphasizing what was *inside* the "box," rather than the box itself. Further studies showed that certain structures are common to plant and animal cells as well as those of microorganisms. Once the contents of cells began to be studied, ideas about the origins of organisms began to change.

For hundreds of years, people had thought that organisms could come from nonliving matter, an idea known as spontaneous generation. Several different studies disproved the idea. In 1855, Rudolph Virchow, a physician and biologist, proposed that all cells produce more cells through time, but his idea was not accepted by everyone. In 1864, Louis Pasteur, a French scientist working with yeast cells, demonstrated that microorganisms cannot arise from completely nonliving matter. By the 1880s, the work of French and German scientists showed how cells divide and produce more cells. Organisms, therefore, come from existing organisms, and their cells arise from existing cells.

Today, the cell theory is summarized in two main ideas: (1) cells are the units of structure and function in living organisms, and (2) all new cells come from cells that already exist. New technologies have made possible extremely detailed studies of cell structure and function, and although there are conflicting ideas about how cells first came to be, the cell theory is still one of the major unifying themes of biology.

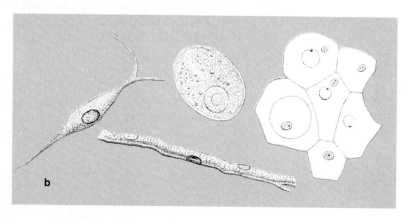

(a) Cork cells as drawn by Hooke, and (b) animal cells as drawn by Schwann.

109

5.2   Biologists Use Microscopes to Study Cells

Biologists began to study cell structure and function in detail as the cell theory developed. To do so, they used different tools and techniques. The light microscope, based on microscopes invented hundreds of years ago, still is used in laboratories for many basic tasks. With a light microscope, light waves pass through a small organism, or thin slices of a larger organism, and the structures are magnified through the lens system.

Tiny organisms, such as those found in a drop of pond water, can be studied in the living state. Colonies of bacteria and many other organisms used in studies of inheritance can be identified or counted using a low-power light microscope. Light microscopes with fine quality lenses can magnify structures up to 1500 times.

Because most cells are nearly transparent, dyes often are used to stain structures within a cell. The structures then can be studied through a light microscope. Dyes, however, may kill or distort cells. To study the smallest structures and events in living cells, biologists must use other methods and instruments, such as the phase-contrast microscope.

The phase-contrast microscope is a special type of light microscope that increases the differences in light and dark areas, thus making cell structures more visible. This instrument makes it possible to study living cells and events that occur within them, such as cell division. Figure 5.2 shows the same cells as seen under an ordinary light microscope and as seen through a phase-contrast microscope.

Electron microscopes show cell parts at very high magnifications. They use an electron beam rather than a light beam to illuminate the object. Modern transmission electron microscopes can enlarge cell structures as much as one million times and provide photographs that show remarkable detail (see Figures 5.8 through 5.12). Scanning electron microscopes allow detailed observations of the surface of biological objects, such as the coleus leaf in Figure 5.3.

5.3   Cells Are of Two Basic Types

Profound differences in the structure of their cells separate all known living organisms into two groups: **prokaryotes** (pro KAIR ee ohtz) and **eukaryotes** (yoo KAIR ee ohtz). Cells of prokaryotes—bacteria—do not have a membrane enclosing their DNA. Cells of eukaryotes—plants, animals, and other

Figure 5.2   Cells from a human cheek lining as seen through (a) an ordinary light microscope (X400) and (b) phase-contrast microscope (X100).

a   b

Figure 5.3 The surface of a coleus leaf, as photographed through (a) a light microscope (X25) and through (b) a scanning electron microscope (X161 000).

familiar organisms—usually have at least one membrane-enclosed structure, the **nucleus**, which contains DNA.

Besides a nucleus, eukaryotic cells also have other membrane-enclosed structures, or **organelles** (or guh NELS), that prokaryotes lack. Compare the prokaryotic and eukaryotic cells shown in Figure 5.4.

Scientists think that prokaryotes represent the earliest type of living cell, yet they are everywhere—in the soil, the air, and the water, and in or on every organism, including humans. They also inhabit extreme environments such as salt flats, hot springs, and thermal vents located deep in the ocean floor. Most prokaryotes are microscopic in size, averaging 1 to 10 μm

Figure 5.4 Electron micrographs of (a) a eukaryotic cell and (b) a prokaryotic cell. Notice the membrane-enclosed organelles and generally more complex structure in the eukaryotic cell.

a   b

Figure 5.5 Prokaryotes are extremely small. Many prokaryotes are visible in this scanning electron micrograph of the point of a pin magnified 383 times.

(micrometers) in length. Up to 35 000 could fit side by side on the head of an ordinary thumbtack, and even the point of a pin can hold hundreds (see Figure 5.5). Chapter 11 describes prokaryotes in more detail.

Cells of eukaryotes are larger (10 to 100 μm) and structurally more complex than those of prokaryotes. The remainder of this chapter deals with the structure and function of eukaryotic cells.

INVESTIGATION 5.1   Observing Cells

Many types of cells have been described and photographed. In this investigation, you will examine some of them.

Materials (per team of 2)

2 pairs of safety goggles	paper towel
2 lab aprons	Lugol's iodine solution in dropping bottle
2 microscope slides	5% salt solution in dropping bottle
2 cover slips	onion
compound microscope	elodea leaf (*Anacharis*)
fine-pointed forceps	prepared slide of stained human cheek
scalpel	cells
dissecting needle	prepared slide of frog blood

SAFETY   **Put on your safety goggles and lab apron**

Procedure

1. Separate one layer from an onion quarter and hold it so that the concave surface faces you. Snap it backwards (see Figure 5.6a) to separate the transparent, paper-thin layer of cells from the outer curve of the scale.

2. Use forceps to peel off a small section of the thin layer and lay it flat on a micro-scope slide. Discard the rest of the onion piece. Trim the piece with a scalpel if necessary and smooth any wrinkles with a dissecting needle. Add 1 or 2 drops of water and a cover slip.

   CAUTION: Scalpel blades and needles are sharp; handle with care.

3. Using your microscope, examine the slide under low power. Look for cell boundaries. Sketch a small part of the field of view to show the shapes and arrangement of the cells.

4. Place a drop of Lugol's iodine solution at one edge of the cover slip. Touch the tip of a piece of paper towel to the opposite edge of the cover slip, as shown in Figure 5.6b. The paper acts as a wick, pulling the Lugol's solution across the slide and into contact with the cells. Record any changes that occur.

   WARNING: Lugol's iodine solution is poisonous if ingested, irritating to skin and eyes, and can stain clothing. Should a spill or splash occur, call your teacher immediately; flush the area with water for 15 minutes; rinse mouth with water.

5. Switch to high power and sketch a single cell. Include as much detail as you can see. Save your sketch for reference in Section 5.3 and Section 5.4.

6. Using forceps, remove a young leaf from the tip of an elodea plant. Place it upside down on a clean slide. Add a drop of water and a cover slip.

Figure 5.6   (a) Removing the thin layer of cells from a section of onion, and (b) adding salt solution under a cover slip.

7. Observe the leaf under low power. By slowly turning the fine adjustment back and forth, count and record the number of cell layers in the leaf.

8. Switch to high power. Select an "average" cell and focus on it. Is there any evidence that the cell is living? If so, what is the evidence?

9. Sketch the leaf cell, including as much detail as you can see. Label any parts you can identify. Keep this sketch for later reference.

10. Place a drop of 5% salt solution at one edge of the cover slip and draw it under the cover slip as you did in Step 4. Observe the cell for any changes and describe what you see. What, if any, structures are evident now that you were not able to see before?

11. Examine a prepared slide of stained human cheek cells. Find several cells separated from others and sketch one or two of them. Include as much detail as you see and label any parts you can identify. What, if any, differences from the onion and elodea cells do you see?

12. Under low power, examine a prepared slide of frog blood. Find an area where the cells are not too crowded and switch to high power. Sketch one or two cells and label any parts you can identify.

13. Wash your hands thoroughly before leaving the laboratory.

Discussion

1. Construct a table in your data book. In the first column, list all the types of cells you observed. Head the other columns with the names of cell parts that you identified. Review your sketches and notes. For each type of cell examined, place an X beneath the name of each cell structure observed.

2. Does the lack of an X indicate that the structure was not present? Why or why not?

3. On the basis of your observations, which kind of cell seems to have less-rounded shapes? Using your observations, which kind of cell has more clearly defined boundaries?

4. What structure may be involved in determining a cell's shape?

5. How did the Lugol's iodine solution help? Why do biologists use different stains to study cells?

Figure 5.7 Generalized animal (left) and plant (right) cells, with enlargements of the major organelles. Some of the terms used here are not explained in this chapter. They are defined in later chapters.

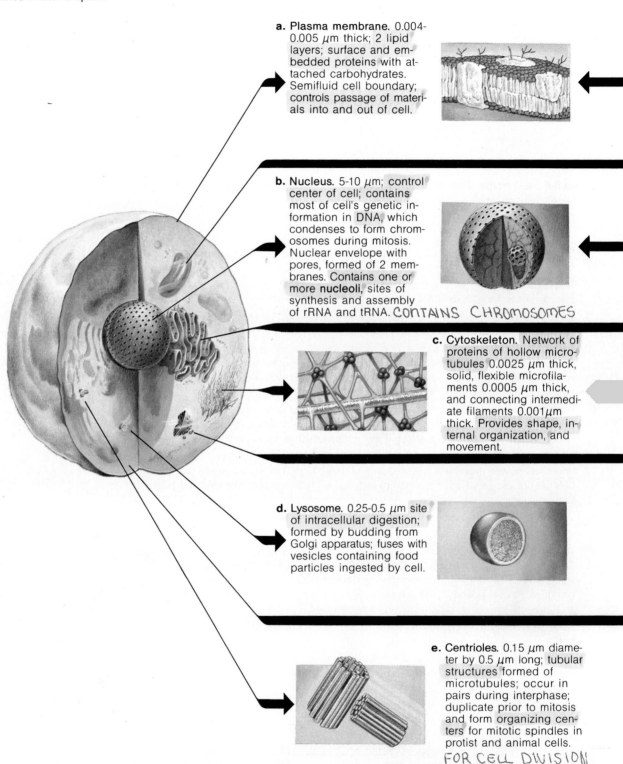

a. Plasma membrane. 0.004-0.005 μm thick; 2 lipid layers; surface and embedded proteins with attached carbohydrates. Semifluid cell boundary; controls passage of materials into and out of cell.

b. Nucleus. 5-10 μm; control center of cell; contains most of cell's genetic information in DNA, which condenses to form chromosomes during mitosis. Nuclear envelope with pores, formed of 2 membranes. Contains one or more **nucleoli**, sites of synthesis and assembly of rRNA and tRNA. CONTAINS CHROMOSOMES

c. Cytoskeleton. Network of proteins of hollow microtubules 0.0025 μm thick, solid, flexible microfilaments 0.0005 μm thick, and connecting intermediate filaments 0.001μm thick. Provides shape, internal organization, and movement.

d. Lysosome. 0.25-0.5 μm site of intracellular digestion; formed by budding from Golgi apparatus; fuses with vesicles containing food particles ingested by cell.

e. Centrioles. 0.15 μm diameter by 0.5 μm long; tubular structures formed of microtubules; occur in pairs during interphase; duplicate prior to mitosis and form organizing centers for mitotic spindles in protist and animal cells. FOR CELL DIVISION

f. Cell wall. 0.1-10 μm thick; formed by living plant cells; made of cellulose fibers embedded in a matrix of protein and polysaccharides; provides rigidity to plant cells and allows for development of turgor pressure.

g. Mitochondrion. 2-10 μm long by 0.5-1 μm thick; enclosed in double membrane; inner membrane much folded; most reactions of cellular respiration occur in mitochondrion; contains small amounts of DNA and RNA; may be several hundred per cell.
POWER HOUSE

h. Endoplasmic reticulum (ER). 0.005 μm diameter; tubular membrane system that compartmentalizes the cytosol; plays a central role in biosynthesis reactions. Rough ER is studded with **ribosomes**, the site of protein synthesis; smooth ER lacks ribosomes.

i. Golgi apparatus. 1 μm diameter; system of flattened sacs that modifies, sorts, and packages macromolecules in vesicles for secretion or for delivery to other organelles.

j. Cytosol. semi-fluid material surrounding organelles; contains enzymes that catalyze cellular reactions.

k. Chloroplast. 5 μm long by 0.5-1 μm thick; enclosed by double membrane; third membrane system forms thylakoids in which light-absorbing pigments are embedded. All reactions of photosynthesis occur in chloroplasts.

l. Vacuole. Variable size; large vesicle enclosed in single membrane; may occupy more than 50% of volume in plant cells; contains water and digestive enzymes; stores nutrients and waste products.

5.4   Membranes Organize Eukaryotic Cells

Each type of cell found in a multicellular organism has a specific role in making life possible for the organism in a changing environment. In a large animal such as a human, there are at least 200 different types of cells. The many types of cells have a similar basic pattern. Depending on their roles, however, they may differ in some details. Figure 5.7 on pages 114 and 115 shows the basic structures in most plant and animal cells. Refer to this illustration as you read the descriptions of these structures.

Every cell has an outer membrane known as the **plasma membrane**. The plasma membrane encloses the cell's contents. It is a thin but active structure that controls the passage of materials into and out of a cell. The plasma membrane is made of two thin layers of lipid molecules arranged in a definite pattern (see Figure 5.7a). Protein molecules on and within the lipid layers perform special functions, such as helping molecules to move in and out of the cell and receiving chemical messages. These proteins and their functions vary in different types of cells.

The cell contents outside the nucleus are known as the cytoplasm (SYT oh plaz um). Extending throughout the cytoplasm is a complex network of membranes that form flattened sheets, sacs, and tubes. Figures 5.7h and 5.8 show a portion of this large network, which is called the **endoplasmic reticulum** (en doh PLAZ mik reh TIK yoo lum), or ER. The ER specializes in the synthesis and transport of lipids and membrane proteins. Still other membranes surround the organelles, which often have their own internal membranes and may be very complex.

Membranes organized as stacks of flattened sacs form the **Golgi** (GOHL jee) **apparatus** (Figures 5.7i and 5.9). This organelle helps to package cell products for export from the cell. Materials synthesized in the ER are then

Figure 5.8   Transmission electron micrograph of rough endoplasmic reticulum in a liver cell (X40 800). What are the round structures along the membranes?

Figure 5.9   Transmission electron micrograph of a Golgi apparatus (X21 930). Part of the nucleus can be seen at lower left. Note the vesicles at the right.

transferred to tiny saclike structures, or vesicles, that are formed by the membranes of the Golgi apparatus. The vesicles move to the plasma membrane and release their products to the outside of the cell.

Clearly, membranes serve an important purpose in cells. The eukaryotic cell may be viewed as a highly organized system divided into distinct parts by membranes. Within each part, a special job is done. Molecules pass back and forth across the membranes within a cell and through the membrane that separates the cell from its outside environment.

5.5 Eukaryotic Cells Contain Various Organelles

In many cells, the most obvious organelle is the nucleus (Figures 5.7b and 5.10), a rounded body surrounded by a double membrane called the **nuclear envelope**. The nucleus contains most of the **genes**, the units of hereditary information that control the basic functions of the cell. For this reason, the nucleus is considered the control center of the cell. Genes are composed of DNA nucleotides. These nucleotides contain information that tells the cell how to function and directs the synthesis of protein molecules. Long strands of nucleotides wrapped around protein molecules form **chromosomes** (KROH moh sohmz), which become visible only during nuclear division.

The powerhouses of the cell are the **mitochondria** (my toh KON dree uh; singular, mitochondrion). They not only help to release energy from food molecules but also help to synthesize building blocks of larger molecules. The chemical reactions that occur in mitochondria provide new supplies of ATP (see Section 4.6), the energy molecules of the cell. The ATP then is used for all types of cell activities. Mitochondria are especially abundant in the parts of an organism that have high energy requirements, such as a muscle in an animal or a growing root tip of a plant. Although they may have different shapes in different cells, all mitochondria consist of two layers of membranes—an outer membrane and a folded inner membrane—as shown in Figure 5.7g.

Along the membranes of the ER may be found tiny bodies called **ribosomes** (RY boh sohmz). Proteins are manufactured on the ribosomes, which are visible in the electron micrograph in Figure 5.8.

Two other types of organelles are common in animal cells. **Lysosomes** (LY soh zohmz), shown in Figure 5.7d, are vesicles containing a variety of digestive enzymes that help break down large molecules and worn-out cell parts. **Centrioles** (SEN tree ohlz) are paired cylindrical structures (Figure 5.7c) that play an important role in cell division.

Many plant cells have **chloroplasts** (KLOR oh plasts). These organelles contain chlorophyll, a pigment that is essential for capturing the energy in sunlight. The chloroplasts (Figure 5.7k), which are the sites of photosynthesis, have a complex arrangement of membranes. The energy captured from the sun and stored in the structure of biological molecules during photosynthesis is released later in the mitochondria. Most plant cells also contain one to several **vacuoles** (VAK yoo ohlz), fluid-filled structures that are surrounded by single membranes (Figure 5.7l). Vacuoles contain enzymes, store nutrients and waste products, and regulate the water content of cells. They may occupy as much as 95 percent of an older cell's volume.

Figure 5.10 Transmission electron micrograph of a mouse liver cell nucleus (X51 300). Note the double membrane that makes up the nuclear envelope. The large dark round body is called the nucleolus.

All the cytoplasmic membranes and organelles are bathed in a gelatin-like substance, the cytosol (SYT oh sol), which is illustrated in Figure 5.7j. Many of the chemical reactions of a cell take place in the cytosol. A complex network of ultrafine protein filaments within the cytosol forms an internal skeleton, the **cytoskeleton** (Figure 5.7c). Constructed of hollow microtubules and solid rods that can be assembled and disassembled rapidly, the cytoskeleton gives the cell its distinct shape and its capacity for movement. The cytoskeleton also may provide internal organization and a framework for the attachment of various enzymes and organelles within the cytosol.

Some types of cells have still other types of organelles, and not all cell structures lie within the plasma membrane. In plant cells and in some other cells, there is a cell wall (see Figure 5.7f) outside the plasma membrane. The cell wall is made of cellulose and other materials that the cell secretes and exports through the Golgi apparatus. The cell wall provides strength and protection. Often when the cell dies, the cell wall remains behind. When Robert Hooke looked at thin slices of cork under his microscope (see Biology Today), he actually was observing the empty cell walls in the bark of a cork oak. For early biologists, the presence of cell walls made plant cells easier to see than animal cells. Most of a large tree's trunk, for example, consists of nonliving cell walls. The living cells of the tree are found in a thin band just underneath the bark of the trunk.

Concept Review

1. What are the main ideas of the cell theory?
2. How did the use of stains advance the study of cells?
3. Compare the advantages of the light, phase-contrast, and electron microscopes.
4. Describe the major differences between prokaryotic and eukaryotic cells.
5. Describe the structure and function of the plasma membrane.
6. Discuss the functions of three cell organelles.
7. How has the meaning of the word *cell* changed since the time of Robert Hooke?

Guidepost: What principles from the physical sciences help explain the functions of organisms?

Cell Functions

5.6 Cell Activities Require Energy

All the different parts of a cell are important to the life of the cell. Each part has a particular role to play, but it is not always possible to tell what that role is just by looking at the part. If all the cogs, wheels, springs, and bearings of a watch were spread on a table, you probably could not understand exactly how the watch worked just by looking at the parts.

Biologists are in much the same situation. They can describe the various parts of a cell and even separate them, but they cannot always figure out how the parts all fit together or how they enable the cell to work. That is the challenge of modern biology. In trying to find out, it usually is more useful to study living cells than nonliving ones.

Scientists, for example, cannot see cells use energy. Instead, it is inferred from what they observe happening in living cells. All cells are able to release energy from complex molecules. The energy in a sugar molecule is released by the mitochondria in small steps. After a cell has taken in energy-rich molecules that can be used as fuel, the molecules are channeled into pathways in the mitochondria where the energy is released. The more active a cell is, the more mitochondria it is likely to contain.

Energy makes it possible for chemical reactions to occur in cells. The sum of all the chemical reactions in a cell or organism is known as its **metabolism** (meh TAB oh liz um). As a part of a cell's or organism's metabolism, biologically important molecules are processed, energy is transferred to ATP, and waste materials are released. Excess carbon atoms, which are released as carbon dioxide (CO_2), are a part of these waste materials. In animals, waste materials include extra nitrogen atoms that can be released in nitrogen-containing molecules such as ammonia (NH_3). These wastes must be removed because they can become toxic to the cell.

Not all energy is used immediately for metabolism. Some cells, for example the cells in green plants that contain chloroplasts, store energy in complex molecules such as sugars. The sugar molecules then can be transformed into other types of molecules such as amino acids, lipids, or nucleotides. These molecules are used to maintain the life of the cell, for growth, and to help build new living substances.

INVESTIGATION 5.2 Diffusion Through a Membrane

How do things get in and out of cells? Using dialysis tubing, you can create a model of a plasma membrane to begin to answer this question.

Materials (per team of 3)

3 pairs of safety goggles
3 lab aprons
2 250-mL beakers
glass-marking pencil
2 15-cm pieces of dialysis tubing

4 10-cm pieces of string
15 mL soluble-starch suspension
15 mL glucose solution
Lugol's iodine solution in dropping bottle
urine glucose test strips

SAFETY **Put on your safety goggles and lab apron.**

Procedure

1. Twist one end of a piece of dialysis tubing. Fold the twisted end over and tie it tightly with a piece of string. Prepare the other piece the same way.

2. Pour soluble-starch solution to within 5 cm of the top of one piece of tubing. Twist and tie the end as in Step 1. Rinse the tubing under running water to remove any starch from the outside.

3. Place the tubing in a beaker of water labeled *A* (see Figure 5.11). Add enough Lugol's iodine solution to give the water a distinct yellowish color.

    WARNING: Lugol's iodine solution is poisonous if ingested, irritating to skin and eyes, and can stain clothing. Should a splash or spill occur, call your teacher immediately; flush the area with water for 15 minutes; rinse mouth with water.

iodine solution

dialysis tubing containing starch solution

water in beaker

Figure 5.11 Setup for diffusion.

4. Repeat Step 2 with the second piece of dialysis tubing, using glucose solution instead of the soluble starch. Place this tubing in a beaker of water labeled *B*.
5. Allow the pieces of tubing to stand for about 20 minutes. Dip a urine glucose test strip into the water in beaker B. Record the color on the strip.
6. Observe the tubing in beaker A. Record any changes, including color, that you see in either the tubing or the water in the beaker.
7. Let beakers A and B stand overnight. Record any changes observed the next day.
8. Wash your hands thoroughly before leaving the laboratory.

Discussion

1. On the basis of the chemical test for starch, what must have happened to the iodine molecules in beaker A?
2. On the basis of the chemical test for glucose, what must have happened to the glucose molecules in beaker B?
3. From the evidence obtained after the beakers stood overnight, what other substances passed through the membrane in beaker B?
4. Which substance did not pass through a membrane? How do you know the substance did not?
5. Physicists can show that all the molecules of a given substance are about the same size but that the molecules of different substances are different in size. Measurements show that iodine molecules and water molecules are very small, glucose molecules are considerably larger, and starch molecules are very large. On the basis of this information, suggest a hypothesis to account for your observations.
6. What assumption did you make about the structure of the dialysis tubing?

5.7 Substances Enter and Leave Cells by Diffusion

Atoms, molecules, and small particles are in constant motion. Although they may move in any direction, in general, molecules tend to move from an area where they are more concentrated to an area where they are less concentrated, until their concentration is the same everywhere (Figure 5.12). This movement is called **diffusion** (dih FYOO zhun). The concentrations of molecules at various points between the high and the low areas form a gradient, which is known as the concentration gradient. Because molecules diffuse from areas of higher concentration to areas of lower concentration, they are said to move *down* the concentration gradient. Many

Figure 5.12 Diffusion is the movement of molecules from an area of high concentration to an area of low concentration until the concentration is the same everywhere. In (a), a crystal of potassium permanganate is dropped into a glass of water. The molecules diffuse through the water (b) until they are evenly distributed throughout (c).

a   b   c

substances can move in and out of cells by diffusion alone. Once a substance has passed into a cell through the plasma membrane, it can continue to diffuse throughout the cytosol down its concentration gradient. Thus, diffusion aids in the even distribution of materials within a cell.

Water moves in and out of cells and diffuses down its concentration gradient in the same manner as other substances. The diffusion of liquids such as water through membranes is called **osmosis** (os MOH sis). For example, suppose a living cell is placed in a salty solution. If the concentration of water molecules is lower in the solution than it is in the cell, water diffuses out of the cell (from an area of higher concentration in the cell to an area of lower concentration), and the cell shrinks (see Figure 5.13a).

If the concentration of water molecules is greater outside a cell, water diffuses inward, and the cell swells (Figure 5.13b). When an animal cell is placed in pure water, osmosis usually causes it to burst. A plant cell swells but does not burst because of the rigid cell walls. (This swelling enables well-watered plants to stand upright and explains why many plants wilt and droop when their cells have lost water.) Figure 5.14 shows how animal and plant cells react to changes in water concentrations as a result of osmosis.

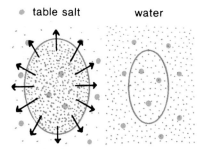

a. if concentration of water is higher inside the cell, water diffuses out of the cell

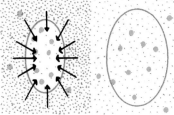

b. if concentration of water is higher outside the cell, water diffuses into the cell

Figure 5.13 A diagram of osmosis. When water concentration is higher *inside* the cell (a), water moves out of the cell, and the cell shrinks. When water concentration is higher *outside* the cell (b), water moves into the cell, and the cell swells. Does this diagram help you understand why supermarkets spray their produce with fresh water?

Figure 5.14 An animal cell (red blood cell) and a plant cell in different solutions. An isosmotic solution contains the same concentrations of dissolved material and water as the cell.

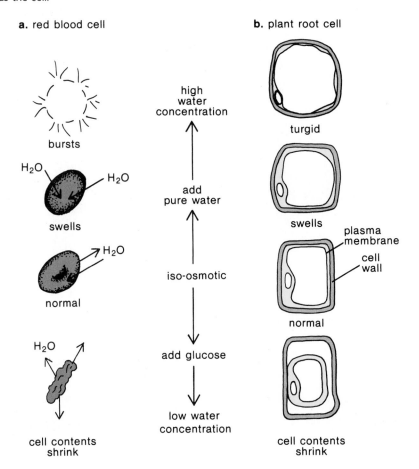

5.8 Cells Move Substances in a Variety of Ways

The cell wall, if it is present, allows free diffusion of most substances. The plasma membrane, however, is differentially permeable. This means that only certain molecules can diffuse through it freely. A small glucose molecule, for example, may pass through the membrane easily, but a large protein molecule cannot. Membranes also may be impermeable to particular substances, which cannot pass through at all. Some substances are too large to pass through a membrane. Others have an electrical charge that makes diffusion impossible because the plasma membrane also is electrically charged. These substances enter or are released from a cell in other ways.

Figure 5.15 illustrates two additional methods by which cells transport substances across membranes. Both methods require special transport proteins in the membrane. In **passive transport**, no energy is expended by the cell. Transport proteins move substances *down* their concentration gradient. Each protein can transport only a specific substance. Cells transport many essential ions by this means. In **active transport**, energy in the form of ATP is used to move substances through the transport proteins. By means of active transport, substances can move across a membrane *against* a concentration gradient. For example, the soil around a plant may contain low amounts of elements necessary for the plant's growth. Ordinary diffusion would not provide the plant with enough of these elements. By means of active transport and other processes, however, the plant's root cells can accumulate these elements in relatively high amounts. The elements then can be transported to all parts of the plant.

In another transport mechanism, the plasma membrane may fold inward, forming a tiny "pocket" that fills with fluid and particles from the cell's surroundings (see Figure 5.16a on page 123). The membrane then closes over the pocket, forming a vesicle that is released into the cell. In this manner, large molecules and particles from the surroundings can be incorporated into the cell. This activity can occur in the reverse direction to expel materials from the cell, as shown in Figure 5.16b.

Figure 5.15   In passive transport (a), substances move through membrane proteins down their concentration gradient. Active transport (b) requires the use of ATP energy as well as membrane proteins. Active transport can move substances against their concentration gradient as well as with their concentration gradient.

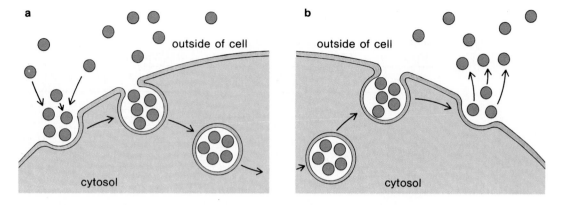

Figure 5.16 The transport of masses of material into a cell and out of a cell. These processes require energy.

Concept Review

1. What activities are involved in metabolism?
2. Why do cells constantly take in and get rid of substances?
3. What is a concentration gradient? How does it affect diffusion?
4. How does osmosis affect an animal cell? How does it affect a plant cell?
5. Why are membranes said to be differentially permeable?
6. Distinguish between passive and active transport. How is each important to a cell or organelle?

Cell Reproduction

Guidepost: How do cells duplicate themselves?

INVESTIGATION 5.3   Cell Size and Diffusion

Why are cells so small? Why do they stop growing after reaching a certain size? What causes cells to stop growing and then to divide? One way to investigate these questions is to build a model. A model is often a small copy of something large. Here you will build a large model of something small.

After reading the procedure, develop a hypothesis that predicts which cell model will do the best job of moving materials in and out of the cell.

Materials (per team of 2)

2 pairs of safety goggles
2 lab aprons
2 pairs of plastic gloves
250-mL beaker
millimeter ruler
plastic spoon

plastic table knife
paper towel
150 mL 0.1% NaOH
3-cm × 3-cm × 6-cm block of
  phenolphthalein agar

SAFETY   Put on your safety goggles, lab apron, and plastic gloves.

TABLE 5.1 Comparison of Agar Cubes

Cube Dimension	Surface Area (cm²)	Volume (cm³)	Simplest Ratio
3 cm			
2 cm			
1 cm			
0.01 cm			

Procedure

1. Using the plastic knife, cut the agar block into three cubes—one 3 cm, one 2 cm, and one 1 cm on a side.

2. Place the cubes in the beaker and cover them with 0.1% NaOH (sodium hydroxide). Record the time. Use the plastic spoon to turn the cubes frequently for the next 10 minutes.

   CAUTION: NaOH is an irritant and can destroy clothing. Avoid skin/eye contact; do not ingest. Should a splash or spill occur, call your teacher immediately; flush the area with water for 15 minutes; rinse mouth with water.

3. Prepare a data table like Table 5.1 and do the calculations necessary to complete it. The surface-area-to-volume ratio is calculated as follows:

$$\text{Ratio of surface area to volume} = \frac{\text{surface area}}{\text{volume}}$$

   This ratio also may be written "surface area : volume." The ratio should be expressed in its simplest form (for example, 3 : 1 rather than 24 : 8).

4. Wear gloves and use the plastic spoon to remove the agar cubes from the NaOH after 10 minutes. Blot them dry. Do not handle the cubes until they are dry. Use the plastic knife to slice each cube in half. Record your observations of the sliced surface. Measure the depth of diffusion of the NaOH in each of the cubes.

Discussion

1. What evidence is there that NaOH diffuses into an agar cube? What evidence is there that the rate of diffusion is about the same for each cube? Explain.

2. List the agar cubes in order of size, from largest to smallest. List them in order of the ratios of surface area to volume from the largest to the smallest ratio. How do the lists compare?

3. Calculate the surface-area-to-volume ratio for a cell model that is 0.01 cm on a side.

4. Which has the greater surface area, a cube 3 cm on a side or a cube 0.01 cm on a side? Which has the greater surface area *in proportion to* its volume?

5. What predictions can you make about the cell model in Question 3?

6. What happens to the surface-area-to-volume ratio of cubes as the cubes increase in size?

7. Most cells and microorganisms measure less than 0.01 cm on a side. What is the relationship between rate of diffusion and cell size?

8. Propose a hypothesis to explain one reason why large organisms have developed from *more* cells rather than *larger* cells.

9. In what other ways can models be used?

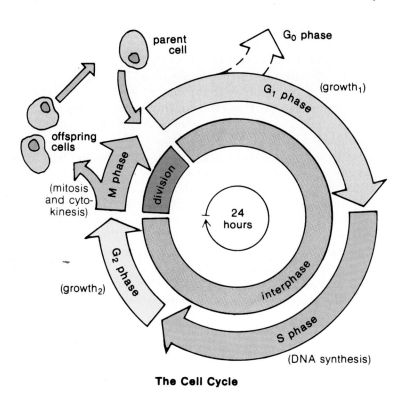

Figure 5.17 The cell cycle includes cell division and interphase. Most of a cell's life is spent in the various phases of interphase.

parent cell

G_0 phase

offspring cells

G_1 phase (growth₁)

M phase
(mitosis and cyto-kinesis)

division

24 hours

G_2 phase
(growth₂)

interphase

S phase
(DNA synthesis)

The Cell Cycle

5.9 The Cell's Life Is a Cycle

Although diffusion supplies cells with needed materials and helps to remove waste products, it takes a long time for substances to diffuse great distances. Therefore if a cell is to function well, it cannot be too large. Usually, when a cell reaches a certain size, it begins a series of changes that permit it to divide into two cells.

The life of a eukaryotic cell consists of a continuous sequence of events called the **cell cycle** (see Figure 5.17). The cell cycle starts with the formation of a new cell and continues until the cell itself has divided into two new cells. Each new cell then begins the cell cycle anew. The dramatic and visible events of cell division are only a brief portion of the cell cycle. A cell spends most of its life between divisions in a stage known as **interphase** (IN ter fayz). Although a cell may *appear* to be inactive in interphase, chemically it is very active indeed. During interphase, all normal metabolic activities take place and, in addition to these activities, intricate preparations for division occur in a precise sequence.

New cells usually contain the same structures as their parent cells. That means the chromosomes, the organelles in the cytosol, and the plasma membrane must be duplicated before a cell divides to form two new cells. If these structures were not duplicated, the offspring cells would be incomplete and unable to survive. Preparations for division take place in the nucleus and the cytosol in distinct phases. The events of the cell cycle occur in fixed order, and each successive step depends on a preceding one.

Just after division, the cell enters the G_1 (Growth 1) phase. During this phase, the cell grows and is metabolically active. New proteins and RNA are made, and other molecules needed for new cells are gathered. Organelles may be duplicated. All these activities, which take place in the cytosol, require energy, raw materials, and enzymes.

Next, the cell enters the S (Synthesis) phase. During this phase, the DNA content of the cell doubles. In the doubling process, the genetic information is copied exactly. This exact copying, or replication, of the DNA ensures that a complete and identical set of genes will be available for each new offspring cell formed during cell division.

Think of this genetic information, which is replicated in each cell generation, as a program that the cell can call on as needed. In a multicellular organism, not all the genetic information is needed or used by every cell, but all of it is replicated in all cells. In unicellular organisms, each new cell generation needs all the information.

When the S phase is complete, the cell enters G_2 (Growth 2), another phase of growth and metabolism. The cell again synthesizes protein, RNA, and other large molecules but often in smaller amounts than in the G_1 phase. A cell in G_2 contains twice as much DNA in its nucleus as a cell in G_1. During G_2, the cell also synthesizes a substance that triggers the beginning of cell division.

During the M (Mitosis) phase, which occurs after G_2, the nucleus reproduces through a series of events called **mitosis** (my TOH sis; from the Greek *mitos*, meaning thread). During this phase, the chromosomes, which usually are stretched out in a threadlike network, become visible with the use of a light microscope. Toward the end of mitosis, the cytosol, with its organelles, also divides. This part of cell division is called **cytokinesis** (syt oh kih NEE sus; from the Greek *cyto*, meaning cell, and *kinesis*, meaning movement).

Thus during its life, a cell passes through two main stages: interphase and cell division. In interphase (G_1, S, G_2), cells carry out normal metabolic activities and preparations for division. During the preparatory phases, the chromosomes, organelles, and cytosol are very active. After a complex series of events, cell division results in two offspring cells. These two new cells then enter interphase again and resume their metabolic activities, including preparations for division.

INVESTIGATION 5.4   Mitosis and Cytokinesis

The growth of an organism is the irreversible increase in the number and size of cells. In other words, a plant or animal grows when it produces new cells that increase in size. When an organism grows, cell parts must be made for each new cell formed. For growth to occur, mitosis and cytokinesis must take place. Not all parts of a plant or an animal grow, and so not all cells of a plant or animal need to carry out mitosis and cytokinesis.

An onion placed in water sprouts slender white roots. This growth occurs partly by repeated duplication of cells. You might expect to see cells undergoing mitosis in a root. Because mitosis is only a small segment of the cell cycle, however, root cells are not always dividing. It is easiest, therefore, to examine the stages of mitosis in stained, prepared slides of root tips that have been sliced very thin longitudinally.

Materials (per person)
compound microscope
prepared slide of *Allium* (onion) root tip
prepared slide of *Ascaris* embryo cells

Procedure

1. Scan the entire length of the *Allium* root tip slide. Study each section of the root tip under low power and then under high power. In which region of the slide do you find many small, tightly packed cells? Are these cells old or new cells? Can you recognize any cells that are undergoing mitosis? How can you tell? Do you see any cells undergoing cytokinesis?

2. The strings, or bars, that appear in the nucleus are chromosomes. Examine these structures in several cells. What happens to the chromosomes during mitosis?

3. Mitosis is a continuous process in which the contents of dividing cells change shape and position. In the region in which you think mitosis is occurring, sketch at least five complete cells that look different from each other. How do the size, shape, and contents of cells undergoing mitosis compare with those of cells in other parts of the root?

4. The continuous process of mitosis often is divided into several stages for ease of study. Compare your sketches with the photographs of a dividing cell shown on the following page and try to identify the stages in your sketches.

5. Compare the length of time for each of the different mitotic stages. To do this, count the number of cells within your field of view that are in each of the 5 stages. As a class, determine the average number of cells found in each of the different stages. What does this suggest about the length of time for each stage? Explain your answer.

6. Examine an *Ascaris* slide and look for stages similar to those you saw in the onion root tip. What differences do you see? Sketch several cells.

Discussion

1. Mitosis produces two nuclei from one nucleus. The number of chromosomes in each new nucleus is the same as that in the nucleus from which they were formed. What does this suggest must happen to the number of chromosomes in the nucleus *before* it divides?

2. In which stage do you think the chromosomes are duplicated? Why does it take so long for this to occur?

3. Based on your observations, describe the differences and similarities of mitosis and cytokinesis in plant and animal cells.

For Further Investigation

Suppose you suspect that the frequency of mitosis in onion roots varies with the time of day. How would you go about getting data to confirm or refute your suspicion?

5.10 The Events of Mitosis Occur as a Continuous Process

Although mitosis is a continuous process, it is convenient to freeze the action at intervals and to describe the process as a series of stages. This is similar to looking at a single frame in a motion picture film. The photographs in Figure 5.18 on page 128 show the stages of mitosis in plant cells. Refer to the photographs as you read the discussion that follows.

During the first stage of mitosis, prophase (see Figure 5.18b, c), the chromosome strands slowly begin to coil like tiny springs. As they coil, they become shorter and thicker. The nuclear envelope begins to disappear. At this time, the chromosomes barely are visible under a microscope. Careful examination shows that each chromosome is double (see Figure 5.19 on

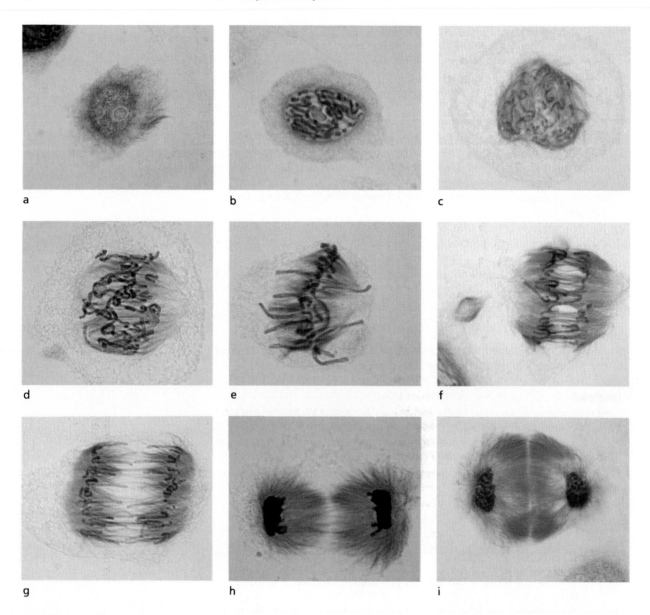

Figure 5.18 Stages of mitosis in the blood lily, *Haemanthus:* (a) interphase, (b) prophase, (c) late prophase, showing disintegration of the nuclear envelope, (d) early metaphase, (e) metaphase, (f) early anaphase, (g) late anaphase, (h) early telophase, (i) late telophase showing formation of the cell plate.

page 129), having replicated during the S (Synthesis) phase. A framework of fibers, known as the spindle apparatus (shown in Figure 5.18d), forms within the cytoskeleton of the cell. The spindle apparatus determines the direction of the cell's division. Some of the spindle fibers attach themselves to the chromosomes.

During the next stage of mitosis, metaphase (Figure 5.18d, e), the chromosomes move to the center of the cell, perpendicular to the spindle. As anaphase begins (Figure 5.18f, g), the chromosome strands separate. Pulled by the spindle fibers in opposite directions toward the two poles of the dividing cell, one strand of each replicated chromosome moves toward either pole.

Now the final events of mitosis occur. During telophase (Figure 5.18h, i), the chromosomes gather at opposite ends of the cell, and a new nuclear

Figure 5.19   Chromosomes in a root tip cell of the hyacinth, *Hyacinthus orientalis*. What indicates that replication already has taken place?

envelope forms around each group of chromosomes. Thus, two new nuclei are formed. Each nucleus contains a set of chromosomes identical to those of the parent cell.

Meanwhile, the other organelles, which also duplicated during interphase, segregate. The cytoplasm divides into two parts, more or less equal in size. In plant cells, a cell plate (Figure 5.18i) begins to form across the middle of the cell, dividing the cell in two.

Mitotic cell division in animal cells is similar to cell division in plants. Animal cells, however, unlike plant cells, contain a pair of centrioles (see Figure 5.7c) that duplicate as mitosis begins. One pair gradually moves toward the opposite side of the cell from the other pair, as shown in Figure 5.20. The pairs of centrioles act as the poles of the cell. Later in prophase as the nuclear envelope breaks down, the spindle apparatus begins to form between the pairs of centrioles. No cell plate forms in animal cells. Instead, the cell constricts (see Figure 5.21) across the middle as cytokinesis begins. A new plasma membrane forms across the constricted portion of the cell. Table 5.2 on page 130 summarizes the events in the different phases of the cell cycle.

Figure 5.20   Before mitosis begins, the centrioles duplicate and move to opposite poles of the cell.

Figure 5.21 A scanning electron micrograph of a dividing animal cell (X42).

TABLE 5.2 Events in the Cell Cycle

Phase		Events	Cell Part Affected
Interphase	G₁ (Growth 1)	Cell makes new proteins, RNA, and other molecules. Organelles may be duplicated. Energy, materials, and enzymes used.	Cytoplasm
	S (Synthesis)	Cell duplicates its DNA. 2 complete, identical sets result. In animals, centrioles duplicate.	Nucleus Cytoplasm
	G₂ (Growth 2)	Similar to G₁, but molecules often produced in smaller amounts. Cell synthesizes substance that triggers the beginning of mitosis.	Cytoplasm
Mitosis and Cytokinesis	Prophase	Chromosomes condense, nuclear envelope disappears. Spindle forms, spindle fibers attach to chromosomes.	Nucleus Cytoplasm
	Metaphase	Chromosomes move to center of cell.	Nucleus
	Anaphase	Chromosome strands pulled toward poles.	Nucleus Cytoplasm
	Telophase	Chromosomes gather at poles, nuclear envelopes form. Division of cytoplasm, cell plate formed (in plants). Cytokinesis, new plasma membrane.	Cytoplasm New nuclei Cytoplasm

5.11 Cells Become Specialized in Development

Complex organisms, such as yourself, develop from the division of a single cell (see Chapter 7). During development, a human may form about 200 different types of cells, each type having a specialized function. As cells become specialized, they take on specific shapes, and their divisions may become more distinctive. For example, different types of cells may require different amounts of time to complete a cell cycle. Some cells divide about every eight hours. Others divide only once every several months. Certain types of cells—such as muscle and nerve cells—almost never divide in adults. In plants, cell division is confined to specialized cells in the tips of stems, branches, and roots, and in bark.

Figure 5.22   Scanning electron micrograph of dividing cancer cells (X6000).

Almost all normal animal cells undergo a certain limited number of divisions—usually no more than 50. There appears to be some type of brake on cell division, although scientists do not know why. Some cells seem to lose control, however, and divide rapidly and abnormally, often at the expense of the organism. These are cancer cells, shown in Figure 5.22. Cancer cells grow and divide uncontrollably. If their growth cannot be controlled or the cells are not removed in time, they may cause the death of the entire organism. In many ways, cancer cells act as rejuvenated, undifferentiated cells. If we can learn why and how cells differentiate, a way may be found to stop the cancer cell's process of undifferentiated growth and reproduction.

Because cells live for a limited time, life on the earth can continue only through repeated cell division. The complex events of cell division ensure accurate reproduction of chromosomes and genes. Although accidents do happen, most cells receive complete and correct genetic information. When the processes of cell reproduction are interrupted or interfered with, abnormal cell activities (such as those in cancer cells) result, or the death of the cell occurs.

Concept Review

1. In what way might diffusion limit the size of cells?
2. Describe the activities in each phase of the cell cycle.
3. What does the term *replication* mean?
4. What is the role of the spindle apparatus in mitosis?
5. In your own words, describe the process of mitosis.
6. What is the biological importance of cell division?

Summary

Cells are the basic units of organisms, and all cells arise from preexisting cells. Different types of microscopes make possible the detailed study of cell structures.

A plasma membrane surrounds each cell and helps to control the passage of materials in and out of the cytoplasm. Materials may move into a cell by diffusion, passive transport, or by energy-requiring active transport. In eukaryotic cells, internal membranes form a variety of structures and enclose organelles, converting the cell into several parts. Each part has a specialized role in the life of a cell. The chemical reactions of a cell—its metabolism—take place in the organelles and in the cytosol, the gelatin-like portion of the cytoplasm.

The continuous sequence of events in the life of a cell is known as the cell cycle. Cells that divide spend most of their life in interphase, the time between divisions. During this phase, cells are metabolizing and preparing for division. Not all cells divide, however. Most plant cells and the nerve and muscle cells in animals remain in a growth phase of the cell cycle.

The cell nucleus contains most of the genes, which control the basic functions of a cell. The genes in the nucleus are located in the chromosomes, each of which replicates itself in the S phase of the cell cycle. During mitosis and cytokinesis, the doubled chromosomes separate, the cytoplasm divides, and two new cells are formed from one existing cell. Thus, each cell receives genetic instructions from its parent cell. In the development of complex organisms, many different types of cells are formed. Each type of cell possesses a specific structure and function. Continued life depends on cell division.

Applications

1. During the late nineteenth century, most of the knowledge of detailed cell structure was gained by studying dead cells. Some biologists of that time objected to many of the conclusions that were drawn from such observations. They argued that the processes of killing, staining, and mounting cells could cause a cell's structure to appear different from what it was in a living cell. What types of evidence are available today to meet at least some of these objections?

2. Which cell parts can you see using the highest magnification of your microscope? What does this indicate about the sizes of these parts compared with all the other cell parts you cannot see?

3. While working in a police laboratory, you are given a tiny sample of material and asked to identify it as either plant or animal matter. How could you decide which it is?

4. Trace the path of a molecule as it enters and moves through a plant cell and eventually enters a vacuole. Through which structures must it pass?

5. On the basis of your understanding of osmosis, describe what would happen to (a) a marine jellyfish placed in a freshwater stream and (b) a frog placed in ocean water. Some fish (for example, shad and striped bass) annually swim from the ocean into freshwater rivers and back. How are they able to do this?

6. Here is a quotation from *The Rime of the Ancient Mariner*, a poem about a sailor.

   Water, water, everywhere,
   And all the boards did shrink;
   Water, water, everywhere,
   Nor any drop to drink.

   Why did the boards of the ship shrink, and why couldn't the mariner drink any water?

7. THC is a chemical that stops chromosomes from separating during anaphase of mitosis.

What do you think the nuclei of the two resulting cells might look like if a cell were sprayed with THC during mitosis?

8. Almost every living thing, including you, starts off as a single cell. When you grow, however, you have cells that are extremely different from each other. Consider the difference between your brain cells, skin cells, and stomach cells—all of which were formed by the genetic information in your chromosomes. What does this suggest about the genetic information found in that first cell?

9. In which type of cells is a cancer more likely to begin—a young cell or a highly specialized cell?

Problems

1. Find out what the oldest known cells on earth looked like. Do they show more resemblance to a modern-day prokaryotic cell or eukaryotic cell?

2. Using a microscope or photomicrographs in books, examine various types of cells from multicellular organisms. Discuss the relationships between the structural forms of the different cells and their functions.

3. A cell can be compared to a small country. A cell has a membrane at its outer edge that screens what comes into the cell. Likewise, a small country may have border guards at its edges to prevent the entry of certain people or objects. Make a list of cell parts. Then, try to think of parts of the country that may have similar functions.

4. What parts and processes do all cells have in common? Consult a cell biology text to determine which organelles and cell structures are found in plant, animal, bacterial, and fungal cells. What does this suggest about the similarity of related functions in all of these cells?

5. There are many plant cells that humans use for things other than food. Consult a book on economic botany to see what these uses are.

6. This chapter describes mitosis in cells that have single, well-defined nuclei. Investigate what is known about what happens to nuclear material during division in (a) a cell that lacks a nucleus (a bacterium, for example), and (b) a cell with more than one nucleus (a paramecium, for example).

7. It may be possible for you to attach a 35-mm camera onto a microscope. Take a series of photographs of cells in different stages of mitosis. Enlarge them and make a display for your entire class. If a camera setup is not available, make large drawings of each stage. For a more difficult assignment, try making all of the drawings in three dimensions.

8. In unicellular organisms, cells usually separate shortly after division. In some unicellular organisms, however, they remain attached and form colonies. In multicellular organisms, cells remain attached, but more strongly in some cases than in others. Investigate the ways in which the cells are held together.

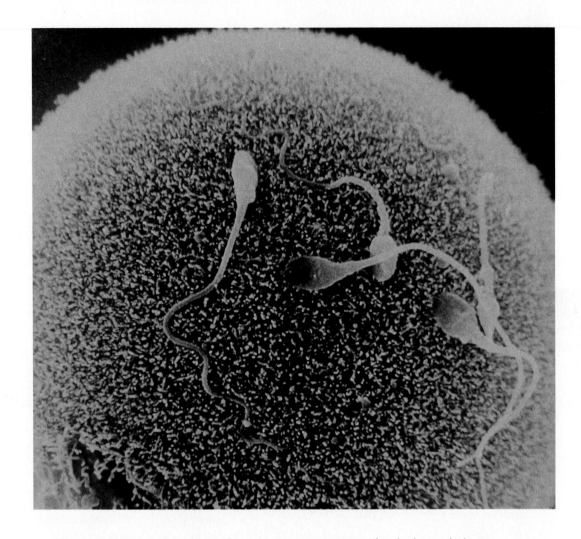

These are human sperm attempting to penetrate an egg. What is the evolutionary advantage of having contributions from both parents during sexual reproduction?

Continuity Through Reproduction 6

Chapter 5 describes how cells reproduce by dividing in two. When that happens, the "parent" cell disappears, but its life continues in its offspring. The process of reproduction for entire organisms, however, usually is not that simple. When entire organisms are produced, the parent usually lives on as an individual and in many cases cares for the offspring.

Producing offspring is not necessary for individual survival. For example, a plant that does not get enough water during a particular year may survive but be unable to flower or produce seeds. The life of the plant, therefore, does not depend on its production of seeds every year. If this entire species is to survive, however, individuals of the species must produce new members. Because individuals die, the species will disappear if new individuals are not formed. The formation of new individuals that are like their parents is called reproduction.

All the organisms alive today arose from preexisting organisms. Although organisms have many different methods of reproduction, every method leads to a marvelous event—the beginning of a new life. This chapter discusses some of these methods.

Reproduction

Guidepost: Why is reproduction important?

6.1 Reproduction Is Essential for the Continuation of Life

A new organism grows, develops, and eventually becomes large enough and mature enough to reproduce. The time when an organism becomes able to reproduce varies among species. Some organisms, such as bacteria, can reproduce after 20 minutes or so of life; other organisms may require months or years to become mature.

For a unicellular organism, reproduction occurs only once. Multicellular organisms, in contrast, may continue to reproduce for days, months, or years. These reproductive periods may form a large part of their life cycle. A **life cycle** is all the events that occur between the beginning of one generation and the beginning of the next. Because life cycles are continuous, they often are shown in diagrams as circles (see Figure 6.1 on page 136).

Some organisms, such as humans, outlive their reproductive periods. Other organisms continue to reproduce until they die. The giant tortoises

Figure 6.1   In the life cycle of the leopard frog (*Rana pipiens*), the animal changes rapidly (in the course of several weeks) from an embryo within an egg, to a tailed tadpole with structures and behaviors suited for life in the water, to a tailless adult frog suited to life on land.

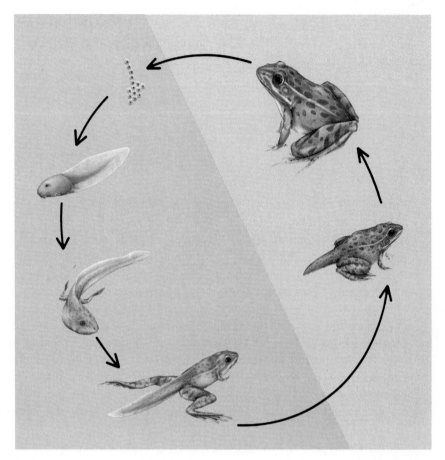

of the Galápagos Islands may live as long as 200 years, but they cannot reproduce until they are about 40 years old. Thereafter, they may continue to reproduce as long as they live.

Even though some organisms live for a long time, no organism lives forever. Although the life span of an organism is part of its genetic program, not all organisms live out their genetic life spans. Some are eaten or die of accidents, infections, or other diseases. The secret of biological success for a species lies in its members' ability to reproduce and leave offspring before death occurs. When organisms do reproduce successfully, their species continues. A species becomes extinct if its members fail to reproduce or if they fail to leave enough offspring to allow for deaths as a result of accidents and disease. Successful reproduction ensures the continued existence of a species.

6.2   Reproduction May Be Sexual or Asexual

In complex organisms, reproduction generally requires two parents and two different special cells, one from each parent. When these cells are united, a new individual begins. This is known as **sexual reproduction**.

Humans, frogs, worms, and flowering plants reproduce sexually. In sexual reproduction, only certain types of cells can unite to form a new individual. These special cells are called **gametes** (GAM eets). The other cells that make up the individual are known as body cells.

Sexual reproduction affects the behavior of organisms, their genetic makeup, and their evolution. The nuclei of gametes, like those of all cells, contain chromosomes. Each chromosome in a gamete carries genes from the parent organism that produced the gamete. When parents reproduce, they pass genetic information on to their offspring, and the new individuals then develop in response to these genetic instructions. Because it combines genetic material from two different individuals, sexual reproduction increases the amount of genetic variation in a population.

In **asexual reproduction**, new individuals originate from a single parent. Either the parent divides into two (or more) individuals, or new individuals arise as buds from the parent's body (see Figure 6.2). No gametes are needed for asexual reproduction. In plants, asexual reproduction sometimes is called vegetative reproduction.

Some animals and plants can reproduce both sexually and asexually. For example, the potato is a flowering plant that can use sexual reproduction to produce seeds, but it also can reproduce asexually. The little "eyes" on potatoes are actually buds—groups of cells that can undergo rapid mitosis and develop into new plants. When potatoes are not used, the buds sprout and begin to grow, as shown in Figure 6.3a.

A sea star is another organism that can reproduce sexually or asexually. To obtain food, sea stars invade oyster beds in coastal waters, open the oyster shells with their powerful arms, and then feast on the oysters inside. Scuba divers used to collect the sea stars to save the oyster beds. In an effort to destroy the sea stars, the divers would cut off the arms and throw the pieces back into the water. A sea star arm, however, can regenerate an entire sea star if part of the central body is attached to it (see Figure 6.3b). In their efforts to eliminate sea stars from the oyster beds, the divers actually increased the sea star population.

Figure 6.2　A budding hydra.

Figure 6.3　Asexual reproduction in a plant and an animal. The sprout on the potato (a) grew from a bud that can produce an entire new potato plant. The sea star can regenerate a new body from one arm (b).

a

b

Concept Review

1. What is the main difference between sexual and asexual reproduction?
2. What do regeneration and vegetative reproduction have in common?
3. How can having two different types of reproduction help some species survive?
4. From the viewpoint of the individual, how does reproduction differ from other life processes, such as digestion and cellular respiration?

Figure 6.4 Eggs and sperm of several organisms. Note that the sperm are enlarged much more than the eggs, and direct comparison between the two is not possible. In the chicken, frog, and fish, the ova are surrounded by other materials (shown in outline).

Eggs (×2.8)	**Sperm** (×602)
chicken	
frog	
fish	
human	
rat	

Formation of Reproductive Cells

6.3 Gametes Are Reproductive Cells

Guidepost: How do gametes differ from other cells?

In most types of sexual reproduction, each parent produces a different type of gamete. For example, one parent may produce large, stationary gametes, and the other parent may produce smaller gametes that can move. In some organisms, the reproductive cells of both parents look similar. Gametes usually differ in appearance from other cells in an organism, and they contain fewer chromosomes.

Males produce gametes called **sperm cells**, which usually are quite small. (The word *sperm* can be used to refer to one or many sperm cells.) Females produce egg cells, or **ova** (singular, ovum). An ovum may be large. Sometimes it is thousands of times larger than a sperm cell. Figure 6.4 compares the ova and sperm of several organisms.

A sperm cell consists of little more than a nucleus in the head, a tail that can move the cell about, and an energy generator—its mitochondria (see Figure 6.5). An ovum, on the other hand, contains, in addition to its nucleus, other organelles, such as mitochondria and ribosomes, that are important for cell function. In many cases, an ovum also contains a reserve supply of food for the new organism. Biologists use the differences in size and function of gametes to define gender. An organism that produces ova is called female. An organism that produces sperm is called male. Some organisms, however, actually change their sex during their lifetime. Such an ability can be extremely useful. An individual can be whatever sex is more productive at a given moment, thus helping to ensure the continuation of the species. For example, the blueheaded wrasse shown in Figure 6.6 lives in tropical reefs and eats parasites off the scales of other fish. It usually lives in groups of one male and five or six females. The male, largest of all, mates with all the females. The females have a social hierarchy based on size. When a female of high status dies, each female of lower status moves up a notch. When the male dies, however, the largest female fish becomes male. Within one hour after the male's death, the dominant female starts to act like a male. Within two weeks, the fish begins to produce sperm.

Figure 6.5 A scanning electron micrograph of human sperm showing the head and the middle piece that contains the mitochondria (X19 500). The mitochondria provide power for the tail.

Figure 6.6 A dominant male blueheaded wrasse (*Thalassoma bifaciaum*). When this male dies, the largest female will assume his role.

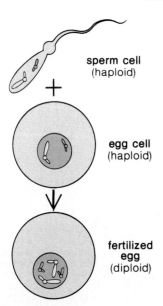

sperm cell
(haploid)

+

egg cell
(haploid)

fertilized
egg
(diploid)

Figure 6.7 Gametes unite in the process of fertilization, restoring the diploid number of chromosomes. What number of chromosomes in the new individual come from each parent?

6.4   The Chromosome Number of a Species Remains Constant

The cells of an organism contain the number of chromosomes that are typical of its species. For example, the body cells of corn plants have 20 chromosomes, the body cells of the common fruit fly, which is used in genetic research, have 8 chromosomes, and the body cells of humans have 46 chromosomes. Each chromosome in a body cell has a partner of the same length and appearance. In other words, the chromosomes occur in matching pairs, and for the most part, each pair has genes that code for the same traits. One member of each chromosome pair originally came from the sperm and the other from the ovum that produced the organism.

A new individual begins at **fertilization**, when the nuclei of two gametes—one gamete from each parent—unite and produce a egg, or **zygote** (ZY goht). When these nuclei unite, the chromosomes from each come together in the nucleus of the zygote. If each of the gametes contained the same number of chromosomes as the rest of the cells of each parent, what would happen from one generation to the next? You would expect the number of chromosomes to double. In humans, 46 chromosomes from each parent would result in 92 chromosomes in the child. Yet, humans normally have 46 chromosomes in each body cell. What makes this constancy possible?

Gametes have only half the number of chromosomes that body cells have—*one* chromosome from each matching pair. Because gametes contain only one chromosome from each matching parental pair, they are said to be **haploid** (HAP loyd) cells. Haploid cells are represented by the symbol *n*. Cells with both chromosomes of each pair are called **diploid** (DIP loyd) and are represented by the symbol *2n*. The union of two haploid gametes restores the chromosomes to the diploid number, as you can see in Figure 6.7. In this illustration, the diploid number is four—two pairs of chromosomes. For humans, normal haploid gametes contain 23 chromosomes. The diploid body cells each contain 46 chromosomes.

Gametes become haploid through a special type of cell division known as **meiosis** (my OH sis). In sexual reproduction, meiosis and fertilization are complementary processes: meiosis produces the haploid (*n*) chromosome number in gametes, and fertilization restores the diploid (*2n*) chromosome number in the zygote.

6.5   Gametes Are Formed by Meiosis

The number of chromosomes is reduced from diploid to haploid when gametes are formed. This reduction takes place during meiosis, which requires two cell divisions in succession. The stages of meiosis are similar to those of mitosis, described in Section 5.9.

During the S phase of the cell cycle preceding meiosis, the chromosomes replicate and produce twin structures. Each structure is called a **chromatid** (KROH muh tid). A specialized area known as the **centromere** (SEN troh meer) connects the twin chromatids, which contain identical DNA. Meiosis consists of one replication of chromosomes followed by two

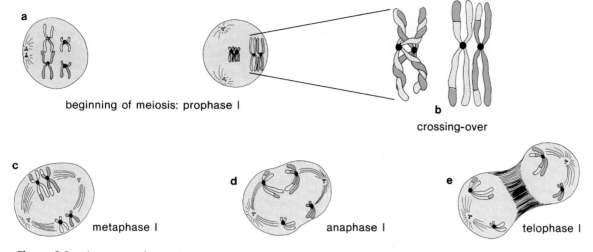

a

beginning of meiosis: prophase I

b

crossing-over

c

metaphase I

d

anaphase I

e

telophase I

Figure 6.8 The stages of meiosis—first meiotic division.

divisions. The first meiotic division is called I; the second is called II. Once meiosis begins in a cell, no more DNA is made—the DNA already present is distributed to the gametes during the two divisions. Figure 6.8a–h illustrates the events of meiosis in a cell with three pairs of chromosomes. Refer to the illustration as you read the description that follows.

As meiosis begins (a), each chromosome is a double structure made of two chromatids joined at the centromere. (The number of centromeres equals the number of chromosomes.) In prophase I, a unique event occurs. Remember that the sperm contributes one chromosome to every pair of chromosomes and the ovum contributes the other. After the chromosomes replicate, the matching pairs come close together and may twist around each other (b). In this position, parts of a chromatid of one chromosome may break off and exchange places with the identical parts of a chromatid of the other matching chromosome. This process is called **crossing-over**, because chromatid segments are exchanged between the two chromosomes of a pair, which results in new combinations of genes.

Toward the end of prophase I, the chromosome pairs move to the equator of the cell. Metaphase I (c) marks their arrival in the center of the cell, where they line up in pairs. In anaphase I (d), the matching chromosomes of each pair separate and begin to move toward opposite poles of the spindle. Each chromosome still consists of two chromatids attached at the centromere.

The matching chromosomes, one from each pair, continue to move in opposite directions. As they approach the poles in telophase I (e), cytokinesis occurs, resulting in two cells. As telophase I ends, the matching chromosomes are in different cells for the first time. Each new cell now has only half as many chromosomes as a body cell and, therefore, contains half the parent cell's total genetic information. Because the amount of genetic information per cell has been reduced by one-half, the first meiotic division often is called *reduction division*.

Figure 6.8   The stages of meiosis (continued)—second meiotic division.

Meiosis continues with a short prophase II in which no additional crossing over occurs. In metaphase II (f), the chromosomes again move to the equator in each of the two cells. This time, however, there are no matching chromosome pairs. The chromosomes line up in single file down the center of each cell, just as in mitosis.

In anaphase II (g), the centromeres divide, and the two chromatids of each chromosome separate and move toward opposite poles. The chromatids (now called chromosomes) gather at the poles and are enclosed by new nuclear envelopes in telophase II (h). Cytokinesis occurs again, resulting in four haploid cells, each containing n single chromosomes. You can become more familiar with the events of meiosis by modeling them in Investigation 6.1.

Although there are similarities between the stages of meiosis and those of mitosis, there are important differences. Figure 6.9 on page 143 compares the two processes.

Male animals, including human males, usually produce four equal-sized sperm from each original cell. Further cytoplasmic changes occur as each sperm develops a tail and the DNA becomes concentrated in the head of the sperm cell. The mitochondria packed into the middle part of the sperm provide energy for the tail, which will move the sperm through the female reproductive tract. When the cytoplasmic changes are complete, the cells are fully formed sperm.

Females usually produce only one ovum from each original cell that begins meiosis. In the first meiotic division, one complete set of chromosomes is cast off in a small cell called a polar body, which consists of little but the chromosomes. Most of the cytoplasm remains in the larger cell. In

Mitosis   **Meiosis**

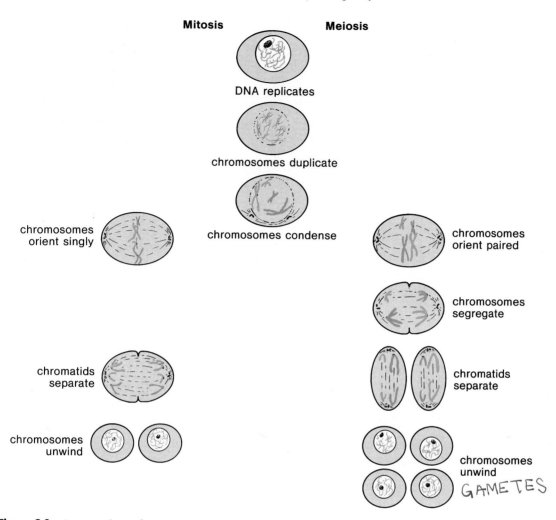

DNA replicates

chromosomes duplicate

chromosomes
orient singly

chromosomes condense

chromosomes
orient paired

chromosomes
segregate

chromatids
separate

chromatids
separate

chromosomes
unwind

chromosomes
unwind

GAMETES

Figure 6.9 A comparison of mitosis and meiosis.

the second meiotic division, another polar body with a haploid set of chromosomes is cast off. In most species of animals, the two polar bodies disintegrate. The remaining cell, which becomes the ovum, contains a haploid set of chromosomes and cytoplasm. Figure 6.10 on the next page compares sperm and egg formation.

Concept Review

1. Compare the two types of gametes.
2. Explain the biological distinction between male and female.
3. How do gametes differ from other cells of the body?
4. What is the significance of the chromosome number in gametes?
5. How does fertilization affect the chromosome number of a cell?
6. How is a zygote formed?
7. Compare and contrast the stages of meiosis and mitosis.

Figure 6.10 In human males, meiotic division usually results in four equal-sized sperm, each with 23 chromosomes. How does the formation of an egg differ from the formation of sperm?

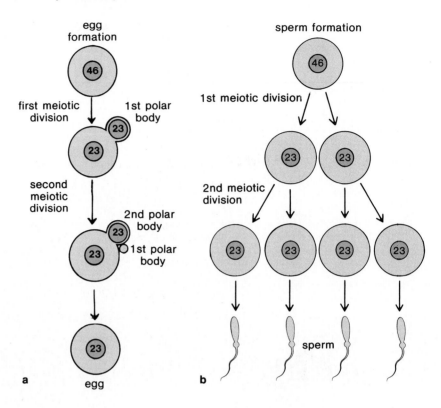

INVESTIGATION 6.1 A Model of Meiosis

Many biological events are easier to understand when they are explained by models. In this investigation, you will use a model to duplicate the events of meiosis.

Materials (per team of 2)
modeling clay, red and blue
4 2-cm pieces of pipe cleaner
large piece of paper

PART A Basic Meiosis
Procedure

1. Use the clay to form two blue and two red chromatids, each 6 cm long and about as thick as a pencil.
2. Form four chromatids, two of each color, 10 cm long.
3. Place the pairs of similar chromatids side by side. Use pipe cleaners to represent centromeres. Press a piece of pipe cleaner across the centers of the two red, 6-cm chromatids. This represents a chromosome that has replicated itself at the start of meiosis. Do the same for the other three replicated chromosomes (see Figure 6.11).
4. On a sheet of paper, draw a spindle large enough to contain the chromosomes you have made. Assume that the spindle and chromatids have been formed, and the nuclear membrane has disappeared.
5. Pair the two 6-cm chromosomes so that the centromeres touch. Pair the two 10-cm chromosomes. Assume that the red chromosome of each pair was derived from

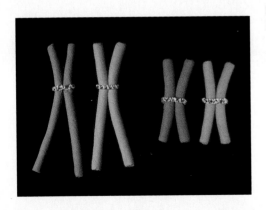

Figure 6.11 Chromosome models.

the female parent. Its matching chromosome, the blue one, came from the male parent.

6. Arrange the two chromosome pairs along the equator (middle) of the spindle so that the red chromosomes are on one side and the blue on the other.

7. Holding on to the centromeres, pull the chromosomes of each matching pair toward opposite poles of the spindle. Once the chromosomes have been moved to the two poles, the first meiotic division is completed.

8. Draw two more spindles on the paper. These new spindles should be centered on each of the poles of the first meiotic division. Both spindles should be perpendicular to the first spindle. Your model is now ready for the second division of meiosis.

9. Place the chromosomes from each pole along the equator of each of the two new spindles. Unfasten the centromere of each chromosome. Grasp each chromatid at the point where the centromere had been attached. Pull the chromatids to opposite poles of their spindles. Try to move all the chromatids at once, as occurs in a living cell. Draw a circle around each group of chromosomes that you have.

Discussion

1. How many cells were there at the start of meiosis? How many cells are formed at the end of meiosis?

2. How many chromosomes were in the cell at the beginning of meiosis? How many chromosomes were in each of the cells formed at the end of meiosis?

3. At the end of meiosis, what are the cells called?

4. How many of your cells at the end of meiosis had only red chromosomes in them? How many had only blue chromosomes in them?

PART B   Effects of Chromosome Position
Procedure

10. A real cell is three-dimensional. Although the red chromosomes (from the female) may be on one side and the blue (from the male) may be on the other when they line up at the equator, there is an equal chance that one red and one blue chromosome will be on the same side. Attach the chromatids as they were at the beginning of the investigation. Go back to Step 6 and arrange the chromosomes so that one blue and one red chromosome are on each side. Now complete meiosis.

Discussion

1. How do these gametes compare with those you made earlier?

2. What difference does this make in terms of genetic variation in the offspring?

3. How many different types of gametes could be made if there were three sets of chromosomes instead of just two?

PART C   Effects of Crossing-Over
Procedure

11. Reassemble your chromosomes. To show crossing-over (Figure 6.8b), exchange a small part of the clay from a chromatid making up one chromosome with an equal part from a chromatid of its matching pair. The colors make the exchange visible throughout the rest of the investigation.

12. Place your chromosome pairs along the equator of the spindle as in Step 6, and complete meiosis.

Discussion

1. How many different types of gametes did you form? Did you form any different ones from those formed by others in your class?

2. In general, how do you think crossing-over affects the number of different types of gametes that are formed?
3. In crossing-over, what actually is exchanged between the chromatids?
4. What are some of the advantages of using a model to visualize a process?
5. How did this model improve your understanding of the process of meiosis?
6. What are some disadvantages of this model?

Guidepost: How does reproduction occur in humans?

The Human Reproductive System

6.6 Gamete Development Differs in Males and Females

In humans and other animals, meiosis and gamete formation take place in special reproductive organs called **testes** (singular, testis) in males and **ovaries** (singular, ovary) in females. In early development, these reproductive organs look alike.

When the developing organism, the **embryo**, is about eight weeks old, these reproductive organs become clearly male or female. Like the ovaries, the testes develop in the abdominal cavity of an embryo. Before birth, however, the two testes move downward and to the outside of the body, where they are housed in a pouch called the scrotum, which is located just below the penis (see Figure 6.12a).

Each testis is made up of thousands of tiny tubules, shown in Figure 6.12b, in which meiosis occurs. Figure 6.12c shows a cross section of one tubule. The diploid cells that will undergo meiosis are located around the outer edge of each tubule. As meiosis proceeds, the new cells are formed toward the center of the tubule. Cells in the various stages of meiosis are shown in Figure 6.12c. The last stages of meiosis occur near the center of the canal, where the dark heads of mature sperm can be seen with their long tails extending into the center of the canal.

The sperm move out of these canals into a special collecting duct near the testis. This curved duct, which is about 50 cm long, can hold billions of sperm. The sperm remain there until they are released in sexual activity, or until they are reabsorbed by the body. Mature human males can produce huge numbers of sperm and may release from 200 million to 500 million at one time.

In humans, meiosis and the formation of sperm cells take about nine weeks. Remember, however, that the development of sperm is a continuous process, and only mature sperm cells are absorbed or released. As some sperm are absorbed or released, others have matured and moved from the tubules in the testis to the collecting duct.

Gametes are not produced until puberty, or sexual maturity, in males, but in females, meiosis begins in the embryo. Millions of potential ova form in the ovaries of the female embryo. Certain diploid cells of the newly formed ovary begin meiosis, but the first meiotic division soon is interrupted. The potential ova stop developing in prophase I and remain in suspended division for many years.

Meiosis in the ovary does not begin again until puberty, when a girl enters adolescence. At this time, about once a month, one cell resumes

a

kidney

ureter

seminal
vesicles

urinary
bladder

prostate
gland

urethra
vas
deferens
scrotum
testes
penis
epididymis

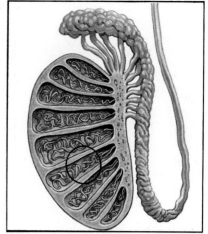

b

Each testis is composed of packed
coils of tubules in which sperm
develop.

Figure 6.12 The human male reproductive system.

c

Cross section through a tubule,
×100. Around the edge of the central
canal are the dark heads of the
sperm, with their tails extending into
the canal.

meiosis. That potential ovum continues into the second meiotic division, and the process stops once more. Later, the cell is released from the surface of the ovary, a process called ovulation (ohv yoo LAY shun). In humans, one ovum usually is released each month, although sometimes two (or more) may be released.

Only about 400 to 500 ova are released during an entire lifetime, and the reproductive span of a female may extend from puberty (which usually begins between ages 9 and 14) until age 45 or 50. Thus, some of the partly mature ova may take as long as half a century to complete meiosis.

6.7 Hormones Control Reproductive Cycles

Both the onset of puberty and the lessening of gamete formation in later life are controlled by changes in the body chemicals that influence the reproductive organs. Many different organs in the body produce these chemicals, called **hormones**. Traveling in the blood to all parts of the body, hormones act as messengers that influence other organs. A hormone from one organ may affect another organ, which can respond in striking ways. The first organ, in turn, can be influenced by hormones from still other organs. This complex set of interactions, a network of stimuli and responses, is called a **feedback system**. This is one way the various parts of the body communicate with each other.

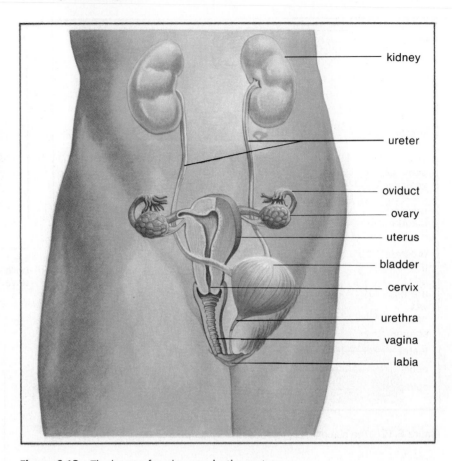

kidney

ureter

oviduct

ovary

uterus

bladder

cervix

urethra

vagina

labia

Figure 6.13   The human female reproductive system.

Hormones produced in the brain stimulate the testes and ovaries to secrete distinctive sex hormones. Although the embryo's sex is determined by the sex chromosome at the time of fertilization, the development of an embryo into a male or a female baby is controlled by the sex hormones. Sex hormones also help bring about some of the noticeable indicators of approaching sexual maturity—changes in voice and body proportions, growth of hair, and increased interest in sex. Although hormones are essential for sexual reproduction in both males and females, the hormone interactions in females often are more complex.

In females, the nervous system, several glands and organs, and a variety of hormones interact to regulate a monthly ovum-releasing cycle, known as the **menstrual cycle** (from the Latin *mensis*, meaning month). During this cycle, usually a single ovum is discharged from one of the ovaries, which are located low in the abdominal cavity (see Figure 6.13). The ovum enters one of two tubular structures called oviducts. The oviducts act as passageways through which the ovum moves into the **uterus**, a muscular, pear-shaped organ. If the ovum is fertilized by a sperm, the fertilized egg develops in the uterus until the time of birth. At that time, strong muscular contractions push the baby out of the uterus through the vagina.

During the menstrual cycle, the inner lining of the uterus builds up in preparation for a fertilized egg. If the egg is not fertilized, the lining, which

Birth Control

The earth's population is increasing rapidly. In an effort at control, some overpopulated countries are trying to limit the number of children born. To do so, they rely on birth control, or contraception.

Although abstinence (refraining from sexual intercourse) is the only *sure* way to prevent pregnancy, contraception can prevent fertilization of an egg or disrupt development of the zygote. It does so by preventing (a) development of gametes, (b) physical contact between two gametes, or (c) implantation of a fertilized egg.

Many women prevent the development of gametes by taking birth control pills containing estrogen and a synthetic hormone called progestin. The combination of the two hormones inhibits the development of egg follicles in the ovaries. The pill is about 99 percent effective, but it may have serious side effects and must be prescribed by a doctor. To date, studies on the side effects have been inconclusive. Women in Sweden, Finland, and Norway can have a hormonal contraceptive implanted under the skin of the upper arm that can prevent pregnancy for up to five years. Several other countries have injectable contraceptives that last for one to two months. Not all these options are yet available in the United States.

Contact between gametes can be prevented by interfering with the movement of ova or sperm cells. A membranous sheath, or condom, can be placed over the penis prior to intercourse, thus preventing sperm from entering the body of the female. Condoms also prevent the spread of AIDS and other sexually transmitted diseases. Another method uses a flexible cap, or diaphragm, over the cervix, the entrance to the uterus, to prevent sperm from entering.

Surgical procedures, or sterilization, also can prevent the movement of ova or sperm. In males, the duct that carries sperm to the penis may be tied and cut. This simple procedure, called a vasectomy (vuh SEK tuh mee), is widely used in some countries. In females, the oviducts may be tied and cut. This method is called tubal ligation (TOO buhl ly GAY shun). Both methods generally are considered irreversible. Currently, sterilization is the most popular form of contraception in the United States.

If pregnancy does occur, it can be terminated involuntarily or voluntarily. Many conceptions and embryos are aborted naturally. If a major chromosomal or genetic abnormality is present, the body may abort the embryo during its first few days of development. Such spontaneous abortions may affect 70 percent of all conceptions.

Many abortions are undertaken because contraceptives failed. Experts blame contraceptive failure for about 2 million unwanted pregnancies a year in the United States. Many of these pregnancies are allowed to continue, but every year about 0.8 million of them are ended by abortion. That represents about half of the 1.5 million abortions occurring annually. The debate surrounding planned abortions is complex and emotionally charged. Because there are no clear-cut answers, the debate is likely to continue for a long time. Many experts feel that the United States lags about 30 years behind other countries in contraceptive development and that new, easier-to-use contraceptives might help reduce our abortion rate, which is one of the highest in the industrialized world.

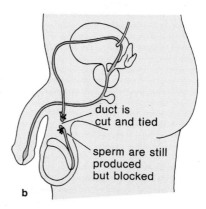

Tubal ligation (a) and vasectomy (b) are permanent methods of preventing pregnancy.

149

develops a rich blood supply, disintegrates and passes from the uterus through the vagina to the outside of the body. Discussions of the menstrual cycle usually take as their starting point the first day of menstrual flow (menstruation, or "period").

The hormonal interactions that regulate the menstrual cycle are shown in Figure 6.14, opposite. Refer to the diagram as you read the descriptions that follow. In the menstrual cycle, the **hypothalamus**, a part of the brain shown at (a), acts in much the same way that a thermostat acts in a house. A thermostat set at 20° C does not keep the room at exactly 20° C. The temperature fluctuates somewhat as the furnace is turned on and off. Just as the thermostat monitors and adjusts temperature, the hypothalamus monitors and helps to adjust the level of hormones in the body. The hormonal fluctuations serve to coordinate the release and fertilization of an egg with the preparations for a receptive environment in the uterus.

At the start of the menstrual flow, two of the hormones secreted into the bloodstream by the ovaries are at low levels. These low levels cause the hypothalamus to secrete a releasing hormone that stimulates a gland, known as the **pituitary** (pih TOO ih ter ee), just underneath the brain. As a result, the pituitary secretes two hormones that act on the ovaries. These hormones are called follicle-stimulating hormone (FSH) and luteinizing hormone (LH). FSH causes an ovum to begin maturing inside a fluid-filled sac, or follicle, on the surface of the ovary (b).

The two pituitary hormones also stimulate the maturing follicle (c) to secrete the hormone **estrogen** (ES troh jen). Estrogen affects the uterus, causing an increase in the blood supply and a thickening of the lining. The effects of estrogen and the two pituitary hormones build for about 14 days. Just before day 14 a sudden increase in LH causes the follicle to burst and release the ovum. This process is known as ovulation (d).

The ruptured follicle is converted to the corpus luteum (KOR pus LOOT ee um). The corpus luteum (e) releases estrogen and another hormone, **progesterone** (proh JES tuh rohn). Progesterone and estrogen act together to stimulate additional buildup of the lining of the uterus.

In the brain, meanwhile, the feedback system enables the hypothalamus to detect the increased levels of estrogen and progesterone in the bloodstream and to slow down secretion of its releasing hormone. This change causes the pituitary to secrete less FSH and LH. If the ovum is not fertilized as it travels down the oviduct toward the uterus, the corpus luteum starts to degenerate (f), and the levels of progesterone and estrogen drop. The drop in estrogen and progesterone levels decreases the blood supply to the lining of the uterus. Menstruation begins as the blood-filled uterine tissues are shed and passed through the vagina. The onset of the menstrual flow marks the beginning of another cycle. Notice the approximate timing of the events in the cycle in Figure 6.14.

The first menstruation indicates that a young female has become capable of producing ova and of having them fertilized. Menstruation alone, however, is an unreliable indicator of fertility. Remember that an ovum first is released from the ovary and begins to travel down the oviduct. If the ovum is *not* fertilized, then the menstrual cycle begins. A young female who has not had her first menstrual period still may have released an ovum that can be fertilized.

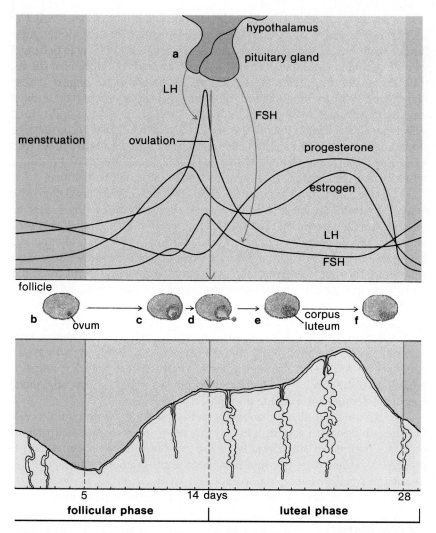

Figure 6.14 The menstrual cycle. Interactions among the nervous system and several organs, glands, and hormones regulate the cycle.

Females do not remain fertile for the rest of their lives. Just as menstruation signals the beginning of the ability to reproduce, **menopause** (MEN oh pahz), the cessation of the menstrual cycle, indicates that a female no longer is capable of reproducing. The hormonal effects of menopause are just as profound as those of the menstrual cycle and vary widely from individual to individual. Menopause is a normal part of the life cycle of a female and usually occurs between the ages of 45 and 50. Males do not have a hormonal cycle similar to the menstrual cycle, nor do they experience menopause, although the number of sperm a male produces may decline with age. If enough sperm are present, however, a male of advanced age is capable of impregnating a female.

6.8 Only One Sperm Fertilizes the Ovum

Before fertilization can take place, sperm cells must enter the body through the vagina. This usually occurs through sexual intercourse. As the male becomes sexually excited, the penis fills with blood and becomes rigid. The sperm cells begin to move through a series of ducts into the penis. After further stimulation, the male reproductive system responds by expelling 3 to 4 mL of **semen** (SEE men). In addition to hundreds of millions of sperm, this whitish fluid contains the secretions of various male glands. These glandular fluids provide food and lubrication for the sperm and counteract the acidity of urine, which may be harmful to sperm. (Urine and semen pass through a common duct in the penis.) Figure 6.12 shows reproductive structures in males. After sperm cells are released into the vagina, they swim in all directions. Many of them swim up along the moist linings of the female reproductive tract, enter the uterus, and swim into the two oviducts.

The released ovum's trip to the nearest oviduct may take two days. The ovum, however, can be fertilized successfully for only 24 hours after its release. Human sperm, on the other hand, may live for 48 hours after they are deposited. If sexual intercourse occurs during the 24 hours after an egg has been released, or if live sperm are present from previous intercourse at the time the egg is released, the ovum may meet many sperm cells, and one sperm may penetrate the outer membranes of the ovum.

The membranes of the ovum immediately react to fertilization and act as a barrier to the entry of other sperm cells. Only one sperm cell fertilizes each ovum, and only the head of the sperm, which contains the genetic material, enters the egg. If only one sperm fertilizes the ovum, why are so many sperm released in semen? Sperm secrete an enzyme that helps to break down the layer of cells that surrounds an ovum. Apparently, many sperm are required to provide enough enzyme.

Unlike the diploid cell that enters meiosis and forms four haploid cells in males, in females each diploid cell that enters meiosis eventually forms only one haploid ovum. Virtually all the cytoplasm of the original cell is preserved in the ovum, where it provides the material and energy needed for the early cell divisions of development. After the embryo is attached to the inner layer of the uterus, it receives nourishment from its mother.

Concept Review

1. Describe the male and female reproductive systems.
2. Describe what happens in the menstrual cycle.
3. If an ovum can be fertilized only during a 24-hour period, how can pregnancy occur even when a female does not engage in intercourse during that period?
4. Trace the movement of sperm from the time they are released to the time they encounter an ovum.

Summary

Reproduction is essential for the continuity of life on the earth. Organisms usually reproduce either sexually or asexually. Sexual reproduction increases genetic variability in a population because genetic material is provided by two different parents.

Sexual reproduction requires two parents. Each parent contributes a gamete to the new individual, forming a zygote. The gametes contain only half the genetic information of the parent. The reduction of genetic material is accomplished by meiosis. Genetic material is exchanged and reduced by half in two successive cell divisions. In males, a single diploid cell gives rise to four haploid sperm, but in females, a diploid cell results in only one ovum. When the gametes combine in fertilization, the chromosome number is replenished.

The menstrual cycle is regulated by a complex feedback system involving hormones produced by the brain and other organs. During the cycle, the lining of the uterus builds up in preparation for receiving a fertilized egg. If fertilization of the ovum does not occur, the blood-rich lining of the uterus disintegrates and passes from the body.

Applications

1. Explain how the behavior of chromosomes during prophase I in meiosis differs from chromosome behavior during prophase in mitosis.

2. Why is crossing over possible in meiosis but difficult in mitosis?

3. Bacteria reproduce by splitting into two different cells. Do you think this is sexual reproduction or asexual reproduction? Explain.

4. Suppose there are two similar plant species—one that reproduces only by sexual reproduction, and one that reproduces only by asexual reproduction. Which species has a better chance of surviving over the next one million years, and why?

5. Mitochondria have their own DNA, which is circular. This DNA differs from the DNA found in the nucleus. Studies show that mitochondrial DNA is derived from the female parent, not the male parent. Use your knowledge of meiosis and fertilization to explain this phenomenon.

6. Suppose you find a cell that contains 15 chromosomes. Is this cell a gamete or a body cell? How do you know?

Problems

1. Most nursery managers use asexual reproduction to propagate some of their plants. Invite a local gardener or nursery manager into your class to demonstrate techniques of cloning, grafting, or vegetative propagation.

2. What effect does the environment have on the development of genetically identical organisms? Take several cuttings from a single plant, making sure that the cuttings come from parts of the plant that are of similar age. After the cuttings have developed their own roots, plant them in identical pots, but expose them to different environmental factors. Observe their growth.

3. Many crop plants have twice the number of chromosomes as their ancestors. These plants are called polyploids—individuals that have more than the normal number of sets of chromosomes. Investigate the evolution of several crop plants and report to the class on the importance of polyploidy in their development. One useful reference is N. W. Simmonds, ed., *Evolution of Crop Plants* (New York: Longman Inc., 1979).

4. Is human reproduction lacking in seasonality? Record by months the birthdays of members of your biology class and as many other classes as possible. Present the data in the form of a bar graph. What do the data indicate?

5. There is some indication that the age at which human females reach sexual maturity, or puberty, is declining. Collect data on this phenomenon and on its possible causes.

6. The environment of the female reproductive tract is slightly acidic, which can be harmful to sperm cells. Investigate the components of semen and the ways in which they help counteract the acidity.

7. There are two types of twins: fraternal (dizygotic) and identical (monozygotic). Explore the cause of twinning in each case.

8. Make a model of the female hormonal interactions that regulate the menstrual cycle. Indicate in your model when pregnancy can occur, and what will happen to the hormonal levels if pregnancy does occur.

These one-day-old white-footed mice (*Peromyscus leucopus*) obviously do not appear similar to their mother. What biological processes must occur from now until the time when they will resemble their mother?

Continuity Through Development

Frogs croaking in a marsh, cats crying in the alley, insects chirping outside your window, all are the sounds of the living world. Reproduction is a fundamental part of the living world, but how does the new life created by this process come about?

In most complex organisms, the fertilization of an egg is the first in a series of intricate steps. The fertilized egg is still a single cell, yet within that one cell is all the genetic information needed to form an organism that has millions or billions of cells. Many of these cells differ greatly in both structure and function. How can one set of genetic instructions determine the fate of so many different types of cells? This chapter describes the series of events that occur after fertilization and result in a multicellular organism.

Development

Guidepost: How does a zygote become a multicellular, complex organism?

INVESTIGATION 7.1 The Yeast Life Cycle

Baker's yeast (*Saccharomyces cerevisiae*) is a unicellular organism that reproduces both sexually and asexually. Because the cells have a characteristic shape at each stage, all the major stages of the life cycle can be seen under the microscope.

Yeast cells may be either haploid or diploid. Haploid cells occur in two mating types ("sexes"): mating-type **a** and mating-type α (alpha). When **a** and α cells come in contact, they secrete hormonelike substances called mating pheromones (FAYR uh mohnz), which cause haploid cells of the opposite mating type to develop into gametes. The **a** and α gametes pair and then fuse, forming a diploid zygote. The fusion of **a** and α gametes is similar to fertilization in animals, except that both parents contribute cytoplasm and nuclei. Yeast zygotes reproduce asexually by budding. When cultured on a solid growth medium, a yeast zygote may grow into a visible colony that contains up to 100 million cells.

Diploid yeast cells do not mate, but in times of stress, such as during a period of unbalanced food supply, the diploid cells may sporulate, or form spores. The spores remain together, looking much like ball bearings, in a transparent saclike structure called an ascus (ASK us).

In this investigation, you will start with two haploid yeast strains of opposite mating types, mate them in order to form a diploid strain, and try to complete an entire yeast life cycle.

Materials (per team of 2)

microscope slide
cover slip
dropping pipet
glass-marking pencil
compound microscope
container of clean, flat toothpicks
bottle of water

6 self-closing plastic bags labeled *waste*
  (1 for each day)
3 agar growth medium plates (YED)
1 "unknown" agar medium plate
agar slant cultures of **a** and α yeast
  strains

Procedure
PART A   Sexual Reproduction: Haploid to Diploid
Day 0

1. Prepare fresh cultures of the **a** and α yeast strains from the agar slants. Colonies of mating-type **a** are red; those of mating-type α are cream colored. Touch the flat end of a clean toothpick to the **a** strain, then gently drag it across the surface of an agar growth medium plate to make a streak about 1 cm long and 1 cm from the edge (see Figure 7.1). Discard the toothpick in the self-closing plastic waste bag, being careful not to touch anything with it. Use a glass-marking pencil to label the bottom of the plate near the streak with an **a**. Use a new, clean toothpick to put a streak of the α strain near the opposite side of the agar on the same plate, and label it α. *Use a new, clean toothpick from the container for each thing you do. Be careful not to touch the ends of the toothpicks to anything except yeast or the sterile agar. Discard used toothpicks in the plastic waste bag. Keep the lid on the plate except when transferring yeast.* Label this plate *I* and add the date and your names. Incubate upside down for 1 day, or 2 days if your room is very cool.

2. Wash your hands thoroughly before leaving the laboratory.

Day 1

3. Use a microscope to examine some yeast cells of either mating type. To prepare a slide, touch a toothpick to the streak of either the **a** strain *or* the α strain in plate I, mix it in a small drop of water on the slide, and place a cover slip over the drop. Discard the toothpick in the waste bag. Examine the cells with the high power lens. In your data book, sketch the cells.

4. Use a clean toothpick to transfer a small amount of the **a** strain from the streak in plate I to the middle of the agar. Use another clean toothpick to transfer an equal amount of the α strain to the same place. Being careful not to tear the agar surface, thoroughly mix the two dots of yeast to make a mating mixture. Discard used toothpicks in the waste bag. Invert and incubate plate I at room temperature for 3 or 4 hours, then refrigerate until the next lab period. (Or, refrigerate immediately, and then incubate at room temperature for 3 or 4 hours before the next step.)

5. Wash your hands thoroughly before leaving the laboratory.

Figure 7.1   Procedure for making a mating mixture from **a** and α yeast strains (a), for subculturing the mating mixture (b), and for inoculating the "unknown" medium plate (c).

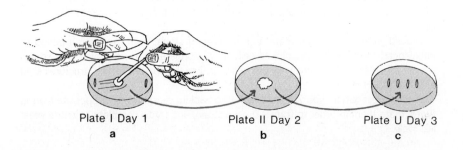

Plate I Day 1   Plate II Day 2   Plate U Day 3
      a   b   c

Day 2

6. Remove plate I from the refrigerator. What color is the mating mixture colony? Use the microscope to examine the mating mixture, as in Step 3. Discard used toothpicks in the waste bag. Sketch what you see and compare it with your earlier drawings. Describe any differences in the types of cells you see. When haploid cells of opposite mating types (a and α) are mixed, they develop into pear-shaped, haploid gametes. Do you see any gametes? A diploid zygote is formed when two gametes fuse. Growing diploid cells are slightly larger and more oval than haploid cells. Do you see any evidence that gametes may be fusing into zygotes? Compared to Day 1, are there more or fewer diploid cells?

7. Make a subculture by transferring some of the mating mixture with a clean toothpick to a fresh agar growth medium plate. Discard used toothpicks in the waste bag. Label this plate *II*. Invert and incubate at least overnight but not more than 2 nights.

8. Wash your hands thoroughly before leaving the laboratory.

Day 3

9. Use the microscope to examine the freshly grown subculture in plate II. Discard used toothpicks in the waste bag. What types of cells are present? Sketch each type. If any of the types seen in Step 6 have disappeared, explain what happened to them.

PART B The Effect of the Environment on the Life Cycle

(Day 3, continued)

10. On a plate of "unknown" medium, make several thick streaks of the freshly grown subculture. Discard used toothpicks in the waste bag. Label this plate *U*. Invert and incubate at room temperature at least 4 days.

11. Wash your hands thoroughly before leaving the laboratory.

Day 7

12. Use the microscope to examine yeast from plate U. Discard used toothpicks in the waste bag. You may need to use the fine adjustment on the microscope to distinguish cells at different levels. What cell types are present now that were not present before? Sketch these cell types, and compare them with the cell types you saw at other stages.

13. Refer to the introduction. How do you think the "unknown" medium differs from the growth medium? If the cells in the sacs are most frequently found in groups of four, do you think they were formed by meiosis or mitosis? Explain. Are the cells in the sacs haploid or diploid? Explain. What part of the life cycle seen on Day 1 do these cells most resemble?

PART C Closing the Cycle: The Next Generation

(Day 7, continued)

14. Transfer some yeast from plate U to a fresh agar growth medium plate. Discard used toothpicks in the waste bag. Label this plate *III*, invert and incubate at room temperature for about 5 hours, and then refrigerate until the next lab period. (Or, refrigerate immediately and then incubate at room temperature for 5 hours before the next step.)

15. Wash your hands thoroughly before leaving the laboratory.

Day 8

16. Use the microscope to examine the growth from plate III. Discard used toothpicks in the waste bag. What life-cycle stages are present? Sketch the cells and compare

them with the stages you observed before. What evidence is there that a new life cycle has started?

17. Discard used culture plates as directed by your teacher, and wash your hands thoroughly before leaving the laboratory.

Discussion

1. In the three parts of this investigation, you have observed the major events of a sexual life cycle. You could readily observe these events in yeast because it is a unicellular organism. In plants and animals, including humans, similar cellular events occur, but they are harder to see. Although the changes were too slow for you to see occurring, your sketches provide a record of the sequences. Think of them as "pauses" in a tape of a continually changing process. Notice in particular how the cycle repeats. Draw diagrams on one page, showing the different shapes of cells you observed in the order in which they appeared. Indicate where you first saw each type and when it disappeared if it did.

2. Compare your sketches with the diagram in Figure 7.2, and try to identify each of the forms you saw.

3. For each of the cell forms you observed, indicate whether it was haploid or diploid.

4. Mark the points in your diagram where cells changed from haploid to diploid and from diploid to haploid.

5. Why do you think the two different mating types are not called "female" and "male"?

6. Can you think of a good argument for calling any particular point in the cycle the "beginning" or the "end"? Why or why not?

7. What would you expect to happen if you allowed the yeast cells in Step 14 to grow for another day and then put them on the "unknown" medium again?

Figure 7.2   Yeast life cycle, showing both asexual and sexual reproduction.

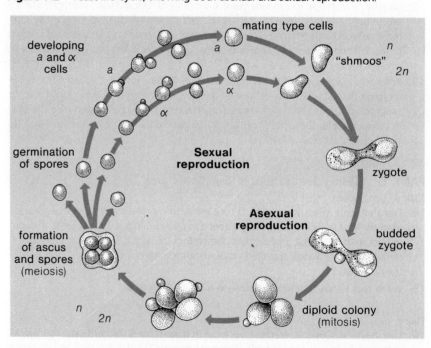

TABLE 7.1   Comparison of the Yeast Life Cycle and Sexual
Reproduction in Animals

Yeast	Animals
Mating pheromones	Sex hormones
Mating type **a** and α gametes	Gametes (ova and sperm)
Fusion	Fertilization
Zygote	Zygote
Asexually reproducing diploid cells	Diploid body cells
Meiosis and formation of spores	Meiosis and formation of gametes

8. Table 7.1 summarizes the similarities between stages in the yeast life cycle and the events in sexual reproduction in animals. For each change you observed in the yeast life cycle, indicate the step in the human sexual reproduction cycle that is most similar.

For Further Investigation

The mating mixture in Step 9 (Day 3) should contain mostly diploid cells, which cannot mate. Use the procedure of Step 4 to test mating ability by mixing samples with each of the **a** and α strains. What would you expect to see?

7.1 A Zygote Gives Rise to Many Cells

At the moment of fertilization, when the haploid (n) set of chromosomes in the sperm nucleus joins with the haploid (n) set in the ovum nucleus, the diploid ($2n$) number of chromosomes is restored. The zygote that results is ready to begin **development**, a series of events that converts a single cell into a fully formed individual. Three basic activities result in embryonic development: cell division, cell movement, and cell **differentiation**—the process by which new cells specialize and become different in appearance and function from their parent cells.

Soon after fertilization, the zygote divides into two cells by mitosis. Each of these two cells divides again to form four cells, then eight cells, and so on. No new cytoplasm is added between the divisions. Soon the tiny embryo is made of dozens of cells. In animals, there is no growth for a while—the cells continue to divide, forming new cells smaller than the parent cells.

These early cell divisions are called **cleavage** (KLEE vaj) because the original mass of the egg is cleaved, or divided, into smaller and smaller cells. Cleavage converts the zygote into a **blastula** (BLAS choo luh), a hollow ball of several hundred cells. Figure 7.3 on page 160 shows some of the stages of cleavage.

Once the blastula has formed, both individual cells and entire sheets of cells begin to change position in an ordered and precise manner. Many cells flow from the surface to the interior of the blastula, so that a new layer of cells lines the original blastula. As a result of these cell movements, the

Figure 7.3 Stages of cleavage in the frog: (a) fertilized egg, (b) 2-cell stage, (c) 4-cell stage, (d) 8-cell stage, (e) early blastula, and (f) late blastula.

blastula becomes a **gastrula** (GAS truh luh). Figure 7.4a illustrates this process, which is known as gastrulation. The gastrula has three distinct cell layers—the original surface layer, or ectoderm (EK toh derm); the new inner layer, or endoderm (EN doh derm), that formed from cells flowing into the blastula; and a third layer, the mesoderm (MEZ oh derm), that forms between those two layers. Figure 7.4b shows these three cell layers, which give rise, either singly or in combination, to all the different tissues of the body. Investigation 7.2 illustrates these developmental stages in a marine worm.

Figure 7.4 Development of a frog gastrula (a). Cross section of a vertebrate embryo, showing the locations of the three primary tissue layers (b).

INVESTIGATION 7.2   Development in Polychaete Worms

The marine polychaete worm, *Chaetopterus variopedatus*, lives in a leathery, U-shaped tube in the sand near the low tide level. Figure 7.5 shows the major features of the worm. The sperm and eggs are visible inside the parapodia (leglike extensions from the body) near the posterior end of the worm. The sperm appear ivory-white and give the male parapodia a smooth, white appearance. The ovaries are in yellow coils, and the eggs inside them give the female parapodia a grainy appearance.

In this investigation, you will observe the worm, remove eggs and sperm, fertilize the eggs, and observe the development of the embryos. The initial stages of development you will observe appear to be the same for nearly all sexually reproducing organisms.

Materials (per team of 2)

microscope slides	dissecting scissors
cover slips	dissecting needle
dropping pipet	cheesecloth
finger bowls	paper towels
microscope	seawater
forceps	fresh water

Figure 7.5 Polychaete worm in a glass tube (top) and the worm's natural tube (bottom). The head is to the right, the parapodia (the fingerlike extensions) at the left.

Procedure

Your teacher will give you specific instructions on obtaining the eggs and sperm.

1. Get a small double-thickness piece of cheesecloth, rinse it in fresh water, and then rinse it in seawater. Place the wet cheesecloth in the bottom of a finger bowl.

2. Using forceps and dissecting scissors, remove 1 parapodium with eggs from the female worm and place it on the wet cheesecloth in the finger bowl. Use the pipet to remove any eggs that spilled from the parapodium and to drop them onto the cheesecloth. Cut open the parapodium and use the dissecting needle and forceps to release the eggs.

   CAUTION: Dissecting needles are sharp; handle with care; replace cork on tip after use.

3. Lift the cheesecloth and pour seawater into the finger bowl to a depth of 1 cm. Holding the cheesecloth by the four corners, gently lift and lower it in the water, and move it around slowly. The movement should allow most of the eggs to filter through the cheesecloth and drop into the seawater while sand and other debris stay on the cheesecloth. The eggs are tiny yellow or yellow-orange dots in the seawater. Note the time in your data book. Discard the cheesecloth in the container designated by your teacher.

4. You must wait 15 minutes from the time you put the eggs in seawater before you add sperm. *Do not add the sperm before 15 minutes have passed.* While you are waiting, add 10 mL of fresh seawater to another finger bowl. Remove a parapodium with sperm from the male worm and quickly place the parapodium into the finger bowl with the 10 mL of seawater. With a dropping pipet, pick up any sperm that were released in the bowl with the worm. (The sperm will appear as a small cloud in the water.) Add these sperm to your finger bowl.

5. Place one drop of the sperm and seawater mixture on a clean microscope slide and add a cover slip. Examine the sperm with low power of a microscope. They will appear as small, moving dots. Switch to high power and observe the sperm. If you see no moving sperm, take another sample from the 10 mL of seawater. If moving sperm still are not evident, remove another parapodium with sperm and repeat Steps 4 and 5.

6. After the eggs have been in the seawater for 15 minutes, add one drop of the sperm and seawater mixture to them. The eggs should all be fertilized within 30 minutes.

After 30 minutes, remove the eggs from the finger bowl with the sperm and place them in another finger bowl with fresh seawater. The development of the embryos will take place in this bowl.

7. While you are waiting for the 30 minutes to pass, obtain fertilized eggs from bowls designated by your teacher and examine them under the microscope. Look for embryos at the 4-, 8-, and 16-cell stages and note any other developmental stages. Your teacher may also direct you to transfer fertilized eggs prepared by other classes to bowls of fresh seawater.

8. Wash your hands thoroughly before leaving the laboratory.

Discussion

1. Examine the illustration of a human sperm in Figure 6.4. How do the polychaete worm sperm compare with the picture of a human sperm? Describe any similarities or differences.

2. Describe the development process you observed with the polychaete worm embryos.

3. The polychaete worm larvae usually develop within 24 hours after fertilization. Based on what you observed and on your knowledge of cleavage and cell division, approximately how many cell divisions have occurred between the 1-cell stage and the swimming larvae?

4. Describe any similarities between the development of the polychaete worm embryos and the development of human embryos.

7.2   Cell Differentiation Is a Major Part of Development

How does a single cell—the zygote—develop muscle tissue, blood tissue, skin tissue, nerve tissue, and all the other tissues necessary to the functioning of a complete organism? Each of these tissues is made of specific types of cells—muscle cells, blood cells, skin cells, nerve cells—and each of these cell types not only looks unlike the others but also produces different proteins. Red blood cells, for example, produce the protein hemoglobin, whereas intestinal cells produce digestive enzymes. Because genes control the production of proteins, scientists once wondered if each cell type contained a different set of genes.

Experiments with amphibian eggs, however, have shown that almost every type of cell in an organism contains a set of genes identical to the set in the zygote. Study the experiment shown in Figure 7.6. In this experiment, an unfertilized egg was exposed to ultraviolet light to destroy its nucleus. Then, the nucleus was withdrawn from a skin cell that had undergone differentiation. When the nucleus was injected into the unfertilized egg, the egg was able to develop into a normal tadpole. This could not have happened unless a complete set of genes was present in the differentiated cell nucleus.

If all cells in an organism contain identical, complete sets of genes, then there must be differences in the genes' activity. As cells begin to specialize, different sets of genes are activated in different cells for varying lengths of time. Thus, after an initial period of activity, one set of genes may stop its activity while another continues. The specific genes and the sequence depend on factors in the genes' environment.

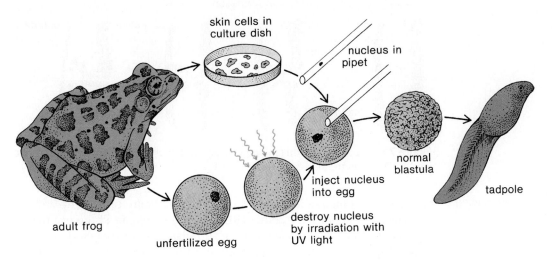

skin cells in
culture dish

nucleus in
pipet

inject nucleus
into egg

normal
blastula

tadpole

adult frog

unfertilized egg

destroy nucleus
by irradiation with
UV light

Figure 7.6 Diagram of an experiment showing that the nucleus of a differentiated cell contains the same set of genes as the zygote.

Study again the experiment shown in Figure 7.6. Note that when the nucleus of the differentiated cell was injected into the egg, the gene action changed to resemble the action of a normal, undifferentiated egg nucleus. Genes that were inactive in the differentiated cell became active again, allowing a normal tadpole to develop. In some way, the cytoplasm of the egg (the environment of the genes) had modified the pattern of gene action in the cell nucleus.

Recall that during cleavage, the zygote divides into smaller and smaller cells, each containing a portion of the egg's original cytoplasm. Experimental evidence suggests that the cytoplasm in an egg is not uniform. Instead, the various molecules that regulate gene activity are located in different portions of the cytoplasm. As cleavage proceeds, the newly formed cells receive only specific portions of the egg's original cytoplasm. As a result, each cell receives different regulatory molecules (see Figure 7.7). These regulatory molecules activate specific genes in each cell, leading to the production of specific proteins. Thus, although the cells all have the same genetic information, they specialize because different genes are activated and then different proteins are produced. Differences in the cytoplasm arising during cleavage, therefore, can play an important role in the differentiation of cells.

regulatory
molecules

Figure 7.7 Regulatory molecules are not uniformly distributed in the cytoplasm of an egg. As the egg divides, different molecules may be incorporated in the offspring cells, resulting in the activation of different genes.

7.3 Interactions Between Cells Influence Differentiation

Interactions between different cell types also can play an important role in differentiation. During the development of the eye, for example, the cells of the embryonic brain divide and expand until they contact the skin ectoderm. The skin ectoderm cells respond to this stimulus by thickening, pushing inward, and pinching off a rounded mass. Genes for specialized lens proteins are activated, causing the mass to become transparent and

Figure 7.8 Neural ectoderm (an outpocketing of the brain) induces formation of the lens of the eye in the skin ectoderm.

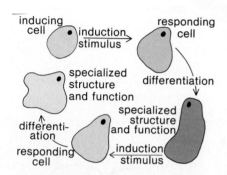

Figure 7.9 A model of cell-to-cell interactions. Chemical signals transmitted from one embryonic cell to another initiate differentiation into specialized types of cells. These signals may occur at different stages of development, bringing about increasing specialization.

Figure 7.10 The neural groove in a frog embryo.

begin functioning as a lens. An extension of the brain forms a cup around the lens that becomes the retina (REH tin uh)—the part of the eye that is sensitive to light. The lens cells then stimulate other skin ectoderm cells to develop into the cornea (KOR nee uh), the transparent covering of the eye (see Figure 7.8). Cells from the brain and the ectoderm must interact to begin lens formation. In experiments in which a cellophane barrier was inserted between the brain and the ectoderm, the lens failed to develop.

How can one tissue or cell type influence the differentiation of other cells? Regulatory molecules, in the form of chemical signals, appear to be responsible. Figure 7.9 suggests a general model for this process. A cell transmits a chemical signal to a second cell, thus activating genes that cause the second cell to differentiate into a specialized cell type. The second cell, in turn, may transmit a chemical signal to a third cell, and so on. Researchers have found, for example, that undifferentiated cells from the ectoderm do not develop into nerve cells unless a special chemical (fibronectin) is present.

Although the nature of most of these regulators is not known, various hormones can affect development dramatically. For example, the conversion of a tadpole to a frog is caused by a hormone from the thyroid gland, and the egg-larva-pupa-adult development in insects is regulated by a group of interacting hormones. Such interactions are a common feature of differentiation in many organisms.

The first organ system to develop in an embryo is the nervous system, which triggers the development of other systems. The ectoderm at the top of the embryo begins to form two long, parallel folds of cells, clearly shown in Figure 7.10. These folds enlarge and become prominent ridges that border a groove. Eventually the ridges join to become a hollow tube, called the neural tube, and other ectodermal cells close over it, as shown in Figure 7.11. The neural tube runs the length of the embryo and eventually gives rise to the brain and spinal cord.

Occasionally the neural tube fails to close completely, and part of the nervous system remains exposed. This condition, called spina bifida (SPY nuh BIH fih duh), often results in abnormalities in the functioning of the nervous system or even in the death of the child. Tests performed before birth can indicate whether a baby will be born with an exposed spinal cord.

The differentiation of cells illustrates the importance of environmental influences, even at the molecular level. Genes are influenced by their cy-

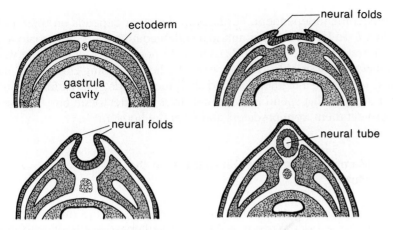

Figure 7.11 Development of the neural tube in a vertebrate embryo.

toplasmic surroundings, and cells in turn are influenced by chemical regulators in their surroundings. In ever-expanding circles, from DNA to whole
organism, each structure is affected by its environment.

Concept Review

1. What is the role of fertilization in triggering development?
2. Explain the difference between a gastrula and a blastula.
3. How can cells affect one another?
4. What cell layers contribute to the formation of the eye?
5. How do regulatory molecules affect cell differentiation?
6. What structures arise from the neural tube?

Animal Development

7.4 Some Animals Develop Independently of Their Parents

Guidepost: What are the
relationships between animal embryos
and their mothers?

Like all cells, fertilized eggs need water to develop normally. Most female
fish, as well as semi-aquatic female animals such as frogs and salamanders,
release their eggs in water. In these cases, the eggs may be fertilized already, or the male may fertilize them after they have been released.

In fact, the need for water even affects mating. When salamanders or
frogs mate, for example, the male awaits the release of hundreds or thousands of eggs by the female. The male then sheds large numbers of sperm
directly over the eggs or nearby. Some of the sperm encounter eggs and
fertilize them, but most of the sperm are lost in the water. Although this
process may seem inefficient, frogs and salamanders have reproduced this
way for millions of years and their survival demonstrates the success of
this method.

Within the egg, young frogs develop as far as the tadpole stage. By then
they can feed, and they have a tail and can swim. At this point, they hatch

Figure 7.12 Section through a human blastocyst. The embryo develops from the inner cell mass, the thicker part on the left.

and become independent. Further development depends on their ability to find food for themselves and to avoid becoming food for predators.

Because reptiles and birds usually do not lay their eggs in the water, their eggs must be able to retain water for the developing embryos. Usually these eggs have a leathery or hard shell that retards evaporation. Reptiles, even reptiles that spend most of their lives in water, often bury their eggs to protect them from predators and from the hot sun.

7.5 Mammalian Embryos Develop Within the Mother

For a short time, a mammalian embryo depends on the small amount of material stored in the ovum. After the blastula becomes implanted in the wall of the uterus, it begins to absorb nutrients brought to the uterus by the mother's blood. The blastula (now called a blastocyst) has a thick mass of cells on one side. As shown in Figure 7.12, this side continues to develop as the embryo. Other cells in the mass develop into embryonic membranes that are important for the developing embryo. One membrane, the **chorion** (KOR ee on), extends fingerlike projections, or villi (VIL eye), into the inner lining of the uterus. The villi are supplied with tiny blood vessels from the embryo's developing circulatory system, as shown in Figure 7.13. The chorionic villi and the uterine lining form the **placenta** (pluh SEN tuh), through which the embryo receives nourishment from its mother and eliminates wastes. The embryo is connected to the placenta by a flexible cord, known as the **umbilical** (um BIL ih kul) **cord**. Gases and other molecules can diffuse from the blood of the mother through the placenta into blood vessels in the umbilical cord and from there into the circulatory system of

Figure 7.13 Cut-away view of a developing human fetus in the uterus. The illustration also shows embryonic membranes and the placental connection. Part of the placenta is enlarged to show the circulation of the mother and fetus. Note that the two circulatory systems do not mix.

Figure 7.14 Prenatal diagnosis of genetic disorders by amniocentesis. Fetal cells and amniotic fluid are withdrawn and cultured for later chromosomal and biochemical studies.

the embryo. Carbon dioxide and other wastes diffuse from the villi into the blood of the mother. The mother's lungs and kidneys dispose of the wastes. During these exchanges, the blood of the embryo and the blood of the mother normally do not mix. About this time, another embryonic membrane, the **amnion**, encloses the body of the embryo and fills with fluid. The embryo then develops in an aqueous environment.

With proper nourishment, the embryo grows, and its cells develop into distinctive types. Similar types of cells are found together in tissues. Tissues perform different functions for the body; muscular tissue contracts, for example, and glandular tissue secretes certain substances. When several tissues are grouped together in a special arrangement for a particular function, an organ is formed. The stomach and heart, for example, are organs. Finally, several organs may be related in general function. These organs form an organ system. The digestive system consists of a whole series of organs that carry food from the mouth through all the steps of digestion and absorption and eliminate the undigested remains.

Often the amniotic fluid contains cells that are cast off from the embryo during its growth. Physicians who wish to test the embryo for biochemical or chromosomal abnormalities without touching the embryo use these cells. A sample of amniotic fluid is removed, and the cells are grown in the laboratory. This process, called amniocentesis (AM nee oh sen TEE sus), is used to test for missing chromosomes, extra chromosomes, and some genetic disorders of a biochemical nature (see Figure 7.14).

A human develops in its mother's uterus for about 40 weeks. Figure 7.15 on page 168 shows several stages of development. After 12 weeks, the embryo is called a **fetus**. At the completion of development (and sometimes earlier), the mother's body undergoes a complex series of hormonal interactions that result in birth. Strong muscular contractions push the fetus from the uterus through the vagina, which usually is able to expand sufficiently to allow the head and body to pass through. After birth, the placenta (afterbirth) is expelled. The vagina returns to normal size, and the uterus shrinks back to the size and shape of a pear.

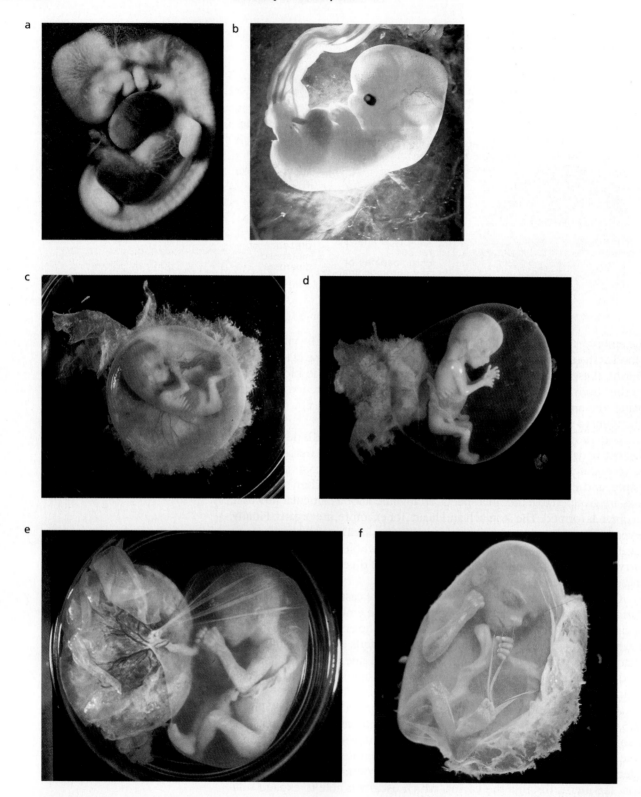

Figure 7.15 Early development of a human embryo: (a) 4 weeks, (b) 6 weeks, (c) 8 weeks, (d) 12 weeks, (e) 14 weeks, (f) 16 weeks.

Progress in Prenatal Diagnosis

Today it is possible to diagnose many genetic diseases and congenital defects while the fetus is still in the mother's uterus. One procedure uses high frequency sound waves, or ultrasound, to determine the overall condition of the fetus and to confirm its age. Because it can be performed without the withdrawal of any fluid or tissue from the uterus or the fetus, ultrasound is called a noninvasive procedure. This test usually is performed during the fourth month of pregnancy. At this time, major developmental abnormalities, such as extra or misplaced limbs, can be detected, and the sex of the fetus can be determined.

Often ultrasound is used in conjunction with more invasive tests. Amniocentesis, which came into prominence in the 1970s, allows physicians to check for chromosomal abnormalities and certain enzyme disorders. For this test, the doctor first uses ultrasound to determine the position of the fetus and the location of the placenta. Next, a long thin needle is inserted through the mother's abdomen into her uterus. About 10 mL of amniotic fluid is withdrawn, along with cells that have been sloughed off from the fetus. These cells are cultured for several weeks and then analyzed. The amniotic fluid is checked for increased levels of a substance called alphafetoprotein (AFP). High levels of AFP are found in such disorders as spina bifida. Amniocentesis cannot be performed until about the sixteenth week of pregnancy because there is not enough room in the amniotic sac to allow the safe withdrawal of fluid. Amniocentesis is quite accurate and safe. The risk of spontaneous abortion after the procedure is less than 0.5 percent (1 in 200).

Chorionic villi sampling (CVS) was first introduced in the early 1980s. CVS can be performed as early as the ninth week of pregnancy. A catheter is inserted through the vagina into the uterus and directed by ultrasound to a region called the *chorion frondosum*, where there are many chorionic villi. Some villi are withdrawn by suction through the catheter. As the cells in the villi are growing and dividing, they can be analyzed quickly. Results from chromosome studies are available in a few days and results from enzyme studies in about one week. Because no amniotic fluid is withdrawn in this procedure, AFP analysis cannot be performed; otherwise, diagnoses from CVS are as accurate as those from amniocentesis. CVS carries a higher risk of spontaneous abortion—about 1 to 2 percent—than amniocentesis does, but it allows earlier diagnosis of many defects.

Prenatal diagnosis is recommended routinely for women 35 years of age and older because the risk of chromosomal disorders increases with advanced maternal age. Prenatal diagnosis also is recommended for women who already have a child with a genetic disorder or who have a history of a genetic disorder in the family of either parent. All detectable chromosomal disorders and more than 200 single gene disorders now are diagnosable using amniocentesis and CVS.

Chorionic villi sampling.

Figure 7.16 Four-year-old child with fetal alcohol syndrome. Note the short nose and the smooth, long, narrow upper lip as well as the narrow eye openings.

Guidepost: How is cancer related to normal developmental processes?

7.6 Embryos Are Affected by Substances in the Mother's Blood

Physicians divide the total time needed for development into three periods, or trimesters. By the end of the first trimester (about 12 weeks), most organs are formed, and the skeleton is clearly visible. The last trimester is a time of rapid growth and the maturation of organs and systems as the fetus approaches the time of birth. Because so much cell differentiation and organ formation occur during the first 12 weeks, embryos in the first trimester are especially sensitive to harmful materials.

The growing embryo shares materials from the mother's blood and therefore can be affected by foods or drugs taken by the mother. For example, if the mother drinks alcoholic beverages, the alcohol in the mother's blood diffuses into the blood system of the fetus, and the fetus may be physically and mentally affected. Some children are born with fetal alcohol syndrome. These children have certain abnormal facial features (see Figure 7.16) and may be mentally retarded. Most doctors recommend that women abstain from alcoholic beverages while they are pregnant or nursing.

Many other substances, including marijuana, cocaine, and heroin, have been shown to cause abnormal fetal development. Smoking marijuana may harm the smoker's fetus, and a mother who smokes cigarettes risks having a child whose birth weight is significantly less than normal, thus threatening its health and development. If a pregnant woman uses a drug such as cocaine or heroin, the baby may be born addicted to the drug. Even caffeine is suspected of damaging the fetus. As you might expect, substances that can harm the mother are likely to harm the fetus. In addition, however, even substances that are harmless or beneficial to the mother may be harmful to the fetus.

Concept Review

1. Are all animals dependent on their parents? Explain.
2. Explain how substances pass from the mother to the embryo in mammals.
3. Explain the relationships between the amnion, the chorion, the umbilical cord, and the placenta.
4. How does the fetus get rid of waste products?
5. Why are proper nutrition and healthful behavior important for a pregnant woman?

Cancer

7.7 Development Usually Is Well Controlled

The entire drama of development depends on a certain number of cell divisions. Fertilization starts these divisions, and the resulting increase in cell number is accompanied by cell movement, changes in cell shape, and differentiation. Ultimately, each cell has a particular task to perform in the development and life of the new individual. Aging is a normal part of development. At the end of the aging process, an organism dies. This process is built into the genetic program and is rather precise for each species.

Figure 7.17 In cancer, abnormal numbers of certain cells are produced. These photomicrographs of stained blood cells show normal human blood (a) and blood from a person with leukemia (b). Note the large number of white blood cells in (b).

When an animal reaches full size, most cell divisions cease. Although some cells can continue to divide once the organism is fully grown, they do so only to replace worn-out or lost cells. For example, blood cells have a limited life span and are replaced regularly by new cells. Cells lining the digestive system also wear away and are replaced continuously. Outer skin cells are lost and are replaced from layers of cells below. Nervous system cells, on the other hand, rarely divide after differentiation. Although generally long-lived, these cells are not replaced. Normal life functions mainly require cell replacement once maturity is reached.

Most cells are able to divide only a certain number of times. One mechanism that controls cell division is called contact inhibition. Contact inhibition causes a cell that is touching another to stop dividing. In cancer, this control over cell division seems to be lost, and cells begin to divide rapidly (see Figure 7.17). These cells have lost their sensitivity to the growth regulation brought about by contact with other cells.

7.8 Cancer Cells Divide Without Limit

Just as organisms have certain life spans, so do the cells that make up the organisms. On rare occasions, however, the genetic controls over normal cell division are altered. Sometimes the cell not only loses control over its division but also loses control over its position in a specific tissue. Normally, cells are able to recognize each other by particular proteins on the surface of each cell type. Cells with the same surface proteins bind together to form tissues and organs. When the genes responsible for those proteins malfunction, the cells may lose their identity and invade surrounding tissues. These cells form a growing mass called a tumor. If the tumor remains in one place near its origin, it usually is benign. Sometimes, small groups of these dividing cells may move to other parts of the body and establish new tumors. Cell masses that continue to divide and to spread (or metastasize) at the expense of other tissues are called malignant tumors.

For a cell to become malignant, certain genes that regulate cell functions must be present. These genes are called **oncogenes** (AHN coh jeenz). Oncogenes first were identified because of their presence in certain viruses that cause cancers in animals. More than 30 oncogenes have been recognized. Oncogenes also are found in normal cells but in small amounts. Thus, oncogenes are considered normal cell-regulating genes that have gone awry.

At least three genes, called tumor-suppressors or anticancer genes, have been implicated in cancer. When these genes function normally, they make proteins that keep a cell's growth in check. If these genes become damaged, however, they are no longer able to produce those proteins, and cells become malignant. Researchers now think that these two types of cancer genes can work together—an oncogene turns dividing cells malignant after the tumor-suppressing gene has been damaged.

Scientists in many labs are working on new cancer drugs based on these tumor-suppressor proteins. The growth-controlling proteins would be made by healthy versions of the genes. These proteins then would be administered to patients to slow or reverse the growth of tumors. Eventually, it may be possible to replace patients' defective genes with healthy genes.

Concept Review

1. For what reasons do cells divide in adults?
2. Explain the differences between benign and malignant tumors.
3. Give a brief definition of cancer.
4. Compare and contrast the actions of oncogenes and tumor-suppressing genes.

Summary

Development is the series of events that begins when a zygote starts to divide and grow. Cells that are similar in early divisions gradually specialize to perform certain tasks. This differentiation appears to result in part from the action of regulator molecules located in different areas of the zygote cytoplasm. As the zygote divides, each cell receives duplicate genetic material but different regulators. These regulators influence the genes that direct cell specialization. Cell-to-cell interactions can determine when these regulators "switch on" specific genes, which in turn direct the production of specific proteins.

Development is aided by cell movements and tissue foldings. In animals, the blastula becomes a gastrula—an embryo that has three distinct layers of cells, each of which gives rise to specific tissues. One or more tissues may make up an organ, and groups of related organs make up organ systems.

Animal development takes place in an aqueous environment. In mammals, the fetus develops within the mother and is surrounded by a fluid-filled amnion. The fetus gets food and other substances from the mother's body systems, and so it is affected by substances in the mother's blood, such as alcohol or other drugs. Developing organisms continue to change throughout life. These changes include aging, which eventually ends the life of the individual. The life span of an organism apparently is a part of the genetic program determined at fertilization.

Normally, development is under strict control, and cells in mature organisms rarely divide. When cells begin to divide without limit and invade other tissues, cancerous growth results. Oncogenes and tumor-suppressing genes are implicated in controlling cell division.

Applications

1. Differentiation leads to specialized cells, tissues, and organs. Initially, in the embryo, it involves formation of specific cell layers. What are these embryonic layers, and what structures in the mature animal are formed from each of them?

2. Cloning is the process of producing several genetically identical individuals from a single organism. This practice is relatively common in research involving plants. How is cloning aided by the fact that almost all cells within an organism contain identical sets of genes?

3. If growth is defined as an increase in the size and number of cells, explain why there is no real growth in the developing organism through the blastula stage.

4. Suppose you take the regulatory molecules that control the production of fingers in humans and inject them into an oak tree. Will the oak tree develop fingers? Explain.

5. Do you think the word *development* applies to single-celled organisms? Do you think the word *differentiation* applies?

6. Water is necessary for the development of animal embryos. Frogs, fish, and many other organisms lay their eggs directly in the water. What are some of the disadvantages of such a reproductive system?

7. In what ways is a bird or reptile egg like the amnion of a human?

8. In what ways is the medical problem of cancer related to an understanding of developmental processes?

Problems

1. Generally, the ability to regenerate missing tissues or organs decreases during embryonic development as specialization increases. Investigate why this is so.

2. Find out about the growth and development of plant embryos. How are such embryos different from, and how are they similar to, the embryos of animals?

3. Using a frog blastula, design a hypothetical experiment that might show the importance of cell position in the blastula for normal tadpole development.

4. Why do cells, and people, grow old? Investigate several hypotheses biologists have made about aging, and identify the strong and weak points of each.

5. Carcinogens are cancer-causing agents. Make a list of potential carcinogens found in and around your home or school. What can be done to reduce your exposure to them, and what ways can you think of to reduce their use?

How closely do these two-year-old Alaskan brown bears (*Ursus arctos*) resemble their mother? How would you explain this phenomenon? Aside from contributing genetic information, how else might parents influence their offspring? What might the mother bear be teaching her offspring here?

Continuity Through Heredity

Bob and Mary were the proud parents of a new baby girl they named Lisa. Several weeks after Lisa came home from the hospital, Mary and her pediatrician noticed that Lisa was not gaining weight as fast as she should. In fact, she was losing weight. After running some tests, the doctor told Bob and Mary that Lisa probably had cystic fibrosis.

Children with cystic fibrosis secrete unusually thick mucus, which accumulates and damages the lungs, digestive tract, and other organs (see Figure 8.1). Untreated, children with this disorder normally die by the age of 4 or 5 from liver damage, pneumonia, or a secondary infection. With appropriate medication and treatment, however, they may live to adulthood.

Bob and Mary were crushed by this tragedy. They also were angry and perplexed. Although cystic fibrosis is the most common genetic disorder among Caucasians, the disorder had not been known to occur in either of their families. How was it possible for them to have a child with cystic fibrosis?

The genetic counselor in the hospital told them the defective gene had been present in both families for many generations. A defective gene is an altered form of a gene that is essential for normal development. The appearance of the disorder was a matter of chance.

How can an individual carry a defective gene without also having the altered trait? What is the chemical basis of genes that gives them the ability to carry information from parent to offspring and to direct the biochemistry of the organism?

This chapter discusses genes and inheritance. It describes the major events that have led to our understanding of how genetic continuity is achieved.

The Dual Roles of Genetic Material

Guidepost: How was the genetic riddle solved?

8.1 Genes Determine Biological Potential

Heredity is the transmission of genetic information from one generation to another. The basic unit of information passed along is a gene, which also controls the basic functions of the cell (see Chapter 5). Geneticists study heredity and gene action together, because the visible effects of gene action in individuals and populations provide clues to the genes that are present in an organism or in a population and how they work.

Figure 8.1 People with cystic fibrosis produce a thick mucus that damages the airways of the lungs.

The information in genes is stored in the sequence of nucleotide bases that make up DNA. In response to environmental influences, this information, a type of molecular code, directs all the cell processes involved in the development and function of the organism, including both the reproduction of genes and the effects of the genes themselves. The genes passed along through reproduction provide the continuity between generations that is essential for the continuation of a species. They also provide instructions for the structure, function, and development of an organism during each generation.

How can we predict whether a particular hereditary trait will appear in an offspring of two organisms? When you predict something, you make a statement that something will or will not happen with a certain amount of confidence. In the study of heredity, this type of prediction is expressed in terms of **probability**, a branch of mathematics that predicts the chances that a certain event will occur. Geneticists use probability to predict the outcomes of matings. You can develop the rules of probability for yourself by doing Investigation 8.1.

INVESTIGATION 8.1   Probability

The probability of a chance event can be calculated mathematically using the following formula:

$$\text{probability} = \frac{\text{number of events of choice}}{\text{number of possible events}}$$

What is the probability that you will draw a spade from a shuffled deck of cards like that shown in Figure 8.2? There are 52 cards in the deck (52 possible events). Of these, 13 cards are spades (13 events of choice). Therefore, the probability of choosing a spade from this deck is 13/52 (or 1/4 or 0.25 or 25%). To determine the probability that you will draw the ace of diamonds, you again have 52 possible events, but this time there is only one event of choice. The probability is 1/52 or 2%. In this investigation, you will determine the probability for the results of a coin toss.

Materials (per team of 2)
2 pennies (1 shiny, 1 dull)
cardboard box

Figure 8.2 The chance of drawing one ace is 1/13. What are the chances of drawing two, three, or four aces?

Procedure
1. Work in teams of two. One person will be student A and the other will be student B.
2. Student A will prepare a score sheet with two columns—one labeled *H* (heads), the other *T* (tails). Student B will toss a penny 10 times. Toss it into a cardboard box to prevent the coin from rolling away.
3. Student A will use a slash mark (/) to indicate the result of each toss. Tally the tosses in the appropriate column on the score sheet. After 10 tosses, draw a line across the two columns and pass the sheet to student B. Student A then will make 10 tosses, and student B will tally the results.
4. Continue reversing roles until the results of 100 (10 series of 10) tosses have been tallied.
5. Prepare a score sheet with four columns labeled *H/H, Dull H/Shiny T, Dull T/Shiny H,* and *T/T* (H = heads; T = tails). Obtain 2 pennies—1 dull and 1 shiny. Toss both

pennies together 20 times, while your partner tallies each result in the appropriate column of the score sheet.

6. Reverse roles once so that you have a total of 40 tosses.

Discussion

1. How many heads are probable in a series of 10 tosses? How many did you actually observe in the first 10 tosses?

2. Deviation is a measure of the difference between the expected and observed results. It is *not* the difference itself. It is the ratio of the sum of the differences between expected and observed results to the total number of observations. Thus:

$$\text{deviation} = \frac{\text{difference between heads expected and heads observed } +}{\text{difference between tails expected and tails observed}}{\text{number of tosses}}$$

Calculate the deviation for each of the 10 sets of tosses.

3. Calculate the deviation for your team's total (100 tosses).

4. Add the data of all teams in your class. Calculate the class deviation.

5. If your school has more than one biology class, combine the data of all classes. Calculate the deviation for all classes.

6. How does increasing the number of tosses affect the average size of the deviation? These results demonstrate an important principle of probability. State what it is.

7. On the chalkboard, record the data for tossing two pennies together. Add each column of the chart. In how many columns do data concerning heads of a dull penny appear?

8. In what fraction of the total number of tosses did heads of dull pennies occur?

9. In how many columns do data concerning heads of a shiny penny occur?

10. In what fraction of the total number of tosses did heads of the shiny pennies occur?

11. In how many columns do heads of *both* dull and shiny pennies appear?

12. In what fraction of the total number of tosses did heads of *both* pennies appear at the same time?

13. To which of the following is this fraction closest: the sum, the difference, or the product of the two fractions for heads of one penny at a time?

14. Your answer suggests a second important principle of probability that concerns the relationship between the probabilities of separate events and the probability of a combination of events. State this relationship.

15. When you toss two coins together, there are only three possibilities—H/H, T/T, or H/T. These three combinations will occur 100% of the time. The rules of probability predict that H/H and T/T each will occur 25% of the time. What is the expected probability for the combination of heads on one coin and tails on the other?

16. When you toss a dull penny and a shiny penny together, what is the probability that heads will occur on the dull penny? What is the probability that tails will occur on the shiny penny? Calculate the probability that the dull penny will be "heads" *and* the shiny penny will be "tails" if you toss the two pennies together. Compare this answer to the answer in Question 15. How do you account for the different answers? Are there other ways than Dull H/Shiny T to get the H/T combination?

17. How many different ways can you get the H/T combination on two coins tossed together? What is the probability of each of those different ways occurring? Is

the probability of getting heads and tails in any combination of pennies closest to the sum, the difference, or the product of the probabilities for getting heads and tails in each of the different ways?

18. Your answer suggests a third important principle of probability that concerns the relationship between (a) the probability of either one of two mutually exclusive events occurring and (b) the individual probabilities of those events. State this relationship.

8.2 Mendel's Work Led to the Concept of the Gene

Today we know that each gene brings about the synthesis of a particular protein, such as an enzyme, a muscle protein, or a protein pigment that may affect skin color. The instructions for the synthesis of a particular protein are stored in the form of specific sequences of DNA nucleotides, which make up the genes. The genes are organized into chromosomes. Through the processes of meiosis and fertilization, chromosomes are passed on to each new generation.

Although these ideas may sound simple, it has taken centuries to assemble the information. Most of our knowledge about genetics has been gained in just the last 100 years. Biologists have joined together various hypotheses and experimental results, as if they were working on a jigsaw puzzle, to give us a broad picture of what happens in inheritance.

The first part of the puzzle was put together in a monastery in Austria. There, in the 1860s, Gregor Mendel (see Figure 8.3), a monk who was trained as a mathematician and natural scientist, set forth the basic principles of heredity. Mendel was interested in the inheritance of animal and plant features, or traits, and he began an eight-year series of experiments with ordinary garden peas. Garden pea flowers contain both male and female reproductive parts, as shown in Figure 8.4. Although under natural conditions the pea plant usually self-pollinates, it is possible to collect pollen grains from flowers of one pea plant and transfer them to the flowers of another plant. This technique, known as cross-pollination, results in seeds that are the offspring of two plants, not just one.

Mendel's work, which we now call Mendelian genetics, provided the basis for the modern study of heredity and variation. His experiments were unique in four important ways. First, he concentrated on one trait at a time. Second, he used large numbers of organisms to minimize the influence of chance on his data. Third, he combined the results of many identical experiments. Fourth, he used the rules of probability to analyze his results. By using these methods, Mendel was able to recognize distinctive patterns of inheritance.

To begin his work, Mendel collected various strains of peas and tested each strain to make certain it was genetically pure, or pure breeding. Pure-breeding plants produce offspring that are identical to themselves generation after generation. Mendel worked with strains that showed one of two different forms of the same trait. For example, plants were either tall or short or produced either green or yellow seed colors. By having distinct and contrasting forms of one trait, Mendel could follow the differences in

Figure 8.3 Gregor Mendel (1822–1884). Experimenting with peas in his monastery garden, Mendel developed the fundamental principles of heredity that became the foundation of modern genetics.

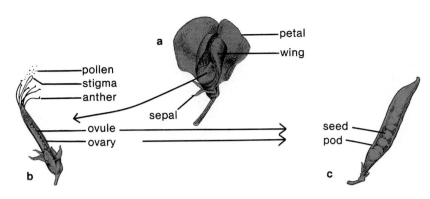

Figure 8.4 Self-fertilization in a pea flower. The petals of the pea flower (a) completely enclose the reproductive organs. As a result, the pollen from the anthers falls on the stigma of the same flower (b). Pollen tubes grow down through the female reproductive organ to the ovules (immature seeds) in the ovary. The ovules develop into seeds, and the ovary wall develops into the pea pod (c).

the offspring. The seven traits shown in Figure 8.5 are the ones Mendel worked with in his experiments.

In his first experiments, Mendel crossed pure-breeding plants that grew from round seeds with pure-breeding plants from wrinkled seeds. Would the offspring produce round seeds, wrinkled seeds, or something in between? Mendel found that *all* the plants from this cross produced round seeds. The wrinkled form of the trait seemed to have disappeared. Today we call the parents of such a cross the P_1 (for parental) generation. The offspring are called the first filial, or F_1, generation.

Mendel then allowed the F_1 seeds to grow into plants and to self-pollinate. The resulting seeds were planted and gave rise to the F_2 (second filial) generation. Of these F_2 plants, approximately 3/4 produced round seeds and 1/4 produced wrinkled seeds. The wrinkled seeds that seemed to have disappeared in the F_1 generation reappeared in the F_2 generation.

seed shape	seed color	seed-coat color	pod shape	pod color	flower position	stem length
round	yellow	colored	inflated	green	axial	long
wrinkled	green	white	constricted	yellow	terminal	short

Figure 8.5 The seven traits of garden peas Mendel studied.

TABLE 8.1   Results from Mendel's Experiments

P₁ Cross	F₁ Plants	F₂ Plants	Actual Ratio
1. round × wrinkled seeds	all round	5474 round <u>1850</u> wrinkled 7324 total	2.96 : 1
2. yellow × green seeds	all yellow	6022 yellow <u>2001</u> green 8023 total	3.01 : 1
3. colored × white seed coats	all colored	705 colored <u>224</u> white 929 total	3.15 : 1
4. inflated × constricted pods	all inflated	882 inflated <u>299</u> constricted 1181 total	2.95 : 1
5. green × yellow pods	all green	428 green <u>152</u> yellow 580 total	2.82 : 1
6. axial × terminal flowers	all axial	651 axial <u>207</u> terminal 858 total	3.14 : 1
7. long × short stems	all long	787 long <u>277</u> short 1064 total	2.84 : 1

Mendel called the round seeds in all the F_1 plants the **dominant** form of the trait. He called wrinkled seeds that had reappeared in the F_2 generation the **recessive** form of the trait.

Mendel repeated this same two-generation cross for six other traits. Table 8.1 shows the data from the F_2 generations in all his experiments. He then calculated the ratio of dominant to recessive forms for each trait. In each case, the dominant form appeared in about 3/4 of the F_2 plants, and the recessive form appeared in about 1/4 of the F_2 plants. All the experiments showed the same 3 : 1 ratio no matter what trait was tested. What was the meaning of this result?

8.3   Mendel Identified the Unit of Heredity

Mendel proposed that each pure-breeding plant had two identical copies of a "factor" for a particular trait. He did not know where the "factors" were located in cells, but he hypothesized that only one copy of a factor went into each sperm or egg cell when gametes were formed. If a parent were pure-breeding "round," for example, all its gametes would have the "round-seed factor." The same would be true for the "wrinkled-seed factor." Thus, the offspring of a round and wrinkled cross would receive a round-seed factor from one parent and a wrinkled-seed factor from the other parent. Mendel called this separation of factors the **principle of segregation**.

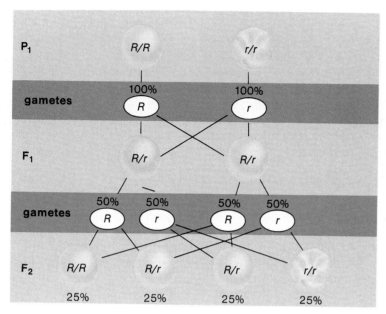

Figure 8.6 One of Mendel's crosses using round and wrinkled peas. Note that the alleles of a gene segregate during gamete formation. What is the probability of *R/r* in the F$_2$?

Today, we call Mendel's "factors" genes, and we call the alternative forms of a gene **alleles** (al LEELZ). The allele for the dominant form is commonly represented by a capital letter (in this cross, *R* for round seeds), and the allele for the recessive form is represented by the same letter in lower case (*r* for wrinkled seeds).

Study Figure 8.6 to help you understand Mendel's results. The symbols in the diagram represent the alleles in three generations of pea plants. The alleles form the genetic makeup, or **genotype** (JEE noh typ), of the plants. The genotype of an individual is responsible for its **phenotype** (FEE noh typ)—its appearance or observable traits. Note that both the *R/R* and *R/r* genotypes produce the same phenotype—round seeds.

In a pure-breeding plant that produces only round seeds, both alleles are the same in every cell: *R/R*. In the same way, pure-breeding plants that produce only wrinkled seeds also have two alleles that are the same, in this case *r/r*. Pure-breeding plants in which both alleles are alike are called **homozygous** (HOH moh ZY gus; from the Greek *homo*, meaning alike, and *zygo*, meaning a pair). When the chromosome pairs containing the alleles separate during meiosis, each gamete receives only one allele. A plant homozygous for a particular trait can produce only one type of gamete (probability = 1, or 100%).

When gametes carrying an *R* allele unite with gametes carrying an *r* allele, all the offspring have the combination *R/r* (probability = 1). The new organisms are **heterozygous** (HET er oh ZY gus; from the Greek *hetero*, meaning different) with respect to the alleles for round and wrinkled seeds. Heterozygous offspring are called **hybrids**. In his experimental crosses, Mendel used *R/R* plants as either the male parent or the female parent. No matter which parent contributed which type of gamete, all F$_1$ plants produced round seeds. Apparently only one *R* allele is needed to direct the plant to produce round seeds, which are the dominant form of the trait for seed shape.

Both the male and female parts of R/r heterozygous F_1 plants produce two types of gametes: 1/2 carry R and 1/2 carry r. If the F_1 flowers self-pollinate, fertilization is random. Any sperm nucleus can fertilize any egg nucleus, regardless of its genotype. Figure 8.6 shows how the probabilities are calculated for an R/r and R/r cross that gives rise to an F_2 generation.

By combining the ideas of probability and random mating as Mendel did, the mathematical regularity in these crosses becomes clear. If more than one trait is involved in the cross, the explanation is more complicated, but the same basic principles apply.

Mendel's experiments defined the basic unit of inheritance, but he could provide no information about its physical or chemical nature. Unable to see such a unit in his studies, he inferred its existence from his experiments. The answer to the riddle of what genes are and how they work could not be known until nearly 100 years later, when techniques for studying biological molecules were developed.

8.4 Genes Direct Biosynthesis

Our understanding of the chemical basis of heredity has come primarily from studies of organisms less complex than garden peas. A giant step toward understanding how genes function came from studying a pink bread mold called *Neurospora crassa* (noo ROS poh ruh CRAS uh). *Neurospora crassa* reproduces both asexually by means of spores and sexually.

Normally, the asexual spores develop into mold that can grow on a nutrient-deficient, or minimal, medium because the mold itself produces most of the nutrients it needs. In the 1940s, George W. Beadle and Edward L. Tatum used X rays to produce heritable changes in the mold. Changes that can be passed on to offspring are called **mutations** (myoo TAY shunz). Beadle and Tatum found that the spores of some mutated molds could no longer grow on a minimal medium. For many of these mutants, however, the addition of a single vitamin or amino acid to the medium permitted the spores to grow. Apparently the radiation-induced mutations prevented the synthesis of some substance the mold normally produced. Although each mutant usually required only a single nutrient, it was not always the same nutrient, and some mutants lacked nearly every vitamin and amino acid the mold normally synthesized. If enzymes control biological reactions, including biosyntheses, as Beadle and Tatum hypothesized, then the defective molds lacked an essential enzyme. Adding a supplement to the simple medium would compensate for the lack of this enzyme and allow the mold to grow. Figure 8.7 illustrates the experimental procedure Beadle and Tatum used.

Beadle and Tatum's work included studies of the inheritance of the deficiencies they induced. It was from these studies that biologists began to understand what genes do. *Neurospora crassa*, like most molds, is able to reproduce sexually. It occurs in two mating types (sexes) that can be crossed to yield offspring in much the same way plants reproduce. Beadle and Tatum found the nutritional deficiencies in the molds were inherited in the same way as the traits Mendel studied. In other words, the X rays

Figure 8.7 Procedure used by Beadle and Tatum. In this case the *Neurospora* spore has lost the ability to synthesize substance c.

produced mutant alleles of normal genes. Other experiments by Beadle and Tatum demonstrated that the forms of the traits produced by the mutant alleles were recessive to the forms produced by the normal alleles.

Eventually, each missing substance in the growth medium was linked to a particular missing enzyme in the mold. Each missing enzyme was the result of a single gene mutation. Beadle and Tatum proposed that each gene specifies the synthesis of one enzyme; they called this the one-gene-one-enzyme hypothesis.

We now know that genes control not only the synthesis of enzymes but also the synthesis of all other proteins. Somehow a gene dictates the way amino acids are combined to form large protein molecules. In other words, genes code for protein synthesis. What in the genes controls the coding? The answer to that question would fill in the next major portion of the genetics picture.

Concept Review

1. What are the chances that a family with three children will have three daughters? After a family has two daughters, what is the probability that their next child will be a daughter?
2. How was mathematics important in Mendel's explanation of his results?
3. Distinguish between the terms *gene* and *allele*.
4. Distinguish between *genotype* and *phenotype*; between *homozygous* and *heterozygous*.
5. Explain how the principle of segregation applies to the movement of chromosomes in meiosis.
6. How do Beadle and Tatum's experiments provide a chemical explanation for dominance and recessiveness?

Guidepost: What are the relationships between chromosome structure and gene function?

Chromosomes and Genes

8.5 Genes Are on Chromosomes

The foundation for Beadle and Tatum's experiments had been established earlier in the century when other biologists had found physical evidence of genes—Mendel's "factors." By 1902, the connection between the activity of chromosomes during meiosis and fertilization and the actions of genes in transmitting hereditary information had become clear. Experiments with egg and sperm cells had shown that the gametes make equal genetic contributions to a new organism. Two biologists—Theodor Boveri and W. S. Sutton—had reasoned independently that if the contributions of sperm and eggs are the same, the genes must be located in corresponding parts of the two types of gametes. They noted that the nuclei of the two gametes are similar even though their cytoplasm differs greatly. Both men concluded that genes are located in the nucleus.

In the 1930s, various studies using the fruit fly *Drosophila melanogaster* (droh SOF il uh mel an oh GAS ter), shown in Figure 8.8, provided experimental support for the conclusions of Boveri and Sutton. Work with *Drosophila* clearly demonstrated the direct relationship that exists between genetic and chromosomal events and firmly established the chromosome theory of heredity, which states that genes are small particles located on the chromosomes.

In recent years, geneticists have learned that each chromosome contains a continuous sequence of DNA, which is associated with specialized proteins that help regulate gene activity. In other words, the genes are integral parts of the chromosomes.

Although the chromosome theory of heredity was established early in the twentieth century, the location of the genes within the chromosomes was less certain. Biologists knew that chromosomes contain both protein and DNA. They did not know, however, which one made up the genes and directed protein synthesis.

In 1952, Alfred Hershey and Martha Chase performed a series of experiments using viruses that attack bacteria. The viruses that Hershey and Chase used have a core of DNA (or sometimes RNA), surrounded by a protein coat. When Hershey and Chase labeled the protein coat with a radioactive marker in order to test for its presence, the bacteria did not show the marker (see Figure 8.9a on page 185). When they labeled the viral DNA with a radioactive marker, however, they found the labeled DNA had entered bacterial cells (see Figure 8.9b). Once inside the cell, the viral DNA took over the cell's machinery and materials and made new virus particles. The new viruses were similar to the original ones and had newly made protein coats. If the viral DNA was labeled with a radioactive marker, the offspring were radioactive also. It became clear to Hershey and Chase that DNA, not the protein coat surrounding the DNA, directed the synthesis of protein and transmitted the virus features from one generation to the next. These and many other experiments convinced scientists that DNA is the genetic material in chromosomes.

Piece by piece, experiment by experiment, biologists were learning how genes are inherited and what they do. These studies set the stage for identifying the molecular structure of the gene and its dual work in heredity and biochemical function.

Figure 8.8 An adult fruit fly, *Drosophila melanogaster.*

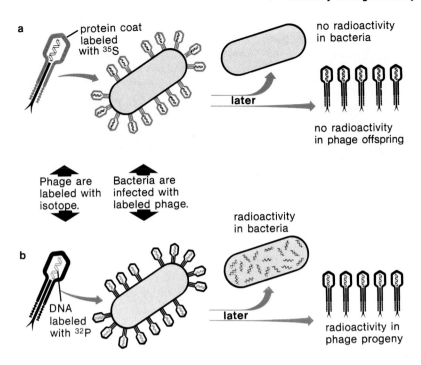

Figure 8.9 The experiments of Alfred Hershey and Martha Chase showed that DNA carries the hereditary instructions of viruses. In (a), phages (viruses that attack bacteria) with protein coats labeled with an isotope of sulfur (^{35}S) were allowed to infect bacterial cells. In (b), phages with their DNA cores labeled with an isotope of phosphorus (^{32}P) were allowed to infect bacterial cells. Later, the bacteria and the phage particles that grew in the cells were tested for radioactivity. The presence of radioactivity inside only the bacteria that were infected with ^{32}P-labeled phage showed clearly that *only* the DNA of the phage entered the bacterial cell.

8.6 Genes Are Long Chains of DNA Nucleotides

Once it was known that genes are composed of DNA and code for specific proteins, molecular biologists could begin to answer more specific questions about the chemical basis of genetics. What is the molecular structure of genes? How do they coordinate their dual roles of transmitting information to the next generation and programming the structure and function of the organism? What is the role of RNA, the other nucleic acid?

In 1953, J. D. Watson and F.H.C. Crick proposed a structure for DNA that explained how DNA could pass information from generation to generation. They hypothesized that a DNA molecule is a long, twisted, double-stranded structure. Each strand consists of a chain of nucleotides, and each nucleotide consists of three smaller units: a sugar, a phosphate group, and a nitrogen-containing base. There are four types of nucleotides in DNA, and each one has a different base: adenine, thymine, cytosine, or guanine. The sugars and phosphates form the sides of the strands and join the nucleotides together. Each base from one strand pairs with a base from the other strand. In this way, thousands of base pairs join the two strands together, as shown in Figure 4.25 on page 98.

According to Watson and Crick, the two strands are complementary because the molecular shape of each base can pair only with a particular complementary base. Adenine (A) can pair only with thymine (T), and cytosine (C) can pair only with guanine (G). These A—T and C—G base pairs occur along the entire length of a DNA molecule. The chemical bonds that hold the bases together are weak, however. When they are broken, two separate, complementary strands result. In DNA replication, each old strand serves as a pattern for the formation of a new, complementary strand. The result is two identical double strands of DNA, each of them exactly like the

Figure 8.10 Replication of DNA. The strands come apart at the bonds between the nucleotides. New nucleotides, which temporarily bear extra phosphates, are added one by one. Eventually two new DNA molecules are produced.

adenine   guanine   cytosine   thymine   deoxyribose   phosphate

original double-stranded molecule. Figure 8.10 shows this process, which you can practice in Investigation 8.2.

The molecular DNA model developed by Watson and Crick explains several essential properties of genes: (1) genes replicate (exactly duplicate) during cell division; (2) genes mutate occasionally; and (3) genetic instructions are passed from generation to generation through meiosis and fertilization.

Still, biologists did not know how the genetic instructions, coded in sequences of nucleotides in nuclear DNA, are translated into the structure and function of proteins in the cytoplasm. What, they wondered, was the link between instructions for a protein and the construction of the protein in the ribosomes found in the cytoplasm.

INVESTIGATION 8.2   DNA Replication and Transcription to RNA

DNA replication is the process by which exact copies are made of the DNA found in prokaryotes and in the chromosomes of eukaryotes. During replication, the genetic code contained within a sequence of nucleotide bases in DNA is preserved. Transcription is the process by which the message in DNA is transferred to RNA. How do replication and transcription take place? In this investigation, you will observe some of the basic steps in these processes.

Materials (per team of 2)

paper clips: 40 black, 40 white, 32 red,   pen
   32 green, 5 silver   string
masking tape

Procedure

PART A   Building the DNA Molecule

1. Use the following key for nucleotide bases and the colored paper clips to build a double-stranded segment of DNA.

   black = adenine (A)   white = thymine (T)
   green = guanine (G)   red = cytosine (C)

2. Construct the first DNA strand by linking the colored paper clips together to represent the following sequence of nucleotide bases:

   A A A G G T C T C C T C T A A T T G G T C T C C T T A G G T C T C C T T

3. Attach a piece of masking tape to the AAA end of the strand, and label the strand with the Roman numeral *I* by marking the masking tape.

4. Now construct the complementary strand of DNA that would pair with strand I. Remember that thymine (T) pairs only with adenine (A), and guanine (G) pairs only with cytosine (C).

5. Attach a piece of masking tape to the TTT end of the complementary strand and mark a Roman numeral *II* on the masking tape.

6. Place strand II beside strand I and check to make certain you have constructed the proper sequence of nucleotide bases in strand II. Green paper clips should be opposite red paper clips, and black paper clips should be opposite white paper clips. Make any correction required in order to have the proper sequence of base pairs.

7. Using a short piece of string, tie a simple knot to join the first paper clips in strand I and strand II, as shown in Figure 8.11.

8. Repeat Step 7 for the paper clips on the opposite ends of strands I and II.

PART B   DNA Replication

9. Place the double-stranded DNA molecule in a straight line on your table, with the AAA end of strand I on your left. (Do not untie the string at the end.)

10. Beginning at the AAA end of strand I, count 22 paper clips from left to right. Tie the 22d pair of paper clips together, as shown in Figure 8.11.

11. Untie the string holding the two strands together at the AAA end of strand I. Separate strand I and strand II so they form a Y.

Figure 8.11 A simple knot of string holds the ends of the two strands together (top); the knot at paper clip 22 represents the temporary end point of DNA replication (bottom).

12. DNA replication on strand I begins at the AAA end and proceeds toward the fork of the Y (the point at which the nucleotide bases are joined at the 22d base pair). Construct the new complementary strand for strand I beginning at the AAA end and working toward the fork of strands I and II. When you have finished, tie another knot at the AAA end of strand I to join it to its new complementary strand.

13. Replication of strand II begins at the fork of the Y and moves outward toward the TTT end. Build the complementary DNA strand for strand II, proceeding from right to left, and then tie the two strands together at the TTT end.

14. Now untie the knots that join strands I and II at the 22d base pair and at the right-hand end of the original double strand.

15. Continue the replication of strand I from left to right until you complete the new complementary strand. Tie another knot to join strand I and its new complementary strand at the right-hand end of the molecule.

16. Continue the replication of strand II from the right-hand end to the left. When you reach the 22d base pair, join the two segments of the newly formed strand (the complement of strand II).

17. Tie another knot in the right-hand end of strand II to join it to its new complementary strand.

PART C   mRNA Formation

18. Repeat Steps 9–11, using one of the two double-stranded DNA molecules you just made. This represents one chromosome. Assume this chromosome has only two genes on it, one 21 and one 15 nucleotides long.

19. RNA is produced by transcription from strand I. The rules for forming mRNA are the same as for DNA except that uracil (silver) is used in place of thymine (white). Beginning at the AAA end of strand I, construct an mRNA molecule for the first gene on strand I.

20. Remove the new mRNA and rejoin the original DNA double strand with a simple knot.

Discussion

1. Compare the two new double-stranded molecules of DNA you made. How are they similar to the original DNA molecule containing strands I and II? How are they different?

2. Describe the differences in the ways strands I and II are replicated.

3. Describe how DNA replication makes it possible to produce two identical cells from one parent.

4. How many nucleotides are in the mRNA molecule? How many amino acids does this mRNA molecule code for?

5. How does the number of nucleotides in the two genes of this model compare with the number of nucleotides in your genes?

6. Would it make any difference in the type of protein formed if strand II were used as a blueprint for the mRNA instead of strand I?

8.7   DNA Makes RNA and RNA Makes Protein

The molecule that carries the information from the chromosomes in the nucleus to the ribosomes in the cytoplasm is RNA. Like DNA, RNA consists of chains of nucleotides. In each RNA nucleotide, however, the sugar is ribose instead of deoxyribose, and thymine is replaced by a similar nitrogen base called uracil. RNA is synthesized by copying one of the complementary DNA strands.

Three different types of RNA occur in cells. One type—ribosomal RNA (rRNA)—along with a number of proteins, makes up the ribosomes. Another type is messenger RNA, or mRNA. It carries the DNA message from the nucleus to the ribosomes. The third type, transfer RNA (tRNA), carries amino acids to the ribosomes. These amino acids are added to a growing chain of amino acids.

As long chains of amino acids are completed, proteins are formed. The 20 different types of amino acids found in proteins can be connected in any order. The structure and function of each protein is determined by the amino acids that are present and by the order in which the amino acids are attached. DNA, then, provides information about the precise sequence of amino acids in the protein chain. How can the DNA instructions guarantee the sequence?

For a moment, imagine the bases in DNA (A, T, C, and G) are like the letters in a sentence written in a secret code (see Figure 8.12). Somehow the code must specify all the arrangements of 20 different amino acids. Does each base in a DNA strand code for one amino acid? No, because the four

Figure 8.12 The DNA genetic code is written in mRNA. Each triplet is a codon for an amino acid. For example, UGG codes for the amino acid tryptophan. The codon AUG signals start, and UAA, UAG, and UGA signal stop. Several amino acids have more than one codon.

First Letter	Second Letter				Third Letter
	U	**C**	**A**	**G**	
U	phenylalanine	serine	tyrosine	cysteine	U
	phenylalanine	serine	tyrosine	cysteine	C
	leucine	serine	stop	stop	A
	leucine	serine	stop	tryptophan	G
C	leucine	proline	histidine	arginine	U
	leucine	proline	histidine	arginine	C
	leucine	proline	glutamine	arginine	A
	leucine	proline	glutamine	arginine	G
A	isoleucine	threonine	asparagine	serine	U
	isoleucine	threonine	asparagine	serine	C
	isoleucine	threonine	lysine	arginine	A
	(start) methionine	threonine	lysine	arginine	G
G	valine	alanine	aspartate	glycine	U
	valine	alanine	aspartate	glycine	C
	valine	alanine	glutamate	glycine	A
	valine	alanine	glutamate	glycine	G

bases could code for only four amino acids. A code of 2 bases could account for 16 amino acids—still not enough. A sequence of 3 bases, however, could provide 64 combinations of the 4 bases, more than enough to code for 20 amino acids. For this reason, scientists assumed the bases are arranged in codons, or "words," of three and that each codon stands for one amino acid. Experiments have supported this assumption. It took several years of work to determine which codon, or triplet of bases, codes for each amino acid. Most of the 64 triplets code for some amino acid; several amino acids are specified by more than one codon, as shown in Figure 8.12.

The steps by which genes coordinate molecular activity are now clear. DNA codes for protein structure. The coded instructions are transcribed when RNA is synthesized along one of the two DNA strands (see Figure 8.13). In its turn, mRNA carries these instructions from the nucleus to the

Figure 8.13 Formation of part of a strand of mRNA along one strand of a DNA molecule.

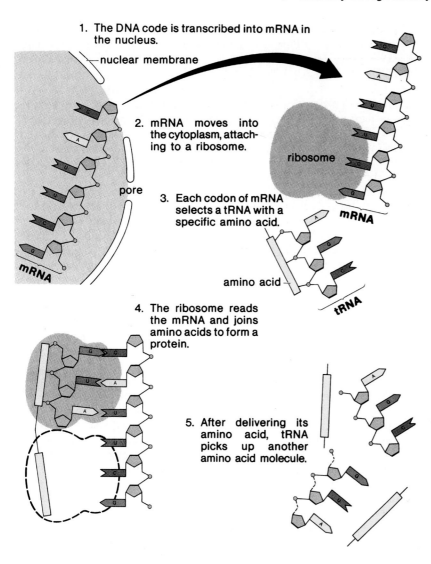

1. The DNA code is transcribed into mRNA in the nucleus.

nuclear membrane

2. mRNA moves into the cytoplasm, attaching to a ribosome.

ribosome

pore

3. Each codon of mRNA selects a tRNA with a specific amino acid.

mRNA

amino acid

tRNA

mRNA

4. The ribosome reads the mRNA and joins amino acids to form a protein.

5. After delivering its amino acid, tRNA picks up another amino acid molecule.

Figure 8.14   How DNA determines the formation of a protein.

ribosomes. In the ribosomes, a triplet of mRNA nucleotides specifies which tRNA will bring in a particular amino acid. At the ribosome, the amino acids are attached, one at a time, to the end of the growing protein chain. The sequence of amino acids determines how the chain can fold into a three-dimensional structure. At the end of the process, a new protein molecule has been formed. Figure 8.14 illustrates these steps.

Concept Review

1. Why would the knowledge that genes are made of DNA rather than protein be helpful to biochemists trying to isolate and study genes?
2. Sketch or describe the structure of the DNA molecule.
3. Describe the evidence for the chromosome theory of heredity.
4. How is the genetic information coded in DNA translated into protein?

Guidepost: How are different patterns of inheritance explained?

Patterns of Inheritance

INVESTIGATION 8.3   A Dihybrid Cross

Sexually reproducing organisms have haploid and diploid stages. In flowering plants and animals, only the gametes are haploid, and traits (the phenotype) can be observed only in the diploid stage. When Mendel made dihybrid crosses to study the inheritance of two different traits, such as seed shape and seed color, he could observe the traits only in the diploid cells of the parents and their offspring. He had to use probability to calculate the most likely genotypes of the gametes.

In the yeast *Saccharomyces cerevisiae*, however, phenotypes of some traits can be seen in the haploid-cell colonies. In Investigation 7.1, for example, you could observe the color trait, red or cream, in both haploid and diploid stages of the yeast life cycle. Not all yeast traits are visible, however, so yeast geneticists use the methods Beadle and Tatum devised for studying mutations. That is, they isolate mutants that cannot make some essential substance. The inability of the mutant and the ability of the normal strain to make the substance define two different forms of a trait.

In this yeast dihybrid cross, you will follow two forms of each of two traits: red versus cream color, and tryptophan-dependent (requires this amino acid to grow) versus tryptophan-independent (does not require this amino acid). In Investigation 7.1, a red strain of yeast of one mating type is crossed with a cream-colored strain of the other type. The diploid strain is cream colored. If there is a single gene for this trait, then there must be one allele that determines cream color in the haploid strain and another allele that determines red color. In the diploid strain, then, there must be one of each of these alleles. Which form of the color trait is dominant?

By observing the color of a colony that a haploid strain forms, you can predict with certainty which allele it carries. In the case of a cream-colored diploid, however, you cannot be sure of the genotype. It could carry either one allele for cream color and one for red, or two alleles for cream color. The trait defined by tryptophan-dependence or independence works in much in the same way. A tryptophan-dependent haploid must carry the allele for tryptophan-dependence (a defective form of the gene) and a tryptophan-independent haploid must carry the allele for tryptophan-independence (the functional form of the gene), but a tryptophan-independent diploid could be either homozygous or heterozygous for the functional allele.

Using symbols for the traits, the allele for the dominant cream form is *R* and the recessive red form is *r*; the normal, tryptophan-independent form of that gene is *T* and the tryptophan-dependent form is *t*. Whereas the color trait is visible, determining tryptophan dependence or independence requires a simple growth test.

Table 8.2 shows how all possible combinations of these two traits can be determined by testing the yeast on two kinds of growth medium: a nutritionally complete medium (COMP), and a medium lacking tryptophan (MIN).

TABLE 8.2   Growth of Yeast on Two Media

Haploid	Diploid	On COMP medium	On MIN medium
RT	*R/R T/T* *R/r T/T* *R/r T/t*	growth, cream	growth, cream
Rt	*R/R t/t* *R/r t/t*	growth, cream	no growth, (cream?)
rT	*r/r T/T* *r/r T/t*	growth, red	growth, cream
rt	*r/r t/t*	growth, red	no growth, (red?)

Materials (per team of 2)
PART C

container of clean flat toothpicks
glass-marking pencil
2 paper templates for streaking strains
3 self-closing plastic bags labeled *waste*
   (1 for each day)
biohazard bag
complete growth medium agar plate
   (COMP)

minimal + adenine growth medium agar
   plate (MIN)
COMP plate with 24-hr. cultures of yeast
   strains **a**1, **a**2, **a**3, **a**4, α1, α2, α3, and
   α4

Procedure
PART A   Predicting the Gametes of the Parents

The diploid parents of this dihybrid cross would have been *r/r T/T* and *R/R t/t*. To predict
the F$_2$ offspring, it is necessary to predict the different types and relative numbers of
gametes that can be produced. The simplest way to make this prediction is to diagram
how the chromosomes separate at meiosis. The diagram in Figure 8.15a shows the gametes predicted from the pure-breeding diploid red parent (*r/r T/T*).

   Two pairs of chromosomes are represented, one carrying the gene for color (in this
case red, *r*) and the other carrying the gene for the ability to make tryptophan (in this
case tryptophan-independent, *T*). At the first meiotic division (M$_I$), chromosome pairs
segregate (separate). At the second meiotic division (M$_{II}$), the two chromatids of each
chromosome segregate into the nuclei of the gametes. In this case, segregation can
occur in only one way because only one allele of each gene is represented. All the gametes from a pure-breeding diploid are the same (*rT*). The probability of this parent
producing a gamete with the genotype of *rT* is 1/1, or 100%.

1. In your data book, copy and complete the diagram shown below in Figure 8.15b
    (or use the diagram your teacher provides) for the cream, tryptophan-dependent
    parent (*R/R t/t*).

2. Use the symbols given to describe for each parent strain the genotypes and phen-
    otypes of all the possible gametes and the relative probabilities of their occurrence.

3. The haploid gametes produced from these two pure-breeding parents mate to form
    the diploid zygotes of the F$_1$ generation. Use the symbols to describe the genotype
    and phenotype of the diploid zygotes that could be formed from the fusion of these
    gametes. At this point your diagrams should show that they all will have the di-
    hybrid genotype *R/r T/t*.

Figure 8.15   Probabilities of gametes that could be formed by yeast organisms that
are homozygous for two traits.

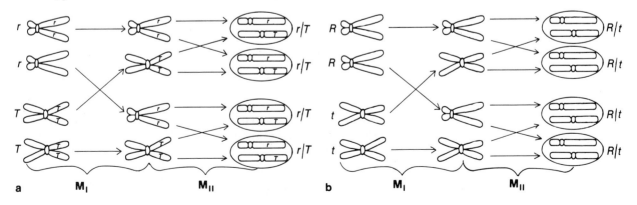

a   **M$_I$**   **M$_{II}$**   b   **M$_I$**   **M$_{II}$**

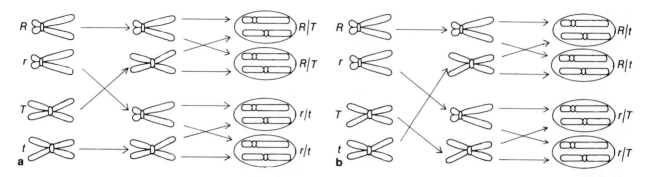

Figure 8.16 Gametes that could be formed by yeast organisms heterozygous for two traits.

PART B   Predicting the Gametes of the F₁ Diploids

Because there are two equally probable ways that the alleles of the two genes can separate at M$_I$, the chromosome diagram for the segregation of the *R*, *r*, *T*, and *t* alleles is more complicated. These are shown in Figures 8.16a and 8.16b. Since these two patterns of segregation are equally probable, the two diagrams together illustrate the relative numbers of all possible genotypes.

4. Copy the diagrams in Figure 8.16 in your data book, or tape in the diagrams your teacher provides. Fill in the missing symbols. Each chromatid should have a symbol.
5. What is the total number of different gametes represented in the diagram?
6. How many different genotypes are represented? Give their symbols.
7. How many times is each genotype represented?
8. What is the probability of occurrence of each genotype among the total number of gametes shown? Compare this prediction with the gametes in Figure 8.19 on page 197.
9. Draw a checkerboard diagram for all the possible crosses among these gametes. This diagram should illustrate the 16 different combinations predicted. Instead of male and female gametes, use mating-type **a** and mating-type *α* haploid strains. In each square of the diagram construct the diploid genotype that would result from the fusion of the corresponding gametes. How many different genotypes are there? How many different phenotypes? Does this diagram predict a 9 : 3 : 3 : 1 ratio? What specific phenotypes would be represented in these ratios?

PART C   Testing the Predicted Phenotypes
Day 0

10. Make two templates for setting up crosses. Copy the pattern shown in Figure 8.17 onto a piece of paper twice, so that each fits the bottom of a petri plate (or use the templates your teacher provides).
11. Tape one template to the bottom of a plate of complete growth medium (COMP) so that you can read it through the agar.
12. Transfer a small sample of each strain onto the agar directly over its corresponding label. To do this, touch the flat end of a clean toothpick to strain **a**1 on the plate your teacher provides. Then, gently drag the toothpick on the COMP agar plate to make a streak about one cm long. Discard the toothpick in the self-closing waste bag, being careful not to touch anything with it. Repeat this procedure for all eight strains, using a new toothpick for each transfer. *Be careful not to touch the ends of the toothpicks to anything except yeast or the sterile agar. Discard used toothpicks in the self-closing plastic waste bag. Keep the lid on the plate at all times except when transferring yeast.*

Mating-type a strains

	a1 *R T*	a2 *R t*	a3 *r T*	a4 *r t*
α1 *R T*				
α2 *R t*				
α3 *r T*				
α4 *r t*				

(Left axis label: Mating-type α strains)

Figure 8.17 Template for dihybrid yeast cross.

13. Invert and incubate at room temperature.
14. Wash your hands thoroughly before leaving the laboratory.

Day 1
15. On the same COMP plate, make a mating mixture for each of the mating-type **a** strains with each of the mating-type α strains. To do this, use the flat end of a clean toothpick to transfer a dot of the **a1** strain of freshly grown cells to each of the boxes below it on the template. Discard the toothpick in the waste bag. Repeat this procedure for **a2**, **a3**, and **a4**, using a new toothpick for each strain. Using the same procedure, transfer a dot of the freshly grown cells of each α strain to each of the boxes to the right of the strain on the template. Place these dots side-by-side, but not touching, as shown in the first box in Figure 8.17. Use a clean toothpick to mix each pair of spots together. Be sure to use a clean toothpick each time you mix pairs together. Discard all toothpicks in the waste bag.
16. Invert and incubate at room temperature.
17. Wash your hands thoroughly before leaving the laboratory.

Day 2
18. Test each mating mixture and parent strain for its ability to grow on MIN agar. To keep track of the tests, tape a copy of the template to the bottom of the MIN plate. Then, use the flat end of a clean toothpick to transfer a small amount of each strain and mixture from the COMP plate to the corresponding position on the MIN plate. Be sure to use a clean toothpick each time you change strains and mixtures. Discard all toothpicks in the waste bag.
19. Invert and incubate both plates until the next day.
20. Wash your hands thoroughly before leaving the laboratory.

Day 3
21. Copy the score sheet in Figure 8.18 on page 196 in your data book (or tape in the copy your teacher provides). Record the color and growth phenotypes of each parent haploid strain and F_2 diploid on the score sheet for both plates.
22. Tabulate the different phenotypes observed among the F_2 diploids and the number of times each one occurred among the 16 crosses.

Figure 8.18 Score sheet for dihybrid yeast cross.

Mating-type a haploid parents

F_2 Diploids

23. Compare the F_2 phenotypes with the predictions you made in Part A. Explain how your results either support or contradict your predictions.
24. Discard all plates in the biohazard bag.
25. Wash your hands thoroughly before leaving the laboratory.

Discussion

Because yeast exhibit most of the same traits in the haploid stage (gametes) and the diploid stage, you knew the precise genotypes of the gametes (haploid strains) that you mated to produce the F_2 diploids. This removed the element of chance at this step. In Part A, however, you had to deal with the role of chance to predict the numbers and genotypes of the gametes from the F_1 diploids. Throughout the entire process, beginning with the parental cross (P) and going through to the F_2 offspring, there are some steps at which chance plays a role, so the results can only be expressed as a probability. In other steps, chance is not a factor, so you can predict the outcome exactly.

1. List the steps in which chance is a factor.
2. List the steps in which chance is not a factor.
3. Explain why the outcome of some steps involves chance, whereas the outcome of others does not.

8.8   Dihybrid Crosses Produce a Distinctive Pattern

What has been described thus far about nucleic acids and chromosomes explains Mendel's results with crossing plants that differed in one trait. Each trait depended on the production of certain proteins, which were specified by certain genes. Mendel, however, also crossed plants that differed in two traits. For example, he crossed plants that were pure breeding for round seed shape and yellow seed color with plants that were pure breeding for wrinkled seed shape and green seed color. This type of cross is called a dihybrid cross because the offspring that result are heterozygous for two (*di*) different traits.

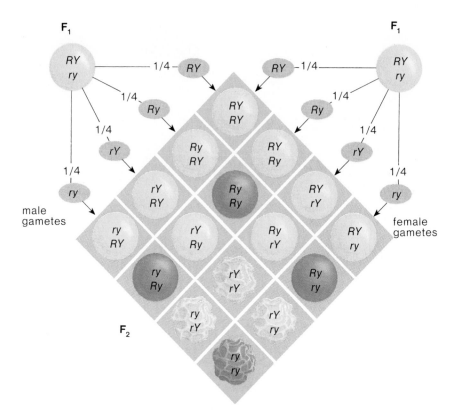

Figure 8.19   A dihybrid cross that illustrates the principle of independent assortment. What phenotypes are present in the F₂ generation, and in what ratios? These results demonstrate that the traits for seed shape and seed color assorted separately. Note that there are 12 round to every 4 wrinkled seeds, and 12 yellow to 4 green seeds. Each trait independently shows the 3 : 1 ratio typical of a monohybrid. R = round seed; r = wrinkled seed; Y = yellow seed; y = green seed.

Figure 8.19 illustrates the four phenotypes Mendel saw in the F_2 generation of a dihybrid cross. Note that the phenotypes do not occur with equal frequency. Instead, the ratio is 9/16 to 3/16 to 3/16 to 1/16—a 9 : 3 : 3 : 1 ratio. How can these results be explained? Remember that during meiosis the chromosome pairs separate independently. When gametes were formed in the F_1 generation, the genes for seed shape and color separated independently, resulting in four different combinations of the alleles of these genes. The phenotypes in the F_2 offspring can be predicted on the basis of their alleles. These results laid the foundation for Mendel's **principle of independent assortment**: Alleles for one trait segregate independently of alleles for other traits during gamete formation.

Although Mendel used garden peas in his experiments, his findings are applicable to humans. For example, think of the little girl with cystic fibrosis discussed in the introduction. Medical geneticists now know that cystic fibrosis is a recessive disorder. What hypothesis about the biochemical basis of the disease does this knowledge suggest?

Assume that F represents the allele for the normal condition and f represents the allele for cystic fibrosis. What are the genotypes of Bob, Mary, and Lisa? What can the genetic counselor tell this couple about the probability that their next child will have cystic fibrosis?

Figure 8.20 Inheritance of flower color in morning glories, an example of codominance.

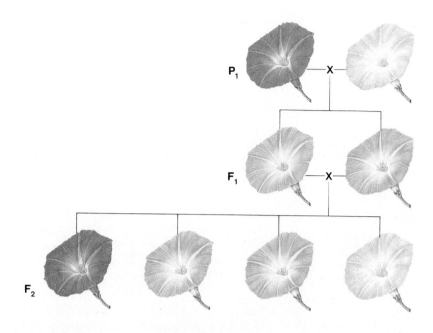

8.9 Gene Expression Can Affect Patterns of Inheritance

Mendel worked with traits that showed clear dominant and recessive inheritance. Complete dominance does not always occur, however. Sometimes the phenotype of hybrid organisms is different from the phenotype of either homozygous parent. When both alleles contribute to the phenotype of a heterozygote, the alleles are said to show **codominance**. The flower color of the morning glories illustrated in Figure 8.20 is an example of codominance.

Sometimes there are more than two alleles for a trait. ABO blood types in humans are determined by three alleles: I^A, I^B, and i. This system of human blood types illustrates the existence of **multiple alleles**, genes that have more than two allelic forms, and offers another example of codominance. These blood types determine whether or not blood from one person can be transfused safely to another person. Allele I^A causes the formation of blood factor A. Allele I^B causes the formation of factor B. These two alleles show codominance; when a person has the genotype $I^A I^B$, both types of factors are produced. Allele i does not cause either factor to form. Table 8.3 shows the genotypes that are responsible for the various phenotypes. Note that an individual has only two of the alleles.

Many traits exhibit more than two phenotypes. These traits vary across a broad range. Suppose you plotted the heights of all the tenth graders in your school on a graph (see Figure 8.21). Between the shortest and the tallest person, there would be many other people who cover the whole range of height. This type of distribution is called continuous. Traits that exhibit more than two phenotypes are different from traits such as cystic fibrosis, which is either present or absent. Such "either/or" traits, which generally are controlled by a single pair of alleles, are called discontinuous, or discrete.

TABLE 8.3 ABO Blood Types

Genotype	Blood Type (Phenotype)
$I^A I^A$ or $I^A i$	A
$I^B I^B$ or $I^B i$	B
$I^A I^B$	AB
ii	O

Figure 8.21 Continuous and discontinuous variations. Most human traits are determined by multiple genes and are influenced by the environment. Some traits (usually controlled by single genes) exist in only two forms. Histogram *a* shows the distribution of the heights of students in a high-school biology class. Histogram *b* shows the distribution of sexes in the same class.

Continuous variability results from **multifactorial inheritance**—the interaction of at least several genes with a large number of possible environmental variables. Most human traits, such as height, weight, intelligence, hair color, skin color, and eye color, do not occur in an either/or condition. Instead, these traits vary from person to person to person. Each trait is controlled by more than one gene, and each gene contributes to the final appearance of the trait. For example, the production of the skin pigment melanin is thought to be controlled by four major genes, and not one of these genes demonstrates simple dominance. The independent assortment of the alleles for melanin production results in a continuous variation in skin color among humans. Skin color, however, also is affected by the amount of exposure to sunlight. Some disorders, such as cleft lip and spina bifida, also may be caused by multiple factors. They are caused by a number of genes interacting with certain environmental factors present in the mother's uterus during pregnancy.

Although these examples do not display the same ratios as Mendel's simple crosses, they still are considered to result from Mendelian inheritance. That is, they can be explained in terms of genes residing on chromosomes transmitted through meiosis during sexual reproduction. Not all genetic information is transmitted this way. In many eukaryotic organisms, some of the genes are associated with cytoplasmic organelles, such as chloroplasts and mitochondria. This genetic information follows a pattern of inheritance unrelated to meiosis. In prokaryotic organisms, which lack a complete sexual cycle and do not undergo meiosis, there are a great variety of non-Mendelian patterns of inheritance.

8.10 X-Linked Traits Show a Modified Pattern of Inheritance

Remember that genes are integral parts of chromosomes, which occur in matching pairs. When chromosomes are stained, the stains concentrate in specific regions and thus create a characteristic banding pattern. The two

(drawing) (photo)
No. 7 showing chromatids

(drawing) (photo)
X showing chromatids

Figure 8.22 Human chromosomes 7 and X. The banding patterns evident in the photographs are shown in the diagrams. These two chromosomes, which have almost the same length and centromere location, can be distinguished by their banding patterns.

chromosomes in each pair show the same banding pattern, and their pattern is different from the patterns of other chromosome pairs. Figure 8.22 illustrates how two similar-sized chromosomes can be distinguished on the basis of their banding patterns. Once the chromosomes are stained, they can be photographed through a microscope. When the photograph is enlarged, individual chromosomes may be cut out and arranged in order by size and shape to form a karyotype similar to the one shown in Figure 8.23. With some experience, counting and identifying chromosomes becomes a relatively simple procedure, and any unusual, missing, or extra chromosomes can be detected fairly quickly.

Using karyotypes, biologists have found that humans have 23 pairs of chromosomes—22 pairs that are the same in both sexes (known as autosomes), and one pair that is different in males and females. This pair determines the sex of an organism, and the chromosomes that make up the pair are called the sex chromosomes. In humans and other mammals, the sex chromosomes are referred to as *X* and *Y* chromosomes. Females have two X chromosomes in each body cell; males have one X chromosome and one Y chromosome. Each ovum produced by a female, therefore, contains an X chromosome. Males, on the other hand, produce two types of sperm— half with an X chromosome and half with a Y chromosome. Thus during fertilization, it is the sperm that determines the sex of the offspring. If an ovum is fertilized by an X-bearing sperm, the zygote develops into a female. If the sperm contains a Y chromosome, the offspring is a male. Because each chromosome contains many genes, and each gene is capable of affecting some part of development, complete chromosomes are essential for normal development and sex determination.

While studying *Drosophila* at Columbia University in the early 1900s, Thomas Hunt Morgan noted some mutant flies with striking differences. Some mutants had white eyes instead of red; others had short wings instead of long; and still others had yellow or black bodies instead of gray ones. Morgan and his colleagues observed hundreds of mutations in fruit flies.

Figure 8.23 A karyotype is a display of human chromosomes arranged as homologous pairs. It is prepared by cutting out individual chromosomes from a photograph and matching them, pair by pair. Is this the karyotype of a male or a female?

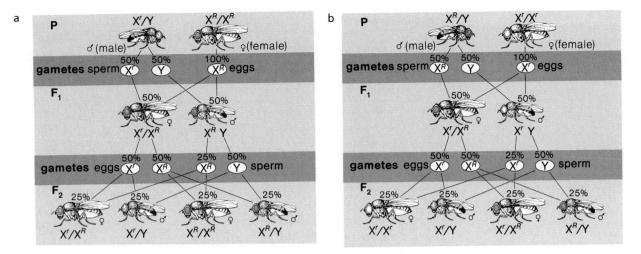

Figure 8.24   X-linked inheritance. When Morgan crossed a white-eyed male with a normal red-eyed female, all the F_1 flies (a) had red eyes. The F_2 flies showed the expected 3 : 1 ratio of red to white eyes, but only male flies showed the recessive trait. A reciprocal cross (b). All male offspring have white eyes because they inherited their X chromosomes from their mother. Compare this cross with the F_1 results of the first cross. The ratio of F_2 flies also is shown.

They carefully tested each mutation by mating mutant with nonmutant flies. In most cases, the mutations were inherited according to Mendelian patterns. One exception, however, was the first mutation Morgan had observed—a male with white eyes.

When Morgan crossed the white-eyed male with normal red-eyed females, the F_1 generation contained only red-eyed flies, supporting his assumption that the white-eye form of the eye color trait is recessive. The F_2 generation had the expected 3 : 1 ratio of red-eyed flies to white-eyed flies, as shown in Figure 8.24a. Morgan noticed, however, that only males showed the white-eye form. Although not all males were white-eyed, this form seemed to be linked to a fly's sex.

Morgan continued breeding his flies and eventually obtained some white-eyed females. He then crossed a white-eyed female with a red-eyed male and studied the inheritance of white eyes. The F_1 offspring of this cross were not all red-eyed, however. Only F_1 females showed the dominant red-eye form. All the males had white eyes. This cross is shown in Figure 8.24b.

Morgan hypothesized that the gene for eye color is on the X chromosome and there is no corresponding gene on the Y chromosome. Thus, a single allele for the recessive form, if on a male's X chromosome, could result in a white-eyed male. White eye color was the first example of an **X-linked trait**, a trait whose gene is on only the X chromosome.

Since then many other X-linked traits have been found in *Drosophila*, and more than 200 have been identified in humans. One well-known X-linked human trait is hemophilia, a disorder in which blood does not clot properly. Others include red-green color blindness and Duchenne muscular dystrophy, a crippling, muscle-wasting disease that results in premature death. The pattern of inheritance of these X-linked traits is the same

Figure 8.25 A pedigree for red-green color blindness. How does the pattern of inheritance compare with the pattern for white-eye trait in *Drosophila?*

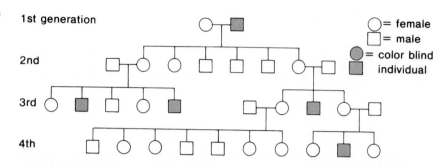

as that of white eyes in *Drosophila*. Study the pedigree shown in Figure 8.25 and try to determine the genotype of each individual on the chart. Charts illustrating family histories of traits are one of the major tools available for studies of inheritance in humans.

8.11 Abnormal Chromosomes Can Affect Patterns of Inheritance

Some types of birth defects are caused by abnormal numbers or types of chromosomes. These chromosomal abnormalities are present from birth and often from the moment of fertilization. One such condition is Down syndrome. (A syndrome is a group of symptoms associated with a particular disease or abnormality.) Individuals with this condition have distinctive features of the eyes, mouth, hands, and sometimes internal organs. All have retarded mental development, though the degree of retardation is highly variable.

Individuals with Down syndrome have 47 chromosomes in each cell instead of 46. The extra chromosome is the tiny number 21, visible in the karyotype shown in Figure 8.26. Down syndrome results from trisomy-21, which means the number-21 chromosome is present in triplicate.

Abnormal numbers of sex chromosomes cause several other syndromes. For example, individuals having only one X chromosome and no Y chromosome (a 45,X karyotype) usually are short, underdeveloped, and sterile females. This condition is known as Turner syndrome. Females who have a 47,XXX karyotype often have limited fertility and may have slight

Figure 8.26 Karyotype in Down syndrome (a); a child with Down syndrome (b).

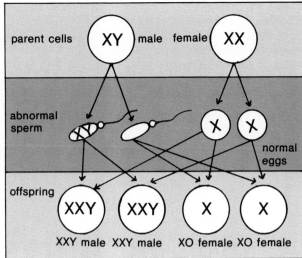

Figure 8.27   Nondisjunction can occur during the formation of either sperm or ova. What combinations of sex chromosomes would result from fertilization between an abnormal sperm and an abnormal egg?

intellectual impairment. A 47,XXY karyotype results in Klinefelter syndrome. These males often are tall and sexually underdeveloped.

These abnormalities arise when nondisjunction occurs—when chromosome pairs do not separate in meiosis. Nondisjunction results in the formation of abnormal gametes, as shown in Figure 8.27. Some sperm or egg cells get extra chromosomes, and some have too few chromosomes. These gametes, when fused with normal gametes, usually result in abnormal development.

Apparently there must be at least two of each type of chromosome in every cell for survival of the embryo. A striking exception is the X chromosome—45,X individuals can survive a missing chromosome. Extra sex chromosomes (X or Y) usually permit a fetus to develop, although the development may be abnormal. Extra autosomes, however, except for trisomy-21, rarely permit development to continue. Examination of cells from spontaneously aborted fetuses indicates that most have abnormal chromosome numbers.

Why do most trisomies of autosomes result in death of the fetus, whereas trisomies of sex chromosomes do not? A possible answer was provided by the British geneticist Mary Lyon in 1961. She suggested that early in the development of a normal female, one X chromosome in each body cell is inactivated. Lyon proposed X-inactivation as a possible explanation for a darkly staining mass, called a Barr body, that normally appears in the cell nuclei of females but not of males (see Figure 8.28). Females with three X chromosomes (47,XXX) have two Barr bodies, suggesting that all but one X chromosome in a cell are inactivated. One reason trisomies of sex chromosomes may not be as disruptive as autosomal trisomies is that the extra X chromosomes are inactivated and condensed into Barr bodies. Apparently a few genes remain active in the condensed X chromosomes, resulting in the abnormalities associated with extra X chromosomes.

Figure 8.28   The darkly staining Barr body in this nucleus is the condensed X chromosome. This cell is not dividing, so the chromosomes are not visible as discrete bodies.

Figure 8.29 If the piece of a chromosome carrying the genes B and C breaks off, four types of chromosomal abnormalities can result: deletion, duplication, inversion, or translocation to another chromosome.

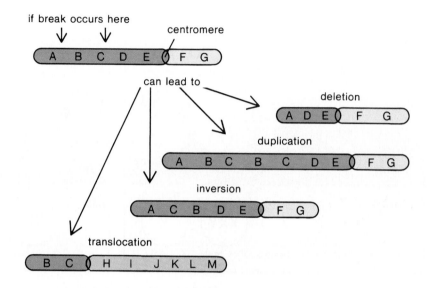

Sometimes abnormalities result from altered chromosome structure. For example, a piece of chromosome may be missing (deletion), or a section may be reversed (inversion). The missing piece of one chromosome may attach to its pairing partner, producing a duplication, or it may attach to an unrelated chromosome, resulting in a translocation. Figure 8.29 shows diagrams of these alterations in chromosome structure. It is clear that the proper number and types of active chromosomes, as well as the proper structure, are essential for normal development. The consequences of abnormal chromosome numbers apparently reflect a delicate balance among the many gene products that participate cooperatively in the total development and function of an organism.

Figure 8.30 Crossing-over frequently occurs between chromosome pairs during the early stages of meiosis.

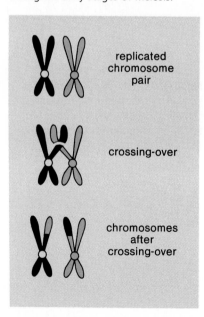

replicated chromosome pair

crossing-over

chromosomes after crossing-over

8.12   Many Genes Are on Each Chromosome

Because there are far more genes than chromosomes, many genes must be located on each chromosome. Genes that are close together on the same chromosome are said to be linked, which suggests that they are inherited together. For example, in humans the gene for the ABO blood group is on chromosome 9. An oncogene that may be involved in certain types of cancer also is on chromosome 9, close to the ABO gene. A geneticist studying those traits would find little or no evidence of independent assortment. Why?

Linked genes do not always remain together. Recall from the discussion in Section 6.5 that chromosome pairs often exchange fragments during the first meiotic division (Figure 8.30). That exchange gives rise to recombinations that increase the variability resulting from sexual reproduction. The farther apart two genes are on a chromosome, the more likely a break will occur between them. If two genes are far enough apart on a chromosome, the breaks between the genes may be so frequent that the principle of independent assortment applies to their inheritance. In other words, these genes exhibit no linkage. Some of the genes in Mendel's experiments sorted independently for this reason.

Figure 8.31 A few of the genes that have been mapped to the human X chromosome and to autosome 9.

The frequency with which two linked traits become separated provides a way to determine the relative distance between the two genes for those traits on the chromosome. With this information, geneticists can construct accurate maps showing the sequence of genes on chromosomes. Genetic maps now exist for chromosomes of humans (see Figure 8.31) and many other organisms.

Concept Review

1. How does the movement of chromosomes during meiosis explain the principle of independent assortment?
2. How does codominance account for the presence of more than two phenotypes of a trait?
3. What are multiple alleles? Can a person have more than two alleles for a single gene? Explain.
4. Describe how multifactorial inheritance explains continuous variability in a trait such as height.
5. How can karyotypes contribute to genetic research?
6. How does the inheritance of X-linked traits differ from that of other traits?
7. Why are the effects of genotypes with extra X chromosomes apparently less severe than those of other trisomies?
8. How does crossing-over affect linkage?
9. Describe how linkage can be used to prepare chromosome maps.

Genetics and Technology

8.13 New Tools Aid Genetics Research

New technology provides the tools for new discoveries that, in turn, lead to additional technology. The relationship between science and technology is illustrated best by two of the most important innovations in molecular biology: the ability to determine the exact base sequence of DNA,

Guidepost: How do technology and science fit together?

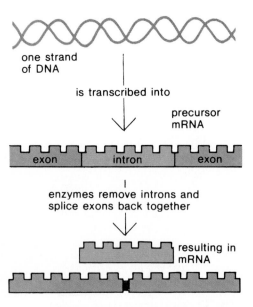

one strand
of DNA

is transcribed into

precursor
mRNA

exon | intron | exon

enzymes remove introns and
splice exons back together

resulting in
mRNA

Figure 8.32   Formation of eukaryote mRNA.

and the ability to transfer DNA from the chromosomes of one organism to those of another.

Studies of the base sequence of genes have allowed evolutionary biologists to compare DNA from different organisms. Through these comparisons, biologists have been able to assess more carefully the degree to which those organisms are related.

These types of studies have led to other discoveries. In the late 1970s, for example, biologists learned that many eukaryotic genes are not continuous. That is, the sequences that code for a protein are separated by noncoding intervening sequences, called introns (IN tronz). After the gene is transcribed, the introns are cut from the RNA strand while it is still in the nucleus (see Figure 8.32). The remaining pieces—called exons (EKS onz)—are spliced together into the mRNA that leaves the nucleus and directs the synthesis of protein on the ribosomes. While biologists had considered the relationship between DNA base sequences and the sequence of amino acids in the protein as simple, this discovery changed their ideas. The function and evolutionary significance of introns still are unknown.

As a result of new techniques available from genetic engineering, the number of genes mapped to human chromosomes is increasing by at least 25 percent per year. This information is valuable in detecting carriers of genetic disorders (see Biology Today). A major project is under way to map and sequence the human genome—the entire complement of human genetic material in a haploid cell. The human genome consists of 3 billion base pairs of DNA. The goal of the human genome project is to identify the exact type and position of each of those DNA base pairs. This information will increase our understanding of the origins, nature, and variability of human traits. It also could open doors to new ways of diagnosing and treating such inherited disorders as cystic fibrosis. One of the expected benefits of the human genome project is a better understanding of the importance of introns as well as other DNA sequences with functions that presently are undetermined.

8.14   Proteins Can Be Manufactured by Genetic Engineering

The genetic code is almost universal—the same genetic code functions in bacteria and in humans. Can the genetic instructions from one type of organism be interpreted by another type of organism?

The answer is yes. The protein-making machinery of a cell reads DNA instructions even when they come from another type of organism. By using special types of enzymes, DNA from one organism can be cut into small pieces and spliced into the DNA of another organism. The resulting DNA is called **recombinant DNA**, because DNA from different organisms has been recombined.

Molecular biologists first used these techniques to study how genes are turned on and off in response to hormones and other physiological and environmental factors. Knowledge of these mechanisms has provided new insights into cancer, AIDS, and many other diseases.

Other applications of transplanted genes also are widespread. For example, individuals who have diabetes produce too little insulin, a protein made by the pancreas. For years, diabetics have depended on insulin extracted from the pancreases of hogs and cattle. Although the animal insulin molecules are similar to human insulin, they are different enough to cause problems for some diabetics.

Today, diabetics can use a molecule that is exactly like the molecule their own pancreases should be making. Human insulin is manufactured in large amounts from recombinant DNA. The technique, represented in Figure 8.33, uses plasmids—small circles of DNA outside the chromosome of bacteria. A fragment of DNA coding for human insulin is inserted into a plasmid, and a bacterial cell takes in the plasmid from a culture medium. The recombinant DNA replicates along with the DNA of the bacterial cell. This process is known as gene cloning. It is analogous to conventional cloning—the reproduction and growth of genetically identical cells or organisms by asexual reproduction. The new cells containing a cloned gene can produce whatever product is coded for by the recombinant DNA, in this case, human insulin. Cloning makes it possible to isolate quantities of a gene for chemical analysis and DNA sequencing.

Medicine is only one area using practical applications of genetic engineering. Genetic engineers also have produced a strain of bacterium that prevents the formation of frost on plants to temperatures as low as − 6° C.

Figure 8.33 Techniques of recombinant DNA. To splice a human gene (in this case, the one for insulin) into a plasmid, scientists take the plasmid out of an *Escherichia coli* bacterium, open the plasmid at a specific site with a cutting enzyme, and splice in insulin-making human DNA. The resulting hybrid plasmid can be taken up by another *E. coli* bacterium, where it replicates together with the bacterium, making it capable of producing large quantities of insulin.

Figure 8.34 Loss of one codon in the gene for CF results in the deletion of one amino acid, causing production of a defective protein.

Calves have been born from fertilized eggs that have had genes to speed growth and make cattle leaner inserted into them. Genetic engineering has been used to improve other food-producing animals as well.

Genetic engineering also has been applied to genetic screening—including prenatal diagnosis—for disorders such as cystic fibrosis, Huntington disease, and muscular dystrophy. In 1989, Lap-Chee Tsui and John R. Riordan of the Hospital for Sick Children of Toronto and Francis Collins of the University of Michigan Medical School announced the identification and cloning of the cystic fibrosis gene. This unusually large gene consists of 250 000 base pairs on chromosome 7. Most of the gene is composed of introns that separate 24 exons. These exons code for a protein made up of 1480 amino acids. The protein's shape suggests it extends through the plasma membrane and thus may have a role in membrane transport. Approximately 75 percent of the mutant genes that cause cystic fibrosis have lost a single codon, resulting in the deletion of a single amino acid from the corresponding protein (see Figure 8.34). Now scientists must determine the function of the protein and how it malfunctions in cystic fibrosis. The search for the remaining 25 percent of the mutations that alter this gene also is under way, but these mutations may be more difficult to find because the gene is so large that many abnormalities can occur.

Genetic engineering provides a good example of the relationship between science and technology. Most of the organisms engineered by geneticists are used in research to demonstrate the organization and function of genetic material, the molecular web that connects all living things. The same techniques and knowledge used to engineer organisms offer potential solutions to many practical and societal problems.

Concept Review

1. How do science and technology interact?
2. How are introns and exons involved in production of mRNA?
3. What is meant by the *human genome project*, and what are some potential benefits?
4. How can bacteria make human insulin molecules?
5. What is genetic engineering, and what are some of its applications?

Summary

Genetic information is passed to the next generation by heredity. The genes provide a set of instructions for this information in the sequence of DNA bases. A triplet of bases codes for one amino acid. RNA copied from the DNA transmits genetic information to the ribosomes, where amino acids are connected in chains to form distinctive proteins. The sequence of DNA bases specifies the sequence of amino acids in a protein.

Because genes are located on chromosomes, patterns of inheritance are explained by chromosome behavior in meiosis and fertilization. Sex determination in many organisms depends on X and Y chromosomes.

Genes exist in different forms called alleles. Normally, only two alleles for the same trait are present in an individual—one on each member of a

Genetic Screening: Prevention with Problems

The isolation of the gene for cystic fibrosis was a major breakthrough in biomedicine. There is new hope now that research will produce a cure for the disorder. The ability to detect the CF gene through DNA analysis also makes prevention of the disease through genetic screening possible. DNA analysis could lead to the detection of carriers (heterozygotes) in the general population. Once recognized, carriers could be advised of the risk of having a child with CF and could choose alternative methods such as adoption, artificial insemination, or *in vitro* fertilization and embryo transfer. Any gametes used in these methods could be screened for the CF gene. DNA analysis also permits the detection of the disorder in developing fetuses (homozygotes). The parents of a fetus diagnosed with CF could choose to abort the pregnancy or to carry it to term.

While the ability to screen a population for a variety of genetic disorders allows for early detection and prevention, it also raises many difficult questions about ethics and public policy. According to Dorothy Nelkin and Laurence Tancredi, sociologists who study the social impact of medicine, DNA analysis in genetic screening "has shifted the focus of the health-care system from looking for *actual* disease in individuals to looking for *tendencies to develop disease*." As a result, there is a danger that people who have no symptoms of disease may be labeled as handicapped and may be discriminated against by health or life insurance companies. Potential employers, for example, may refuse to hire such individuals because they do not want to risk increased costs for medical benefits and do not want to spend time and money training an employee whose life expectancy might be limited.

Neil A. Holtzman, a pediatrician/geneticist at Johns Hopkins, claims the problems raised by DNA analysis are intensified because the tests may not always predict with certainty whether a person will develop a particular disorder. This is especially true for multifactorial disorders such as heart disease, cancer, and schizophrenia, which result from a complex interaction of genes and environmental conditions. Furthermore, says Holtzman, there are too few laboratories that can perform DNA analysis with precision and not enough trained genetic counselors to interpret the results.

Holtzman, Nelkin, and Tancredi recommend the development of new regulations to control how genetic screening is done and what is done with the results. Without regulation and greater public education about genetics, there is danger that some individuals may lose some control of their lives. Carried to extremes, some people might be forbidden to marry, to have children, or to do certain kinds of work.

The ethical, legal, and policy issues raised by DNA analysis and genetic screening must be weighed against the potential benefits to society. Such value conflicts often accompany advances in science and technology and affect individuals, families, and society at large.

Method of identifying the CF gene.

chromosome pair. In meiosis, these paired alleles segregate into different gametes. Random fertilization brings alleles together in ratios that can be predicted by the rules of probability. Mendel used these rules to describe the behavior of alleles in gamete formation and fertilization. The principles of dominance, segregation, and independent assortment are the basis of Mendelian genetics.

Other patterns of inheritance are variations of Mendelian genetics. Alleles for some traits show codominance, and the inheritance of other traits, such as the ABO blood types, is controlled by multiple alleles. Both of these hereditary patterns result in several possible phenotypes. Multifactorial inheritance, which involves several genes interacting with environmental variables, results in continuous variability. X-linked traits and differences in the number or structure of chromosomes also produce unusual patterns of inheritance. The relatively mild effects of sex chromosome trisomies are explained by the hypothesis that all X chromosomes except for one are inactivated early in embryonic development.

Genes on the same chromosome are linked. The frequency of crossing-over between linked genes is used to develop chromosome maps, which are helpful in understanding and screening for genetic disorders. The human genome project will increase our knowledge of genetic disorders and provide information about human variability and the function of introns and other DNA sequences that do not code for proteins.

Using recombinant DNA techniques, biologists can splice DNA from one species into the DNA from another species. Although the benefits from genetic engineering technology already are great and will increase, society is faced with many ethical and legal problems raised by applications of this technology.

Applications

1. Cystic fibrosis is a common genetic disorder among Caucasians. The disorder is recessive; the gene is not X-linked. A young woman has a brother who has cystic fibrosis. The woman visits a genetic counselor to find out whether she carries the allele for cystic fibrosis. Before performing any tests, what can the counselor tell her about the probability that she carries the allele? (Hint: What are the genotypes of the woman's parents?)

2. Suppose you were flipping a coin, and the first nine times it came up "heads." What is the probability that it will be "heads" on the next flip?

3. For each characteristic you have, there is at least one special enzyme that controls the chemical reaction responsible for that characteristic. This is true for all living organisms, which means there must be millions of different types of enzymes. How is it possible that

only four types of nucleotides can produce the codes for all these different enzymes?

4. Suppose one of your chromosomes was damaged so that just one nucleotide of one gene was missing. Explain how the mRNA formed along this gene and the protein formed along the mRNA would be affected.

5. Most human traits are multifactorial. How does the distribution of such traits differ from either/or traits?

6. Barton Childs, a medical geneticist at Johns Hopkins School of Medicine, has said the following about the relationship between genes and the environment: "Genes propose; environments dispose." What does this statement mean?

7. How do Mendel's various principles explain his results?

8. How does the behavior of chromosomes during the first meiotic division explain the ratios Mendel observed?

9. Most flowering plants have both male and female reproductive structures on the same individual. Would you expect these plants to have sex chromosomes? Explain.

10. In a dihybrid cross, how will the type of F_2 offspring differ if the two genes are on the same chromosome (linked together) rather than on separate chromosomes?

11. Baldness is an X-linked trait. If you are a male, can you tell by looking at your father whether or not you will become bald?

12. Genetic engineering involves the ability to transfer genetic material from one organism to another, even from one species to another. What does this tell you about the importance of DNA in the biosphere? What does it tell you about the continuity of life in the biosphere?

Problems

1. Investigate two discontinuous human traits. Determine what is known about the genetics of those traits. Indicate how the environment influences the expression of those traits.

2. Some viruses contain RNA but no DNA. Find out how these viruses take control of human cells and reproduce other viruses.

3. "Wisconsin fast plants" are a species of mustard plant that grows from seed and that flowers within two weeks. Obtain some seeds and conduct crossing experiments with some of the many varieties.

4. Explore the relationship between advancing maternal age and the occurrence of chromosomal disorders such as Down syndrome.

5. There are a great many opportunities for employment in the field of genetics. Investigate the educational requirements for a particular job, such as a genetic counselor. Find out about the day-to-day tasks of a person employed in such a position.

6. Build a three-dimensional model of DNA.

7. Debate the merits and possible negative aspects of genetic engineering.

8. Many human genes have been mapped to specific locations on individual chromosomes. Using a genetics textbook, investigate a current human gene map and locate the genes for the following traits: (a) Rh blood group; (b) ABO blood group; (c) hemophilia; (d) major histocompatibility complex (MHC) (these genes control the acceptance or rejection of tissues received in transplant operations); and (e) Huntington disease.

9. Human DNA can be inserted into a bacterium, and the bacterium will express the human genes. Because eukaryotic genes have introns and prokaryotic genes do not, bacteria do not have the enzymes necessary to cut the introns out of human DNA. Investigate how genetic engineers get around this problem. (Hint: The process involves an enzyme called reverse transcriptase.)

10. It is now possible to determine whether an individual carries the gene that causes Huntington disease. Find out about the disease and the method of detection. If you were at risk to develop the disease, would you want to know if you carried the fatal gene? What are the pros and cons about knowing?

a

b

The organism in (a) died more than 150 million years ago, and its remains were preserved in rock. What does it look like to you? What familiar group of modern organisms does it most closely resemble? What does it have in common with these modern organisms? How does it differ? How would you account for these differences?

The organism shown in (b) is a modern bird, the anhinga (*Anhinga anhinga*). How do you know it is a bird? How does it resemble the organism in (a)? How does it differ? How are these two organisms, one extinct and the other very much alive, examples of *unity* of pattern and *diversity* of type? Discuss your responses with a classmate.

Continuity Through Evolution

9

All things are affected by time. Organisms are born, grow old, and die over periods of time. Plants sprout, grow, and bloom in some way before they die back with the changing seasons. Even the transmission of genetic material happens through time.

For an individual organism, 100 years is a long time, but some changes require longer periods of time than that. Each species alive today is only the tip of a branching tree that extends far back in time to an ancient earth inhabited by gradually changing populations of organisms. How did today's species come to be? Evolution—change through time—is the biological process that links all species, no matter how they differ.

Evolution is the source of all biological diversity and the one process that explains the unity in life. This chapter examines the evidence for ways in which evolution occurs.

Diversity, Variation, and Evolution

Guidepost: What does diversity of type and unity of pattern mean?

INVESTIGATION 9.1   Variation in Size of Organisms

Within a population of organisms, small or large variations make each individual different from the others. Variations among familiar organisms are fairly obvious. For example, you are very aware of differences among humans, but all bluebirds may appear alike to you. In this investigation, you will study individual variations in size among members of a sample population, draw graphs of the data, and interpret them.

Materials (per team of 3)
metric ruler
graph paper
50 objects of one type—dried bean
   seeds, peanuts in the shell, leaves
   from the same type of tree

PART A   Line Graph
Procedure
1. Choose one object and measure its length in millimeters. How "typical" do you think this object is in terms of its length? Guess how many millimeters larger the longest object will be. Then, guess how much smaller the smallest object will be. Record the object's length and your two guesses in your data book.

2. Choose two team members to measure the objects. The third member will record the data. Using a metric ruler, measure the length of the other 49 objects to the nearest millimeter. (If an object measures 11.5 mm, for example, round up to 12.0 mm.) Record the measurements in your data book with your first measurement.

3. Construct a grid. Label the x (horizontal) axis *Length*. The size of your largest object will determine the length of this axis. Divide the axis into 10 or 20 equal parts and label each mark (*10 mm, 20 mm*, etc.). Label the y (vertical) axis *Number of Objects*. Divide this axis into 20 equal parts and label each mark (*1, 2, 3*, etc.).

4. Draw a line graph on the grid. Group the objects according to their length. Then, count the number of objects in each group and mark this number on the graph. For example, suppose there were 3 objects that were 20 mm long. Find the 20 mm mark on the x axis and trace an imaginary line up from this point. Trace an imaginary line from the third mark along your y axis across the graph. At the point these lines meet, draw a large dot. Continue to construct the entire graph.

5. Connect all of the dots with one continuous line.

Discussion

1. Is there variation within your population? What is the shape of your line? Does it resemble those of your classmates?

2. What would be the shape of the line if you pooled all of the data in the class?

3. How close were your guesses about the largest and smallest objects? Do you think that using a small sample is a good way to predict the characteristics of a large population? Explain.

PART B   Frequency Distributions and Histogram
Procedure

6. Using your data from Part A, construct a frequency distribution table as follows:

   a. Determine the range of length in the sample by finding the difference between the largest and smallest objects.

   b. Divide the range into 8 to 10 intervals, selecting a convenient size. For example, if the range of the sample is 23 (the smallest, 16 mm; the largest, 39 mm), it could be divided into 8 intervals of 3 mm each, as follows: 16–18, 19–21, 22–24, 25–27, 28–30, 31–33, 34–36, and 37–39.

   c. Assemble your data in a frequency distribution table with the intervals listed in the X column and the number of individuals falling into each interval (frequency) listed in the F column, as shown in Table 9.1.

7. Construct a histogram similar to Figure 9.1 from your frequency distribution table. A histogram is simply a bar graph with the intervals on the horizontal axis and the frequency on the vertical axis.

TABLE 9.1

x	f
16–18	2
19–21	7
22–24	4
25–27	9
28–30	14
etc.	etc.

Figure 9.1   A histogram.

8. Calculate the mean, or average, of your data. The mean is the sum of all measurements divided by the number of individuals you measured.

9. What is the mode, or high point, on the histogram? The value of the mode usually is given as the midpoint of the interval having the highest bar. If the high point falls in the 16 to 18 range for example, choose 17 as the mode. In Figure 9.1, the mode is 29.

10. Find the median, which is the value for the middle of the sample, when the values (lengths) are arranged in order. If a series of measurements is 2, 2, 3, 4, 7, 8, 9, 9, and 11, the median would be 7.

Discussion

1. Look at the data and your histogram. What is the difference in length between the longest and the shortest objects in your sample?

2. Given the overall size of the objects, do you think this difference is important? What might be the advantage or disadvantage of being smaller or larger than average?

3. Would you have noticed the differences if you had not measured the objects?

4. Do you think there would be any size difference if the objects were of different ages or from different plants?

5. When do you think a histogram is a better, or more accurate, way to present data than a line graph?

9.1   Living Organisms Are Both Similar and Varied

The same types of cells and molecules can form different organisms. A daffodil is obviously different from an oak tree, and a horse does not look much like a pig. Even though these organisms look quite different, however, they share many similarities. Every type of organism also must have some way of obtaining energy, reproducing, and exchanging substances with the environment. Because there are a limited number of ways in which these processes can be accomplished, organisms that appear quite different may have similar ways of accomplishing the processes.

Biologists have observed a unity of pattern in the structures and functions different organisms have in common. The most striking example of this unity of pattern is the genetic code (the triplet sequence of bases in DNA that specifies which amino acids are incorporated into a protein) discussed in Chapter 8. There apparently is only one basic genetic code for all organisms, including humans. The cell theory discussed in Chapter 5 describes a unity of pattern at the cellular level.

This unity of pattern also can be seen in the structure of various organisms. Vertebrates, animals having backbones (a column of vertebrae), provide a good example of structural unity, as you can see in Figure 9.2 on the following page. Even the most diverse types of vertebrates, such as whales and bats, have related limb patterns. Organisms also may share many functional features, such as ways of using food and eliminating wastes. Because of these similarities, the results of research on one group of organisms often are applicable to other groups, including humans. The nerve cells in a squid, for example, function much like the nerve cells in a human. Therefore, scientists can study squid nerve cells to gain an understanding of how human nerve cells function.

a

Figure 9.2 (a) Cross sections of the leaves of a fern, *Cyrtomium balcatum*, (×10), Douglas fir, *Pseudotsuga taxifolia* (×20), and a corn plant, *Zea mays* (×35). Notice the similarities in tissue structure. (b) Bones of the forelimbs of seven vertebrates. Use the color key to find comparable structures.

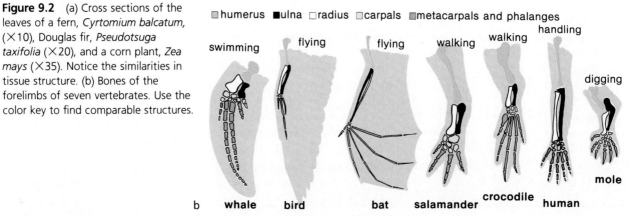

□humerus ■ulna □radius □carpals ■metacarpals and phalanges

b

Reproductive and developmental processes also have a unity of pattern. The life cycles of all animals contain the same sequence of events—meiosis produces gametes, fertilization forms zygotes, and mitosis eventually produces adults. As Figure 9.3 shows, for several vertebrates, the early stages of embryologic development are remarkably alike. These similarities do not mean that a human passes through fish, amphibian, or reptile stages

Figure 9.3 Embryologic development of some vertebrates. Zygotes are shown on the left, adults on the right, and comparable developmental stages are shown in between. Drawings are not to the same scale.

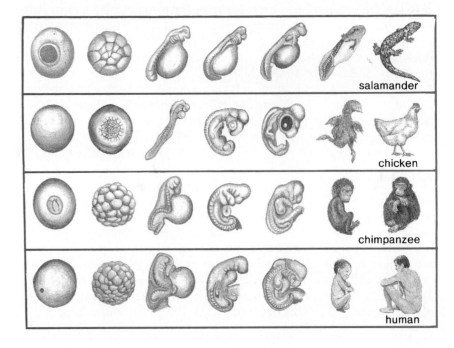

during development. Rather, the similarities show that the same fundamental processes occur in the development of many different structures found in vertebrates.

Even with these similarities, however, there is amazing diversity in the types of organisms on the earth. Humans, squirrels, cows, frogs, spiders, trees, orchids, mushrooms, and bacteria are only a few of all the different forms of life. There are millions of different organisms, each with unique structures, behaviors, and ways of life. How can there be both great unity and great diversity in life?

9.2 All Organisms Can Be Grouped into Species

Although every organism is unique, some have more characteristics in common than others have. Organisms that are so closely related they can mate and produce fertile offspring are grouped in the same species.

The distinctness of a species is maintained in three basic ways. First, behavioral, geographical, or physical differences may prevent one species from mating with another. The inability of one group to interbreed successfully with any other groups is known as **reproductive isolation**. Clearly, an elephant and a crab cannot mate and produce offspring. Alaskan brown bears and polar bears, however, have mated successfully in zoos, but no such cross has been discovered in the wild. Because brown bears live in forests and polar bears live on snowfields and ice floes, they rarely, if ever, encounter one another in nature. They have become separate species.

Second, mating may result in offspring, but the offspring may not live. If the eggs of a bullfrog are fertilized by a leopard frog, they develop for a short time, but soon die.

Third, organisms of two species may mate and produce live offspring, but the offspring are sterile, unable to reproduce. Mules, the offspring of a female horse and a male donkey, almost always are sterile (see Figure 9.4). Exceptions, such as the Indian paintbrush shown in Figure 9.5, often occur in plant species, which can form fertile hybrids with other species. Generally, however, the term *species* refers to a group of actually or potentially interbreeding populations that are reproductively isolated from other groups.

Figure 9.4 Krause (center), a mule, with her two colts. Technically a mule is a cross between a male donkey and a female horse; normally mules are sterile. Sometimes strange things happen, however. In 1984, Krause gave birth to a colt. Geneticists determined the number of chromosomes in both Krause and the new colt. The tests showed that Krause and her offspring were mules. Horses have 64 chromosomes, donkeys 62, and mules 63. In 1987, Krause gave birth to another colt. Examples such as these illustrate the difficulty in defining a species.

Figure 9.5 Indian paintbrush: *Castilleja rhexifolia* (a) and *C. sulphurea* (b), two different species, can form a fertile hybrid (c).

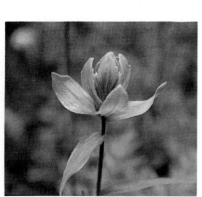

a   b   c

Figure 9.6 Five breeds of dogs. Beginning with the Great Dane, breeds are represented by every other dog. Mongrel offspring are intermediate between the breeds. What types of variations can you see?

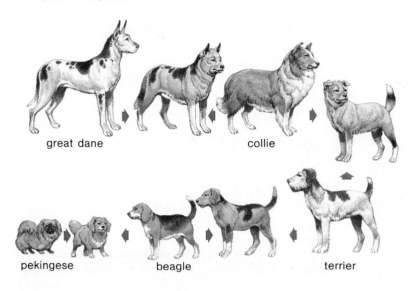

great dane   collie

pekingese   beagle   terrier

Sometimes, organisms in the same species may not look like one another. For example, not all dogs look alike. There are many intermediate breeds between a Great Dane and a beagle (see Figure 9.6). Middle-sized dogs can interbreed easily, but extremely different dogs, such as a Great Dane and a Pekingese, may have difficulty interbreeding because of physical differences. They are grouped within the same species because they can mate with intermediate breeds.

Differences among individuals of a species are called **variations**. Sometimes, as Investigation 9.1 demonstrated, members of the same species may look very similar but are quite different. The grass blades in a lawn, for example, may look identical. Closer examination, however, reveals that some blades are long, others are short, some broad, others thin. Variations occur among individuals of all sexually reproducing populations. Variations may be small or major. Humans, for example, do not all look alike. Many populations vary from one end of their geographic range to the other. For example, the flickers shown in Figure 9.7 belong to the same species even

Figure 9.7 The red-shafted flicker and the yellow-shafted flicker have large geographic ranges. They once were classified as separate species. Because they can and do interbreed, forming fertile hybrids, they now are thought of as one species.

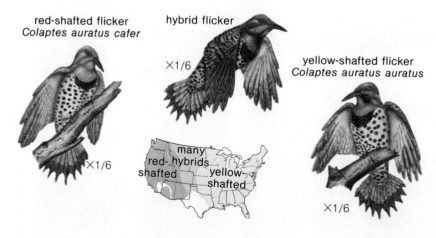

red-shafted flicker
Colaptes auratus cafer

hybrid flicker

×1/6

yellow-shafted flicker
Colaptes auratus auratus

×1/6

many
red-hybrids
shafted   yellow-
shafted

×1/6

Figure 9.8 Snow geese come in two color forms: all white and mainly dark. Colonies often contain birds of both color forms, but geese usually pair with others of the same color.

though the populations on the West Coast are red-feathered and those on the East Coast are yellow-feathered. The hybrid individuals are orange-feathered. Some populations consist of two or more distinct varieties of individuals, such as the snow geese and blue geese shown in Figure 9.8.

9.3 Darwin Observed Variation Among Organisms

How does such diversity and variation come about? In the nineteenth century, Charles Darwin offered one answer to this question. In 1831, at the age of 22, Darwin signed on as a naturalist with the H.M.S. *Beagle* for a five-year voyage around the world (see Figure 9.9). During this trip, Darwin

Figure 9.9 Young Charles Darwin (left) sailed around the world on the *Beagle*. The route taken (right) included stops in South America and the Galápagos Islands. During the trip, Darwin found evidence that would later support his theory of evolution by natural selection.

Figure 9.10   The Galápagos finches were able to adapt to many different food sources because few species competed with them. Note the variations in the sizes and shapes of the beaks. The six species on the left are tree species feeding on various insects or on fruit and buds. The six dark species on the right feed on seeds of different sizes or on cacti. The finch at the top right uses cactus spines as tools to extract insects from under the bark of trees.

studied the natural history and biology of South America and various islands in the Pacific.

The Galápagos (guh LAH puh gohs) Islands, off the coast of Ecuador, particularly interested Darwin. As he explored these islands, Darwin encountered a fascinating assortment of plants and animals. He collected many biological specimens, including marine and land iguanas and various types of birds.

One day, an island resident told Darwin that he could tell from which island any of the tortoises came by the size and shape of their shells and the length of their legs and necks. Although this observation meant little to Darwin at the time, years later it led him to wonder whether each of the islands was somehow producing its own forms of these tortoises and other creatures.

When Darwin finally returned to England, he continued studying the specimens he had collected during the long voyage. His collection of Galápagos finches was examined by specialists who determined that the specimens represented 13 species, differing primarily in the size and shape of their beaks (see Figure 9.10). Darwin surmised that these finches must originally have come from the South American mainland, but why, he wondered, were these island birds so different from those finches to be found on the mainland? Also, why did the assortment of finches differ so much from one island to the next?

During his studies, Darwin came across an essay by Reverend Thomas Malthus (1766–1834) that warned about the dangers of human overpopulation. Malthus pointed out that populations tend to increase and that if humans continued to reproduce at the same current rate, they eventually would outstrip their food supply. Darwin applied this idea to his own work and concluded that most species have a high reproductive potential but not all individuals reproduce. Populations, he argued, are kept in check in part because some organisms fail to reproduce. If this was true, then what was it that affected reproductive output, and did it affect all individuals in the same way?

Concept Review

1. What does "unity of pattern" mean in evolution?
2. What is a species?
3. How do species maintain their distinctness?
4. How was Darwin influenced by the variety of organisms he observed on the Galápagos Islands?
5. Distinguish between variation and diversity.

Evolution and Natural Selection

9.4 Darwin Identified Selection as a Force in Evolution

Darwin knew that breeders could accentuate desirable characteristics by carefully selecting animals for mating. For example, by mating only the offspring of the greatest milk producers, breeders could develop high-yield dairy cattle. The roosters in Figure 9.11 also are the results of choosing animals for breeding. This process is known as **artificial selection**.

Guidepost: How does evolutionary change occur?

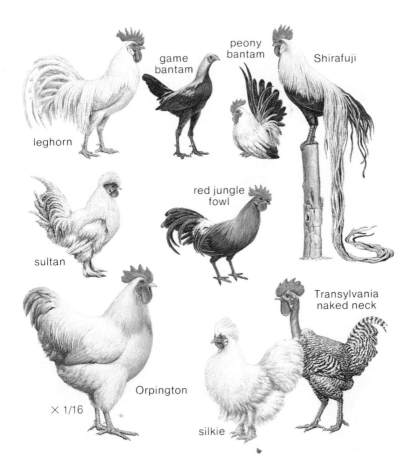

Figure 9.11 An example of artificial selection. The red jungle fowl, a bird of Southeast Asia, is thought to be the species from which the many breeds of domestic chickens have been developed. Which of the breeds shown retains the largest number of the wild bird's traits?

game bantam

peony bantam

Shirafuji

leghorn

sultan

red jungle fowl

Transylvania naked neck

× 1/16

Orpington

silkie

Darwin thought that some process similar to artificial selection might affect organisms in their natural environment. He suggested that in nature, the animals that reproduced were somehow *selected* by the environment. Individuals with characteristics that helped them survive in their environment could produce more offspring than individuals that did not have such characteristics. He called this process **natural selection**. Scientists now define natural selection as the process by which those characteristics that permit survival and reproduction are continued and eventually replace less advantageous characteristics. The characteristics that enable some members of a species or population to survive and reproduce are called **adaptations**. Adaptations are part of the variations in the group.

If natural selection acts on variations among members of a population, what accounts for these variations? Darwin proposed that variations appear randomly and without design. If a new variation allows its bearer to produce more offspring, it will spread through future generations.

Darwin was not the only nineteenth-century scientist trying to explain how species change and variations are passed on. In France, Jean Baptiste Lamarck suggested that one species could give rise to another. He could not explain changes, but he surmised that some force causes an organism to generate new structures or organs to meet its biological needs. Once formed, he argued, these structures continue to develop through use, and the development attained by the parents is inherited by the offspring.

Lamarck's idea of how giraffes developed their long necks is the most famous example of his hypothesis. He maintained that as giraffes stretched their necks in an effort to reach the topmost branches of trees, the effort somehow altered the animals' hereditary materials, and the longer neck was passed on to the offspring. Today we know that the characteristics an organism develops during its life, called acquired characteristics, are not passed on to offspring. Consider, for example, the Doberman pinschers in Figure 9.12. If characteristics acquired by the parents during their lives, such as short ears and bobbed tails, are passed on to their offspring, why do young Dobermans have floppy ears and long tails?

Figure 9.12   Do these photos of an adult Doberman and its offspring support Lamarck's idea that acquired characteristics are inherited?

In contrast to Lamarck, Darwin suggested that in any population of giraffes, some animals would have longer necks than others. These giraffes would be able to reach more leaves and would be better fed. These better-fed animals would be more likely to bear young and rear them to reproductive age. Thus, because of natural selection, longer necks would become common in future generations. These ideas of Darwin's came to be known as the theory of evolution by means of natural selection. Investigation 9.2 illustrates how natural selection operates in predator-prey relationships.

INVESTIGATION 9.2   Natural Selection—A Simulation

Biologists consider natural selection to be the chief mechanism of evolutionary change and the process responsible for the diversity of life on earth. This investigation illustrates one way in which natural selection operates.

Materials (per team of 6)

2 1-m × 2-m pieces of fabric with different colors and patterns

100 paper chips of assorted colors in plastic bag

small bowl

1 set of colored pencils with colors similar to chip colors

5 sheets of graph paper

PART A   Natural Selection
Procedure

1. Spread out the fabric "habitat" on a table top. Examine the paper chips and record the colors of the chips.
2. Appoint one team member as the keeper of the bag of paper chips. All other members are predators, whose prey are chips. The keeper will keep track of the number of turns each predator takes and the number of prey remaining.
3. Turn your backs to the table and allow the keeper to spread the chips uniformly over the fabric, making sure no chips stick together.
4. Imagine yourselves as predators, the paper chips as your prey, and the fabric background as your habitat. One at a time, turn around and select a paper chip using only your eyes to locate it. Do not use your hands to feel the chips. When you have selected a chip, place it in the bowl and turn around. Continue taking turns until only 25 paper chips remain on the fabric and the keeper signals you to stop.
5. Carefully shake the fabric to remove the survivors.
6. Group the survivors according to color by placing chips of the same color together. Arrange them in a row. Record the number of each color that survived.
7. Assume each survivor produces three offspring. Using your teacher's reserve supply, place three chips underneath each survivor.
8. Mix the survivors and their offspring thoroughly and distribute them as in Step 3. (Do not use the chips that were "eaten.")
9. Repeat the entire process of selection (Steps 3–7) four more times. Be sure to record the total number of chips of each color at the start of each round.

Discussion

1. Prepare a histogram for each generation (five histograms) using colored pencils that match as closely as possible the chip colors. Study the histograms of each generation. Was one color chip represented more than others in the first generation of survivors? Why do you think these chips "survived"? What, if any, change occurred between the first and second generation? The first and fifth generation?

2. Compare the original and the last survivor population. Which, if any, color from the original population is not represented in the survivor population? Why?

3. Examine the colors of chips in the fifth generation and the fabric habitat. How do the colors of the survivors relate to the colors of the habitat?

4. How are your results related to the process of natural selection?

5. Assuming no new individuals migrate into the habitat, how will the population change with time? How are your results related to evolution?

PART B   Variation and Changes in the Environment
Procedure

10. Using a piece of fabric (habitat) with colors different from the first, begin a new series of predator-prey interactions with the 100 paper chips used in Part A.

11. Repeat Steps 3–8 five times.

Discussion

1. An organism is adapted to its environment if it can grow and reproduce in it. In Part A, which color of chip was "best adapted" to the habitat? Which color of chip was "least adapted"?

2. In Part B, which color of chip was "best adapted" to the new habitat? Which color was "least adapted"? How do these compare to the chips from Part A?

3. Changing colors of habitat is similar to a change in the environment. Name as many types of changes in the environment as you can.

4. How are species with genetic variation in their offspring able to survive in environments that change through time?

5. What do you think would happen to a population if it produced only offspring that were all alike?

For Further Investigation

If new individuals of different colors immigrate into your population, what will be the effect on the population, assuming that habitat and predators remain the same?

9.5   Darwin's Theory Changed Biology

Darwin developed his theory during a time when the prevailing scientific opinions were quite different. He lacked conclusive proof for his ideas. Although the field of genetics might have offered support, little was known about it then. Not having definite proof and expecting a storm of controversy, Darwin was reluctant to present his ideas. He did discuss them with biologists who also were close friends, however, and they urged him to publish his theory.

In 1858, Darwin received an unfinished paper from Alfred Russel Wallace, a young man who was working in Indonesia. Wallace's paper was, in essence, a sketchy outline of the principles of natural selection. Darwin had planned to set forth his findings at a scientific meeting, but now he insisted that Wallace disclose his ideas and share the credit. Both papers were read at the meeting, but Darwin's was presented first because he had extensive supporting evidence from his travels and observations. He subsequently published his ideas in a book entitled *On the Origin of Species*, in 1859.

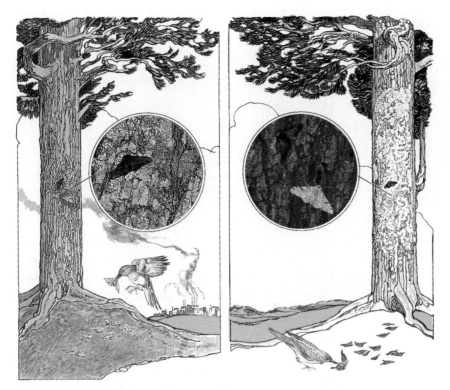

Figure 9.13 The basis for selection in the peppered moth (*Biston betularia*). Dark and light forms on a tree with light-colored lichens (a). The two forms of the moth on a tree blackened with soot (b). Which moth in each photo is most likely to be eaten by predators?

As Darwin had feared, his theory of evolution by natural selection met with protest and opposition. His data came from his years of collecting on the *Beagle* journey, and his ideas were developed carefully. Nevertheless, the theory required the support and defense of some the best scientists of his time.

Theories usually are modified as our knowledge increases. This process of continuous, minor modification is an example of how science works. Occasionally, as in the case of Darwin's theory, a major theory is proposed that alters the way scientists look at their data. After that, the process of modification resumes in the light of the new theory.

Darwin's theory of natural selection has been supported by the findings of thousands of other biologists. A few "natural experiments" even have demonstrated evolution occurring. One example concerns the peppered moths found in the British countryside. In the 1700s, many trees in Britain were covered with pale gray-green lichens. Light-colored moths, like those shown in Figure 9.13, could be found on these lichens. Darker moths were rare, and they usually were eaten by birds because they were so visible against the light-colored trees. Later in the century, as the Industrial Revolution progressed, coal-burning factories spewed out clouds of black smoke. As soot darkened the tree bark and covered the lichens, the light-colored moths became easier to see and darker moths became less visible.

Figure 9.14 Natural selection explains the increase of penicillin-resistant bacteria.

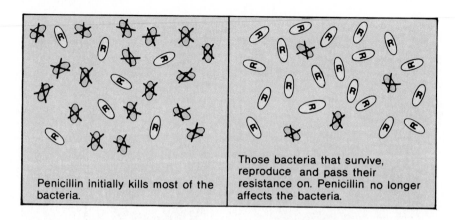

Penicillin initially kills most of the bacteria.

Those bacteria that survive, reproduce and pass their resistance on. Penicillin no longer affects the bacteria.

Birds began to eat the conspicuous light-colored moths, and their numbers dwindled. By the late 1840s about 98 percent of the moths were dark. Today, as pollution controls curb the production of soot, more light-colored moths can be found. This change in color is the result of a change in the frequency of alleles for color in moths.

Other natural experiments have shown the same process of natural selection in bacteria (see Figure 9.14). About 30 years ago, penicillin killed most types of bacteria, but now many strains are unaffected by the antibiotic. This has come about because a few bacteria always had a genetic resistance to the penicillin. As nonresistant individuals were killed, the penicillin-resistant strains were left to reproduce. Therefore, there were a greater number of resistant individuals in the next generations. To avoid having this happen again, antibiotics are used more carefully today.

Extensive field work and experimentation continue to support Darwin's theory of natural selection as one evolutionary force. Mechanisms other than natural selection also have been proposed and have supporting evidence. Thus, now the question is not *whether* evolution occurs, but *how* it occurs, and how rapidly it occurs.

Concept Review

1. Compare and contrast artificial selection with natural selection.
2. What are adaptations and how are they important to the survival of an individual and a population or species?
3. Explain the difference between Lamarck's view of evolution and Darwin's view.
4. What is meant by evolution through natural selection?

Guidepost: How do genetic mechanisms influence evolution?

Evolution and Genetics

9.6 Evidence from Genetics Supports Natural Selection

In this century, although they are the framework for much of today's research, Darwin's original ideas have been modified and expanded greatly. Evolution may be defined as change in a species or population through time, but what type of change? Mendel's discoveries, which led to the chromosome theory of heredity, showed the way in which variations could be inherited and maintained in a population. Darwin clearly understood that only hereditary variations could have any meaning in the evolutionary process. Variations that are not passed on do not lead to changes in a species or a population. Hereditary variations, however, are products of an organism's genes. There are at least two major sources of the variability required for evolution. One of these sources is mutations and another is genetic recombinations.

Mutations, which are discussed in Chapter 8, are changes in DNA molecules. The rate at which mutations occur is not known precisely. Some scientists maintain that the average mutation rate in humans is about one mutation per 100 000 gametes per generation. Others contend that the figure is lower—about one mutation per one million gametes per generation. The mutation rate is difficult to determine because many mutations cannot be detected. A mutation in a gene for the recessive form of a given trait, for example, will not be exhibited in a heterozygous individual. Furthermore, there may be many mutations in introns—the pieces of DNA that are not translated into protein. At this time, in fact, biologists do not know if mutations in introns have any evolutionary significance. As DNA analysis improves, the ability to detect mutations also will improve, and a more accurate estimate of mutation rates will be possible. Whatever the actual mutation rate may be, however, it has operated in large populations through long periods of time, producing enough variation for the process of natural selection to operate.

Genetic recombinations that result from meiosis and sexual reproduction increase the genetic variability in a given population by producing different combinations of genes (see Figure 9.15). A human, for example,

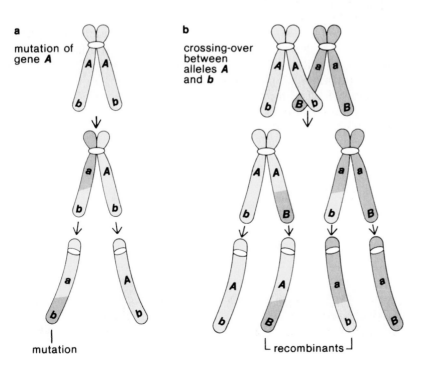

a

mutation of gene **A**

b

crossing-over between alleles **A** and **b**

mutation

recombinants

Figure 9.15 Mutation and crossing-over are the major sources of genetic variation. Mutations (a) give rise to new alleles, such as when *A* mutates to *a*. Before mutation, only the combination *Ab* was possible. Crossing-over (b) provides new combinations. Before meiosis, only the combinations *Ab* and *aB* occur. Crossing-over between *A* and *b* results in the recombinants *AB* and *ab* in the gametes.

possesses 23 pairs of chromosomes. Even if each chromosome pair contained only one pair of heterozygous alleles, independent assortment would produce more than 8 million different sorts of gametes. Thus, more than 60 trillion (6×10^{13}) genetic recombinations are possible in one child. Considering that there are a great many heterozygous pairs of alleles, the actual number of possible recombinations is enormous. This process of genetic recombination contributes to the **gene pool**—the sum of all the alleles of a population.

In any species, the variety of possible hereditary types is almost limitless because of genetic recombination. New mutations continue to occur and form new genotypes that produce new phenotypes. These phenotypes are subjected to natural selection, which eliminates certain gene combinations from the population's gene pool. Natural selection, however, is not the only evolutionary mechanism at work.

9.7   Several Evolutionary Mechanisms Can Affect Populations

In a plant population made up entirely of homozygous red-flowered plants, the gene pool for flower-color alleles consists of only the allele for red flowers—a frequency of 100 percent. In that homozygous population, the frequency of alleles for white flowers is zero. On the other hand, if the population consists entirely of *heterozygous* red-flowered plants, then the allele frequencies are 50 percent red-flower alleles and 50 percent white-flower alleles.

Study the cross shown in Figure 9.16, and determine the allele frequencies for each generation. In general, the frequency of alleles in a population's gene pool remains relatively stable, particularly if the population is large and well adapted to its environment. This type of population is said to be in equilibrium.

A number of factors, however, can change the frequency of alleles in a population. The first factor is mutation. A mutation results in the introduction of a new allele and causes an immediate, but small, shift in equilibrium. Populations of organisms (bacteria, for example) that produce a large number of offspring within a short time are influenced more by mutations than populations of organisms that produce a few offspring over a relatively long time span.

Figure 9.16   In this cross of two heterozygous red-flower plants, what are the allele frequencies in the parental generation? In the offspring generation?

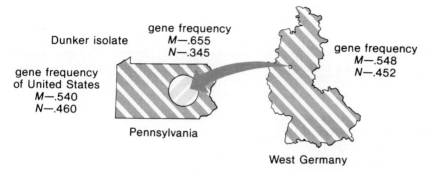

Figure 9.17 Genetic drift in *MN* blood groups. A Dunker group that had a relatively high frequency of the allele *M* immigrated to Pennsylvania. Descendants of these Dunkers continue to exhibit a higher frequency of the alleles for *M* and a lower frequency of the alleles *N* than others in West Germany or the rest of the United States.

The second factor is migration, the movement of organisms into or out of the population and therefore the gene pool. If organisms entering or leaving the population have a genotype different from that of the rest of the population, a change in equilibrium occurs. The movement of genes into or out of a population because of migration is called gene flow.

The third factor is a *random* change in allele frequencies that results in a population with distinct characteristics. This random change, which usually occurs in small populations, is called **genetic drift**. The Dunker population of West Germany (see Figure 9.17) is a good example of genetic drift. The Dunker population arose from a small group of people that had a higher frequency of a particular blood type allele (and a lower frequency of a second allele) than the surrounding population. Succeeding generations retained this frequency differential because of religious beliefs against marriage outside the group. When some of the group emigrated to the United States, the differential frequency increased and then was maintained by the same marital bias. A random change in allele frequencies may be caused by other factors as well. If a population is small, a storm or an attack by predators could kill a large proportion of the population, and the survivors may have a different allele frequency than that found in the original group.

The fourth factor is artificial or natural selection. These types of selection determine which individuals in a population will reproduce and pass on their genes. If there are strong selection pressures on the population, the equilibrium of the gene pool will change rapidly. Weak selection pressures will change the population more slowly.

The fifth factor that can influence the equilibrium of a gene pool is the nonrandom mating of individuals within the population. If certain individuals in a population show a preference for mating with individuals that have a particular phenotype, the mating pattern ceases to be random. Such nonrandom mating is known as **sexual selection** (see Figure 9.8). Sexual selection influences the allele frequencies of a population in future generations. If these allele frequencies lead to reproductive isolation, the population may become a new species. Investigation 9.3 considers some of the factors that can lead to **speciation**, the development of new species.

INVESTIGATION 9.3 A Step in Speciation

The small salamanders of the genus *Ensatina* are strictly terrestrial. They even lay their eggs on land. Nevertheless, these salamanders need a moist environment and do not thrive in arid regions. In California, *Ensatina eschscholtzii* has been studied by R. C. Stebbins at the University of California (Berkeley). This investigation is based on his work.

Materials (per team of 1)
outline map of California
8 different colored pencils

PART A Collection Areas
Procedure
1. Imagine that you are working with Stebbins's salamander specimens, some of which are pictured in Figure 9.18. In the following list, salamanders are identified by subspecies. (A subspecies is a geographically restricted population that differs consistently from other populations of the same species.) The parentheses after each subspecies contain a number and a color. The number is the total of individuals Stebbins had available for his study. The color is for you to use in designating the subspecies. Following this is a list of collection areas. They are indicated by a code that fits the map of California in Figure 9.19. For example, 32/R means that one or more *E. e. croceator* specimens were collected near the intersection of Line 32 and Line R.
   a. *E. e. croceator* (15; brown): 32/R, 32/S, 30/T, 31/T
   b. *E. e. eschscholtzii* (203; red): 30/M, 32/O, 34/S, 35/V, 36/W, 35/Z, 38/Y, 40/Z
   c. *E. e. klauberi* (48; blue): 36/Z, 38/a, 40/a, 39/a
   d. *E. e. oregonensis* (373; purple): 9/B, 7/E, 6/E, 13/C, 10/C, 7/D, 15/D
   e. *E. e. picta* (230; yellow): 2/B, 2/C, 3/C, 4/C
   f. *E. e. platensis* (120; green): 8/J, 10/J, 11/M, 13/M, 15/M, 15/O, 17/M, 15/P, 20/Q, 24/S, 21/R, 25/T, 26/U
   g. *E. e. xanthoptica* (271; orange): 17/G, 17/F, 19/H, 19/O, 20/I, 20/J, 21/I
2. Plot each collection area by making a small X mark on an outline map that has a grid like the one in Figure 9.19. Use the colors indicated for each subspecies population to make a distribution map of *Ensatina eschscholtzii* in California.

Discussion
1. Is the species uniformly distributed? Use your knowledge of the species' ecological requirements to offer an explanation of its distribution. Are there any other factors that might affect distribution?
2. Consider the physiography of California in Figure 9.19. Does the species seem more characteristic of mountain areas or of valley areas?
3. Do you expect any pattern in distribution of subspecies? Why or why not?
4. Examine the salamanders in Figure 9.18. Note that some subspecies have yellow or orange spots and bands on black bodies. Some have fairly plain, brown-orange bodies. One has small orange spots on a black background. There are other differences as well. For example, some of them have white feet. Now refer to your distribution map. Does there appear to be any order to the way these color patterns occur in California? For example, do the spotted forms occur only along the coast? Do spotted forms occur in the north and unspotted ones in the south?
5. Subspecies *E. e. eschscholtzii and E. e. klauberi* are different from each other. What relationship is there between their distributions?

Figure 9.18 Specimens of the salamander *Ensatina eschscholtzii*.

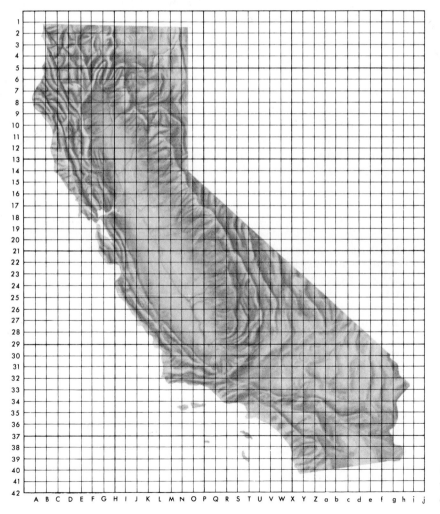

Figure 9.19 Map of California, with the grid to be used in plotting distributional data.

PART B   Additional Collections
Procedure

3. You may wonder if there are salamanders in some areas for which you have no records. You also may wonder if there might be additional subspecies for which you have no specimens. A biologist faced with these questions would leave the laboratory and go into the field to collect more specimens. Imagine that you have done so and returned with the following data:

   b. *E. e. eschscholtzii* (16; red): 36/Z, 41/Z, 33/M, 34/W, 34/U

   c. *E. e. klauberi* (23; blue): 40/b, 40/Z, 36/a

   h. Unidentified population 8 (44; black and green): 4/I, 5/H, 7/H, 7/F, 6/J, 9/F

   i. Unidentified population 9 (13; black and red): 28/T, 27/T, 26/T, 28/S, 29/T

   k. Unidentified population 11 (131; black and blue): 23/J, 24/K, 24/I, 29/M, 25/J, 25/I

   l. Unidentified population 12 (31; black and yellow): 6/C, 7/C, 6/B

4. Mark with a 0 the following places that were searched for *Ensatina* without success: 11/I, 14/I, 17/K, 19/K, 22/N, 26/Q, 5/M, 32/U, 32/A, 35/F. Specimens of populations 8 and 9 are shown in Figure 9.18. There are no illustrations for populations 11 and 12.

Discussion

1. According to Stebbins, the unidentified populations are not additional subspecies. What, then, is the probable genetic relationship of populations 8, 9, and 11 to the subspecies plotted on the map?

2. On this basis, describe (or make a colored drawing of) the appearance you would expect specimens of population 11 to have.

3. Why is it unlikely that you would ever find individuals combining characteristics of *E. e. picta* and *E. e. xanthoptica*?

4. Look at the distribution of the original collections of *E. e. eschscholtzii* and *E. e. klauberi*. What reasons were there for trying to collect additional specimens from extreme southwestern California?

5. How do the results of the additional collections differ from the results in other places where two different populations approach each other?

6. Bear in mind the biological definition of a species and also the appearance and distribution of the named populations of *Ensatina*. Which one of these populations could be considered a species separate from *E. e. eschscholtzii*? (This population was indeed once considered by biologists to be a separate species.)

7. Now imagine that, while examining salamanders in another collection, you find specimen j from population 10 shown in Figure 9.18. Compare its characteristics, especially the spotting pattern, with those of the named populations. Also consider the distribution of these populations. Between which two is this specimen most likely a hybrid? On your map, draw a line along which you might expect to collect other specimens like this one.

8. In a brief paragraph, explain why Stebbins concluded that there is only one species of *Ensatina* in California.

9. Suppose volcanic activity in northern California should become violent and completely destroy all the salamanders in that region. How would this event affect the species *Ensatina*?

For Further Investigation

What accounts for the one record of *E. e. xanthoptica* in the Sierras, with the rest of the subspecies occurring along the coast?

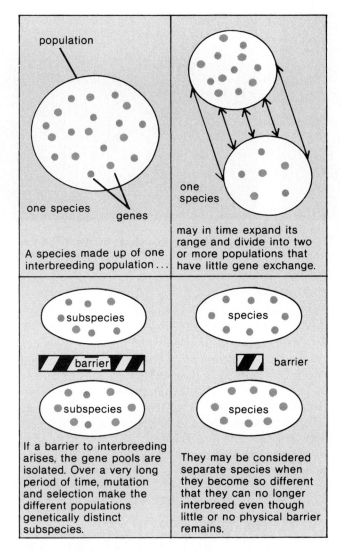

Figure 9.20   A simple model of how two populations of a species may become isolated and evolve to produce two species.

9.8   Isolation Is Needed for Speciation

Reproductive isolation maintains the distinctness of a species. It also is necessary for the development of new species. The evolution of a new species through time, known as speciation, is shown in Figure 9.20. Speciation usually does not occur within a population in which individuals are free to interbreed. If a population separates into two small groups that remain isolated, however, changes between the two groups through time may become so great that interbreeding is impossible even if the groups meet. When the two groups can no longer interbreed, speciation has occurred. For example, the Kaibab and Abert squirrels of the northern and southern rims of the Grand Canyon probably once belonged to the same species but diverged when they became separated by the canyon (see Figure 9.21).

Figure 9.21   The Kaibab (a) and Abert (b) squirrels are related. When they became isolated on opposite rims of the Grand Canyon, they began to develop different characteristics. How many can you see?

The tendency of a population to disperse into new areas, which was discussed in Chapter 2, also may lead to speciation. As the parent population grows, offspring populations move off into new areas that offer different opportunities and different problems. Each population then gradually adapts to its new environment or dies out. In time, these adaptations render the populations so genotypically different that they do not interbreed even when they inhabit the same area. They have become separate species. This process of dispersal, adaptation, and subsequent speciation is called **adaptive radiation** (Figure 9.22). The Galápagos finches are thought to be descendants of a mainland finch that dispersed to the islands and underwent adaptive radiation.

Figure 9.22   Adaptive radiation in mammals. As the first mammals dispersed to new areas, they adapted to the new environmental conditions. Gradually, these adaptations produced different species and, eventually, all the different mammals alive today. Representative animals from the orders included are (1) pangolin, (2) squirrel, (3) bat, (4) lion, (5) buffalo, (6) horse, (7) elephant, (8) armadillo, (9) rabbit, (10) monkey, (11) mole, (12) whale, (13) aardvark, (14) manatee, (15) kangaroo, (16) mastodon, (17) litoptern, (18) creodont, (19) *Brototherium*, (20) *Uintatherium*.

Figure 9.23 Two models for the rate of speciation. The traditional model (a) contends that new species branch off the parent species as small populations. Species descended from a common ancestor become more and more unlike as they acquire new adaptations. According to the model of punctuated equilibria (b), a new species changes the most at the time it branches from a parent species and changes little for the rest of its existence.

The development of species may not always take a long time. Most experts agree that the different organisms represented in the fossil record changed little for millions of years. Occasionally, however, many new organisms, not represented in the earlier fossil record, appear. These periods of long stability, followed by the sudden appearance of new forms, are called **punctuated equilibria**. According to the punctuated equilibria hypothesis, major genetic changes may have caused new species to evolve in relatively few steps (see Figure 9.23). Although biologists debate whether the development of most new species has been relatively quick or a process of gradual change, even the punctuated equilibria model requires thousands of years for the formation of new species.

Debates over the mechanisms of evolution are good examples of the refinement of scientific theories. New information leads to new questions and hypotheses. Each new hypothesis then must be tested to determine its validity.

9.9 Various Evolutionary Patterns Occur

A variety of evolutionary patterns can be found in the living world. Some of these patterns occur between groups of closely related organisms. Divergence occurs when organisms within a species become so different that they can no longer interbreed. Even after divergence occurs, however, closely related groups still can evolve in the same direction. When this happens, it is known as the process of parallel evolution. For example, the African ostrich and the South American rhea, shown in Figure 9.24, probably are descended from a fairly recent common ancestor. Even though the two groups of birds have diverged, they have evolved in a similar manner because they were subjected to the same type of environmental pressures.

Other patterns of evolution occur between more distantly related groups of organisms. In convergence, different groups have developed the same

The Cheetah—On the Way to Extinction?

Plant and animal extinction has many causes. Humans often cause changes in the environment that reduce or eliminate food, living space, and breeding grounds, thus modifying or destroying habitats. In the case of the cheetah, which the Convention of International Trade in Endangered Species has listed as threatened by extinction, other factors may be more important. Estimates of the cheetah population vary from 1000 to 25 000 animals. Recent studies have shown that a "population bottleneck" has limited the population. At some time in the cheetah's recent evolutionary history, an infectious disease or a natural catastrophe must have reduced its population. This in turn caused a reduction in the variety of alleles in its gene pool.

To test the hypothesis that a population bottleneck had reduced the genetic variation in the cheetah and was an important factor in the decline in the cheetah population, scientists used their knowledge of transplant and skin graft rejections in humans. Frequently, an organ transplant or a skin graft is rejected because the donated organ or skin is recognized as foreign by the immune system of the recipient. The greater the genetic variation between donor and recipient the greater the possibility of rejection. In genetically similar individuals, the chances of rejection are greatly reduced.

The researchers exchanged skin grafts between six pairs of unrelated cheetahs. They also grafted patches from one part to another of each cheetah. As a second control, the researchers grafted skin patches from another species, the domestic cat, onto each cheetah; the bottom photo on the right shows one of these grafts after 12 days. The grafts between unrelated cheetahs (on the left) and from one part to another of the same cheetah (in the middle) are healing. There was no rejection of the graft between unrelated cheetahs or between parts of the same cheetah. The graft from the domestic cat, however, shows acute rejection.

The rejection of the grafts from the domestic cat indicated that the cheetah's immune system was responsive to genetically different transplanted tissues. Therefore, if there were a normal variation of alleles in the gene pool, one would expect grafts between unrelated cheetahs to be rejected after 7 to 13 days. The nonrejection of these grafts supported the hypothesis that the population bottleneck reduced the allelic variation in the gene pool of the cheetah.

The cheetah population is an example of a genetically uniform species that could be wiped out by a deadly infectious disease. Should such a disaster happen, evolution of the cheetah species would stop, and it would become one of the many animal species that are known only through books and museums.

Young cheetah (top); skin grafts (bottom).

a

b

Figure 9.24 An ostrich (a) and rhea (b). What similarities and differences can you see?

adaptations because they live in the same type of environment. Compare the spiny, leafless euphorbia (yoo FOHR bee uh) of the African desert with a cactus from the Mexican desert. Their body forms are similar, but their flower structure shows that they are members of two different families (see Figure 9.25).

Figure 9.25 Convergence in plants. Cactus is on the top, euphorbia on the bottom. How are these plants similar? How are they different?

Figure 9.26   A yucca, *Yucca glauca*, and a Pronuba moth carrying a pollen ball on a yucca flower.

In coevolution, two unrelated groups become uniquely adapted to one another. One of the most common examples of coevolution is the adaptation that has developed between certain flowers and their pollinators. One such relationship is that of the yucca plant and the pronuba moth of the southwestern United States, shown in Figure 9.26. When the yucca flowers open, several small pronuba moths fly into the flowers and mate there. Each female moth collects pollen from that flower and rolls it into a ball. She then flies to another yucca flower, bores into its ovary, and deposits her fertilized eggs there. Afterward, she pollinates the flower with the ball of pollen. Later, the moth eggs hatch into larvae that feed on some, but not all, of the yucca seeds that are maturing in the ovary of the yucca blossom.

Both the yucca and the pronuba moth benefit from this relationship, and neither can complete its life cycle without the other. Other insects that are attracted to the yucca blossom may derive benefit for themselves but are unlikely to pollinate the flowers.

Concept Review

1. How does genetics help explain evolution?
2. Explain how different factors can affect a population's gene pool.
3. Explain how reproductive isolation is necessary for the formation of a new species.
4. Compare and contrast the evolutionary processes of convergence and divergence.
5. Explain the difference between parallel evolution and coevolution.

Summary

Although all organisms are similar in some basic ways, they also exhibit a remarkable amount of diversity. A unity in pattern is evident from the genetic code common to all organisms and from similar developmental stages and structures. The process of evolution explains how these similarities and differences arise.

Charles Darwin developed a theory of evolution through natural selection, which explains how adaptations to particular environments allow organisms to survive and produce offspring. Our current knowledge of genetic mechanisms, such as mutation and recombination, explains how these adaptive variations are inherited by some of the offspring.

Populations are the units of evolution. Populations are affected by mutation and recombination, migration, genetic drift, natural and artificial selection, and nonrandom mating. When populations of a species become isolated from each other, changes can accumulate until they become separate species.

Although biologists do not completely agree about the mechanisms of evolution, the overwhelming majority agree that diversity of type and unity of pattern are best explained by evolution. Evolution pervades all aspects of biology and provides a foundation for understanding many observations.

Applications

1. How do you explain that the variability in a domesticated species is greater than in the same species, or a similar one, in the wild? (For example: dogs vs. wolves; chickens vs. red jungle fowl.)

2. Polydactyly (more than the normal number of fingers or toes) is caused by several different genes. One type of polydactyly is a dominant form of the trait, yet the phenotype is rare in humans. Type O blood is the recessive form of the blood type trait, yet in some populations of North American Indians, as many as 97 percent of the individuals may have type O blood. Explain these two situations.

3. In many species of birds, populations that live at higher altitudes lay more eggs per clutch than do those that live at lower altitudes. Would you expect the former gradually to replace the latter? Why or why not?

4. How does Mendel's work, of which Darwin was unaware, provide support for Darwin's theory of evolution by natural selection?

5. What effects may modern medicine have on the future evolution of humans? Distinguish carefully between the data and your interpretations of them.

6. Suppose there are two different populations of flowering plants living in fields next to one another. The plants of the two populations look different, but where the populations meet they form fertile hybrid offspring. Do you think these different populations belong to the same species?

7. Genetic variation is important for the long-term survival of a population, but how might it be considered a problem for individuals?

8. What is the difference between natural selection and evolution?

9. If there were no mutation, do you think that evolution still would occur?

10. Is the rate of evolution likely to be higher or lower in a small population compared with a large population? How about the likelihood of extinction?

11. Many people believe, incorrectly, that the phrase "survival of the fittest" means that "only the strong survive." What does "survival of the fittest" really mean?

Problems

1. Suppose no data existed to support the theory of evolution until the twentieth century. Using only data compiled since 1900, how would scientists develop the theory of evolution today? In what ways would it resemble Darwin's presentation? How would it differ?

2. Suppose species of organisms were not related to one another and did not change through time. How would our view of biology change? Would it be easier or harder to study?

3. Using an imaginary population of organisms, detail how it could evolve into two distinct species. Describe the environment of the population and the factors that led to its separation into two species.

4. Endemic species are those that are restricted to certain areas—not widely found. Identify some of the endemic plant and animal species of your community or state and try to learn what might have allowed them to become endemics.

5. What is the evidence for the punctuated equilibria theory of evolution? Present the evidence to your class.

6. Charles Darwin was a remarkably versatile scientist and natural historian. While most associate his name with natural selection and evolution, he was a prolific writer on a number of diverse subjects. Find a book other than *Origin of Species*, read it, and write a report on it.

7. Time is an important factor in the evolution of organisms. Make a time scale and place the major events in the evolution of life on earth on this scale. First, get a very long piece of butcher paper that covers the entire length of the longest wall of your classroom. Divide the scale into five equal parts—each of these parts is equal to 1 billion years of time. Divide each of these sections into 10 equal parts of 100 000 000 years. Now indicate on the time scale major events beginning with the formation of the earth some 4.6 billion years ago and ending with the first fossil evidence of humans.

Section Three

Diversity and Adaptation in The Biosphere

Biologists have described an incredible diversity and number of organisms, and more are discovered every year, primarily in the tropical rain forests. Why are there so many types of living things? In what ways are all living things related? What adaptations have allowed them to survive? Section Three describes some representative forms of organisms and examines how each is uniquely adapted to its environment. It also explains how biologists name organisms and how they keep track of the different forms and bring order to the diversity of life.

A summer field in Maine. How many different species of flowers can you count? Check your estimate with a classmate and try to resolve any differences you may have. What important biological phenomenon is illustrated?

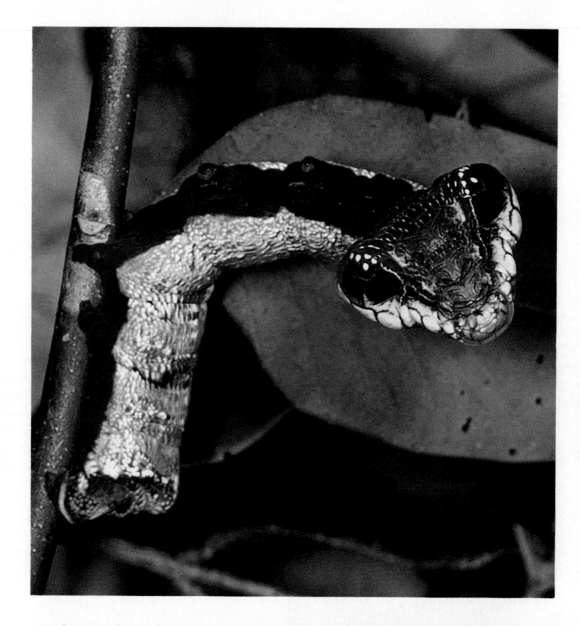

When you observe an organism, it is not always apparent what type of organism it is. What is the organism in this photo? Although it may appear to be a snake, it is, in fact, a caterpillar. What type of organism is a caterpillar? How can you tell it is a caterpillar? What possible advantages would there be for a caterpillar to look like this?

Ordering Life in the Biosphere 10

How many different types of organisms can you recognize? A hundred? A thousand? Biologists have described more than 1.4 million different species of organisms. Millions more have never been named or completely described. Some biologists estimate that we share this planet with as many as 10 million species, and that number does not include organisms that are extinct.

Evolution has produced overwhelming diversity among organisms. In fact, evolution has produced diversity in every aspect of life—in size, from the smallest microorganism to the largest whale; in habitat, from the deepest ocean bottom to the top of the highest mountain; and in way of life, from growing on a rooted tree to flying high above land or water. To make sense of the relationships that exist among organisms, we must have some method of arranging the organisms in a logical and meaningful way. For example, it is possible to classify birds according to the color of their feathers: all blue-colored birds in one group, gray birds in another group, and black birds in a third group. This type of system could be useful, but it has limitations. The type of information it conveys about an organism it does not take into consideration where the organism lives, what it eats, or how it may be related to other organisms. The classification system most widely used today considers how organisms are related and groups them on the basis of what is known of their evolutionary history. This chapter discusses how biologists classify organisms and examines some current ideas about the origins of this great diversity of life.

Biological Classification

10.1 Classification Is a Way of Seeing Order in Diversity

Guidepost: Why do biologists classify organisms, and how do they do it?

The diversity of the living world is awesome. Even the molecules and structures that make up the great variety of living organisms are diverse. From earliest times, however, humans have searched for order in the diversity of nature. They have tried to find ways to distinguish the many types of living things (see Figure 10.1 on page 244). From these efforts has grown the science of classifying organisms, or **taxonomy** (tak SAHN uh mee).

The basic grouping used in biological classification today is the species. Species differ from one another in at least one characteristic, and they

Figure 10.1 One similarity among these organisms is obvious—their color. How else are they alike?

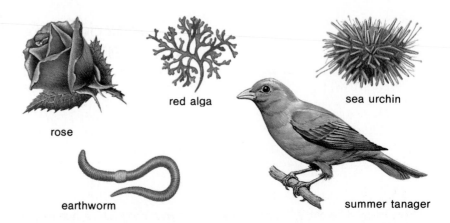

rose

red alga

sea urchin

earthworm

summer tanager

usually do not interbreed freely with other species. Section 9.2 discusses the nature of species and the problems with defining a species. Because some groups of organisms, bacteria and many eukaryotes, for example, reproduce asexually, they cannot be classified according to their interbreeding abilities. Biologists agree, in general, on the types of organisms they classify as species, but the criteria they use when classifying these organisms may differ.

10.2   Classification Is Based on Homologies

To classify organisms, specialists in taxonomy use a variety of characteristics. These characteristics include structure, function, biochemistry, behavior, nutrition, embryonic development, genetic systems, cellular and molecular makeup, evolutionary histories, and ecological interactions. The more constant a characteristic is, the more valuable it is in determining classification. For example, characteristics such as the size or color of organisms may vary greatly, but structures such as skeletal form, reproductive parts, and internal anatomy are less variable. Structure, therefore, generally provides a consistent and useful basis for classifying organisms.

Similarities of structure that indicate *related ancestry* are especially important in classification. For example, fish, amphibians, reptiles, birds, and mammals all share the same limb pattern, as discussed in Chapter 9. The limbs have the same relationship to the body, and they develop in the same way in the young. These types of relationships are called structural **homologies** (hoh MOL uh jeez).

Similarities between body substances (blood, for example) and between molecules (proteins, DNA, or RNA) are called biochemical homologies. Biochemical homologies have become increasingly important in determining relationships. Sometimes comparisons of DNA or RNA sequences have changed our understanding of the relationship between organisms. The greater the similarities in DNA sequences, the more closely related two organisms are thought to be. It also is possible to compare the nucleic acids or the resulting sequences of amino acids. For example, cytochrome *c* is a protein that plays an important role in respiration in animal and plant cells. Between species, the differences in cytochrome *c* are differences in amino acid sequence. If two species are closely related in their evolutionary history, there will be few amino acid differences between their comparable proteins.

Biochemical techniques have helped to clarify some classification problems. For example, for many years experts could not agree on the classification of the giant panda (see Figure 10.2). Some experts grouped pandas with bears; others grouped them with raccoons. New techniques for studying the homologies in DNA and chromosomes, however, have led to a greater understanding of the evolutionary relationships between bears, raccoons, and pandas. Now, the giant panda is classified with the bears, and the evidence suggests that the raccoon and bear families diverged from a common ancestor between 35 and 40 million years ago.

In Investigation 10.1, you can use DNA sequences to determine relatedness. Sometimes studies of DNA sequences offer surprising results. For example, traditional studies of structural characteristics led scientists to think that the relationship between chimpanzees and gorillas was closer than the relationship between chimps and humans. DNA studies, however, have indicated that the relationship between chimpanzees and humans is closer than that between chimps and gorillas (Figure 10.3).

Still, structural characteristics remain particularly important in classification schemes for several reasons. First, biochemical techniques cannot be used on most **fossils**—the remnants or traces of organisms from past

Figure 10.2 Giant panda, *Ailuropoda melanoleuca.*

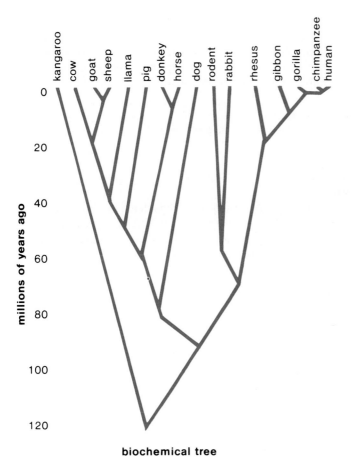

biochemical tree

Figure 10.3 Biochemical trees represent the degree of relatedness of organisms based on analysis of biochemical data, such as amino acid sequences in proteins or base sequences in DNA. In this tree, goats and sheep are closely related while kangaroos are fairly distant relatives of all other organisms shown.

Figure 10.4 What type of organism may have made these trace fossils?

geologic ages. In some fossils, such as fossil bones, the living material has been replaced by minerals. Second, because these organisms no longer exist, scientists only can guess about their behavior. Using preserved specimens or geological records, however, a taxonomist can observe structural characteristics and make guesses about an organism's behavior, based on structural evidence. Sometimes, an organism has left behind tracks, such as the ones shown in Figure 10.4. The organism's behavior must be inferred from the pattern of tracks it leaves. Because organisms of the past may have been ancestors of living organisms, knowledge of past organisms can help to determine the relationships between living organisms.

INVESTIGATION 10.1   DNA Sequences and Classification

Until the mid-1970s, taxonomists usually classified organisms by comparing observable structures in a given organism with those of another organism. For example, a taxonomist might compare the structure of forelimbs in mammals. In recent years, taxonomists also have been able to compare the structure of certain proteins in different organisms.

New techniques now allow biologists to compare the DNA that codes for certain proteins. A hypothesis known as the molecular clock hypothesis uses the comparison of DNA sequences to make predictions about the relatedness of the organisms from which the DNA was taken. According to the molecular clock hypothesis, any changes that occur in the genetic materials of isolated populations are due to mutations in the DNA that are passed from one generation to the next. The rate at which the mutations accumulate can be measured. On the basis of this information, the approximate point when two species diverged from a common ancestor can be determined. This investigation shows you a model of this technology. It may be helpful for you to review the relationship between DNA and proteins discussed in Chapter 8.

Materials (per team of 4)
paper clips: 50 black, 50 white, 50 green, 50 red

PART A   Comparing DNA Strands
Procedure

1. Consult Table 10.1. Which organisms share the most groupings? On the basis of the data in the table, which organisms are least closely related?

2. Synthesize the DNA strands indicated below. Allow the paper clips to represent the four bases of DNA according to the following key:
   black = adenine (A)
   white = thymine (T)
   green = guanine (G)
   red = cytosine (C)
   Hook together paper clips of the appropriate colors. Stretch the string of paper clips out on the lab table with position 1 on the left. Tape a small piece of paper next to the strand and label it as indicated. Each strand will represent a small section (20 bases) of a gene that codes for the protein hemoglobin.

3. Team member 1 will synthesize the following piece of DNA by hooking together paper clips in this sequence:

pos. 1   pos. 20
A G G C A T A A A C C A A C C G A T T A

TABLE 10.1   Examples of Classification of Animals

	Human	Gorilla	Southern Leopard Frog	Katydid
Phylum	Chordata	Chordata	Chordata	Arthropoda
Subphylum	Vertebrata	Vertebrata	Vertebrata	
Class	Mammalia	Mammalia	Amphibia	Insecta
Subclass	Eutheria	Eutheria		
Order	Primates	Primates	Salientia	Orthoptera
Suborder	Anthropoidea	Anthropoidea		
Family	Hominidae	Pongidae	Ranidae	Tettigoniidae
Subfamily			Ranidae	
Genus	*Homo*	*Gorilla*	*Rana*	*Scudderia*
Species	*Homo sapiens*	*Gorilla gorilla*	*Rana pipiens*	*Scudderia furcata*
Subspecies			*Rana pipiens sphenocephala*	*Scudderia furcata furcata*

Source: From C. P. Hickman, L. S. Roberts, and F. M. Hickman, *Integrated Principles of Zoology,* 7th ed. (St. Louis: Times Mirror/Mosby College Publishing, 1984), p. 148.

Label the strand *human DNA*. This represents a small section (20 bases) of the gene that codes for the protein hemoglobin in humans.

4. Team member 2 will synthesize the following strand:

    pos. 1   pos. 20
    T C C G G G G A A G G T T G G C T A A T

Label this strand *chimpanzee cDNA*. cDNA stands for complementary DNA. cDNA is a single strand of DNA that will match up with its partner strand. Remember, the bases in DNA are complementary. That is, adenine (A) always pairs with thymine (T), and cytosine (C) always pairs with guanine (G). This cDNA was made from the gene that codes for chimpanzee hemoglobin.

5. Team member 3 will synthesize the following piece of cDNA:

    pos. 1   pos. 20
    T C C G G G G A A G G T T G G T C C G G

Label this strand *gorilla cDNA*. This cDNA strand was made from the gene that codes for gorilla hemoglobin.

6. Team member 4 will synthesize the following piece of DNA:

    pos. 1   pos. 20
    A G G C C G G C T C C A A C C A G G C C

Label this strand *hypothetical common ancestor DNA*. This strand will be used in the second part of the investigation.

7. Compare the sequences of the human DNA and the chimpanzee cDNA. Match the human DNA and the chimpanzee cDNA base by base (paper clip by paper clip). Remember, black (adenine) always must pair with white (thymine), and green (guanine) always must pair with red (cytosine). Where the bases are complementary (that is, where the colors match correctly), allow the clips to remain touching, as shown in Figure 10.5. Where the bases are not complementary, separate the clips slightly to form a loop, as shown in Figure 10.5. Count the number of loops. Also

Figure 10.5   Paper-clip DNA sequences. Position 1 is at the top.

TABLE 10.2

Human DNA Hybridized to:	Chimpanzee cDNA	Gorilla cDNA
Number of loops		
Number of differences		

count the total number of bases that do not match. Copy Table 10.2 in your data book and record the data in the appropriate columns.

8. Repeat Step 7, using the gorilla cDNA and the human DNA. Enter the data in the appropriate columns in the table in your data book.

Discussion

1. Based on the data you have collected for this one protein, is the gorilla gene or the chimpanzee gene more similar to the human gene?

2. Again, based on the data you have collected for this one protein, does the gorilla or the chimpanzee seem more closely related to humans?

3. Do the data prove that your answer to Question 2 is correct?

PART B   An Evolutionary Puzzle
Procedure

Scientists have determined that mutations in DNA occur at a regular rate. They use this rate to predict how long ago in evolutionary history two organisms began to separate from a common ancestor. In this part of the investigation, you will use your paper-clip model to provide data in support of one of two hypotheses about a common ancestor for humans, chimpanzees, and gorillas.

9. Read the following information about a current debate among scientists who study human evolution:

> Most scientists agree that humans, gorillas, and chimpanzees shared a common ancestor at one time in evolutionary history. However, one group thinks the fossil record shows that gorillas, chimpanzees, and humans split from one common ancestor at the same time. Their model for this split is shown in Figure 10.6a.
>
> A second group thinks the fossil record shows there were two splits. In the first split, gorillas split from the common ancestor. Humans and chimpanzees then shared another common ancestor for perhaps 2 million years. They then split again and evolved into their present states. The model for this pattern of splitting is shown in Figure 10.6b.

10. Use your DNA model and the DNA sequences from Part A to investigate this debate. First, use your human DNA as a guide to synthesize a human cDNA. Do that step as a team. Make certain that the human cDNA is complementary to the human DNA strand.

11. The hypothetical common ancestor DNA synthesized in Part A is DNA for hemoglobin extracted from a hypothetical common ancestor. Now, match all three samples of cDNA (gorilla, human, and chimpanzee) with the common ancestor DNA, one sample at a time. Again, allow the paper clips to touch where the bases match correctly. Form loops where the bases do not match. Copy Table 10.3 in your data book and record your data.

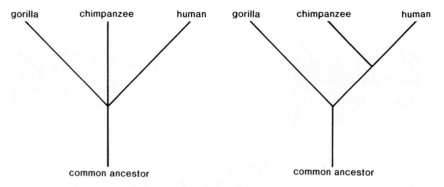

Figure 10.6   Hypotheses of primate divergence.

TABLE 10.3   Hybridization Data: Common Ancestor

Common Ancestor DNA to:	Human cDNA	Chimpanzee cDNA	Gorilla cDNA
Number of loops			
Number of differences			

Discussion

1. Which cDNA is most similar to the common ancestor DNA?
2. Which cDNAs give the most similar patterns of matching and looping when matched to the common ancestor DNA?
3. Which model in the evolutionary debate described above do your data support?
4. Do your findings prove that this model is the correct one? Explain.

10.3   Species Are Grouped into Broader Categories

When relatively few plants and animals were known, there was little need to group them into larger, more inclusive units that reflected common characteristics. Once thousands of plants and animals are known, however, larger groupings become necessary in order to distinguish specific organisms within this bewildering array. In a universal system of classification, organisms of different species are grouped into larger, more general categories based on homologies. For example, dogs, coyotes, and wolves are considered separate species, but they are similar in many ways. Species that share many similar characteristics are grouped into the same **genus** (JEE nus; plural *genera*). The genus for these doglike animals is *Canis*.

In much the same way, similar genera are grouped together in a **family**. Although foxes share many similarities with members of the genus *Canis*, there are also many differences. For example, foxes usually have a lighter build and are more catlike than members of the genus *Canis*. Therefore,

Figure 10.7 Some animals in the order Carnivora. Which two look most alike? Which two look least alike? In what ways?

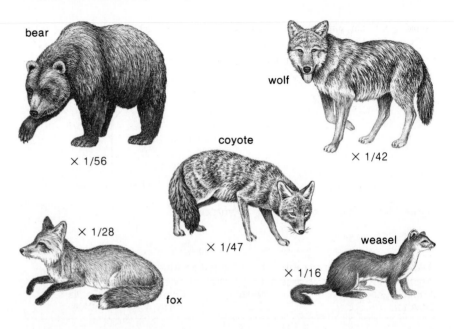

bear

× 1/56

wolf

× 1/42

coyote

× 1/47

× 1/28

fox

weasel

× 1/16

taxonomists place foxes in a separate genus (*Vulpes*), yet because of the similarities that exist, include them in the same family as dogs, coyotes, and wolves—the family Canidae (see Figure 10.7).

Weasels also resemble dogs and wolves in some ways, but they are less like them than are foxes. Taxonomists express this difference by placing weasels in a separate family, Mustelidae. Bears, which eat meat and fruit, have many structural differences from weasels or foxes. Taxonomists place

Figure 10.8 Classification of several animals.

Common Name	Scientific Name	
Human	*Homo sapiens*	
Lion	*Panthera leo*	
Coyote	*Canis latrans*	⎫
Wolf	*Canis lupus*	⎬
Dog	*Canis familiaris*	⎭
Gopher	*Thomomys bottae*	
Ground squirrel	*Spermophilus tridecemlineatus*	
American robin	*Turdus migratorius*	
European robin	*Erithacus rubecula*	
Gopher turtle	*Gopherus polyphemus*	
Green frog	*Rana clamitans*	⎫
Bullfrog	*Rana catesbeiana*	⎬

bears in still another family, Ursidae. These three families (Canidae, Mustelidae, Ursidae) are all meat-eaters. Therefore, they are grouped with other families of meat-eaters into the same **order**, Carnivora.

Because monkeys appear to have little in common with wolves, weasels, and bears, they are placed in a different order, Primates. Monkeys (and many other organisms), however, do have some characteristics in common with wolves, weasels, and bears. These characteristics include hair and the production of milk to feed their young. Therefore, monkeys, wolves, weasels, and bears, as well as other animals that have hair and produce milk to feed their young, are put together in the next larger grouping, the **class**—in this case, the class Mammalia.

Continuing with this method of grouping, taxonomists combine classes containing birds, frogs, fish, and snakes with the class Mammalia to form the **phylum** (FY lum) Chordata. Animals in this phylum all have a primitive spine, or notochord. (Botanists group classes of plants as **divisions** instead of as phyla.) Finally, chordates, snails, butterflies, and thousands of other organisms are grouped into the **kingdom** Animalia. This kingdom contains all the living things we think of as animals.

Notice how the characteristics the organisms have in common become more general at each succeeding level from species to kingdom. At the species level the individuals are so much alike they can interbreed. At the kingdom level only a few characteristics are shared among all the individuals. Figure 10.8 shows the classification of twelve organisms.

By placing the dog family, the bear family, and the weasel family in the same order, taxonomists imply that all these animals long ago descended from a common ancestral group. The farther up the list of levels, the more distant the relationships become. When taxonomists place a dog and a

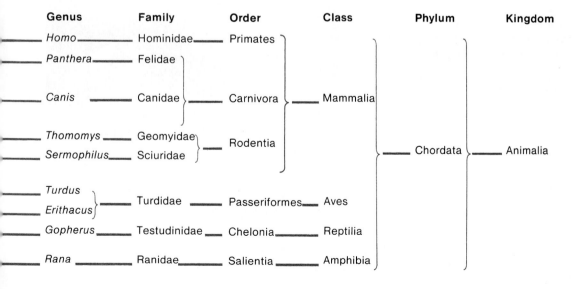

Reproductive Physiologist

Plants and animals are disappearing today at such an unprecedented rate that some researchers estimate more than one million species may become extinct by the end of this century. The primary cause of this decay in diversity is the destruction of habitats due to the expansion of the human population and its demand for space and goods. In some cases, zoos may be a species' last refuge.

Dr. Betsy Dresser has dedicated her life to preserving endangered wildlife. Currently she is the director of the Cincinnati Zoo's Center for Reproduction of Endangered Wildlife (CREW). CREW is known internationally for developing techniques that allow interspecies embryo transfer—the transfer of an embryo from an endangered species to a surrogate parent of a different, but closely related, species. For example, a domestic cat acting as a surrogate mother carried to term an Indian desert cat embryo. CREW's goals are the preservation and propagation of endangered species through the use of embryo transfer techniques and *in vitro* fertilization.

From an early age Dr. Dresser was fascinated with wildlife. After graduating from high school, she took a job in animal research. Although the job gave her hands-on experience with animals, it was not enough. She became a volunteer at the zoo and helped teach junior-high-school biology classes. Eventually she attended college where her interests in wildlife and biology soon centered on reproduction in exotic animals.

Dr. Dresser has worked with cats, sheep, mice, rhinos, and antelope. The common antelope she works with is the eland, and the exotic species is the bongo. Only about 60 bongos exist in captivity. To increase their numbers, seven-day-old bongo embryos were removed nonsurgically from a bongo at the Los Angeles Zoo and transferred into surrogates at the Cincinnati Zoo that same day. Twin bongo calves were born one week apart, one to a surrogate eland mother and one to a surrogate bongo mother.

The techniques Dr. Dresser uses require a great deal of skill. Once the sperm or embryos have been collected, they are frozen until needed. The water in the cells is replaced with glycerol, which, unlike ice, does not pierce the membranes and rupture the cells. The sample is placed in a plastic straw

and the ends are heat sealed. The sample then is cooled to below freezing and plunged into a liquid hydrogen bath.

Primarily because of Dr. Dresser's hard work and dedication, CREW has received funding to build a new center equipped with the latest technology. At the new center, new storage methods will be devised, and the technology will be expanded to the embryos of other species.

Dr. Dresser's long-term goal is to make herself obsolete by creating a world where her line of work is no longer needed. "It won't happen in my lifetime, I know that. The animals are dying so fast, and there is so much to be done and so much we don't know. All the technology in the world isn't the answer to conservation problems. CREW is just one of the tools."

Betsy Dresser

goldfish in the same phylum but in different classes, they imply a very distant evolutionary relationship. Each level in the universal classification system deals with three things: characteristics that pertain to each group; organisms that belong to each group; and varying degrees of evolutionary relatedness among the group.

A taxonomic classification is not necessarily permanent. It is the result of the interpretations of data made by individuals developing the system of classification. Although it is possible to prove that cats and eagles and alligators have claws, whether these three organisms should be grouped together is subject to debate. Biologists do not totally agree about where many organisms fit into the classification scheme, especially at the species level. The more evidence taxonomists obtain, the more complex the relationships of organisms appear to be. Taxonomists also differ in how they interpret the evidence—different inferences may be drawn from the same set of observations. For these reasons, there are several schemes of classification, all of which are designed within the same framework of levels. Investigation 10.2 acquaints you with some of the characteristics taxonomists use to classify organisms.

INVESTIGATION 10.2   Levels of Classification

You will investigate some of the structural characteristics that taxonomists use in separating animal groups at different classification levels. Because you will use the observations of other people (recorded as drawings), your conclusions can be no more valid than those drawings.

Materials (per person)
none

Procedure

1. In your data book, prepare four forms similar to Table 10.4, or tape into your data book the forms your teacher provides. Label the forms *Table A, Table B, Table C,* and *Table D.*

TABLE 10.4   Structural Characteristics

Characteristics	Animals		
	1.	2.	3.
a.			
b.			
c.			
d.			
e.			
Classification level _____			

Figure 10.9

2. For Table A, use the information in Figure 10.9. In the spaces under *Animals*, write *human, chimpanzee,* and *gorilla*. In the spaces under *Characteristics*, copy the italicized key words in each of the following questions. These words should remind you of the full questions when you review the table.

   a. How does the *length of the arms* of the animal compare with the length of its legs?
   b. Are the *canine teeth large*, or are they *small* as compared with other teeth of the same organisms?
   c. How many *incisor teeth* are present in the upper jaw?
   d. Is the *brain* case of the skull *large*, or is it *small* as compared with the overall size of the skull?
   e. Is there an *opposable first toe on the foot*? (An opposable toe is one that can be pressed against all the others, just as your thumb can press against your other fingers.)

3. After studying Figure 10.9, fill in all the spaces in Table A with your answers for each animal. Then write *Family* in the space following *Classification level*. Refer to Appendix 4 to find the family into which each of these organisms has been placed. Write this information in the spaces at the bottom of the table.

4. For Table B, use the information in Figure 10.10. Under *Animals*, write *human, dog,* and *cat*. Under *Characteristics,* copy the italicized words in each of these questions:

   a. How many paired *appendages* (arms and legs) does the animal have?
   b. Are *nails or claws* present on the toes of the foot?
   c. How does the size of the *canine teeth* compare with that of other teeth in the lower jaw?
   d. How many *incisor teeth* are present in the lower jaw?
   e. How does the size of the *collarbone* compare with that of the other organisms?

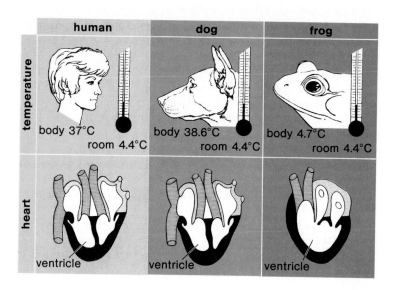

Figure 10.10

5. After studying Figure 10.10, fill in the spaces in Table B with your answers for each animal. Write *Order* in the blank space following *Classification level*. From Appendix 4, select the order into which each of these organisms has been placed. Enter this information in the table.

6. For Table C, use the information in Figure 10.11 and the following questions:

   a. What type of *body covering* (hair, feathers, scales, none) does the animal have?
   b. How many paired *appendages* (arms and legs) does the animal have?
   c. Do the *ears project from* the surface of *the head*?
   d. Is the *body temperature* similar to the temperature of the environment or not?
   e. How many *ventricles* are *in the heart*?

7. The classification level for this table is *Class*. Determine the class for each organism in Figure 10.11 and write it in the table.

Figure 10.11

Figure 10.12

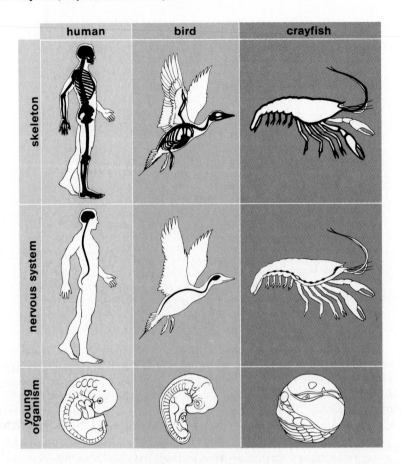

	human	bird	crayfish
skeleton			
nervous system			
young organism			

8. For Table D, use the information in Figure 10.12 and the following questions:

   a. What type of *skeleton* (internal or external) does the animal have?
   b. Is the *position of* the *nerve cord* along the back or along the belly?
   c. Compared with the rest of the nervous system, is the *brain* large or small?
   d. Are paired *appendages* present, or are they absent?
   e. Are there *grooves behind* the *head region* of the very young organism?

9. Write *Phylum* in the space following *Classification level* and add the name of the phylum into which each animal in Figure 10.12 is placed.

Discussion

1. There are more structural similarities between chimpanzees and gorillas than between chimpanzees and humans. How does the classification system you used express this fact? Focus on the levels in the classification system into which these organisms are placed together.

2. How does the classification system you used express the following?

   a. There are more structural similarities between dogs and cats than between dogs and humans.
   b. There are more structural similarities between humans and dogs than between humans and frogs.
   c. There are more structural similarities between humans and birds than between humans and crayfish.
   d. There are more structural similarities between humans and chimpanzees than between humans and dogs.

3. You are told that species A and B are classified in the same kingdom but different phyla. You also are told that species C and D are classified in the same phylum but different classes. What general statement can you make about similarities among species A, B, C, and D?

10.4 Biologists Use a Binomial System

Probably the first plants that were named were useful to humans as spices, medicines, and foods or for religious ceremonies. These useful plants were not given elaborate or formal names. In fact, the common names for many organisms are still important today. If, for example, you ask for some *Lactuca sativa* in a supermarket, you are not likely to get what you want. A request for lettuce works much better. Why, then, do biologists need special names?

Different languages usually have different names for the same organisms. For example, a potato is *papa* in Spanish and *Kartoffeln* in German. To compound the problem, the *same* word in a language may refer to different organisms. In Florida, *gopher* refers to a turtle; in Kansas, to a rodent. Biological names, then, are necessary for scientific exactness. There is no other single, agreed on set of names available for all organisms.

During the Middle Ages, scholars looking for a universal naming system attempted to fit the plant and animal names used by the ancient Greeks to the plants and animals of the rest of Europe. In order to recognize the differences in the plants and animals of England, Germany, and other northern lands from those of Greece, they attached new adjectives to the old name of a similar plant or animal. As new places and new organisms were discovered, the names became more complicated. By the eighteenth century, the name used for the carnation plant shown in Figure 10.13 was *dianthus floribus solitariis squamis calycinis subovatis brevissimis corollis crenatis*, which meant "the pink [a general name for the carnation] with solitary flowers, the scales of the calyx somewhat egg-shaped and very short, the petals scalloped."

In 1753, a young Swedish botanist named Carolus Linnaeus (lih NAY us) developed the solution to this problem of scientific naming. Linnaeus used a system of **binomial nomenclature** (by NOH mee ul NOH men klay chur), or two-word naming, to identify each species. The first word in each name indicates the genus. Thus, all species of pinks are in the genus *Dianthus*. The second word in each name is the species name, and it indicates a group of similar individuals. For the common carnation, Linnaeus picked the species name *caryophyllus*. Neither the genus name nor the species name is, by itself, the scientific name. The scientific name consists of both genus and species names—*Dianthus caryophyllus*. Scientific names always are italicized (or underlined), and the first letter of the genus name always is capitalized.

Linnaeus established two rules that have made his system succeed. The first rule is that the genus name can never be used for another group. *Dianthus*, for example, can be used only to describe pinks. The second rule is that the species name cannot be used for any other species within the same

Figure 10.13   A carnation.

有些蕈菌含有毒性化合物，稱為瓢菌素，若誤食後會引起呼吸及循環的失常，對人類是有害的。蕈類約有70種，其中以瓢菌 *Amanita verna* 最毒（圖 12—6），它有潔白的子實體，雖僅食用少許的蕈帽，也必在一日之內致死，所 以俗稱 "死神蕈"，真是名符其實。通常這類毒蕈，柄的基部有一"杯狀物"（圖12—6），這杯狀物常深藏土中不易發現，毒蕈與可食蕈的區別，普通傳述有如下的幾種謬見："銀匙與 毒蕈共煮則呈黑色"；"可食的蕈帽易於剝落，毒 蕈則不然"；"若昆蟲或其他動物吃過的蕈，人亦 可食之" 等等，其實皆不可信。惟一可靠的選擇方 法是由市場購買人工栽培的食蕈。如從野外探來的 ，就應先經專家檢定後方可食用。

圖 12—6 瓢菌 *Amanita verna* 圖示蕈托 杯狀物位在柄的基部（應藏地下）。這是許多頭毒蕈的特徵。蕈帽下放射狀的蕈褶是蕈的特徵。

Figure 10.14   A short section on poisonous mushrooms from a Chinese BSCS biology book. Note the scientific name.

genus. For example, *caryophyllus* can never be used for any other species of *Dianthus*. The species name may be used with some other genus, but because a genus name cannot be repeated and because the scientific name of an organism always has two words, no other species can be called *Dianthus caryophyllus*.

Although Latin and Greek word roots frequently are used, the scientific name of an organism may come from any language, or it may be invented. For example, *Tsuga* (the hemlocks) comes from Japanese, and *Washingtonia* (a genus of palms) obviously does not come from Latin. All names are given Latin endings, however, and the names always will appear the same. Thus, in a Russian or Chinese biology book, biological names are printed in the same form as they appear in this book, though the rest of the printing is different (see Figure 10.14).

Concept Review

1. What is the basis for biological classification? What advantages does it have?
2. How does the number of characteristics shared by all members of a classification level change as you progress from species to kingdom?
3. Why can equally experienced taxonomists disagree about the classification of a particular species?
4. What are the basic rules of the binomial nomenclature system, and how is the system useful in science?

The Kingdoms of Organisms

Guidepost: What characteristics determine how organisms are grouped into kingdoms?

10.5   Cell Structure Is Evidence for Relatedness

As the universal classification system progresses from species to kingdom, each category contains more organisms. The more organisms or species in a category, the more difficult it is to find homologies among them. At the same time, the more variations there are within a group, the more difficult it becomes to find definite characteristics that distinguish one group from other groups at the same taxonomic level.

As discussed in Chapter 5, all organisms are made up of cells—the basic units of structure and function—and cell products. These cells contain clues about the relationships between organisms. Based on their cell structure, all living organisms are either prokaryotes or eukaryotes (see Figure 10.15).

Prokaryotes (bacteria) are thought to represent the earliest type of living cell. They lack membrane-enclosed internal structures. This means they do not have a nucleus or cytoplasmic organelles such as mitochondria or chloroplasts. Their cellular material does not appear to move. Their rigid cell walls are formed from proteinlike chains and sugars. Their genetic material consists of a single circular thread of DNA, although some also contain several plasmids. Mitosis does not occur. Prokaryotes move by gliding or by means of long, hairlike projections called **flagella** (fluh JEL uh).

Eukaryotic cells, which usually are larger than those of prokaryotes, have a membrane-enclosed nucleus and other membrane-enclosed organelles (see Figure 10.15). The cellular material often appears to move, or stream, within the cell. If cell walls are present, they usually are composed

Figure 10.15   A prokaryotic cell (top) compared with a eukaryotic cell (bottom). Note the greater complexity of the eukaryotic cell (an amoeba) and the presence of many membrane-enclosed organelles.

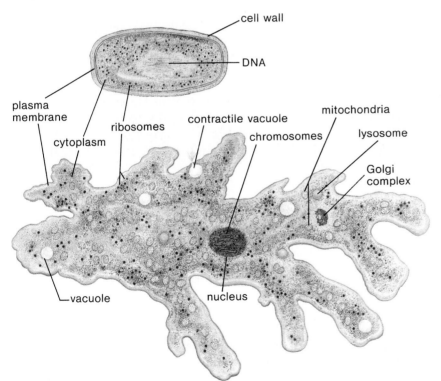

TABLE 10.5   Some Differences Between Prokaryotic and Eukaryotic Cells

Eukaryotes Have	Prokaryotes Have
■ nucleus	■ no nucleus
■ membrane-enclosed organelles	■ no membrane-enclosed organelles
■ chromosomes in pairs	■ single chromosome
■ streaming in the cytoplasm	■ no streaming in the cytoplasm
■ cell division by mitosis	■ cell division without mitosis
■ complex, tubular flagella	■ solid flagella
■ larger ribosomes	■ smaller ribosomes
■ complex cytoskeleton	■ simple cytoskeleton
■ cellulose in cell walls	■ no cellulose in cell walls

of cellulose or another carbohydrate, chitin (KYT in). The DNA of eukaryotes is organized into chromosomes, and there are no plasmids. Cell division includes mitosis and cytokinesis. Table 10.5 summarizes the major differences between prokaryotes and eukaryotes.

10.6   Organisms Are Grouped into Five Kingdoms

Because organisms may look very unalike at the kingdom level of classification, biologists consider several questions about the evolutionary history of an organism before classifying it. Is it a prokaryote or a eukaryote? Is it a producer, or **autotroph** (AWT oh trohf)? Is it a consumer, or **heterotroph** (HET eh roh trohf)? Does it reproduce asexually or sexually, and does it develop from an embryo? Finally, biologists consider the general structure and function of the organism. None of these questions, however, resolves all the problems of classification, and the systems continually change as new knowledge emerges. Today, most biologists favor the five-kingdom classification system shown in Figure 10.16.

At present, all prokaryotes are grouped in one kingdom called the **Prokaryotae** (proh kayr ee OH tee). Prokaryotes display a greater variety of chemical and functional patterns than do eukaryotes. Like green eukaryotes, many prokaryotes produce food through photosynthesis, but the prokaryotes use a greater variety of substances as raw materials. The energy source of prokaryotes differs, also. Some prokaryotes use energy obtained from inorganic chemicals (chemicals that do not contain carbon chains or rings) rather than light to produce food, and many prokaryotes are heterotrophs. Although most prokaryotes are unicellular, there are some prokaryotes that are multicellular. Reproduction usually is accomplished through asexual cell division. When sexual reproduction is used, the parents do not contribute equal amounts of genetic material. Figure 10.17 shows several prokaryotes.

Currently, the eukaryotes are divided into four kingdoms. Most autotrophic, multicellular eukaryotes that produce their own food through photosynthesis belong to the kingdom **Plantae** (PLAN tee). This kingdom includes all organisms developing from an embryo that lacks a blastula stage. Plants have cellulose-containing cell walls; their cells contain chloroplasts;

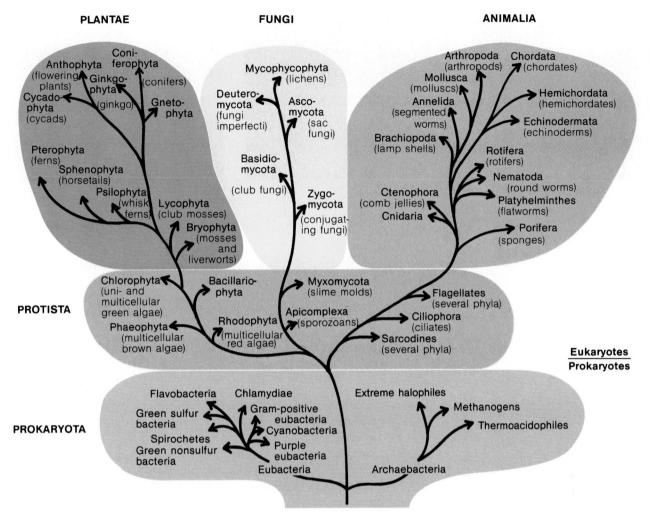

Figure 10.16 Currently all organisms can be classified in the five kingdoms shown in this diagram. However, increased understanding of the prokaryotes may lead to that group being divided into more kingdoms.

Figure 10.17 Representative prokaryotes include (a) cyanobacteria, *Anabaena circinalis;* (b) purple photosynthetic bacteria, *Rhodospirillum rubrum* (stained); and (c) nitrogen-fixing bacteria.

Figure 10.18   Members of the kingdom Plantae: (a) moss, (b) ferns, *Cystopteris fragilis,* (c) Indian paintbrush, *Castilleja.*

they store food as starch; and most reproduce sexually. The bulk of the world's food and much of its oxygen are derived from plants. Figure 10.18 shows a variety of plants.

All organisms developing from an embryo that has a blastula stage are members of the kingdom **Animalia** (an ih MAYL yuh). All animals are heterotrophic and multicellular. Ranging in size from microscopic forms to giant whales, animals are the most diverse in form of all the kingdoms (see Figure 10.19). Most animals have senses and nervous systems that aid in locomotion. In most animals, reproduction is sexual. Members of kingdom Animalia include vertebrates (animals with backbones), such as birds, frogs, and humans; and invertebrates (animals without backbones), such as sponges, worms, and insects.

Organisms of the kingdom **Fungi** (FUN jy) develop directly from spores. Fungi, which vary in size from microscopic species to large mushrooms, are heterotrophs that absorb small molecules from their surroundings through their outer walls. Many are decomposers and play an important

Figure 10.19   Animals exhibit great diversity of type. Shown here are (a) blacktailed prairie dog, *Cynomys ludovicianus;* (b) dragonfly, *Libullaria;* and (c) sea slug, *Hermissendra crassicornis.*

Figure 10.20 Typical fungi include (a) collard stinkhorn, *Dictyophora;* (b) jelly fungus, *Dacrymyces palmatus;* and (c) hat thrower fungus, *Pilobolus.*

role in breaking down organic material. With the exception of yeasts, most fungi are multicellular and have chitinous cell walls. Reproduction may be either sexual or asexual; most fungi are haploid rather than diploid. The kingdom Fungi includes yeasts, molds, bracket fungi, and mushrooms (see Figure 10.20).

All the remaining eukaryotes are grouped in the kingdom **Protista** (pro TIST uh). This kingdom includes algae (plantlike protists), protozoa (animal-like protists), slime molds (funguslike protists), and other organisms. Many protists are microscopic and unicellular, but others are large multicellular organisms that may reach 10 m or more in length. Protists also vary in cellular organization, methods of reproduction, and life-styles (see Figure 10.21). They may be producers, consumers, or decomposers; some even switch from one form of nutrition to another, depending on conditions in their environments. Despite their differences, these organisms probably are more closely related to each other than they are to members of the other kingdoms. The classification of organisms representing each kingdom is shown in Figure 10.22 on page 264.

Figure 10.21 Examples of Protista: (a) single-celled alga, *Chlorella,* (b) protozoa, *Vorticella,* and (c) red raspberry slime mold, *Tubifera ferruginosa.*

a

b

c

Figure 10.22 Classification of an organism from each kingdom.

	Anabaena cyanobacteria	*Amoeba*	*Rhizopus* bread mold	*Quercus alba* white oak	*Homo sapiens* human
kingdom	Prokaryotae	Protista	Fungi	Plantae	Animalia
phylum/division	Cyanobacteria	Sarcodina	Zygomycota	Anthophyta	Chordata
class	Eubacteria	Lobosa	Phycomycetes	Dicotyledoneae	Mammalia
order	Oscillatoriales	Amoebina	Mucorales	Fagales	Primates
family	Nostocaceae	Amoebidae	Mucoraceae	Fagaceae	Hominidae
genus	*Anabaena*	*Amoeba*	*Rhizopus*	*Quercus*	*Homo*
species	*circinalis*	*proteus*	*stolonifer*	*alba*	*sapiens*

10.7 Classifications Can Change

The more information we obtain, the more complex the relationships between organisms appear to be. New knowledge often requires changes in classification, especially at the kingdom level. Any classification system should be exclusive as well as inclusive. That is, the characteristics used should allow us to form groups of similar organisms that are different from other groups. For example, members of the kingdom Prokaryotae are prokaryotes: all other organisms are eukaryotes. Plants are autotrophic; animals and fungi are heterotrophic. These characteristics serve to *include* similar organisms in one kingdom and *exclude* them from others.

At one time it appeared that all organisms could be classified in two kingdoms—animals and plants. As improved microscopes increased our knowledge of microscopic organisms, scientists realized that many did not fit well into either kingdom, and taxonomists added other kingdoms for the prokaryotes and protists. Because the fungi share few characteristics with the other groups, a separate kingdom was proposed for them.

The five-kingdom system of classification presented here is not accepted by all biologists. Some people place multicellular algae with the plants and heterotrophic protists with the animals. Other specialists divide the bacteria into two different kingdoms. Further studies of bacteria may result in the addition of new kingdoms.

Species, genera, families, orders, classes, and kingdoms do not exist in nature; only individual organisms exist. Classification systems are simply a means to think more easily about the great diversity of the world. The changing systems of classification reflect how science works—it is open to modification on the basis of new data. Appendix 4, *A Catalog of Living Things,* shows only a few examples of the diversity of life.

Concept Review

1. Compare and contrast eukaryotic and prokaryotic cells.
2. What is the most fundamental characteristic that separates organisms?
3. What characteristics do biologists use to sort organisms into kingdoms?
4. Describe the distinguishing characteristics of each kingdom.
5. Why do classification systems change?

The Origin of Diversity

10.8   All Species May Have Come from One Ancestral Species

Guidepost: How did life originate and evolve into the many forms found today?

Although organisms arise only from other organisms of the same type, individual organisms vary. The process of natural selection, then, may lead to an increase in the number of species. For example, all of the Galápagos finches probably are descended from the same ancestors. If many types of organisms evolved from a few, the farther into the past scientists look, the fewer species they should find. Ultimately, it is possible that all of today's species came from one ancestral species.

Studies of fossils indicate that many forms have evolved from a few ancestral forms. As scientists have studied fossil evidence, they have found that one group of organisms would flourish, then die out—perhaps because of changes in the environment or because of competition from other groups. As one group faded, another group of organisms would increase in number and diversity. During the age of reptiles, for example, there was a greater variety of reptile species than there is today (see Figure 10.23),

Figure 10.23   The age of reptiles produced tens of thousands of species, from small lizards to flying reptiles and giant dinosaurs. Turtles, alligators and crocodiles, lizards, and snakes are the only surviving orders of reptiles today.

first reptile
(over 280
million
years ago)

reptiles today
(four living orders
and 6000 species)

Age of Reptiles
(16 or more orders and tens of thousands of species,
extending through almost 200 million years)

Figure 10.24   A 3.5 billion-year-old prokaryote fossil (a) in a stromatolite from western Australia. Modern living stromatolites (b) in Shark Bay, Australia, are built by cyanobacteria.

but there were only a few mammal species and almost no birds. When the number of reptiles declined, the number and types of mammals and birds increased.

This pattern appears in each age in the fossil record, and it appears in the history of each group of organisms. Throughout the history of life, new species have evolved and older species have become extinct. Again and again, more species have evolved from a smaller number of species that preceded them.

How far back can scientists trace life? Rocks in South Africa and western Australia have yielded the oldest known microfossils—microscopic pro-karyotes. Because geologists have estimated that the rocks holding these fossils are between 3.0 and 3.5 billion years old, the fossils themselves must be that old. The Australian fossils were found associated with domelike structures called stromatolites (stroh MAT uh lyts). See Figure 10.24a, b. The domes, about 30 cm high by 1.5 m across, are constructed of many wafer-thin layers of rock. So far, scientists have been unable to identify any fossils older than these.

10.9   Stars Provide Evidence of Earth's Early History

How did these organisms evolve? How can the evolution of the first life forms be studied? It is impossible to recreate the earth in a laboratory in order to study how it was formed and how life appeared. To study the earth of the past, therefore, scientists make inferences based on their observations of stars. Many stars develop from revolving clouds of gas and dust. Giant clouds, often with trailing arms of gases, are visible in space today. From their observations, scientists infer that the sun condensed from the solar system's early gas cloud. They further infer that the cloud had trailing arms of gases, some of which may have condensed to form the planets. Because these are only inferences based on observations, however, little about the evolution of the earth or other planets is unquestioned.

Geological evidence suggests that after it first formed approximately 4.6 billion years ago, the earth grew very hot. Because heat speeds up

Figure 10.25   A sequence of events that might have led to the first life.

chemical activity, atoms must have been constantly combining and recombining to form many types of molecules. Even after the earth cooled greatly, it was still much warmer than it is today.

Geological evidence also indicates that the early atmosphere of the earth was composed of volcanic gases, such as methane (CH_4) and carbon dioxide (CO_2). It also included ammonia (NH_3) instead of present-day nitrogen (N_2). Some hydrogen (H_2) and a great deal of water vapor (H_2O) also were present, but free oxygen (O_2) was absent.

With the energy of heat and lightning, the gases of the early atmosphere may have combined to form substances such as amino acids, which are found in all organisms today. Through millions of years, organic compounds may have accumulated in the warm oceans, lakes, and pools, forming an "organic soup." Along the shores, in tide pools, or on particles of clay, these simple compounds may have combined to become more complex molecules. Finally, these more complex molecules may have united somehow to form a simple type of reproducing "living thing." Figure 10.25 summarizes these ideas.

In 1952, to test these speculations, Stanley Miller and Harold Urey at the University of Chicago decided to find out what would happen if a simulated primitive atmosphere were exposed to an energy source. Using apparatus like the one shown in Figure 10.26 on page 268, they passed electric sparks (simulating lightning) through ammonia, methane, water, and hydrogen. When the resulting substances were analyzed later, simple amino acids had been produced.

Other scientists have repeated this experiment. Some investigators have used ultraviolet light instead of electric sparks and obtained the same type

Figure 10.26 A drawing of Miller's apparatus in which conditions thought to exist in the primitive atmosphere were reproduced in the laboratory. Does this experiment prove how life originated on earth?

of results. Since Miller and Urey's first experiments, researchers have synthesized many other types of organic molecules, including nucleotides and carbohydrates. Although these experiments suggest a way in which life might have originated, it is still a long way from complex molecules to even the simplest of known organisms.

There are more questions about the origin of life than there are answers, but it is important to remember that the conditions under which life originated are not the same as conditions today. If life on earth originated in the way that Miller and Urey have suggested, why has it not originated this way again and again during the history of earth? If organic compounds cannot be spontaneously generated today, then where did they originally come from? Why are conditions on earth different today? What has happened to change them? How might the first organisms have arisen?

10.10   The First Cells Probably Were Heterotrophs

The first life forms must have had two essential characteristics of life: first, a means of self-maintenance (maintaining their organization and order); and second, a means of self-replication (reproducing themselves). An early microorganism could not have been immortal. Sooner or later it would have been killed by storms or waves. Hence, for life to evolve successfully, it had to be able to reproduce. There must have been some way to store coded information about how to make new copies of itself.

How could these abilities have developed? Scientists have two major hypotheses; one focuses on self-replication and the other on self-maintenance. According to the first hypothesis, the first life forms were not enclosed in a membrane but consisted of molecules or crystals that could self-replicate. Following this hypothesis, minerals dissolved in clay could have crystallized to form layers with a definite, repeating structure (see Figure 10.27). These layered clays could have acted as "crystal genes" able to store and transmit information in much the same way as DNA.

The second hypothesis suggests that the first life forms were cell-like structures enclosed within a membrane. Through time, some amino acids in the "organic soup" may have formed polypeptides and proteins. Other simple organic molecules also may have formed larger, more complex molecules. Eventually, some of the larger molecules combined into clusters, and the clusters merged to form a primitive cell.

Scientists have proposed different models for these precells. A Russian scientist, A. I. Oparin, suggested that precells might have been like coacervates (koh AS er vayts). Coacervates are clusters of proteins or proteinlike substances held together in small droplets within a surrounding liquid, as shown in Figure 10.28. Sidney Fox of the University of Miami has proposed a second model. He thinks precells were more like microspheres—cooling droplets from a hot water solution of polypeptides. Each microsphere may have formed its own double-layered boundary as it cooled (see Figure 10.29 on page 270). Carl Woese at the University of Illinois has presented a third model for the formation of the first cells. According to his model, the early atmosphere had a high level of carbon dioxide that produced a greenhouse effect. Meteorites that hit the primitive earth provided large quantities of dust that were spread by the wind. Water vapor may have condensed around these dust particles, forming droplets. Each droplet, warmed by the atmosphere, then acted as a primitive cell in which chemical reactions occurred, allowing life to begin. All three models of cell formation have proponents and opponents. If a membrane enclosed the first life form, any one, or none, of the three models could account for the first cell.

What did the first living things use for food? It is tempting to suggest that they made their own food, but food-making is a complex process. The

Figure 10.27   Repeating structures in kaolinite clays. The first chemical systems to self-replicate may have been clays. Minerals in the clay may have formed a crystalline pattern that served as a blueprint for assembling more minerals into layers of similar patterns.

Figure 10.28   This drawing shows how a complex coacervate can form when a water layer surrounds a cluster of protein molecules.

Figure 10.29 Microspheres are formed when dry mixtures of amino acids are heated and then placed in water. This microsphere is 2 μm in diameter. The dark material, including the boundary, is proteinoid.

first cells probably could not have synthesized all the organic compounds they needed. If evolution requires the fewest possible steps, the first living things would have been heterotrophs that used the supply of naturally occurring organic compounds for food. This is known as the **heterotroph hypothesis**. Heterotrophs would have required fewer evolutionary steps to develop than would have more complex autotrophs. Unless evolution took a great many complex steps almost simultaneously, the heterotroph hypothesis seems the most reasonable explanation of the first cells. Thus, the earliest life had appeared and was well established by the time the earth was only one third of its present age (see Figure 10.30).

10.11   Several Types of Photosynthesis Evolved

Continuing the heterotroph hypothesis, as organic compounds were used for food by the primitive cells and as environmental conditions changed, the supply of compounds diminished. Under these conditions, some of the primitive cells must have begun to use visible light from the sun as a source of energy for making food.

    The first organisms able to use sunlight may have acted partly like heterotrophs, taking in organic compounds, and partly like autotrophs, making other organic compounds with energy from sunlight. Certain present-day bacteria have this ability. Gradually the ability to live entirely by means of photosynthesis evolved, although the photosynthesis did not produce free oxygen. Many types of bacteria living today use light energy to synthesize

Figure 10.30 Moving clockwise, in this calendar 4.6 billion years of the earth's history are compressed into one year. Each day on the calendar is equal to almost 13 million years on earth.

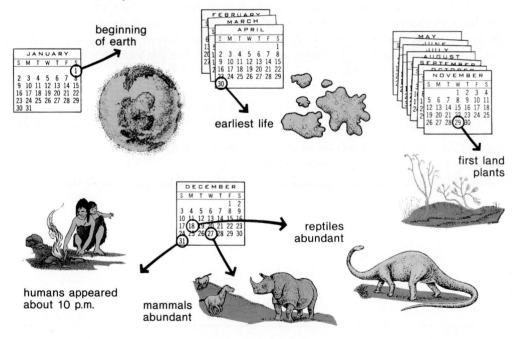

RNA as an Enzyme

Like all other sciences, biology is an incomplete body of knowledge. New information is added constantly to the existing store of knowledge. A dramatic example of the revision of a long-held assumption in biology occurred in the early and mid-1980s, when a young University of Colorado biochemist began to question the role of RNA.

RNA is one of the two nucleic acids that store genetic information. For many years, biologists believed that information molecules and enzymes were mutually exclusive. That is, an enzyme could not serve as an information molecule, and an information molecule could not serve as an enzyme.

In 1981 and 1982, Thomas R. Cech and his colleagues, while working with the protist *Tetrahymena thermophila*, discovered that RNA can indeed act as an enzyme. Like all other eukaryotes, *T. thermophila* contains portions of DNA that are not translated into messenger RNA or protein. These segments, called introns, are cut out of the precursor mRNA that is made from the DNA in the nucleus. The remaining segments of mRNA, the exons, then are spliced together to form a shorter piece of mRNA. This mRNA is released from the nucleus into the cytoplasm, where it directs the synthesis of protein on the ribosomes.

The removal of introns and splicing together of exons requires energy and enzymes. Cech and his colleagues, Paula J. Grabowski and Arthur J. Zaug, conducted splicing experiments. Assuming extracts from *T. thermophila* nuclei would supply the necessary enzymes, they used unspliced RNA with extracts and unspliced RNA without extracts. They found that a particular *Tetrahymena* intron was removed even without the extract from *T. thermophila* nuclei.

Thinking that a nuclear protein might be bound tightly to the RNA, the researchers synthesized artificial *Tetrahymena* RNA, using recombinant DNA techniques. As Cech explained in a 1986 *Scientific American* article, "the resulting RNA had never been near a cell and therefore could not be contaminated by splicing enzymes." Yet, when Cech and his team tested the newly synthesized RNA, the intron was still removed. The only possible conclusion was that the RNA itself acted as an enzyme.

In 1985, Zaug added another startling piece of information. He showed that *Tetrahymena*'s self-spliced intron could organize short pieces of RNA into longer segments. In other words, RNA not only can act as a splicing enzyme but also can organize its own replication in certain situations. Because a system for the replication of information is required for the origin of life, this discovery has implications for evolution theory. For this work, Cech shared the 1989 Nobel Prize in chemistry with Sidney Altman of Yale University, who conducted similar, yet independent, research.

Tetrahymena RNA acts as a catalyst.

Figure 10.31 Mitochondria, chloroplasts, and cilia may have originated as free-living cells.

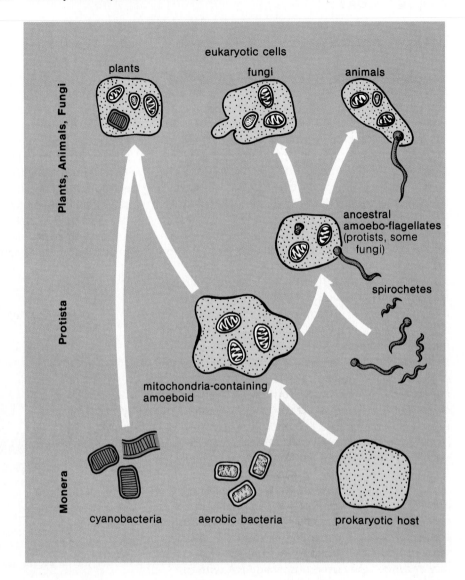

organic compounds from hydrogen sulfide or other compounds. These types of bacteria are **anaerobic** (an eh ROH bik)—able to live in environments in which oxygen is absent or present only in low concentrations. At some point, some primitive cells developed a form of photosynthesis that used water and released oxygen. Today, this type of photosynthesis is found in all autotrophic eukaryotic cells as well as in some modern photosynthetic bacteria.

Although this explanation may account for the evolution of autotrophs from heterotrophs, it does not clarify the evolutionary connection between prokaryotes and eukaryotes. Lynn Margulis of the University of Massachusetts has hypothesized that the mitochondria and chloroplasts found in eukaryotic cells originated as free-living prokaryotes (Figure 10.31). Margulis proposes that small, specialized prokaryotes may have formed close associations with larger anaerobic prokaryotes. The larger cells could have taken in the smaller ones as foodstuffs. Instead of being digested, however, the smaller prokaryotic cells continued their specialized activities in the host cell. The resulting increase in energy would have been beneficial to the host cell. Those prokaryotes that used oxygen to produce ATP could have been the ancestors of mitochondria in eukaryotic cells. Photosynthetic prokaryotes could have been the ancestors of chloroplasts. A large body of evidence supports this hypothesis, which is widely accepted.

With the advent of oxygen-producing photosynthesis, a great change took place in the atmosphere of the earth. Free oxygen began to accumulate in significant amounts, and **aerobic**, or oxygen-requiring, life became possible. Because oxygen molecules react readily, some of them probably combined to form ozone (O_3). As a layer of ozone developed in the upper atmosphere, most of the ultraviolet rays from the sun were blocked from the earth. Without that energy source, the synthesis of organic compounds in the seas was no longer possible. Heterotrophic cells became dependent on autotrophic cells for their source of food and energy.

Life has greatly altered, and continues to alter, the atmosphere of the earth. James Lovelock, a British scientist, has developed a hypothesis to explain the changes in earth's atmosphere. This hypothesis, known as the Gaia (GY uh) hypothesis, suggests that the earth, including all of its abiotic and biotic components, consists of a huge, self-regulating system. The sum of all the living organisms on the earth controls various aspects of the atmosphere, oceans, and land. Is this system responsible for the advent of the current oxygen-rich atmosphere and the changes from the earlier, more hydrogen-rich atmosphere? If evidence for the Gaia hypothesis can be found, then all organisms on earth, including humans, would be functioning components of a larger system. While some data from various disciplines seem to support the hypothesis, it is not well accepted and makes for heated scientific debates.

10.12 Matter Is Organized on Several Levels

At what point did the clusters of proteins discussed in Section 10.10 become alive? It is difficult to give a simple answer. Obviously, a cow is alive and a stone is not. We can say that an amoeba is living and a coacervate is not. Difficulties arise, however, when biologists try to set up an exact classification system for living versus nonliving things.

Sometimes nonliving things appear lifelike. If a salt crystal is added to a concentrated salt solution, for example, it will grow and start the formation of other crystals. There is a difference, however, between this type of growth and the growth of organisms. The salt crystal can grow only by taking from the environment material of the same composition as itself. The primitive cells proposed in the heterotroph hypothesis also may have grown by taking materials like themselves from the environment. The environment eventually became depleted of those materials, however. In the new environment, other organisms were favored. Because they can take in and use materials from the environment that are significantly different from themselves, modern organisms show few superficial resemblances to crystal growths.

Viruses provide a biological puzzle in this connection. Like salt, some viruses can be crystallized (Figure 10.32) and even stored. Yet when placed

Figure 10.32 This single crystal of polio virus, magnified 600 times, appears hollow because the crystal is transparent. The crystal contains about 200 million viruses.

Figure 10.33 The organization of matter on earth forms a continuum. The least complex is at the bottom; matter then increases in complexity to the top of the chart.

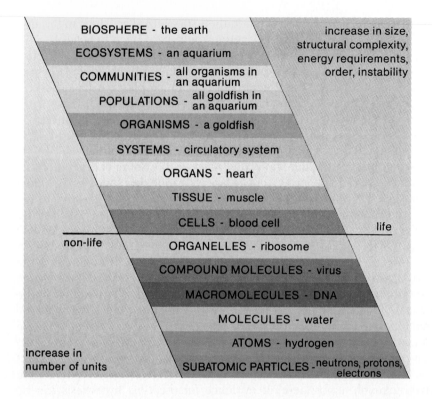

inside a living cell, these viruses can take over the cell and cause it to destroy itself, reproducing more viruses. Outside a cell, viruses are inactive. They cannot take in and use materials from the nonliving environment. Are viruses alive?

The organization of matter, from the smallest subatomic particles to the biosphere, forms a continuum like the one shown in Figure 10.33. Each level is dependent, to some degree, on the levels above and below. Where do we draw the line between living and nonliving? At present, there is no way to decide how complex a system must be before it is called living. The difference between a cow and a stone is obvious. The difference between the simplest living organism and the most complicated nonliving system is not so obvious.

Concept Review

1. What evidence do we have that many species evolved from a few species?
2. Compare environmental conditions on the earth during its early history with conditions today.
3. How can experiments such as those of Miller and Urey provide evidence for the origin of life?
4. How do scientists think primitive cells might have formed?
5. Why do scientists think it likely that the first cells were heterotrophs rather than autotrophs?
6. What changes did oxygen-producing photosynthesis cause on earth?
7. Why is it so difficult to draw a line between life and nonlife?

Summary

Classification systems enable biologists to study the great variety of organisms that exist in the biosphere. Taxonomists have developed a system that names organisms and indicates their relationships. That system is accepted throughout the world and allows biologists from different places to understand one another when discussing a particular organism. Classification systems may be based on many characteristics, but the most important ones are structural and biochemical homologies. Currently all organisms are classified in five kingdoms—Prokaryotae, Protista, Fungi, Plantae, and Animalia. Classification systems will continue to change as biologists learn more about the natural world.

The theory of evolution led to many predictions of what the first life was like. Based on those predictions, scientists have studied the evolution of the earth itself. A variety of experiments has provided evidence that life may have originated in a series of steps beginning with the formation of simple organic molecules. Later, those molecules may have assembled in a variety of aggregates, some of which had the ability to reproduce themselves. Those first cells probably were heterotrophs that used the organic molecules around them as sources of energy. Later, some cells developed the ability to use light energy. Fossil evidence supports the view of very simple beginnings; the oldest fossils resemble the prokaryotes of today.

Applications

1. Compare your name to the scientific name of a plant or animal. How are they similar or different? Is your first name more like a genus or species name? What about your last name?

2. Suppose you joined an expedition into a tropical rain forest and found an organism that was previously unknown to modern science. How would you go about determining its closest relative? Do you think you should start with breeding experiments using different organisms to determine its relationship with them?

3. The species portion of an organism's scientific name usually helps to describe some characteristic of that organism. Find out what each of these frequently used words means: (a) *vulgaris;* (b) *chinensis;* (c) *alba;* (d) *edulis;* (e) *vomitorius;* (f) *tripetalus.* Can you find plants with these words in their names? How appropriate are their names?

4. What part of the earth has the greatest variety of plants and animals? What is happening to this area, and what effect will this have on the rest of the earth?

5. What characteristics do viruses lack that would suggest they are not living organisms?

Problems

1. Morphological analogues are structures having similar functions but different origins. For example, the wing of an airplane and the wing of a bird are analogous but certainly have different origins. List the ideas that humans have borrowed from plants and animals to use in building structures and making materials.

2. What would the atmosphere of earth be like if there were no life on it? Look at the Gaia hypothesis proposed by James Lovelock.

3. Write a report on *The Panda's Thumb* by Stephen Jay Gould. What does his first essay illustrate about evolution? About homologies?

4. You can increase your awareness of the different plants and animals in your area by consulting lists of local flora, field guides, and books on natural history. Set a goal for yourself of becoming familiar with several species of plants and animals previously unknown to you by the end of the year. Find out their names, where they grow, and how they live.

5. Some scientists place the archaebacteria in a separate kingdom. Read about these organisms, and determine what characteristics might warrant their separation.

These rocks are part of a beach on Prince William Sound, Alaska, the site of the *Exxon Valdez* oil spill. What might cause the rocks on one side of the photo inset to be clean, whereas the rocks on the other side still are covered with oil?

Prokaryotes and Viruses 11

Members of the kingdom Prokaryotae—commonly known as bacteria—are the most ancient organisms on the earth. Prokaryotes are microscopic organisms that lack a nucleus and other membrane-enclosed organelles. Found in every available habitat, they affect the living world dramatically. They may be producers, consumers, or decomposers. Because they play essential roles in chemical cycles such as the nitrogen cycle, life on the earth would be totally different without them. Without prokaryotes, there would be few decomposers to break down the bodies of dead plants and animals, the oxygen supply would be decreased, and large oxygen-using animals such as humans would find life difficult. Although some prokaryotes can cause human diseases such as strep throat, rheumatic fever, and skin infections, most are beneficial. They help digest food, provide certain vitamins, and are used in manufacturing such foods as cheese and yogurt.

This chapter introduces a few of the many members of the kingdom Prokaryotae and examines their important roles in the biosphere. The chapter also discusses viruses—microscopic agents of disease that are not truly living but that dramatically affect the lives of many different organisms.

Prokaryotes

Guidepost: What are prokaryotes, and what are their roles in the biosphere?

INVESTIGATION 11.1   Distribution of Microorganisms

Just where and in what abundance are prokaryotes and other microorganisms found? In this investigation, you will discover some of the answers to these questions.

Materials (per team of 3)
4 50-mL beakers
6 sterile swab applicators
glass-marking pencil
transparent tape

stereomicroscope or 10X hand lens
self-closing plastic bag labeled *waste*
4 sterile nutrient agar plates

Procedure
1. Do not open the nutrient agar plates. Label the bottom of the four plates with your team symbol and the date. Number the plates from 1 to 4.
2. Do nothing to plate 1.

Figure 11.1 Inoculating a plate.

3. Uncover plate 2 and expose it to the air for 15 to 20 minutes. Each team should select a different area of the laboratory or the school building, including quiet corners and traffic areas. In your data book, note the exact location. Cover the plate after exposure.

4. Using plate 3, draw a line across the middle of the bottom of the plate. Label one side *Clean* and the other *Dirty*. Carefully remove a sterile swab applicator from its paper wrapper and rub the swab over part of your lab table. Without completely removing it, carefully lift the cover of the plate, as shown in Figure 11.1. Slip in the swab and *gently* rub it over the agar on the side labeled *Dirty*. Be sure the swab does not cut the agar surface or touch the clean side. Replace the used swab in its paper wrapper and discard in the self-closing plastic waste bag.

5. Use soap, or detergent, and water to wash carefully a small area of your lab table. Rub the clean area with a second swab, and then rub the swab over the side labeled *Clean*. Discard the swab in the waste bag as in Step 4. (Instead of wiping the swab over your lab table, you can compare freshly washed laboratory glassware with glassware that has been on the shelf several days.)

6. Divide plate 4 into equal sections. Allow one section to act as a control. Collect leaves, bark, pebbles, soil, or other materials. Using a clean beaker each time, mix each material with a small amount of sterile water, dip a fresh swab into the mixture, and rub over the agar surface of one section of the plate. As a control, dip a swab in plain sterile water and rub over one of the sections. Discard all swabs in the waste bag. Label the sections according to the material used.

7. Tape all plates closed and invert to prevent the water droplets that may condense inside the cover from dripping onto the agar. Incubate at room temperature for 3 or 4 days. In your data book, predict which plates will have the greatest number of microorganisms, which will have the fewest, and why.

8. Wash your hands thoroughly before leaving the laboratory.

9. Observe all plates daily, *without opening them*, and record your observations. A stereomicroscope or 10X hand lens may help in your observations.

    CAUTION: Do not remove the cover of any plate. Harmful microorganisms can be picked up and cultivated even on this simple nutrient agar.

10. On the fourth day, sketch the different types of colonies and write a description of each. There may be mold colonies present; they are fuzzy and larger than bacterial colonies. Record the number of each type of colony per plate.

11. Discard all plates in the biohazard bag designated by your teacher.

12. Each day, wash your hands thoroughly before leaving the laboratory.

Discussion

1. What are the results in plate 1? What was the purpose of plate 1?

2. What are the results in plates 2 through 4?

3. How do you account for the presence or absence of microorganisms on the examined surfaces?

4. Coughing or sneezing may spread droplets from mouth and nose to a distance of 3 m or more. The water in these droplets evaporates rapidly, leaving microscopic bits of dry matter that contain bacteria. Microorganisms from many other sources also may be carried on dust particles. How can you use this information to help interpret the class data?

5. Are microorganisms transmitted by touch alone?

6. In your daily living, how can you protect yourself and others from contamination by microorganisms?

7. Were any microorganisms found on the clean side of plate 3? What is the difference between clean and sterile conditions?

8. Were your predictions about the growth of microorganisms correct?

For Further Investigation

Does the type of medium used in the plates affect the count obtained at any one location? Does the temperature at which the plates are incubated affect the count? Design and carry out experiments that test hypotheses based on these questions.

11.1 Prokaryotes Are Structurally Simple but Biochemically Complex

Prokaryotes are extremely small—a spoonful of garden soil may contain 10 billion of them, and the total number in your mouth is greater than the number of humans who ever have lived on the earth. Prokaryotes cover the skin, line the nose and mouth, live in the gums and between the teeth, and inhabit the digestive tract of humans.

Most prokaryotes are one-celled organisms. The cell may be shaped like a straight rod, a sphere, a spiral, a filament, or a short, curved rod, as shown in Figure 11.2. Some prokaryotes join together in clusters and chains, forming multicellular structures. Just as the shapes of prokaryotes may differ, their reactions to certain stains and the size, shape, and color of their colonies also differ. Many prokaryotes move by means of flagella, but others glide over surfaces. Spiral-shaped prokaryotes move by using a corkscrew motion that involves flagellalike structures between the layers of their cell walls. Although all prokaryotes are microscopic, some are smaller than others. Rickettsia are among the smallest known cells, ranging from 0.3 to 0.7 μm in diameter. These tiny organisms, however, have been responsible for many human deaths from diseases such as Rocky Mountain spotted fever, in which the rickettsia are transmitted by ticks.

Most prokaryotic cells have rigid cell walls made of lipids, carbohydrates, and protein, but no cellulose. Figure 11.3 is a diagram of the structures of a prokaryote. Inside the cell wall is a plasma membrane that encloses the cell contents. An infolding of the plasma membrane forms the mesosome (MEZ oh sohm). The mesosome may aid in secretion, the movement of particles out of the bacterium. It also may have some role in copying the chromosome before cell division. Prokaryotes have one main chromosome, which is made of a continuous, circular molecule of double-stranded DNA. Because there is only one chromosome, the genes are not present in

Figure 11.2 Prokaryotes may be round, rod-shaped, spiraled, or filamentous. Many have flagella; others form colonies.

Figure 11.3 The structure of a typical prokaryotic cell.

Figure 11.4 Fossil prokaryote *Eoastrion* (a), this section through Gunflint chert. These fossils are about 2 billion years old. Modern prokaryotes (b) very similar to *Eoastrion,* from a microbial mat. Organism (c) shown in (b), magnified 100 times.

pairs as they are in eukaryotes. The chromosome is attached to the plasma membrane and located in an area of the cell known as the nucleoid (NOO klee oyd). In addition to one chromosome, prokaryotes usually contain one or more smaller circular DNA molecules known as plasmids. Each plasmid consists of a few genes. As discussed in Section 8.14, plasmids are important in genetic engineering and gene cloning.

Although many of the metabolic processes of prokaryotes are similar to those of eukaryotes, others are unique. As a group, prokaryotes can digest almost anything—even cellulose and petroleum. Prokaryotes are the major decomposers in most ecosystems. Because they break down complex organic compounds into simple inorganic materials used by plants, prokaryotes are essential to all food webs. They also transform inorganic materials into complex organic compounds and serve as food for protists and other microorganisms. Many prokaryotes are photosynthetic; still others are the source of antibiotics.

The fossil record of prokaryotes goes back 3.5 billion years (see Figure 11.4), which leads scientists to think they probably were the first living organisms. By comparison, records of animals date back about 700 million years and records of land plants, 470 million years. Some biologists think that about 2 billion years ago certain photosynthetic prokaryotes started a metabolic revolution that led to the increase of oxygen concentration in the atmosphere of the earth from less than 1 percent to the present-day 20 percent. Humans and other animals could not have evolved without this increase in oxygen.

Prokaryotes, which are among the most hardy organisms known, can survive extremely low temperatures, even freezing, for many years. Some species live in boiling hot springs and others in hot acids. They can survive at high pressure in great oceanic depths and at low pressure high in the atmosphere. They can tolerate total drying by forming endospores (EN doh sporz), complex thick-walled structures that contain DNA (see Figure 11.5). They are among the first organisms to inhabit new environments such as burned or volcanic areas, and they help transform these environments into areas in which other organisms can survive.

Figure 11.5 Some prokaryotes can survive harsh conditions by producing a thick-walled endospore (the dark body in the middle of this cell.) Transmission electron micrograph (X100 000).

All prokaryotes can reproduce asexually. A form of sexual reproduction known as conjugation may occur, but the genetic contribution of the parents is not equal. In conjugation, one prokaryote injects a portion of its DNA through a tube into another, as shown in Figure 11.6. Genetic material also can be transferred between prokaryotes by viruses.

Figure 11.6 Visible evidence for the exchange of cell materials between prokaryotes. This electron micrograph reveals a tube between prokaryotes of two different strains. A plasmid can move through the tube.

More than 10 000 species of prokaryotes have been described, but the vast majority have not been identified. Studies in the molecular biology of these organisms indicate there are two fundamentally different groups of prokaryotes: the **archaebacteria** (AR kee bak TIR ee uh; *archae* meaning ancient) and the **eubacteria** (YOO bak TIR ee uh; *eu* meaning true).

11.2   Archaebacteria Are Different from All Other Organisms

Although they share some characteristics with the eubacteria, archaebacteria differ in basic ways from all other forms of life. Their biochemistry and their adaptation to extreme environments suggest they were among the earliest organisms on the earth. The first archaebacteria may have lived 3.5 billion years ago.

The cell walls of archaebacteria lack a large molecule known as peptidoglycan (PEP tid oh GLY kan), which is made of sugar subunits cross-linked by peptides. This molecule is present in the cell walls of all walled eubacteria. Plasma membranes of archaebacteria contain lipids that are not like those found in the membranes of either eubacteria or eukaryotes. The transfer RNA found in archaebacteria also is different from that found in eubacteria and eukaryotes. Ribosomes in archaebacteria are unique in shape, and the nucleotide sequence of the ribosomal RNA is different from that of eubacteria.

All organisms have ribosomal RNA (rRNA), and the study of an organism's rRNA can reveal its evolutionary history. The sequence of the nucleic acid bases within rRNA changes little through time. By comparing the rRNA of two different organisms, a scientist can estimate the relationship of the organisms to each other. If the sequence of rRNA bases is very similar, the organisms are thought to be closely related. The greater the differences in

Figure 11.7 Relationship of the archaebacteria, eubacteria, and eukaryotes, based on ribosomal RNA sequence data.

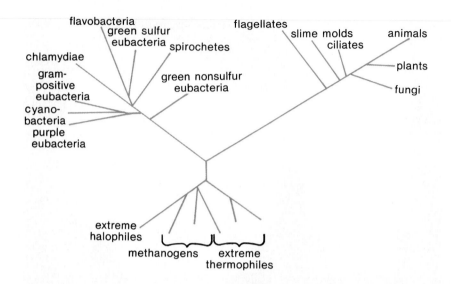

base sequence, the less closely related the organisms are thought to be. Figure 11.7 shows relationships among archaebacteria, eubacteria, and eukaryotes based on rRNA analysis.

11.3 Three Types of Archaebacteria Exist

There are three distinct types of archaebacteria. One type, the thermo-acidophiles (THER moh a SID uh fylz), lives in hot, acidic environments in which temperatures average 70° to 90° C and the water may have a pH of 2 or less (see Figure 11.8). These environments include hot sulfur springs

Figure 11.8 Thermoacidophilic archaebacteria. *Pyrodictium occultum* (a) can grow at temperatures up to 100° C. *Pyrococcus furiosus* (b) grows at temperatures up to 103° C. *Acidianus infernus* (c) grows at temperatures up to 95° C and at pH 1.0 (the pH of concentrated hydrochloric acid). In all three transmission micrographs, the bar = 10 μm.

a

b

c

and smoldering piles of coal-mining debris. Despite the low external pH, the organisms maintain an internal pH close to neutral (pH 7).

Another type, the extreme halophiles (HAL uh fylz), shown in Figure 11.9, requires a high concentration of salt to survive. Halophiles can grow in salt brine. In their presence, salted fish may become discolored and spoiled. The extreme halophiles are found in salty habitats along ocean borders and in salty inland waters such as the Great Salt Lake and the Dead Sea. Although some are photosynthetic, they do not contain chlorophyll. Instead, they absorb light energy through the pigment bacterial rhodopsin (roh DOP sin), which is similar to a visual pigment found in human eyes.

The third type, the methanogens (meh THAN uh jenz), shown in Figure 11.10, is the best studied and most widely distributed of the archaebacteria. Because methanogens are killed by oxygen, they live only in anaerobic conditions. They produce methane (CH_4) gas from hydrogen and carbon dioxide. Methanogens often are found with the eubacteria that decompose organic matter and release hydrogen gas. They are common in stagnant water, sewage treatment plants, the ocean bottom, and hot springs. They play an important role in the carbon cycle by removing organic compounds from the sediments and releasing methane to the atmosphere, where it reacts with oxygen to form water and carbon dioxide. In this way, they increase the carbon dioxide level of the earth's atmosphere. Methanogens also live in the digestive tracts of animals, including cattle and humans.

Cows, goats, deer, sheep, and antelope, unlike other herbivores, have rumens (ROO mens), special enlargements of the digestive tract. Within the rumen, protozoans, eubacteria, and archaebacteria live in an unusual ecosystem. The protozoa and eubacteria contain enzymes that break down the cellulose in the grass and other plant material a cow or goat eats. The methanogens use the carbon dioxide and hydrogen released by those organisms as their source of food and produce methane gas.

Methane gas can be used as a fuel if enough is produced and captured. Some scientists have suggested using methanogens to convert the growing amount of garbage, sewage, agricultural waste, and manure to methane gas. Organic waste would be placed in a methane generator—an airtight, anaerobic container—along with methanogens and various eubacteria, which would grow and reproduce. Methanogens produce large amounts of methane gas, which can be collected and used to heat houses. Although presently rather unpredictable, improved generators may help solve two problems—waste and garbage buildup, and the energy crisis.

Figure 11.9 Halobacteria require a high concentration of salt to survive. This organism is growing beside a salt crystal (the large rectangle).

Figure 11.10 Two methanogens. *Methanobacterium ruminatum* (a), from cow rumen. The bacterium has nearly finished dividing. *Methanospirillum hungatii* (b). Transmission electron micrographs; bar = 1 μm.

a

b

gram-positive cell wall (cross-section)

10–80 nm

a   cytosol

cell wall
(peptidoglycan layer)

plasma membrane

gram-negative cell wall (cross-section)

9–11 nm

b   cytosol

cell wall
(lipoprotein layer)

plasma membrane

Figure 11.11 Diagrammatic representation of cell wall cross sections of a gram-positive (a) and a gram-negative (b) eubacterium showing the differences in complexity.

11.4   Eubacteria Are a Diverse Group

The eubacteria may be anaerobic, photosynthetic, or thermophilic (heat-loving). In fact, their most prominent characteristic is their metabolic diversity. Autotrophs and heterotrophs are intermixed within the various eubacterial phyla. Although structure and function do not provide much information about their evolutionary relationships, a growing body of ribosomal RNA analysis is adding information rapidly. Until more data are assembled, however, classification of eubacteria at the phylum level remains uncertain.

Three major groups can be defined by their cell walls. Some lack a rigid cell wall; some have a thick peptidoglycan wall without an outer lipoprotein layer; and some have a thin peptidoglycan wall that has an outer lipoprotein layer. Figure 11.11 shows diagrams of the two wall types. The thick-walled types are known as gram-positive, because of their reaction to Gram's stain (a widely used bacterial stain); the thin-walled types are gram-negative.

Eubacteria without cell walls are surrounded by a triple-layered membrane composed of lipids. These eubacteria are resistant to penicillin, because penicillin works by inhibiting the growth of the cell wall. Commonly known as mycoplasmas (my koh PLAZ muhs), these eubacteria are probably the smallest (0.2–0.3 μm) organisms capable of independent growth. They are of special evolutionary interest because of their extremely simple cell structure. Although certain species of mycoplasmas (see Figure 11.12) cause diseases such as pneumonia in humans and cattle, most are harmless.

Gram-positive eubacteria, which have thick peptidoglycan cell walls, are widespread in soil and air. Some gram-positive eubacteria produce lactic acid (a 3-carbon compound) in their metabolism. These are used in the preparation of products such as sauerkraut, buttermilk, and yogurt. Others are the source of antibiotics, such as streptomycin, tetracycline, and erythromycin. Still others form endospores that enable them to survive unfavorable growth conditions.

The remaining eubacteria are gram-negative and have a wide variety of metabolic styles. Although photosynthesis is widespread among the gram-negative eubacteria, many photosynthetic forms are anaerobic—they do not use water in photosynthesis and thus do not produce oxygen gas. Based on the pigments present in their cells, these producers may be called green

Figure 11.12 *Mycoplasma pneumoniae* lives in human cells and causes a type of pneumonia.

Figure 11.13   Anaerobic photosynthetic eubacteria. Transmission electron micrograph of *Rhodomicrobium vanielli* (a), a purple nonsulfur eubacterium. Note the photosynthetic membranes around the periphery of the cell; bar = .05 μm. *Rhodospirillum* (b), another purple nonsulfur eubacterium, can live also as a heterotroph. Transmission electron micrograph of *Ectothiorhodospira mobilis* (c). P is the photosynthetic membrane system, R indicates ribosomes, PM is the plasma membrane, CW is the cell wall, and N is the nucleoid. Three species of purple sulfur bacteria: *Thiopedia, Thiospirillum,* and *Chromatium* (d).

eubacteria and purple eubacteria (see Figure 11.13). Many use hydrogen sulfide (H_2S) and release sulfur gas in photosynthesis; others use hydrogen gas (H_2) or compounds such as lactic acid. Like plants, however, they synthesize organic compounds from carbon dioxide and convert light energy to chemical energy.

The cyanobacteria (SY an oh bak TIR ee uh), or blue-green eubacteria, shown in Figure 11.14, are a photosynthetic group with structures and functions similar to those of algae and plants. Cyanobacteria inhabit a wide variety of environments, including hot springs and under the ice of antarctic

Figure 11.14   Cyanobacteria are abundant in all aquatic ecosystems. *Oscillatoria* (a) (X100); *Gleocapsa* (b) (X250), *Calothrix* (c) (X150).

a   b   c

lakes. They are responsible for the "algal" blooms on freshwater streams and lakes that have been polluted with phosphates. Some of the cyanobacteria grow in colonies that can be seen with the naked eye, although individual cells are microscopic.

Cyanobacteria contain chlorophyll associated with membranes, but the membranes are not organized into chloroplasts as they are in eukaryotic cells. Their photosynthesis, like that of plants and algae, uses water and results in the release of oxygen gas. Cyanobacteria are thought to be primarily responsible for the start of the oxygen revolution that changed the earth's atmosphere, making possible the evolution of large oxygen-using organisms.

There are thousands of species of cyanobacteria, and thousands more existed in the ancient past. Remains of these ancient communities are found in fossil stromatolites. Stromatolites still are being produced in the salt flats along the Persian Gulf, the Bahamas, western Australia, and the west coast of Mexico.

11.5   Eubacteria Make Possible the Nitrogen Cycle

Although cyanobacteria make their own food using energy from the sun and carbon from carbon dioxide, most eubacteria get their energy from foods produced by other organisms, either living or dead. If the food source is living, the eubacteria are called parasites. If the food source is dead, they are called decomposers. Some eubacteria neither photosynthesize nor use organic molecules as their source of energy. Instead, they obtain energy from inorganic molecules in the environment. These eubacteria are called chemosynthetic (KEE moh syn THEH tik). Both chemosynthetic and decomposer eubacteria play important roles in the nitrogen cycle.

Nitrogen is essential to all living things. In fact, it is found in two of the four types of biological molecules needed for life—proteins and nucleic acids. Just as carbon cycles through a community and back into the physical environment from which it came, nitrogen also cycles from the biotic to the abiotic parts of the ecosystem. Study the nitrogen cycle shown in Figure 11.15 as you read the remainder of this section.

Although nitrogen gas (N_2) makes up about 78 percent of the atmosphere, neither plants nor animals can use that form of nitrogen. Instead, all consumers get nitrogen-containing compounds in the things they eat, which, of course, can be traced to producers. Producers get their nitrogen-bearing compounds from the soil (or water) in which they grow. Nitrogen gas dissolved in water is present in the soil. Also living in the soil or in the roots of certain plants are nitrogen-fixing eubacteria. These chemosynthetic eubacteria can "fix" nitrogen, converting it to a form they or other organisms can use. Nitrogen fixers convert nitrogen gas to ammonia (NH_3), which can be used to synthesize proteins and other nitrogen-containing compounds.

Farmers discovered long ago that soils in which clover, peas, beans, alfalfa, or other members of the legume family had been grown produced better crops than did other soils, although they did not know why. Today, scientists know the crops grow better because of an increase in the nitrates in the soil. The increased nitrate level in the soil is the result of nitrogen

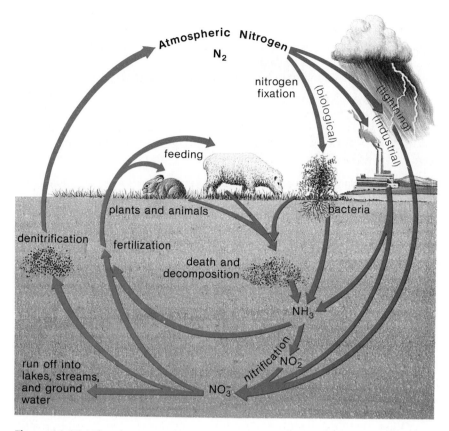

Figure 11.15 The nitrogen cycle.

fixation and chemical reactions following fixation. The plants, however, do not fix the nitrogen. Eubacteria that live in the roots of these and many other legumes are responsible for nitrogen fixation. The eubacteria induce the plant to form swellings on its roots called nodules (NOD yoolz), shown in Figure 11.16. Under favorable conditions, eubacteria in these nodules can fix as much as 22.5 g of nitrogen per square meter every year. Some

a

b

Figure 11.16 Soybean roots with nodules formed by a species of *Rhizobium* (a). Alfalfa nodule cells packed with actively nitrogen-fixing *Rhizobium* (b). Scanning electron micrograph (X640); bar = 10 μm.

Figure 11.17 Nitrifying eubacteria. *Nitrosomonas* (a) converts ammonia (NH_3) to nitrite (NO_2^-). *Nitrobacter winogradskyi* (b) converts nitrite to nitrate (NO_3^-). Transmission electron micrographs, (a) X121 300; (b) X213 400.

free-living eubacteria and cyanobacteria also have the ability to fix nitrogen. The ammonia produced by nitrogen-fixing eubacteria can be used by the legume plants, by eubacteria, or by other plants.

Decomposer eubacteria in the soil break down the complex organic compounds in dead plants and animals. They use the energy in these compounds and convert the compounds to simpler substances, including the gas ammonia. Some of the ammonia escapes into the atmosphere, but much of it dissolves in soil water, where it reacts chemically with hydrogen ions to form ammonium ions (NH_4^+). In the form of ammonium ions, nitrogen may be absorbed by the roots of plants. It then is built into living material again by the plants.

Other chemical reactions also may occur in the soil. Two groups of chemosynthetic nitrifying (NY trih fy ing) eubacteria (Figure 11.17) continue the nitrogen cycle. One group changes the ammonium ions (NH_4^+) to nitrite ions (NO_2^-). The other group changes the nitrite ions to nitrate ions (NO_3^-). Plants can absorb the nitrates from the soil and can use them to build proteins.

Nitrifying eubacteria operate only under aerobic environmental conditions. Oxygen from the air spaces found between soil particles dissolves in soil water. Sometimes, all the spaces fill up with water, leaving no room for air. When that happens, the soil environment becomes anaerobic, and nitrifying eubacteria cannot carry on their activities. Denitrifying (DEE ny trih fy ing) eubacteria, however, thrive in anaerobic environments. They change any remaining nitrates to nitrogen gas. The nitrogen gas gradually escapes into the atmosphere, and the cycle of nitrogen is complete.

Concept Review

1. What characteristics distinguish prokaryotic cells?
2. In what ways are prokaryotes important in the biosphere?
3. In what ways do archaebacteria differ from eubacteria?
4. How can methanogens help with the energy crisis?
5. What characteristics do microbiologists use to distinguish eubacteria?
6. How might cyanobacteria have contributed to the evolution of animals?
7. How does nitrogen fixation take place?
8. How is nitrogen returned from organisms to the atmosphere?

Disease

Guidepost: What is the relationship between a pathogen and a disease?

11.6 Diseases Result from Interrelationships

Disease is a condition that interferes with an organism's ability to perform a vital function. An infectious disease involves a disease-causing agent, or **pathogen**. The disease results from the interaction between an infected organism, or **host**, and the pathogen. The pathogen infects the host. If the pathogen harms the host, symptoms of disease may appear.

How serious a disease becomes depends on the characteristics of both the host and the pathogen. The ability to cause disease is called **virulence** (VIR yuh LENTS). The ability of an infected host to cope with a pathogen is called **resistance**. A pathogen with high virulence may cause death in a host with low resistance. A host with high resistance, however, may show only mild symptoms of the same disease. Sometimes, a moderately virulent pathogen may produce serious illness. During a famine, for example, more people die from disease than from starvation. This is because a poorly nourished host may have much less resistance than a well-fed host.

Although genetics does play a role in the ability of an organism to resist disease, much of an individual's resistance to disease is acquired during the lifetime of the individual, rather than inherited. Resistance, whether acquired or inherited, is called **immunity**. When a pathogen infects a human host, the host produces proteins called **antibodies** that can destroy the pathogen. In addition to antibodies, certain cells of the body's defense, or immune, system have a chemical memory. (The immune system is discussed in Chapter 16.) If the host survives the initial infection, these cells retain the ability to produce the same antibodies again. If the same pathogen infects the host a second time, the host's body can act against it more rapidly, preventing symptoms from developing or reducing their severity.

Each type of antibody is effective against only the particular pathogen that brought about its production, or pathogens that are very similar. For example, a person can have chicken pox only once, because antibodies produced during the first infection usually prevent symptoms from developing after a second infection. These antibodies are not effective against measles, however. Because there are many types of pathogens that cause colds, immunity to one type does not confer immunity to all the others.

Many diseases can be prevented by **vaccines**. A vaccine contains only enough of the killed or weakened disease-causing agent to stimulate production of antibodies by the immune system, usually without producing symptoms of the disease. Memory cells then act to keep later infections by the same pathogen from causing disease. Vaccines first were used by Edward Jenner, an English physician (see Figure 11.18).

Disease may result from a variety of causes. The pathogens involved in infectious diseases may be viruses or eubacteria or other organisms. For example, athlete's foot, ringworm, potato blight, and corn smut are caused by fungi. Protists cause malaria, African sleeping sickness, and amoebic dysentery. Many worms and insects cause diseases of plants and animals, and insects themselves are important vectors, or carriers, of disease. Other diseases, such as scurvy, are the result of diet deficiencies; still others are a result of advancing age; and some, such as asthma, are brought on by reactions to substances or pollutants in the environment. Finally, some disorders, such as cystic fibrosis or Huntington disease, are hereditary.

Figure 11.18   The English physician Edward Jenner demonstrated in 1796 that by inoculating patients with a vaccine made of cowpox viruses, he could protect them from the more serious disease of smallpox.

Figure 11.19 Fire blight is a disease of apples, pears, and related trees and plants. It is caused by the bacterium *Erwinia amylovora.*

11.7 Some Eubacteria Are Pathogens

Although the vast majority of prokaryotes are beneficial or at least harmless to other organisms, some eubacteria are pathogenic. At the present time, however, no archaebacteria are known to cause disease.

Eubacterial diseases may be spread through water, food, or air. Many plant diseases are caused by eubacteria, and almost all types of plants are susceptible to one or more types of eubacterial disease. Fire blight, shown in Figure 11.19, is a common eubacterial disease that can destroy fruit trees. In Florida in 1984, an outbreak of a eubacterial disease called citrus canker led to the destruction of more than 4 million citrus seedlings in four months in an effort to halt the spread of the disease. Citrus canker is caused by one of more than 100 distinct varieties of the gram-negative eubacterium *Xanthomonas campestris.* Other varieties of this eubacterium cause diseases with similar symptoms in beans, cabbages, peaches, and other plants. Citrus canker now threatens the existence of the Florida citrus industry—an annual $2.5-billion business.

Eubacteria also cause human diseases such as cholera, leprosy, tetanus, eubacterial pneumonia, whooping cough, and diphtheria. Many eubacteria multiply in the human digestive tract and leave the body in the feces. If untreated feces enter the water supply, as discussed in Chapter 3, the eubacteria may enter another organism when it drinks the water. Waterborne diseases such as typhoid kill 6 million children each year. Most cases of food poisoning are due to eubacteria. In some cases, the eubacteria produce poisonous substances, or toxins, that are released in the food. These toxins can cause nausea, vomiting, and diarrhea within a few hours of ingestion. In the case of botulism, even a small amount of the toxin can be fatal. In other cases, eubacteria ingested with food multiply in the intestine, causing symptoms even several days after ingestion.

Many eubacteria are airborne. Whenever you open your mouth and exhale, you release droplets of moisture, each of which may contain one or two eubacteria. With a sneeze, you release many thousands of droplets at great speed. Droplets from a sneeze, visible in Figure 11.20, have been clocked at 200 miles an hour. Droplets from a cough travel at about half that speed. Each time you sneeze, you can expel from 10 000 to 100 000 individual eubacteria. The eubacteria travel through the air without being killed and may infect another person standing close to you.

Figure 11.20 Droplets from a sneeze.

Toxic shock syndrome, which is characterized by a sharp drop in blood pressure followed by fever, vomiting, and diarrhea, is caused by certain strains of the eubacterium *Staphylococcus aureus.* Although about 85 percent of the cases of toxic shock syndrome reported in the United States have occurred in menstruating women who use tampons, both men and women can contract the disease. Researchers have found that two fibers contained in super-absorbent tampons absorb strong concentrations of magnesium from the surrounding blood and tissue. These high magnesium concentrations enhance the production of toxins by the eubacteria. As the eubacteria multiply, they produce more toxins, which are released into the body, killing cells and causing other symptoms of the disease. Most people have developed a resistance to the toxin by the age of 20.

Even dental caries, or tooth decay, is caused by eubacteria. The decay begins on the surfaces of the teeth in a film known as dental plaque. This film, which consists of large eubacteria in a complex sugar matrix (shown in Figure 11.21), builds up on unbrushed teeth and in crevices not reached by a toothbrush. Diets high in sugars are especially harmful because eubacteria break down the sugars to lactic acid, which causes the loss of calcium from the teeth. Enzymes released by eubacteria then begin to break down the tooth enamel, allowing other eubacteria to penetrate the teeth. Fluoride makes the teeth more resistant to decay because it retards the loss of calcium.

Figure 11.21 *Streptococcus mutans* is a lactic acid bacterium associated with tooth decay. (X8000)

INVESTIGATION 11.2   Control of Eubacteria

How effective are antibiotics, antiseptics, and disinfectants in controlling the growth of eubacteria? In this investigation, you can test a variety of substances with two common eubacteria.

 After reading the procedure, construct hypotheses that predict how the substances you select will affect the eubacteria.

Materials (per team of 3)

3 pairs of safety goggles
18 construction-paper disks
glass-marking pencil
stereomicroscope or 10X hand lens
2 pairs of forceps
2 sterile swab applicators
transparent tape
metric ruler

self-closing plastic bag labeled *waste*
sterile alcohol pad
4 sterile nutrient agar plates
small containers of distilled water, antiseptics, and disinfectants
6 antibiotic disks
broth cultures of *Micrococcus luteus* and *Escherichia coli*

SAFETY   **Put on your safety goggles.**

Procedure

1. Using a glass-marking pencil, label the bottoms of the four plates *A, B, C,* and *D.* Divide each plate into six equal, pie-shaped sections. Label the sections *1* through *6.* Label plates B and D *Control.*

2. Carefully remove a sterile swab applicator from its paper wrapper and use the swab to streak the entire surface of plates A and B with *Micrococcus luteus,* as shown in Figure 11.22. Replace the used swab in its paper wrapper and discard in the self-closing plastic waste bag.

3. In the same manner, streak the entire surface of plates C and D with *Escherichia coli.* Discard the swab in the waste bag as in Step 2.

4. Use forceps to place one plain disk each near the outside edge of sections 1, 2, and 3 in plates B and D. Be sure the disk adheres to the agar surface but does not break through it.

5. Using forceps, pick up a clean disk and dip it in distilled water. Gently tap the disk against the side of the container to remove excess water and place it near the outside edge of section 4 in plate B. Repeat until you have placed a disk in sections 4, 5, and 6 of plates B and D. Why do you dip these disks in distilled water before placing them on the agar?

6. Your teacher will provide disks that have been soaked in three different antibiotics, *I, II,* and *III.* Place one disk of antibiotic I near the outside edge of section 1 in plate A and one disk near the outside edge of section 1 in plate C. In the same way, place disks of antibiotic II in section 2 of plates A and C, and disks of antibiotic III in section 3 of plates A and C.

Figure 11.22 Two patterns for streaking culture plates.

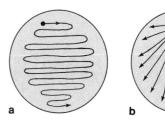

a   b

7. Select any three of the antiseptic and disinfectant solutions your teacher provides. Using forceps, dip a clean disk into the first solution. Place one disk each in section 4 of plates A and C. Clean the forceps with a sterile alcohol pad before taking another disk.

8. Repeat Step 7 with two other solutions, placing them in sections 5 and 6 of plates A and C.

9. Cover and tape all four plates. Invert them and incubate at 37° C.

10. In your data book, record the substance into which you dipped each disk and the section in which you placed it.

11. Wash your hands thoroughly before leaving the laboratory.

12. After one or two days, observe all plates, *without opening them*, and record your observations. A stereomicroscope or 10X hand lens may help your observations. Use a metric ruler to measure the width of any clear areas around the disks. Record your measurements.

 CAUTION: Do not remove the cover of any plate. Harmful microorganisms can contaminate the plates when you inoculate them.

13. Discard all used plates in the biohazard bag designated by your teacher.

14. Wash your hands thoroughly before leaving the laboratory.

Discussion

1. What do the clear areas indicate? What do differences in the widths of the clear areas indicate?

2. What evidence do you have that the inhibition of these eubacteria is due to the chemicals on the disks and not the disks themselves?

3. Which antibiotic was most effective against *Micrococcus luteus*? Against *Escherichia coli*? What might account for the difference in effectiveness?

4. Which of the two species of eubacteria is more sensitive to all the chemicals?

5. Which antiseptic or disinfectant would you use to control *M. luteus*? To control *E. coli*?

6. *E. coli* are normally harmless eubacteria that live in the human intestines. Does the reaction of *E. coli* to antibiotics suggest that antibiotics should be used only when necessary? Explain.

7. How does an antibiotic differ from a disinfectant? From an antiseptic?

For Further Investigation

1. Test different concentrations of the same antiseptic or disinfectant.

2. Test the effects of heavy metals such as copper, lead, or zinc.

3. Check the antiseptics and disinfectants you used to determine the chemicals they contain. Use a reference such as *The Merck Index*, 11th ed. (Rahway, NJ: Merck, 1989) to determine which chemicals are likely to be the active ingredients.

11.8   Viruses Are Unusual Pathogens

What do AIDS, hepatitis, measles, mumps, influenza, colds, and polio have in common? These diseases all are caused by viruses such as those shown in Figure 11.23. A virus is an infectious agent that contains a nucleic acid (either DNA or RNA) and a protein coat. Viruses are so small that they can be seen only with an electron microscope. In fact, viruses can pass through

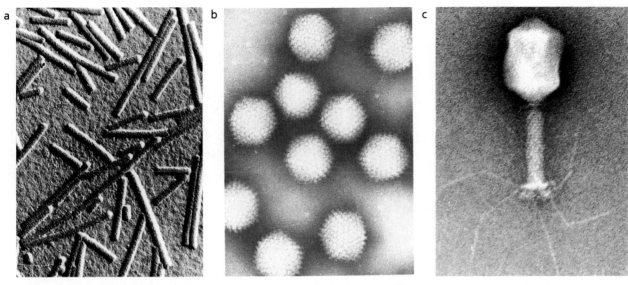

Figure 11.23 All viruses are pathogenic. Tobacco mosaic virus (a) causes a disease of plants (X185 000); adenovirus (b) causes respiratory illnesses in humans (X11 000); T4 bacteriophage (c) (X240 000).

most bacteriological filters. Viruses may play one of two roles when they enter a host cell. As agents of disease, viruses enter the host cells, disrupt their normal functioning, and sometimes kill them. As agents of heredity, viruses can enter cells and cause permanent, inheritable changes. Usually these changes are not harmful and even may benefit the host. Often, the role the virus plays depends on the host cell and environmental conditions.

Viruses differ from living things in several important ways. They are able to produce copies of themselves only inside a living organism. Outside the host cell, viruses neither reproduce, feed, nor grow. They have no metabolism of their own. They do not take in and use energy. They do not have cell parts. Some can be crystallized and survive for years in that state. Only when they enter the appropriate host cells can they resume reproduction. Viruses infect plants, animals, fungi, and prokaryotes (see Figure 11.24). The virus first attaches to the host cell by means of its protein coat and then injects its DNA or RNA into the host cell. The viral nucleic acid takes over the machinery of the host cell, commanding it to make more viral protein and viral nucleic acid. The proteins and nucleic acids then are assembled into new virus particles, and the infected cell ruptures, releasing hundreds of newly made viruses. Each new virus has the ability to infect a single new host cell. Figures 11.25 and 11.26 on the following page show these events for a virus that has entered a eubacterium.

Usually, genetic information is stored in DNA, transcribed into mRNA, and then translated into protein. In most cases, RNA viruses transcribe their genetic information from RNA directly into mRNA. Certain RNA viruses, known as retroviruses, must go through an additional step before they can reproduce. Retroviruses must make copies of DNA from their RNA, the reverse of the normal flow of stored information in biological systems. Like other viruses, a retrovirus binds to the surface of a host cell and injects its RNA into the cell. Reverse transcriptase, a special enzyme associated with

Figure 11.24 TYMV (turnip yellow mosaic virus) in a young cabbage plant.

1. virus approaches eubacterium
2. virus becomes attached to cell
3. viral DNA is injected into cell
4. protein coat remains outside
5. new viral DNA molecules are made
6. viral protein coats are made
7. complete viruses are assembled
8. eubacterium bursts, releasing new viruses

Figure 11.25   A virus attacking a single eubacterium.

Figure 11.26   The eubacterium *Escherichia coli* (a) with T4 phage viruses attached to the surface and inside (X50 000). T4 phage (b) attached to the eubacterium after it has injected its DNA (X200 000).

the RNA, allows the retrovirus to make a complementary DNA copy of its RNA (see Figure 11.27). The newly formed viral DNA then can be integrated into the host cell DNA, replicated at the same time, and transmitted to offspring cells. When the cell produces RNA from its own DNA, it also produces viral RNA, which becomes the source of new viral particles, thus continuing the infection.

Retroviruses are known to cause cancer in some animals and have been associated with certain types of cancers in humans. Human oncogenes have similarities to genes in these retroviruses—both can result in cancerous growth when they are somehow disturbed. The virus that causes AIDS is a retrovirus that attacks and kills certain cells of the immune system. Once these cells are killed, the immune system is unable to perform its normal function of defending the organism against disease.

Figure 11.27   Retroviruses use the enzyme reverse transcriptase to make a DNA copy of their RNA. The viral DNA then is incorporated into the host cell DNA, and the host cell directs production of new viruses.

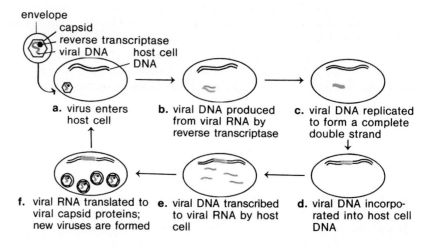

envelope
capsid
reverse transcriptase
viral DNA
host cell DNA

a. virus enters host cell

b. viral DNA produced from viral RNA by reverse transcriptase

c. viral DNA replicated to form a complete double strand

d. viral DNA incorporated into host cell DNA

e. viral DNA transcribed to viral RNA by host cell

f. viral RNA translated to viral capsid proteins; new viruses are formed

Slow Infections of the Brain

Since 1976, 22 cases of a rare and fatal brain infection have been diagnosed in a sparsely populated area of Czechoslovakia. This disease, identified as Creutzfeldt-Jakob disease, or CJD, kills its victims within seven months after the first symptoms appear. Normally, CJD occurs in one person per million every year throughout the world. The incidence in Czechoslovakia is several hundred times that.

CJD resembles kuru, a fatal infection that was common among Papua, New Guinean tribes who ate and handled the brains of their dead during mourning rituals. Kuru victims first lose their coordination and their ability to walk and then become insane. Now that the rituals have ended, the number of kuru victims has decreased. CJD also is similar to "mad cow disease" (bovine spongiform encephalopathy, or BSE), which has caused 15 000 cattle to be destroyed in Great Britain since 1986. Most investigators believe that CJD, kuru, and mad cow disease are unconventional slow infections that are variants of scrapie. Scrapie affects the central nervous system of sheep and goats. Animals infected with scrapie also lose their coordination and become unable to walk. Scrapie and scrapielike diseases are called slow infections because of the long incubation period—from several months to several years.

At least 25 cases of CJD in Western countries can be traced to technological misfortune. Injections of growth hormone from human pituitary glands are responsible for 14 cases. Because of the long incubation period of the disease, more cases caused by technological misfortune probably will appear. Three people who received skin grafts of the tissue that lines the inside of the skull developed CJD. The use of unsterile instruments and electrodes during neurosurgery has produced six cases. One case was transmitted during a cornea transplant and another by root-canal surgery.

The nature of the agent that causes CJD and other slow infections of the brain is controversial. The victims exhibit no immune response, and the infection is resistant to treatments that normally kill infectious agents. Even ultraviolet radiation, which normally destroys nucleic acids, appears to have no effect on the agent.

Some researchers believe the cause is a prion, a deviant piece of pure protein. If this abnormal protein can somehow catalyze its own formation from a healthy version of the protein present in all brains, it may not need any nucleic acid. Other researchers believe the protein-only hypothesis is implausible. Analyses of experiments dealing with radiation and heat lead some researchers to conclude the causative agent probably is a small virus. The most common view, however, is an intermediate one: The agent is a protein with a tiny amount of nucleic acid. The amount is too small to code for more protein but large enough to carry a scrap of genetic information. One researcher has found a fragment of mitochondrial DNA in scrapie proteins extracted from infected brains, but its role is unclear. Until researchers finally can identify the cause of scrapie and related diseases, treatment will remain impossible.

Slow Infections of the Brain

Infection	Affected Organisms
Scrapie	Sheep, goats
Bovine spongiform encephalopathy— mad cow disease	Cattle
Transmissible mink encephalopathy	Mink
Creutzfeldt-Jakob disease Kuru Gerstmann-Straussler syndrome	Humans
Chronic wasting disease with spongiform encephalopathy	Captive Rocky Mountain elk and mule deer
Spongiform encephalopathy	Nyala Gemsbok Domestic cat

Source: From "Oravske Kuru" by Tim Beardsley. Copyright © August 1990 by Scientific American, Inc. All rights reserved.

Concept Review

1. Describe the interaction between host and pathogen that may result in disease.
2. Distinguish between infection and disease.
3. What is immunity? How can a vaccine bring about the development of immunity?
4. Describe, using examples, the major ways by which eubacterial diseases can be spread.
5. What is the role of eubacteria in dental caries?
6. What is the structure of a virus and how does it infect a host cell?
7. How does a virus direct the formation of other virus particles?
8. How does a retrovirus differ from a virus?

Guidepost: What are the characteristics and dangers of sexually transmitted diseases?

Sexually Transmitted Diseases

11.9 Eubacteria Cause Several Sexually Transmitted Diseases

A number of diseases are transmitted by sexual contact with infected persons. These diseases include gonorrhea, syphilis, chlamydia, herpes, and AIDS (*A*cquired *I*mmune *D*eficiency *S*yndrome). Gonorrhea, syphilis, and chlamydia are caused by eubacteria; herpes and AIDS are caused by viruses.

Gonorrhea is one of the most widespread human diseases. It is caused by the eubacterium shown in Figure 11.28, which thrives in the human body but can be killed by drying, sunlight, or ultraviolet light. An infected person can transmit the eubacteria to his or her sexual partners through the moist membranes of the sex organs. In females, symptoms of the infection may be only a mild irritation of the vagina, which often goes unnoticed. In males, however, the organism causes a painful infection of the urethra that results in burning during urination and sometimes the discharge of pus. Gonorrhea also can cause eye infections in adults and in newborns of mothers with the disease. Most gonorrhea responds well to treatment with antibiotics, but the number of infections reported each year continues to increase for many reasons. First, because the symptoms can be mild in females, one female unknowingly can infect many men. Second, oral contraceptives, which are widely used today, alter the pH of the vagina and allow the eubacteria to grow more readily. Third, for unknown reasons, an individual does not develop immunity to the eubacteria, so repeated infections are possible. And fourth, because gonorrhea usually is treated with large doses of penicillin, penicillin-resistant strains are becoming more common. Left untreated, gonorrhea can result in sterility.

Less common but more serious is syphilis, which is caused by the spiral eubacterium shown in Figure 11.29. Syphilis can be fatal. The incubation period—the time between infection by the eubacterium and the appearance of symptoms—for syphilis is about three weeks. The first symptom is a hard, painful ulcer called a chancre (SHAN ker), which appears at the site of infection. The chancre disappears in about one to four weeks, and the infected person may experience no other symptoms for two to four

Figure 11.28 The eubacterium that causes gonorrhea, *Neisseria gonorrhoeae* (X1000).

months even if the disease is left untreated. After this time, the disease enters the second stage, which is marked by a skin rash and infections in various organs. The disease then may become latent, or hidden. Although the disease may remain latent for the remainder of an individual's life, in many people the disease enters a third stage. This stage can produce severe nervous system or circulatory system damage, insanity, and even death. Syphilis is treated with penicillin, but as with gonorrhea, many penicillin-resistant strains have developed, and recent reports indicate an increase in the number of cases.

More common than either gonorrhea or syphilis are chlamydial infections, which can be controlled easily with tetracycline. In females, chlamydial infections usually affect the cervical canal and the vagina and may cause pelvic inflammation. The symptoms may go unrecognized, but if not treated, they may result in sterility. Infected females may pass the disease to their children at the time of birth, resulting in eye and lung infections. Chlamydial infections in males cause irritation of the urethra and produce a discharge from the penis. Because chlamydial infections are difficult to diagnose, they are the most prevalent sexually transmitted disease today.

Figure 11.29 A scanning electron micrograph of *Treponema pallidum*, a spirochete that causes syphilis (X48 000).

11.10 AIDS Is a Deadly Disease

The virus responsible for AIDS was identified in 1983 and named *H*uman *I*mmunodeficiency *V*irus, or HIV. HIV primarily infects those cells of the immune system known as helper *T* cells, but it also may infect brain or nerve cells. Helper *T* cells, brain cells, and nerve cells all have a particular cell-surface molecule called CD4 that serves as the site of attachment for the virus. Inside the host cell, the viral RNA is copied into DNA, which may be integrated into the host cell genome. Each infected cell then can serve as a factory for the production of HIV. When the cell eventually ruptures, hundreds of viruses, each able to infect another cell, are released, as shown in Figure 11.30. Reproduction of HIV in helper *T* cells leads to a gradual decline in the number of helper *T* cells, weakened function of the immune system, and increased susceptibility to opportunistic infections and cancers. Opportunistic infections are infections a normal, fully functioning immune system would be able to repel. Although HIV can damage organs directly, it is the opportunistic infections that account for up to 90 percent of deaths from AIDS.

The incubation period for HIV infection is two weeks to three months. If infection has occurred, HIV antibodies can be detected in the blood at that time. The incubation period for AIDS is much longer. An individual infected with HIV may not develop AIDS for several months or even as long as 12 years. The variation in the AIDS incubation time is related to the length of time the virus remains inactive in the cells of the immune system. Symptoms vary widely, depending on the type and number of opportunistic diseases that occur, and often alternate with periods of relatively good health.

AIDS, the most severe stage of HIV infection, usually leads to death within two years of diagnosis. Diagnosis is based on the presence of certain infections or cancers in a person who tests positive for HIV antibodies. Among those conditions are a variety of infections caused by fungi, protozoans, eubacteria, and other viruses seldom seen in the population (or

Figure 11.30 The several hundred small particles are HIV releasing themselves from a helper *T* cell, which they infect and ultimately destroy. Scanning electron micrograph (X124 000).

only in much milder form), and a type of skin cancer known as Kaposi's sarcoma. Besides Kaposi's sarcoma, the most common opportunistic infection associated with AIDS is a pneumonia caused by a protozoan parasite, *Pneumocystis carinii*. Tuberculosis, which is caused by a eubacterium, also is prevalent among HIV-infected individuals.

HIV usually is introduced into the body through sexual contact, although infection also can occur through contaminated blood. AIDS can occur in both homosexuals and heterosexuals, and heterosexual transmission is becoming more common. Abstaining from sexual activity is one way to reduce the risk of infection. Limiting sexual activity to one noninfected individual and using condoms are two other methods that reduce the risk of acquiring AIDS from sexual contact.

Many individuals with AIDS are drug addicts who use intravenous injections and have shared hypodermic needles with infected people. Some AIDS patients are hemophiliacs and other recipients of blood transfusions that contained HIV. Since 1985, however, all donated blood in the United States has been tested for HIV. Because the virus can cross the placenta and infection can occur at birth or through the mother's milk, an increasing number of AIDS cases are appearing in babies born to HIV-infected mothers. Although there is no evidence that AIDS can be transmitted through casual contact such as shaking hands or swimming in the same pool, several health-care workers exposed to HIV by way of accidental cuts or injections have contracted AIDS.

INVESTIGATION 11.3   Screening for AIDS

The public policy issues surrounding the detection, control, and treatment of AIDS are complex and open to debate. This investigation examines several of these issues.

Materials (per team of 3 or 4)
none

PART A   Screening Donated Blood and Plasma for HIV Antibody
Procedure

1. For this investigation, you will be working in small discussion groups of 3 or 4 students. Read the following paragraphs before forming your small groups.

   HIV infection has been associated with several well-defined high-risk groups, including recipients of blood and blood products. Screening blood for contamination by the causative agent, HIV, has become one of the highest public health priorities and currently is in place nationwide.

   When your body is infected by a virus, it produces antibodies that help destroy the virus. The blood-screening test currently used on all blood detects HIV antibody, not the virus itself. Any blood that tests positive is tested again, using a different, more specific test. If both tests are positive, the person has been infected with HIV. That does not necessarily mean that the person has AIDS or even that he or she will develop AIDS at a later date. It does mean, however, that these possibilities exist, and that the donated blood from that person should not be transfused into other individuals.

   Unfortunately, the current tests sometimes produce *false positive* results. False positive results indicate that a person *has* HIV antibody when in fact the person *does not* have the antibody in his or her blood. False positives are rare, but they do occur. Adverse publicity and personal trauma may result for a donor who receives a false positive report from a blood donation. Fear of

AIDS and concern with the screening process could decrease the pool of volunteer blood donors.

For the person who is truly positive for HIV antibody, notification and elimination from the donor pool is, indeed, appropriate. However, for donors falsely labeled as antibody positive, these results may do needless harm to their social and personal relationships.

2. If your small group is designated group A, your task is to provide arguments in favor of the nationwide screening program for HIV antibody in blood banks. If you are group B, your task is to provide arguments against the nationwide screening program for HIV antibody in blood banks.

Discussion

1. Group A: During your discussion to develop arguments *for* the nationwide screening program, you might wish to consider the following questions:
   a. Of what value is the screening program for the general public?
   b. How do hemophiliacs, who may require frequent blood-cell products, benefit from the program?
   c. Of what value is the screening to blood recipients in general?
   d. Of what value is screening to a healthy donor infected with HIV?
   e. What rights do blood donors have?
   f. What rights do blood recipients have?

2. Group B: During your discussion to develop arguments *against* the nationwide screening program, you might wish to consider the following questions:
   a. What problems might arise from screening for HIV and getting a false negative test, results that indicate a person *has not* been infected with HIV when in fact he or she *has* been infected?
   b. What problems might arise from a false positive result? Consider employment and medical insurance, for example.
   c. How might a false positive test for HIV antibody affect the life of that individual? Consider interpersonal relationships, friends, and family.
   d. In the use of screening tests, there is concern both for the public at large and for the individual. What types of conflicts might arise as a result of these concerns? How might society reach some compromise?

PART B   Coping with a Positive Result
Procedure

3. Remain in your groups and read the following scenario:

   Jean and Roger were married for three years before they decided to begin a family. When Jean's doctor determined that she was pregnant, she ordered an ultrasound to confirm the gestational age of the fetus. The ultrasound revealed that Jean was not carrying one child, but twins. The delivery was uncomplicated and both children, one girl and one boy, were quite healthy.

   Two months later, Roger's boss was admitted to the hospital for surgery, and the employees decided to donate blood to the blood bank to help defray the costs of the surgery. All donated blood was screened for HIV antibody. Roger's test was positive, and the hospital performed the second screening test. That test also was positive, indicating that Roger had been infected with HIV.

Discussion

1. What problems does the positive result pose for Roger?
2. What might this result mean for his wife and children?
3. Do Roger's coworkers and his boss have a right to know about the test results? Why or why not?

4. Using what you know about HIV and AIDS, should Roger and Jean be concerned about any relationships they may have had before they were married? What should they do?

5. Assume that one child tests positive for HIV antibody and the other child tests negative. What rights does each child have?

6. If the children live to attend school, what rights do the other children have?

11.11   Prospects for Treatment of AIDS Are Improving

The efforts to understand AIDS have greatly increased our knowledge of how the immune system works and hold promise for better treatment in the near future. Although it was once thought impossible to vaccinate against AIDS because the protein coat of the virus can change, thus appearing to the immune system as a different virus, prospects for a vaccine are improving. A study of pregnant women infected with HIV found HIV antibodies present in women who had uninfected babies, but not in women whose babies were infected. These results indicate that antibodies can protect against infection. A number of vaccines are being tested and others are in various stages of research and development. Of the trial vaccines under study, some protect chimpanzees from infection and some stimulate the production of antibodies in AIDS patients as well as in healthy people.

Researchers are trying to design specific drugs to block the action of HIV. For example, in test-tube studies, when CD4 (the cell-surface molecule to which HIV attaches) is introduced into the bloodstream, it binds with HIV. Thus HIV is no longer free to bind to CD4 on the cells. Experiments are under way with CD4 molecules attached to toxins that may kill the virus (see Figure 11.31). Three-dimensional computer models are being

Figure 11.31   Linking a toxin to a molecule that recognizes glycoproteins on an infected cell's surface should only destroy sick cells.

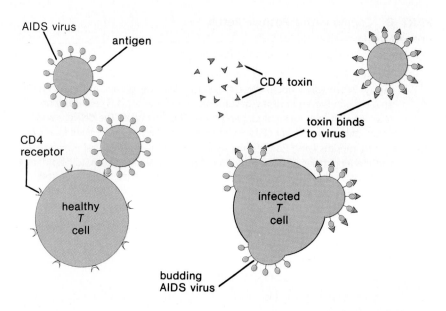

used to screen other drugs for the same binding capability. AZT, the first drug used against AIDS, interferes with HIV's ability to reproduce. Cytokines—molecules that act as immune-system regulators—have been able to turn off production of HIV in laboratory studies and to delay the onset of AIDS in people infected with HIV. These and other drugs under investigation have serious side effects, however, and many researchers think combinations of drugs will prove the most effective therapy. Although initial studies of various combinations are promising, particularly in treating and preventing recurrence of opportunistic infections, much remains unknown.

There is, however, new hope for treating Kaposi's sarcoma. Recent evidence indicates that HIV stimulates tumor development by causing the production of a growth-stimulating factor on which the sarcoma is dependent. These findings may lead to new treatments with drugs that block the action or synthesis of the growth factors, thus stopping or slowing the tumor growth.

Concept Review

1. What are some of the reasons for the increase in gonorrhea despite the effectiveness of penicillin in treating it?
2. Describe the symptoms of each stage of syphilis.
3. What characteristics of chlamydial infections increase the spread of this disease?
4. Distinguish between HIV and AIDS.
5. What is meant by *opportunistic* infections?
6. Describe some of the problems involved in developing treatments for HIV infection and AIDS.

Summary

Prokaryotes are the most ancient, most widespread, and most metabolically diverse of any organisms. They also are the most important ecologically. They are major decomposers, indispensable members of all food webs, and essential to the cycling of matter. They include two distinctly different groups—archaebacteria and eubacteria. Both groups lack a nucleus and other membrane-enclosed organelles. Archaebacteria have distinctive cell walls and other characteristics that distinguish them from eubacteria. Apparently they are as closely related to the eukaryotes as to the eubacteria. A large percentage of archaebacteria inhabit extreme environments like those that may have been typical on the early earth—temperatures near or above boiling, brine ponds saturated with salt, waters with very high or very low pH, and the extreme pressures of the ocean depths. Among the archaebacteria are the methanogens, which can produce methane gas from organic matter such as sewage, garbage, agricultural waste, manure, and other waste materials.

Among the eubacteria are many photosynthetic types, some of which are anaerobic, using substances other than water as a source of hydrogen. The cyanobacteria are aerobic photosynthesizers that most likely were responsible for the initial buildup of oxygen gas in the early atmosphere. Numerous types of eubacteria are involved in the nitrogen cycle.

Disease results from the interaction between a host and a pathogen. Pathogenic eubacteria are responsible for many infectious diseases, including whooping cough, typhoid, eubacterial pneumonia, syphilis, and numerous plant diseases. Viruses, which have a simple structure of DNA or RNA with a protein coat, also cause many plant and animal diseases, including tobacco mosaic, colds, influenza, chickenpox, and AIDS. Although AIDS continues to be a deadly disease, prospects for treatment are improving.

Applications

1. If the first organisms on earth were heterotrophic prokaryotes, from where did they get their food if there were no other living things?

2. After an atomic blast that wipes out all living things in a local area, which type of microorganism might you expect to colonize the blast site first? Why?

3. The adaptations of archaebacteria to extreme modern-day environments suggest they were among the earliest organisms on earth. Explain this statement.

4. In what way are microorganisms economically important to humans? In other words, how do we use microorganisms for our own good?

5. How does the nitrogen cycle illustrate the interdependence of organisms in the biosphere?

6. Will a vaccine for a particular disease help an individual if he or she already has the disease? Explain your answer.

7. What characteristics of sexually transmitted diseases make them particularly difficult to combat?

8. During famines, diarrhea is one of the major killers, especially of infants. Explain why this might be.

9. How does cancer fit into the classification of diseases?

10. What distinguishes viruses from living organisms?

11. HIV does not kill people—the cause of death usually is an opportunistic infection. Explain what is meant by an opportunistic infection and why these occur in AIDS patients.

12. Why might a cure for AIDS be found if scientists could block the activity of reverse transcriptase?

Problems

1. Both DNA and RNA are used to study the evolutionary relationships of organisms to each other. Compare the techniques. What are the advantages of each?

2. Design a methane generator for a home that uses food waste as an energy source. How would you control the smell of rotting food and prevent contamination? How would you collect the methane gas without allowing it to build up to dangerous levels? How would you provide the appropriate environment for the methanogens?

3. Draw a model of the nitrogen cycle based on the information in your text—try not to refer to your book's illustration. You also may want to refer to the textual material on the nitrogen cycle in a general ecology book, but do not look at the illustrations. After you have drawn your model, compare yours to the illustrations in any book you used. How are the illustrations similar or different? Which do you think is better?

4. Start a small experimental compost pile at your school. What type of materials should you add to your compost? Which materials decompose and which do not? How fast do materials change and how do they change? What environmental conditions favor the development of a compost pile?

5. Investigate an infectious disease of a plant or animal using the following outline of topics: history, symptoms, pathogen, vector (where applicable), treatment, epidemiology. Suggested diseases: anthrax, bacterial meningitis, Dutch elm disease, filariasis, tsutsugamushi fever, hoof-and-mouth disease, black stem rust of wheat, brucellosis, fire blight of pears, or schistosomiasis.

6. Using the same method as in problem 5, study a noninfectious disease.

7. In recent years epidemics of virus-caused influenza have occurred and spread over many parts of the world. Yet vaccination against influenza viruses is common. Why are the vaccines ineffective against new epidemics? Use your understanding of pathogens, resistance on the part of both host and pathogen, and immunity to answer the question.

8. Scientists are aware that some infectious diseases have not been identified. First cases or infrequent cases may go unrecognized. The first epidemic draws attention to the disease. This was the case with the human disease now called Legionnaire's disease. It is named after the 1976 American Legion convention in Philadelphia at which an epidemic occurred. A number of people died. Consider such an epidemic, its victims, and their autopsies (studies of blood and other body parts after death). If you were attempting to identify the cause of the disease, how would you use people in each of the following careers: biochemists, food inspectors, medical doctors, metallurgists, microbiologists (virologists and bacteriologists), pathologists, and toxicologists? If necessary, do library research on the nature of these professions.

9. In developed countries, infectious diseases are no longer the leading causes of death as they are in many parts of the Third World. Choose one of the most common infectious diseases in the Third World and report on the causative agent, the frequency of the disease, and attempts to control it.

10. In the early 1980s, the World Health Organization reported that smallpox had been eliminated worldwide. Report on the campaign to eradicate this viral disease.

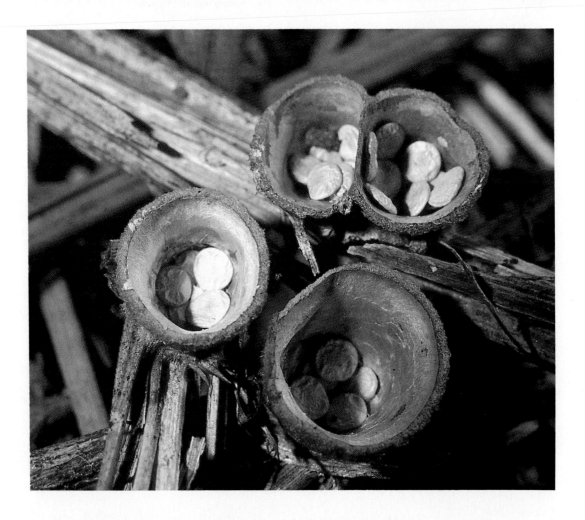

These four structures are called white-egg bird's nest fungi. Of what possible adaptive value are these odd structures? Share your idea with a classmate, and ask for the classmate's explanation.

Eukaryotes: Protists and Fungi 12

Protists and fungi are eukaryotes. They possess organized nuclei and membrane-enclosed organelles. Evidence suggests that some of these organelles originated from free-living prokaryotic ancestors.

The kingdom Protista includes a diverse group of organisms that have little in common. Its members are eukaryotes, but not plants, animals, or fungi. Protists may be unicellular, colonial, or multicellular. They may be consumers or producers, and some switch from one form of nutrition to the other, depending on conditions. All protists carry out aerobic cellular respiration in mitochondria. Most have flagella at some stage of their life cycle. Producer protists carry out photosynthesis in chloroplasts.

All organisms in the kingdom Fungi are multicellular heterotrophs that absorb molecules from their surroundings. Many fungi are decomposers and play an important role in food chains by making nutrients available to plants. This chapter describes several species of protists and fungi and examines their roles in the larger community.

Autotrophic Protists

Guidepost: What characteristics distinguish the various groups of algae?

INVESTIGATION 12.1 Variety Among Protists and Cyanobacteria

Many protists live in fresh water as do many cyanobacteria. In this investigation, you will use a dichotomous key to make a series of choices between alternative characteristics and identify a number of common protists and cyanobacteria.

Materials (per team of 2)

2 microscope slides
2 cover slips
flat toothpick
compound microscope
protist/cyanobacteria key

A Catalog of Living Things
Detain™
protist/cyanobacteria survey mixture
pond, stream, or puddle samples

Procedure

1. Place a drop of Detain™ (protist slowing agent) on a clean slide. Add a drop of either the survey mixture or a sample and use a toothpick to mix the two drops. Place a cover slip on the mixture.

2. Use the low power (100X) of a microscope to examine the mixture. For more detailed observations, switch to high power (400X).

3. Use the protist/cyanobacteria key in Appendix 3, *A Catalog of Living Things* (Appendix 4), and other references to identify as many organisms as possible.

4. Wash your hands thoroughly before leaving the laboratory.

Discussion

1. Choose one of the specimens you observed. Write a short paragraph describing the organism's environment, other organisms that might live in the same environment, and their possible interactions with the protist you observed.

2. Using the key's descriptions, summarize the distinguishing characteristics of cyanobacteria and the major groups of protists included in the key.

12.1 Algae Are Photosynthetic Protists

Algae, perhaps the most familiar protists, are aquatic and have relatively simple structures. They can be found just about anywhere you find tiny droplets of water—in the air, on tree trunks and branches, in the bottoms of streams, on soil particles, and on rocks at the seashore. They may be unicellular, colonial, or multicellular (see Figure 12.1). Multicellular species may form either long filaments or thin plates of similar cells or plant-like structures that have division of labor.

Although all algae are photosynthetic, not all of them appear green. Both unicellular and colonial algae float near the surface of bodies of water, using energy from sunlight and carbon dioxide from the water to carry out photosynthesis. Multicellular algae, on the other hand, are adapted to living along shores and in shallow water, where nutrients are generally abundant but living conditions are difficult. These algae are subject to periodic drying and to variations in temperature, light, and other factors. As a result, different groups of algae have become adapted to particular niches.

The term *algae* is one of convenience. It is no longer used in formal classification because in recent years biologists have found differences in the biochemical characteristics of the various groups. Algae vary greatly in the compounds that make up their cell walls, in their food storage compounds, and in the pigments present in their cells. Like the cell walls of plants, the cell walls of most algae contain cellulose, but they may contain a variety of other compounds as well. Some algae store food as starch, but others store it as carbohydrates that are found only in algae.

In addition to chlorophyll, algae may contain various other pigments. Often, these pigments mask the bright green of the chlorophyll and produce the characteristic colors for which some algae are named. The variety of pigments suggests that the various groups may have evolved from symbiotic relationships with different photosynthetic prokaryotes.

12.2 Green Algae May Have Been the Ancestors of Plants

The majority of green algae (phylum Chlorophyta; kloh ROF it uh) live in fresh water. You can find them in your aquarium at home or on the shady side of trees. Many green algae are unicellular and microscopic, but they may be so abundant that they color the water of ponds and lakes green or

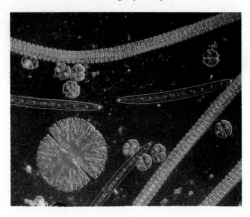

Figure 12.1 Mixed algae: *Micrasterias*, large two-lobed alga; *Cosmarium*, small two-lobed alga; *Closterium*, oblong alga with conspicuous vacuoles; and *Dismidium*, filamentous alga (X100).

Figure 12.2 Green algae: (a) *Acetabularia,* a large one-celled alga (X1); (b) water net, *Hydrodictyon reticulatum,* a filamentous form (X100); and (c) sea lettuce, *Ulva,* a multicellular alga (X1).

form huge, floating mats. Some marine species form large, multicellular seaweeds. Figure 12.2 shows a few members of this large group.

Green algae reproduce both asexually and sexually. Some sexually re-producing species produce flagellated gametes that look identical (see Figure 12.3). Other species produce flagellated sperm and larger, non-motile ova that are retained in the parent algae. Many of the multicellular green algae have complex life cycles, similar to those in plants, during which they produce both gametes and spores.

Biologists think plants originated from green algae. Several types of evidence support this hypothesis. Green algae, but not other algae, contain the same combination of photosynthetic pigments as plants do. Both store reserve foods as starch, and both have cellulose-containing cell walls. Like plants, several species of green algae have developed true multicellularity with some division of labor. Other species show intermediate steps be-tween unicellularity and multicellularity. Finally, many species of green algae have patterns of reproduction and life cycles in common with plants.

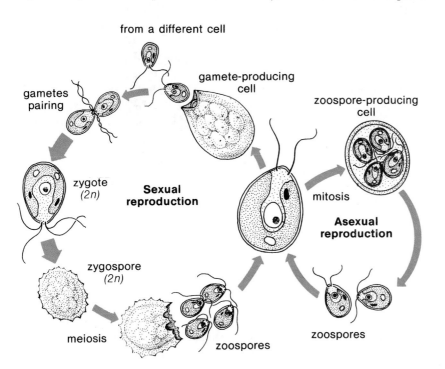

Figure 12.3 Life cycle of *Chlamydomonas,* a unicellular green alga, showing both sexual and asexual reproduction. Notice that the zygote is the only diploid stage in the life cycle.

Figure 12.4 Diatoms: (a) Strained mixture photographed through a light microscope (X200); and (b) scanning electron micrograph showing the double silicon shell and the pattern of pores (X4000).

a   b

12.3 Diatoms Are Important Producers

Diatoms (phylum Bacillariophyta; buh SIL er ee OFF ih tuh) are some of the most intricately patterned creatures on earth, as shown in Figure 12.4. Their cell walls are formed by a double shell, the two halves of which fit together like a box. The shells are made of organic materials impregnated with silica. Electron microscope studies have shown that the patterns on the shells are created by the pores that connect the inside of the cell with the outside environment. Diatoms are microscopic, occuring as single cells, simple filaments, or colonies; some move by gliding. Abundant in both fresh and marine waters, they are the principal producers in many food webs and provide much of the world's oxygen. Diatoms also grow in damp places on land or even on moist flowerpots in greenhouses. Usually brownish in color, diatoms contain various yellow pigments that mask their chlorophyll. They store food in the form of oils, and they reproduce either asexually by cell division, or else sexually by producing gametes. The zygote resulting from fertilization grows to the full size for the particular species of diatom and undergoes cell division. The offspring cells then produce new silica-containing shells.

Silica shells remain intact for many years, even after the organisms inside have died. Great masses of shells that have slowly accumulated through many years (diatoms date back to the Cretaceous period, 144 million years ago) now are mined as diatomaceous earth. That material is used many ways—for example, as a filter (both for beer and for swimming pools), as an abrasive in silver polish, and as a brightener for the paint used to mark highways.

12.4 Brown and Red Algae Are Multicellular

The major seaweeds of the world are the multicellular brown and red algae. The brown algae (phylum Phaeophyta; fay OFF it uh) inhabit rocky shores in the cooler regions. Kelps, for example, form extensive offshore beds, where they are the primary producers for many communities of animals and microorganisms. Brown algae have cellulose in their cell walls. They store their food as a unique carbohydrate or as oil, rather than as starch, which is present in most other algae. Other pigments usually mask the chlorophyll in the brown algae.

Brown algae (Figure 12.5a, b) may be very large, sometimes reaching as much as 100 m in length. Some are unspecialized, but others are differentiated into areas for anchorage (holdfast), support (stipe), and photo-

a b c

Figure 12.5 (a) A brown alga, *Macrocystis pyrifera,* showing air bladders that help keep it afloat; (b) a brown alga, *Postelsia,* showing holdfast, stipe, and blade; (c) a coralline red alga, *Bossiella.*

synthesis (blade), as seen in Figure 12.5b. Some kelps have conducting cells in the center of the stipe that resemble the sugar conducting cells of plant stems.

Most seaweeds of warm oceans of the world are red algae (phylum Rhodophyta; roh DOF it uh). They generally grow attached to rocks or other algae (Figure 12.5c). Because they absorb blue light, which penetrates farther into the water than other light wavelengths, red algae can grow at greater depths in the ocean than any other algae. Biologists think the chloroplasts of red algae may have evolved from ancestral cyanobacteria because the two groups contain similar pigments. The cell walls of red algae contain an inner layer of cellulose and an outer layer of other carbohydrates such as agar, used for culturing bacteria in the laboratory.

In many parts of the world, humans eat red and brown algae as part of their regular diet. Although algae do not provide many kcals, they do contain some vitamins and trace elements necessary to human health. Red and brown algae produce compounds used in the manufacture of food and goods that require a smooth texture. When you eat marshmallows, ice cream, or certain cheeses, and when you use hand lotion, lipstick, or paint, you probably are using products derived from algae.

Concept Review

1. What characteristics of green algae make them probable ancestors of all plants?
2. What makes diatomaceous earth useful, and how is it used?
3. How are brown and red algae important to humans?
4. Which algae group can live in deeper water than any other? Why?

Heterotrophic Protists

Guidepost: What characteristics distinguish the various heterotrophic protists?

INVESTIGATION 12.2 Life in a Single Cell

The relationship between structure and function in plants and animals is always significant but not always obvious. Careful and detailed observations of living organisms, sometimes over long periods of time, will reveal which structures are performing the life functions. In this investigation, you will observe structures of the protozoan *Paramecium* and form hypotheses about their possible functions.

Materials (per team of 2)

2 18-mm × 150-mm test tubes
2 microscope slides
2 cover slips
dropping pipet
compound microscope
rubber stopper to fit test tube
test-tube rack
incandescent light source
forceps

aluminum foil
6 flat toothpicks
Detain™ in squeeze dropping bottle
alizarin red stain in squeeze dropping bottle
India ink in small bottle
yeast-brilliant cresyl blue preparation in
    squeeze dropping bottle
cotton thread wetted with 0.1*M* HCl
paramecium culture

PART A   Structure and Function

Procedure

1. Place a drop of Detain™ in the center of a clean slide. Add a drop of paramecium culture and use a toothpick to mix the two drops together. Place a cover slip on the mixture.

2. Use the low power (100X) of a microscope to examine the slide. Locate at least one paramecium.

3. In your data book, make a simple drawing of the paramecium that you have seen through the microscope. Include in your sketch only the parts of the organism you observe.

4. What structure performs what function? Label the structures in your drawing according to the following scheme. (You may not have observed all the structures listed below.)

    a. Structure(s) associated with movement.
    b. Structure(s) associated with food intake.
    c. Structure(s) associated with digestion.
    d. Structure(s) associated with elimination of undigested food.
    e. Structure(s) associated with oxygen and carbon dioxide exchange.
    f. Structure(s) associated with excretion of fluid wastes.
    g. Structure(s) associated with removal of excess water.

    It is not necessary at this time to know the names of the parts you have labeled.

5. Carefully raise the cover slip and add a drop of alizarin red stain to your slide. Mix with a toothpick and gently replace the cover slip.

6. Use low power to examine the slide. Can you see structures in the paramecium that you did not see before? Add these structures to your diagram, using a colored pencil to distinguish them from your first observations.

7. Use the list in Step 4 to develop hypotheses about the function of any newly visible structures.

Discussion

1. What difference do you observe in the paramecium after adding the alizarin red stain?

2. What structures could you see before adding the alizarin red? What structures could you see afterwards?

3. What functions of a living organism are not represented in your seven hypotheses?

4. Now look at Figure 12.6. What clues to functions do you find in the names of the structures?

5. With this new information, change any of the structure-function relationships you think need changing.

6. What should you do to determine if your hypotheses are good?

Figure 12.6   Structures in a paramecium.

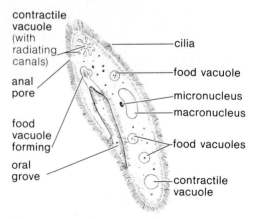

contractile vacuole (with radiating canals)

anal pore

food vacuole forming

oral grove

cilia

food vacuole

micronucleus

macronucleus

food vacuoles

contractile vacuole

PART B Movement
Procedure

8. Prepare a fresh slide as in Step 1.

9. When the paramecia begin to slow down, switch to high power (400X) to observe movement in detail. Which end of the paramecium is usually in front as the organism moves?

10. The front end of an organism is referred to as *anterior*, the back end as *posterior*. Using the diagram you made in Part A, label what you believe to be the anterior and posterior ends. Can you identify a top (*dorsal*) and a bottom (*ventral*) side?

11. Diagram the motion of a paramecium as it moves in one direction. Use arrows to show the directions of body motions.

12. Carefully raise the cover slip and use the tip of a toothpick to add a small amount of India ink to the paramecium culture on the slide. Mix well; the culture should be light gray. Gently replace the cover slip.

13. Use high power to observe the surface of one of the organisms. The hairlike structures you see are cilia. Notice carefully what happens to the India ink particles that touch an organism. Adjust the light until you can see individual cilia.

Discussion

1. What appears to be the function of the cilia?

2. What does the pattern of movement of the India ink particles tell you about the activities of the cilia?

3. On the basis of your observations in Part B, review and revise (if necessary) the hypothesis you formed in Part A about how a paramecium moves from one place to another.

PART C Ingestion, Digestion, and Elimination
Procedure

14. Prepare a fresh slide as in Step 1, but add a drop of yeast-brilliant cresyl blue preparation. Observe what happens to the yeast cells. Notice their color when they are outside the paramecium. Then, find some that have been ingested and observe their color. Yeast cells that have been ingested will be in small, round structures called food vacuoles. What color are the yeast cells when they are outside the paramecium? What happens to the yeast cells that are encountered by the paramecium? What parts of the paramecium are involved first in the ingestion of the yeast cells? Where are the food vacuoles formed in the paramecium?

15. Carefully watch a newly formed food vacuole and note any changes in it. Within 2 or 3 minutes you should begin to notice a color change in the food vacuoles. Describe the color change. How do you account for it?

16. Continue to observe the food vacuoles. By means of circles representing food vacuoles and arrows representing their movement, diagram the path a food vacuole takes from its formation to its disappearance. How do the food vacuoles move through the organism?

17. Watch carefully to see how undigested food in a food vacuole is eliminated from the paramecium. From what part of the body does this happen?

Discussion

1. Refer to the hypothesis you formed in Part A regarding the structures in the paramecium responsible for food intake, digestion, and the discharge of undigested food. Were you correct?

2. On the basis of your observations in Part C, summarize how you now think a paramecium ingests and digests food and eliminates wastes.

PART D Exchange of Materials
Procedure

18. Like most living things, a paramecium needs a constant supply of oxygen and must eliminate carbon dioxide, liquid wastes, and excess water. To discover the structures associated with these functions, prepare a slide as in Step 1 and observe under high power.

19. Considering the environment of the paramecium, through what structure would you expect oxygen and carbon dioxide to move? How do you think this movement occurs?

20. Concentrate your observation on either end of the paramecium. Locate a structure that contracts and then seems to disappear. It becomes visible again and repeats this process. The canals leading into this structure give it a star-shaped appearance. The structure is called a contractile vacuole. Is fluid entering or leaving the vacuole when it contracts? If you identified any contractile vacuoles in Part A, what hypothesis did you form about their function?

21. How many contractile vacuoles are there in a paramecium? Where are they located? Add them to your drawing. Do the contractile vacuoles move through the cell as food vacuoles do, or do they stay in one place? Is the fluid within the enlarged vacuole clear or granular? How many contractions does the vacuole make in one minute?

Discussion

1. Students who investigated the question, "What is the function of a contractile vacuole?" made the following observation: Urea (an end product of nitrogen metabolism) has not been found in any great amount in the fluid of a contractile vacuole. What effect does that observation have on a hypothesis that the contractile vacuole is an excretory organ used for eliminating nitrogenous wastes?

2. Two further observations were: (a) If vacuoles of freshwater paramecia do not function, the body swells; and (b) Injection of distilled water into the organism increases the rate of contraction of the vacuole. Considering these observations, what probably is the function of a contractile vacuole in the paramecium?

3. List the ways in which water might enter a paramecium.

4. The students also observed that water intake along with food in food vacuoles is only a fraction of the amount of water that is expelled by the contractile vacuole. What would you suggest is the major method of water entry?

6. Develop a hypothesis that accounts for the function of a contractile vacuole.

PART E Behavior
Procedure

22. Fill two test tubes 1/4 full with paramecium culture. Place one test tube upright in a test-tube rack so that it receives uniform light from all directions.

23. Stopper the second test tube. Wrap the top half of the test tube with aluminum foil. Lay the test tube on its side and evenly illuminate the bottom half.

24. Now observe the upright test tube. In a rich paramecium culture, a concentration of the organism looks like a white cloud. Examine the tube for any concentration. What evidence do you find that the paramecia are or are not moving toward one part of the tube? In this controlled situation, is the tendency of organisms to concentrate in one part of the tube a response to gravity? Why do you think so? What other factors might account for this concentration? What additional investigations would you devise to check all variables, including gravity?

25. Observe the partially covered test tube placed on its side. Has the paramecium culture concentrated at one end of the test tube? Is there a response to light? How

do you know? Why could this investigation not have been done with the test tube in an upright position? How might this response to light act as an advantage for a paramecium?

26. Place a drop of paramecium culture on a slide. Use forceps to place a small piece of thread wetted with 0.01*M* HCl across the middle of the drop. Add a cover slip. Using low power, focus on the piece of thread. In what way does the paramecium respond to the acid?

 CAUTION: HCl is an irritant; do not allow the thread to touch your skin. Should contact occur, call your teacher immediately; flush the area with water.

27. Each day, wash your hands thoroughly before leaving the laboratory.

Discussion

1. Summarize your observations of a paramecium's behavior.

12.5 Flagellates May Be Consumers or Producers

Protozoa are animal-like protists that are mostly heterotrophic and unicellular. The term *protozoa*, like *algae*, is one of convenience. As a group, the flagellates could be considered both protozoa and algae because some are heterotrophic and others are autotrophic. Most flagellates are unicellular organisms that move by means of flagella. Some have only one flagellum; others have thousands. Flagellates are abundant in damp soil, fresh water, and the ocean. Most species reproduce asexually by dividing in two, but a few reproduce sexually by fusion of identical-appearing gametes. Among the several phyla of flagellates, some contain chlorophyll and synthesize food when light is present. Others are heterotrophic and capture smaller microorganisms, which they digest internally. Some flagellates can be either consumers or producers, depending on conditions.

Euglenoids are a small group of flagellates similar to the organisms in the genus *Euglena*, which are commonly found in fresh water (see Figure 12.7a). Many species contain chloroplasts and therefore manufacture their own food. The nonphotosynthetic forms absorb dissolved organic substances or ingest living prey. Most species orient to light by means of a light-sensitive eyespot. Euglenoids lack rigid cell walls but have supporting strips of protein inside their plasma membranes.

Dinoflagellates, in contrast, are marine flagellates armored in stiff cellulose walls. Two beating flagella cause the cells to spin like tops as they move through the water (Figure 12.7b). Many dinoflagellates are red in color, and some produce powerful toxins that affect the human nervous system. If these reproduce in great numbers, they cause what are known as "red tides." During red tides, the toxins may become concentrated in food chains. Clams and other organisms that eat dinoflagellates may become poisonous to humans at that time.

Many flagellates live in symbiotic relationships with other organisms. For example, the parasitic flagellate *Trypanosoma gambiense* (Figure 12.7c) causes African sleeping sickness in humans. Symbiotic flagellates live in the digestive tract of termites, where they break down the wood eaten by the insects. Without the flagellates the termites would starve to death because termites cannot digest cellulose any more than humans can.

a

b

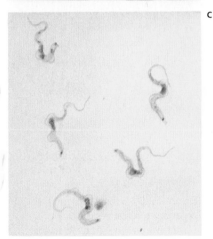
c

Figure 12.7 Flagellates: (a) *Euglena* (X150); (b) a dinoflagellate, *Gymnodinium* (X12 000); and (c) *Trypanosoma gambiense*, one of the parasites that causes African sleeping sickness.

Figure 12.8 An amoeba (a) is a sarcodine that surrounds its prey with pseudopods and engulfs them (X180). Foraminiferans (b) build shells of calcium carbonate (X200).

12.6 Many Sarcodines Use Pseudopods

Amoebas and their relatives are protozoan protists called sarcodines. The sarcodines are thought to have evolved from flagellates, and many develop flagella during some stage of their life cycle. They reproduce asexually by cell division. Some amoebas live in ponds, puddles, and damp soil, whereas others live within larger organisms in symbiotic relationships that may be parasitic. The organism that causes amoebic dysentery in humans is one example of a sarcodine parasite.

The cytoplasm of a sarcodine is a granular, grayish or transparent mass that has no definite shape. In contrast to the fast-moving flagellates, sarcodines ooze or float about. They move and feed by means of pseudopods (SOO doh podz), fingerlike extensions of cytoplasm. The cytoplasm continuously flows into one or several pseudopods, and the organism flows in the direction of whichever pseudopod is longest. Pseudopods also surround and engulf bacteria, protists, algae, or other aquatic organisms that the sarcodines use as food, as shown in Figure 12.8.

Most shell-bearing sarcodines live in the sea. Radiolarians, for example, have an outer shell of silica, a sandlike substance. Many long, stiff pseudopods radiate from a radiolarian's shell. The foraminiferan (FOR am ih NIF er un) shown in Figure 12.8b builds a shell of calcium carbonate, which is the same material clam shells are made of. During millions of years, great numbers of foraminiferan shells have accumulated at the bottom of seas and gradually have solidified into rock. Chalky deposits, such as the white cliffs of Dover in England, result from these accumulations of shells.

12.7 Sporozoans Are All Parasites of Animals

Sporozoans (phylum Apicomplexa; ap ih kum PLEK suh) are spore-forming protozoa that are parasites of animals. Unlike the sarcodines or flagellates, these protozoa lack any means of self-propulsion. Their complex life cycles involve more than one host. Sporozoans reproduce sexually, and their zygotes form thick-walled cysts that can be passed to new hosts through infected blood. Inside the cysts, the zygotes divide rapidly, forming large numbers of tiny spores. When the cysts break open, each one of the spores can infect a cell within a new host.

In the human bloodstream, sporozoans like those shown in Figure 12.9 cause serious and even fatal diseases such as malaria. In the past, malaria has been controlled by drugs that inactivate the parasitic sporozoan and

Figure 12.9 Sporozoans: (a) the organism that causes malaria, *Plasmodium vivax*, showing several different stages of the life cycle in a blood smear; and (b) *Gregarina*, a gregarine (X75).

a   b

a b

Figure 12.10 Ciliates: (a) *Vorticella campanula* (X125); (b) cilia of *Paramecium caudatum*. Scanning electron micrograph (X1500).

pesticides that kill the mosquitoes that carry infected blood from one host to another. Unfortunately, both sporozoans and mosquitoes have developed resistance to the chemicals used to control them. Today, malaria remains a principal cause of human death in the world.

12.8 Ciliates Have Two Types of Nuclei

The most specialized and complex of the protozoan protists are ciliates (phylum Ciliophora; sil ee OFF er uh). Most are unicellular, free-living organisms found in marine or freshwater habitats. They have definite, semi-rigid shapes, as shown in Figure 12.10a. Active swimmers such as *Paramecium* have distinct anterior and posterior ends. All ciliates move by means of **cilia** (SIL ee uh)—short, whiplike extensions that beat in rhythm, driving the organism through the water (see Figure 12.10b). In some species, the beating of cilia causes a current that sweeps food particles into the organism. In others, bundles or sheets of cilia function as mouths, paddles, teeth, or feet.

Ciliates have two types of nuclei, macronuclei and micronuclei. The macronuclei are required for growth, reproduction, and cellular functions. Macronuclei do not divide by mitosis. Instead, when the cells divide, the macronuclei are distributed between the offspring. Micronuclei are exchanged by two ciliates during conjugation (see Figure 12.11).

12.9 Slime Molds Resemble Both Fungi and Protozoa

Slime molds (Myxomycota; mik soh my KOH tuh) are funguslike heterotrophs that usually grow among damp, decaying plant material. If you search through such material soon after a heavy rain, you may find a slime-mold plasmodium (plaz MOHD ee um)—a glistening sheet or net with no definite form. The plasmodium may be more than one meter in width and brightly colored. As it moves from one place to another, the plasmodium engulfs bacteria, spores, and small bits of dead organic matter.

As long as there is food and water, a plasmodium continues to grow. When food becomes depleted, the plasmodium may clump into mounds and form **sporangia** (spoh RAN jee uh)—stalked structures that produce reproductive cells or spores (see Figure 12.12a). When a sporangium breaks

Figure 12.11 Conjugating *Paramecium multimicronucleatum*, showing the macronuclei (X100).

a

Figure 12.12 (a) Slime mold sporangium, *Hemitrichia clovata* (X60). The plasmodium stage of *Physarum polycephalum* also is visible. (b) Life cycle of a slime mold. Which part of the cycle is best adapted to a dry environment? Which parts of the cycle would require abundant moisture?

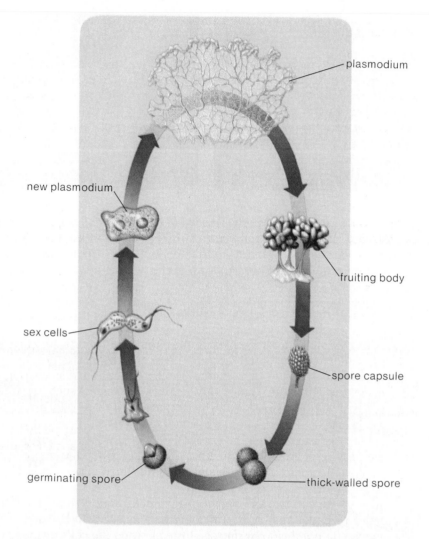

open, a large number of spores are released and carried away by wind, insects, or raindrops. The spores can survive unfavorable environmental conditions, but whenever they land in suitable spots with enough moisture, they can germinate. Each spore gives rise to a tiny organism with one or two flagella. After swimming about for a short time, these organisms lose their flagella and fuse in pairs. By feeding and growing, each pair develops into a new plasmodium. Figure 12.12b shows the life cycle of a slime mold.

Concept Review

1. Are flagellates algae or protozoa? Explain.
2. What roles do flagellates play in the biosphere?
3. Describe how a sarcodine moves and obtains food.
4. How do sporozoans differ from other protozoa?
5. In what way are ciliates considered complex?
6. How do spores of slime molds and sporozoans differ?
7. Which characteristics of slime molds are animal-like and which are funguslike?

Models of Eukaryotic Origins

Two of the major ideas about the origin of eukaryotic cells are known as the *autogenesis hypothesis* and the *endosymbiosis hypothesis*. According to the autogenesis model, eukaryotic cells evolved through the specialization of internal membranes that originally were part of the plasma membrane of a prokaryote. The internal membrane system, including the nuclear envelope, Golgi apparatus, endoplasmic reticulum, and other organelles enclosed by a single membrane may have resulted from the plasma membrane folding in on itself. Mitochondria and chloroplasts may have developed their double membranes through more complex foldings.

According to the endosymbiosis model, the ancestors of eukaryotic cells were symbiotic prokaryotic cells, with certain species living within larger prokaryotes. One modern example of endosymbiosis is found in the digestive cells of *Hydra viridis*, which contain algal cells of the genus *Chlorella*. These digestive cells recognize the appropriate algae and eject any others. Accepted algae are moved to the base of the digestive cell where they reproduce by mitosis until the normal algal population is reached. Thereafter the algae reproduce when the host cell reproduces. The presence of the algae enables the hydra to survive when food is scarce.

Many biologists think mitochondria originated as symbiotic aerobic bacteria. The bacteria that became mitochondria were engulfed by ancestral eukaryotic cells. Before these cells acquired the bacteria, they lacked the enzymes that would allow them use oxygen in respiration. The engulfed bacteria, however, were able to perform these reactions, and they became the inner part of mitochondria. All the mitochondria within a eukaryotic cell arise from the division of the existing mitochondria, just as new bacteria are produced through cell division. Mitochondria divide by simple fission, splitting into two, and they apparently replicate and divide their circular DNA molecule as do bacteria. Mitochondrial replication, however, is impossible without a cell nucleus because most of the genes required for mitochondrial division are located in the nucleus and translated into proteins by ribosomes in the cell's cytosol. Moreover, mitochondria cannot be grown in a cell-free culture.

During the time that mitochondria have existed as endosymbionts in eukaryotic cells, most of their genes have been transferred to the chromosomes of the host cell. Mitochondria, however, still have genes of their own contained in a circular molecule of DNA, similar to that of bacteria. This mitochondrial DNA contains several genes that produce some of the proteins essential for the functioning of mitochondria in aerobic cellular respiration.

Symbiotic events similar to those proposed for the origin of mitochondria could explain the origin of chloroplasts, which are characteristic of photosynthetic eukaryotes. Some ancestral eukaryotic cells may have ingested photosynthetic bacteria, which evolved into chloroplasts. The incorporation of photosynthetic bacteria would have conferred one great advantage on the cell: It then could manufacture its own food. Like mitochondria, chloroplasts have their own circular DNA, but many of the genes that code for necessary chloroplast components are located in the nucleus.

It is possible that the evolution of eukaryotic cells required both autogenesis and endosymbiosis. Endosymbiosis may be responsible for mitochondria and chloroplasts, and autogenesis may be responsible for other organelles.

A hydra, *Chlorahydra viridiscima*, with buds.

Guidepost: What are the
characteristics of the fungi?

Fungi

INVESTIGATION 12.3 Growth of Fungi

Mushrooms and molds are common fungi. In previous investigations some of your cultures may have been contaminated by molds, which may have given you the idea that molds are everywhere. If this is true, what features of molds make them so plentiful?

Materials (per team of 3)

glass tumbler	masking tape
microscope slide	transparent tape
cover slip	15-cm metric ruler
dropping pipet	glass-marking pencil
compound microscope	self-closing plastic bags labeled *waste*
stereomicroscope or 10X hand lens	glycerine-water solution (l : l)
sheet of white paper	sterile potato-dextrose agar plate
scalpel	fresh mushrooms
flat toothpick	+ strain of *Rhizopus stolonifer*
3 sterile swab applicators	− strain of *Rhizopus stolonifer*

Procedure
PART A   Observing Mushrooms
Day 1

1. Examine a fresh mushroom. The mushroom itself is the reproductive body of this particular fungus. It consists of a stalk and an umbrellalike cap.

      CAUTION: Do not eat the mushrooms.

2. Use the scalpel to cut the cap from the stalk.

      CAUTION: Scalpels are sharp; handle with care.

   Look at the underside of the cap and find the gills. Place the cap, gill side down, on a sheet of white paper.

3. Label a piece of masking tape with your team name and place it on the bottom of the tumbler. Cover the cap with the tumbler to protect it from air currents.

4. Set the cap, paper, and tumbler in a place where they will not be disturbed.

5. Wash your hands thoroughly before leaving the laboratory.

Day 2

6. Remove the tumbler and lift the cap straight up from the paper. Replace the tumbler. What do you observe on the paper? What is the relationship between what you see on the paper and the structure of the mushroom cap?

7. Remove a sterile swab from its wrapper. Tilt the tumbler and use the swab to transfer a small amount of this material to a slide. Add a drop of glycerine-water and a cover slip. Replace the swab in its wrapper and discard in the waste bag.

8. Examine the material under both low and high power of the compound microscope. Approximately how many particles are present in your sample? Using this number, estimate the total number of such particles coming from the mushroom.

9. Wash your hands thoroughly before leaving the laboratory.

PART B   Growing *Rhizopus*
Day 1

10. Draw a line across the middle of the bottom of a potato-dextrose agar plate. Label one side + and the other −.

11. Carefully remove a sterile swab applicator from its paper wrapper and transfer a few spores from the + culture of *Rhizopus stolonifer* to the + section of your plate. Make a 1-cm streak about 1 cm from the outside edge. Be careful not to dig into the surface of the medium. Replace the used swab in its paper wrapper and discard it in the waste bag.

12. Using a fresh sterile swab, repeat Step 11 with the − strain and place the spores in the − section of your plate.

13. Cover the plate and tape it closed. Invert and incubate at room temperature for 10 days.

14. Wash your hands thoroughly before leaving the laboratory.

Days 2–10

15. Observe the plate daily using a stereomicroscope or a 10X hand lens. As the mold starts to grow on the plate, you can see that it is composed of many branching threads called hyphae. Hyphae perform various functions. Some support the sporangia, and others anchor the mold to the agar and absorb water and nutrients.

16. Measure the growth of the mold daily and record the data in your data book. What difference is there in the growth rate of the two strains?

17. As the two strains grow toward the center of the plate, observe them carefully for changes, particularly where hyphae from the two strains have touched. The enlarged cells near the tips of the hyphae are sex cells. How do the sex cells of the two strains differ from each other? What occurs when the pair of cells touch one another?

18. Note carefully the zygospore formed by fusion of the two sex cells. How does it seem to be adapted to a land environment?

19. Each day, wash your hands thoroughly before leaving the laboratory.

Day 10

20. Unseal and partially open your plate. Use a flat toothpick to transfer a bit of material from the mature culture to a glass slide. Reseal the plate. Add a drop of glycerine-water solution and a cover slip.

21. Observe under the low power of a compound microscope. Where are the sporangia produced? How might the position of the sporangia affect the distribution of spores? Does each sporangium produce a large or a small number of spores? Why is the number of spores produced important in the reproduction of *Rhizopus*?

22. Using high power, observe a few spores. How is their structure adaptive to a land environment?

23. Discard your plate in a biohazard bag as instructed by your teacher.

24. Wash your hands thoroughly before leaving the laboratory.

Discussion

1. Why is the world not covered with mushrooms? How numerous would mushrooms be if they produced only a small fraction of the number of spores you calculated for one mushroom in Part A?

2. Would the process you observed in Part B occur if both sides of the plate had + or both had − strains of *Rhizopus* on them? Explain.

3. How does the number of spores produced by *Rhizopus* compare with the number you calculated for the mushroom in Part A?

4. Which type of reproduction, sexual or asexual, appears likely to produce the greatest number of molds?

5. Which type of reproduction would be most advantageous to the mold in adjusting to changing environmental conditions?

Figure 12.13 Fungal mycelium growing on the inner surface of the bark of a log (X1).

For Further Investigation

With your classmates, conduct a mold-growing race. Select a medium, a source of spores, and an environment for growth. Keep a daily record of the mold growths. At the end of the race, present your procedures and results to the class.

12.10 Fungi Are Important Decomposers

Fungi are nonmotile, heterotrophic eukaryotes. They include mushrooms, molds, mildews, rusts, smuts, and many other, less familiar, organisms. Many fungi cause diseases, especially plant diseases, but many more form essential symbiotic relationships with plants. Others are used for baking breads, making alcoholic beverages, and producing drugs. Fungi reproduce by means of spores that germinate and grow slender tubes called **hyphae** (HY fee). A large, tangled mass of hyphae called a **mycelium** (my SEE lee um), shown in Figure 12.13, makes up the body of a fungus. Each fungal spore is capable of producing an entire mycelium. A mycelium will grow on virtually anything from which the fungus can obtain food. It grows especially well in moist soil that contains large amounts of dead plant and animal matter. Sometimes growth can be very rapid: A single fungus can produce a kilometer of hyphae in just 24 hours. The cell walls of fungi contain a carbohydrate known as **chitin** (KYT in). In some cell walls, the chitin is combined with cellulose, the same material found in plant cell walls. Chitinous walls are hard and resistant to water loss, which allows fungi to live in some extreme environments.

Along with bacteria, fungi are the most important decomposers in the biosphere. Digestion begins outside a fungus. It secretes digestive enzymes that break down food into small organic molecules. After the food is broken down, the fungus absorbs the small molecules. In this way, fungi decompose dead organisms.

Fungi reproduce both asexually and sexually. In asexual reproduction, the reproductive part of the fungus may appear above ground. A mushroom is an example. It is formed of tightly packed hyphae and may produce one million spores. Most spores are surrounded by tough, resistant walls that enable them to survive drying and extreme temperatures. Spores usually are dispersed by wind and air currents. Because their small size keeps them suspended in air for a long time, fungal spores are widely distributed. Fungi also may reproduce sexually by conjugation. In these fungi, the hyphae of different mating types come together and fuse. Because the hyphae of most fungi look very much alike, classification of fungi is based on differences among their sexual reproductive structures. The next sections describe some phyla of fungi.

12.11 Conjugating Fungi Can Destroy Foods

The common black bread mold, *Rhizopus stolonifer*, is probably the most familiar example of conjugating fungi (phylum Zygomycota; ZY goh my KOH tuh). Without preservatives or refrigeration, bread will develop a black, fuzzy growth like that in Figure 12.14 in a few days. The growth begins as

Figure 12.14 Bread covered with mycelium of *Rhizopus*.

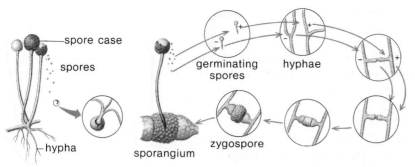

Figure 12.15   Asexual and sexual reproduction in *Rhizopus*.

Asexual reproduction

Sexual reproduction

a spore that germinates on the surface of the bread, producing hyphae. Then, some of the hyphae push up into the air and form sporangia. Mature sporangia are black, giving the fungus its color. The sporangia eventually break open, releasing many spores, each of which can germinate and spread the mold.

Sexual reproduction in *Rhizopus* occurs when nuclei from the hyphae of two different mating types fuse (see Figure 12.15). The resulting zygotes develop into thick-walled zygospores (ZY goh sporz), structures that give the phylum its name. Zygospores can survive periods of extreme temperature or dryness. When they eventually germinate, they give rise to sporangia that release spores and begin the cycle again.

Black bread mold can appear on many foods, including fruits such as grapes, plums, and strawberries. There also are parasitic species in this phylum, some of which attack crop plants. Most conjugating fungi, however, live in the soil and decompose dead plant and animal material.

12.12   Sac Fungi Have a Special Reproductive Structure

Phylum Ascus

The sac fungi (phylum Ascomycota; AS koh my KOH tuh) include yeasts, powdery mildews, and many common blue-green molds, as well as morels and truffles. Some sac fungi cause diseases such as chestnut blight and Dutch elm disease, both of which have destroyed many trees in the United States. Sac fungi are distinguished from other fungi by a microscopic sexual reproductive structure called an ascus (AS kus), which gives this phylum its name. Figure 12.16 shows several examples of sac fungi.

a   b   c

Figure 12.16   Sac fungi: (a) leaves covered with a powdery mildew; (b) a morel, *Morchella augusticeps;* and (c) red cup fungus, *Peziza*.

fruiting body
of sac
fungus

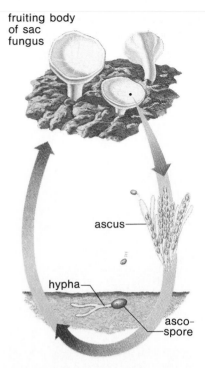

ascus—

hypha—

asco-
spore

Figure 12.17 Sexual reproduction in a sac fungus.

When sac fungi reproduce sexually, two hyphae conjugate and form an ascus, illustrated in Figure 12.17. Inside the ascus, nuclei from the hyphae fuse and become ascospores. The ascospores eventually are released from the ascus and may travel long distances on wind currents. Each ascospore can germinate and produce hyphae. In asexual reproduction, an individual hypha segments into huge numbers of spores that are dispersed by wind, water, or animals.

Yeasts are single-celled ascomycetes. They are small, oval cells (see Figure 2.1 on page 29) that reproduce asexually by budding and sexually by the production of ascospores. Yeasts are essential to the production of breads and alcoholic beverages, such as beer and wine. Under anaerobic conditions, yeasts convert the sugars found in bread dough or fruit juices into carbon dioxide and ethyl alcohol. The carbon dioxide bubbles cause bread dough to rise and beer to foam. When bread dough is baked, the alcohol evaporates.

12.13   The Reproductive Structures of Some Club Fungi Are Edible

Of all the fungi described here, club fungi (phylum Basidiomycota; buh SID ee oh my KOH tuh) are probably the most familiar. The mushrooms you buy in a grocery store are the sexual reproductive structures of club fungi. The phylum also includes smuts, rusts, jelly fungi, puffballs, and stinkhorns (see Figure 12.18).

Most species of mushrooms, both edible and poisonous, are club fungi. A mushroom is a spore-producing body composed of masses of tightly packed hyphae. Under a mushroom cap, you can see structures called gills radiating out from the center. Between these gills many spores are produced on tiny clubs, or basidia (buh SID ee uh), which give this phylum its name (see Figure 12.19).

A toadstool is a poisonous mushroom. One of the most poisonous is the "death cap," *Amanita*. One bite of certain species of *Amanita* can be fatal. Some species of toxic mushrooms have hallucinogenic effects and sometimes are eaten deliberately. Other club fungi are plant parasites, such as the organisms causing wheat rust and corn smut.

Figure 12.18   Club fungi: (a) puffballs, *Lycoperdon perlatum;* (b) bracket fungus from an Australian rain forest; (c) the poisonous *Amanita muscaria.*

c

×1300

basidium

×1/2

×250

Figure 12.19   Reproductive structures of a field mushroom.

12.14   Imperfect Fungi Have No Known Sexual Reproductive Structures

The classification of fungi is based principally on their sexual reproductive structures. Some fungi are unclassifiable because they seldom, if ever, show reproductive structures. Biologists have investigated certain fungi for years without discovering sexually produced spores. Some of these fungi, however, do resemble species whose sexual reproductive structures are known, and they can be classified accordingly. For example, if a fungus produces a large number of asexual spores, it can be classified on the basis of similarities in sporangia. Once the sexual structures of such fungi are discovered, they are reclassified on that basis.

If no sexual reproductive structures ever have been discovered for a fungus, it is called "imperfect." For convenience only, taxonomists place imperfect fungi in a separate group. Figure 12.20 shows two examples. Some imperfect fungi are important in making certain cheeses and antibiotics. For example, *Penicillium roquefortii* is responsible for the distinctive flavor

a

b

Figure 12.20   Imperfect fungi: (a) a species of *Penicillium* growing on an orange; and (b) asexual reproductive structures (conidiospores) of *Aspergillus* (X90).

and appearance of Roquefort cheese, *P. camembertii* is used to manufacture Camembert cheese, and *P. notatum* produces the drug penicillin. Other imperfect fungi are parasites of crop plants and animals. For example, athlete's foot is caused by an organism that for years was thought to be an imperfect fungus. Recently, sexual spores have been discovered, showing athlete's foot fungus (and a number of related skin disease fungi) to be a sac fungus.

Concept Review

1. How are hyphae related to a mycelium?
2. What adaptations to a terrestrial environment do fungi show?
3. What characteristics are used to distinguish fungi from each other?
4. How are fungi important to humans?
5. What is meant by the term *imperfect fungi*?

Guidepost: What roles do fungi play in a community?

Fungi in a Community

12.15 Fungi Support Many Food Chains

A soil ecosystem is like a city in that its food supply comes from the outside. Most soil organisms are consumers and live below the surface of the soil. In soil, food made during photosynthesis comes down into plant roots from the leaves and stems. Plant roots can serve as food for some consumers, but for most of them, the food supply comes from the remains of other organisms. Therefore, decomposers such as fungi and bacteria are very important in soil communities. Figure 12.21 shows a decomposer food chain.

A dead root in the soil still contains a large amount of starch; likewise, a dead animal still contains fats and proteins. Beetles and other small animals living on or just below the soil surface use these substances as food. Other substances, such as cellulose in plant bodies and chitin in insect bodies, can be broken down only by microorganisms and fungi. Decomposers that use cellulose and chitin as food leave simple organic substances as waste products. The wastes still contain energy that other fungi and bacteria can use. These fungi and bacteria, in turn, leave simple inorganic substances in the soil, which still other decomposers can use. Thus, one decomposer depends on another for its food supply.

This type of food chain is like an assembly line in reverse. Instead of building up step-by-step from simple to complex compounds, the decomposer food chain breaks down complex organic substances. At the end of the chain, only inorganic substances such as carbon dioxide, nitrates, ammonia, and water remain. Plants then can assemble these inorganic materials into organic compounds and begin the cycle again.

Some soil organisms produce substances that harm other organisms. These substances may accumulate in the soil around the organism that formed them, reducing the growth of competing organisms. We use these substances as antibiotics to combat bacterial infections in animals. Aureomycin, for example, is derived from a bacterium, and penicillin is derived from a fungus.

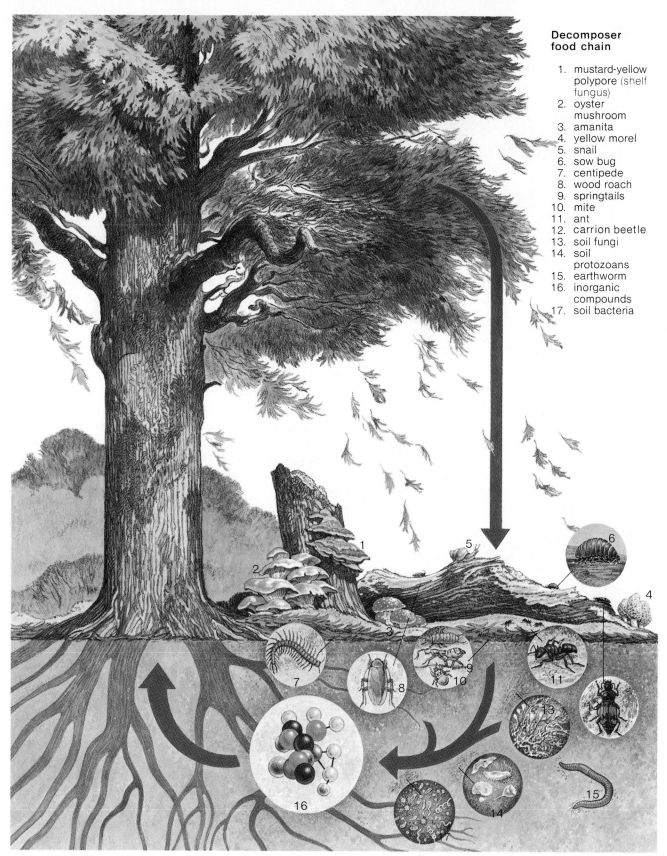

Decomposer food chain

1. mustard-yellow polypore (shelf fungus)
2. oyster mushroom
3. amanita
4. yellow morel
5. snail
6. sow bug
7. centipede
8. wood roach
9. springtails
10. mite
11. ant
12. carrion beetle
13. soil fungi
14. soil protozoans
15. earthworm
16. inorganic compounds
17. soil bacteria

Figure 12.21 A decomposer food chain. Identify each organism and the role it plays in the process of decomposition.

Figure 12.22   Predacious fungi capture small roundworms present in the soil. *Dactylella drechsleri* traps the worms with small adhesive knobs (X400).

Fungi even can act as predators (see Figure 12.22). Several species of soil fungi form hyphae with tough branches that curl into loops. The loops from adjacent hyphae interlock, forming a network and producing a sticky fluid. Small worms living in the soil, called nematodes, are caught in the loops and held fast despite violent struggles. Other fungal hyphae then invade the bodies of the captive nematodes and digest them.

12.16   Mycorrhizal Fungi Help Many Plants to Grow

If you trace the networks of fungal hyphae down through loose soil, you will find some leading to plant roots. The relationship between the roots and the fungi, which causes slight modifications of the roots, is known as **mycorrhizae** (my koh RY zee).

Mycorrhizae are a symbiotic relationship. The fungi either form a sheath around the root or penetrate the root tissue. Some plants, such as orchids and pines, do not grow, or else grow poorly, if their mycorrhizal fungi are not present. The mycorrhizae increase the surface area of the root through outward extensions (Figure 12.23). The fungi absorb soil nutrients and secrete an acid that makes the nutrients readily usable by the plant. Thus, the fungi transfer nutrients to the plant, and photosynthetic products from the plant nourish the fungi. Mycorrhizae also absorb water and protect the plant against certain pathogens in the soil.

Mycorrhizae occur in all plant families. They are found in some of the oldest plant fossils, dating from the Devonian period. Mycorrhizae even may have made it possible for plants to adapt to a terrestrial existence. Scientists speculate that the fungi may have transported nutrients to the roots of primitive plants and prevented them from drying out long enough for the plants to establish themselves on land.

12.17   Lichens Are Symbiotic Pairs of Organisms

Figure 12.23   A scanning electron micrograph of mycorrhizae between a *Boletus* mushroom and an aspen, *Populus tremuloides* (X50).

You can find lichens on the bark of pine trees, on tombstones in New Hampshire, buried under arctic snow, or on rocks in Arizona. The shape and color of lichens are so characteristic that for several hundred years biologists described them as if they were single organisms. Actually, a lichen is made up of a fungus and either an alga or a cyanobacterium in a symbiotic relationship. More than 25 000 "species" of lichens have been described. They are named according to the fungus part of the lichen, because the fungus is the component that reproduces.

A lichen has the structural framework of fungal hyphae. Many small producers, usually green algae or cyanobacteria, live in the branches of the hyphae. The producers can live alone, but the lichen fungi do not grow well when separated from their partners. Many lichen fungi are classified as sac fungi, but they are unlike any of the sac fungi that do live alone.

Lichens secrete acidic compounds that break down rocks. Thus, they often are the first organisms to colonize bare rock, bare soil, or ice (see Figure 12.24). Lichens can survive in these extremely dry areas because they have the ability to dry out completely and become inactive, or dormant. When dormant, lichens can tolerate temperature extremes that would

Figure 12.24 Fruticose lichens (a) on a tree trunk; crustose lichens (b) colonizing on rock.

kill other organisms, even hardy ones. However, they do not grow until they become wet again. In rain, fog, or dew, a lichen may absorb up to 30 times its own weight in water; then the producer in the lichen quickly begins to photosynthesize. Growth continues until the lichen dries out again. Because lichen growth occurs in spurts, the average growth rate is extremely low—sometimes just 0.1 mm per year. Thus some small lichens may be several hundred years old.

One common lichen, reindeer moss (Figure 12.25), covers great areas of the Arctic. Although it is called a moss, it is actually a lichen that reindeer and caribou eat. Reindeer moss, like all lichens, absorbs some nutrients from the soil or rock and from rainfall, then concentrates the nutrients in its cells. Toxic compounds also may be absorbed and concentrated in the lichen bodies. If the air is extremely polluted, the lichens will die, thus providing an indication of how polluted the air is in a given area. Some reindeer moss has accumulated radioactive materials from aboveground atomic bomb testing in the Arctic. Reindeer and caribou that have eaten

Figure 12.25 Reindeer moss, a type of shrubby lichen.

the lichens have taken in the radioactive materials. When Eskimos have eaten these animals, they also have absorbed the radioactive materials. Thus materials may be passed along unexpectedly through a food chain to a consumer at the top of the chain.

Concept Review

1. Compare and contrast a food chain involving only decomposers with one that includes plants and animals.
2. How are some fungi predators?
3. Describe mycorrhizae. What makes this relationship symbiotic?
4. What characteristics make lichens good pioneer organisms?
5. Why do lichens make good indicators of air quality?

Summary

Protists and fungi are two kingdoms of eukaryotes. Algae are autotrophic protists and may be distinguished from each other by their dominant pigments—red, brown, yellow, or green. Heterotrophic protists include protozoa and slime molds. Most protozoa can be distinguished by their method of movement. Some do not move at all, whereas others move by using cilia, pseudopods, or flagella. Some protozoa are parasites that cause diseases of humans and other animals. Slime molds are multicellular, heterotrophic protists that exhibit animal-like and funguslike characteristics.

Fungi are all heterotrophic and can be distinguished from each other by their sexual reproductive structures. Whereas some fungi are edible, others destroy foods, and still others are parasites. In a community, fungi play a major role as decomposers. Fungi also may be part of symbiotic relationships with other living organisms, as they are in mycorrhizae and in lichens.

Applications

1. What characteristics of algae make them more advanced than the cyanobacteria discussed in Chapter 11?
2. For an aquatic organism, what are the advantages of storing energy reserves as lipids, starch, or some other carbohydrate, and *not* as sugar?
3. Is color a good characteristic to use to distinguish between autotrophic and heterotrophic organisms? Explain.
4. Suppose you moved a freshwater paramecium into salt water. What do you think would happen to the rate of contraction of the contractile vacuole? Explain.

5. *The Blob* was a movie about an alien organism that lands on earth and consumes all living things in its path. Which organism in this chapter does the Blob resemble and in what ways?
6. Many fungi are decomposers of plant tissues. Considering this, what is the significance of the fact that fungi have cell walls made up mostly of chitin?
7. The bodies of many fungi are located underground, but the reproductive structures are above ground. What is the advantage of this arrangement?
8. In which of the ecological relationships between organisms described in Chapter 3 are fungi involved?

9. How do the characteristics of lichens cause a taxonomic problem?

10. Mycorrhizal fungi use sugars from the plants to which they are attached. Why are these fungi not considered to be parasites of the plants?

Problems

1. Start reading the labels of the foods you eat. Make a list of those foods that include algae or algal products. Look for carrageenan or algin, two algal products commonly used.

2. Draw or paint a picture depicting an ecosystem in which algae play an important role. Show the algae and the organisms that depend on these producers. You might consider a freshwater pond or lake with different types of algae and fish or a coastal ecosystem that includes sea otters, brown algae, and sea urchins. Try to depict through your illustrations the relationships between the organisms in the ecosystem.

3. Algae are good indicators of water quality—many species are found only in standing, polluted waters. Ask a local phycologist (a person who studies algae) or a local expert on water quality to visit your class and speak on the types of organisms normally found in your water reservoirs, the problems they can cause, and how the problems are resolved.

4. Mitochondria have their own circular DNA that is different from the DNA found in the nucleus of the cell. Studies in humans have shown that mitochondrial DNA is of maternal origin, that is, it is inherited from the mother only. How might you explain this? (Hint: refer to the discussion of fertilization in Chapter 6.)

5. Investigate how fungi and bacteria are used to help fight disease—especially in the production of antibiotics.

6. Make a collection of mushrooms and lichens from areas where collecting is permitted. *Do not eat any of them.* Describe their size, shape, and color; if possible, note to what each was attached. What role do you think each played in the community?

7. How many of the groups of organisms discussed in this chapter can be found in your locality? Take into consideration wild and cultivated, indoor and outdoor, and aquatic and terrestrial organisms.

This is a curious-looking plant. What type of plant is it? Use *A Catalog of Living Things* (Appendix 4) to find its name. What are the structures at the top of the plant? Check your answer with a classmate.

Eukaryotes: Plants 13

Several hundred million years ago, the land surface of the earth was barren. Life existed only in the water. One-celled algae colored the surface waters of the oceans shades of green. Some algae joined together into multicellular organisms and evolved a degree of cell specialization. These organisms probably were the ancestors of modern plants (see Chapter 12).

This chapter examines several of the characteristics that enabled plants to invade the land and describes a few examples of modern plants. Today, plants adapted to life on land are found in most habitats, from prairies to tundra and in the forests and deserts of the world. They come in the widest possible range of sizes. The largest organisms in the world are the giant redwoods in California, such as those shown in Figure 13.1. Some of these trees are more than 100 m tall and 7 m in diameter. Between a simple multicellular green alga in the ancient sea and a giant California redwood tree with billions of cells, the difference in cellular coordination is tremendous. The evolution of such a complex land organism from a relatively unspecialized multicellular ancestor did not occur overnight. It took millions of years.

The Evolution of Land Plants

Guidepost: What adaptations allowed plants to colonize land?

13.1 Two Major Groups of Land Plants Evolved

Plants are multicellular, photosynthetic organisms adapted primarily for life on land. The evolution of multicellular green algae that had some division of labor among their cells produced many adaptations that allowed plants to live in this environment. The oldest known plant fossils are simple, branched structures that had several important adaptations.

Because the climate on land may vary from hot to freezing and from wet to dry, water is the major limiting factor for any organism. On land, the only reliable source of water is underground where it is too dark for photosynthesis. What possible advantages of life on land could overcome such a major disadvantage?

Without the presence of other organisms, plants that were able to tolerate the dry conditions that exist on land would have had little competition for food, living space, and the other necessities of life. Another advantage would have been the abundance of carbon dioxide and oxygen

Figure 13.1 The giant redwoods, *Sequoia sempervirens,* are the largest organisms in the world.

present in the air. A third advantage for these plants would have been increased light levels, because the air does not absorb as much light as water does. Finally, there would have been plenty of space, and nutrients would have been readily available in the soil.

A multicellular alga tossed onto the shore would have had a better chance of surviving than would a one-celled alga. The outer cells of the organism might have protected the inner ones from drying out rapidly, and the inner cells might have been efficient at photosynthesis. Other cells of the same organism might have specialized in collecting water or nutrients from the environment. The resulting division of labor would have enabled the alga to exploit the resources in its new environment. A specialized, multicellular alga, somewhat like the modern *Chara* in Figure 13.2, might have been able to live and reproduce on land if ocean spray or tides kept it moist.

Two plant groups apparently evolved from relatively complex multicellular green algae. The more complex group, which includes fossils of the oldest land plants, has many adaptations to life on land, including **vascular tissue**—cells joined into tubes that transport water and nutrients throughout the body of the plant. These plants are called vascular plants. The less complex group, the **bryophytes** (BRY oh fyts), are not aquatic, yet they possess few adaptations to life on land. This group includes the true mosses, hornworts, and liverworts. Bryophytes are called nonvascular plants because they do not have vascular tissue.

Vascular plants can be divided into two groups: those that produce seeds, and seedless plants such as ferns that reproduce with spores. Vascular plants that produce seeds can be divided further into two groups: those that produce naked seeds, such as pine trees, and the flowering plants, which produce seeds enclosed in a fruit. Figure 13.3 summarizes the known evolutionary history of plants.

Figure 13.2 The multicellular green alga *Chara.* The fossil record of *Chara*-like algae extends back about 400 million years. Why is *Chara* considered to be a model of the ancestors of land plants?

Figure 13.3 The evolutionary history of plants. What evidence was used to construct this history?

INVESTIGATION 13.1 Increasingly Complex Characteristics

Biologists sometimes use pairs of terms such as *less complex* and *more complex* when discussing diversity among organisms. A species that has changed little from its ancestors is said to be less complex than organisms that differ greatly from their ancestors. Conversely, a species that has few of the characteristics of its ancestors is said to be more complex. After studying many types of evidence in the fossil record and in living organisms, scientists have reached fairly general agreement about which characteristics have been in existence for a long time and which are more recent. Table 13.2 is based on such studies. Because there may be many degrees of complexity, the terms *less complex* and *more complex* are not absolute, but they are useful for making the types of comparisons you will make in this investigation.

Materials (per team of 6)

10 labeled specimens of organisms of
    various kingdoms and divisions
compound microscope

stereomicroscope or 10X hand lens
microscope slides
cover slips

Procedure

1. In your data book, prepare or tape in a table similar to Table 13.1, with enough lines for all 10 specimens.

TABLE 13.1

Name of Organism	Numerical Values of Choices Made	Total Complexity Score	Rank
1.			
2.			
etc.			

2. Determine the complexity score for each of the labeled specimens. Start at the left of Table 13.2. Arrows from the starting point lead to two descriptions. Choose the one that fits the organism you are scoring. More complex organisms such as plants are represented at each station as well as some less complex organisms from other kingdoms.

3. Proceed across Table 13.2 by following the arrows and choosing in each column the description that best fits each organism. Continue as far as the arrows go.

4. With each description there is a number. As you proceed, record the numbers of your chosen descriptions in the second column of your data table. The complexity score for the organism is the sum of all the numbers appearing after the descriptions you used in working through Table 13.2. The more alike two organisms are, the more alike their scores will be. The greater the difference between two organisms, the greater will be the difference in their scores. More complex organisms, such as plants, will have high scores (maximum 26), and less complex organisms, such as prokaryotes, will have low scores (minimum 3).

5. When you have the complexity score for each of the organisms, give the organism with the lowest score a rank of 1 and the organism with the highest score a rank of 10. Then rank the rest of the organisms according to their scores. Record the rankings in the column at the right side of your data table.

Discussion

1. Basing your conclusions on the way the complexity score key was designed, list some of the most important differences among the organisms you observed.
2. What are some of the less important differences?
3. Using the information in the table, list the characteristics you would expect to find in one of the less complex organisms.
4. Do the same for one of the more complex plants.
5. Today's land plants evolved from water-dwelling ancestors. Plants that live on land are in constant danger of drying out. Suggest how each of the following characteristics found in a more complex plant would help the plant live on land: (1) roots; (2) stems that contain vascular tissues; (3) seeds; and (4) flowers.

TABLE 13.2   Key for Determining the Complexity Score for an Organism

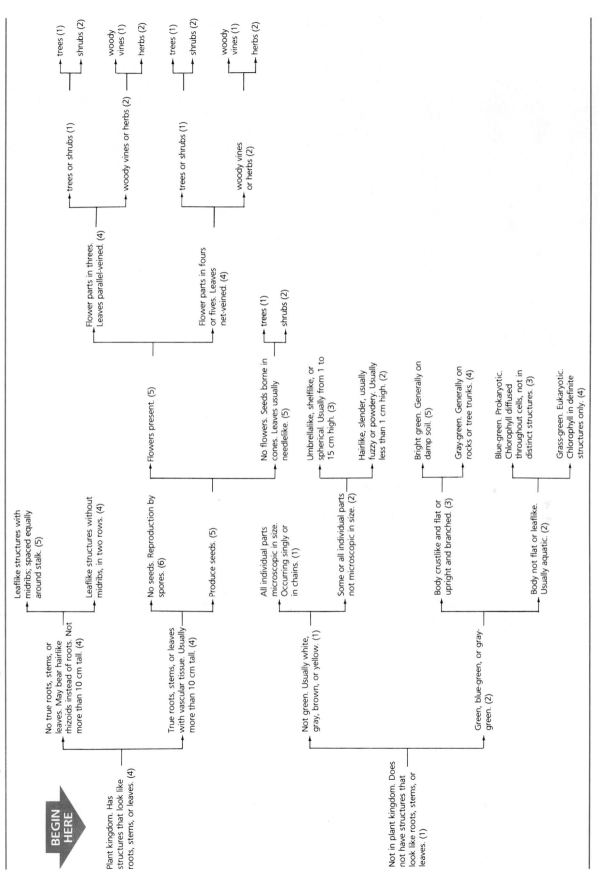

13.2 Vascular Plants Have Adaptations That Conserve Water and Permit Gas Exchange

The success of plants on land depends largely on their ability to absorb and hold water. Structures that enable them to do so include roots, vascular tissue, and an outer covering that retards water loss.

Vascular plants such as ferns, conifers (cone-bearing plants), and flowering plants have well-developed root systems that penetrate into many parts of the soil. An extensive root system provides an efficient way to collect water and nutrients from the soil and bring them to the main body of the plant. Rooted plants offer a good example of division of labor—the cells in the root collect water and nutrients, and the aboveground cells absorb sunlight and produce food through photosynthesis.

Tall plants require cells that can transport water from the roots to the leaves and cells that can support an upright body. Because vascular tissue serves both these functions, vascular plants can grow taller than bryophytes and thus capture more sunlight. Bryophytes never grow very tall because they lack an efficient water transport system.

In bryophytes, water can evaporate from the entire surface, so a moss plant, for example, dries out quickly. Vascular plants produce a waxy covering, or **cuticle**, that covers the plant body above ground, reducing the amount of water that can evaporate from its surface. The cuticle is often thick on the leaves of plants living in dry places, such as the sedum in Figure 13.4. The covering does not prevent gas exchange with the environment because vascular plants have slitlike openings, or **stomates** (STOH maytz), in the surface of their leaves. Stomates (Figure 13.5), found in the oldest fossil plants, permit carbon dioxide and oxygen to enter or leave the plant. Roots that absorb water, vascular tissue that supports the plant and conducts water, a cuticle that prevents evaporation, and stomates that permit gas exchange are characteristics that enable plants to live on land.

Figure 13.4 Plants such as this *Sedum adolphic* have very thick cuticles. How are thick cuticles adaptive?

13.3 Bryophytes Require Water for Reproduction

Reproductive adaptations also enabled plants to survive on land. Bryophytes are restricted to moist areas because their sperm are flagellated, as are those of animals and algae. For sexual reproduction to occur, the plants must be bathed in water so the sperm can swim to the egg. Thus, bryophytes reproduce sexually only where water sprays them or after they are wet with dew or rain. Like all sexually reproducing plants, bryophytes have a life cycle in which a haploid (*n*) phase alternates with a diploid (*2n*) phase. This type of life cycle is called **alternation of generations**.

The carpet of moss shown in Figure 13.6a actually is many individual plants. These small plants are haploid, and each is called a **gametophyte** (guh MEET oh fyt). Gametophytes produce gametes, usually in special structures near the tips of the plants. In wet conditions, sperm are produced in male gametophytes. The sperm swim in a film of water to the egg cell in the female gametophyte, and a diploid zygote is produced.

The zygote divides by mitosis and develops into a diploid embryo. Eventually, the embryo grows out of the female gametophyte into a stalk-like structure, the **sporophyte** (SPOR oh fyt). The sporophyte is visible above the small haploid individuals shown in Figure 13.6b. Meiosis occurs within

Figure 13.5 A single stomate in the surface of a leaf.

Figure 13.6 Mosses (a) thrive in moist conditions such as streambanks. Moss sporophytes (b), *Polytrichium,* growing from the female gametophyte plants.

the capsule at the end of the sporophyte, and haploid spores are formed. The capsule helps to protect the spores. Once the spores are released, they are carried to wherever wind or water transports them. Most spores fall onto unfavorable habitats and die. If a spore reaches a favorable environment—usually a moist soil surface—its wall can burst open. The cell within begins to divide by mitosis, producing long green threads that resemble the filaments of many aquatic algae. The gametophyte moss plant develops from these threads. When a spore germinates and grows into another gametophyte plant, the life cycle diagrammed in Figure 13.7 is complete.

Figure 13.7 The life cycle of a moss. Which generation is predominant?

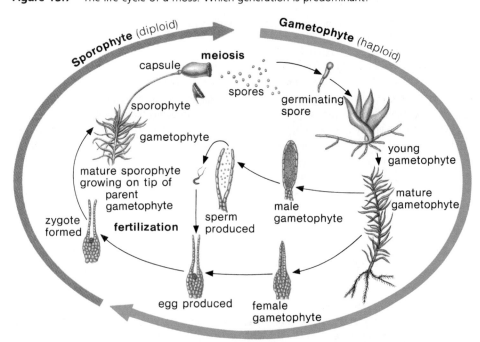

INVESTIGATION 13.2 Reproductive Structures and Life Cycles

Although their reproductive organs differ as do the environments in which they live and reproduce, the basic principles of sexual reproduction are the same in a moss, a flower, a bee, and a human. In this investigation you will learn how the structures of a moss and a flower serve reproductive functions in their respective environments.

Mosses form mats on logs and on the forest floor, growing best in damp, shaded environments. Sporophytes grow out of the tops of the gametophytes and often look like hairs growing out of the mat of moss. Mosses cannot reproduce unless they are wet. Flowering plants, on the other hand, are found in many different environments and climates. They need water to live but not to reproduce.

Materials (per team of 2)

2 microscope slides	prepared slide of moss male and female
2 cover slips	reproductive organs
dissecting needle	15% sucrose solution
scalpel	moss plant with sporophyte
forceps	fresh moss
compound microscope	gladiolus flower
stereomicroscope or 10X hand lens	other simple flowers for comparison
modeling clay	fresh bean or pea pods
prepared slide of filamentous stage of	
moss	

Procedure
PART A Moss

1. Examine a moss plant with sporophyte attached. The sporophyte consists of a smooth stalk terminated by a little capsule. Separate the two generations by pulling the sporophyte stalk out of the leafy shoot of the gametophyte.

2. Using a dissecting needle, break open the capsule of the sporophyte into a drop of water on a slide.

    CAUTION: Needles are sharp. Handle with care.

    Add a cover slip and examine under the low power of a compound microscope. What are the structures you observe? How are these structures distributed in nature? How are they adapted for life on land?

3. Most moss spores germinate on damp soil and produce a filamentous stage that looks like a branching green alga. Examine a prepared slide of this stage.

4. The filamentous stage gives rise to the leafy shoot of the gametophyte. Using forceps, carefully remove a leafy shoot from the fresh moss. How does this shoot obtain water and nutrients for growth?

5. The reproductive organs of the gametophyte are at the upper end of the leafy shoot. Examine a prepared slide of these organs under the low power of a compound microscope. The male sex organs are saclike structures that produce large numbers of sperm cells. The female sex organs are flask-shaped and have long, twisted necks. An egg is formed within the base of the female organ. How does a sperm reach the egg? Would you expect to find moss plants growing where there was little or no water? Explain. The union of the egg and sperm results in a cell called the zygote. Where is the zygote formed? What grows from the zygote?

PART B Flowers

6. Examine the outside parts of a gladiolus flower. The outermost whorl of floral parts may be green and leaflike. These green sepals protected the flower bud when it was young. In some flowers, such as lilies, the sepals look like an outer whorl of

petals. Petals are usually large and colored and lie just inside the sepals. Both sepals and petals are attached to the enlarged end of a branch. These parts of the flower are not directly involved in sexual reproduction. What functions might petals have?

7. Strip away the sepals and petals to examine the reproductive structures. Around a central stalklike body are 5 to 10 delicate stalks, each ending in a small sac, or anther. These are the male reproductive organs, or stamens. Thousands of pollen grains are produced in the anther. The number of stamens varies according to the type of flower. How many stamens are present in the flower you are using? How is pollen carried from the anthers to the female part of the flower?

8. If the anthers are mature, shake some of the pollen into a drop of 15% sucrose solution on a clean slide. Add a cover slip and examine with the low power of a compound microscope. What is the appearance of the pollen? How is the pollen adapted for dispersal?

9. Make another pollen preparation on a clean cover slip. Use modeling clay to make a 5-mm-high chamber, slightly smaller than the cover slip, on a clean slide. Add a small drop of water to the chamber and invert the pollen preparation over it. Examine after 15 minutes and again at the end of the lab period. What, if any, changes have occurred? (If no changes have occurred, store the slide in a covered petri dish containing a piece of cotton moistened with water, and examine it the next day.)

10. The central stalk surrounded by the stamens is the female reproductive organ, or carpel. It is composed of an enlarged basal part, the ovary, above which is an elongated part, the style, ending in a stigma. How is the stigma adapted to trap the pollen grains and to provide a place for them to grow?

11. Use a scalpel to cut the ovary lengthwise.

    CAUTION: Scalpels are sharp; handle with care.

    Using a hand lens or stereomicroscope, look at the cut surface. How many ovules can you see? Each ovule contains one egg. To what stage of the moss life cycle is the ovule comparable? How close to the egg can the pollen grain get? If the pollen grain cannot get to the egg, how do the sperm produced by the pollen reach it? To what stage of the moss life cycle is a pollen grain comparable?

12. The union of egg and sperm causes extensive changes in the female reproductive parts. Fertilization of the egg stimulates the growth of the ovary and the enclosed ovules. Carefully examine a fresh bean or pea pod. Open the pod to find the seeds. What part of the female reproductive apparatus is the pod of a bean or pea? What is the origin of a seed? If you plant ripe bean or pea seeds and water them, to what will they give rise? What can you conclude develops within a seed as a result of fertilization?

13. If time permits, examine other types of flowers. Compare the numbers of various parts and the ways the parts are arranged with respect to each other.

14. Wash your hands thoroughly before leaving the laboratory.

Discussion

1. In alternation of generations in a moss, which is the predominant, independent generation? Which is the less conspicuous generation?

2. Compare the life cycle of a moss (with alternation of generations) with your life cycle (with no alternation of generations).

3. Would you expect the most variation in flowering plants or in those reproducing by asexual means? Explain.

4. Compare and contrast the life cycle of a moss with that of a flowering plant.

5. Do flowering plants represent more or less adaptation to a land environment than mosses? Explain.

Figure 13.8 Scanning electron micrograph of a pollen grain from the cactus *Opuntia polyacantha* (X525).

13.4 Flowering Plants Have Special Reproductive Adaptations

Unlike bryophytes and seedless vascular plants, which require water to be present for fertilization, seed plants produce special structures called **pollen grains** (Figure 13.8), in which the sperm develop. The pollen grains may be blown by wind or may be accidentally carried by animals from one plant to another. This efficient means of transferring sperm to the egg under dry conditions is most highly developed in the flowering plants. Brightly colored, scented flowers attract insects and contain nectar the insects drink. As an insect drinks, it picks up pollen from one flower and carries it to another flower while searching for more nectar. Because seed plants are not restricted to moist conditions, sexual reproduction can occur whenever the sperm and egg are fully developed. Like bryophytes, flowering plants have a life cycle with alternation of generations, but the gametophyte generation of flowering plants is smaller and protected, at least for a while, by the sporophyte.

Flowering plants are considered to be the most complex of the vascular plants. Their reproductive structures are found in flowers. A flower is actually a short branch bearing groups of specialized leaves. Some of these leaves may resemble ordinary leaves in many ways, but others are so different in structure that it is hard to think of them as leaves at all. If you closely examine a flower such as the buttercup illustrated in Figure 13.9, you will see a number of green, leaflike structures called sepals (SEE pulz) on the underside of the flower. Before the bud opens, the sepals cover and protect the other parts of the flower. The most conspicuous parts of a flower are the colorful petals. Although petals are often leaflike in shape, they are not usually green.

Just inside the circle of petals of a typical flower is a ring of male reproductive structures, the **stamens** (STAY menz). In the center of the flower is the **carpel** (KAR puhl), the female reproductive organ. Although most plants have both male and female organs within the same flower, a few plants produce flowers with only female parts or only male parts. Stamens usually have an enlarged tip, the anther, whereas the carpel tip (the stigma) is more pointed. Despite their shape, however, both stamens and carpels are thought to be modified leaves that have been adapted for reproductive roles.

Figure 13.10 shows the relationship between reproductive structures and the life cycle of one type of flowering plant. At the base of the carpel is an enlarged portion, the ovary, that contains one or more small structures called **ovules** (OHV yoolz). Meiosis occurs in a special cell in each ovule, resulting in the formation of four haploid cells, the female spores. These spores do not separate from the sporophyte as they do in mosses. Instead, three of the spores disintegrate. The fourth spore divides three times by mitosis, forming eight nuclei. The nuclei, with their surrounding cytoplasm, form seven cells, one of which is the egg cell. (One of the cells contains two nuclei, the polar nuclei.) Figure 13.11a shows these seven cells, which constitute the female gametophyte.

In the stamens, cells in the anthers undergo meiosis, each giving rise to four haploid cells, the male spores. Each spore contains one haploid nucleus that divides by mitosis, forming two nuclei. A spore wall thickens around each nucleus, forming a pollen grain. A single stamen may contain

Figure 13.9 Diagram of the flower structure of a buttercup flower.

petal —
stamen —
carpel —

sepal —

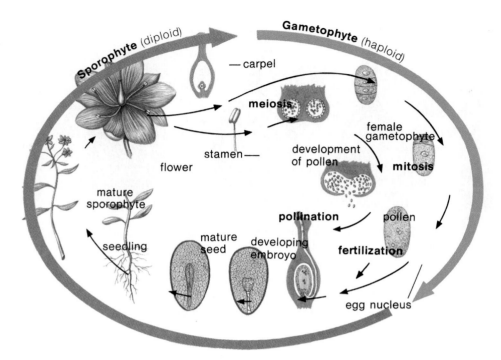

Figure 13.10 The life cycle of a flowering plant. The parts are drawn to different scales. Which generation predominates?

thousands of pollen grains. Each mature pollen grain is a male gametophyte (see Figure 13.11b).

Pollination is the transfer of pollen from the stamens to the carpel, either within a flower, between flowers of the same plant, or between plants of the same species. The sticky stigma at the top of the carpel can trap pollen grains carried to it by wind, water, or a visiting animal. Only pollen grains from flowers of the same species are useful in fertilization.

The hard wall of the pollen grain protects the haploid cell until it lands on a stigma. Once there, a thin finger of tissue, the **pollen tube**, grows from the grain into the carpel. Within the pollen tube, one nucleus, the tube nucleus, leads the way. The other nucleus divides to form two sperm nuclei. The pollen tube grows down the carpel, transporting the sperm nuclei to the ovule. Fertilization occurs when one sperm nucleus unites with the egg, forming a zygote. This diploid cell gives rise to the embryo. The other sperm nucleus unites with the polar nuclei, leading to the formation of **endosperm** (EN doh sperm), a mass of food-storing cells. The endosperm is triploid—it has three sets of chromosomes, one set from the sperm and two from the polar nuclei. Figure 13.12 on page 342 summarizes the stages of fertilization and embryo development in a flowering plant.

The life cycle of a flowering plant is similar to that of a moss in two ways. First, meiosis occurs just before spore formation. Second, there is alternation of generations between the sporophyte and gametophyte portions of the life cycle. There are, however, several differences. First, abundant surface water is not necessary for fertilization in flowering plants. Second, the gametophytes are smaller than the sporophytes. Third, the gametophyte that produces an egg and the spore that produces the female gametophyte do not separate from the sporophyte plant. Thus, these structures

Figure 13.11 Female gametophyte of a lily (a). Mature ovule ready to be fertilized; the egg nucleus is one of those on the right. Mature pollen grain with two nuclei (b). A pollen grain is the male gametophyte of a seed plant.

a

b

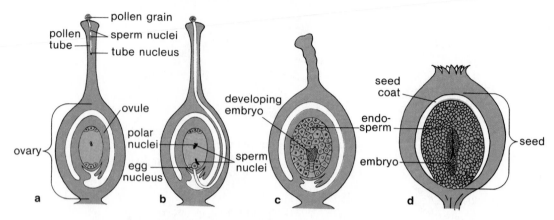

Figure 13.12 Pollination, fertilization, and development of the embryo, endosperm, and seed in a flowering plant.

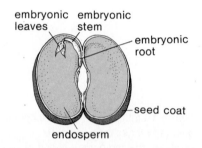

Figure 13.13 A bean seed, showing the embryo and endosperm.

Figure 13.14 Evolution and specialization in sporophytes and gametophytes.

are better protected from the environment than are their counterparts in a moss. Finally, the embryo (young sporophyte) grows for a short time, then becomes dormant. The embryo and its endosperm are surrounded by a protective coat formed from ovule tissues. This package is a **seed**, diagrammed in Figure 13.13. The seed protects the young sporophyte, which remains dormant until environmental conditions are suitable for germination. Moss embryos, on the other hand, cannot tolerate dry conditions. Also, moss spores contain no embryos and only a small amount of food.

Land plants show a trend toward increasing specialization of the sporophyte and decreasing specialization of the gametophyte, as shown in Figure 13.14. Seed plants have succeeded because of their adaptations to the changing environment on land. Cuticle, vascular tissue, root systems, sperm carried in pollen, spores protected on sporophytes, and embryos protected in seeds makes flowering plants well suited to life on land.

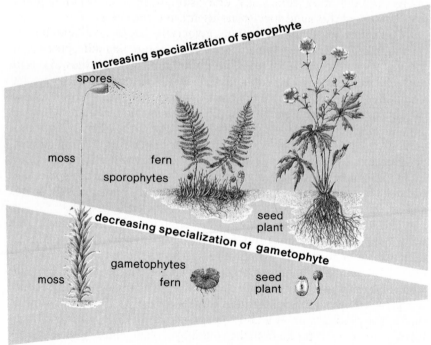

Concept Review

1. Describe the limiting factors and advantages of life on land for the first land plants.
2. What types of adaptations were necessary for plants to live on land, and why were they important?
3. Briefly distinguish between the different types of land plants.
4. Compare and contrast gametophytes and sporophytes.
5. Compare and contrast the life cycles of bryophytes and flowering plants.
6. Explain the roles of the reproductive structures of flowering plants.
7. Describe the origin of the embryo, the endosperm, and the seed.

Bryophytes and Seedless Vascular Plants

Guidepost: How do bryophytes and seedless vascular plants differ in structure and habitat?

13.5 Bryophytes Have No Roots, Stems, or Leaves

Most bryophytes (division Bryophyta; bry OFF ih tuh) are relatively small. Few of them exceed 20 cm in height. Although they have structures resembling stems and leaves, these terms are not used in describing bryophytes because they lack the vascular tissue of other land plants. There are three classes of bryophytes: true mosses, the largest class; liverworts; and hornworts. Figure 13.15 shows several bryophytes.

True mosses often grow in clumps or small clusters in rock crevices and on the shady side of trees. An individual moss plant from such a clump is simply an upright, green, stemlike stalk with threadlike structures called **rhizoids** (RY zoidz), which perform the function of roots by aiding in absorption and helping to hold the plant in place. Many flat, green, leaflike structures are attached spirally along the stalk. Water and nutrients are absorbed throughout the body of the bryophyte, so most grow in fairly damp places, and a few grow in water. Under dry conditions many mosses become dormant. When dormant, the life processes slow down, and the plant appears dead. Normal activities resume when the plant comes in contact with water. Because many bryophytes can photosynthesize in limited light, they

Figure 13.15 Bryophytes: (a) moss, *Polytrichium;* (b) liverwort, *Marchantia polymorpha;* (c) hornwort, *Anthoceros punctatus.*

a   b   c

often are found on the ground in forest ecosystems where other plants cannot grow.

One important group of mosses is found in boggy places in the cold and temperate parts of the world. These plants, from the genus *Sphagnum*, form peat bogs—small lakes and ponds completely filled with living and dead mosses. *Sphagnum* produces a very acidic condition in the water that keeps decomposers such as bacteria and fungi from growing, thus allowing these plants to build up through time, layer on layer. People in Ireland and other countries cut blocks of peat, dry it, and use it for fuel or to build small enclosures. Dry peat absorbs water quickly and holds the water well, characteristics that make peat attractive to gardeners, who add it to their soil to lower the pH and to increase the water-holding capacity.

Liverworts and hornworts grow in very moist areas, such as on the banks of streams where water spray keeps the soil wet. Some liverworts have flattened, lobed bodies, and other species may look "leafy," although they do not have true leaves. Liverworts have distinct top and bottom surfaces, with numerous rhizoids projecting from the bottom surface into the soil. The life cycle of liverworts is much like that of true mosses, and they can reproduce both sexually and asexually.

Hornworts look much like liverworts but are distinguished by their sporophytes, which are elongated capsules that grow like horns from the matlike gametophyte. Each photosynthetic cell of a hornwort contains a single large chloroplast, in contrast to the many smaller chloroplasts typical of the cells of most plants.

Although bryophytes are less complex than vascular plants, there is no evidence that they are the ancestors of vascular plants. The fossil record for bryophytes is small. The earliest bryophyte fossils are about 350 million years old. By that time, vascular plants were already established on land. The first fossils of vascular plants appear 50 to 100 million years *earlier* in the fossil record.

13.6 Club Mosses and Horsetails Are Seedless Vascular Plants

The roots, stems, and leaves of vascular plants make up the vascular system through which water, sugar, and dissolved nutrients move from one place to another in the plant.

Rhynia major (RY nyuh MAY jer), shown in Figure 13.16, best represents the oldest land plants. *Rhynia* had an underground stem that probably served to anchor the plant and to absorb water. From this underground stem grew upright branched stems that had stomates. At the tips of the stems were sporangia, which split open to release thick-walled spores. The closest living relative of *Rhynia* is the whisk fern, *Psilotum* (sy LOH tum), shown in Figure 13.17.

The club mosses (division Lycophyta; ly KOF ih tuh) are evergreen plants that seldom grow more than 40 cm tall. Although the word *moss* is a part of their name, they are not true mosses. Their branching, horizontal stems grow on the surface of the soil or just below it. The most noticeable part of a club moss plant is an upright branch growing from one of these stems (see the chapter opening photo). Club mosses reproduce by spores, which are produced on modified, specialized leaves. In many species, these

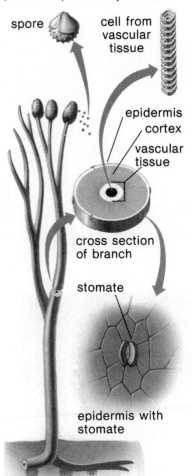

Figure 13.16 A reconstruction of the extinct plant *Rhynia*. It had no leaves or roots and stood about 30 cm high. Sporangia were produced at the tips of upper branches. The presence of stomates indicates that photosynthesis occurred in the branches. The cells of vascular tissue (upper right) were oblong and sometimes hollow. Thick-walled spores were produced in fours in the sporangia, an indication that the spores were produced by meiosis.

spore

cell from vascular tissue

epidermis
cortex
vascular tissue

cross section of branch

stomate

epidermis with stomate

Figure 13.17 The whisk fern, *Psilotum*. Note the sporangia along the stems.

leaves form club-shaped cones at the tips of short, upright stems. The name *club moss* is derived from this feature. Club mosses are rather common in the eastern and northwestern United States and often are used to make Christmas wreaths. They rarely grow in the dry states of the Southwest.

Horsetails (division Sphenophyta; sfen OFF ih tuh) have hollow, jointed, upright branches that grow from horizontal underground stems. Their small, scalelike leaves grow in a circle around each stem joint. Spores are produced in conelike structures at the tips of some of the upright branches, such as those shown in Figure 13.18. In middle latitudes, horsetails rarely reach a height of two meters, but in the American tropics one species may grow several meters tall. They are found in moist places, such as along streams. Horsetails are harsh to the touch; their tissues contain silica, a compound present in sand. Because American Indians and the pioneers scrubbed pots and pans with them, they are commonly called scouring rushes.

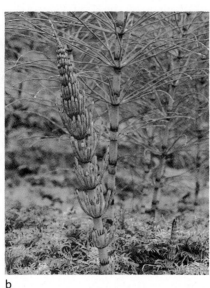

Figure 13.18 Spores of *Equisetum* are produced in these conelike structures (a). The vegetative stage of *Equisetum* (b).

a

b

×1/350

×1/250

×1/350

×1/77

×1/151

Lepidodendron   Sigillaria   Cordaites   seed fern   Calamites

Figure 13.19 Some trees of the coal-age forests: *Lepidodendron* and *Sigillaria* were club mosses. *Cordaites* were primitive cone-bearing plants. The seed ferns have no living species. *Calamites* were horsetails.

Relatives of the club mosses and horsetails can be traced back about 430 million years. During the Coal Age, about 300 million years ago, great parts of North America were covered by shallow swamps and seas. The warm and wet environment allowed plants to grow year round. Under these conditions giant relatives of today's club mosses, horsetails, and ferns, as well as seed-producing plants, covered the land. Some of these plants were more than 20 m tall (Figure 13.19). As they died, their large stems were covered with mud and soil before they completely decayed. A tremendous number of plants from the Coal Age were compressed over long periods of time and under high temperature and great pressure. Eventually, they became fossil fuels, mainly coal and some natural gas.

13.7   Fern Leaves Grow from Underground Stems

Ferns (division Pterophyta; ter OFF ih tuh), like the club mosses and horsetails, reproduce by spores. At certain times of the year, small brown spots develop on the undersides of fern leaves, as shown in Figure 13.20. Each spot consists of a cluster of sporangia. Each sporangium produces a large number of spores, which are almost microscopic in size.

When a sporangium is mature, it opens and the spores are thrown out into the air. Spores are very light and can be carried for incredible distances by wind. If a spore falls in a suitably moist place, it germinates and develops rapidly into a thin, green, heart-shaped plant that is rarely over one centimeter in diameter (see Figure 13.21a). This small gametophyte plant, which is completely different from the familiar fern with its large leaves, is seldom noticed in the woods. The gametophyte produces the sperm or eggs, or in

Figure 13.20 Sporangia on the underside of a licorice fern leaf, *Polypodium glycyrrhiza.*

a

b

Figure 13.21   Fern gametophyte (a) and sporophyte (b).

some cases, both. Like the moss sperm, the fern sperm must swim in a film of water to fertilize the egg. The zygote produced by that fertilization eventually grows into the conspicuous spore-bearing fern plant that most people recognize. This large plant is the sporophyte generation of the fern, seen in Figure 13.21b.

The ferns native to most of the United States are shade-dwelling plants with underground roots and stems. From these stems, roots grow downward and new sets of upright leaves appear above ground each spring. In Hawaii and elsewhere in the tropics, many species of ferns have stems that grow upright. These tree ferns may reach a height of 20 m with leaves 5 m in length, as Figure 13.22 illustrates. Most fern species are found in the tropics, but many can be found in forest ecosystems around the world.

Figure 13.22   Tree fern on the island of Sumatra.

Concept Review

1.   How do bryophytes differ from vascular plants?
2.   How does reproduction in seedless vascular plants differ from that in bryophytes?
3.   What events must occur for a fern sporophyte to be produced from a gametophyte?
4.   How is the reproductive strategy of ferns and horsetails adaptive?

Seed Plants

Guidepost: How are seed plants adapted to life on land?

13.8   Many Conifers Are Evergreens

Humans have used plants for food, clothing, shelter, and medicines. The conifers (division Coniferophyta) are woody vascular plants with seeds borne in cones. Conifers provide all of the paper pulp and most of the lumber used in home construction and furniture. They include pines, firs, spruces, and junipers, among many other species. Figure 13.23 on page 348 shows several conifers and related plants.

Almost all conifers are trees or shrubs, and all are at least somewhat woody. Many have leaves that are like needles or scales, such as those in

Figure 13.23 Conifers and related plants: (a) Ponderosa pine, *Pinus ponderosa;* (b) cycad, *Dioon edule;* and (c) maidenhair tree, *Ginkgo biloba.* Seeds of these plants are not enclosed in tissues as are those of flowering plants.

Figure 13.24, and most of these plants are evergreen. An evergreen tree or shrub appears green throughout the year because it always maintains most of its leaves. A few leaves die at different times of the year and drop to the ground. Although the number of conifer species is small, the number of individual conifers is enormous.

Many common conifers are well adapted to life in dry habitats. For example, although pine trees may grow where there is much snow, the snow is really frozen water and is not available for growth. In the spring, much of the snow evaporates, and the melted snow may run off into streams before it soaks into the soil. The leaves of pines are well adapted for growth in dry places. The long, narrow needles reduce the amount of water lost by evaporation. In addition, a pine needle often is covered by a thick, waxy cuticle that further reduces water loss.

Figure 13.24 Many conifer leaves are needlelike or scalelike. (a) Single needles of Douglas fir, *Pseudotsuga menziesii;* (b) scalelike leaves of juniper, *Juniperus chinensis "pfitzerii."*

Figure 13.25 Pine cone and seed.

winged seed detached
from scale

cone scale bearing seed
on its upper surface

Conifers reproduce by seeds that are attached to the upper surface of cone scales. A seed developing in a cone may be protected by the scales, somewhat as a small coin may be concealed between the pages of a book. If two scales are separated slightly, you can see the seed between them, as shown in Figure 13.25. Thus the seeds in cones are not completely covered as they are in the fruits of flowering plants.

Conifer spores are of two types and are produced in different cones, illustrated in Figure 13.26. Typically, pollen develops from spores in the small male cone, and pollination occurs in the spring when the pollen is blown onto a female cone. The larger, more familiar female cones contain the ovules. Pollen grains, the male gametophytes, lodge in a sticky substance secreted by the ovule. Once lodged, the grains develop pollen tubes within which the sperm are formed. Within the ovule, the female gametophytes develop and produce eggs. Fertilization occurs approximately a year after pollination, and the seed requires an additional year to mature. Table 13.3 summarizes these events.

Figure 13.26 Male cone (a); female cone (b) of the piñon pine, *Pinus edulis.*

a

b

TABLE 13.3   Seed Development in a Pine

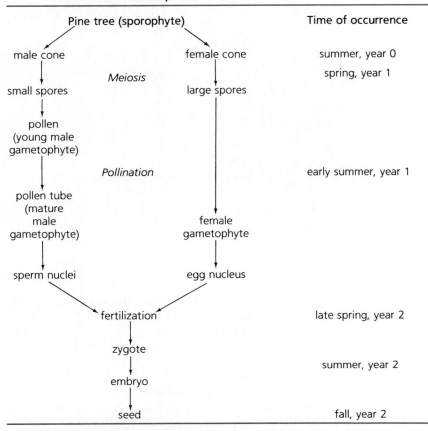

	Time of occurrence
Pine tree (sporophyte)	
male cone → female cone	summer, year 0
Meiosis	spring, year 1
small spores large spores	
pollen (young male gametophyte)	
Pollination	early summer, year 1
pollen tube (mature male gametophyte) female gametophyte	
sperm nuclei egg nucleus	
fertilization	late spring, year 2
zygote	
embryo	summer, year 2
seed	fall, year 2

From Weier, Stocking, Barbour, Rost, eds., *Botany: An Introduction to Plant Biology,* 6th ed. (NY: John Wiley, and Sons), copyright © 1982. Reprinted by permission of John Wiley, and Sons, Inc.

13.9   Many Flowering Plants Have Special Pollinators

Flowers are the distinguishing feature of the most successful division in the plant kingdom, the Anthophyta (an THOF ih tuh). Although we commonly appreciate flowers for their beauty, their major role is in reproduction. When a flower opens, it reveals the reproductive structures, the stamens and carpels. Insects, birds, bats, or other animals that visit the flower may pick up pollen from the anther and carry it to the next flower they visit. When pollen is transferred from a flower of one plant to a flower of another, the process is called cross-pollination. The main pollinators of flowers are insects. Pollen also may be transported from flower to flower by wind. Some flowers, however, seldom or never open. In this case, the pollen falls on the stigma of the same flower. These flowers are self-pollinating. Flower and vegetable gardeners say they "self." Most plants, however, have devices that prevent self-pollination.

The sepals and petals are not directly involved in seed formation, so a flower can function without them. In fact, in a few plants a flower may consist of only a single stamen or a single carpel. Petals and their adaptations,

Figure 13.27 A few examples of the diversity of flowers. (a) Colorado blue columbine, *Aquilegia caerulia;* (b) sunflower, *Helianthus annuus;* (c) poinsettia, *Euphorbia pulcherrima;* and (d) brome grass, *Bromus.*

however, usually play a major role in flower pollination. Much of the diversity among flowering plant species lies in their flowers (see Figure 13.27). This diversity usually is related to the way pollination occurs. If pollen is transferred from stamen to carpel by insects, the petals of the flower often are large and brightly colored, as in Figure 13.28a. The petals often have small glands that produce a sugar solution called nectar. These adaptations attract pollinating insects. On the other hand, flowers in which pollen is transferred by wind usually have small sepals and petals or none at all. These flowers often are located high on the plant and produce an abundance of pollen (Figure 13.28b). Their carpels commonly have large, long, or feathery structures at the tips, which are covered with a sticky fluid. These adaptations increase the likelihood that some pollen will stick to the carpels.

The great variety of flowers has come about, in part, by the coevolution of flowers and pollination agents. For example, hummingbirds need lots of nectar to supply their energy needs. Although they do not have a good sense of smell, they can see the color red very well. Flowers pollinated by hummingbirds usually are well adapted to their pollinators, as you see in

Figure 13.28 Insect-pollinated flowers such as the aster in (a) usually are brightly colored. Wind-pollinated flowers often lack petals and sepals and produce an abundance of pollen. Shown in (b) is the male catkin of a willow, *Salix.* Note the many stamens.

spurs of flower

Figure 13.29   The columbine and hummingbird are uniquely adapted to each other.

Figure 13.29. The columbine flowers shown in the illustration are red, have little or no scent, and produce copious amounts of nectar. The nectar is found at the bottom of a long tube formed by the red petals, a shape that makes it difficult for other organisms to rob the flower of nectar. Hummingbirds, however, have long beaks that can probe the flower and reach the nectar. The stamens stick out in such a position that the hummingbird's head is dusted with pollen when it visits these flowers. When the hummingbird flies to another flower of the same species, the tip of the carpel is in a perfect position to have pollen from the hummingbird's head scraped onto it. Thus, the flower is pollinated while the hummingbird drinks nectar. Interactions of this type, in which both organisms become uniquely adapted to each other, have helped shape the way flowering plants have evolved.

INVESTIGATION   13.3   Fruits and Seeds

The survival of plants depends on their ability to reproduce. In seed producers, the most complex plants, reproductive ability is enhanced by mechanisms that protect and disperse the seeds (and fruits) so the seeds do not compete for nutrients, light, and water with the parent plants. Dispersal increases the likelihood that some seeds will not be eaten. In flowering plants, seeds are protected by the tissues of the mature ovary, or fruit. (Conifers, which lack an ovary, produce naked seeds on cone scales.)

In this investigation, you will try to determine how the seeds of various fruits are dispersed. You also will use the fruits to make a dichotomous key, which can help you distinguish between objects by focusing on their similarities and differences.

Materials (per team of 3)
set of fruits

Procedure
PART A   Seed Dispersal
1. Blow gently at the fruits. What happens? Why? What parts of the fruits are important in allowing them to be wind-dispersed?
2. Gently rest the sleeve of your blouse or shirt on the fruits and then lift up your arm. Which fruits were lifted? What parts of these fruits are important in allowing them to be dispersed by animals?
3. Some fruits attract birds and other animals that eat the fruits but do not damage the seeds inside. What characteristics of the fruits might serve as attractants?
4. Some fruits contain a chemical that acts as a laxative. How might this function in seed dispersal? How might the contents of bird droppings assist in survival of the new seedlings?

PART B   Dichotomous Key
5. Design a dichotomous key using the fruits from Part A. A dichotomous key should separate all the available fruits into individual categories. Assemble the fruits and review the characteristics you observed in Part A.
6. To form a dichotomous key, use characteristics that some fruits have and some do not, rather than characteristics that all share. For example, *fruits with spines* versus *fruits without spines* might be a good characteristic to use in building your key. On the other hand, *fruits that contain seeds* would not be a good characteristic because almost all fruits contain seeds.
7. To divide the fruits into two groups, choose one major characteristic *not* shared by all the fruits. Separate your fruits into two groups—one group that possesses the

characteristic and another group that does not. In your data book, prepare a table similar to Table 13.2 on page 335 but with the lines only, or tape in the table your teacher provides. Write the characteristic on the table as shown below:

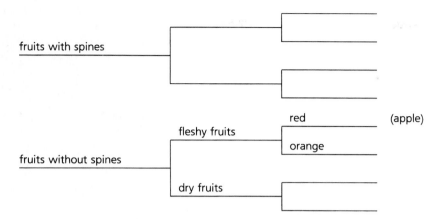

8. Now focus on all the fruits in just one of your two groups. Choose another characteristic that separates those fruits into two groups. Write this characteristic in the second column of the key.

9. Continue to select characteristics and write them on the key until you have produced a key that separates all the fruits into individual categories. In other words, your key should split the fruits into smaller and smaller groups until each fruit is in a group by itself. Once this is done, place the name of the fruit (maple, ash, etc.) next to the appropriate line on the right side of the key.

10. Exchange keys with another team. See if you can follow their key and they can follow yours.

11. Wash your hands thoroughly before leaving the laboratory.

Discussion

1. What are some of the ways seeds/fruits are dispersed?
2. Describe how a dispersal mechanism that relies on the presence of other organisms might develop in a plant species.
3. What would happen to the distribution of plants that produce cockleburs if they lived on an island having no animals?
4. Explain how poplar trees might come to inhabit an island in the middle of a large lake.
5. What is the purpose of a dichotomous key?
6. What characteristics did you use to develop your dichotomous key? What characteristics did your class as a whole use?
7. Could you develop a dichotomous key for organisms that looked identical to each other? Explain.

13.10 Flowering Plants Produce Fruits with Seeds

After pollination and fertilization, seeds begin to develop. The carpel, often with other parts of the flower, develops into a protective fruit around the seed, as shown in Figure 13.30. There may be many seeds in a fruit. Each seed began its development when an egg cell in one ovule was fertilized

Figure 13.30 Stages in the development of tomato fruit from flowers.

by a sperm from one pollen grain. In flowering plants, then, an embryo is protected within a seed, and seeds are protected within a fruit.

Part of the embryo in the seed consists of one or two modified leaves, called **cotyledons** (kot ih LEE dunz). Another part is a beginning of a root. Each seed also contains a supply of food that is used when the embryo starts to grow. The food may be stored in the endosperm, or it may be stored in the embryo itself, usually in the cotyledon.

Beans and peas are examples of seeds. They are enclosed in protective pods, which are just one of many types of fruits. Each bean or pea contains a small embryo and a supply of stored food for its early development. The entire pea or bean can act as food for humans. Peas and beans can be germinated easily, and each embryo gives rise to a plant that, in turn, gives rise to new flowers and fruits. Apples and oranges are examples of fruits that contain a number of seeds. Under natural conditions, these fruits eventually decay, leaving their seeds behind to germinate and give rise to the next generation.

Fruits show as much diversity as flowers, as indicated in Figure 13.31. This diversity is related to the method of dispersal of the fruits and their seeds—that is, how they are scattered from the parent plant. In Investigation 13.3 you observed some of the structures that aid in seed dispersal. In many cases, part of the carpel becomes thick and fleshy, as in the fruits of peach, plum, and tomato. Fleshy fruits, often red in color, may be eaten by birds or mammals. The seeds in many such fruits have thick coats that permit them to pass through an animal's digestive system unharmed. They are dropped later at some distance from the parent plant. Many fruits are not fleshy but have other adaptations that aid in scattering their seeds. These fruits may have spines that catch on the fur of an animal that brushes up against the plant. The fruit is carried from the plant and later falls off or is brushed off by the animal. Many fruits and seeds are lightweight and have

a

b
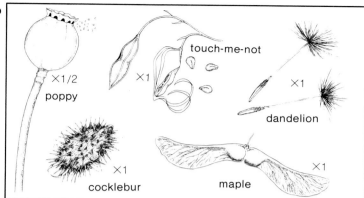

touch-me-not

×1/2
poppy

×1

×1
dandelion

×1
cocklebur

maple
×1

Figure 13.31 Each edible fruit is a mature ovary containing the seeds of the plant (a). Many fruits have structural adaptations for seed dispersal (b). Can you describe them?

special winglike projections that help them to be carried away from the plant by wind. An entire plant, such as the tumbleweed, can be broken off near the ground and blown about by the wind. As the tumbleweed bounces about, it drops its fruits (and the seeds within) all along its path.

13.11 Flowering Plants May Appear Very Different

There is great diversity in the size of flowering plants and in the life span of their shoots—the parts that appear above ground. Many flowering plants are trees. A tree bears leaves well above the ground where they are likely to receive more light than do the leaves of shorter plants. Because of their size, trees can store large reserves of food in trunks and roots and can survive through bad years. Trees have relatively long life spans. A tree species may survive even if a year's seed crop is destroyed. Most species of flowering plants are not trees, however. Some, such as roses and raspberries, are woody shrubs. Others, such as ivy, grapes, and hundreds of tropical species, are woody vines that grow on rocks, walls, or other plants. Most, however, are neither trees, shrubs, nor vines. Instead, they are nonwoody, or herbaceous (her BAY shus), plants such as those in Figure 13.32.

Flowering plants are divided into two large classes—the **monocots** (MON oh kotz) and the **dicots** (DY kotz). In monocots, (monocotyledons), the embryo contains a single cotyledon. The monocots include grasses and grain-producing plants such as wheat, rice, and corn—the chief food plants of the world. The pasture grasses that feed cattle, another source of human food, are also monocots. Without monocots, the human population never could have reached its present state.

The seeds of the dicots (dicotyledons) have two seed leaves. This class is larger than the monocot class. Most fruits and vegetables, such as carrots, lettuce, apples, and grapes, are dicots. In addition, the so-called hardwoods used in furniture, flooring, hockey sticks, and baseball bats come from dicot trees. Almost all shade trees are dicots, also. Figure 13.33 on page 356 illustrates the major differences between monocots and dicots.

Figure 13.32 Herbaceous plants include marigolds, common garden flowers.

Monocot

Dicot

sepals

stamen

carpel

petals

pith

conducting tissue

embryonic shoot

cotyledon

embryonic root

Figure 13.33 Comparison of monocot and dicot characteristics.

Biological Illustrator

Marjorie Leggit is a freelance biological illustrator. Biological illustrators illustrate reports, journal articles, and books prepared by researchers. Many illustrators work in natural history museums, painting and constructing the background scenery for different types of displays. These artists may recreate Alaskan ice floes to show off a group of polar bears or a South American hillside for a display of llamas.

Marjorie's family is composed of writers, sculptors, painters, and woodcarvers, so an interest in art comes naturally to her. Her interest in science, Marj thinks, may have developed because she could do so much drawing in science classes.

Although a biological illustration course in college sparked her interest, Marjorie continued to study fine arts. After studying for a year in Europe, however, she realized she missed science and the technical art that goes with it. She returned to college and designed her own program of independent studies that led to a degree in scientific art. Her program included an internship at the Denver Museum of Natural History in Denver, Colorado. While searching for a full-time job after graduation, Marj illustrated a professor's botanical field guide. Her persistence was rewarded eventually with a position at the Field Museum in Chicago, Illinois.

Marj later returned to Denver to be married. Reestablishing her career there proved to be difficult because opportunities for biological illustration were far fewer than in very large cities where more research is conducted. For seven years she worked at art-related jobs, including geological drafting and computer-aided graphics. Although she learned a whole new field of art, Marj was not satisfied with the direction of her career.

To keep in touch with the field of biological illustration, she contacted the Guild of Natural Science Illustrators (GNSI) in Washington, DC. Knowing Colorado had nothing similar, she founded the Colorado Chapter of GNSI and made friends with a group of local illustrators.

In 1986, Marj went into business for herself as a biological illustrator and graphic artist. She contacted old and new local clients as well as publishers throughout the country. She spent many hours researching the possibilities for assignments and marketing her skills.

Marj did many of the illustrations used in this textbook, including Figure 13.33. To ensure accuracy in her work, Marj does a great deal of research and uses many references and photos. She consults extensively with authors when planning an illustration. It is essential that she understand the author's intent, what the illustration is meant to portray, and how it relates to the text and to other illustrations. Proper interpretation depends on good communication between author and artist. Marj strongly believes that the hard work involved in creating illustrations has its rewards. At times the hours are long, the deadlines close to impossible, and the challenges seemingly insurmountable. However, she is doing exactly what she wants to be doing—full-time biological illustration.

Concept Review

1. How are conifers adapted to dry conditions?
2. How do conifers differ from flowering plants?
3. What role does flower color play in the reproduction of flowering plants?
4. How are fruits related to seed dispersal?
5. In what ways do monocots and dicots differ?

Summary

The ancestors of plants were probably multicellular algae. A multicellular organism has many advantages over a single cell, including size, division of labor, and ability to conserve water. Other adaptations that permit plants to absorb and hold water are roots, vascular tissue, and the cuticle. All plants have a life cycle that alternates between two different generations, a gametophyte and a sporophyte. In flowering plants, however, the egg cell, the gametophyte that produces the egg, and the spore that produces the gametophyte all remain protected in the flower. Mosses are nonvascular plants that lack roots, stems, and leaves. They possess some adaptations to life on land, but they still require a moist environment. Club mosses, horsetails, and ferns are seedless vascular plants, and many of their ancestors produced today's fossil fuels. Conifers and flowering plants are seed-producing vascular plants, but flowering plants have a fruit that encloses their seeds. The great diversity of flowers and fruits is the result of the co-evolution of plants and their agents of pollination or dispersal.

Applications

1. Which organisms in this chapter would you consider to be less complex? Which would you consider to be more complex? Explain your answer.
2. Could descendants of the first land plants still be in existence today? Explain.
3. Would you expect to find a cuticle around the tissues of a plant's roots? Why or why not?
4. What advantages and disadvantages does a seed-producing plant have compared with one that produces only spores? How do you interpret the words *advantage* and *disadvantage* here?
5. Suppose you found a population of ferns in which the young sporophytes appeared only after a rainstorm. How would you explain this occurrence?
6. Why are nonvascular plants short?
7. Considering only spores, what reproductive advantage do flowering plants have over ferns?

8. What advantage do wind-pollinated plants gain by having flowers with very small sepals and petals?
9. The fossil record reveals a parallel increase in the diversity of flowering plants and the diversity of certain insect groups. How might you explain this?
10. Explain the following statement: The plant parts that furnish the greatest amount of human food are either seeds, roots, or underground stems.

Problems

1. Desert plants are land plants adapted to extremely dry conditions. What adaptations allow them to live with little water? Make a collection of desert plants or find out about them from a plant ecology text.
2. Fruits come in many sizes and shapes and with diverse appendages. Collect different fruits and try to identify their plants.

3. Bring to the classroom different pieces of furniture, tools, sculptures, or other objects made from wood. What qualities of the wood make it useful for each object? Can you identify which tree provided the wood? Are some woods better than others for certain tasks?

4. Throughout history, plants have played religious and cultural roles in different societies. Many of these roles have been captured in paintings, sculptures, and tapestries. Find artwork that contains a plant or plants as an important part of the production. What does the plant signify?

5. Find or make some drawings of plants that were important during the time of the dinosaurs. Hang the drawings in your classroom. What happened to these plants? Did any of their descendants survive until today?

6. In what ways do plants that live entirely in water differ in structure, reproduction, and limiting factors from plants that live entirely on land?

7. Observe the plants that grow without human help in a city. Try to discover the characteristics that enable them to live successfully in an urban environment.

8. How many of the different types of plants discussed in this chapter can be found in your locality? Consider wild and cultivated plants, indoor and outdoor plants, aquatic and terrestrial plants.

9. Choose some cultivated plants. Investigate the history of their domestication. Examples: wheat, apple, potato, cotton, corn, cabbage, sugarcane.

Frequently, bright colors on animals are warnings to potential predators. What might the bright colors on this frog signal? From what you know about how frogs have sometimes been used by the natives of Central and South America, suggest a name for this frog.

Eukaryotes: Animals

The animal kingdom presents an enormous diversity of organisms, adapted to a wide variety of habitats and life-styles. Although they display many structural and functional differences, all animals share certain characteristics. All are essentially multicellular in organization and heterotrophic. Most are capable of locomotion for some part of their lives, and most are able to respond rather quickly to changes in their environments.

Most animals share a common evolutionary origin. Thus, in spite of the great diversity among animals, their basic features were developed in ancestral forms. Given similar habitats, similar solutions to common problems may have evolved even in unrelated groups. This chapter explores some of the problems animal groups encounter and some of the adaptations that have evolved to solve these problems.

The Animal Way of Life

Guidepost: What are the problems of animal life, and what are some of the solutions?

INVESTIGATION 14.1   Diversity in Animals: A Comparative Study

Animals are similar in many ways but differ in other ways. In this investigation, you will sharpen your observation skills by studying several different organisms. You also will examine the relationship between structure and function through observing the adaptations of different organisms to their environments.

Materials (per team of 6)

Station 1
3 stereomicroscopes or 10X hand lenses
3 Syracuse watch glasses
3 dropping pipets
3 compound microscopes
3 No. 2 paint brushes
3 prepared slides of longitudinal sections of hydra
5 to 10 live hydras
daphnia culture

Station 2
6 stereomicroscopes or 10X hand lenses
3 Syracuse watch glasses
5 to 10 small pieces of raw liver

3 No. 2 paint brushes
3 compound microscopes
3 prepared slides of cross sections of planarians
3 whole mounts of carbon-fed planarians
5 to 10 live planarians

Station 3
3 10X hand lenses
moistened paper towels
3 compound microscopes
6 boxes containing damp soil
3 prepared slides of cross sections of earthworms

(continued)

Materials (continued)

5 to 10 live earthworms

Station 4

3 No. 2 paint brushes

3 shallow boxes or pans

1 small bowl containing lettuce and fruit pieces

3 land hermit crabs

Station 5

3 4-qt battery jars

aquarium

3 live frogs

Procedure

1. Review the guidelines in Appendix 1 concerning the proper care and handling of laboratory animals. Follow these guidelines carefully.

2. Make an enlarged copy of Table 14.1 (or use the table provided by your teacher). It should extend across two facing pages in your data book. Each of the 13 spaces should allow for several lines of writing.

3. In the *Characteristics* column, copy the italicized words from each of the following questions. The 13th space is for any other observations you make. All the specimens of one animal species and the materials and equipment needed for observing them are arranged at one station. Each team will have a turn at each station. Record only your *observations*, not what you have read or heard about the organism.

   a. What is the *habitat* of the animal? Does it live in water, on land, or both?
   b. Is *body symmetry* radial or bilateral?
   c. Does the animal have a *skeleton*? If it does, is it an endoskeleton or an exoskeleton?
   d. Is the animal's body *segmented* or is it *unsegmented*?
   e. Which type of *digestive cavity* does the animal have, a sac or a tube?
   f. Does it have *paired appendages*?
   g. How does the animal *obtain oxygen*? Through lungs, gills, skin, or a combination of these?
   h. Are any *sense organs* visible? If so, what types and where?
   i. How does the animal *move* from one place to another?
   j. Does it make any types of *movement* while it remains more or less in one spot?
   k. How does the animal *capture* and take in *food*?
   l. How does it *react* when touched lightly with a small brush?

Station 1 Observing Hydras

4. Place a single hydra in a small watch glass with some of the same water in which it has been living. Wait until the animal attaches itself to the dish and extends its tentacles, then slowly add a few drops of a daphnia culture with a dropping pipet.

5. Touch the hydra gently with the brush and observe its reactions.

6. Examine a prepared slide of a longitudinal section of a hydra under a compound

TABLE 14.1

Characteristics	Hydra	Planarian	Earthworm	Hermit Crab	Frog
1.					
2.					
3.					
13.					

microscope. Try to determine the presence or absence of a skeleton and of a digestive system.

Station 2 Observing Planarians

7. Place 1 or 2 planarians in a watch glass containing pond or aquarium water. Add a small piece of fresh raw liver. Observe using a stereomicroscope or hand lens.

8. Use a compound microscope to examine cross sections of a planarian. Examine whole mounts with a stereomicroscope. Determine the presence or absence of a skeleton and a digestive system.

Station 3 Observing Earthworms

9. Pick up a live earthworm and hold it gently between your thumb and forefinger. Observe its movements. Do any regions on the body surface feel rough? If so, examine them with a hand lens.

10. Place a worm on a damp towel. Watch it crawl until you determine its anterior and posterior ends. Use a hand lens to see how the ends differ. Describe.

11. Place an earthworm on the loose soil and observe its movements as it burrows.

12. Use a compound microscope to examine cross sections of the body under low power and high power. Does it have a skeleton?

Station 4 Observing Hermit Crabs

13. Observe the movements of the appendages and the pattern of locomotion of a live land hermit crab. Observe the antennae. Touch them gently with the brush. Note the animal's reaction.

    CAUTION: Keep your fingers away from the crab's pincers.

14. Place a small piece of food from the food dish in with the hermit crab. Observe how the hermit crab eats.

Station 5 Observing Frogs

15. Your teacher will show you how to catch and hold a frog so as not to injure it. Place a frog in a battery jar and study its breathing movements while it is not moving about. To do this, observe from the side, with your eyes at the level of the animal.

    CAUTION: Do not rub your eyes while handling live frogs. Wash your hands immediately after handling.

16. Remove the frog from the jar and place it in an aquarium. Observe its movements. How do they compare with the movements of a frog hopping and moving about on a laboratory table?

17. If a hungry frog is available, your teacher may be able to show you how it captures food.

18. Wash your hands thoroughly before leaving the laboratory.

Discussion

Review what you have learned about each of the organisms in Table 14.1. By reading across the table, you should be able to compare and contrast the characteristics of the five animals you have studied. For each animal, select five functions it performs as part of its way of life and describe how its structure enables it to perform these functions.

14.1 Animals Adapt to the Demands of Their Environments

Animals are heterotrophic. Therefore, they usually must seek food. As a result, adaptations that allow animals to be **motile** (MOH tuhl), or capable of movement from place to place, have evolved. Motility is easier if the organism is elongated in the direction of movement. Fish, for example, are

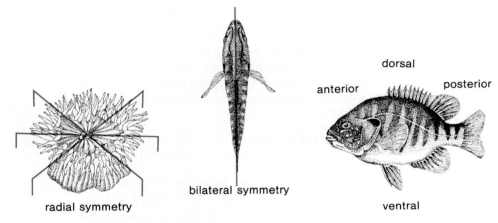

Figure 14.1 Body symmetry.

streamlined in a manner that minimizes water resistance as they swim. Motility is helped further if the sensory organs—organs that detect food, light, and other stimuli—are concentrated in the end that meets the environment first, usually the head. An organism can move more easily if it has a balanced body, with neither side heavier or larger than the other.

Animals have the type of body plan that is best suited to their life-style (see Figure 14.1). For motile animals, the term **bilateral symmetry** (by LAT er ul SIM eh tree) is used to describe the body plan. In bilateral symmetry, the right side is an approximate mirror image of the left side. Motile animals have an **anterior**, or head end, a **posterior**, or tail end, and **dorsal** (top) and **ventral** (bottom) surfaces.

As active movement requires coordination, evolution has produced an area in which sensory organs and nerve cells are gathered—a head—which usually is found in the anterior end. Structures that help a motile animal capture prey, such as claws and teeth, usually are located toward this end. Digestive, excretory, and reproductive structures usually are located toward the animal's posterior.

Organisms that are **sessile** (SES il), or nonmotile, often have a body plan best described by the term **radial** (RAYD ee ul) **symmetry**. Their body parts radiate from a center, much as spokes radiate from the hub of a bicycle wheel. Radial symmetry allows sessile or drifting **aquatic** (uh KWAH tik) (water-dwelling) animals to interact with the environment from all sides. Sessile organisms, however, must deal with certain limitations. They must depend on food that floats by them, or they must create water currents to bring food to them. They cannot actively seek mates and, therefore, must rely on the chance meeting of eggs and sperm released into the water.

Every environment imposes special demands on the organisms living there. Sea water is the most uniform and least stressful environment for animal life. Oxygen is usually adequate, temperatures and salt content are fairly constant, and there is little danger that the organism will dry up. In contrast, the salt and oxygen contents of fresh water vary greatly.

Organisms that live in water have special adaptations. Gills, for example, allow the organisms to make use of the oxygen found in water, and circulatory systems (systems that transport materials) help to maintain the organisms' water and salt balance. On land, oxygen is plentiful, but the organisms that live there must protect themselves from the dangers of drying up—dangers that are increased greatly because air temperatures fluctuate daily and seasonally. Air does not provide the same buoyancy as water, so large **terrestrial** (ter ES tree uhl), or land-dwelling, animals require good supportive structures. On the other hand, there is less resistance to movement in air than in water. **Appendages** (uh PEN dij ez), such as arms and legs, which would hinder an animal's movement in water, may help on land. Thus, long appendages specialized for locomotion have evolved in terrestrial animals.

a

14.2   As Size Increases, Complexity Increases

As shown in Figure 14.2, organisms exhibit an enormous range in size. During the course of evolution, most animal groups have shown a trend toward increasing size, a result of their being multicellular. Although larger animals may be able to obtain more food than smaller ones and may get away from predators faster, increased size imposes special problems. Larger animals require an increased division of labor among their body parts, which leads to greater complexity in the organism. Consider how many different parts your body is made of and all their different functions.

No matter what the body size, individual cells are remarkably consistent in size among all organisms. Large animals have *more* cells than small animals but not larger cells. The presence of more cells usually leads to increased specialization of tissues, which in turn requires more coordination. Through diffusion, small animals, for example, can exchange materials easily with their environment, but within large animals diffusion is too slow to move materials. These animals require specialized circulatory systems to obtain oxygen, remove wastes, and transport substances. These systems must be finely coordinated. In general, the larger the animal, the more complex are the controls necessary to coordinate its activities.

b

Figure 14.2   Animals vary greatly in size. The microscopic *Daphnia* (a) contains perhaps a few hundred cells (X40); the elephant (b) millions.

All animals face common problems of feeding, self-maintenance, and reproduction. In members of the same phylum, the solutions to these problems often are similar. The next section examines eight important animal phyla, ranging from the least complex to the most complex. It describes the special features and modifications that improve the adaptation of each phylum to its environment. (Further descriptions of animal phyla are in Appendix 4, *A Catalog of Living Things.*)

Concept Review

1.   Compare and contrast bilateral and radial symmetry.
2.   Compare the advantages and disadvantages of aquatic and terrestrial environments for animals.
3.   Describe how increased size leads to increased complexity.
4.   How does an animal's structure help it survive in different environments? Give some examples.

Diversity and Adaptation in Animals

14.3 Sponges and Cnidarians Are the Least Complex Animals

Sponges (phylum Porifera; poh RIH fer uh) are an ancient group with an extensive fossil record that dates back to the early Precambrian period, about 550 million years ago (see Figure 14.3). The several thousand species living

Figure 14.3   The evolution of animals throughout the earth's history.

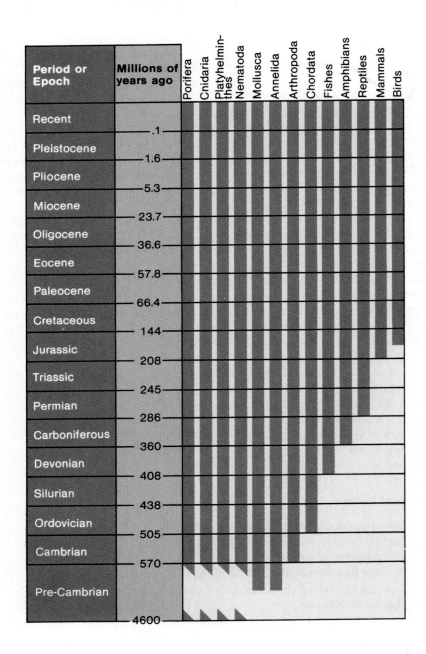

today are all aquatic, occupying a variety of marine and freshwater habitats. They show considerable diversity of form, and many are brightly colored, as shown in Figure 14.4.

Sponges are sessile animals without any apparent symmetry. The name Porifera, pore-bearer, aptly describes the body of a sponge, which consists of a bag pierced by many pores and canals. The cells are loosely organized and do not form true tissues or organs. If a sponge were forced through a sieve, its cells would break apart and then flow back together again, re-forming the sponge on the other side. The pores and canals function as a water-filtering system. Water containing oxygen and microscopic food is pulled through pores in a current created by the beating flagella of specialized cells that line the inner body. Water containing carbon dioxide and other cellular wastes flows back out of the body cavity through an opening called the mouth. The body wall is supported by interlocking particles of a tough material that form a type of skeleton. Figure 14.4c shows the sponge body plan.

Cnidarians (nih DAR ee unz), members of the phylum Cnidaria, are aquatic animals usually found in shallow, warm marine habitats, although some are found in freshwater and cold marine habitats as well. The fossil record of the cnidarians, like that of the sponges, goes back more than 600 million years. Cnidarians include corals, jellyfishes, sea anemones, and freshwater hydras (see Figure 14.5 on page 368). Many of these organisms harbor symbiotic algae within their cells, which allow them to thrive in nutrient-poor waters. The algae produce some sugars useful to cnidarians, and the cnidarians provide protection for the algae. Coral reefs—limestone ridges just under the surface of the sea—are formed by the secretions of several species of cnidarians and are home to the greatest diversity of animal life in the water. Corals have produced coral island groups such as the Bahamas and Bermuda.

Cnidarians are radially symmetrical and, thus, are adapted to a sessile existence, although some are motile. They have two basic body forms, tubular and bowl-shaped. Reef-building corals and freshwater hydras have

Figure 14.4 Sponges: (a) brown volcano sponge, *Hemectyon ferox;* (b) purple vase sponge, *Dasychalina cyathina;* (c) body plan of a sponge.

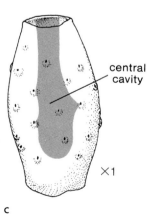

central cavity

×1

a

b

c

Figure 14.5   Cnidarians: (a) cnidarian body plan; (b) night-blooming orange cup corals, *Tubastraea coccinea;* (c) jellyfish, *Aequorea aequorea;* (d) burrowing sea anemone, *Pachycerianthus.*

tubular bodies that usually are attached to some surface. Free-swimming or drifting jellyfishes have inverted, bowl-shaped bodies that have dangling tentacles.

The body of either type of cnidarian is little more than a thin-walled bag with an outer and inner tissue layer, as shown in Figure 14.5a. Sandwiched between the layers is a jellylike material that contains a network of nerve cells and contractile fibers. Digestion occurs in the central cavity of the bag; there are no internal organs. Food is taken in and undigested particles are released through the same opening. The opening is surrounded by tentacles containing specialized cells that sting prey on contact. These cells give the phylum its name, Cnidaria (nettles).

14.4   Flatworms Have True Organs

Flatworms (phylum Platyhelminthes; plah tih hel MIN theez) represent the next evolutionary step in complexity. They are the least complex of the bilaterally symmetrical animals that have organs, organ systems, and a distinct head at the anterior end. They also have greater organ specialization and division of labor than are found in either sponges or cnidarians.

Flatworms (Figure 14.6a) are adapted to a great variety of habitats. Many, including tapeworms and flukes, are parasitic, and some are parasites of humans. Free-living forms, such as planarians, may be marine or may live in fresh water or moist soil.

Figure 14.6   Flatworms: (a) marine flat worm, *Alleena mexicana;* (b) flatworm body plan. Note the much-branched digestive cavity.

Figure 14.7 Nematodes: (a) in garden soil; (b) *Trichinella spiralis,* a parasite (X65).

Flatworms lack circulatory systems. The typical flatworm digestive system is diagrammed in Figure 14.6b. It is a branched sac with only one opening, which serves to take in food and to eliminate wastes. The branches bring food directly to the cells or within diffusible distance of them. Parasitic forms often have no digestive system. Instead, they absorb digested food directly from the host organism. Oxygen and carbon dioxide easily diffuse in and out of all cells because of the flat body plan.

14.5 Roundworms and Mollusks Have a Body Cavity

Roundworms (phylum Nematoda; nee muh TOH duh) are slender cylindrical animals that are present in abundance in almost every type of habitat (see Figure 14.7). There may be one million roundworms (also called nematodes) in a shovelful of good garden soil. Many are efficient predators, and some are parasites that are able to infect almost all types of plants and animals, often causing serious disease.

Roundworms have a complete digestive system with two openings. This adaptation permits continuous processing of food. The digestive system is usually tubelike and is suspended in a fluid-filled body cavity within the body wall. The internal body cavity represents an evolutionary advance in animal body plans, and the resulting tube-within-a-tube shown in Figure 14.8 is the basic body plan of animals classified in all the more complex phyla.

Figure 14.8 Roundworm body plan, a tube within a tube.

mouth   digestive cavity   anus

Figure 14.9   Mollusks: (a) octopus, *Octopus vulgaris;* (b) rock scallop, *Hinnites giganteus;* (c) molluscan body plan.

An internal body cavity is also typical of mollusks (phylum Mollusca; muh LUS kuh), which include snails and slugs, clams, oysters, and scallops, as well as squid and octopuses (see Figure 14.9a, b). Most mollusks live in a variety of aquatic habitats, from the shallows along the shores to the ocean depths. Their diversity of form, feeding methods, and life-styles results from their adaptation to these very different habitats.

Mollusks have soft bodies in which the ventral wall is modified into a head-foot region adapted for feeding, sensory reception, and locomotion, as shown in Figure 14.9c. They have a circulatory system and structures specialized for exchanging gases with the environment. Their most distinctive characteristic is a mantle, a modified body wall that forms a cavity and encloses the internal organs. In many mollusks, the mantle secretes substances that create a shell. For these mollusks, the mantle forms the lining of the shell. Many mollusks protect themselves by withdrawing the head or foot into the cavity that the shell surrounds. In aquatic forms, muscular pumping of the mantle maintains a current of water that brings in food and oxygen and carries out wastes. The jet propulsion of squid and octopuses also is accomplished in this way.

14.6   Annelids and Arthropods Are Segmented

The basic adaptive feature of annelids (AN ih lidz; phylum Annelida; an NEL ih duh), or segmented worms, is their body segmentation, which shows clearly in Figure 14.10a. The annelid body consists of a series of ringlike

Figure 14.10   Annelids: (a) night crawler, *Lumbricus;* (b) marine annelid, *Sabella.*

a   b

compartments, each resembling the next. Many internal body organs are repeated through several compartments. The fluid-filled body cavity also is segmented. The contraction of muscles surrounding each segment allows the animal to burrow and swim.

Although terrestrial annelids such as earthworms are most familiar to us, most annelids are aquatic. Marine annelids are abundant, diverse, and important in food chains. They live under rocks or in coral crevices, burrow in sand and mud, or secrete tubes from which they extend feathery tentacles for feeding (see Figure 14.10b). Scientists have found worm burrows made during the Precambrian period. Leeches, another type of marine annelid, are adapted to bloodsucking and have supporting adaptations such as cutting jaws, suckers, and the production of substances that inhibit blood clotting.

Three-quarters of all known species of animals are arthropods (phylum Arthropoda; ar THRAP uh duh), the most numerous animals on earth. The arthropods include crustaceans and insects as well as centipedes, millipedes, and arachnids. They are found in every type of habitat, have a wide variety of feeding habits, and are uniquely able to adapt to changing conditions. In fact, their adaptability has contributed to their great evolutionary success.

Arthropods have segmented bodies, usually with three major divisions: a head, a thorax (chest), and an abdomen. Many of the segments have pairs of jointed appendages, an adaptation that is efficient for locomotion. Appendages often are modified for specialized functions such as food getting, sensory reception, swimming, or flying. Figure 14.11 shows the basic arthropod body plan.

There are three major groups of living arthropods, the chelicerates (chih LIS er ayts) the crustaceans, and the uniramia (yoo nih RAY me uh). Chelicerates include horseshoe crabs, spiders, harvestmen, scorpions, and ticks and mites. Figure 14.12 shows examples of chelicerates. Unlike the separate head and thorax of many insects, the head and thorax of chelicerates are joined. They have four pairs of legs and two pairs of chelicerae—appendages adapted for feeding. Spiders are the most familiar chelicerates. Most spiders make webs, and all have poison glands that are used to immobilize their prey. Only a few, such as the brown recluse and black widow, are dangerous to humans. Many ticks and mites transmit pathogens that

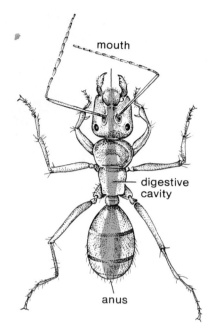

Figure 14.11 Arthropod body plan. Note the appendages.

Figure 14.12 Chelicerates: (a) horseshoe crab, *Limulus polyphemus;* (b) tick, *Dermacenter;* (c) nursery web spider, *Pisaurina mira.*

a

b

c

Figure 14.13   Crustaceans: (a) spiny lobster, *Panulirus argus;* (b) sow bug, *Oniscus.*

cause several serious diseases, including Rocky Mountain spotted fever and Lyme disease, which is characterized by fevers, chills, muscle aches, and pains.

Crustaceans are mostly aquatic, and many of their appendages are adapted for swimming. They breathe by means of gills and have three distinct body parts with two pairs of antennae or "feelers" (see Figure 14.13). Lobsters, crabs, and shrimp are among the larger and more familiar members of this group, but most crustaceans are small, even microscopic. Tiny crustaceans such as krill are the basic food supply for saltwater fishes and whales. A number of crustaceans have adapted to a life on land. Sow bugs and pill bugs, which live in moist places such as under rocks, are common terrestrial animals.

Uniramia include the millipedes, centipedes, and insects. Insects (Figure 14.14) are the most numerous and diverse arthropods. There are perhaps one million species, and about one billion individual insects for every human on earth. Insects have three body parts, one pair of antennae

Figure 14.14   Insects: (a) glorious beetle, *Plusiotis gloriosa;* (b) yellow jacket wasp, *Vespidae;* (c) butterfly, *Graphium weiskei.*

a   b   c

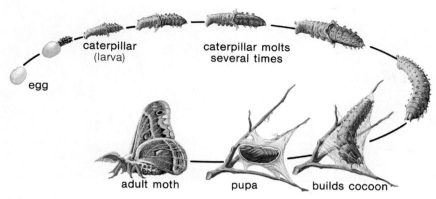

caterpillar
(larva)

caterpillar molts
several times

egg

adult moth   pupa   builds cocoon

Figure 14.15   The life cycle of the Cecropia moth, or American silkworm. An egg develops into a small caterpillar in 10 days. The caterpillar, or larva, eats leaves, molts 4 times, and grows up to 10 cm in length in several weeks. It then spins a cocoon around itself, and the pupa develops within the cocoon. The adult moth emerges in the spring.

on the head, and three pairs of legs on the thorax. The abdomen usually is visibly segmented. Most adult insects have one or two pairs of wings attached to the thorax. Many insects undergo **metamorphosis** (met uh MOR phuh sis), a change in body form and function that occurs during development. Moths and butterflies are examples of insects that undergo metamorphosis. Figure 14.15 illustrates the life cycle of a moth. In its immature, or larval (LAR vuhl), stage as a caterpillar, it feeds continually and may do great damage to trees, crops, and ornamental plants. Many adult insects live only long enough to mate and lay eggs.

14.7   Chordates Have Internal Skeletons

At some time in their lives, all chordates (phylum Chordata; kohr DAH tuh) possess three characteristics: (1) paired gill slits; (2) a dorsal, tubular nerve cord; and (3) a **notochord** (NOH toh kord)—a flexible, rodlike structure that extends the length of the body. In the vertebrate class, the notochord is replaced by a backbone made up of bony or cartilaginous pieces called vertebrae (VER teh bray) that surround the nerve cord. Vertebrates have paired appendages such as legs, wings, and fins. They also possess a well-developed brain that provides good coordination. Because chordates are highly adaptable, they have succeeded in occupying most types of habitats. Fish, amphibians, reptiles, birds, and mammals are all members of phylum Chordata.

Cartilaginous fishes, a class that includes sharks and rays, have an internal skeleton made up entirely of **cartilage**—a tough, elastic connective tissue that is stiff enough to provide support but is more flexible than bone. The rays shown in Figure 14.16, for example, have oddly flattened and flexible bodies that are adapted to gliding along on the sea bottom.

a

b

Figure 14.16   Cartilaginous fishes: (a) nurse shark, *Ginglymostoma cirratum;* (b) sting ray, *Dasyatis americana.*

Bony fishes have skeletons made of bone, and most have an outer covering of scales. Almost all fishes obtain their oxygen through covered gills, although lungs that carry out gas exchange have evolved in air-breathing lungfishes. There is great diversity among bony fishes. In fact, of all the vertebrate classes, this one has the most species, occupying almost all the waters of the earth (see Figure 14.17).

Amphibians include frogs, toads, and salamanders, shown in Figure 14.18. Most live on land as adults but return to water to reproduce. Amphibians lay their eggs in water, where the young develop before returning to land. The limbs of amphibians, which are poorly adapted for walking, may have evolved from the lobed fins of the lungfishes. This group is of special importance in biology because of its evolutionary position between water- and land-dwelling animals.

Reptiles (Figure 14.19) have completely adapted to terrestrial life, although some spend at least part of their lives in water. Their embryos develop in eggs with leathery shells that are laid on land. Reptiles have an outer covering of scales, which prevents excessive water loss, and they breathe by means of lungs all their lives. Dinosaurs, which were early reptiles, are thought to be the ancestors of birds.

Figure 14.17   Bony fishes: (a) grunt, *Haemulon flavolineatum;* (b) a male Siamese fighting fish, *Betta splendens.*

a

b

Figure 14.18   Amphibians: (a) leopard frog, *Rana pipiens;* (b) newt, *Pseudotriton ruber.*

a

b

Figure 14.19 Reptiles: (a) Western long-nosed snake, *Rhinocheitus lecontei lecontei;* (b) iguana lizard, *Iguanidae.*

All birds (Figure 14.20) have feathers, and all animals with feathers are birds, although some feathers are highly modified. No other class of animals is so easy to characterize. All birds have wings, and although not all can fly, those that can have the greatest motility of any animal. In addition, all birds hatch from eggs that have hard shells. Unlike most other chordates, birds have a nearly constant body temperature, an adaptation that enables them to maintain a high rate of metabolism at all times.

Bird feathers show many structural adaptations. Feathers form the soft down of ducks, the long, beautiful plumes of ostriches, and the waterproof coat of penguins. Bird wings also show many structural adaptations. The wings of penguins are not used for flying, but rather for swimming. Albatrosses, which have long, slim wings, spend almost all their lives gliding on air currents. Modification of other structures, such as feet and beaks, also has been important to the ability of birds to adapt to many types of ecosystems. Figure 14.20c illustrates several adaptations.

From the tiny shrew, which measures about 15 cm, to the giant blue whale, which is 30 m long and the largest animal that has ever lived, all mammals share two characteristics. First, all have hair, although it sometimes is not very evident. Even whales have hair at birth, though they lose

Figure 14.20 (a) Canada warbler, *Wilsonia canadensis;* (b) pelican, *Pelecanus occidentialis carolinensis;* (c) diversity of types and unity of pattern. Although species of birds vary greatly, they share characteristics that group them together. Can you determine which feet belong to each bird? How do these differences adapt each bird to its life-style?

a

b

c

a   b   c

Figure 14.21   (a) buffalo, *Bison bison;* (b) porpoise, *Tursiops;* (c) northern sea lion, *Eumetopias jubala*. What similarities can you see?

it soon after. Second, all species of mammals feed their young with milk, a fluid secreted from mammary glands in the skin. This adaptation allows a long period of development under parental care and increases the chances that the young will survive. Like birds, mammals maintain a constant body temperature regardless of the environmental temperature.

Although they have many characteristics in common, mammals show great diversity in size, structure, and color, some of which is evident in Figure 14.21. Diverse modifications for motility and for eating have adapted mammals to function in a variety of specific ecological niches. Modifications for reproduction permit mammals to be grouped into three categories: monotremes, which lay eggs; marsupials, which have pouches in which the young develop; and placentals, which have special internal structures that nourish the young until birth.

Concept Review

1. What characteristics do sponges and cnidarians share, and how are they different?
2. What adaptations of evolutionary significance are shown by flatworms? by roundworms? by annelids?
3. What is the distinctive structure found in mollusks and what functions does it perform?
4. Describe characteristics that distinguish a spider, a crustacean, and an insect.
5. What characteristics distinguish chordates from other phyla?
6. What structural adaptations distinguish the vertebrate classes?

Life Functions in Animals

14.8   Digestion May Be Intracellular or Extracellular

A variety of mechanisms for getting food have evolved in animals. Sponges demonstrate a rather primitive method. Individual cells with a single flagellum keep a current of water moving through the sponge. When a food particle comes by, the cell may engulf it and draw it into a vacuole, just as a paramecium does. The process of taking food particles into a body cavity is called ingestion.

Marsupials and Monotremes

All animals of the class Mammalia have hair—at least at some point in their life—and produce milk for their young. They also maintain a constant internal temperature. On the basis of reproductive modifications, the animals in this class can be divided into three groups: the monotremes, the marsupials, and the placental mammals.

Monotremes lay eggs as birds do, but fertilization occurs internally, as it does in other mammals. In contrast to marsupial and placental mammals, monotremes have not developed a great diversity of species. Today, only two members of this group survive: the platypus and the echidna, or spiny anteater. One possible explanation for this scarcity of species is that the early monotremes were displaced by the more rapidly evolving marsupials.

Because monotremes are not well represented in the fossil record, most information about them is derived from the two living species. These animals resemble reptiles in their skeletal structure and in their method of reproduction. The eggs of the duck-billed platypus, for example, are fertilized in the oviduct. As the eggs continue down the oviduct, gland secretions add albumin and a thin, leathery shell. The platypus lays its eggs in a burrow and incubates them for about 12 days. Thus, monotremes have no period of pregnancy. The developing embryo is nourished entirely by nutrients stored in the egg. Because it has a bill, a newly hatched young platypus cannot suck milk from its mother—instead, it must lick milk from the mother's fur.

Marsupials are mammals, such as opossums, kangaroos, wombats, and koalas, that give birth to immature young, then carry them in protective pouches. There is more diversity among marsupials than there is among monotremes. The koala, for example, is specialized for life in the treetops. Its long arms, curved claws, and hands with a viselike grip permit the koala to climb very smooth tree trunks. Unfortunately, the destruction of forests, the killing of thousands of koalas for their fur, and now epidemic venereal disease among them have reduced their numbers to a dangerously low level.

The kangaroo family contains a great many species, but all kangaroos have elongated hind feet and a long, heavy tail that serves as a balance for jumping or as an extra leg for standing or walking. The young of the largest marsupial, the red kangaroo, are only 1.9 cm long at birth but can grow to over 1.65 m and can weigh as much as 81 kg. The first pregnancy of the season lasts 33 days. When the young "joey" is born, it crawls unaided to its mother's pouch, where it attaches to a nipple. Although the mother immediately becomes pregnant again, the presence of the joey in the pouch arrests the development of the new embryo in the uterus. When the joey leaves the pouch, the embryo resumes development and is born about a month later. Though the developmental pattern among other marsupials may differ somewhat, in all marsupials, the young are born after a very short pregnancy.

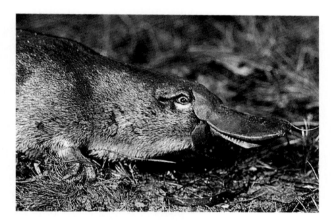

A platypus, *Ornithorhynchus anatinus*. Unlike the bill of a duck, the bill of the platypus is flexible and leatherlike. They also have sharp venom spurs on their hind legs.

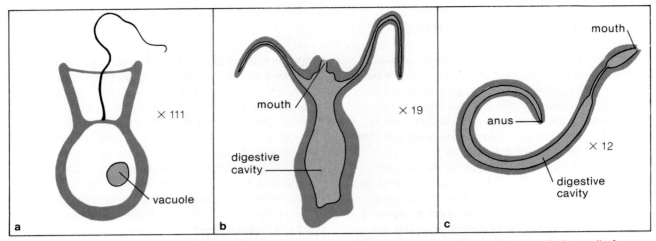

Figure 14.22   Types of digestive cavities: (a) intracellular (vacuole) in a cell of a sponge; (b) extracellular with one opening (sac) in a hydra; (c) extracellular with two openings (tube) in a roundworm.

In contrast to sponges, most aquatic and terrestrial animals actively pursue their food and have means to capture it. Some predators poison their prey. For example, the tentacles of some cnidarians have stinging capsules. Within each capsule is coiled a long, hollow barbed thread. When a food organism brushes one of the tentacles, the thread shoots out with such force that it pierces the body of the victim. A paralyzing poison is injected into the prey, which then is drawn into the cnidarian's body cavity by the tentacles.

More complex food-getting structures also have evolved. Among vertebrates, jaws, beaks, and teeth aid food getting. Special adaptations of these structures often are present. A snake can unhook its lower jaw from its upper jaw and move the two independently, permitting it to swallow prey larger than its own head.

Once ingested, food enters a digestive cavity inside the animal's body. Figure 14.22 shows three types of digestive cavities: a vacuole within an individual cell, a sac, and a tube.

Most nutrient particles that animals ingest are too large to pass through plasma membranes. Even the microscopic particles that sponges take in are much too large to pass directly into cells. Food ingested by animals must be broken down into small molecules. The physical and chemical processes of this breakdown are known collectively as **digestion**. Digestion may be classified as either intracellular or extracellular, according to where it occurs.

The breakdown of large pieces of food into smaller ones is known as physical digestion. Most mammals have teeth that cut or grind food into smaller pieces. For others, movements of the digestive cavity itself break down food. The gizzard of a bird, for example, is a specialized part of the stomach in which food is ground up (see Figure 14.23). Some birds swallow sand and small pebbles that aid the grinding. Physical digestion, whether in the mouth or the stomach, increases the surface area of food, so that the chemical part of digestion can take place more quickly. At this point, enzymes take over the task of breaking down food.

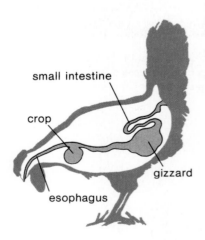

Figure 14.23   A bird swallows food without chewing and stores it temporarily in the crop. Sand and small pebbles in the gizzard aid in grinding the food.

In sponges, and to some extent in cnidarians and flatworms, chemical digestion takes place in a vacuole inside a cell. This process is called intracellular digestion. The vacuole is formed when a food particle is surrounded by an infolding part of the plasma membrane. After the vacuole breaks free, it fuses with a lysosome, which contains digestive enzymes but is not digested by them. Chemical digestion occurs within this new structure. In this way the cell itself avoids being digested. Small molecules produced by chemical digestion pass out of the vacuole into the cell's cytoplasm, making the nutrients available to the cell.

In most animals, digestion is extracellular, meaning that it takes place in a digestive cavity into which enzymes are secreted. In the digestive sac of a cnidarian, for example, some of the cells lining the cavity secrete enzymes. Likewise, in the simple digestive tube of a roundworm, digestive enzymes are produced by cells in the lining of the tube. In more complex animals, however, digestive enzymes are secreted by tissues specialized as glands. The enzymes may be specialized for a particular type of food. Some herbivorous vertebrates, such as cows, have a special chamber (the rumen), which contains cellulose-digesting microorganisms that enable the animal to utilize the cellulose fibers. Most of the fiber in the rumen is digested by numerous anaerobic prokaryotes—up to one billion per milliliter of rumen contents. Presumably the prokaryotes benefit, and the cow obtains much of its nutrition from prokaryote products.

The molecular products of digestion are absorbed into the animal's cells. Often the food molecules diffuse into an animal's blood and are carried to cells throughout the body. The rate of food absorption depends on the surface area of the digestive cavity lining—the greater the surface area, the faster the rate of absorption. In complex animals, the lining has many folds, an adaptation that increases surface area.

14.9   Animals Have Several Ways of Obtaining and Transporting Materials

Figure 14.24   (a) Gas exchange in a hydra. (b) In fishes, gills are thin filaments supported by bony structures and richly supplied with blood vessels. Each filament is made of disks that contain numerous capillaries. Water flows past these disks in directions opposite the flow of blood through the capillaries. A covering over the gills, called the operculum, protects the delicate filaments.

To derive energy from cellular respiration, which requires oxygen, all animal cells must be able to obtain oxygen. Cells also must be able to get rid of carbon dioxide, the waste product of respiration.

In small animals, such as the hydra, gas exchange can occur entirely through the body wall (Figure 14.24a) because the surface area is large enough to allow sufficient amounts of oxygen to enter the organism through

a   b

diffusion. As body size increases, however, the ratio of surface area to volume becomes smaller. Large animals usually do not have enough body surface to allow sufficient gas exchange. Instead, they have organs that increase the surface area through which respiration can occur. In aquatic animals, these organs are usually gills (Figure 14.24b), which often have a feathery or platelike structure. Dissolved oxygen diffuses in and carbon dioxide diffuses out through the gill's cells. Gills are remarkably similar in a wide variety of animals.

Terrestrial animals lose a great deal of water through their respiratory organs because water can diffuse through any surface that also permits diffusion of respiratory gases. Also, cell surfaces must be kept moist so that gases can diffuse through them. For this reason, terrestrial animals that breathe through gills, such as sow bugs, or take in oxygen through their skin, such as frogs and earthworms, must live in places where the air is moist and evaporation is slow.

Many terrestrial animals live where the air is dry. These animals need an extensive surface *inside* the body where air can be kept moist. There are two principal ways of meeting this requirement. Insects and some other arthropods have a complicated system of air tubes that carry oxygen to all parts of the body. Body movements move air through the tubes. Air-breathing vertebrates have lungs, which are divided into a large number of tiny air sacs. Respiratory gases diffuse through the enormous moist surface area of these sacs.

In a single-celled organism, movement of the cytosol provides an adequate system for transporting materials. In some multicellular organisms such as sponges, cnidarians, and free-living flatworms, almost every cell has part of its surface exposed to the environment. Each cell can obtain its own oxygen and get rid of its own wastes. Although every cell may not take in food directly, no cell is very far from others that do. Diffusion and active transport are sufficient to move substances into and out of cells.

In roundworms, transport is complex. These organisms have a fluid-filled body cavity surrounding the digestive tube. As the roundworm wriggles, the fluid within the body cavity is squeezed about from one place to another. In this way substances dissolved in the fluid are carried to and from the body cells.

Most animals that have body fluid have a circulatory system, either open or closed. Both types of circulatory systems include a pump (the heart) and a series of tubes through which nutrients can be delivered and waste products removed from the cells. In arthropods and most mollusks, a heart pumps blood through blood vessels that empty into body spaces. The blood moves through these spaces, eventually emptying into another set of vessels that carry it back to the heart. This incomplete system is called an open circulatory system.

Annelids, on the other hand, have a closed circulatory system. In a closed system, blood always flows within vessels. An earthworm has five pairs of hearts and a complex set of finely branched vessels. These vessels empty into a large dorsal vessel, which returns the blood to the hearts. Valves in the dorsal vessel keep the blood flowing in only one direction through the system. Vertebrates, including humans, also have closed circulatory systems. Figure 14.25 illustrates transport in several animals.

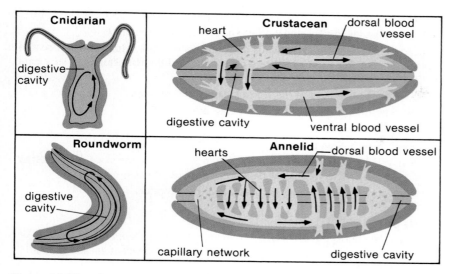

Figure 14.25 Diagrams of fluid transport in four invertebrate animals.

INVESTIGATION 14.2 Temperature and Circulation

Circulation in a multicellular animal is affected by many environmental factors. You might ask, for example, how increasing temperature would affect the rate at which an animal's heart beats. This investigation examines the effects of environmental temperature on the pulse rate of the dorsal blood vessel of an earthworm.

▶ Construct a hypothesis about the effects of various temperatures on the pulse rate of a worm's dorsal blood vessel. Record this hypothesis in your data book.

Materials (per team of 3)

50-mL beaker of tap water	2 damp paper towels
500-mL beaker	ice
nonmercury thermometer	warm water (not more than 40° C)
10X hand lens	watch or clock with a second hand
clean, chemical-free dissecting tray	large earthworm

Procedure

1. Review the guidelines in Appendix 1 for the proper care and handling of laboratory animals. Follow these guidelines carefully.

2. Place an earthworm in the pan, dorsal side up. The dorsal side is much darker and rounder than the ventral side.

3. Moisten the earthworm with a few drops of water. Place a damp piece of paper towel gently over the head region so the worm will wiggle and move very little. Touch the worm as little as possible.

4. Locate the dorsal blood vessel running directly down the center of the dorsal surface. Find the part where you can most easily see the actual pulsing of the vessel. (It may be helpful to use a 10X hand lens.) Make a trial count of the number of beats in 1 minute. You will expose the worm to five different temperatures from cold (ice water) to warm but not hot. Do not expose the worm to temperatures higher than 40° C—it may die.

5. Add just enough ice water to the tray to submerge your worm (about 1 cm).

6. Lay the thermometer in the pan so the bulb is submerged.
7. Wait 3 minutes to allow the worm to adjust to the temperature.
8. Read the temperature of the water and immediately begin counting the pulse rate for 1 minute. Record the beat rate per minute and the temperature in your data book.
9. Pour off the water and let the worm breathe for 2 minutes.
10. Add water that is about 10° C to the tray, as in Step 5.
11. Repeat Steps 6–8 with the 10° C water.
12. Repeat Steps 5–8 with water at or close to 20° C, 30° C, and 40° C.
13. When you are finished collecting data, return the worm to the place designated by your teacher. These worms will be retired to a nice compost pile.
14. Wash your hands thoroughly before leaving the laboratory.

Discussion

1. In your data book, prepare a graph from your data that shows the effect of temperature on the beat rate per minute. Plot the independent variable on the horizontal axis.
2. Why is it important to wait a few minutes after adding water to the pan before counting the pulse rate?
3. Why must the water be poured off to allow the worm to breathe?
4. What would be the experimental advantage of starting either with the coldest or warmest temperatures and working gradually in the other direction instead of skipping around to different temperatures?
5. Describe any patterns revealed by the curve on your graph. Do the data support or refute your hypothesis? Explain.
6. To how many other types of organisms can you generalize your data?
7. What are some flaws or sources of error in the experimental procedure used here? Be very critical.
8. If you were to test your hypothesis again and had no limit on equipment or materials, how would you design the new experiment?

14.10   Excretory Systems Maintain Water and Chemical Balance

Wastes are removed from cells by a process called **excretion** (ek SKREE shun). What are wastes? Some products of metabolism are poisonous—ammonia, for example, which is formed in the breakdown of proteins. Other substances, however, are toxic only if large amounts accumulate. Sodium chloride is a normal cell substance that is toxic in large amounts. Ordinarily this salt is constantly diffusing into cells from the fluid surrounding them, and so the normal proportion of sodium chloride in cells can be maintained only by continuous excretion. In general, any substance can be called "waste" if an organism has too much of it.

In small aquatic animals, wastes may simply diffuse through plasma membranes to the outside environment. Sponges and cnidarians also can excrete wastes directly through their body surfaces, because all their cells contact the water environment or are very close to it. Most animals, however, have special systems for ridding their bodies of wastes.

Animal cells constantly produce water through cellular respiration. What happens to that water depends on the type of environment in which an animal lives. For example, the cells of a jellyfish contain a particular mixture of substances in water, whereas seawater contains a different mixture of substances in water. Normally the concentration of water molecules inside the jellyfish equals the concentration outside it. As fast as metabolic water is produced, it diffuses into the environment. This occurs in many other marine invertebrates as well.

Fresh water, on the other hand, contains few dissolved substances, and so the concentration of water molecules is high. In the cells of a freshwater planarian, the concentration of dissolved substances is high, and the water concentration is low. Therefore, water is always diffusing *into* a planarian, and metabolic water is constantly added to this excess. If water was not somehow expelled from the cells, eventually the cells would burst. Planarians and other freshwater animals excrete water by active transport.

Terrestrial animals have the opposite problem. As a human, for example, you take in a great deal of fresh water, and your cells also produce water by metabolism. As a land animal, however, your greatest danger is of drying out. Your kidneys reabsorb water that would otherwise be lost from the body with wastes.

Within the body, animals maintain water balance in a variety of ways. In planarians, for example, this function is performed by a system of flame cells spread throughout the body and connected by small tubes, or tubules. Each flame cell has cilia that project into the cell cavity. Wastes enter the flame cells from the tissues. The waving action of the cilia moves fluid and wastes out into the tubules. Many tubules join together and eventually empty wastes into the environment through a pore. In this system, there are many flame cells and many pores. Figure 14.26a shows the excretory system of a planarian.

Most annelids, mollusks, and crustaceans have a somewhat different excretory system. The functional unit for these organisms is a tubule leading from the body cavity to the outside (see Figure 14.26b). There are many of these tubules. Wastes from within the body cavity, and from the network of capillaries surrounding each tubule, diffuse into the tubules and are

Figure 14.26   (a) The excretory system of a planarian. A flame cell and an excretory pore are shown enlarged. Wastes and excess water removed by flame cells are eliminated through pores in the body wall. (b) Excretory tubes in the earthworm. The near side of the body, including the digestive tube, is cut away. Most of the segments of the body have a pair of excretory tubes, one on either side. Capillaries from the circulatory system surround the excretory tubes.

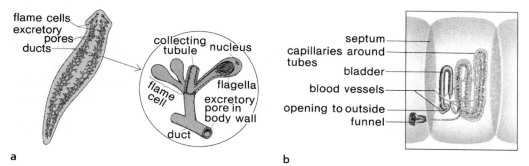

a   b

emptied outside. At the same time, useful materials may be absorbed back into the blood or body fluid from the tubules. Among vertebrates, a somewhat similar system of tubules is found in the kidneys, which are described in Chapter 16.

The excretory organs evolved chiefly as water-regulating devices. In most animals, however, they also regulate the excretion of wastes, especially those that contain nitrogen. Amino acids contain nitrogen and are used by cells to build proteins, but an animal may take in more amino acids than it can use. Some proteins are continually broken down to amino acids. Unlike carbohydrates and fats, amino acids and proteins cannot be stored in large amounts. As a result, there is usually a surplus of amino acids. This surplus cannot be used for energy until the amino group ($-NH_2$) is removed from each amino acid.

In vertebrates, amino groups are removed mainly in the liver. As they are removed, ammonia (NH_3) is formed. Ammonia is toxic but also soluble. If a large supply of water is available, the ammonia can be carried out of the body in solution. For example, in freshwater fishes, nitrogenous wastes may be excreted through the gills as ammonia.

In other vertebrates, the ammonia is converted to other forms, and the kidneys are the main route for excretion. In birds, reptiles, and insects, amino groups are excreted as uric (YOOR ik) acid. Uric acid is almost insoluble and is excreted with only a small loss of water. In most adult amphibians and mammals, amino groups are converted to a waste called urea (yoo REE uh). Unlike uric acid, urea is soluble. It diffuses into the blood, from which it is removed by the kidneys and then excreted in solution in the urine.

14.11 Nervous Systems Control Responses to Stimuli and Coordination of Functions

For the most part, animals can make rapid adjustments in response to changes in their environment. Even though sponges have no true nervous systems, they are able to adjust to environmental changes, or **stimuli** (STIM yoo ly), such as changes in temperature or chemical concentration. Nerve cells, alone or in nervous systems, make these adjustments possible. Cnidarians have nerve cells, some of which are specialized to receive only certain stimuli. The nerve cells are connected in a network that permits simple coordinated responses.

Flatworms have nerve networks, but they also have a centralized system of two nerve cords that extend along the length of the body. At the anterior end is a large mass of nerve tissue, known as a ganglion (GANG glee un), where nerve impulses are exchanged. Flatworms also have cells specialized for receiving a variety of stimuli. For example, the eyespots of planarians cannot form images, but they can detect the direction and intensity of light.

Annelids have well-developed nervous systems. A main nerve cord extends along the ventral side of the body. Many ganglia occur along the nerve cord. A large ganglion, sometimes called a brain, is found at the anterior end of the body. Although earthworms have no obvious sense organs, they

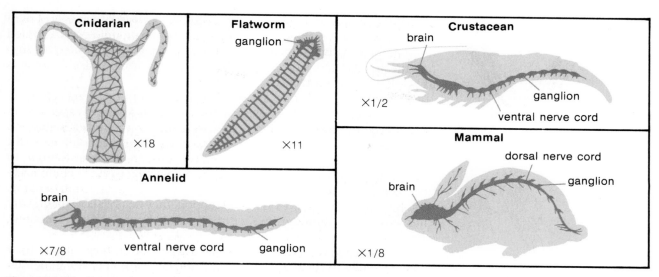

Figure 14.27 Nervous systems in five diverse types of animals.

can detect many stimuli. Other annelids have specialized sense organs, including eyes. Mollusks and arthropods also have nervous systems, basically of the annelid type.

A dorsal, tubular nerve cord, usually called a spinal cord, is a distinctive characteristic of the chordate phylum. In vertebrates, an anterior enlargement of this nerve cord—the brain—dominates the rest of the nervous system. Figure 14.27 compares the nervous systems of several animals.

The ability to receive and react to stimuli from the environment is one of the basic characteristics of living things, although it is developed to different degrees in different organisms. Most animals detect different types of stimuli by using specialized receptors at the nerve endings. In humans, for example, receptors in the skin of the fingertips are sensitive to pressure but not to light. Receptors in the eyes are light sensitive, but they are not sensitive to sound waves. In most animals, some receptors are concentrated in special organs, such as eyes, ears, and nose. Other receptors, such as those sensitive to pressure, touch, cold, and heat, are widely distributed in the skin.

14.12 Muscles and Skeletons Provide Support and Locomotion

In animals, motion usually involves muscle cells. Sponges are an exception. They have no muscle cells, although individual cells are capable of movement. Sponge larvae swim by means of cilia, just as many protists do.

Cnidarians have specialized cells that allow them to move their tentacles. Flatworms have cells organized into definite muscle tissues, but locomotion still is accomplished largely by cilia. In all other animal phyla, muscle tissues are organized into bundles (muscles) controlled by nervous systems.

a

b

humerus

radius

ulna

Figure 14.28 Comparison of exoskeletal structure (a) with endoskeletal structure (b).

The skeletons or support systems of animals consist of a variety of mineral compounds. A sponge's skeleton is made of small, rigid parts scattered through the soft, living tissues. The stony skeletons built up by corals support the animals and protect them from predators. A mollusk's skeleton is its shell, which is protective.

Support and protection are the functions of arthropod and chordate skeletons. In these phyla, skeletons also aid in movement, especially in locomotion. Arthropod skeletons are external, with the muscles attached to the inner surfaces, as shown in Figure 14.28a. These **exoskeletons** (EK soh SKEL eh tunz) are composed of chitin, a tough, flexible material. In many arthropods, calcium compounds in combination with the chitin make the exoskeleton hard and strong. Thin, flexible chitin joints permit bending. Although exoskeletons provide good support, they do not allow for growth. Thus, arthropods periodically shed their exoskeletons and produce new, larger ones.

Vertebrates have **endoskeletons** (EN doh SKEL eh tunz) inside their bodies (see Figure 14.28b). Muscles are attached at points along the outside of the endoskeleton. The skeletons may consist of either cartilage or bone, both of which contain cells that secrete new material. Thus the skeleton can grow as the animal does.

14.13 Animals Are Adapted for Reproductive Success

Reproduction is primarily sexual in the animal kingdom. Even in species that ordinarily reproduce asexually, such as many of the cnidarians, sexual reproduction does take place. The alternation of haploid with diploid generations present in all plants is unknown in animals, although some animals have haploid and diploid groups. Cnidarians alternate a sexual generation with an asexual generation, but both are diploid. Male and female gametes are different in animals, and meiosis occurs just before or during gamete formation.

In most adult animals, an individual has either ovaries or testes, not both. A difference in reproductive organs usually is accompanied by differences in appearance (see Figure 14.29), behavior, or other characteristics. In some animals, an individual has both ovaries and testes. This is true of many annelids, most flatworms, and some crustaceans and mollusks. Among vertebrates, this type of individual occasionally occurs but is not considered normal.

Cnidarians reproduce both sexually and asexually. Hydras, for example, reproduce asexually by budding through much of the year. Under certain conditions, however, they produce ovaries and testes—an adaptation that increases the chances of fertilization in sessile or slow-moving animals. Eggs in the ovary are fertilized by sperm shed into the water. Figure 14.30 shows both types of reproduction in a hydra.

Flatworms have a complex reproductive system. Most individuals have both ovaries and testes, but they do not fertilize their own eggs. Instead, two worms exchange sperm. Free-living forms also reproduce asexually by fragmentation, and even a small piece can regenerate into an entire worm.

Figure 14.29 Some secondary sexual characteristics in animals. Can you think of other instances where the male and female of a species do not closely resemble each other?

Parasitic forms such as *Schistosoma* (shis toh SOH muh) have complex life cycles involving many larval forms and two or more hosts, as you can see in Figure 14.31 on page 388. Parasitic forms produce huge numbers of eggs, an adaptation that increases the chances for survival of the species.

In roundworms, the sexes are separate, and asexual reproduction does not occur. The females may release as many as 200 000 eggs a day, and the egg cases and developing embryos can survive for many years before resuming growth. As with flatworms, parasitic roundworms have complex life cycles with several hosts.

Earthworms also have both ovaries and testes, but the ova are not fertilized by sperm from the same individual. When two earthworms mate, sperm from one individual are deposited in a special sac of the other. Likewise, sperm from the second worm are deposited in the first. As ova move from the ovary, they pass the sperm-storage sac. Sperm are released, and the ova are fertilized.

In fishes and amphibians, fertilization occurs externally, and the zygote develops in the water. Fertilization is internal in all other vertebrates. The zygote may develop in an egg laid by the female, as in birds and most reptiles, or within the body of the female, as in mammals.

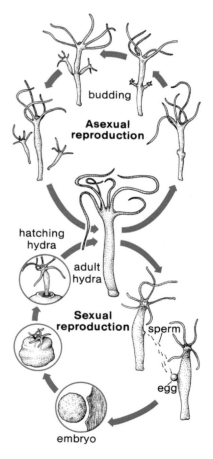

Figure 14.30 A hydra can reproduce either asexually or sexually.

Figure 14.31 Life cycle of *Schistosoma haematobium.* Adult flukes in the veins of a human host produce fertilized eggs that are expelled with the urine. In the water, the larval stages, or miracidia, hatch from the egg and infect snails, the intermediate hosts. Further changes take place in the snail, and immature worms, or cercaria, emerge. The cercaria penetrate the skin of a human, migrate into the small blood vessels, and are carried by the circulatory system to small veins in the bladder, where they mature. Adult worms may live 30 years and produce thousands of eggs each day. This parasite is widespread in many parts of the world and causes serious disease.

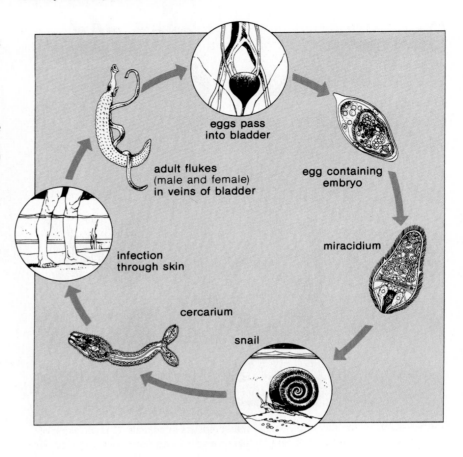

eggs pass
into bladder

adult flukes
(male and female)
in veins of bladder

egg containing
embryo

miracidium

infection
through skin

cercarium

snail

Concept Review

1. How do the various food-getting devices of animals illustrate structural diversity?
2. In what ways do digestive cavities differ among animals?
3. Why does a very small aquatic animal require no breathing system?
4. How do gills and lungs function in gas exchange?
5. What types of animals can survive without circulatory systems?
6. Distinguish between open and closed circulatory systems.
7. What are the basic functions of excretory systems?
8. Compare the nervous system of a vertebrate with that of a flatworm.
9. How do skeletons differ in vertebrates and arthropods?
10. Under what circumstances is internal fertilization necessary?

Summary

All types of animal forms and functions have resulted from their heterotrophic way of life. Evolutionary development favored motility, which improved animals' ability to obtain food. Motility, in turn, called for streamlining and led to the evolution of bilateral symmetry and a head where sensory organs and nerve cells are concentrated. Larger animals have more choice of food sources but also more cells, requiring more coordination and greater specialization. Systems evolved to handle basic functions such as digestion, gas exchange, transport, excretion, and reproduction. Evolution, through natural selection, has resulted in a variety of adaptations that enable animals to solve the problems of life that different environments impose on them.

Applications

1. Why do plants lack heads?

2. What is the advantage of having organs and of having a division of labor? In other words, what advantages do animals with organs have over those without?

3. What are the advantages and disadvantages of having an exoskeleton?

4. Many insects undergo metamorphosis. What are the advantages of having such different body forms within the same life cycle?

5. The importance of surface area to the function of a structure or system has been emphasized in this chapter. How is a greater surface area important to digestion? To respiration? Can you think of any other examples where surface area plays an important role?

6. Birds and mammals maintain a constant body temperature. In what ways is this adaptation an advantage?

7. What features of arthropods and vertebrates enabled them to adapt to terrestrial life?

8. One of the major principles of biology is that an organism's structure is related to its function. Choose one organism from any phylum and explain how it illustrates this principle.

9. Another of the major principles of biology is that an organism and its environment are complementary—the environment shapes the organism and vice versa. Choose one organism from any phylum and explain how that organism illustrates this principle.

Problems

1. Humans are carnivores. They eat other animals to obtain their nutrients and energy. Make a list of the types of animals people around the world eat—arranging the food animals by phyla and classes.

2. In 1990, the 17-year cicadas emerged all over the Midwest after a long developmental period. Investigate these unusual organisms and report on their development. What other organisms show unusual developmental processes?

3. One of the major limitations to life on land is the lack of water. What adaptations do desert-dwelling animals have that allow them to tolerate the hot and dry conditions?

4. Learn more about animals in their native habitats. To find out more about birds, contact a local Audubon Society chapter. Join members in one of their Christmas counts of birds in your area.

5. Take a field trip to your local zoo or aquarium. Compare the food, behavior, artificial habitats, and the form and structure of different animals there.

6. List the principal life functions of animals, then compare the structures by which each is accomplished in a sponge, a planarian, an insect, and a mammal. Because all these animals exist in large numbers, we must conclude that despite the diversity of their structures they all successfully carry on their life functions. Explain how this can be.

7. Make a list of animals from a number of phyla and arrange it in order of the care given the young—from least care to most care. List the same animals in order of numbers of young produced—from most to least. Explain any relationships you find between the two lists.

Organisms vary greatly in size. A redwood tree may be 100 m tall and a blue whale may weigh 150 metric tons. A bacterium, on the other hand, may be only 0.4 μm wide and weigh a small fraction of a milligram. No matter how different they appear on the outside, all organisms must accomplish the same basic tasks, such as acquiring food and eliminating wastes, to remain alive. The mechanisms and processes they use, however, may vary. Section Four examines how two different types of organisms go about the day-to-day business of life.

Women's hurdles. How many different organs must work together in this athlete as she jumps these hurdles? Must she think about all of these organs to make them work? How can a complex multicellular organism, such as a human, coordinate so many different parts of the body?

A typical combination pizza. What probably are the most abundant nutrients in this meal? The least abundant? How do you think two slices of pizza compare in fat content to a chicken thigh and leg? To two beef tacos?

The Human Animal: Food and Energy 15

Humans consume an amazing variety of foods. Your lunch may include a hamburger and fries or several pieces of pizza and a soft drink. Later in the afternoon, you may add a candy bar, an ice cream cone, or popcorn and more soft drinks to your digestive tract. How can your body use the protein in hamburger and the potato starch in french fries to grow and provide your muscles with energy? How do the plant and animal tissues you eat become human cells and tissues?

More and more these days, processed and convenience foods are replacing foods rich in vitamins and nutrients. How does what you eat affect your health? Can you control your diet and still enjoy dessert? This chapter examines how the human digestive system operates and the importance of maintaining a complete and well-balanced diet.

Ingestion and Digestion

15.1 Specific Actions and Reactions Take Place Along the Digestive Tract

Ingestion—the process of taking food into the entrance of the digestive tract—begins with the first bite of a hamburger. In humans, the entrance to the digestive tract is bounded by teeth, tongue, and palate. Once food has entered the mouth, the chewing action of the teeth begins the mechanical breakdown of food. This grinding action breaks the large pieces of hamburger and bun into smaller pieces, much like chopping a log into kindling to help start a fire.

As you chew, your highly muscular tongue keeps the food in contact with the teeth, moves it to mix it with saliva, and manipulates it into position for swallowing. Saliva, a watery secretion containing digestive enzymes, begins chemical digestion. Taste buds (Figure 15.1), sensory nerve-endings in the tongue, tell you of the food's flavor. The tongue then moves the moistened and ground hamburger to the back of the mouth, where it is swallowed.

Connecting the mouth to the stomach is the **esophagus**, a tube made of two layers of muscles—an outer layer running lengthwise and an inner circular layer. Together, these muscles move the hamburger to the stomach by a process known as peristalsis (per ih STAWL sis), as shown in Figure

Guidepost: How is food prepared for absorption?

Figure 15.1 A cross section of a taste bud.

393

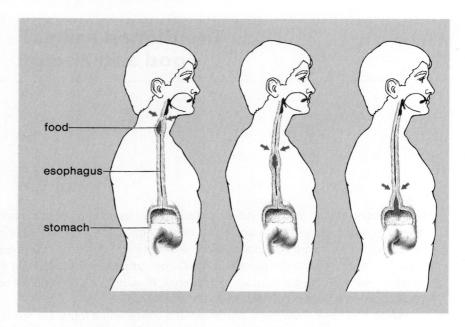

Figure 15.2   Peristalsis in the wall of the esophagus forces food through the esophagus and into the stomach.

15.2. Contraction of the circular muscle immediately above the swallowed food pushes the food down as the muscle in front of the food relaxes. Peristalsis proceeds in a wavelike sequence down the length of the esophagus.

Normally the food's journey is a one-way passage through the digestive tract, or gut, which may be described as a tube that begins at the mouth and ends where the undigested material exits the body. Figure 15.3 shows the human digestive tract and summarizes the activities that take place in it. As the food moves along approximately 9 m of digestive tract, it goes through various sections that function as a disassembly line. The **stomach**, which lies between the esophagus and the intestine, is an enlargement of the gut. A circular muscle, or sphincter (SFINK ter), at either end of the stomach closes the opening. Often thought of as a storage organ to be packed with food, the stomach actually holds food only long enough to prepare it for entrance into the small intestine.

15.2   Chemical Digestion Reduces Large Food Molecules to Smaller, Absorbable Molecules

In the stomach, food is churned and diluted, the proper pH is established, and enzymatic action continues. The kneading and churning action of the stomach muscles continues the mechanical breakdown of food. This motion also helps mix in a watery fluid called gastric juice, which is secreted into the stomach by glands in the stomach wall. Gastric juice contains digestive enzymes and hydrochloric acid, which aids in the breakdown of proteins into polypeptides. The digestive enzymes that catalyze the breakdown require a low (acidic) pH to function effectively. The hydrochloric acid in the gastric juice provides an appropriate environment for enzymatic action.

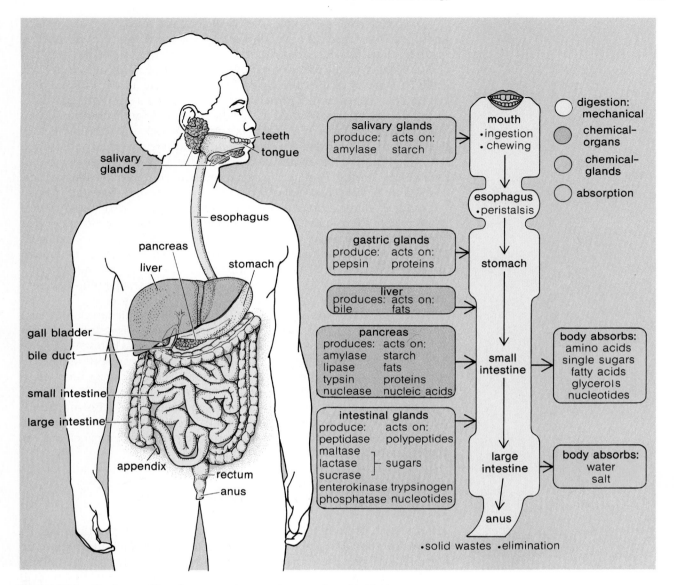

Figure 15.3   The human digestive system is a continuous tube with highly specialized organs and tissues along its length. Mechanical and chemical digestion are followed by absorption.

Churning combined with the action of gastric juice and digestive enzymes changes the contents of the stomach into a thick liquid. The partially digested food is held in the stomach by the contraction of the pyloric valve, the sphincter between the stomach and small intestine. The pyloric valve relaxes irregularly, allowing the partially digested food to spurt into the small intestine. It takes about four hours to empty the stomach of food.

Chemical digestion is completed and absorption of food molecules occurs within the **small intestine**, a tube approximately 6 m long. Absorption occurs when the molecules move through the intestinal walls and enter the bloodstream. The blood then carries the molecules to the cells, where the food finally becomes a part of the body.

Food molecules go through their final chemical breakdown in the small intestine. Three types of molecules are used in all living cells as building materials and as sources of energy—carbohydrates, proteins, and fats (lipids). Although the food you eat must supply these molecules, your body cannot use the molecules directly from the food. Just as an old brick building can be demolished and the bricks used to construct a patio or a fireplace, so the human body tears food molecules apart and forms useful building blocks. Of all the molecules that make up a food, only the small molecules that make up carbohydrates, proteins, and lipids are absorbed and used to synthesize the specific molecules that make up the human body. Carbohydrates are strings of single sugars; proteins are composed of amino acids; and simple fats are made up of a glycerol molecule and three fatty acids.

Once the large molecules enter the small intestine, digestive enzymes begin to split them apart. Enzymes or digestive juices enter the upper part of the small intestine from the pancreas, the liver, and the intestine itself. The digestive juices all have a high (basic) pH, which helps to neutralize the hydrochloric acid of the soupy mixture that enters from the stomach. This is important because the enzymes of the small intestine require a neutral environment.

The pancreas delivers pancreatic (pan kree AT ik) juice through a duct leading into the small intestine. Pancreatic juice contains several enzymes that catalyze digestion of polypeptides, fats, and carbohydrates. See Figure 15.3 on the preceding page. Intestinal juice, secreted by glands in the wall of the small intestine, contains other enzymes. These enzymes function in the digestion of dipeptides and double sugars.

Fats must be pretreated before they can be digested. This function is performed by bile, a substance secreted by the liver and stored in the gall bladder. Bile breaks the fat droplets into fine particles that pancreatic enzymes then convert to fatty acids and glycerol.

15.3   Absorption Takes Place in the Small Intestine

Almost all absorption of nutrients occurs in the lower part of the small intestine. The process of absorbing food molecules is much like wiping up a spill—the larger the sponge, the more that is absorbed. The small intestine is like a large sponge—if every wrinkle and bulge of the human small intestine were flattened, the total surface area would be about 300 m², about half the size of a basketball court. This enormous surface area is possible because of the presence of folds in the intestinal wall that are covered by tiny bulges known as villi (see Figure 15.4). These villi are covered by cells that have surface membranes folded into fine projections called microvilli. As proteins and carbohydrates are broken down, amino acids and single sugars are produced. These end products are absorbed into the blood by diffusion or active transport through the walls of the small intestine.

Fatty acids with 12 or fewer carbons are absorbed directly into the blood capillaries and carried to the liver for processing. Fatty acids with more than 12 carbons are absorbed by the cells of the intestinal wall, where they recombine with glycerol and are packaged in droplets. These droplets are absorbed by the **lymphatic** (lim FAT ik) **system**, a system of vessels that transports fluid called lymph (LIMF) from the body tissues to the blood.

Figure 15.4 A cross section of the small intestine. Intestinal villi are shown enlarged in the micrograph and drawing. Digested foodstuffs enter the blood and the lymph through the villi.

The absorption processes are selective, but they are not infallible. There is no control mechanism to prevent absorption of excess food, and many harmful molecules can masquerade as useful molecules and be absorbed. For example, DDT, an insecticide that was used heavily on food crops in the United States until 1972, is absorbed and stored in fatty tissue and in human milk.

15.4 The Large Intestine Absorbs Water

Normally, digestion and absorption are completed in four to seven hours. Substances left in the small intestine then pass into the large intestine, where more water is absorbed. In a diet consisting of approximately 800 g of food and 1200 mL of water per day, the following approximate amounts of digestive secretions are added:

1500 mL saliva
2000 mL gastric secretions
1500 mL intestinal secretions
 500 mL bile
1500 mL pancreatic secretions.

Of this total—approximately 9000 mL including food and water—8500 mL are absorbed in the small intestine and 350 mL in the large intestine. The remaining 100 mL of water and 50 mL of solids are expelled from the body as solid wastes.

No further absorption of food substances occurs in the large intestine. Undigested foods, indigestible substances, mucus, dead cells from the digestive tube lining, and bacteria are concentrated by water removal to form waste matter, or feces (FEE seez). If the fecal matter remains in the large intestine too long, almost all the water is reabsorbed and very dry feces are passed. The feces exit the body through a sphincter muscle called the **anus** (AY nus).

What happens to the 8850 mL of absorbed solids and liquids? The next section traces the fate of molecules that are intended as fuel for the energy-producing processes.

Concept Review

1. Distinguish between mechanical and chemical digestion.
2. Distinguish among ingestion, digestion, and absorption.
3. Trace the movement of a hamburger and bun through the digestive system, indicating what happens in each part.
4. Discuss the digestion of carbohydrates, including where it takes place and the substances involved.
5. What is the role of the small intestine in the digestion of proteins and fats?

Guidepost: How do cells obtain energy from food molecules?

Cellular Respiration

INVESTIGATION 15.1   Food Energy

All foods contain energy, but the amount of energy varies greatly from one food to another. This investigation uses a calorimeter (Figure 15.5) to measure the amount of energy, in calories, in some foods. Recall that a calorie is the amount of heat required to raise the temperature of 1 g (1 mL) of water 1° C. Caloric values of foods in diet charts are given in kilocalories (1000 calories) or kcals. Table 15.1 lists accepted caloric values for some common foods.

Using the LEAP-System™ with a thermistor, or a thermometer if the LEAP-System is unavailable, you can measure the change in temperature (ΔT) of a known volume of water. The temperature change is caused by the absorption of the heat given off by burning a known mass of food. Based on the change in temperature, you can calculate the amount of energy in the food. Instructions for use of the LEAP-System are provided in the LEAP-System BSCS Biology Lab Pac 2.

TABLE 15.1   Energy in Kcals of Common Foods Burned in a Highly Efficient Calorimeter

Food	Kcal/100 g
Almonds	640
Bacon	626
Brazil nuts	695
Bread, white	261
Butter	733
Cashews	609
Cornflakes	359
Dried coconut	570
Dried lima beans	341
Kidney beans	350
Lard, shortenings	900
Mayonnaise	720
Peanut butter	619
Pretzels	362
Puffed rice	363
Roasted peanuts	600
Salad oil	900
Walnuts, black	672
Walnuts, English	702

Materials (per team of 3)

3 pairs of safety goggles
3 lab aprons
100-mL graduated cylinder
250-mL Erlenmeyer flask
Apple IIe, Apple IIGS, Macintosh, or IBM computer
LEAP-System™
printer
thermistor
extension cable for thermistor
15-cm nonroll, nonmercury thermometer (optional)

balance
forceps
tin can with cutout air and viewing holes
cork with sample holder
kitchen matches
20-cm × 30-cm piece of extra heavy aluminum foil
2 pot holders
small container of water
3 whole peanuts
3 walnut halves

SAFETY   Put on your safety goggles and lab apron. Tie back long hair and roll up long, loose sleeves.

Procedure

1. Decide who in your team will be the experimenter, who will be the recorder/computer operator, and who will be the safety monitor to assure correct safety procedure is followed.
2. Copy Table 15.2 in your data book or tape in the table your teacher provides.

TABLE 15.2   Energy Content of Walnut and Peanut Samples

| | Temperature of Water (°C) | | | | Food Energy | | |
	Mass of sample	Before burning	After burning	Change	Calories	Kcal	Kcal per gram
Walnut sample 1							
Walnut sample 2							
Walnut sample 3							
Average							
Peanut sample 1							
Peanut sample 2							
Peanut sample 3							
Average							

3. Using the balance, determine the mass to the nearest 0.1 g of each peanut and each walnut half. Record each mass in the table.

4. Obtain a 250-mL flask, a tin can, a cork with sample holder, and a piece of extra heavy aluminum foil. This equipment may be used to make a calorimeter like the one shown in Figure 15.5. Practice assembling and disassembling the equipment.

5. With the calorimeter disassembled, measure 100 mL of tap water and pour it into the flask.

6. Bend the thermistor wire at a right angle about 3 cm from the tip of the sensing element. Insert the thermistor into the flask. The sensing element should be submerged in the top of the water and not touching the glass. Bend the thermistor

Figure 15.5   The calorimeter setup.

wire at the top of the flask to hold the sensing element in position. If not using the thermistor, set the thermometer in the flask.

7. Measure the temperature of the water and record in the table.

8. Place a peanut in the wire holder anchored in the cork. Then, place the cork on the piece of aluminum foil.

9. Carefully set fire to the peanut. This may require several matches. Discard burned matches in the container of water.

    WARNING: Matches are flammable solids. In case of burns, call your teacher immediately; place burned area under *cold* running water.

10. Place the tin can over the burning sample with the viewing hole facing you. Place the flask of water on top of the tin can.

11. Begin recording data. Continue recording, even when the peanut has burned completely, until the water temperature begins to decrease. (It will continue to rise after the sample has burned out as the water absorbs heat from the tin can.) Record the maximum water temperature. If using a thermometer, take readings as soon as the sample has burned out and then at 30-second intervals.

12. Allow the calorimeter to cool about 2 minutes before disassembling.

    CAUTION: The flask and tin can will be hot; use pot holders to handle and place on the aluminum foil *only*. The sample holder also will be hot; use forceps to remove burned sample. In case of burns, call your teacher immediately; place burned area under *cold* running water.

13. Repeat Steps 7 through 12 until you have data for three samples each of peanut and walnut. Change the water in the flask each time.

14. Wash your hands thoroughly before leaving the laboratory.

Discussion

1. Run stripcharts to obtain graphs of the data. Determine the average change in temperature for each sample. Calculate the number of kcals produced per gram. To do this, multiply the increase in water temperature (average change) by 100 (the number of mL of water used). This step will give you the number of calories. To convert to kcals, divide by 1000. To calculate kcals produced per gram of food, divide this number by the number of grams of food burned. Enter all data in the table.

2. How do your data (adjusted for 100 g) compare with the values for 100 g of the same or similar food in Table 15.1? (The kcals listed in most diet charts are per 100 g, per ounce, per cup, or per serving. To compare your results, you may need to convert to common units.)

3. How do you account for any differences?

4. If the same amount of each food were completely burned in the cells of the human body, would you expect the energy release to be greater or less than your results? Than published results?

5. Which of the two foods you tested seems to be the better energy source?

6. Why might some foods with fewer kcals be better energy sources than other foods with more kcals?

7. What was the original source of energy in all the foods tested?

For Further Investigation

Determine the kcals in snack foods such as potato chips, corn chips, and cheese snacks.

15.5   Cellular Respiration Releases Energy from Food Molecules

Cells require a continuous supply of energy to maintain order, move substances, build organic molecules, grow, and carry on all their other activities. Because living things can use only chemical energy, the energy for these activities must come from food. As a result of the digestive processes described in the previous sections, amino acids, single sugars, fatty acids, and glycerol are delivered to your cells. How do your cells obtain energy from these molecules?

Cells release energy from the basic molecules of food through the process of cellular respiration—a type of controlled burning. When something is burned, a great deal of energy is liberated all at once in the form of heat and light. Cells release the same amount of energy gradually in a stepwise series of reactions that are controlled by enzymes. As a result of this controlled release, the energy present in these basic molecules is freed in small amounts that can be captured and stored in molecules of ATP. It might be said that the energy stored in food is like money in the bank, whereas energy stored in ATP is like "small change." It is readily available for use in cellular work.

Cellular respiration in humans generally is aerobic; that is, it requires oxygen, which is transported along with small food molecules in the blood. The oxygen and small food molecules move across the plasma membranes of every cell by diffusion, passive transport, or active transport and can be used by the cell in a variety of ways. The primary food molecule utilized in cellular respiration is glucose, a single sugar that has the formula $C_6H_{12}O_6$. During respiration, the energy stored in a glucose molecule is released slowly as the molecule is broken down. Its gradual breakdown yields six carbon dioxide molecules. The overall reaction may be described in the following equation:

$$C_6H_{12}O_6 \; + \; 6\,O_2 \xrightarrow{\text{enzymes}} 6\,CO_2 \; + \; 6\,H_2O \; + \; \text{energy}$$
$$\text{(glucose)} \quad \text{(oxygen)} \qquad \text{(carbon} \qquad \text{(water)}$$
$$\text{dioxide)}$$

This equation summarizes a complex series of chemical reactions that involve many steps. Figure 15.6 provides a simplified diagram of this series of reactions.

The steps of cellular respiration can be divided into three stages. In the first stage, known as **glycolysis** (gly KAWL uh sis), glucose is split into two 3-carbon molecules and some of the released energy is transferred to ATP. Glycolysis takes place in the cytosol. The second stage is the **Krebs cycle**. During the Krebs cycle each of the 3-carbon glucose fragments is disassembled and six carbon dioxide molecules are formed. Hydrogen atoms also are released. Special molecules carry the hydrogen atoms to the third stage, the **electron transport system**. In this system, the relatively large amount of energy present in a glucose molecule is stored in several smaller amounts in ATP. Each hydrogen atom is separated into an electron and a proton and transferred through many small steps to oxygen, eventually forming water. The Krebs cycle and the electron transport system operate in the mitochondrion (see Figure 15.7 on page 402).

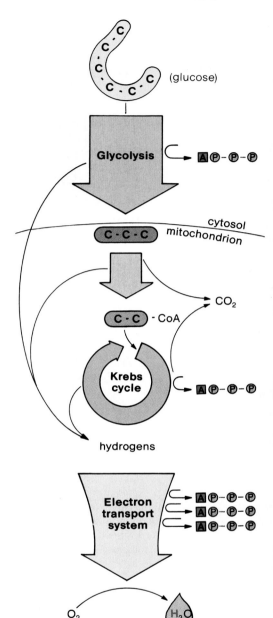

Figure 15.6   Cellular respiration occurs in three stages. Glycolysis occurs in the cytosol; the Krebs cycle and the electron transport system take place in the mitochondrion.

a

b

outer membrane

inner membrane

intermembrane space

inside of mitochondrion

Figure 15.7 (a) Electron micrograph of a mitochondrion in the pancreas of a bat. (b) A three-dimensional drawing of the mitochondrion shown in (a). Note the projections formed by the inner membrane. Embedded in these projections are all the enzymes of the electron transport system and ATP formation. Some enzymes of the Krebs cycle are on the projections; most are in the inside area of the mitochondrion. Such organization makes possible the efficient function of this organelle.

15.6 Glycolysis Begins the Energy-Yielding Process

As a single glucose molecule proceeds through the reactions of glycolysis, three important things happen. First, the carbon chain is broken into two pieces. Second, some ATP is formed. Third, hydrogen atoms from glucose become available for use in the electron transport system. Figure 15.8 diagrams the steps of glycolysis. Refer to the figure as you read the description that follows.

The glucose molecule is a relatively stable molecule that contains a fairly large amount of chemical energy. To start the energy-releasing process, some energy must be expended, just as energy (in the form of a match) is required to make a fuel burn. The starting energy is provided by the enzymatic transfer of phosphates from two ATP molecules, converting them to ADP, as shown at (a). The resulting molecule enters the next reaction (b) in which the 6-carbon chain is split into two 3-carbon molecules. In the remaining reactions of glycolysis (c), the 3-carbon molecules are rearranged to form two molecules of pyruvic acid. No further energy is expended in these rearrangements, each of which is catalyzed by a specific enzyme. Instead, these reactions yield energy.

Four ATP molecules are produced in glycolysis, which is a net gain of two ATP molecules. In addition, the hydrogen atoms released during the

Figure 15.8 In glycolysis, glucose breaks down in small steps to two 3-carbon molecules of pyruvic acid. ATP and NADH form in these reactions, which involve many rearrangements.

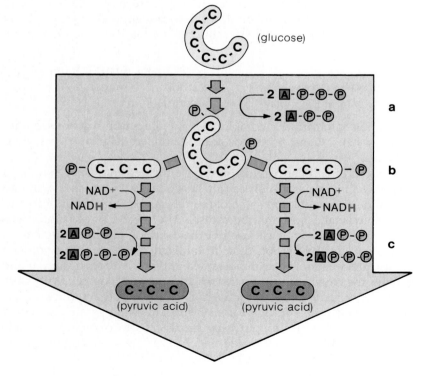

reactions combine with a hydrogen-carrier molecule called **NAD⁺**, or nicotinamide adenine dinucleotide (nik uh TEE nuh myd AD uh neen DY NOO klee oh tyd). A carrier molecule transfers substances from one reaction to another, much as a mail carrier transfers letters from the post office to a person's home. The molecule that is formed when NAD⁺ picks up a hydrogen molecule is NADH. NADH then transfers hydrogen atoms (electrons and protons) from the reactions of glycolysis to the electron transport system.

The end products of glycolysis are two molecules of pyruvic acid, two ATP molecules, and two NADH molecules. Most of the chemical energy in the original glucose molecule is retained in the pyruvic acid molecules, which are transported into the mitochondrion. There, a complex enzyme system converts each pyruvic acid molecule into a molecule of acetic acid, a 2-carbon compound. Each acetic acid molecule combines with a carrier molecule, **CoA** (coenzyme A), forming acetyl CoA. CoA delivers the acetyl group to the Krebs cycle, where respiration continues. Figure 15.9 shows these reactions, which also release two molecules of carbon dioxide and two hydrogen atoms that combine with NAD⁺, forming NADH.

15.7 The Krebs Cycle Completes the Breakdown of Glucose

Figure 15.10 illustrates the second stage of cellular respiration, the Krebs cycle. Study the diagram as you read the description of this stage, which takes place in the mitochondrion.

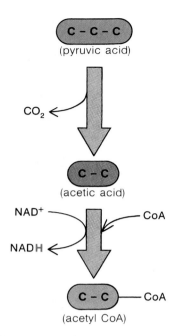

Figure 15.9 The conversion of pyruvic acid to acetyl CoA in a mitochondrion. This step prepares the pyruvic acid formed during glycolysis for entrance into the Krebs cycle.

Figure 15.10 The reactions of the Krebs cycle complete the breakdown of glucose to carbon dioxide. In addition, ATP and NADH form in these reactions. Two molecules of acetyl CoA enter the cycle for each molecule of glucose.

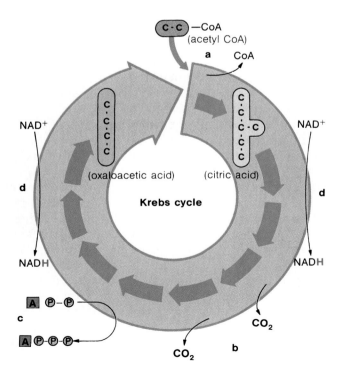

The acetyl CoA formed after glycolysis enters the cycle at (a). At this point, an enzyme combines the acetyl group with a 4-carbon compound (oxaloacetic acid) and releases CoA. The 6-carbon compound that results is citric acid. Citric acid then undergoes many rearrangements, each catalyzed by a specific enzyme. These rearrangements release the remaining two carbons of glucose as carbon dioxide (b). During one of the rearrangements, (c), enough energy is released to combine one phosphate with ADP, forming a molecule of ATP. Many hydrogen atoms also are released in the Krebs cycle. These are picked up by NAD^+ or another carrier (d) and transferred to the electron transport system. Eventually oxaloacetic acid is regenerated, allowing the cycle to continue.

The Krebs cycle completes the breakdown of the carbon chain that makes up glucose. As a result of glycolysis and the Krebs cycle, one 6-carbon glucose molecule has been broken down to six 1-carbon molecules: the carbon dioxide molecules you exhale. These reactions often are called the carbon pathway. The direct energy gain has been two ATP molecules produced in glycolysis and another two ATP molecules formed during two turns of the Krebs cycle (one turn for each molecule of pyruvic acid formed from one glucose molecule in glycolysis). Most important, both processes have released hydrogen atoms that are transported to the electron transport system by carrier molecules. The complete aerobic respiration of 1 molecule of glucose to 6 molecules of carbon dioxide can produce up to 38 molecules of ATP. Stage three of respiration, the electron transport system, requires oxygen to produce most of that ATP.

15.8 The Electron Transport System Packages Energy from Glucose as ATP

Each electron transport system consists of a series of electron carriers embedded in the inner membrane of a mitochondrion. During the final stage of respiration, illustrated in Figure 15.11, hydrogen atoms are accepted by the electron transport system and separated into electrons and protons. The electron carriers transfer the electrons step by step through the system to a terminal carrier. The terminal carrier combines the electrons with protons and molecular oxygen, forming water. Only this final step of respiration uses oxygen.

As the electrons move from one carrier to the next (central part of the arrow in Figure 15.11), they release energy. Some of the energy is used to transport protons across the inner membrane of the mitochondrion. The protons become concentrated in the space between the two membranes (right side of the arrow), and then they diffuse through an enzyme complex in the inner membrane, where ATP is synthesized from ADP and phosphate. In this way, much of the energy from the electrons is transferred to ATP.

A maximum of 34 molecules of ATP can be formed in the electron transport system for each molecule of glucose. With the 4 ATP molecules formed during glycolysis and the Krebs cycle, a total of 38 ATP molecules can be formed in aerobic respiration. That represents approximately 44 percent of the energy available in a molecule of glucose, a remarkable efficiency. In comparison, an automobile engine converts only about 25 percent of the energy in gasoline into a usable form.

Figure 15.11 The electron transport system consists of a series of enzymes in the inner mitochondrial membrane through which electrons are transferred. Energy released from electron flow is used to build up a high concentration of protons in the intermembrane space. As protons move down their concentration gradient, they pass through an enzyme complex in the membrane that synthesizes ATP. Most of the cell's ATP is made in this way.

15.9 The Krebs Cycle Plays a Central Role in Cellular Metabolism

Carbohydrates, fats, and proteins all may be used as sources of energy in the process of respiration. Breakdown products of all three eventually enter the Krebs cycle. Carbohydrates can enter the carbon pathway as glucose or as any 3-carbon sugar. Fats first are separated into glycerol and fatty acids. Glycerol is converted into one of the 3-carbon intermediates of glycolysis before entering the carbon pathway. Fatty acids first are broken down to acetic acid and form acetyl CoA in the mitochondria. The acetyl group then is decomposed in the Krebs cycle and gives off two molecules of carbon dioxide.

When proteins are used in respiration, they first are decomposed to amino acids. The amino group then is removed, and the remaining carbon skeletons may be broken down to the same 2-, 3-, 4-, and 5-carbon molecules formed in glycolysis and the Krebs cycle. These compounds then enter the carbon pathway at various points and are decomposed by the Krebs cycle, giving off carbon dioxide.

Figure 15.12 Respiratory and biosynthesis pathways. The Krebs cycle plays a central role in both.

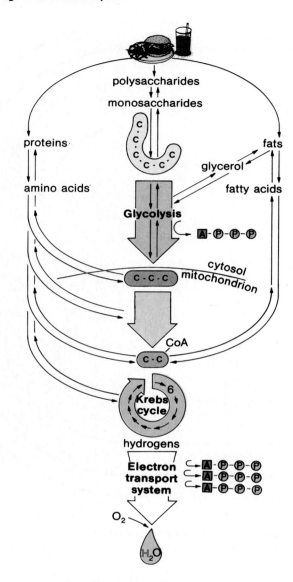

As Figure 15.12 shows, the Krebs cycle plays a central role in the carbon pathway. The intermediate compounds of the Krebs cycle and glycolysis also provide carbon skeletons that serve as building blocks for the synthesis of other compounds. These biosynthesis (by oh SIN thuh sis) reactions provide the cell with enzymes and all other materials needed for maintenance, cell repair, and growth.

INVESTIGATION 15.2 The Marathon

A marathon covers a distance of 42 195 meters. Athletes train for years to build up enough endurance to run this grueling race. Some of the many factors involved in running a marathon are diet, conditioning, race strategy (when to run fast or slow), and weather. Internal body conditions, such as body fluid levels, levels of glycogen in the liver, levels of glucose and lactic acid in the blood, heart rate, and breathing rate, are

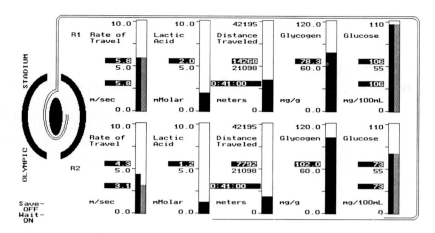

Figure 15.13   Appearance of the computer screen for two marathon runners, one across the top, the other across the bottom. From left to right, the bars show the rate of travel (requested, a solid bar; attained, a hatched bar), lactic acid, distance traveled, glycogen, and glucose (requested in solid; attained in hatching).

extremely important. These internal conditions are regulated by homeostatic mechanisms and vary from one person to another.

This investigation uses the LEAP-System™ to provide a computer simulation of a marathon. The simulation considers only a few of the factors involved. Figure 15.13 shows what appears on the computer screen. The marathon starts and ends in the "stadium" and is "run" on a track around the perimeter of the screen. The distance to the finish line is the same for all lanes of the track. Using the dials, you can control two factors for each runner—the rate of travel and the level of blood glucose. How fast a runner can go (rate of travel) depends on how much glucose and lactic acid are in the blood. The level of lactic acid is related to the rate of travel, and the level of glucose depends on the rate of travel and the amount of glycogen stored in the liver. Your task is to decide on a race strategy, based on regulating the rate of travel and blood glucose levels, that will result in the best possible marathon time. A good time is 2:40; an excellent time is 2:32. If you run the race in less than 2:30, please share your strategy with BSCS. Instructions for use of the LEAP-System are provided in the LEAP-System BSCS Biology Lab Pac 2.

Materials (per team of 3)

Apple IIe, Apple IIGS, Macintosh, or
   IBM computer
LEAP-System™
2 dials

Procedure
PART A   Trial Run

1. Decide which team member will control the runner's rate of travel (dial 1); which one will control the level of blood glucose (dial 2); and which one will watch the lactic acid and glycogen bars and advise the dial operators. The three of you must work together to achieve a good race time while keeping the glycogen, glucose, and lactic acid under control.

2. *For this trial run, the teams at each computer should work with one runner only.* Use the LEAP-System "run" procedure to select **MARTHNC1.SPU** for one runner.

3. Follow the prompt on the screen to start the marathon. Experiment with different levels of the variables to determine the effect on the time of the race. You will need to cooperate with your partners to regulate the optimum levels for blood glucose and rate of travel. (Normal blood glucose range is 80 to 120 mL/100 mL blood.) Try several combinations of settings to see what happens as the variables change. Record these settings in your data book.

4. Use the LEAP-System "reinitialize" procedure to restart the race. Repeat as often as time permits, recording the settings for each run.

PART B   Planning the Race

5. As a class, review the relationships that exist between glucose, glycogen, lactic acid, and exercise.

6. Decide on several race strategies that account for the level of blood glucose and the rate of travel at the beginning of the race, at various points in the race, and for the final kilometer. What assumptions are you making in your strategy? If, for example, you sprint the first 400 meters, what are you assuming is the effect of the sprint? Record your assumptions in your data book.

7. In your data book, write "if/then" hypotheses that include the strategies and assumptions from Step 6. The "if" part of each hypothesis should contain the conditions specified in your strategies. The "then" part of each hypothesis should be your predicted outcome, assuming your conditions are correct. For example, "If lactic acid rises to 8.0 mM, then rate of travel will slow to 3.5 m/sec."

PART C   The Marathon

8. Start the race as in Step 2, but select **MARTHNC2.SPU** for two runners or **MARTHNC3.SPU** for three runners. Follow your strategies carefully. Record any changes in strategy made during the race. Record the results of the race in your data book. Include your finishing time if you finished, how far you ran, and so forth.

9. Based on the results in Step 8, decide with your partners on a strategy that will improve your runner's time in the marathon. Develop a hypothesis about how your runner's time could be improved and by how much. Record your hypothesis in your data book.

10. Use your new strategies and rerun the race.

Discussion

1. Compare how your strategies turned out. Were the prediction parts of your hypotheses correct? What changes did you make in the second trip?

2. Compare your strategies with those of other teams in the class. How were the strategies similar? How were they different?

3. Describe the strategy that produced the best running time among all the runners. Tell why it was better than another strategy that was used.

4. What effect did the rate of travel have on your runner's performance?

5. What is glycogen? Where is it formed? How and when is it utilized? Why is glycogen supply important for a marathon runner?

6. What effect did the level of glycogen have on your runner's marathon performance?

7. What is lactic acid and when is it formed? What effect does an increasing level of lactic acid have on a marathon runner's performance?

8. What is glucose? How is it utilized in exercise? How is the supply replenished?

9. Describe how cooperation between the two controlled variables influences the outcome of the marathon.

10. Marathon runners frequently "hit the wall" late in a race and have difficulty continuing. Based on your experience with this simulation, what might produce the effect of "the wall"?

11. Why is carbohydrate loading important before a long race such as a marathon?

12. What important factors for a marathon runner were ignored in this simulation?

15.10 Energy Releasing Processes Are Essential to Life

Without the presence of oxygen, the entire process of cellular respiration stops because all the electron carrier molecules become "loaded" and cannot transfer their loads. In fact, as the level of oxygen decreases, the process slows down. Oxygen is the final acceptor of electrons. If enough oxygen is not present, the electron flow stops, and the electron transport system backs up like a plugged drain. NADH is unable to transfer the hydrogen it is carrying, and therefore, it is unable to pick up more hydrogens. As a result, the Krebs cycle also stops.

When muscles are severely taxed, as they are during strenuous exercise, for example, the lungs and circulatory system may not be able to meet the oxygen demands of the muscle cells. If this condition occurs, the cells carry out **fermentation** (fer men TAY shun). Glycolysis continues, producing two ATP molecules, pyruvic acid, and NADH. The pyruvic acid does not proceed to the Krebs cycle, however. Instead, two hydrogen atoms are transferred from NADH to the pyruvic acid. This converts the pyruvic acid to the 3-carbon molecule lactic acid (Figure 15.14). Thus NAD$^+$ is freed to pick up more hydrogen, and glycolysis can continue, although it provides only limited amounts of ATP. The lactic acid accumulates in the muscles, eventually reducing the ability of muscle fibers to contract and causing muscle fatigue.

To illustrate the vital nature of these processes, consider what happens when they are stopped—for example, by the action of cyanide, one of the fastest acting poisons. Cyanide combines chemically with the final carrier in the electron transport system and blocks the production of ATP by the mitochondria. Without ATP, cellular activities in nerve cells and other cells cannot continue, and a person becomes unconscious within three to six minutes.

The hamburger you had for lunch may serve one of two purposes. First, it can provide usable energy for life functions. Second, the breakdown of the hamburger can provide molecules to be reassembled into larger molecules that build or repair cells in the body. This second process, known as biosynthesis, is dependent on the presence of appropriate building materials. These include the carbon skeletons formed in the Krebs cycle as well as various elements and vitamins that must be obtained in the diet. The next section examines some of the requirements for biosynthesis and some of the ways that today's environment may make it difficult to obtain these requirements.

pyruvic acid (from glycolysis)   lactic acid

Figure 15.14 Conversion of pyruvic acid to lactic acid in muscles. This reaction, in which NADH also is converted to NAD$^+$, enables glycolysis to continue, supplying energy in the form of ATP to the muscle cells.

Concept Review

1. Why can food be considered a fuel?
2. Compare the release of energy in burning and the release of energy in cells.
3. What role does glucose play in cellular respiration?
4. How is oxygen involved in cellular respiration?
5. How are fats and proteins used in cellular respiration?
6. In what ways does the Krebs cycle play a central role in the process of cellular metabolism?

Guidepost: What nutrients do humans need for optimal growth and health?

Nutrition

15.11   Dietary Habits Affect Health

During the approximately 2 million years the human genus has existed, the human digestive system has evolved in accordance with a dietary selection that remained fairly unchanged until this century. From the available evidence, early diets were high in fiber and vegetable matter and low in fat, sugar, and salt. During the last 50 years, however, diets have become high in fat, refined sugar, and salt and low in fiber. During these 50 years, there also has been a rise in disabilities and illnesses that may be linked to new dietary habits.

The function of the digestive system is to break down large molecules and to absorb the resulting smaller molecules that can meet the needs of the body's cells. The digestive system can use only the materials supplied to it, however. If you decided to have a hamburger, fries, and a soft drink for lunch, will this meal provide the necessary nutrients for your body to continue growing and functioning? Research has indicated that eating too much processed food, eating too much saturated fat (animal fats, tropical oils, and cottonseed oil), not eating enough complex carbohydrates (starches), and eating more kcals than are needed or expended may damage the human body.

Although Americans now consume about 3 percent fewer kcals than they did in 1910, they get so little exercise that two out of every five are overweight. Furthermore, because their intake of refined sugars and animal fats has increased, some overweight individuals actually are malnourished due to the absence of essential nutrients in their diets. The average American takes in more food than his or her body requires but even so does not ingest the proper amounts of basic and essential nutrients.

Does your diet supply you with all the nutrients you need? Look at Table 15.3, an analysis of a fast-food lunch. When he ate that lunch, a male between the ages of 15 and 18 years would have consumed 34.5 percent of his daily recommended kcal allowance. A female in the same age group, however, would have taken in 46 percent of her kcals for the day. These kcal recommendations are based on activities for the average high-school student and include walking to school and some daily physical activity, such as physical education class. A student on one of the athletic teams who works out regularly would have an increased allowance. A less active student would need to decrease the number of kcals to avoid gaining weight.

Of the carbohydrates in the fast-food lunch, only 28 g, or 23 percent, are complex carbohydrates. The 92 g of refined carbohydrates represent 77 percent of the total carbohydrate intake, well above the recommended 10 percent. All 378 kcals of fats eaten probably are saturated fats. Current recommendations state that only 10 percent of the total daily intake of fats should be saturated fats.

In 1988, the first *Surgeon General's Report on Nutrition and Health* singled out saturated fat as the number one dietary problem in the United States. Five of the 10 leading causes of death in the United States—cardiovascular disease, certain types of cancer, stroke, diabetes mellitus, and atherosclerosis—are diseases in which the diet plays a role. According to the report, less fat and more fiber in the diet should reduce the risk of contracting many of these conditions.

TABLE 15.3 Analysis of a Fast Food Lunch

	Kcals	Protein (grams)	Carbohydrates (grams)	Fats (grams)
Hamburger	606	29	51	32
Fries	215	3	28	10
Soft drink	145	0	41	0
Totals:	966	32	120	42

Age	Sex	Recommended Daily Kcal Intake	Kcals in This Meal		Percentage of Total Kcals
15–18	Male	2800	966	=	34.5
15–18	Female	2100	966	=	46.0

Nutrient (energy content)	Grams × Kcal/g	=	Total Kcals	Actual Percentage of Daily Kcal Intake	Recommended Percentage of Daily Kcal Intake
Protein (4 kcal/g)	32 × 4	=	128	4.6 (male) 6.1 (female)	12 12
Carbohydrates (3.8 kcal/g)	120 × 3.8	=	456	17.1 (male) 22.9 (female)	58* 58*
Fats (9.1 kcal/g)	42 × 9.1	=	382	13.5 (male) 18.0 (female)	30** 30**

*Only 10% of the total carbohydrate intake should be refined carbohydrates—77% of the carbohydrate in this lunch is refined.
**Only 10% of the total fat intake should be saturated fats—all of the fat in this lunch is saturated.

The report recommends that most people should:

1. Reduce their intake of fat, especially saturated fat, by eating leaner meats and fish, and more whole grains, vegetables, fruits, and low-fat dairy products.
2. Get sufficient exercise to burn off the calories they consume.
3. Eat more whole grains and dried beans to increase the amount of fiber in their diet.
4. Limit their intake of sodium and sugar.
5. Limit their consumption of alcohol; pregnant women should not drink at all.

Many questions about diet and nutrition remain to be answered through research. Until more information becomes available, the most reasonable recommendation is to eat a variety of foods in moderation and to reduce intake of processed and convenience foods.

INVESTIGATION 15.3 Assessing Risk for Cardiovascular Disease

Several factors are known to contribute to cardiovascular disease, a leading cause of death in the United States. We can control some of these factors but not others. In this investigation, you will assess your chances of acquiring cardiovascular disease. Once you know your chances, if you wish, you can modify your life-style to reduce the risks of cardiovascular problems.

Materials (per team of 2)

centimeter tape measure   sphygmomanometer (optional)
bathroom scale   calculator

Procedure
PART A   Calculating Your Ideal Weight

1. Work with your partner to make the measurements. Measure your wrists in centimeters at their *smallest* circumferences. Add both measurements and divide them by 2 to get an average.

$$\frac{\text{left wrist} + \text{right wrist}}{2} = \text{average}$$

2. Measure your forearms and calves at their *largest* circumferences. Measure your ankles over the ankle bones. Calculate an average for each as in Step 1.

3. Add the four averages and divide by 17.07 for males or 16.89 for females.

$$\frac{\text{wrist} + \text{forearm} + \text{calves} + \text{ankles}}{17.07 \text{ or } 16.89} = \text{quotient}$$

4. Square the quotient.

$$\text{quotient} \times \text{quotient} = \underline{\hspace{1cm}}$$

5. Measure your height (in centimeters) without shoes.

6. Multiply your answer in Step 4 by your height.

$$\text{height} \times \text{quotient}^2 = \underline{\hspace{1cm}}$$

7. Multiply the answer in Step 6 by 0.0111 to obtain your "ideal" mass in kilograms. To convert this value to weight in pounds, multiply by 2.2.

$$\text{height} \times \text{quotient}^2 \times 0.0111 \times 2.2 = \text{ideal weight in pounds}$$

8. Determine your real weight by weighing yourself on a bathroom scale. Compare your real weight with your ideal weight. Is your real weight over, under, or equal to your ideal weight?

9. Find the difference between your real weight and ideal weight.

PART B   A Self-Check

10. Take the self-check below. If a sphygmomanometer is not available, omit item 6.

Self-Check

Directions: Read each item and score the appropriate number in the space to the right of the number.

1. _____ Statistics show that males are more likely to suffer heart attacks than females. If you are (a) male, score 1 point, or (b) female, score 0.

2. _____ Heredity can influence your chances of heart disease. If one or more of your parents, grandparents, or siblings have suffered a heart attack, you may have an inherited tendency toward this condition. If one or more of your parents, grandparents, or siblings have suffered a heart attack (a) before age 60, score 12 points, (b) after age 60, score 6 points, and (c) neither (a) nor (b), score 0.

3. _____ A person with diabetes is likely to build up fatty deposits in the arteries. If you have (a) diabetes and are now taking insulin or pills, score 10 points, (b) diabetes in your immediate family, score 5 points, or (c) no diabetes or can control it with diet, score 0.

4. _____ Smoking has been shown to contribute to cardiovascular disease. Substances inhaled during smoking damage the lining of blood vessels. If you (a) smoke 2 packs or more per day, score 10 points, (b) smoke 1–2 packs per day or quit less than a year ago, score 6 points, (c) smoke less than 1 pack per day or quit 1–10 years ago, score 3 points, (d) never smoked, score 0.

5. _____ High amounts of cholesterol in the diet can clog or narrow the arteries. This places added stress on the heart and arteries. If you eat (a) 1 serving of red meat per day, more than 7 eggs per week, use butter, whole milk, and cheese daily, score 8 points, (b) red meat 5–6 times a week, 4–7 eggs per week, use margarine, low-fat dairy products, and some cheese, score 4 points, (c) poultry, fish and little or no red meat, 3 or fewer eggs per week, use margarine and skim milk, score 0.

6. _____ High blood pressure increases the heart's work and places wear and tear on blood vessels. If you are able to measure your blood pressure, score this item. Otherwise, omit score. If your blood pressure at rest measures (a) 160/105, score 8 points, (b) between 160/105 and 140/90, score 4 points, (c) less than 140/90, score 0.

7. _____ Overweight people run a higher risk of heart disease than those not overweight. (Refer to the calculation of the difference between your ideal and real weight in Step 9, Part A.) If you are (a) 25 pounds overweight, score 4 points, (b) 10–25 pounds overweight, score 2 points, (c) less than 10 pounds overweight, score 0.

8. _____ Aerobic exercise is that type of activity that temporarily increases your heart rate, stimulates sweat production, and causes deep breathing (jogging, bicycling, swimming, and other similar activities). If you engage in aerobic exercise (a) less than once per week, score 2 points, (b) 1–2 times per week, score 1 point, (c) 3 or more times per week, score 0.

_____ Total Score

11. After completing the self-check, use the self-check score interpretation to determine your risk of cardiovascular disease.

Self-Check Score Interpretation	
Total Score	**Risk of Heart Attack, Stroke, or Cardiovascular Disease**
26 pts. or more	high risk
25–14 pts.	medium risk
13 pts. or less	no risk

Use the following information if item 6 (blood pressure analysis) was omitted.

Self-Check Score Interpretation	
Total Score	**Risk of Heart Attack, Stroke, or Cardiovascular Disease**
22 pts. or more	high risk
21–12 pts.	medium risk
11 pts. or less	low risk

Discussion

In Part A, you calculated your ideal weight. This weight is only an approximation of what your body weight should be. It is normal for your real weight to vary 1 to 10 pounds from your ideal weight. Part B in this investigation allowed you to assess your chances of acquiring cardiovascular disease. The information obtained in Part B is only a guide, but it can be useful in examining your life-style. Using this information, answer the following questions:

1. Based on your self-check score, do you appear to have a high, medium, or low risk of cardiovascular disease?
2. What aspects of your life that could increase your chances of cardiovascular disease are you incapable of changing?
3. What modifications, if any, can you make in your life-style to reduce your chances of having cardiovascular disease?

For Further Investigation

Does your daily diet have the proper number of kcals to maintain your ideal weight? To find out, keep a record of everything you eat for five days. After obtaining an average of your kcal intake for the five days, determine whether your average kcal intake is appropriate for your weight. A quick calculation to determine your approximate kcal needs per day follows:

$$\text{ideal weight} \times 15 = \text{kcal needs per day.}$$

15.12 Lipids Are Concentrated Sources of Food Energy

Your body needs several types of nutrients. Carbohydrates and fats are the primary sources of energy. Proteins supply the body with building materials and enzymes. They also may be used to provide energy. Vitamins and several elements are essential substances that form cofactors in cellular respiration and other cellular activities. As cofactors, they enhance the activity of many enzymes, and they help to release the energy of carbohydrates, fats, and proteins and to maintain normal body function. The tables entitled *Elements Important in Humans* and *Vitamins Important in Humans* in Appendix 3 detail the major functions of the vitamins and essential elements. Figure 15.15 shows several foods that are high in vitamins.

Lipids, or fats, are the most concentrated source of food energy available. They release approximately 9 kcals for every gram of fat. Proteins and carbohydrates release only 4 kcals per gram. Lipids aid in the metabolism of cholesterol, a lipid associated with animal products and linked to heart disease, and they aid in the absorption of vitamins A, D, E, and K. They are used structurally in cellular membranes and in the production of hormones and hormonelike substances. Lipids also supply essential fatty acids. Essential nutrients, such as fatty acids, are those the body cannot produce at all or cannot produce in adequate amounts and, therefore, must be obtained from ingested foods.

Although fats are important nutritionally, most Americans today consume far too many and the wrong type. Two types of fats are of particular concern: saturated fats, which include animal fats such as lard and butter,

Figure 15.15 Examples of foods high in vitamins.

and unsaturated fats, which include most vegetable oils (except cottonseed oil and tropical oils such as palm oil and coconut oil). Fats containing fatty acids that do not have double carbon bonds (two carbon atoms sharing two chemical bonds) are known as saturated fats. Saturated fats tend to be solid when left at room temperature. Fats that contain one or more double carbon bonds in their fatty acids tend to be liquid at room temperature. These fats are known as monounsaturated or polyunsaturated fats, depending on the number of double carbon bonds present. (Review Figure 4.17, page 91.)

Reducing fat intake, particularly the intake of cholesterol and saturated fat, is important for individuals because of the apparent link between high blood cholesterol levels and cardiovascular disease (disease of the heart and blood vessels). During the past 30 years, scientific research has indicated that a high dietary intake of both cholesterol and saturated fat tends to raise blood cholesterol level in individuals. On the other hand, a diet that is rich in polyunsaturated fat may lower blood cholesterol levels for some people. In the disease atherosclerosis (ATH uh roh skleh roh sis), excess cholesterol is deposited on the inner lining of the arteries. The bulky fat streaks are called plaques. An accumulation of plaques, like that shown in Figure 15.16, can inhibit the flow of blood and increase blood pressure. Eventually a clot may form, obstruct the artery, and cause a heart attack or stroke. Table 15.4 lists the factors that increase the risk of heart disease. Data show that half of all the deaths in the United States are caused by this type of disease.

In the last 30 years, autopsies of individuals between the ages of 18 and 30 have revealed significant development of cholesterol deposits in the arteries. A recent study has shown fatty streaks in arteries of children as young as two, indicating that atherosclerosis may begin in childhood. As a result

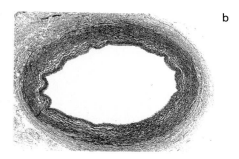

Figure 15.16 Atherosclerosis occurs as a fatty deposit on the walls of an artery (a). Note the small opening that remains. A healthy artery with large opening is shown in (b).

TABLE 15.4 Some Risk Factors Associated with Heart Disease

Risk Factors	Physiological Result	End Result
eating & drinking too much / not exercising enough	overweight	
high total fat consumption / high saturated fat consumption / low poly-unsat.: sat. fat ratio / high cholesterol consumption	elevated blood cholesterol	higher risk of heart disease
no exercise / overweight	elevated blood pressure	
high fat diet / diabetes / smoking	accelerates the atherosclerotic process	

Figure 15.17 Cholesterol content of various foods, per 250 mL (1 cup) serving.

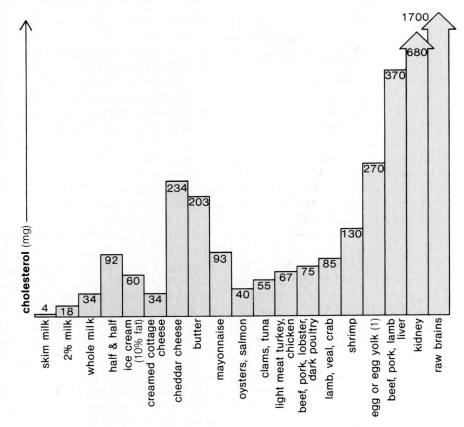

of these studies, scientists think it is important to adopt moderate-fat and moderate-cholesterol diets early in life, before life-style patterns have been too firmly established. Figure 15.17 shows the cholesterol content of various foods.

15.13 Carbohydrates Provide Energy, Nutrients, and Fiber

Fruits, vegetables, and whole grains, such as those shown in Figure 15.18, are important because they provide the essential elements and vitamins necessary for good health, as well as carbohydrates, the body's source of energy. Refined sugars, which contain no other nutrients, provide only kcals—sometimes called "empty" kcals. Reports indicate that three of every five kcals ingested by Americans come from fats or added sugars. In the American diet, processed and convenience foods, which have a high refined sugar content, are replacing more nutritious foods that supply vitamins and essential elements. Refined and processed sugars are added to almost every type of processed and convenience food. The food industry has many terms for the sugars and sweeteners it uses, such as sucrose, raw sugar, turbinado sugar, brown sugar, total invert sugar, corn syrups, honey, fructose, levulose, dextrose, lactose, and others. All of these, however, can

Cardiovascular Diseases

Cardiovascular diseases—diseases of the heart and blood vessels—are the leading cause of death in the United States. About one person in five has some form of cardiovascular disease. Heart attacks are the main cause of cardiovascular deaths and account for about one-fifth of all deaths. A heart attack occurs when some area of the heart muscle does not receive enough blood. Sometimes a blood clot forms in the blood vessels of the heart, causing a heart attack. The clot blocks the passage of blood through those vessels, so blood does not reach a portion of the heart muscle. Heart attacks also may result if a vessel is blocked by deposits of materials. If the damaged area of the heart is fairly small, recovery can occur in the manner illustrated in the figure. Sometimes the amount of damage is slight and thus difficult to detect. Because they reveal abnormalities in the timing of the heart contractions, electrocardiograms are useful for evaluating the overall condition of the heart. Abnormalities in the timing of heart contractions may be associated with damaged heart tissue.

Strokes are caused by an interference with the blood supply to the brain. They often occur when a blocked vessel bursts in the brain. Strokes also may be associated with a blood clot or a coagulation of other materials in a blood vessel. The overall effects of a stroke depend on how severe the brain damage is and where the stroke occurs in the brain.

Atherosclerosis, which often is a factor in heart attacks and strokes, is a buildup of deposits within the arteries. The accumulation of fatty materials and deposits of cholesterol or other cellular debris within the arteries can impair the proper function of the arteries. When this condition is severe, the arteries can no longer expand and contract properly, and the blood moves through them with difficulty. The accumulation of cholesterol is thought to be the prime contributor to atherosclerosis. Diets low in cholesterol frequently are prescribed for this condition.

Arteriosclerosis, or hardening of the arteries, occurs when calcium is deposited in arterial walls. This usually occurs when atherosclerosis is severe. When atherosclerosis and arteriosclerosis occur, the flow of blood through the arteries is restricted, and the arteries cannot expand enough to handle the volume of blood pumped out by the heart. Thus, the heart must work harder.

At least three factors are linked to cardiovascular disease: smoking, high blood pressure, and high blood cholesterol. The current advice for reducing cardiovascular risk is simple: Do not smoke; have your blood pressure checked and regulated, if necessary; and follow a diet that lowers the levels of cholesterol in your blood. Finally, try to be aware of new information as it becomes available. Research on nutrition and cardiovascular disease is very active, and new data often result in changed dietary and activity recommendations. Look for valid reports based on controlled experiments and avoid popularized claims of quick fixes and miracle remedies.

thrombus

scar tissue

a. The hearts own system of blood vessels supplies the muscular walls of this powerful pump with food and oxygen. Sometimes blood flow may be blocked (circled area). The muscles are damaged by lack of oxygen and foods.

b. With smaller blood vessels taking over from the blocked artery, scar tissue begins to form in the dead muscle cells.

c. If the heart attack is not too severe, the scar tissue will shrink over several months—eventually normal heart function may be established.

Events in a heart attack and recovery.

Figure 15.18   A variety of high-fiber foods.

Figure 15.19   Foods rich in protein come from a variety of sources.

contribute significantly to obesity and to the national $10 billion-a-year dental bill for cavities. Because of the current labeling laws, sometimes it is impossible for consumers to determine how much refined or processed sugar they actually consume.

Fruits, vegetables, and whole grains also are sources of dietary fiber, or roughage, as well as vitamins, essential elements, and carbohydrates. Dietary fiber consists of cellulose and other carbohydrates that are abundant in plants and in unprocessed plant foods. Humans lack the enzymes necessary to digest these substances. Instead, fiber absorbs water rather like a sponge, causing bulkier feces that move more quickly and easily through the large intestine.

The current emphasis on fiber intake began with some observations made by a number of British physicians. They noted the reduced incidence of diseases such as cancer of the colon and atherosclerosis in certain African rural villages. The diet of these villagers is low in processed food and high in fiber. When these villagers moved to urban areas and began eating the low fiber, high fat diet of Western cultures, they suffered the same diseases as longtime urban dwellers. The physicians proposed the hypothesis that a high fiber diet would protect against the diseases of Western cultures. This hypothesis is still being tested, and caution is advised in adding large amounts of fiber to the diet. In large amounts, bran can interfere with absorption of calcium and other essential elements.

15.14   Protein Provides the Body Framework

Protein is needed daily for the repair and maintenance of body tissues, as well as for normal growth and development. Protein also supplies the amino acids needed to make the type of protein required by the body. Figure 15.19 shows some foods rich in protein. Humans can synthesize only 8 to 10 of the 20 amino acids they need. The others—called essential amino acids—must be obtained from foods. Eggs, milk, fish, soybeans, cheese, meat, and poultry provide complete proteins—proteins that contain all the essential amino acids in the proportions needed by the body. Although grains, nuts, and seeds are good sources of protein, they usually do not provide all the essential amino acids or the required amounts or proportions. It is possible, however, to obtain complete proteins by combining these foods so that a missing amino acid in one is provided in the other. Grains and legumes (beans, peas, and peanuts), and legumes and nuts are examples of such combinations. Vegetarian diets should contain a balance of soybean products, leafy dark green vegetables, and some dairy products to supply the necessary balance of amino acids.

Most Americans exceed the protein requirement of 0.5 to 0.8 per kilogram of body weight per day. Contrary to popular claims, consuming excess protein does not result in "muscle building." Instead, the protein is respired to provide energy or converted to fat and stored.

15.15 Eating Disorders Are Widespread

Dieting has become almost epidemic in the United States, but in a growing number of cases, dieting becomes an obsession that leads to a potentially fatal eating disorder called anorexia nervosa (an oh REX ee uh ner VOH suh). A person suffering from anorexia literally starves. Bulimia (buh LEE mee uh), another eating disorder, also is increasing in the United States. A bulimic individual indulges in eating binges, usually involving large quantities of carbohydrate-rich "junk" food, and then induces vomiting or uses laxatives in an effort to purge the body of the food and thus not gain weight.

Since 1980, these disorders have increased at an unparalleled rate. For reasons that are poorly understood, 90 to 95 percent of the individuals suffering from eating disorders are female. Researchers now estimate that 5 to 10 percent of the adolescent girls and young women in the United States suffer from eating disorders. As many as one in five female high-school and college students may suffer from bulimia.

People suffering from either of these diseases usually have an irrational fear of being fat, have low self-esteem, and are unable to see themselves as they actually are. Even in advanced cases, where the individuals may weigh only half of their normal weight, they still perceive themselves as fat. The disease can result in the cessation of the menstrual cycle, skin rashes and dry skin, loss of hair and nail quality, dental cavities, and gum diseases. In its extreme form, affected individuals may die from damage to vital organs, heart failure, rupture of the esophagus, or other causes.

Although anorexia and bulimia are believed to be psychiatric disorders, scientists also are searching for physiological explanations for these conditions. Changes in the levels of certain hormones are involved, but whether these changes cause the conditions or result from the conditions has not been established. The mortality rate of these disorders is perhaps the highest of any condition classified as a psychiatric disorder. In treating these conditions, it is important to recognize the shame often associated with the condition as a problem. Professional help should be sought promptly. Both a psychiatrist or psychologist and a physician should be involved in treatment, and the family as a whole must play a strong, supportive role.

Concept Review

1. What are the three major causes of dietary damage to the human body?
2. Describe the role of each of the basic food molecules in normal functioning of the human body.
3. Distinguish between saturated and unsaturated fats. How are these differences important in nutrition?
4. What health problems may be related to excess fat intake?
5. Explain the importance of dietary fiber.
6. What problems are associated with taking in too many or too few kcals?

Summary

Digestion is the process by which food is broken down into small molecules. The process begins with mechanical breakdown in the mouth and continues in the various compartments of the digestive tract. Most of the chemical breakdown occurs in the upper part of the small intestine in the presence of specific digestive enzymes. The end products of digestion are absorbed from the small intestine and delivered to the cells by the blood and lymph. Inside the cell, the food molecules may be stored, respired to obtain usable energy, or used to synthesize other necessary molecules.

In cellular respiration, the energy stored in food molecules is converted to ATP. The reactions of respiration take place in the cytosol and mitochondria and, in addition to providing usable energy in the form of ATP, provide carbon skeletons for biosynthesis. Carbohydrates, especially glucose, are the major source of energy in respiration, but fats and proteins also can be respired. These nutrients, as well as vitamins and essential elements, which are required in small amounts, must be supplied in the diet. Increased intake of processed foods, fats, and refined sugars in the United States and other industrialized nations appears to be related to major health problems such as obesity and heart disease.

Applications

1. How is chemical digestion related to the various types of chemical syntheses carried on by the cells?
2. Research the various types of teeth in humans. How do these different teeth help in the digestion of the food you eat?
3. What changes occur in the pH of the human digestive system as food passes through it?
4. Based on comparative length, why does it make sense that most food is digested and absorbed in the small intestine and not in the large intestine?
5. Why are feces semisolid although digested food in the small intestine is semiliquid?
6. Compare the major events of glycolysis and the Krebs cycle with the electron transport system. Which reaction yields more energy?
7. Explain the central role of the Krebs cycle in both cellular respiration and cellular synthesis reactions.
8. If brain death usually occurs in a person who is without oxygen for 5 minutes, how is it possible for a child to survive for 20 minutes after falling into cold water under the ice of a lake?
9. Why do you lose weight rather than remaining the same weight if you do not eat?

10. A quotation from the Bible says that "all flesh is grass." In terms of the food you eat, what does this mean?

11. Why do we need complete proteins in our diet?

12. In what ways are vitamins and essential elements important to health?

13. Cellulose (fiber) and starch are both carbohydrates, yet humans can digest starch but not cellulose. Why is this, and why are people supposed to eat fiber?

Problems

1. Write a play to illustrate how sugar is completely broken down in cellular respiration. Have members of the class take different roles in the play and then videotape the play.

2. What are some of the factors in today's society that might cause a high-school student to become anorexic or bulimic?

3. People living in different parts of the world have very different diets. Most of the population in Africa, Asia, and Australia cannot drink milk. Investigate why this is so.

4. Suppose, because of the success of the artificial kidney, a biomedical engineer wished to design an artificial digestive system. What major functions would you have to build into the artificial digestive system?

5. What percentage of the calories you eat are fat? It may be difficult for you to measure what you eat, but try to keep a diary of the foods you eat that contain fat. Divide this list into fat or oils from plants and fat from animals. If the labels on processed foods provide nutrition information, record the percentage of fat in the various foods.

6. Many diet plans actually can be harmful. What is the strangest diet plan you have heard about? What is the evidence that is presented to potential dieters about its ability to help a person lose weight?

7. What did our ancestors eat? Use anthropology books to learn about the diet of hunter-gatherers of today and of many thousands of years ago.

8. Compare the diet and health of ethnic people in their native land and after they move to the United States. For instance, Japanese and Japanese-Americans used to eat different foods, and the incidence of diet-related health problems in the two countries was different. Today, many Japanese people have adopted a Western diet, and as a consequence, their health has declined. Read about this problem.

The clotting of human blood, showing both red blood cells and the fibrin network (×10 000). Use a metric ruler to measure the size of one of the red blood cells. Use the ratio of the magnification to this measurement to determine the *actual* size of this red blood cell. Check your answer with a classmate. Approximately how large is the widest opening in the fibrin? Develop a hypothesis based on these observations that explains how blood clots.

Once a meal is eaten, digested, and absorbed through the cells of the digestive tract, the food molecules must be distributed throughout the body. The absorbed food molecules must be delivered, and they must be controlled carefully so the supply of food suits the needs of the cells. Excesses of certain nutrients can be as harmful as shortages. The balance of supply and demand is part of the homeostasis of the body.

This chapter examines how cells are supplied with the building materials and energy provided by food, as well as oxygen and other necessary substances, and how wastes are removed. The chapter also considers the role of the immune system in defending the body from invasion and how the various systems interact to regulate the internal environment of the body.

Circulation

Guidepost: How does the circulatory system distribute materials to all functioning cells?

INVESTIGATION 16.1 Exercise and Pulse Rate

You probably have experienced the sensation of your heart beating strongly inside your chest when you participated in a physical activity such as running, aerobic exercise, or athletic competition. The heartbeat rate increases in response to signals from the nervous and endocrine systems, which are monitoring the entire body. Pulse rate, as measured by using a pulse probe with the LEAP-System™, or by taking the pulse rate manually at the wrist if a LEAP-System is not available, is a measure of how fast the heart is beating. The LEAP-System BSCS Biology Lab Pac 2 includes instructions for use of the system.

 Develop a hypothesis about how exercise (both mild and vigorous) affects heart rate and recovery time. Recovery time is the length of time it takes for the heartbeat to return to the normal resting rate. Design an experiment to test your hypothesis using the materials listed below.

Materials (per team of 2)
Apple IIe, Apple IIGS, Macintosh, or IBM
    computer
LEAP-System™
pulse probe
watch (or clock) with second hand (optional)

Figure 16.1   Taking a pulse with (a) the pulse probe and the LEAP-System or by (b) the wrist method.

Procedure

Use the LEAP-System with a pulse probe (see Figure 16.1a) to carry out the experiment you designed. In your data book, record the pulse rate from the screen. Organize your data in a table, and construct graphs at the end of the experiment to display the data. If a LEAP-System is not available, use the wrist method (see Figure 16.1b). Count the pulse directly, organize your data in a table, and construct graphs to display the data.

Discussion

1. What was the control in your experiment?
2. What label did you put on the *y* axis of your graph? On the *x* axis?
3. What is the influence of exercise on heartbeat rate?
4. Are there any differences in heartbeat rates and recovery times for the different types of exercise among the students in your class? Explain.
5. Are there any differences in heartbeat rates and recovery times between males and females in your class?
6. Note any other conclusions you can make based on your data.

16.1   The Heart Pumps Blood Through a Series of Tubes

Humans, like most animals that have body fluid, have a system of tubes or vessels that circulate fluid and transport materials throughout the body. These tubes or vessels make up the circulatory system, which carries raw materials to the cells and removes wastes from the cells' environment. Human circulation occurs in a closed system. A single, muscular heart with four chambers pumps blood through three types of blood vessels. **Arteries** have thick, muscular walls. They carry blood away from the heart. **Veins** have thin walls with little muscle. They carry blood toward the heart and have valves that prevent backflow. **Capillaries** (KAP ih layr eez) are thin-walled, narrow tubes that connect arteries and veins. Figure 16.2a shows the human circulatory system.

In the seventeenth century, William Harvey, a British physician, showed through a remarkable series of experiments that blood leaves a vertebrate's heart through arteries and returns to the heart through veins. In 1628, Harvey published a paper on the subject. Because microscopes were not yet widely used, Harvey never actually saw blood passing from arteries to veins. Later

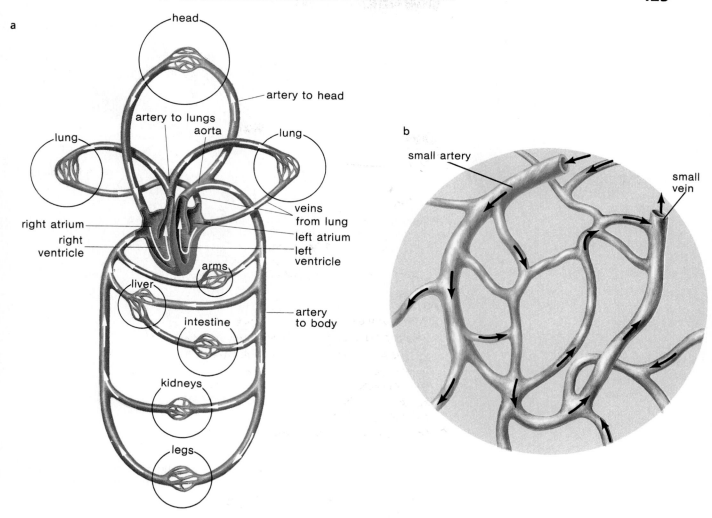

Figure 16.2 (a) A simplified drawing of the human circulatory system. Oxygen-rich blood is shown in red; oxygen-poor blood is shown in blue. (b) A typical capillary bed.

in the century, however, another scientist was able to observe capillaries, like those shown in Figure 16.2b, connecting the arteries and veins. These observations provided further evidence for Harvey's ideas about the circulation of the blood.

In humans, as in all mammals and birds, the heart is separated into right and left sides. Each side of the heart is connected to its own set of veins and arteries. The right side of the heart receives oxygen-poor blood from almost all parts of the body and sends it to the lungs. The left side receives the newly oxygenated blood from the lungs and sends it to all parts of the body. Each side of the heart has two chambers. The top chamber, or **atrium** (AY tree um), forms the upper part of the heart and receives incoming blood from the tissues of the body. The thin-walled atrium bulges with this blood, and as the lower heart muscle relaxes, the blood flows into

a

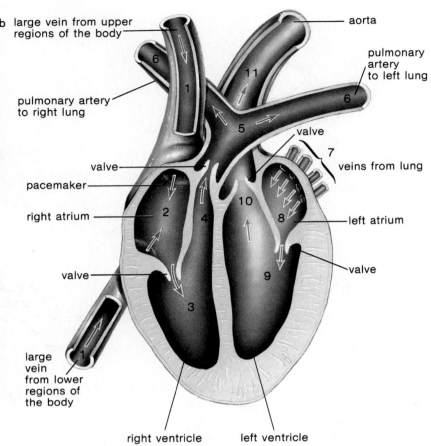

b large vein from upper regions of the body

aorta

pulmonary artery to left lung

pulmonary artery to right lung

valve

veins from lung

valve

pacemaker

right atrium

left atrium

valve

valve

large vein from lower regions of the body

right ventricle   left ventricle

Figure 16.3 (a) Valves of the human heart. (b) A drawing of a cross section through a human heart and the blood vessels leading to and from it. Trace the flow of blood into and out of the heart by following the numbers and arrows.

a second chamber, the thick, muscular **ventricle** (VEN trih kul). Figure 16.3 shows these structures clearly.

On each side, the atrium and the ventricle are separated by valves, flaps of tissue similar to those shown in Figure 16.3a. The structure of these valves prevents the blood from flowing back into the atrium when the ventricle contracts. Ventricles, atria, and associated structures of the heart are shown in cross section in Figure 16.3b. As the ventricle contracts, the muscular wall squeezes the blood, forcing shut the valves between the two chambers and preventing backflow into the atrium. As the blood is pumped out of the heart, pressure forces open the valves at the entrance to the artery, and blood flows through the vessel away from the heart. When the ventricles relax between heartbeats, back pressure from the blood forces the valves shut, thus blocking the artery and preventing blood from flowing back toward the heart. Figure 16.4 summarizes the pumping action of the heart. The contraction of muscles surrounding the veins helps to push blood through the body. Valves in the veins prevent backflow, so the blood flows only toward the heart (see Figure 16.4d).

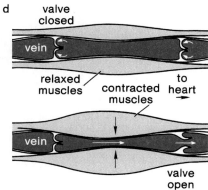

Figure 16.4 The cardiac cycle. Blood enters the atria (a), which contract, forcing blood into the ventricles. The atria relax and fill (b), and the ventricles contract (c), forcing blood into the pulmonary artery and the aorta. Then, the ventricles relax and the atria contract, repeating the cycle. Movement of blood in veins (d) is brought about by pressure from adjacent muscles. Compression forces blood in both directions, but valves keep it from flowing backward and away from the heart.

Capillary walls are only one cell thick. As blood flows through the capillaries, some substances move from the blood through the thin walls of the capillaries into the body tissues. Other substances move from the tissues into the blood. This exchange of materials occurs by diffusion because substances in the tissues and substances in the blood are present in unequal concentrations.

16.2 Plasma Contains Blood Cells and Other Substances

Blood is composed of cells and other substances suspended in **plasma** (PLAZ muh), a clear straw-colored liquid that is 90 percent water and 10 percent dissolved substances. Most of the dissolved substances are proteins called plasma proteins. The other dissolved substances include essential elements, absorbed food molecules (such as single sugars, amino acids, and fatty acids), respiratory gases (oxygen and carbon dioxide), waste products, and regulatory substances (hormones and enzymes). Some of the plasma proteins play important roles in homeostatic functions, such as blood clotting and maintenance of blood osmotic pressure. Other proteins are essential products of the immune system, which protects the body against invasion by organisms or foreign particles.

Red blood cells and white blood cells, the two types of human blood cells, are manufactured in the bone marrow. Red blood cells are specialized for the transport of oxygen. As a red blood cell matures, it loses its nucleus and other cell structures and becomes filled with **hemoglobin** (HEE moh gloh bin), an iron-containing pigment that combines readily with oxygen. Because a human red blood cell lacks a nucleus and because it undergoes a great deal of wear and tear, it lives only 110 to 120 days. After this period of time, the red blood cells are removed from circulation and destroyed in the liver and spleen. Iron from the hemoglobin is salvaged in the liver and used by bone marrow cells to make new red blood cells.

White blood cells, shown in Figure 16.5, play a major role in defending the body against invading pathogens. Although there are several types of

Figure 16.5 Human blood smear showing red blood cells, two large white blood cells, and a platelet (the small particle in the right half of the photo) (X500).

white blood cells that differ in size and function, all white blood cells contain a nucleus. Because they lack hemoglobin, white blood cells are colorless. White blood cells can move about like amoebas, slipping through the thin walls of capillaries and wandering among cells and tissues. Some white blood cells engulf pathogens in the same way as an amoeba engulfs food. Others synthesize antibodies, complex proteins that react with pathogens and other foreign substances.

16.3 Clotting Is an Interaction Between Platelets and Plasma Proteins

Normally, when you have a cut or scrape, the blood at the skin surface hardens, or clots. Blood clotting comes about through a complex sequence of events that involve some 30 proteins called clotting factors, as well as small cell fragments called **platelets** (PLAYT lets). Clotting begins when plasma and platelets come in contact with a rough surface, such as a torn tissue. The platelets gather on the rough surface, become sticky, and attract more platelets, forming a plug that partially seals the wound. They also release substances that act with clotting factors in the plasma to begin a chain of chemical reactions (see Figure 16.6). As a result of these reactions, a substance called prothrombin (proh THROM bin) activator is formed. In the presence of calcium ions, the prothrombin activator catalyzes the conversion of prothrombin, a plasma protein, to thrombin. Acting as an enzyme, thrombin converts the soluble plasma protein fibrinogen (fy BRIN oh jen) to its insoluble form, fibrin (FY brin). The fibrin forms a network of threads that trap platelets and other materials and form the clot (see the chapter opening photograph).

Sometimes a clot may form when there is no injury. For example, a clot may form within a blood vessel, blocking circulation. If the clot blocks one

Figure 16.6 A blood clot is the result of a cascade of enzymatic reactions that end when the soluble protein fibrinogen is converted to insoluble fibrin strands. The fibrin strands provide a network in which platelets are trapped, forming a blood clot (see the chapter opener).

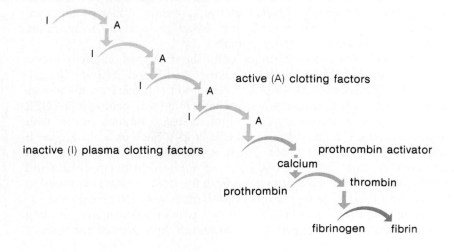

of the arteries supplying blood to the heart, blood flow to that area of the heart is cut off, resulting in a heart attack. A clot blocking an artery in the brain results in one type of stroke. Strokes usually result in paralysis of part of the body and may result in death. (See Chapter 15, Biology Today.)

16.4 Some Cells and Materials Move Between Blood Vessels and Tissue

Normally the water and nutrients in plasma pass through capillary walls to and from body tissues. White blood cells also move freely between blood vessels and tissues. Red blood cells do not leave the blood vessels, and only small amounts of plasma proteins are found in tissues. Thus, the fluid that bathes tissue cells is lower in protein content than the blood plasma. This difference in concentration is important in regulating the exchange of materials, illustrated in Figure 16.7, between the blood and the cells.

Some of the fluid that bathes the tissues may ooze back into the blood capillaries, but most of it collects in a different set of vessels. This tissue fluid is called lymph, and the tubes that carry it are called lymph vessels. Muscle contractions caused by movement squeeze the vessels and move the lymph along. In the walls of the small intestine, lymph vessels in the villi absorb fats. Many of the metabolic wastes of cells also pass into the lymph. For these reasons, lymph has a higher fat content and a higher waste content than does blood.

The vessels of the lymphatic system divide into tiny, twisted passages that make up the lymph nodes. Lymph flows slowly through these nodes. Pathogens and other foreign materials that have entered the body can be engulfed by white blood cells in the lymph nodes. The lymphatic system, shown in Figure 16.8 on page 430, is a part of the body's defense system.

Eventually all lymph vessels join, forming a duct in the region of the left shoulder. The duct empties into a vein, and the fluids that moved into the tissues at the capillaries return to the blood before it enters the heart.

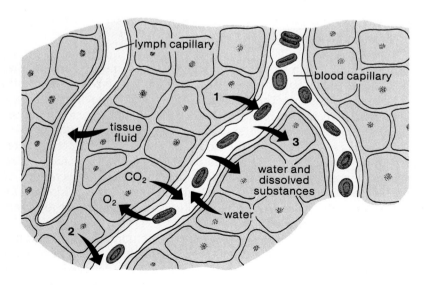

Figure 16.7 The exchange of materials at a capillary. What do you think is happening at points 1, 2, and 3?

Figure 16.8 The human immune system. In addition to the lymphatic system, the bone marrow, thymus, tonsils, and spleen are important components of this system.

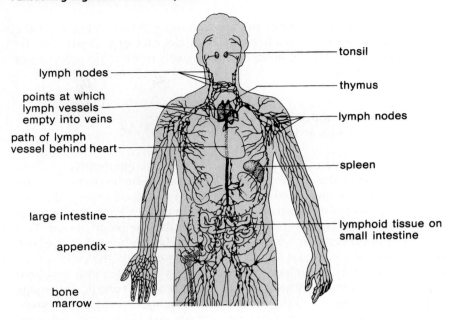

lymph nodes

points at which lymph vessels empty into veins

path of lymph vessel behind heart

large intestine

appendix

bone marrow

tonsil

thymus

lymph nodes

spleen

lymphoid tissue on small intestine

Concept Review

1. Compare and contrast arteries, veins, and capillaries.
2. Explain how a four-chambered heart functions in circulation.
3. How is blood kept flowing in one direction?
4. Compare and contrast plasma and whole blood.
5. Why is it good to have a long chain of events leading to clot formation?
6. Describe the role of the lymphatic system in circulation.

Guidepost: How does the immune system protect the body from invasion?

Immunity

16.5 The Body Has Several Defenses Against Foreign Invaders

Immunity is the capacity of the human body to resist most pathogens that might damage tissues and organs. The body has two types of immunity against foreign substances. The first type is nonspecific protection. This type involves intact skin and mucous membranes and the inflammatory process. The second type is the immune system's specific immune response.

The layers of skin covering the body and the layers of mucous membranes lining the digestive and respiratory tracts provide a protective barrier against invasion. Mucous membranes are protected by secretions of mucus. Some membranes consist of ciliated cells that sweep foreign objects away. Other surfaces are washed by fluids such as saliva or tears that contain substances active against pathogens.

If the barrier formed by the skin or mucous membranes is broken, substances in the circulating blood and lymph initiate an inflammatory response. During an inflammatory response, injured cells release a chemical called histamine and other substances. These cause nearby capillaries to swell and become "leaky." Various types of white blood cells then pass through the capillary walls and gather at the site of injury, engulfing pathogens that may have entered through the cut.

16.6 Protection by the Immune System Is Specific

In contrast to the nonspecific defenses, the response of the immune system to foreign invaders is very specific. The immune system includes the structures shown in Figure 16.8 and a variety of interacting white blood cells.

The primary cells of the immune system are small white blood cells known as **lymphocytes** (LIM foh syts). Lymphocytes are specialized for the recognition of foreign substances. Like other blood cells, they develop in the bone marrow. Some lymphocytes also mature in the bone marrow. These types of lymphocytes are called *B* **cells** (*b*one-marrow-derived lymphocytes). Their function is to produce antibodies that act against foreign substances in the body. Other lymphocytes migrate to the thymus gland, where they mature into *T* **cells** (*t*hymus-derived lymphocytes). There are at least two types of *T* cells. Killer *T* cells attack infected host cells directly. Helper *T* cells interact with *B* cells to make antibodies. In addition, amoebalike white blood cells called **macrophages** (MAK roh fayj ez) act like processing plants that engulf and disassemble complex materials such as pathogens. Figure 16.9 shows the cells involved in the immune response.

Lymphocytes are activated by antigens—substances the body recognizes as foreign. Antigens (*anti*body *gen*erators) usually are large proteins or carbohydrates that make up the cell walls or other parts of viruses, fungi, bacteria, or other pathogens. Other large molecules or parts of molecules also can serve as antigens. Surface receptors on the lymphocytes bind to the antigen that has a shape that best fits the receptor. The fit between receptor and antigen is like a lock and key. Each lymphocyte may have several thousand receptors on its surface, but all the receptors are identical, so each cell can recognize only one foreign shape. To be able to recognize all potential antigens, humans have millions of different lymphocytes. To start an antibody response, macrophages, *B* cells, and helper *T* cells must cooperate, as shown in Figure 16.10 on page 432. When a particular pathogen enters the body, it may bind to a *B* cell surface receptor that fits its shape,

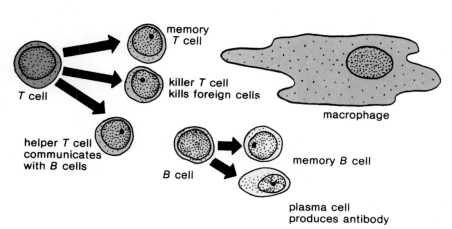

Figure 16.9 Major cells of the immune system. *B* cells, when stimulated by an antigen, differentiate and become plasma cells that secrete antibodies. *T* cells are of two types. Killer *T* cells attack infected cells and invading pathogens by secreting toxic chemicals. Helper *T* cells secrete lymphokines, helper molecules that enhance the immune response. Both *B* cells and *T* cells give rise to long-lived memory cells. Macrophages ingest pathogens and display their antigens to *T* cells, triggering the specific immune response.

memory
T cell

T cell

killer T cell
kills foreign cells

macrophage

helper T cell
communicates
with B cells

B cell

memory B cell

plasma cell
produces antibody

as illustrated in (a). At the same time, a macrophage may engulf another pathogen of the same type and disassemble it, which is illustrated in (b). After the pathogen is disassembled, antigens from the pathogen appear in receptors on the plasma membrane of the macrophage (c). A helper *T* cell that binds to the antigen in these receptors (d) becomes activated.

The activated *T* cell begins to produce helper molecules known as lymphokines (LIM foh kynz). Lymphokines stimulate pathogen-bound *B* cells to enlarge and divide rapidly (e). This action gives rise to a clone of identical plasma cells, which secrete thousands of antibodies into the bloodstream (f). These antibodies are the same as the surface receptors on the original *B* cell. In the bloodstream, the antibodies bind to the antigens that stimulated their production.

Figure 16.10 Production of antibodies in the specific immune response.

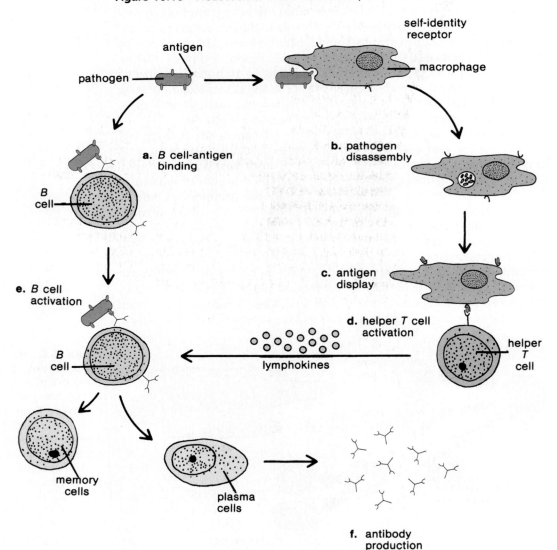

When antibodies have bound to the antigens on a pathogen, it becomes easier for macrophages to engulf and destroy the pathogen. Antibodies bound to antigens on a pathogen also can activate the **complement system**, a complex group of blood proteins. Once activated, the complement system can form holes in the plasma membrane of the pathogen, causing it to burst, as shown in Figure 16.11.

Lymphokines also stimulate the division of helper *T* cells, which produce still more lymphokines and ensure the continued stimulation of *B* cells and the continued secretion of antibodies. The invading antigen-bearing pathogens usually are destroyed within 7 to 14 days by the overwhelming numbers of antibodies released into the bloodstream. As the numbers of antigens decline, the stimulus that began the antibody response is removed from the bloodstream, and the process gradually comes to a halt. The antibody response, however, also includes the production of long-lived *B* and *T* cells known as memory cells. These memory cells remain in the circulation for many years—even a lifetime—and protect against repeated invasions by the same pathogen.

Killer *T* cells act against the cells of the body that are infected by a virus or other pathogen. Although antibodies cannot act on a virus inside a cell, as the virus multiplies, viral antigens appear in the self-identity receptors on the cell's surface. Killer *T* cells then bind tightly to these antigens and form pores in the cell's plasma membrane (see Figure 16.12 on page 434). As a result, the cells burst, exposing the viruses to antibody action.

The first time a pathogen enters the body, it takes 7 to 14 days for the production of mature plasma cells that then can secrete antibodies at the

Figure 16.11 Activation of the complement system. When two antibody molecules bind to adjacent sites on a pathogen, they can activate the first factor in the complement system, which, in turn, activates other complement factors. Activated complement factors aggregate within the plasma membrane of a pathogen, forming a small pore. Fluid rushes in through the pore, causing the pathogen to burst.

1. An antibody binds to antigen on the surface of a pathogen.

2. Another antibody binds, sending chemical signals that begin the process of activating complement factors.

3. The first activated complement factor binds to the antibodies and activates other complement factors.

4. Activated complement factors gather in sequence on the surface, forming a pore in the membrane.

5. Fluid rushes in through the pore, . . .

6. . . . causing the pathogen to burst. Macrophages then dispose of the pathogen.

Figure 16.12 A killer *T* cell (a) destroys a tumor cell by secreting a protein that makes holes in the tumor cell's membrane. As a result, the tumor cell (b) becomes leaky, takes in water, and bursts. Scanning electron micrographs, X5000 and X9200.

maximum rate of several thousand molecules per cell per hour. During this time, the antigen-bearing pathogens (for example, measles viruses) also are multiplying, producing the symptoms of disease. The second time measles viruses invade the body, however, memory cells immediately begin large-scale production of antibodies, and the infection is overtaken before any symptoms appear. Thus we say the individual is immune to measles. Figure 16.13 shows the different immune responses to primary and secondary infections.

This immunity, brought about by the rapid response of memory cells, is the basis for vaccination against many infectious diseases. Vaccines are prepared from weakened, killed, or closely related or modified pathogens. When injected into the body, the vaccine stimulates production of plasma and memory cells, usually without producing disease symptoms. The immunity is specific. Immunity to measles does not provide immunity to other diseases such as chicken pox or mumps. Because there are so many similar, but different, viruses that can cause diseases such as the common cold and

Figure 16.13 Response of the immune system to antigens. At the first exposure to an antigen, it takes 7 to 14 days to build a large clone of antibody-producing plasma cells. After the second exposure, however, the clone of memory cells immediately begins large-scale production of antibodies and also multiplies rapidly, further increasing antibody production.

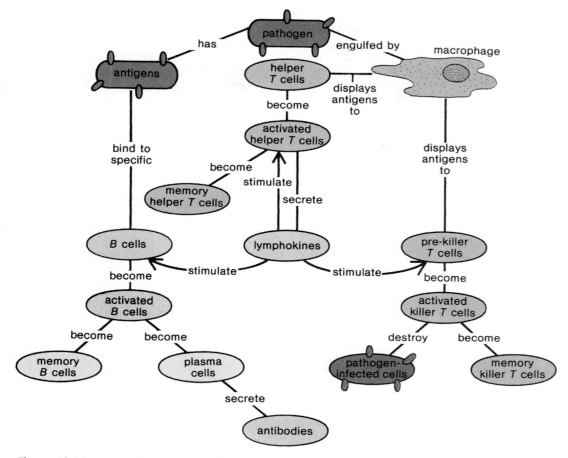

Figure 16.14 Interactions in the specific immune response.

influenza, immunity does not develop. For the same reasons, attempts to produce vaccines have not been successful. Figure 16.14 summarizes the steps involved in the specific immune response.

INVESTIGATION 16.2 The Alpine Slide Mystery

In this investigation you will develop hypotheses that might explain the severe allergic reaction experienced by several people after riding on a summertime amusement ride called an alpine slide.

Materials (per team of 3)
none

Procedure
1. Read the following information:

    During the two-week period from 22 July to 8 August 1982, five young men vacationing in Vermont experienced severe allergic reactions. Certain areas of their skin developed slightly raised patches that were inflamed and itched. They had symptoms like those of hay fever in the nose and eyes. They also had some difficulty breathing and blood pressure in the low range.

The doctors who treated these young men recognized that they were reacting in a highly sensitive manner to the presence of some foreign substance in their bodies. They treated them with injections of antihistamines and epinephrine, drugs that prevent the production of the agents that cause the reactions. All five young men recovered.

The doctors were puzzled by these medical events. As they checked the histories of the five young men, they found a common experience. All their allergic reactions occurred within a 3- to 20-minute period after each one of the five sustained some skin abrasions while riding on an alpine slide. These slides are common in ski areas, where they are built on the ski runs. They generate income for the ski areas during the spring, summer, and fall. The slide consists of a free-running sled that glides down a mountainside on an asbestos-cement track. The abrasion injuries occurred as a result of falling off moving sleds.

A check of 27 similar alpine slides in the United States revealed 24 similar cases of severe allergic reactions, 22 of which followed skin abrasions. The evidence seemed to indicate a relationship between allergic reactions and skin abrasions sustained while riding the alpine slide.

2. What do you think might have caused the severe allergic reactions in these people? Make a list of possibilities.

3. Consider the following additional information:

Data from the Alpine Slide in Vermont	
Rides per day	approximately 1400
Rate of injury (abrasions or friction burns)	1–2% (14–18) per day
Rate of allergic reaction	0.0002%

Do you think that abrasions alone are sufficient to cause severe allergic reactions? Explain.

4. If abrasions alone are not the cause, what other factors might play a role in the apparent interaction between skin abrasions and severe allergic reactions? Discuss possible environmental factors and list them in order of likelihood.

5. Now consider the following historical data on the occurrence of severe allergic reactions from other alpine slide operators across the United States.

Time Period	Rate of Allergic Reaction
23 May–27 June (1976–1981)	0 cases
28 June–4 August (1976–1981)	1.71 cases/10^5 rides
5 August–6 November (1976–1981)	0 cases

Consider the symptoms that describe severe allergic reactions: itching and inflamed skin, watery and reddened nose and eyes, difficulty in breathing. What factors cause allergies in people you know? Are any of these allergies seasonal?

6. Look at the alpine slide environment shown in Figure 16.15. Describe what you see. Do you see any factors that should be included in the list of possible causes of the severe reactions?

Figure 16.15 An alpine slide.

Based on the list your team has developed, what might be the most likely cause of the reactions? Begin by eliminating proposed causes that do not vary with the calendar, and focus on those that consider the entire alpine slide environment.

7. Suppose you suspected that a skin abrasion alone could not cause the allergic reaction, that some other factor also was necessary. How could you demonstrate convincingly that the young men had to have an abrasion *and* had to be exposed to another factor for the allergic reaction to occur? In other words, how do you show an association between the allergic reactions in the young men, the abrasions they received, and any other factors?

Discussion

1. Compare your hypotheses about the causes of the allergic reactions with those of the other teams. Are they similar?
2. Compare the ways in which the different teams proposed to demonstrate the association between the allergic reactions and the factors that caused them. Are they similar?
3. If you were the operator of the alpine slide and knew that skin abrasions during certain seasons could lead to severe allergic reactions, what steps could you take to protect your customers?

16.7 Many Problems Can Arise with the Immune System

The primary role of the immune system is to distinguish "self" from "nonself." In the developing embryo, the immune system learns to recognize substances that are present as "self." Later in life, this recognition may break down, and the immune system may produce antibodies or *T* cells that act against the individual's body cells. Disorders resulting from the action of the immune system against self are known as **autoimmune** diseases. In one autoimmune disease, antibodies are formed that interfere with the nerve

Monoclonal Antibodies

Antibodies are highly specific proteins produced by the immune system to help fight infection. Each type of antibody is effective against only one type of antigen. In 1975, biologists developed a technique using two different types of cells to produce a given antibody artificially.

Antibodies are produced by *B* cells, but *B* cells do not grow well in culture. They tend to die off after a few generations. Cancer cells, however, will grow in culture almost indefinitely. The technique for producing antibodies fuses *B* cells with cancer cells. First, mice are injected with a specific antigen—for example, the virus that causes hepatitis. Certain *B* cells in the mice have surface antibodies complementary to the virus antigens; those *B* cells multiply. Then, the *B* cells are collected and fused in culture with a special type of cancer cell called a myeloma cell. The resulting hybrids, known as hybridomas, are grown in a special culture that eliminates both the parent *B* cells and the myeloma cells. Only the hybridomas remain, and these cells possess the characteristics of both parent cells—they produce the specific antibody of the parent *B* cell, and they have the longevity of the parent myeloma cell. These genetically identical, or cloned, hybridomas are grown in culture to produce quantities of the desired antibody. The resulting product is called a monoclonal antibody.

This method of making monoclonal antibodies has several drawbacks. It usually takes several months, and it is difficult to make human monoclonal antibodies. In addition, this method, which is meant to yield large quantities of antibodies, will sometimes yield only one, or at most a few, of the desired antibodies.

A new technique being developed may solve these problems. It is comparatively fast, taking only a few days to get specific antibody clones, and it can be used to make human monoclonal antibodies. It also avoids the use of mammalian myeloma cells and uses instead the eubacterium *Escherichia coli*, which can be grown in enormous quantities.

Essentially, this genetic engineering technique calls for inserting into *E. coli* those genes from an animal cell that code for antibody production. The engineered eubacteria then would produce the complete range of antibodies, instead of just a specific one, and researchers would screen the eubacteria for the desired antibody.

The technique, however, has not been perfected. One problem is the structure of the antibody itself. An individual antibody consists of four proteins arranged into a Y-shaped molecule. The arms of the Y form the sites at which an antigen binds to the antibody. Initially, the engineered *E. coli* was able to produce only a portion of the antibody molecule—the antigen-binding sites at the arms of the Y. Some of these fragments, however, can bind antigen almost as well as a whole antibody molecule can. As long as the fragments bind antigens specifically, they can be used in research. As the technique is improved, researchers expect to insert longer gene segments into *E. coli*, making it possible to produce full-sized antibodies that can help fight infection.

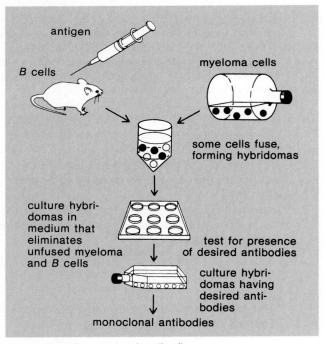

Production of monoclonal antibodies.

stimulation of muscles, resulting in muscular weakness. In another, antibodies are produced against the individual's own DNA, thus disrupting many body functions. An autoimmune response appears to be involved in some other disorders, including rheumatoid arthritis and multiple sclerosis.

Allergies are apparently a poorly adapted response of the immune system to foods or substances commonly present in the environment, such as pollen and dust. Sensitive individuals produce a class of antibodies that combine with specialized cells found in the skin and membranes of the eye, nose, mouth, respiratory tract, and intestines. These cells release histamine and other substances that cause the congestion, sneezing, and itching typical of "hay fever" or cramps and diarrhea in the case of food allergies. Treatment involves the use of antihistamines, which suppress some of the allergic symptoms but also may cause side effects such as drowsiness.

Some of the blood group systems present in humans can cause problems under certain circumstances. The blood types that arise from the ABO blood group system are not always compatible. This incompatibility comes about because antibodies in the plasma of one blood type (for example, type A) react with antigens on the red blood cells of another blood type (for example, type B). The reaction causes the red blood cells to clump, or agglutinate (uh GLOO tin ayt). Because of this clumping, transfusing blood of one type into blood of another type can cause a fatal reaction.

Unlike other antibodies, which are produced only in response to an antigen, ABO blood group antibodies develop spontaneously and normally are present in the blood. Each blood group is characterized by the presence of a particular antigen and a different antibody, as shown in Table 16.1.

Rh factors are antigens sometimes found on the surface of red blood cells that can cause problems in transfusions and during pregnancy. Rh positive individuals have Rh antigens on their red blood cells; Rh negative individuals do not. Normally, antibodies to Rh antigens are not present in plasma. Antibodies may develop, however, if Rh positive blood is transfused into an Rh negative individual or if an Rh negative woman has an Rh positive baby. There usually is no problem for the fetus in the first pregnancy, unless the woman has been sensitized previously by a transfusion of Rh positive blood. During birth, however, red blood cells from the infant may enter the mother's blood, stimulating formation of Rh antibodies in her plasma. During a later pregnancy, these antibodies may pass into the fetal blood, causing the destruction of fetal red blood cells (see Figure 16.16). These problems are prevented if an Rh negative mother is injected with Rh antibodies within 72 hours of delivery of an Rh positive child. The Rh antibodies destroy the fetal Rh positive red cells in the mother's body, preventing her immune system from forming antibodies.

Figure 16.16   The events leading to Rh disease. Fetal red blood cells may enter the maternal blood during childbirth, stimulating formation of Rh antibodies in the mother's plasma. These antibodies may pass into the fetal blood during a subsequent pregnancy, destroying the fetal red blood cells.

Rh⁺ red blood cells

Rh⁻

Rh⁺

development of anti-Rh anibody

Rh⁺

Rh⁻

fetal red blood cells with Rh antigen

maternal antibodies against Rh antigen

TABLE 16.1   ABO Blood Group Antigens and Antibodies

Blood Group	Antigen on Red Blood Cell	Antibody in Plasma
0	none	anti A, anti B
A	A	anti B
B	B	anti A
AB	A, B	none

The immune system also is responsible for organ transplant rejections. Rejection is brought about by killer *T* cells. These cells identify the transplanted organ as a foreign antigen. A few days after an organ transplant, they invade the tissue and begin to destroy it. To prevent this reaction, physicians use drugs to suppress the immune response. Unfortunately the suppression of the immune response leaves the transplant recipient vulnerable to infection, the major cause of death among organ transplant recipients. Efforts to control the immune response and to improve matching of donor and recipient tissues are active areas of research.

Concept Review

1. Distinguish between specific and nonspecific immunity.
2. Define the term *antigen* and provide several examples of antigens.
3. Describe the steps involved in antibody production.
4. Why does the second exposure to a pathogen usually not produce disease?
5. In what way are allergies examples of an inappropriate immune response?
6. Describe some problems that can arise when the immune system malfunctions.

INVESTIGATION 16.3   Exercise and Carbon Dioxide Production

When you exercise, you breathe faster than when you are at rest. Does exercise also affect the concentration of carbon dioxide in the air you exhale?

You can test for the presence of carbon dioxide by bubbling your breath through water. Carbon dioxide in your breath combines with water to form carbonic acid ($CO_2 + H_2O \rightarrow H_2CO_3$), which lowers the pH of the solution. Using a pH probe and the LEAP-System™, or a pH meter or wide-range pH paper if the LEAP-System is not available, you can measure the amount of carbon dioxide in the water by detecting a change in pH. The LEAP-System BSCS Biology Lab Pac 2 includes instructions for use of the system.

▶ Read the procedure for the investigation and complete the statement: If carbon dioxide is a waste product of cellular respiration, then. . . .

Materials (per team of 2)

100-mL graduated cylinder
2 250-mL flasks
Apple IIe, Apple IIGS, Macintosh, or IBM
  computer
LEAP-System™
printer

2 wrapped drinking straws
2 pH probes
pH meter or wide-range pH paper
  (optional)
forceps (optional)
watch or clock with a second hand

Procedure

Each member of the team should do the investigation. One person should be responsible for the computer while the other keeps track of his or her own time breathing into the flask. Then switch responsibilities.

1. In your data book, prepare a table similar to Table 16.2.
2. Label two flasks *A* and *B*.
3. Place 100 mL of tap water in each flask.
4. Record the initial pH in both flasks by using two pH probes with the LEAP-System (Figure 16.17), a pH meter, or by dipping small strips of pH paper into the water and comparing the color to a standard color chart.

TABLE 16.2 pH Readings

Experimental Condition	Flask A		Flask B	
	Initial	Final	Initial	Final
At rest				
Mild exercise				
Vigorous exercise				

5. Put a drinking straw in the water in Flask B. Sit quietly for 1 minute and then gently blow air from your lungs through the straw into the water for 1 minute. Do not suck any material through the straw. Swirl the flask.

6. Determine the pH in both flasks and record the readings in the table.

7. Discard the water in both flasks. Rinse the flasks and the straw.

8. Repeat Steps 3–7, but this time, before blowing through the straw, exercise for 1 minute by walking in place.

9. Repeat Steps 3–7, but this time, exercise more vigorously for 1 minute (run in place, dance, or do jumping jacks). Discard the straw.

10. Wash your hands thoroughly before leaving the laboratory.

Discussion

1. Run stripcharts to obtain graphs of the data. Observe the graphs carefully. In which flask was the pH decrease greatest? Why? In which flask was the pH decrease smallest? Why?

2. If the pH changed in Flask A, how would you use that information to explain the results in Flask B?

3. How is the production of carbon dioxide related to activity?

4. What factors other than exercise might play a part in determining the amount of carbon dioxide given off?

5. Compare your results with the results of other students in your class. Try to explain any differences.

6. What is the source of carbon dioxide in your breath?

7. What was the purpose of Flask A in this experiment?

Figure 16.17 LEAP-System setup for measuring carbon dioxide production.

Guidepost: What role do the respiratory and excretory systems play in maintenance of homeostasis?

Gas Exchange and Excretion

16.8 Cellular Respiration and Gas Exchange Are Not the Same

Figure 16.18 Movements of breathing in humans.

inhalation

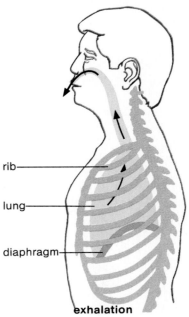

rib

lung

diaphragm

exhalation

Frequently, cellular respiration is confused with the process that exchanges vital gases with the environment. Cellular respiration is primarily a process of energy release. Because cellular respiration uses oxygen and releases carbon dioxide, gas exchange also is involved. As carbon dioxide accumulates, it produces acidic conditions that are poisonous to cells and must be removed. Respiration is the overall process of exchanging oxygen and carbon dioxide with the environment and the blood. The process is accomplished by the respiratory system. Respiration may be studied in three stages. The first stage is the process of breathing, which involves the movement of air in and out of the lungs. The second stage is the exchange of gases between the internal surface of the lungs and the blood. The third stage is the exchange of gases between the blood and the tissue cells.

The respiratory organs function to exchange gas molecules between the inner surface of the lungs and the blood. Breathing moves the air that has been inhaled by the mouth and nose through a series of air tubes into the lungs. Movement of the air is accomplished by the action of two groups of muscles. The first group makes up the **diaphragm** (DY uh fram), a muscular wall that separates the body cavity into two parts. The second group is the rib muscles. Working together, the diaphragm and rib muscles act to change the size of the chest cavity, as shown in Figure 16.18. When you inhale, your diaphragm moves down, your ribs move up and out, and the cavity enlarges. When the chest cavity expands, the pressure within the chest falls. Because the pressure within the chest cavity is lower than the atmospheric pressure outside, air rushes in. When you exhale, the volume of the chest cavity is reduced, the internal pressure becomes greater than the atmospheric pressure, and air is forced out. The rhythmic increase and decrease in the volume of the chest cavity is the mechanical pump that drives air in and out of the lungs.

Atmospheric air is dry, sometimes cold, and often dirty. The air you breathe passes through your nose and down the windpipe, or **trachea** (TRAY kee uh), into passageways that enter the lungs and end in air sacs, or **alveoli** (al VEE oh ly). The air is moistened, warmed, and cleaned by cells lining these air passageways. Alveoli are grapelike clusters of cavities formed by single-layered sheets of cells. Each lung has millions of these cavities, and the walls of these cavities are covered with a network of capillaries. Figure 16.19 illustrates the human respiratory system and details of the alveoli. Oxygen diffuses into the bloodstream through the alveolar walls. Because the many alveoli provide an enormous amount of surface area, the lung can exchange a large volume of gases in a short time. If all the alveoli of the human lung were spread out flat, the surface would cover an area of about 70 m²—the size of about five parking spaces.

Some diseases affect the normal functioning of the respiratory system. In people with emphysema (em fuh SEE muh), for example, the alveolar walls lose elasticity, making it difficult for the lungs to remove air low in oxygen and high in carbon dioxide. As a result, less fresh air, which contains more oxygen and less carbon dioxide, can be brought in for air-blood

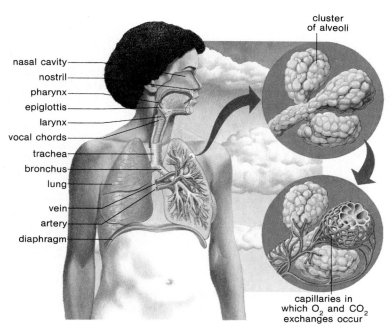

nasal cavity
nostril
pharynx
epiglottis
larynx
vocal chords
trachea
bronchus
lung
vein
artery
diaphragm

cluster
of alveoli

capillaries in
which O_2 and CO_2
exchanges occur

Figure 16.19 The parts of the human respiratory system involved in breathing. The lung has been cut away to expose the branching system of bronchial tubes. Part of the lung has been enlarged to show alveoli and their relation to capillaries. Millions of alveoli in each lung give the tissue a spongelike appearance.

gas exchange. Emphysema is associated with smoking or living in areas of high air pollution. The condition becomes progressively worse, and sufferers may become completely incapacitated.

16.9 Respiratory Gases Are Transported in the Blood

Once oxygen has diffused from the alveoli into the blood, it must be transported to the tissues. Because oxygen does not dissolve readily in plasma, special oxygen-transporting molecules are required. In humans and all other vertebrates, the red blood cells are packed with hemoglobin molecules. When the concentration of oxygen is relatively high, as in the alveoli, each hemoglobin molecule combines with four oxygen molecules. When the red blood cells reach the tissues, where the concentration of oxygen is relatively low, the hemoglobin gives up its oxygen. Hemoglobin enables blood to carry about 60 times more oxygen than could be transported by plasma alone. Figure 16.20 on page 444 shows the steps in the transport of oxygen and carbon dioxide.

Carbon dioxide is more soluble than oxygen in the blood, and a small amount dissolves in plasma. The hemoglobin in the blood transports approximately 25 percent of the carbon dioxide in the human body. The remaining 75 percent is carried in the blood as bicarbonate ion (HCO_3^-). To create this ion, carbon dioxide first reacts with water in the blood to form carbonic acid. The carbonic acid then ionizes to form hydrogen and bicarbonate ions. As with oxygen transport, the amount of carbon dioxide in the alveoli and the tissues determines the direction of the reaction. Carbon dioxide usually diffuses into the lungs because it is more concentrated in the blood than in the alveoli.

Figure 16.20 Oxygen and carbon dioxide transport. Oxygen from the alveolus enters red blood cells and combines with hemoglobin (Hb), forming oxyhemoglobin (HbO$_2$). The oxygen is carried in this form to body cells where it is released (see bottom of illustration). Carbon dioxide diffuses from body cells into the red blood cells. There, it combines with water to form carbonic acid (H$_2$CO$_3$), which ionizes to bicarbonate ions (HCO$_3^-$) and hydrogen ions (H$^+$). The bicarbonate ions diffuse into the plasma. The hydrogen ions combine with hemoglobin (H-Hb). Some of the carbon dioxide combines directly with hemoglobin (HbCO$_2$). In the lung capillary, the reverse reactions occur. Bicarbonate ions from the plasma are converted to carbon dioxide and diffuse into the alveolus.

16.10   The Kidneys Are Major Homeostatic Organs

The body's internal chemical environment must be regulated at all times. Regulation involves excreting cellular wastes, controlling concentrations of ions and other substances, and maintaining water balance. Carbon dioxide and nitrogen compounds are the chief cellular wastes. In humans, the lungs excrete carbon dioxide. The kidneys excrete nitrogen compounds, regulate ion concentrations, and maintain water balance.

Figure 16.21 shows the human urinary system. The kidneys are dark red, bean-shaped organs about 10 cm long, located on each side of the back wall of the body cavity, just above waist level. Each kidney contains about one million working units or **nephrons** (NEF rahnz). A nephron is a long, coiled tubule. One end of a nephron opens into a duct that collects urine.

Figure 16.21 The human urinary system.

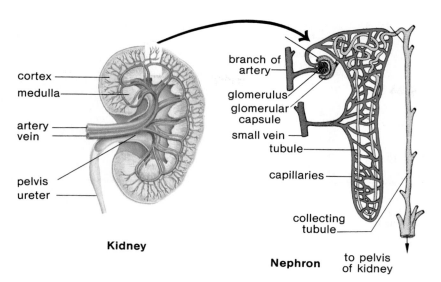

cortex
medulla

artery
vein

pelvis
ureter

Kidney

branch of
artery
glomerulus
glomerular
capsule
small vein
tubule
capillaries

collecting
tubule

Nephron to pelvis
of kidney

Figure 16.22 A section through the human kidney and an enlarged view of one nephron with its surrounding capillaries.

The other end forms a cup that encloses a mass of capillaries. Other capillaries form a network that closely surrounds the entire nephron, as shown in Figure 16.22. This intimate relationship between the blood and the kidneys is essential to kidney function.

The ball of capillaries is called a **glomerulus** (glah MER yoo lus) and the cup of the nephron is called the glomerular capsule. The wall of a nephron is just one cell thick and is in direct contact with a capillary wall, also a single cell thick. Often the tubule and capillary walls appear merged into one undivided structure.

All the collecting ducts of the nephrons empty into the **ureter** (YOOR ee ter), a large tube that leads from each kidney to the urinary bladder. Urine is stored in the bladder until it is discharged to the outside through another tube called the **urethra** (yoo REE thruh). In males, the urethra also carries semen out of the body.

INVESTIGATION 16.4 The Kidney and Homeostasis

The cells of the human body are surrounded by a liquid that is remarkably constant in its properties. The continuous regulation of the many dissolved compounds and ions in this internal environment is referred to as homeostasis.

The kidneys play an important role in homeostasis by regulating blood composition and by regulating the levels of many important chemicals and ions. The production of urine and its elimination from the body are critical functions of the kidneys and the urinary system.

Materials (per team of 3)
none

PART A   Blood versus Urine
Procedure
1. The relationship of structure and function in the kidney is illustrated in Figure 16.22 at the top of this page. Use this illustration and the data in Table 16.3 to answer the discussion questions.

TABLE 16.3 Comparison of Materials in Blood and Urine

	% in Blood as It Enters Kidney	% in Urine as It Leaves Kidney
Water	91.5	96.0
Protein	7.0	0.0
Glucose	0.1	0.0
Sodium	0.33	0.29
Potassium	0.02	0.24
Urea	0.03	2.70

Discussion

1. What do the data for water indicate?
2. Protein molecules are not normally found in the urine. Give some possible reasons for this.
3. The information for glucose is similar to that for protein. Can you explain these data?
4. Based on what the sodium data indicate, what do you think may happen to the sodium content in the urine of a person who increased his or her intake of sodium chloride?
5. How do the data for potassium differ from those for sodium?
6. How would you interpret the data for urea?
7. Summarize the functions that take place between blood and urine and identify the structures where these functions occur.

PART B Filtration, Reabsorption, and Secretion
Procedure

2. The micropuncture method was used in a second study of the six materials presented in Table 16.4. Under a microscope, a very fine pipet was used to withdraw samples of fluid at four points along the nephron. Study Table 16.4, which shows the data that were collected using this technique. Use the data to answer the discussion questions.

TABLE 16.4 Comparison of Materials at Four Points Along a Nephron

	In Blood Entering Glomerulus	In Tubule from Glomerulus*	In the Urine	In Blood Leaving Nephron**
Water	100	30	1	99
Protein	100	0	0	100
Glucose	100	20	0	100
Sodium	100	30	1	99
Potassium	100	23	12	88
Urea	100	50	90	10

*The numbers in this column represent proportions, not actual numbers of molecules. For every 100 molecules of water in the blood, 30 are found in the tubule.
**The numbers in this column were obtained by subtracting the proportionate number of molecules of the substance in the urine from the proportionate number of molecules of the substance originally in the blood (100).

Discussion

1. Which function, secretion or reabsorption, is greatest as far as the movement of water in the kidney is concerned?
2. Proteins are involved in which of the three kidney functions?
3. Compare the protein data with the glucose data. What is the difference? Explain the difference.
4. In some samples, glucose is found in the urine. What do you think might cause this condition?
5. Why are excess glucose molecules in the blood excreted?
6. The data tell us that the concentration of sodium in the blood is greater than in the urine, yet most of the sodium ions in the urine move back into the blood. What process makes this movement possible?
7. Urea is a by-product of amino acid metabolism. Next to water, urea is the most abundant material found in urine. If urea were allowed to accumulate in the blood, what might happen?
8. Homeostasis is the maintenance of a relatively stable internal environment in an organism. Discuss your ideas about how the kidney functions as a homeostatic organ.

16.11 Nephrons Filter the Blood

Three processes are involved in the function of the nephrons: filtration, secretion, and reabsorption. Figure 16.23 summarizes these processes. Filtration occurs in the glomerulus of the nephron, where the fluid portion of the blood is forced into the glomerular capsule. Blood cells and most of the plasma proteins are retained in the glomerulus. The filtrate in the glomerular capsule contains nitrogenous wastes, ions, and much of the blood's water content. Approximately 150 to 180 L of fluid enter the nephrons each day, yet only about 1.5 L of urine are eliminated from the bladder. More than 99 percent of the fluid arriving in the glomerular capsule is returned to the blood.

Figure 16.23 Major processes in the formation of urine.

Secretion and reabsorption take place along the nephron tubule. As the filtrate from the glomerulus moves through the nephron, the cells of the tubular wall selectively remove substances that are left in the plasma after filtration has taken place. The cells then secrete these substances into the filtrate. Substances such as penicillin may be removed from the blood in this manner.

Cells of the tubular walls also reabsorb useful substances from the filtrate and return these substances to the blood. Glucose, amino acids, essential ions such as sodium and potassium, and most of the water are conserved in this manner. Secretion and reabsorption require active transport for this exchange of materials.

The nephrons maintain appropriate blood-sugar levels by the process of glucose reabsorption. If the blood-sugar concentration rises too high, not all of the glucose can be reabsorbed, and some of it is excreted in the urine. If the blood-sugar level is low or normal, almost all the glucose in the nephrons is reabsorbed into the blood, and no glucose is present in the urine.

Water-salt balance also is regulated in the nephron and the nephron's collecting duct. Here, about 99 percent of the water that has left the capillaries and entered the nephron is reabsorbed into the blood. The mechanism is complex and involves the maintenance of a high salt concentration in the tissues surrounding the nephron and its collecting duct. As a result, water leaves the collecting duct by osmosis. The reabsorption of water is controlled by hormones, and the final concentration of the urine occurs in the collecting duct.

Although the kidneys can regulate blood-sugar levels, remove nitrogenous wastes from the blood, regulate water-salt balance, and excrete excess salt within limits, they cannot excrete high concentrations of salt. In fact, perspiration and the kidneys working together cannot remove high concentrations. People shipwrecked on the ocean can easily perish from dehydration by drinking the salt water. Instead of replenishing the body's water supply, they will lose more water than before drinking as the kidneys attempt to excrete the salt.

The kidneys have still other functions, such as pH regulation, but blood-sugar regulation, water-salt balance, and nitrogenous waste excretion are their major functions. If the kidneys fail, vital or poisonous blood components are not maintained at normal concentrations, homeostasis fails, and death follows.

Concept Review

1. Distinguish between respiration and cellular respiration.
2. Describe the mechanism that operates to move air into and out of the human lungs.
3. Describe how oxygen and carbon dioxide are carried in the blood.
4. List the major regulatory functions of the kidneys.
5. Describe the structure of the nephron.
6. Explain how filtration, secretion, and reabsorption contribute to formation of urine.

Temperature Regulation

Guidepost: How do humans maintain a constant internal body temperature?

16.12   The Rate of Chemical Reactions Is Influenced by Temperature

Metabolism is the sum of all the chemical reactions that occur within a cell or an organism. Like other chemical reactions, these enzyme-controlled reactions are influenced by changes in temperature. Within limits, they slow down at low temperatures and speed up at high temperatures. Therefore, maintaining a constant internal temperature allows for more efficient chemical processes to take place. It also permits an animal to be active when environmental temperatures are low.

In humans and other animals that are able to maintain a constant internal temperature, the temperature of the skin and the tissues just beneath it may fluctuate. Internal body temperature in most individuals, however, changes little in the course of a day. On waking up, a human's temperature is usually about 36.2° C, and it increases to perhaps 37.6° C by late afternoon, averaging about 37° C. It is important to maintain the body's temperature within these narrow limits.

To accomplish this regulation of internal temperature, the body must balance the amount of heat it produces with the amount it loses. Therefore, when you become overheated, during strenuous exercise for example, the extra heat produced must be dispelled from the body. Otherwise, the body temperature could increase to the point at which enzyme functions would be impaired.

One of the by-products of cellular respiration is heat. It is this heat, produced by the breakdown of foods, that maintains the body's temperature. Because muscles and organs are the most active tissues in the body, they carry on more respiration and thus produce more heat than any other body tissues. The activity of the muscles determines the rate of heat production. When you sleep, there is little muscular movement, and heat production decreases. As you exercise or shiver, your active muscles produce more heat.

16.13   Major Heat Loss Occurs Through Evaporation and Radiation

Most of the body's heat loss, 80 percent or more, occurs through the skin. The remainder takes place through the normal function of the respiratory, digestive, and excretory systems.

Evaporation at average temperatures and normal rates of activity is only one of the methods of cooling the body. At high temperatures or during vigorous activity, however, it is the only method by which heat can be lost through the skin. How can you explain that the same temperature on two different days may feel hotter on one of those days? At a temperature of 38° C (100° F) and a humidity of 90 percent, a person would feel hotter than at the same temperature with a humidity of 45 percent. That is because evaporation takes place more rapidly in a drier environment, thus cooling your skin more efficiently.

Figure 16.24 Lizard absorbing radiated heat. Animals such as this do not have internal mechanisms that regulate body temperature.

Radiated heat is transferred from one object to another even when the two objects do not touch. You can feel the radiant heat of the sun without actually being in contact with the sun's surface. The skin gains or loses heat through radiation, depending on the environment. If the environment is cool, body heat is lost through radiation. Conversely, in a hot environment heat is gained by radiation from hot surfaces (see Figure 16.24).

16.14   Internal Temperature Is Controlled by the Brain

Heat loss or gain also may be affected by the amount and type of clothing worn, the ingestion of hot or cold food or liquid, or even by standing in a breeze. Internal body temperature can remain constant only if the rate of heat loss is the same as the rate of heat production, as shown in Figure 16.25. Receptors that respond to temperature changes are necessary to keep body temperature constant. In an area of the brain called the hypothalamus, there are receptors that detect changes in blood temperature. Much as a thermostat controls a furnace, air conditioner, or water heater, the receptors in the hypothalamus act as a thermostat to regulate body temperatures.

The body thermostat is set at 37° C. When the temperature of the blood increases by as little as 0.01° C, the cells of the hypothalamus send messages to the sweat glands and blood vessels of the skin. In response to these messages, the sweat glands increase their rate of secretion. The blood vessels in the skin begin to become larger, or dilate, allowing for greater blood flow to, and heat loss from, the body surface. If body temperature continues to increase, this process will accelerate the rate of sweating and dilation, causing more heat loss from the skin.

In a cold environment, the reverse of this process occurs. When the temperature of the blood reaching the hypothalamus drops, messages are sent that slow production of sweat and that narrow, or constrict, the blood vessels in the skin. If this reaction does not raise the temperature sufficiently, a second mechanism is initiated. Shivering and skeletal muscle contractions cause an acceleration of cellular respiration and heat production. When you are very cold you use more oxygen and tire more quickly.

The deliberate use of hypothermia (lowering of body temperature) often is called for in open-heart surgery and other prolonged surgical procedures. The mechanisms involved in survival of extremely low body temperatures, however, are still poorly understood.

Concept Review

1. What is the advantage of maintaining a constant body temperature?
2. How does the body prevent damage from overheating as you are vigorously exercising?
3. What are some of the mechanisms by which your body attempts to keep you warm in very cold weather?
4. How does the human body regulate its body temperature?
5. How is body temperature related to homeostasis?

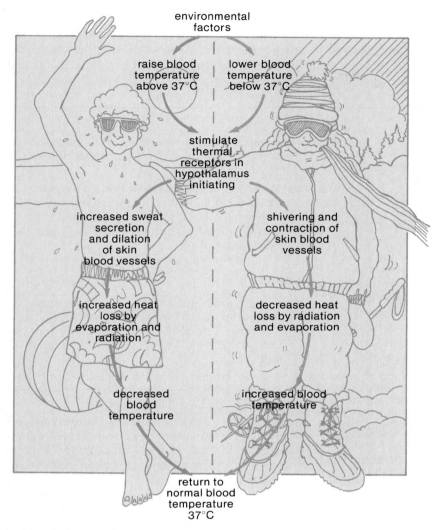

Figure 16.25 How heat-dissipating and heat-conserving mechanisms operate to maintain normal body temperature.

Summary

For its cells to function effectively, the body requires a constant internal environment. Homeostasis, the maintenance of this stable environment, is achieved by cooperative action of all body systems. The heart and blood vessels that make up the circulatory system are specialized to deliver raw materials to the cells of the body and to remove wastes from the cells' environment. The immune system serves to protect the body from foreign substances and pathogens through both nonspecific and specific types of protection. The respiratory system supplies the body with oxygen, and it aids the excretory system in eliminating waste substances. In addition, the excretory system plays a central role in regulating the concentration of many essential substances in the blood. A constant internal temperature enables chemical reactions to proceed at optimum rates. All of these systems are coordinated by the nervous and endocrine systems, the subject of the next chapter.

Applications

1. Describe the route followed by a molecule of oxygen as it moves from the air in the external environment to a mitochondrion in a muscle cell.

2. A temporary reddening of the skin surface sometimes is called a "flush" and sometimes a "blush." The first term often is used in cases of fever. The second is used to describe a reaction to some situation in the external environment. Do you think the body mechanism is the same in both cases? If it is the same, how does it operate? If it is not the same, what are the differences?

3. Using chapters 15 and 16, trace the pathways of the hamburger, fries, and soft drink you might have had for lunch. Describe what happens to the food molecules from the time they are absorbed in the intestine until their remains are excreted in the urine or eliminated through the anus.

4. Distinguish among excretion, secretion, and elimination. Explain which system(s) of the body are involved in each.

5. Construct a diagram or concept map showing the relationship between the circulatory and respiratory systems.

6. What is the effect of the presence of respiratory pigment such as hemoglobin on the transport of oxygen?

7. How are breathing and gas transport affected by climbing to high altitudes?

8. If the human immune system has a long-term memory against foreign invaders, why is it necessary for people to get vaccinated each year in order to protect themselves from contracting influenza?

9. How is the body's allergic reaction to pollen similar to how the body develops immunity to a disease?

10. Suppose you needed an organ transplant. Who would be the best choice as the donor of the organ—your spouse, one of your parents, an identical twin, or your fraternal twin? Who would be the worst choice as the donor, and why?

11. Some people used to be called "universal blood donors," and some used to be called "universal blood recipients." What was meant by this, and why are these terms no longer used?

Problems

1. In cases of kidney failure, the patient must undergo periodic hemodialysis on a kidney machine. How does this machine function? Compare it with the functions of the human kidney.

2. Why do surgeons place their patients in hypothermia before procedures such as open-heart surgery? What are the advantages?

3. Because the heart is often irreparably damaged by disease, much work has been done on designing artificial hearts. What problems must be overcome in order to design an effective artificial heart?

4. Under what circumstances would you apply cardiopulmonary resuscitation (CPR)? What are the precautions you must observe in exhaled air ventilation and external cardiac compression?

5. Take a field trip to a local heart clinic and meet with some of the technicians who operate the equipment. An opportunity to view such high-tech marvels as the ultrasound machine with Doppler image capabilities that can produce an echocardiogram is an experience to remember.

6. Ask a representative of the American Lung Association to visit your class and to bring pictures, films, and slides of diseased lungs. Ask him or her to speak about the causes and prevention of lung disorders.

7. Your book has mentioned the importance of surface area to the functioning of structures in the body; find the references in this chapter

and the previous chapter. Find references in other chapters to the relationship between surface area and function in various organisms. Why is surface area so important, and what types of structures often serve to increase it?

8. In a complete physical examination by your physician, samples of urine and blood are analyzed. What type of information about your health can be determined by analyzing these fluids?

9. Thermography is a technique used to detect diseases by checking the temperature patterns in the surface of a person's skin. Find out more about this technique, what diseases might be discovered by its use, and why the process works.

10. Hank Gathers was a very talented young college basketball player at Loyola Marymount College in California. He had a heart condition that he knew about, and still he decided to play. During a basketball game his heart began to beat arrhythmically, and he died. What was this heart condition, and what are other life-threatening heart conditions? If you loved to participate in some activity, yet knew this activity was potentially life-threatening, would you continue? Make a list of reasons for and against.

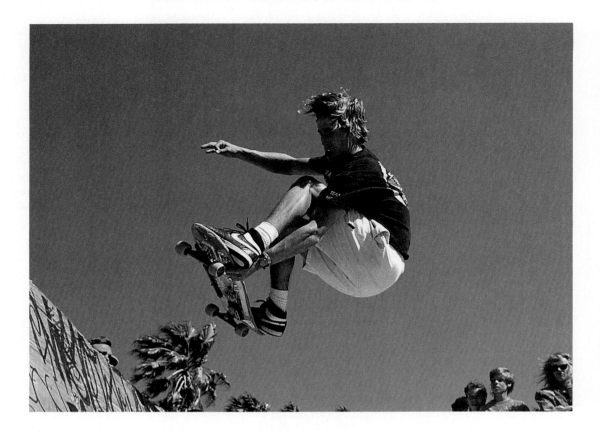

This skateboarder is performing a neuromotor task that requires a high level of concentration, training, and coordination. Name three sensory organs he is using to perform this task. What other organs is he using? Describe at least four separate functions the brain is carrying out. How is the skateboarder able to do so many things at once?

The Human Animal: Coordination 17

In performing any activity involving movement, the body must integrate several systems. A skateboarder's execution of a half-pipe, a musician's playing of a violin concerto, or an actor's speaking of the lines in a play all depend on the function and integration of the skeletal, muscular, nervous, and endocrine systems. Specialized cells in the nervous system transmit information from muscles, organs, and sensory receptors to the brain. The brain uses the information to determine the nature of electrical signals and chemical messages to be sent to muscles, organs, or glands. This chapter examines some of the mechanisms that make this integration possible and that maintain homeostasis in the body.

Human Movement

Guidepost: How do muscles, bones, and joints enable humans to move?

17.1 Muscle Contraction Depends on Energy from ATP

Humans and other vertebrate animals have three types of muscles, as shown in Figure 17.1 on page 456. Each of these three types has a different structure and function. Skeletal muscle moves the bones of the skeleton in response to conscious control. Because it is controlled by the individual, it is called voluntary muscle. Each skeletal muscle is composed of muscle tissue, which is made up of numerous individual cells acting together. Skeletal muscle is used in making adjustments to the external environment, such as raising your arm to swat an annoying mosquito. Figure 17.2 shows a skeletal muscle at successively greater magnifications. As shown in (a), the muscle consists of individual fibers that run its entire length. Each muscle fiber, the equivalent of a muscle cell, is made up of many parallel myofibrils (b), and each myofibril is made up of protein organized into thick and thin filaments (c). The thick filaments are composed of a protein called myosin. The thin filaments are made of another protein called actin. The thick and thin filaments are arranged in an overlapping pattern that gives skeletal muscle its characteristic striated appearance under the microscope. In muscle contraction, portions of the thick filaments attach to the thin filaments and pull the thin filaments toward each other.

Smooth muscle is found in the walls of blood vessels, the digestive tract, and portions of the respiratory tract. Organized into sheets of cells,

Figure 17.1 Types of muscle tissue. (a) Skeletal muscle cells fuse end to end, forming a long fiber with characteristic striations. (b) Smooth muscle is made of long, spindle-shaped cells arrayed in parallel. (c) Cardiac muscle cells are organized into interconnecting chains that form a lattice.

smooth muscle functions involuntarily in the operation of body organs and the regulation of the internal environment. In some tissues, the muscle cells contract only when they are stimulated by a nerve impulse or hormone, and then all of the cells contract as a unit. The smooth muscles in blood vessels operate in this fashion. In other body organs, such as the wall

Figure 17.2 The structure of skeletal muscle. See text for details.

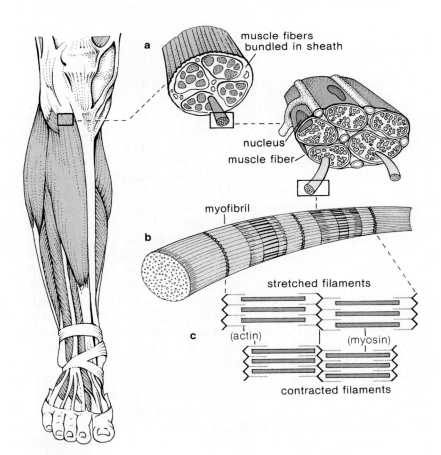

of the intestine, the individual cells of smooth muscle may contract spontaneously, creating a slow, steady contraction of the tissue.

Cardiac (KARD ee ak) muscle is found only in the heart, and its contraction is involuntary. Cardiac muscle is striated. It is composed of chains of single cells, each with its own nucleus. These chains are organized into branches that interconnect and form a lattice. This latticelike structure is crucial to how heart muscle contracts. Heart contraction begins when channels across the plasma membrane of the muscle cells open, admitting ions that change the electrical charge of the membrane. When two cardiac muscle fibers touch one another, their membranes make an electrical junction. As a result, the change in electrical charge of one fiber initiates a wave of contraction throughout the heart, with the change in electrical charge passing rapidly from one junction to the next. Thus, a mass of cardiac muscle tends to contract all at once, rather than a bit at a time.

Muscle contraction, whether voluntary or involuntary, is initiated by a nerve impulse. The nerve impulse causes the nerve endings on the muscle (Figure 17.3) to secrete chemicals that stimulate the muscle to contract. Muscle contraction also requires the presence of calcium ions and energy in the form of ATP (first discussed in Section 4.6). The supply of ATP in a single muscle fiber, however, can sustain full contraction for only about one second. For a muscle fiber to continue contracting, the ATP must be regenerated rapidly. This is made possible by using three energy sources. The first is creatine phosphate (KREE uh tin FOS fayt), a substance in the muscle cell that can pass its energy to ADP. Creatine phosphate provides *immediate* regeneration of ATP. The second energy source is the ATP formed during normal cellular respiration, and the third is glycogen, the form in which glucose is stored in the liver and in muscles (discussed in Section 4.8). In the muscle cells, the glycogen is broken down to glucose, which then is respired to produce more ATP. In the liver, glycogen is broken down to glucose, which is released into the blood in the presence of certain hormones. When an athlete is mentally preparing for exertion, the brain triggers the release of these hormones, which prepare the body for action.

Figure 17.3 Nerve endings on skeletal muscle fibers (X60). Nerve impulses stimulate the nerve endings to secrete their chemical messengers, initiating the events that result in contraction of muscle fibers.

INVESTIGATION 17.1 Muscles and Muscle Fatigue

How do muscles move bones? Do muscles always work efficiently and at the same rate, or do they suffer from fatigue? In this investigation, you will observe the relationship between muscle movement, exercise, and muscle fatigue.

Materials (per team of 2)

watch or clock with a second hand	brass brad
sheet of thin cardboard	tape
scissors	metric ruler
string	

PART A   Muscles and Movement
Procedure

1. Using the materials provided and the figure on page 460 as a guide, construct a working model of the upper and lower arm that shows the attachment of the biceps and triceps. Use cardboard for bones, string for muscles, tape for tendons, and a brad for the elbow. (Hint: When cutting the string for the biceps and triceps muscles, be sure you have enough to compensate for the movement of the joint.)

2. Place the model on the table. Grasp the biceps string just below the upper arm attachment site. Gently pull the string. What do you observe? Release the string, but do not reposition the lower arm.

3. Now grasp the triceps string just below its upper attachment site and gently pull. What do you observe? How does this movement differ from that in Step 2?

Discussion

1. Many muscles in the body work in pairs. Some of these muscles are called flexors and extensors. How do such muscle pairs result in movement?

2. Is the biceps a flexor or an extensor? Is the triceps a flexor or an extensor?

3. What other joints of the body are moved by such muscle pairs?

4. Would movement be possible without flexors? Without extensors? Explain.

5. How do animals with external skeletons, such as lobsters and beetles, move?

PART B   Exercise and Muscle Fatigue
Procedure

4. Rest your forearm on the table, palm up, and grasp your upper arm with your other hand. Have your partner rest his or her forearm on the table with the hand over your hand and wrist and grasp his or her own upper arm with the free hand.

5. Try to bend your arm up as your partner exerts pressure to keep your hand on the table. Describe what happens in your arm.

6. Switch roles with your partner and repeat Steps 4 and 5. Record any differences.

7. Make a tight fist with one hand, and then extend the fingers as far as you can. Do this as fast as possible for 30 seconds. Your partner should keep track of the elapsed time on a watch or clock with a second hand and the number of times you make a fist. Record the data. Do you feel any muscle fatigue? Where?

8. Rest one minute and repeat Step 7.

9. Repeat Steps 7 and 8 using your left hand. Compare the performance of your left hand with that of your right hand. Is there any difference?

10. Change roles with your partner and repeat Steps 7–9.

Discussion

1. How many times were you able to make a fist the first time? After the rest period?

2. What differences did you notice between when you started and when you finished? What might have caused those differences?

3. Account for any performance differences between your right and left hands.

17.2   Interaction of Bones, Joints, and Muscles Produces Movement

Vertebrate movement is accomplished through the coordinated interaction of the bones of an animal's skeleton, the joints between the bones, and the contraction of muscles that span the joints. Figure 17.4 shows the human skeletal and muscular systems. The skeleton, consisting of numerous bones and cartilage, is the internal framework of the body of a vertebrate. Most bones develop as cartilage, which gradually is replaced by bone through growth and further development. The hardness of bone results, in part, from

skull

cervical
vertebrae
clavicle

scapula

sternum

ribs
thoracic
vertebrae
humerus

lumbar
vertebrae
innominate
ulna
radius
carpals
metacarpals
phalanges

femur

patella (knee cap)

tibia

fibula

tarsals
metatarsals
phalanges

trapezius

deltoid

pectoralis
major

triceps

biceps

rectus
abdominis

external
oblique

sartorius

rectus
femoris

Figure 17.4   Human skeletal and muscular systems.

deposits of calcium and magnesium compounds. In certain places of the body, for example in a portion of the rib cage and at the ends of bones, the cartilage is not replaced by bone. Retention of cartilage in these areas permits greater flexibility.

Joints are the places where two bones come together and body movement can occur. The structural features of the bones at the joint determine the degree of movement possible. Cords of connective tissue called ligaments hold together the bones that move at a joint. A joint capsule that surrounds the joint and encloses a lubricating joint fluid provides further support.

Figure 17.5 A flexor-extensor muscle pair in the human arm. Flexors bend a limb at a skeletal joint; extensors straighten the limb again.

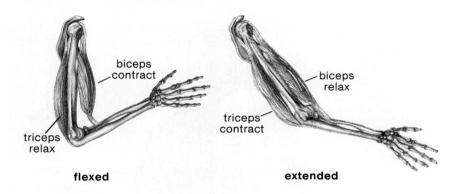

biceps contract

triceps relax

flexed

biceps relax

triceps contract

extended

Muscles are attached to bones by flexible cords of connective tissue called tendons. Tendons span joints so that, when a muscle contracts, the force is applied at the joint and movement occurs. A muscle attached to a bone is like a rope attached to a wagon. You can pull a wagon with a rope, but you cannot push it. To move the wagon back to its original position, you must attach a rope to the other end and pull it. Muscles act in the same way—in opposing pairs (see Figure 17.5). To touch your hand to your shoulder, you contract one set of muscles, producing the flexed position. When you return your hand and arm to the extended position, you contract another set of muscles and relax the first set. These two sets of muscles, called flexors and extensors respectively, work in opposition to each other. All your skeletal movements are performed by contraction and relaxation of opposing sets of muscles.

17.3 Cardiovascular Fitness Depends on Regular Exercise

Even when an animal appears to be at rest, its muscles are not completely relaxed. The muscles of any healthy organism are always in a state of partial contraction called muscle tone. This produces the firmness that can be felt even in relaxed muscles. Today, physical fitness is a household word and a large business. For many, fitness means losing weight, whereas for others it means building muscles or improving muscle tone. The primary benefit of the effort to be physically fit, however, is to improve cardiovascular fitness—the ability of the lungs, heart, and blood vessels to deliver oxygen to the cells.

The American Heart Association recommends exercising 20 to 30 minutes, three times a week, as one way of preventing coronary problems. The heart, like any muscle, gets stronger when it is exercised. The average person's heart pumps approximately 5 L of blood per minute. During intense exercise the amount of blood pumped may rise to between 15 and 30 L per minute. At this rate, a marathon runner's heart can deliver as much as 4000 L of blood to the body during a 43-km race—enough to fill a 12-gallon gas tank 80 times.

Cardiovascular fitness is important for everyone's health. The performance of an Olympic-caliber athlete, however, requires a great deal more than ordinary fitness. Centers such as the U.S. Olympic Training Center use the latest in modern electronic technology to help athletes improve their performances. Because of the progress that has been made in sports medicine, experts believe the 1996 Olympics will see athletes reach the limits of human performance in those events requiring a short burst of energy. Researchers also predict that within a decade or two, women will equal or surpass men in marathon events because of their more efficient method of utilizing stored energy.

Concept Review

1. Describe the structure and action of the different types of muscle tissues.
2. How is the supply of ATP in muscle regenerated?
3. Explain how the contraction of muscles moves bones at joints.
4. How does cardiovascular fitness differ from physical fitness?

The Nervous System

Guidepost: How does the brain coordinate body movements and determine behavior?

INVESTIGATION 17.2   Sensory Receptors and Response to Stimuli

How do we know about the world in which we live? We receive information through receptors. The activities of these receptors, coordinated by the brain, are called the senses: touch, sight, hearing, smell, and taste. Sensory receptors respond to stimuli. In this investigation, you will test some aspects of the senses of touch and sight.

 In areas of the skin that are especially sensitive, are the touch receptors more numerous or simply more sensitive? Read the procedure for Part A, and then construct a hypothesis about the relationship between the sense of touch and the number and/or sensitivity of the touch receptors.

Materials (per team of 2)
centimeter ruler
probe or blunt dissecting needle

colored pencils (red, blue)
meter stick

PART A   Touch
Procedure

1. In the center of a 4-cm square sheet of notebook paper, trace with a ruler a 1-cm square. Cut out the 1-cm square to form a hole in the larger square 1 cm² in area. You will use this paper template to find the number of touch receptors in 1 cm² of skin surface.
2. In your data book, trace six or seven of these templates in which to record data.
3. Work in pairs, with one partner applying the stimulus while the other reports the sensation felt.
4. With the subject's eyes closed, place the template on the underside of the subject's wrist. Using the probe, touch the skin *very lightly*, trying not to depress the skin surface. The subject should respond by saying yes whenever he or she feels the touch.
5. In one of the templates in your data book, map the subject's touch receptors, using a red dot to represent a touch receptor (a yes response) and a blue dot to represent an area touched without sensation.
6. Try three different areas on the wrist. Record responses in the same way as in Step 5.

7. Proceed in a similar manner to locate touch receptors on the upper arm, back of the neck, and the fingertips. Record your data as before.

Discussion

1. What was the average number of touch receptors per cm² of wrist surface? What was the number on the upper arm surface? On the back of the neck surface? On the fingertip surface?
2. Do your data support or refute the hypothesis you developed before performing Part A?
3. Suggest other ways that you could test your hypothesis.
4. Which area had the most touch receptors? The fewest? Construct a hypothesis that accounts for the differences in receptor frequency in these various areas.
5. What is the relationship, if any, between hairs on the skin and touch receptors?

PART B   Sight
Procedure

8. On a sheet of paper, place a cross (+) and a circle (O) 6.5 cm apart. Make the cross and circle about the size of the ring holes in your notebook paper. Enclose these marks within a 5-cm × 10-cm rectangle.
9. Working individually, hold the paper about 60 cm away from your eyes and close one eye. Slowly bring the paper closer while keeping your open eye fixed on the cross. In your data book, record your observations.
10. Draw a line from the cross almost to the circle, interrupting it in the vicinity of the circle and continuing it again on the other side of the circle.
11. Repeat Step 9. Record your observations.

Discussion

1. What happens to the circle and the cross as you bring the paper nearer your eye (Step 9)?
2. What happens to the line in the vicinity of the circle (Step 11)?
3. What do these experiments indicate about the sense of sight?

PART C   Response to Stimuli
Procedure

12. The seated subject should rest an arm on the desk so the hand is extended past the edge of the desk. The last three fingers of the extended hand should be curled into a fist, leaving the thumb and forefinger free, about 3–4 cm apart, and parallel to the ground.
13. The experimenter should stand and hold a meter stick so its lowest number is between the thumb and forefinger of the subject.
14. Without any warning, the experimenter will drop the stick. The subject will try to catch it with just the thumb and forefinger of the extended hand.
15. Record the point on the meter stick at which the subject caught it.
16. Repeat the test four more times, recording this distance each time. Determine the average for the five trials.
17. Repeat the exercise while the subject counts by threes to 99. Average the time (distance on the meter stick) it takes the subject to catch the meter stick while counting.
18. Repeat the procedure while the subject's eyes are closed. The experimenter will snap his or her fingers to signal when the stick is released. Repeat four more times and calculate the average.

Discussion

1. In the first test, describe the events in your body that enable you to respond and catch the meter stick.
2. What receptors are involved?
3. In the second test, what effect does thinking have on the response time?
4. Compare the response time in the first test with that in the third test. In which situation was response time shortest?
5. What can you conclude about the responses to different signals?

17.4   Neurons Transmit Nerve Impulses

An animal's rapid adjustments to the environment are brought about primarily by the nervous system. These adjustments or movements are initiated by an internal or external signal known as a **stimulus** (STIM yoo lus). The information is carried through the nervous system by **neurons** (NYOO rahnz), or nerve cells (see Figure 17.6). Each neuron has an enlarged region, the cell body, that contains the nucleus and other organelles. Most cell bodies are located in the brain or spinal cord. Two types of nerve fibers, dendrites and an axon, extend from the cell body. In humans, these fibers may be almost one meter long. Dendrites receive stimuli and conduct impulses *toward* the cell body. The axon carries impulses *away* from the cell body. Nerves are bundles of nerve fibers.

Neurons transmit **nerve impulses**. A stimulus occurring at the end of a nerve fiber starts a process of chemical and electrical change that travels like a wave over the length of a neuron. This wave of change is the nerve impulse. The chemical reactions that occur during the nerve impulse require oxygen and energy and produce carbon dioxide and heat. Electrical changes also take place. The plasma membrane of a resting neuron (one not transmitting an impulse) is electrically more positive on the outside than on the inside, as shown in Figure 17.7a on page 464. This difference in electrical charge is due to the differential permeability of the membrane to sodium and potassium ions.

Stimulation of a neuron causes a small segment of the plasma membrane to become more permeable to sodium ions. As positively charged sodium ions rush into the neuron, the electrical charge of the membrane suddenly changes (see Figure 17.7b). This local change is the nerve impulse, and it starts a similar change in the next small segment. In this manner, the impulse moves from one end of the neuron to the other in a wave of local chemical and electrical changes as sodium and potassium ions move across the membrane (Figure 17.7c, d, e, f).

One neuron is separated from another neuron by a small space called a **synapse** (SIN aps). When an impulse reaches the end of the axon of one neuron, it causes the release of a chemical substance known as a neuro-transmitter (NYOOR oh trans MIT er). The neurotransmitter diffuses across the synapse to the membrane of the next neuron, where it causes a new nerve impulse to be initiated. Because transmission across a synapse takes place in only one direction, neurons conduct impulses in only one direction, from the axon of one neuron to the dendrites or cell body of the next

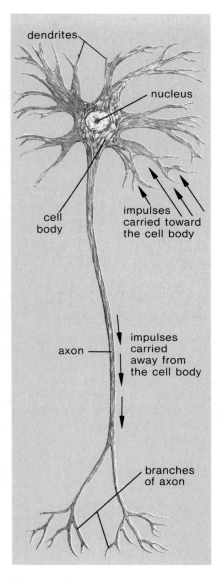

Figure 17.6   In this typical neuron, note that the axon always carries impulses away from the cell body.

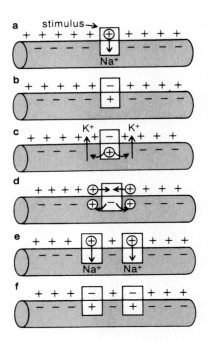

Figure 17.7 The plasma membrane of a resting neuron is more positive on the outside than on the inside (a). A stimulus changes membrane permeability, Na⁺ rushes in (b), and K⁺ diffuses out (c). The impulse moves as a wave of such changes in electrical changes (d, e, f).

Figure 17.8 Types of neurons.

neuron. Neurotransmitters also cause muscles to contract and certain glands to secrete hormones.

Nerve impulses may travel at speeds greater than 100 m per second in large fibers. The impulses are of short duration—usually a few ten-thousandths of a second. The neuron, however, cannot carry another impulse until it has had time to recover. To restore the resting state, sodium ions must be pumped back outside the plasma membrane by means of active transport. Only after recovery can the neuron respond to a new stimulus.

In humans, the three types of neurons shown in Figure 17.8 can be distinguished by their functions. Sensory neurons receive information from the internal and external environments by means of specialized nerve endings called receptors. (Receptors also may be specialized neurons such as those in the retina of the eye that react to light.) They transmit impulses toward the brain or spinal cord. Interneurons transmit impulses from one neuron to another. Motor neurons transmit impulses that originate in the brain or spinal cord to an effector—a muscle or a gland. Effectors respond to an impulse with movement or secretion of some chemical.

17.5 The Central Nervous System Coordinates Body Functions

Nervous systems have two basic functions: to control responses to the external environment and to coordinate the functions of internal organs and thereby maintain homeostasis. Because the vertebrate nervous system is complex, it is convenient to divide it into parts that differ in function. The two main parts are the central nervous system, or CNS, and the peripheral nervous system, or PNS. Composed of the brain and spinal cord, the CNS is specialized to receive and process incoming information and to determine and initiate the appropriate responses. Communication between the CNS and the rest of the body is provided by the PNS, which consists of the

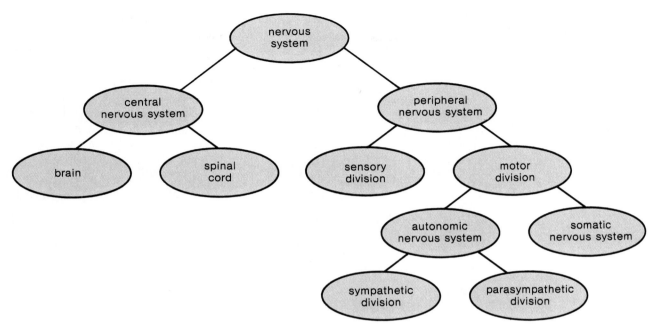

Figure 17.9   Organization of the human nervous system.

sensory and motor neurons. Figure 17.9 shows the relationships between the parts of the nervous system.

Two major activities are carried out by the spinal cord: integration of simple responses to certain types of stimuli and transfer of information to and from the brain. Spinal integration normally is in the form of a reflex—an automatic, involuntary response to a stimulus, such as withdrawal from a pinprick (see Figure 17.10). Most reflexes are fairly complex and sometimes involve many interneurons between the sensory and motor neurons.

Personality, consciousness, learning, and memory are all functions of the **cerebrum** (seh REE brum), in which the senses of seeing, hearing, tasting, smelling, and feeling also are experienced. The cerebrum is made

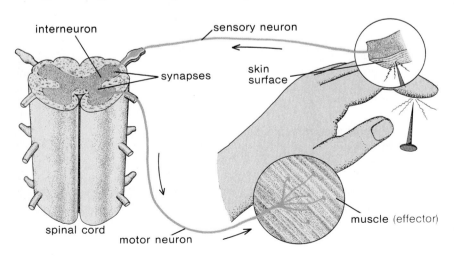

Figure 17.10   A reflex is an automatic response to a stimulus that occurs before the stimulus is received and interpreted by the brain, thus reducing the reaction time to a potentially life-threatening situation. Getting jabbed with a pin causes an impulse to travel up a sensory neuron to the spinal cord. The impulse travels through an interneuron to a motor neuron and on to the arm muscles. The arm immediately jerks away from the stimulus. At the same time, the stimulus is traveling through the same interneuron to the brain, where it is received and interpreted as pain.

Figure 17.11 The major parts of the human brain are associated with specific senses and abilities.

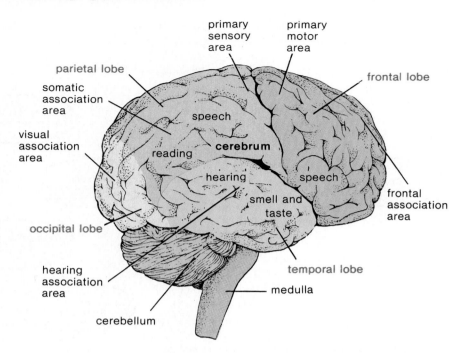

up of several lobes and is divided into a left and a right hemisphere, each having different functions. Verbal and analytical skills such as reading, speech, writing, mathematics, and logic generally are associated with the left hemisphere, whereas spatial relationships and artistic skills such as music and art are associated with the right hemisphere. The right hemisphere controls functions on the left side of the body and, conversely, the left hemisphere controls functions on the right side of the body. Figure 17.11 illustrates the structures and functional areas of the human brain.

Located under the cerebrum, the **cerebellum** (ser eh BEL um) coordinates the contractions of skeletal muscle. Maintenance of posture and muscle tone as well as coordination of movements such as walking and running are under the influence of this part of the brain. The **medulla** (meh DUL lah) regulates automatic, unconscious activities, such as the heartbeat rate, contractions of the blood vessels, and the depth and rate of breathing. Other functions include the control of coughing, sneezing, swallowing, and vomiting. The medulla keeps several of the body's vital functions operating properly without any conscious intervention.

17.6 The Peripheral Nervous System Transmits Information

The peripheral nervous system (PNS) includes all the nerve pathways through which the central nervous system communicates with the rest of the body. Sensory neurons carry signals from the external and internal environments *to* the CNS. Motor neurons, by contrast, carry signals *from* the CNS to muscles and glands. The motor division of the PNS can be further

The Human Eye and Vision

All the input from sensory neurons to the central nervous system arrives in the same form—as nerve impulses. To the brain, every arriving nerve impulse is identical to every other one. The information the brain derives from sensory input is based on the frequency with which these impulses arrive and on the neuron that transmits the input. If the optic nerve, for example, is stimulated, the impulses generated are perceived as a flash of light. If the auditory nerve is stimulated, the stimulation is perceived as a noise. Most receptors are either chemical or mechanical, except for the receptors in the eye. Vision, the perception of light, is dependent on receptors that can detect the electromagnetic energy present in light.

The human eye is sensitive to almost 7 million shades of color and about 10 million variations of light intensity. What makes this sensitivity possible? Light waves that enter the eye through the cornea (shown in a) are bent toward the lens. The iris (the colored part of the eye) consists of smooth muscles that can contract and relax involuntarily, changing the size of the pupil and thus regulating the amount of light that enters the eye. The waves pass through the lens and are focused on the retina, which has several different layers of tissue (b). Specialized sensory cells called rods and cones receive stimuli resulting from the incoming light waves (c). Rods are responsible for peripheral vision (the edges of the field of vision), and they respond to dim light for black-and-white vision. Cones are responsible for daylight color vision. Each cone contains pigments that are most sensitive to wavelengths corresponding to one of the colors red, green, or blue.

The structure of the pigments in both rods and cones is altered by light, producing a change in the electrical charge of the cells. This change in charge is a signal to neighboring neurons of the presence of light. Signals converge and diverge among neurons of the retina. They even may spread out and inhibit activity in nearby cells. As a result, the frequencies at which signals are transmitted to the brain change in response to changes in the size, location, and intensity of light stimuli. The neurons of the retina eventually relay the signal to the optic nerve fibers. From there, the signals travel to a part of the brain called the thalamus and then on to visual processing centers in the cerebrum.

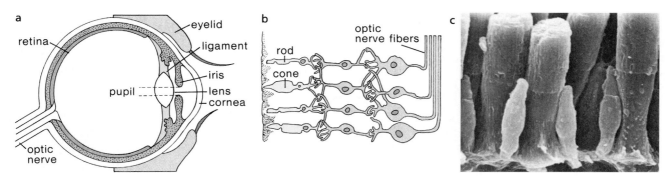

A cross-section diagram of the human eye (a); layers of tissue in the retina (b); rods and cones (c).

subdivided into the somatic and the autonomic nervous systems (as diagrammed in Figure 17.9). Motor neurons in the somatic nervous system carry signals to skeletal muscles. The somatic nervous system often is considered "voluntary" because the skeletal muscles are subject to conscious control (although much skeletal movement is the result of reflexes). Motor neurons in the autonomic, or "involuntary," nervous system act to coordinate organ function and help to maintain homeostasis.

The autonomic nervous system itself is made up of two divisions—the sympathetic and the parasympathetic. These two divisions generally have opposite effects on an organ, acting much like an accelerator and a brake pedal. For example, the sympathetic division speeds up heart rate in emergencies, and the parasympathetic division slows it down again. For the most part, the sympathetic division serves as a stimulator and plays an essential role in preparing the body to respond to emergency situations. Figure 17.12 shows some of these contrasting actions.

The autonomic nervous system helps to control blood pressure, stomach movements and secretions, urine formation and excretion, body temperature, and many other functions necessary to maintenance of a steady state. It also plays a major role in emotions. The functions of the autonomic nervous system, in turn, are controlled by the hypothalamus, a part of the brain located under the cerebrum. In addition to receiving and sending nerve impulses, the hypothalamus has specialized cells that produce hormones, and thus it serves to coordinate the nervous system with the endocrine system.

Figure 17.12 The autonomic nervous system is composed of two divisions that act like a brake and an accelerator. The parasympathetic division generally acts to maintain normal body function. The sympathetic division acts to prepare the body for emergency. The two divisions have opposing effects on a variety of organs and body functions.

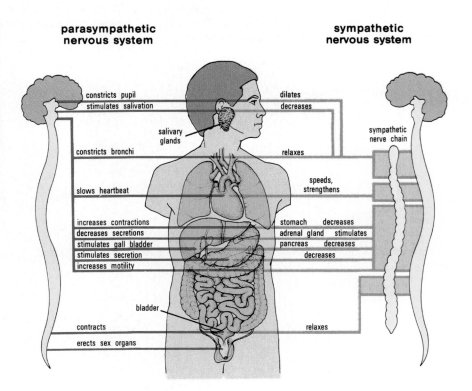

Concept Review

1. What is the relationship between a stimulus and a nerve impulse?
2. Briefly describe the electrical and chemical changes involved in a nerve impulse.
3. How does the central nervous system operate to coordinate body functions?
4. How does the autonomic nervous system function in maintenance of homeostasis?

The Endocrine System

17.7 Hormones Are Chemical Messengers

Even before you were born, the endocrine (EN doh krin), or ductless, glands and the hormones they secrete were functioning in the homeostatic control of your body. The amount of growth you experience, the regulation of your blood sugar levels, the regulation of the calcium level in your bones and blood, and the rate at which you metabolize food are only a few examples of the endocrine glands at work. Endocrine glands are capable of sensing the level of specific hormones in the blood and responding with an increase or decrease in their own secretions. This network of glands, illustrated in Figure 17.13, makes up the endocrine system, which acts with the nervous system to coordinate body functions. Because these systems work together so closely, they sometimes are referred to as the neuroendocrine system. Endocrine response, however, is slower and longer lasting than that of the nervous system.

Although hormones are secreted directly into the bloodstream and carried all through the body, they produce their effects only on specific target

Guidepost: How are the nervous and endocrine systems coordinated?

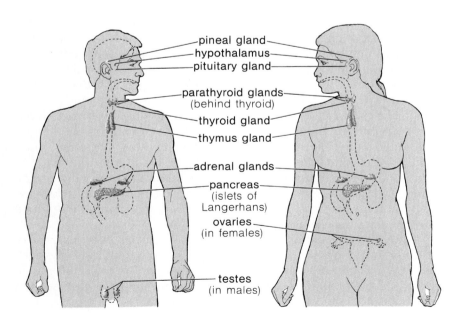

Figure 17.13 The endocrine system. The dotted lines show organs such as the stomach and kidneys.

pineal gland
hypothalamus
pituitary gland
parathyroid glands
(behind thyroid)
thyroid gland
thymus gland
adrenal glands
pancreas
(islets of Langerhans)
ovaries
(in females)
testes
(in males)

organs, such as the ovaries or testes of the reproductive system discussed in Chapter 6. This specific action is possible because the cells of the target organ have receptors that fit the shape of a particular hormone. When molecules of the hormone reach the cells of the target organ, they bind to the receptors, much the way an antigen binds to an antibody. This binding begins a series of events that cause the target organ to take particular actions.

A hormone can be effective for as short a time as two minutes or as long a time as several hours. Hormones constantly are being removed from the blood by the target organs, excreted by the kidneys, or broken down to simpler compounds inside the liver. The effect of a hormone on the cells of its target organ depends to a large degree on the level of the hormone concentration in the blood. Feedback systems maintain the concentration of most hormones within a predetermined range. A feedback system operates as a cycle in which the last step affects the first step. Each endocrine gland continuously receives information, or feedback, through the blood in the form of chemical signals about the cellular process its hormones affect. If conditions change, the gland can act on the information and adjust its rate of secretion.

17.8 Several Hormones Act Together to Control Blood Glucose Levels

Control of blood glucose demonstrates how a feedback system works. Endocrine tissues embedded in the pancreas produce two hormones: **insulin** (IN suh lin) and **glucagon** (GLOO kuh gahn). These are the primary regulators of glucose levels in the blood, although several other hormones are involved in different ways. Low levels of blood glucose provide feedback to the pancreas, which causes its cells to secrete glucagon. Glucagon increases the conversion of glycogen to glucose in the muscles and liver, thus increasing blood glucose levels. The feedback is negative because the release of glucagon reverses the low level of blood glucose, raising it toward an average level. Insulin has the opposite effect of glucagon and is secreted in response to high levels of blood glucose. Insulin enhances the uptake of glucose by muscle and liver cells, thus removing glucose from the blood. Insulin also is regulated by a negative feedback system. The release of insulin reverses the high level of blood glucose, lowering it toward an average level. The negative feedback systems that control the production of insulin and glucagon, as well as other negative feedback systems that regulate most endocrine glands, thus work to maintain homeostasis.

Insulin and glucagon have opposite effects in the homeostatic control of the level of blood glucose. After a meal, glucose levels rise as carbohydrates are digested. As a result, insulin secretions increase and glucagon secretions decrease. Conversely, during fasting, insulin secretions decrease and glucagon secretions increase. Figure 17.14 summarizes these blood glucose control mechanisms.

Maintenance of appropriate glucose levels is crucial. In the absence of glucagon, blood glucose levels may drop too low, resulting in hypoglycemia (hy poh gly SEE mee uh), or low blood sugar. Hypoglycemia is dangerous because brain cells must obtain a constant supply of glucose from the blood, and if the supply is cut off, the person may lose consciousness.

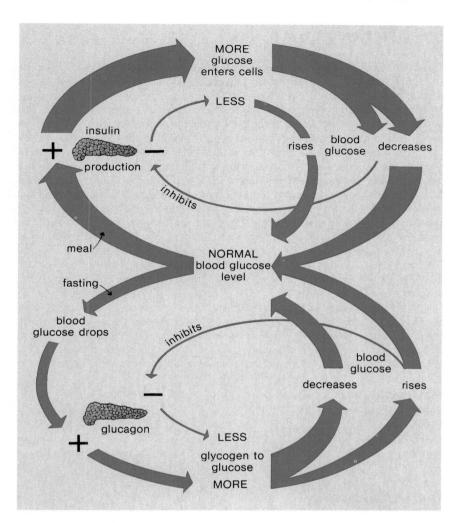

Figure 17.14   Insulin and glucagon function together to maintain a fairly stable level of blood glucose.

In the absence of insulin, blood glucose levels tend to remain elevated, causing changes in many body systems. Complications resulting from the high blood glucose levels may include circulatory disorders and vision problems that can lead to blindness. In addition, the kidneys are unable to reabsorb all the glucose and, as a result of increased glucose in the collecting ducts, less water is removed and returned to the blood. Both glucose in the urine and thirst are symptoms of insulin deficiency. The lack of insulin also prevents glucose from entering body cells, which then metabolize fats and proteins instead. Products of fat and protein metabolism enter the blood and lower the pH, interfering with many body functions. At high levels, these products can cause coma and death.

The disease diabetes mellitus (MEL luh tus) results from insufficient insulin secretion. There are two major forms of diabetes mellitus, both apparently with a hereditary basis. In type I diabetes, the immune system attacks and destroys the insulin-secreting cells of the pancreas; type I diabetics are therefore insulin dependent. The causes of type II diabetes are not well understood, but type II often can be controlled by diet and drugs that stimulate the pancreatic cells to secrete insulin.

INVESTIGATION 17.3   A Bike Trip

During summer, an all-day bike trip can be fun, but it needs planning. In this computer simulation, planning is especially necessary for rider 1, an insulin-dependent diabetic. Among the factors to be considered are conditioning, diet, pretrip meals, breathing rate, heart rate, and levels of liver glycogen, blood glucose, body fluid, and blood lactic acid. These internal conditions are regulated by homeostatic mechanisms and vary from one person to another. This simulation considers only a few of these factors.

Figure 17.15 shows the conditions of the trip as they appear on the computer screen. The bike trip starts and ends at home; midway is a forest park where you can stop as long as you like. The temperature will rise to 95° F by midafternoon, and the trip covers hills as well as level stretches. The left bar on the screen shows the grade of the road at each point of the trip. Uphill is a positive percentage, with a maximum grade of 6 percent. Downhill is a negative percentage; a flat road shows as 0 percent. You control the speed of travel by using the dial to change the bicycle gears. This simulation assumes a constant rate of pedaling. The bikes can cover more distance in tenth gear but climb hills more easily in lower gears. You must watch the demand and level of both glucose and fluid. If either level drops too low, you will "crash." If one rider crashes, the other(s) must stop and the trip is over.

Cooperation is essential as you develop strategies and test your hypotheses about how to complete the bike trip, beginning with the pretrip selections—insulin dose for the diabetic rider, breakfast for all riders, and the food and drinks to pack in the saddlebags. Your task is to cooperate with your team in planning the best strategy for the trip. This is not a race—the goal is to complete the trip and return home.

Materials (per team of 7)

Apple IIe, Apple IIGS, Macintosh, or IBM
    computer
LEAP-System™
2 dials

Procedure
PART A   Practice Trip

1. In each team, a group of three controls each rider, and one team member is the keyboard operator. Decide which group will control rider 1 (diabetic) and which

Figure 17.15 Appearance of the computer screen for the bike trip with two riders, one across the top, the other across the bottom. From left to right, the bars show grade percentage, glucose demand, blood glucose, fluid demand, and body fluid level. The box to the left of the bars shows the gear, which is controlled by the dial.

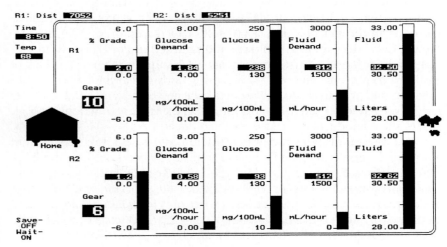

TABLE 17.1   Pretrip Selection Record

Menu: Biketrip

	Trip No.	1	2
Saddle bag weight allowed (grams)		_____	_____
Current saddle bag weight (grams)		_____	_____

Rider Startup Information

	Trip No.		
R1 Insulin dose:		_____	_____
R1 Breakfast:		_____	_____
R2 Breakfast:		_____	_____
R3 Breakfast:		_____	_____

Meals/Drinks

	Trip No.		
Peanut butter & jelly sandwich (120g)		_____	_____
Ham & cheese sandwich (9g)		_____	_____
Tuna fish sandwich (90g)		_____	_____
Pizza slice (120g)		_____	_____
Hog dog in bun (90g)		_____	_____
Water (250g)		_____	_____
Sport drink (250g)		_____	_____
Orange juice (250g)		_____	_____
Milk (240g)		_____	_____
Diet soda pop (360g)		_____	_____
Soda pop (360g)		_____	_____
Vegetable juice (180g)		_____	_____

Snacks

	Trip No.		
Hard boiled egg (60g)		_____	_____
Granola bar (30g)		_____	_____
Chocolate bar (45g)		_____	_____
Potato chips (30g)		_____	_____
Carrot sticks (30g)		_____	_____
Apple (120g)		_____	_____
Orange (120g)		_____	_____
Peanuts (30g)		_____	_____
M & M's (45g)		_____	_____
Canned pudding (120g)		_____	_____
Cheese (30g)		_____	_____

rider 2. Each group includes one gear shift dial operator and two advisers—one to watch glucose demand and level, and one to watch fluid demand and level. Designate one of the advisers to be a recorder also. The dial operator changes gears according to the grade and input from the advisers. Each group must cooperate to enable its rider to complete the bicycle trip. Each group also must cooperate with the other rider's group, because if either rider crashes, both riders stop and you must start again with another strategy.

2. In your data book, prepare a pretrip selection record like Table 17.1 and a trip selection record like Table 17.2, or tape in the tables your teacher provides.

3. *For this practice trip, the team should work with the diabetic rider only.* Use the LEAP-System "run" procedure to select **BIKETRP1.SPU** for one rider. The menu shown in Figure 17.16 will appear on the screen.

TABLE 17.2   Trip Selection Record

Trip No. _____

Rider No.	Stop No.	Distance Covered	Computer Time	Meals/Drinks/ Snacks Taken	Consequences/ Conclusions

Figure 17.16 Menu screen for the bike trip.

Menu: BIKETRIP
Saddle Bag Weight Allowed (grams) 5000
Current Saddle Bag Weight (grams) 1935

RIDER STARTUP INFORMATION	
R1 Insulin Dose:	Low
R1 Breakfast:	Bacon and Eggs
R2 Breakfast:	Captain Krunch
R3 Breakfast:	Not going today

MEALS/DRINKS	
Peanut Butter Jelly Sandwich (120g)	2
Ham & Cheese Sandwich (90g)	0
Tuna Fish Sandwich (90g)	0
Pizza Slice (120g)	0
Hot Dog in Bun (90g)	0
Water (250g)	2
Gatorade (250g)	2
Orange Juice (250g)	2
Milk (240g)	0
Diet Soda Pop (360g)	0
Soda Pop (360g)	0
Vegetable Juice [V8] (180g)	0

SNACKS	
Hard Boiled Egg (60g)	0
Granola Bar (30g)	1
Chocolate Bar (45g)	0
Potato Chips (30g)	0
Carrot Sticks (30g)	0
Apple (120g)	1
Orange (120g)	0
Peanuts (30g)	0
M & M's (45g)	1
Canned Pudding (120g)	0
Cheese (30g)	0

F10-Exit Menu   ↑ ↔ ↓   Select   RET-Enter Box

4. Follow the prompts at the bottom of the screen to make pretrip decisions about insulin dose, breakfast, and what to pack in the saddlebags. Each rider is allowed 2500 g. As you make selections, the computer displays total grams in the BIKETRIP box. If you overload the saddlebags, the computer will beep. Adjust the contents as needed. Record all selections in Table 17.1.

5. Follow the prompt on the screen to start the bike trip. To stop for a snack or a meal, press your rider number. A window will appear showing the items and the quantity of each remaining in the saddlebags. Follow the prompts to select items for your rider. Use Table 17.2 to record for each stop the rider number, stop number, distance covered, time from the screen, the snack/drink taken by the rider, and the resulting changes in glucose and/or fluid levels.

6. Watch the changing levels of glucose and fluid on the screen as the bike trip proceeds. Use the LEAP-System "reinitialize" procedure to restart the trip. The menu will reappear.

7. Experiment with different doses of insulin, with different breakfasts, and with different selections of food and drinks for the saddlebags. In your data book, record what problems you encounter and what steps you can take to overcome them.

8. Repeat the trip as often as time permits, recording the selections for each run.

PART B   Planning the Trip

9. As a class, discuss the relationships between glucose, insulin, and exercise, and the effects of exercise on body fluid and fluid demand.

10. Decide on several trip strategies that account for all the variables you encountered in your practice trips. This could include not only what to eat for breakfast and what foods to take along but when to stop for food and/or drink. What assumptions are you making in your strategy? Record these assumptions in your data book. For example, if you stop for orange juice, what are you assuming about the effect of the orange juice on your ability to continue?

11. In your data book, write if/then hypotheses that include the strategies and assumptions from Step 10. The "if" part of each hypothesis should contain the conditions specified in the strategies. The "then" part of each hypothesis should contain the predicted outcome, assuming your conditions are met. For example, "If breakfast is skipped, then the rider will need to stop for a snack at 9:00 AM."

PART C The Trip

12. Start the trip as in Step 3, but select **BIKETRP2.SPU** for two riders or **BIKETRP3.SPU** for three riders. (All riders share the same saddlebags.) Follow your strategies carefully. Record any changes in strategies made during the trip. Record snacks/meals for all riders in the trip selection record.

13. Revise your strategies, if necessary. Develop a new hypothesis about how the trip will be completed, including insulin dosage, breakfast, and saddlebag contents. Record your hypothesis in your data book.

14. Use your new strategies and rerun the bike trip.

Discussion

1. Compare the results of your strategies. Were the prediction parts of your hypotheses correct? What changes did you make in the second trip?

2. Compare your strategies with those of other teams in the class. How were they similar? How were they different? Which strategies resulted in successful completion of the bike trip?

3. Describe the trip strategy that produced the most successful result in the bike trip. Tell why it was better than another strategy used.

4. What effect did the insulin dose have on rider 1's performance in the trip?

5. What effect did the food you selected for breakfast have on the outcome of the trip?

6. What effect did the type of food and drink you packed in the saddlebags have on the outcome of the trip? What effect did the time at which you stopped to eat or drink have?

7. Describe how cooperation between riders influences the outcome of the trip.

8. What effect does temperature have on the outcome of the trip? What body functions are influenced by temperature?

9. Now that you know the effects of selecting the proper food and drink for a bike trip, describe how what you eat and drink can be important in your everyday activities.

10. How does the diet of an insulin-dependent diabetic differ from the diet of another bike rider on the trip?

11. Write a short paragraph describing what you have learned about the type of activities in which an insulin-dependent diabetic can participate and what limitations diabetes can place on the person's life.

17.9 The Hypothalamus and Pituitary Control Other Endocrine Glands

Chief among the endocrine glands is the pituitary, which secretes hormones that control the rate at which several other endocrine glands function. The pituitary, in turn, is controlled by hormones from the hypothalamus. Some of these hormones pass through blood vessels into the pituitary, causing the pituitary to release its own hormones. Other hypothalamic hormones pass through secretory neurons and are stored in the pituitary for later release. Pituitary hormones control the thyroid and the adrenal glands and, through the reproductive organs, regulate reproductive activities such as the menstrual cycle and ovulation, sperm production, embryonic development, and the production of the secondary sex characteristics. Figure 17.17 illustrates the relationship between the hypothalamus

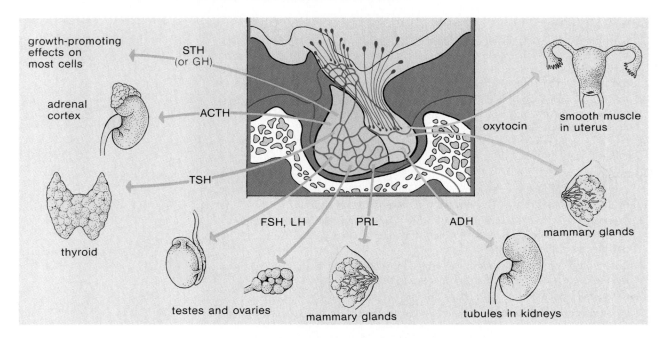

Figure 17.17 Connections between the hypothalamus and the two lobes of the pituitary. Under the influence of hormones from the hypothalamus, the anterior pituitary secretes six hormones, which affect the target organs shown. The hypothalamus also secretes two hormones that are stored in the posterior pituitary and released at a later time, affecting the target organs shown.

and the pituitary. The secretions of the hypothalamus and the pituitary are regulated by feedback mechanisms among these two glands and their target tissues. Table 17.3 on page 478 presents a summary of hormones produced in humans and their roles in coordination and homeostasis.

Located at the base of the neck, the thyroid gland secretes the hormone **thyroxine** (thy ROK sin), which regulates the rate of cellular respiration. If too little thyroxine is produced, the rate of cellular respiration decreases. Food is stored rather than used in energy release, resulting in an increase in weight. Other effects are sluggishness, sleepiness, and lowered body temperature. Too much thyroxine, on the other hand, increases the rate of cellular respiration, so little food is stored. The individual usually loses weight, has excess energy, and is nervous. The thyroid is controlled by the hypothalamus and pituitary in the feedback system shown in Figure 17.18.

There are two adrenal glands, one on top of each kidney. Each adrenal gland has two parts that function as two separate endocrine glands—the outer, larger adrenal cortex and the inner, smaller adrenal medulla shown in Figure 17.19. The adrenal cortex secretes several different steroid hormones that supplement the action of the sex hormones secreted by the reproductive organs, regulate salt and water balance, fight inflammation, or help the body respond to stress. Release of some of these hormones is regulated by a feedback system involving the hypothalamus and pituitary.

The adrenal medulla secretes two major hormones in response to nerve signals from the sympathetic division of the autonomic nervous system. The hormones, **epinephrine** (ep ih NEF rin, also known as adrenalin) and **norepinephrine**, help increase mental alertness, raise blood pressure, speed

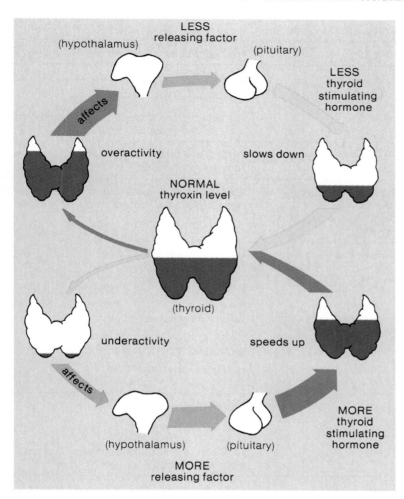

LESS
releasing factor
(hypothalamus)
(pituitary)

LESS
thyroid
stimulating
hormone

overactivity

slows down

NORMAL
thyroxin level

(thyroid)

underactivity

speeds up

affects

(hypothalamus)   (pituitary)

MORE
thyroid
stimulating
hormone

MORE
releasing factor

Figure 17.18   The feedback mechanism between the pituitary, hypothalamus, and the thyroid glands. This mechanism controls the level of thyroxine in the blood.

Figure 17.19   The structure of the adrenal gland, showing the cortex and the medulla.

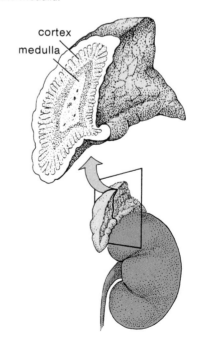

cortex
medulla

up heart rate, dilate coronary blood vessels, increase respiratory rate, and speed up the rate of metabolism. These effects—often called the fight-or-flight response—supplement the immediate response of the sympathetic division and allow the body to react to threats or emergencies. Epinephrine also can enhance an athlete's performance by stimulating the breakdown of glycogen, thus providing additional energy sources to the cells.

Concept Review

1. If hormones are transported throughout the body in the circulatory system, how are they able to act on a specific target organ?
2. Describe how a simple feedback system operates.
3. Compare the functions of insulin and glucagon in the control of blood glucose.
4. Describe the interrelationships of the hypothalamus and the pituitary.
5. Describe the feedback loop that controls secretion of thyroxine by the thyroid.
6. How does the adrenal gland function in the fight-or-flight response that enables the body to respond to emergency situations?

TABLE 17.3   Origins and Effects of Hormones

Gland/Organ	Hormone	Target	Principal Action
Adrenal cortex (outer portion)	aldosterone	kidneys	affects water and salt balance
	androgen, estrogen	testes, ovaries	supplement action of sex hormones
	cortisol	general	increases glucose, protein, and fat metabolism; reduces inflammation; combats stress
Adrenal medulla (inner portion)	epinephrine, norepinephrine	general (many regions and organs)	increase heart rate and blood pressure; activate fight-or-flight response
Heart	atrial natriuretic factor (ANF)	blood vessels, kidneys, adrenal glands, regulatory areas of the brain	regulates blood pressure and volume; excretion of water, sodium, and potassium by the kidneys
Hypothalamus	corticotropin-releasing hormone (CRH)	anterior pituitary	stimulates secretion of ACTH
	gonadotropin-releasing hormone (GnRH)	anterior pituitary	stimulates secretion of FSH and LH
	prolactin-inhibiting hormone (PIH)	anterior pituitary	inhibits prolactin secretion
	somatostatin	anterior pituitary	inhibits secretion of growth hormone
	thyrotropin-releasing hormone (TRH)	anterior pituitary	stimulates secretion of TSH
Hypothalamus (via posterior lobe of pituitary)	antidiuretic hormone (ADH, vasopressin)	kidney	controls water excretion
	oxytocin	breasts, uterus	stimulates release of milk; contraction of smooth muscle in childbirth
Pancreas (endocrine tissues)	glucagon	liver	stimulates breakdown of glycogen to glucose
	insulin	muscles, liver cells	lowers blood sugar level; increases storage of glycogen

Stress, Drugs, and the Human Body

17.10   Stress Is the Body's Response to Change

The physiological changes that characterize the fight-or-flight response affect the entire body and are known collectively as stress. Changes in the environment that cause stress are called stressors. Stressors may be physical, such as stepping in front of an oncoming car, exposure to extreme heat or cold, infections, injuries, prolonged heavy exercise, lack of sleep, or loud sounds. Stressors also may be emotional, for example fears about real or imagined dangers, personal losses, anger at a boss or teacher, or unpleasant social interactions (or lack of social interactions). Even pleasant stimuli, such as feelings of joy or happiness, can act as stressors.

To understand what happens to the body during a stress reaction, imagine a stressful situation in your life. First, you experience a change in

TABLE 17.3   (continued)

Gland/Organ	Hormone	Target	Principal Action
Parathyroid	parathyroid hormone (parathormone)	intestine, bone	stimulates release of calcium from bone; decreases excretion of calcium by kidney and increases absorption by intestine
Pineal	melatonin	unknown in humans— perhaps hypothalamus and pituitary	regulation of circadian rhythms (day/night cycles)
Pituitary (anterior lobe)	adrenocorticotropic hormone (ACTH)	adrenal cortex	secretes steroid hormones
	follicle stimulating hormone (FSH)	ovarian follicles, testes	stimulates follicle; estrogen production; spermatogenesis
	growth hormone (GH)	general	stimulates bone and muscle growth, amino acid transport, and breakdown of fatty acids
	luteinizing hormone (LH)	mature ovarian follicle, interstitial cells of testes	stimulates ovulation in females, sperm and testosterone production in males
	prolactin	breasts	stimulates milk production and secretion
	thyroid stimulating hormone (TSH)	thyroid	secretes thyroxine
Reproductive organs ovaries	estrogen	general	development of secondary sex characteristics; bone growth; sex drive
	progesterone	uterus (lining)	maintenance of uterus during pregnancy
testes	testosterone	general	development of secondary sex characteristics; bone growth; sex drive
Thymus	thymosin	lymphatic system	possibly stimulates development of lymphatic system
Thyroid	calcitonin	intestine, kidney, bone	inhibits release of calcium from bone; decreases excretion of calcium by kidney and increases absorption by intestine
	thyroxine	general	stimulates and maintains metabolic activities

your environment that you consciously or unconsciously decide is a stressor. Then, your brain sends signals to the hypothalamus, which activates stress responses in the adrenal gland and other body tissues. These responses help your body prepare for vigorous muscular activity—to fight or flee. If your stressor is like most in today's society, however, neither action is appropriate, and the net result to the body is like stepping on the accelerator of a car while the brake is on. A lot of fuel is burned, but you do not get anywhere.

Long-term stress lasting days or weeks can cause considerable damage to the body. People under constant stress have increased levels of adrenal hormones in their blood and urine. One of these hormones (cortisol) depresses the activity of the immune system, resulting in a lowered resistance to infectious diseases. Long-term stress is related to higher rates of high blood pressure, atherosclerosis, and ulcers in people who are relatively

Figure 17.20 Changes in the internal or external environment lead to physiological changes in the body that prepare it for a fight-or-flight response. These changes are initiated by impulses arising from the hypothalamus that directly affect body tissues and by impulses directed toward the adrenal medulla that cause it to release epinephrine, which affects the same target tissues.

young. Psychological problems such as insomnia, anxiety, and depression also are associated with prolonged stress. Figure 17.20 summarizes the effects of stressors on the human body. Stress management offers ways to reduce stressors in one's life and get rid of the extra energy the body builds under stress. It is healthy to learn how to manage long-term stress and reduce its effect on the body. Table 17.4 suggests ways to manage stress.

TABLE 17.4   Recommendations for Managing Stress

1. Do not take on too many responsibilities. The most common stressor is trying to do too much with the time and mental and physical resources you have.

2. Balance your time. Make sure you have time to work and to play. Learn how to work effectively but also teach yourself how to relax—through relaxation techniques, music, hobbies, and/or exercise. The word *recreation* is just that—a "re-creation" of energies for your mind and body. It can stop the stress response.

3. Get enough rest, sleep, and exercise. Allow your body and mind to renew.

4. Work off the energy created by stress. Take a break and blow off steam by exercising regularly. Do something you enjoy, such as playing tennis, biking, walking, swimming, or playing basketball.

5. Avoid self-medication. Drugs, including alcohol and tobacco, might seem to produce a *temporary* reduction of stress, but drugs do not cure the causes, and they have harmful side effects. In addition, they can be addictive.

6. Talk to someone you trust about the stressors in your life. Your parents, a friend, or a school counselor might help you see how you could avoid some stressors.

7. Check with your doctor or school nurse if you suspect you have a stress-related medical problem.

17.11   Psychoactive Drugs Affect the Central Nervous System

For some people, the effort to manage stress is difficult and confusing. They seek to escape stress and to find a short cut to confidence and satisfaction through the use of chemical substances that affect the nervous system. The functions of the central nervous system can be modified in a variety of ways, particularly by psychoactive (sy koh AK tiv), or mind-altering, drugs. Drugs are nonnutritive substances that change the way the body or brain functions. Used properly, certain drugs are important in maintaining health. Drugs include medicines, such as antibiotics and aspirin, that are used to cure or treat diseases or to relieve minor discomforts. Psychoactive drugs, which include tranquilizers, opiates, barbiturates, and alcohol, may be prescribed to relieve severe pain, aid sleep, and calm nervousness. Because these substances work on the brain, they can change the way a person thinks, feels, or acts. Many of them are used illegally, without a prescription, for their pleasurable side effects.

Any drug can be abused, that is, used in a way that causes personal or social problems. For example, overuse of aspirin can lead to severe abdominal bleeding. Consumption of too much caffeine in coffee, tea, cocoa, or cola drinks can lead to irritability and sleeplessness. Abuse of psychoactive drugs, in addition to causing a variety of physical effects, can lead to more serious problems. Many of the psychoactive drugs induce tolerance, so that larger and larger amounts are required to produce the same effect. Most also are addictive, producing physical or psychological dependence. In physical dependence, the body becomes unable to function normally without the drug. Withdrawal may produce symptoms ranging from discomfort to convulsions, which can be fatal. In psychological dependence, the user exhibits an intense craving for the drug and its effects. Sources, Uses, and Effects of Some Psychoactive Drugs in Appendix 3 summarizes information about the more commonly used psychoactive drugs.

Psychoactive drugs can be classified as depressants, stimulants, or hallucinogens. Depressants reduce the activity of the central nervous system, some of them by blocking nerve transmission. Alcohol, the most widely used and abused depressant, releases inhibitions and may produce feelings of happiness. Tranquilizers dull the emotions and reduce anxiety by calming and relaxing the user, but overuse can lead to psychological or physical dependence. Opiates (opium, morphine, heroin) are used to relieve pain and relax muscles; they also produce effects similar to tranquilizers. Opiates induce tolerance and are physically addictive. Barbiturates dull the mind and are used to induce sleep and relax tensions. Like opiates, they induce tolerance and are addictive. Barbiturates are especially dangerous when used with alcohol, opiates, anesthetics, or tranquilizers.

Stimulants increase activity of the central nervous system and usually induce a feeling of well-being. Examples include caffeine, the nicotine in tobacco, amphetamines, cocaine, and crack, a form of cocaine. Stimulants increase alertness and reduce appetite. Cocaine and crack, in addition, create intense feelings of happiness and can block pain. All stimulants can induce tolerance and produce powerful psychological and physical dependence. Nicotine and crack are two of the most addictive drugs known. Figure 17.21 illustrates the distorting effect of amphetamines on the building of a spider's web.

Figure 17.21   Many of the complex behavior patterns of animals can be disturbed by chemical substances. A female spider builds webs in the characteristic regular form (a). Several hours after feeding on sugar water containing a small dose of amphetamine, the spider built the unusually small, irregular web (b). Although we cannot compare directly the behavior patterns of invertebrate animals and humans, experiments such as this provide important information about the effects of chemical substances on brain function.

a

b

Hallucinogens are drugs that alter thinking, self-awareness, emotion, space-time perception and the perception of incoming sensations. Marijuana, the most widely used of all hallucinogens, contains substances that affect transmission of nerve impulses in the brain. Long-term use may lead to loss of motivation and interest in daily life. LSD blocks nerve impulses and distorts sensory perceptions, producing hallucinations and mood swings. Peyote and psilocybin are naturally occurring hallucinogenic substances that have been used widely in religious ceremonies. The effects of all hallucinogens are related to the expectations of the user, the setting in which use takes place, and the potency of the drug.

Concept Review

1. How does long-term stress affect the body, and what can you do to reduce its effects?
2. What is meant by drug abuse?
3. Contrast physical and psychological dependence.
4. Compare and contrast the general effects of depressants, stimulants, and hallucinogens.

Summary

The nervous and endocrine systems work in concert to bring about movement and homeostasis of internal systems. Muscle contractions are initiated by nerve impulses triggered by external and internal signals; bones, joints, and muscles work together to bring about the desired movements. A nerve impulse is a wave of chemical and electrical changes that travels over the length of a neuron, crossing the synapse between neurons by means of specialized neurotransmitters. Interpretation and coordination of nerve impulses resides in the central nervous system, which receives information about the external and internal environments through sensory neurons and sends messages about the action to be taken through motor neurons to muscles and endocrine glands. Most integration occurs in the brain, which is the seat of conscious activities such as speech, learning, memory, and the senses, as well as unconscious activities such as breathing and heartbeat rates. In addition to carrying information to and from the brain, the spinal cord integrates reflexes such as withdrawal from a pinprick.

The hypothalamus integrates the network of endocrine glands that constitutes the endocrine system, which secretes hormones directly into the bloodstream. The hypothalamus secretes hormones that control the secretions of the pituitary and other endocrine glands through negative feedback mechanisms. Based on messages from sensory receptors throughout the body, the hypothalamus influences a variety of body functions and is especially important in the fight-or-flight response. Stress is the sum of the physiological changes that occur in response to stressors, which may be physical and/or emotional. Psychoactive drugs, which affect the central nervous system, cannot cure the causes of stress. Long periods of sustained stress do major harm to the body, so learning to manage stress is an important skill.

Applications

1. When and how would food eaten 24 hours before exertion affect the amount of energy available for the muscles? What would be the effect of food eaten 4 hours before?

2. Explain the action of the muscles of the neck as you bring your left ear to your left shoulder and then move the right ear to the right shoulder.

3. Do all organisms need muscles to move? Explain your answer.

4. Why do muscles get tired and cramped? Relate physical activity to the physiology of the muscle.

5. A thermostat in your house works on a feedback system. Explain how this feedback system works, and how it resembles several systems in your body.

6. How do the structures of the eye relate to color blindness?

7. Do you get more muscle fibers when you pump iron, or do your muscle fibers just get bigger?

8. Beginning with a stimulus and ending with a movement, place the following in order: synapse, effector, receptor, sensory neuron, motor neuron, interneuron, muscle.

9. Drinking too much coffee, tea, or cola may cause the heart to skip a beat or muscles to twitch involuntarily. What is the explanation?

10. Describe how a pituitary tumor in a teenager could affect the onset of sexual maturity.

11. Select one endocrine gland and describe how a negative feedback loop works with the secretion of that gland and the hypothalamus and pituitary gland.

12. Do you have the same kind of stressors your parents have? How are they similar, and how are they different?

13. How does being able to express your feelings to someone help relieve stress symptoms?

14. Explain how having a hobby outside of an occupation can help reduce the effects of long-term stress.

15. What would be the effects of a drug that heightens perception but depresses the function of the hypothalamus? How would it affect the performance of a high-wire walker, a musician, or a basketball player?

Problems

1. Make a working model of an arm (different from the one you made in Investigation 17.1) showing the relationships between bones, muscles, and joints.

2. Compare the skeletons of several different types of animals, including birds, four-legged animals, and snakes. Can you determine where the muscles were attached to the bones or how the bones helped each animal move?

3. Make a graph showing the fastest recorded times in the marathon through history and another graph of the fastest recorded times in the 100-meter dash. What is the trend in times? Do you think there are any limits to how fast people can run?

4. Research any one physical fitness fad, such as jogging, racketball, or aerobics, and determine its relative value for cardiovascular fitness.

5. Invite a speaker from Al-anon or Alateen to speak to your class about alcohol dependence.

6. What are some different methods to help people to stop smoking? Why do you think some methods are successful and others are not? Develop a program to be used in your school involving teachers, students, staff, and administrators to help people quit smoking. Challenge each other to stop.

7. Biofeedback has been used to control functions such as blood pressure and pulse. Research the validity of claims to control autonomic or subconscious functions.

8. In 1986, two young, strong, apparently well-conditioned athletes died from heart attacks caused by cocaine. What might be the physiological mechanism?

9. Study the use of hormones in the raising of beef cattle, poultry, and dairy animals, as well as in the training of athletes.

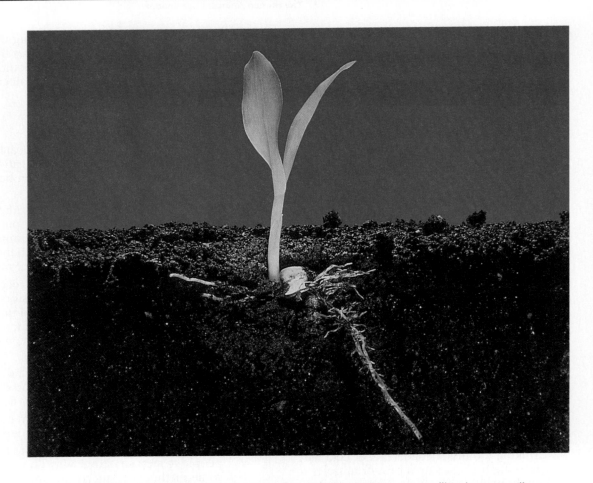

How would you design an experiment to learn if this young corn seedling is responding to a source of light? What other forces are acting on the growth of this plant?

The Flowering Plant: 18 Form and Function

Most plants live in two different environments at the same time—air and soil. Leaves and stems usually are above ground, where they are surrounded by air and where light, carbon dioxide, and oxygen are available. Roots are in the soil, which supplies water and nutrients. Each part of the plant is adapted for its environment and for a specific function in the life of the entire plant. Most leaves carry out photosynthesis, stems support the leaves and transport materials to other plant parts, and roots anchor the plant and absorb water and soil nutrients. The plant uses sugars produced during photosynthesis to build its structure.

How is the plant's structure produced, and how does each part function? How are the activities of all the parts coordinated as a whole? How are the various parts adapted for their environments? This and the next chapter begin to answer these questions for flowering plants, the most familiar and important plants for humans.

Leaf Structure

Guidepost: How are the parts of a leaf adapted for photosynthesis?

18.1 Leaf Structure Allows for Movement of Carbon Dioxide

Usually the main function of a leaf is to produce food through photosynthesis. How is its structure adapted for that function? The leaf of a flowering plant consists of an enlarged flat portion, the blade, connected to the plant stem by a petiole (PET ee ohl). The blade may be in one piece or divided into separate leaflets (a compound leaf), as shown in Figure 18.1. The large surface area of the blade permits maximum exposure to light.

Figure 18.1 Simple leaves (a) and a compound leaf (b). How does leaf structure enhance a plant's ability to carry out photosynthesis?

a   b

Figure 18.2   A portion of a leaf blade showing internal structures. Colors are diagrammatic only.

Figure 18.3   Scanning electron micrograph of a leaf cross section. Note that air can move freely throughout the interior of the leaf.

Figure 18.4   Scanning electron micrograph of the surface of a leaf showing open stomates.

Figure 18.2 illustrates the cellular structure of a leaf blade, which is made of several different tissues. Covering both the upper and the lower surfaces of the leaf is a transparent layer of cells called the epidermis. Epidermal cells secrete a waxy substance that forms a waterproof coating, the cuticle. Photosynthesis occurs in the **mesophyll** (MEZ oh fil) cells, which contain numerous chloroplasts. There are two types of mesophyll cells. In the upper, palisade layer, the mesophyll cells are elongated and packed together tightly. This space-saving arrangement exposes a maximum number of cells to light in a minimum amount of space. In the lower, spongy layer, the cells are rounder and packed loosely with many empty spaces between them. These empty spaces, which are clearly visible in the scanning electron micrograph in Figure 18.3, allow gases to move around freely and diffuse into all the mesophyll cells.

Substances enter and leave the leaf by two different routes: veins and stomates. Veins, which are continuations of the vascular tissues of the stem and root, supply the leaf cells with water and nutrients. Gases move into and out of a leaf by diffusion through millions of stomates (slitlike openings) in the leaf surfaces (see Figure 18.4). Because many of the stomates open into the spongy mesophyll's air spaces, the carbon dioxide used in photosynthesis and the oxygen used in respiration can reach all the cells of the leaf.

The concentration of carbon dioxide in the air is relatively constant. Within the leaf, however, the concentration varies. Cellular respiration goes on at all times in leaf cells, producing carbon dioxide and water as by-products. During the day, mesophyll cells use the carbon dioxide in photosynthesis. As a result, the concentration of carbon dioxide becomes lower in the leaf than it is in the air, and carbon dioxide diffuses in through the stomates. At night, photosynthesis does not occur. As a result of continuous respiration, the level of carbon dioxide builds up in the leaf. Eventually, it becomes higher than the level of carbon dioxide in the air and diffuses out.

INVESTIGATION 18.1   Water and Turgor Pressure

Some plants resist the forces of gravity, wind, rain, and snow because of the rigidity of the plant body. In herbaceous plants, however, many cell walls are thin and structurally weak. Their support depends on the turgor pressure the contents of individual cells exert against their walls. This turgor pressure varies as the volume of the cell contents changes when water is taken in or lost. In these plants, maintaining position is directly related to maintaining shape, which depends on turgor pressure.

▶ After reading the procedure, construct a hypothesis that predicts the change for each potato core.

Materials (per team of 3)

3 pairs of safety goggles	dissecting needle
3 lab aprons	test-tube rack
50-mL graduated cylinder	aluminum foil or plastic wrap
3 25-mm × 200-mm test tubes	paper towels
15-cm metric ruler	10% sucrose solution
balance	20% sucrose solution
glass-marking pencil	distilled water
cork borer (5- to 10-mm diameter)	white potato
scalpel	

SAFETY   **Put on your safety goggles and lab apron.**

Procedure

1. Label the test tubes A, B, and C.

2. Using a cork borer, cut three cores from a potato. With the scalpel, trim each core so it is at least 30 mm long. Make all cores as nearly the same length as possible. Place one core in each labeled test tube.

    CAUTION: Cork borers and scalpels are sharp; handle with care.

3. Copy Table 18.1 in your data book, or tape in the table provided by your teacher.

TABLE 18.1   Potato Core Data

Measurements	Core A (distilled)			Core B (10% sucrose)			Core C (20% sucrose)		
	1st day	2nd day	gain or loss (+ or −)	1st day	2nd day	difference (+ or −)	1st day	2nd day	difference (+ or −)
Length (mm)									
Diameter (mm)									
Mass (g)									
Rigidity (+,++,+++,++++)									
Volume (mL)									

Figure 18.5 (a) Read water volume from the bottom of the meniscus. (b) Units of measure for mass changes can be entered on the left and a graph plotted for changes in mass of potato cores at different water concentrations.

4. Measure the length and diameter of each core to the nearest millimeter and record the measurements in the table.

5. Determine the mass of each core to the nearest 0.1 or 0.01 g. Record the data in your table.

6. Determine the rigidity of each core by holding it at each end between your fingertips and gently bending it. Record the rigidity, using + + + + for very rigid, + + + for moderately rigid, + + for slightly rigid, and + for soft.

7. Measure the volume of each core by the following method. Pour water into the graduated cylinder until it is about half full. Hold the cylinder at eye level and read the line on the level with the lower part of the curved surface of the water, as in Figure 18.5a. This curved liquid surface is called the meniscus (men IS kus). Record this exact amount of water.

   Holding the core by a dissecting needle, sink it just under the water's surface and record the new water level. The difference between the two water levels represents the volume of the core in milliliters. Record the volume of each core in your table.

   CAUTION: Dissecting needles are sharp; handle with care.

8. Pour distilled water into test tube A until the core is covered. Pour 10% sucrose solution into test tube B until the core is covered, and pour 20% sucrose solution into test tube C in the same way. Cover each test tube with foil or plastic wrap, and store them in a test-tube rack until your next class period. What is the water concentration of the 10% sucrose solution? of the 20% solution?

9. Wash your hands thoroughly before leaving the laboratory.

10. The next day, remove the cores from the test tubes, blot them dry with paper towels, and repeat Steps 4–7.

11. Wash your hands thoroughly before leaving the laboratory.

Discussion

Review the discussion of diffusion in Investigation 5.2 and Section 5.7. Plasma membranes of potato cells are permeable to water and impermeable to sucrose molecules. The thin cell walls are permeable to both. The walls of these cells can be stretched or contracted only to a limited extent. Their shape when contracted, however, can be changed readily.

1. Complete the data table by calculating the differences, if any, between the initial and final measurements for each core. Use a + to indicate increase and a − to indicate decrease. What is the relationship between the concentration of water and the change in mass of the potato cores?

2. In your data book, prepare a graph such as the one in Figure 18.5b and plot the changes in mass of the three cores.

3. Predict the water concentration at which a potato core would not change in mass.

4. Describe an experiment that would test your prediction.

5. How does the change in volume compare with the change in mass for each core?

6. Which of the cores showed the greatest change in rigidity? Which showed the least change?

7. How does the change in rigidity compare with the change in mass for each core?

8. Explain the differences in rigidity among the cores placed in various concentrations of sucrose.

9. Do your data support the hypothesis you constructed at the beginning of this investigation? Explain.

10. What sources of error or other variables may have affected your results?

For Further Investigation

1. Using the same experimental approach, investigate the effects of 30% and 40% sucrose solutions. Compare the data obtained with the data derived from your original experiment.

2. Investigate the effects on elodea leaves of sucrose solutions of the same range of concentrations as used in this investigation. Mount detached leaves in the sucrose solution on a microscope slide, and add a cover slip. Locate several cells under low power and then observe under high power for 5 minutes.

18.2 Guard Cells Regulate the Rate of Water Loss

Although water molecules are present in the air as water vapor, they are never as abundant in the air as they are in the leaf. Thus, the plant loses water as it diffuses into the air through the stomates. This water loss is called **transpiration** (trans pih RAY shun).

When plant cells have adequate supplies of water, the water exerts a pressure, known as turgor (TER ger) pressure, against the cell walls. Turgor pressure supports the stem and leaves. If more water is lost from a plant by

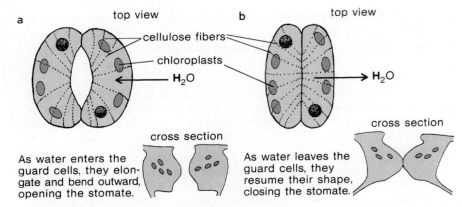

Figure 18.6   Guard cells control the opening of stomates. Because they are attached to each other at both ends and encircled by rigid cellulose fibers, the guard cells elongate and bend outward when they take up water.

transpiration than is replaced through the roots, the cells lose turgor pressure. As a result, the stem and leaves no longer are held upright and the plant wilts. If water is not replaced, the plant cells, and the plant, will die.

How can a plant conserve water and still allow carbon dioxide to enter? Each stomate is surrounded by a pair of specialized cells called guard cells. When the guard cells fill with water, they bend outward, as shown in Figure 18.6a. As they bend, the stomate opens and carbon dioxide can diffuse into the leaf. When a plant loses more water than it can replace, the turgor pressure of the guard cells decreases. The guard cells no longer bend outward, and as a result, the stomates close (Figure 18.6b). In this manner, the plant is able to reduce the loss of water under dry conditions. During these times, however, little photosynthesis takes place.

18.3   Some Leaves Have Functions Other Than Photosynthesis

Although the major function of most leaves is photosynthesis, many leaves photosynthesize little, if at all. These leaves have other functions that are adapted to the environmental conditions in which the plants live. For example, the spines on cacti (Figure 18.7) are modified leaves that are hard, nonphotosynthetic, and often dangerous to touch. Because water loss occurs mostly from leaves, plants with smaller leaves (and, therefore, less surface area) lose less water than do plants with larger leaves. In the dry habitats in which cacti are found, spines reduce water loss, and they help to protect the cactus from herbivores. In cacti, photosynthesis takes place in the stem rather than the spines.

Other plants in dry habitats have modified leaves that store water. These plants, known as succulents (SUK yoo lents) have tissues that are so juicy that water can be squeezed from them. After a rainstorm, the roots of succulents quickly absorb more water than the plant can use immediately. The water is stored in the leaves of succulents (and in the stems of cacti) until the plant needs it. A waxy cuticle so thick it can be scraped off with a fingernail also helps the leaves to hold in water.

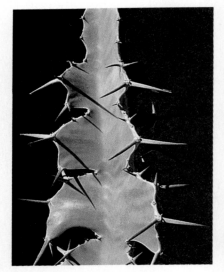

Figure 18.7   Cactus spines are modified leaves. How might this modification be adaptive to a cactus?

Although bogs are watery habitats with soil that often lacks nitrogen, most bogs are populated with great numbers of plants—including insectivorous plants. These plants also have leaves adapted for their environment. Venus's flytrap, the sundew, and the pitcher plant all have modified leaves that help attract and capture insects. The insects are digested within the leaf, and their nutrients, including nitrogen-containing compounds, are absorbed by the plant. In this way, insectivorous plants can live in the nitrogen-poor soils of bogs.

Concept Review

1. How is the arrangement of the two types of mesophyll cells in the leaf important to photosynthesis?
2. Explain why carbon dioxide moves into a leaf during the day.
3. How does turgor pressure provide support for a leaf?
4. How do guard cells function to prevent water loss through the stomates?
5. What advantage do small leaves have over large leaves in a dry habitat?
6. How do leaves of insectivorous plants help them live in nitrogen-poor soils?

Stems and Conduction

Guidepost: How are stems adapted for conduction and support?

18.4 Stems Support Leaves and Conduct Materials

The plant stem connects the leaves and roots and supports the leaves and reproductive organs in the light and air. Stems also carry water and nutrients from the roots to the leaves and sugars from the leaves to nonphotosynthetic parts of the plant. Air diffuses into stems through lenticels (LENT ih selz), small openings on the stem's surface, thus supplying oxygen to stem cells for cellular respiration. Stems may be either herbaceous or woody. Herbaceous stems are rather soft and rely on turgor pressure for support. Woody stems, on the other hand, have cells with thick, stiff walls that provide support.

Although stems differ greatly in structure, all have buds. A bud is a miniature shoot with a short stem, small immature leaves, and sometimes flowers. Buds on the bare twigs of deciduous trees are quite conspicuous, as shown in Figure 18.8 on the following page. Growth from a terminal (tip) bud lengthens the twig. Growth from a lateral (side) bud starts a new branch. During the winter, a bud usually is covered with tough, protective scales that are modified leaves. Bud scales fall off when growth resumes in the spring, leaving a ring of scars where they were attached. Each ring of scars indicates where growth began for that year. Because these bud scales are formed only in the fall, the age of a branch can be determined from the number of scars.

Some plants have stems modified for storage. White potatoes are underground stems modified for storage; their "eyes" are buds. (Sweet potatoes, which do not have buds, are roots modified for storage.) An onion bulb is a small stem with modified leaves attached. The upper part of each leaf is green and produces sugars through photosynthesis. The bottom part of each leaf is white, lies underground, and stores the sugars in the form

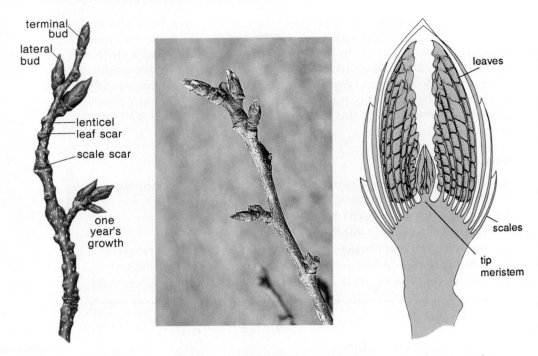

Figure 18.8 A diagram (a) and photograph (b) of a dormant woody twig, and a diagram of a longitudinal section of a terminal bud (c).

of starch. Figure 18.9 illustrates different types of underground stems. Most stems, however, grow above ground.

Figure 18.10 shows the microscopic structure of a young woody dicot stem in three dimensions. The stem is composed of several tissues. Around the outside is the epidermis, a single, protective layer of cells that will be replaced by bark as the stem grows older. Just inside the epidermis is a layer of thin-walled cortex cells that carry on photosynthesis or function as

Figure 18.9 Structural adaptations of underground stems. What do you think is the principal function of each type?

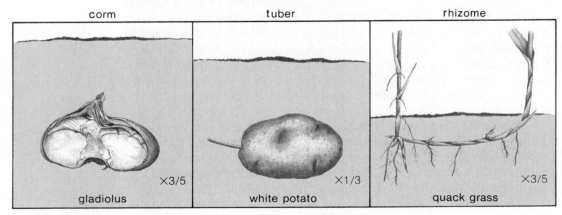

epidermis ——
cortex ——
cambium ——
pith ——
phloem ——
xylem ——

Figure 18.10   Diagram of a young woody stem showing microscopic structures. Colors are diagrammatic only.

storage areas, depending on the stem. Next to the cortex are the vascular tissues. **Phloem** (FLOH em) is the vascular tissue that transports sap containing dissolved sugars and amino acids from the leaves throughout the plant. **Xylem** (ZY lem) is the vascular tissue that helps support the plant and transports water and dissolved nutrients from the roots throughout the plant. Figure 18.11 illustrates these two tissues. At the center of a young woody stem there may be a group of loosely collected cells called the pith.

a

b

Figure 18.11   Stained, longitudinal section of xylem tissue (a) and phloem tissue (b). How do these tissues help a plant survive on land?

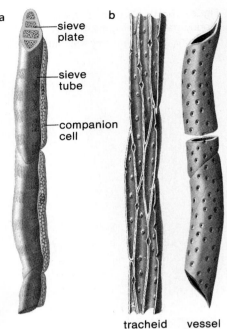

Figure 18.12 Cells from vascular tissues of flowering plants. Phloem cells (a) and xylem cells (b). What adaptations do these cells demonstrate?

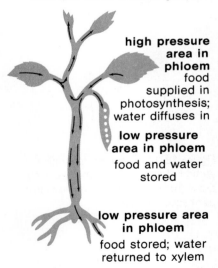

Figure 18.13 The pressure-flow hypothesis, based on fluid pressure, for transport of food and water in phloem.

The pith stores food in young stems and usually is destroyed as the stem grows older.

Phloem is composed of tubes formed from sieve cells. As each sieve cell develops, many small holes form in both end walls. The cytoplasm of one cell connects with the cytoplasm of adjacent cells through these holes, making a continuous cell-to-cell system of sieve tubes. As a sieve cell matures, its nucleus disintegrates. Located beside each sieve cell is a smaller, companion cell that has a nucleus. Companion cells appear to regulate the activity of sieve cells. Figure 18.12a illustrates the structure of phloem cells.

Lying between the phloem and xylem cells is the vascular **cambium** (KAM bee um). The cells of this tissue divide by mitosis and form other plant tissues. When a cambium cell divides into two new cells, one of these cells eventually becomes either a conducting or a supporting cell. The other cell remains a part of the cambium tissue and divides again. Phloem cells are formed continually at the outer surface of the cambium; xylem cells are formed at its inner surface. The cambium also forms fiber cells, which are interspersed with the vascular xylem and phloem and strengthen the stem.

Eventually, most of a woody stem consists of xylem. Wood, which is found in older stems, is made completely of xylem cells. Xylem usually is composed of two types of cells: tracheids (TRAY kee idz) and vessels. Tracheids are long cells with thick walls that grow to their full size and then die. The walls of these dead cells contain thin areas called pits. Pits in one tracheid are lined up with those in adjacent tracheids, allowing water and dissolved substances to pass easily from one to another. Vessels are made of long, thick-walled cells joined end to end to form miniature pipes that extend throughout the stem. The end walls of vessel cells disintegrate, and they, like tracheids, die. Both tracheids and vessels are arranged to form a continuous conducting system from the root, up through the stem, and out into the veins of the leaf. Figure 18.12b illustrates both types of xylem cells. You will be able to observe these structures in Investigation 18.2.

18.5   Pressure Within Phloem Cells Helps Move Sugars Down a Plant

Sugars move from the leaves to roots and other nonphotosynthetic parts of the plant through the phloem cells. Phloem cells contain living cytoplasm through which water and dissolved sugars must pass. The rate at which the fluid moves through phloem is thousands of times faster than diffusion alone could account for. What is the mechanism of phloem transport?

The best explanation of the movement of sugars through the phloem is the **pressure-flow** hypothesis. According to this hypothesis, water and dissolved sugars flow through the sieve tubes from an area of higher pressure to an area of lower pressure, as shown in Figure 18.13. Sugars secreted by mesophyll cells in the leaf move by active transport into the phloem cells. The resulting high concentration of sugar causes water to diffuse into the phloem cells, increasing the turgor pressure there. This area of higher pressure forces the sugar-water solution to move into the next phloem cell. In this manner, sugars are moved from cell to cell—from higher-pressure area to lower-pressure area—until they reach a cell where they will be used or stored. There, sugars are removed from the phloem by active transport. At

City Forester

Ron Morrow is the city forester for Colorado Springs, a job that includes caring for the city's trees, flower beds, and greenhouses. He also is a private forestry consultant who shows home-owners how to care for their trees.

Pursuing a lifelong interest in the natural world, Ron majored in forestry science in college, focusing on forest pathology—the study of insects and diseases that affect trees. After obtaining his master's degree, he worked in the Coronado National Forest in Arizona as a timber management specialist and as a guide and interpreter for visitors. Since becoming city forester 14 years ago, he has been responsible for maintaining the trees and flowers in new and existing parks and other city-owned land. He also has developed and implemented many new and innovative plans.

The care of trees involves both maintenance and disease control. Recently trimmed trees are more vigorous and less susceptible than untrimmed trees to damage from wind and snow loads. It also is necessary to ensure that the trees do not obstruct street signs or interfere with power lines. About 80 000 trees along the streets of Colorado Springs and 20 000 trees in the parks are individually maintained. If diseased trees are found, and pesticide spraying is ineffective, the trees (both living and dead) must be cut down and buried. Any trees that

die naturally are examined for signs of wildlife nesting. If such signs are found and the tree is not a hazard to people, it is tagged as a "wildlife tree" and allowed to remain in place. Every tree on city property is recorded on a computer. Information about the type of tree, its height, diameter, condition, and value is readily available. The computer also stores information about the history of the tree and any treatment it has received.

Twenty-two full-time employees and 10 to 20 seasonal employees for the forestry and greenhouse divisions work under Ron's direction. Four city greenhouses produce about 120 000 plants and flowers for more than 195 flower beds throughout the city. The greenhouse division also is responsible for the planting and maintenance of a formal garden. In cooperation with a local horticulture group, Ron has initiated a demonstration garden of native and drought-resistant plants.

Ron has helped to draft several new ordinances, including one that requires developers to maintain a certain amount of "green space" in each new project. Another ordinance requires that trees be planted and maintained in parking lots—the number of trees must be in proportion to the number of parking spaces.

Urban forests and parks, Ron observes, not only add to the beauty of a city but provide valuable habitat for many different animals, and the trees also serve as windbreaks. He believes, too, that trees and parks are important to mental health: "You get out, see open space and wildlife, and you feel better." A well-managed city forestry program can help to ensure that as a city grows, it can still be attractive to people and other animals.

495

this time, water also leaves the phloem cell. Thus, the flow of sugar and water from the leaves to the root can continue. The entire process is dependent on an uptake of water by phloem cells in the leaves and a loss of water by phloem cells in the storage areas or in areas of rapid growth.

18.6 Transpiration Pulls Water Up Through the Xylem of a Plant

Water must travel more than 100 m to reach the top of the tallest trees. Many experiments have shown that water from roots rises through root xylem to stem xylem to the xylem in the veins of leaves. This movement is against the force of gravity. Imagine you are standing with a long soda straw on top of a building three stories tall. The straw reaches into a bottle of root beer on the ground. No matter how hard you try, you cannot suck the root beer up from the bottle, not even with a vacuum pump. How do water and materials dissolved in the water move to the top of a tree?

Water is thought to move by a process known as **cohesion-tension**, illustrated in Figure 18.14. Under certain conditions, water can be pulled up a narrow tube if the water is in a continuous column. One condition is that the tube must have a very small diameter. A second condition is that the tube must be made of a material to which water molecules will adhere.

Figure 18.14 The principal water-conducting structures in a tree. Trace the path of water from the root hairs to the leaves.

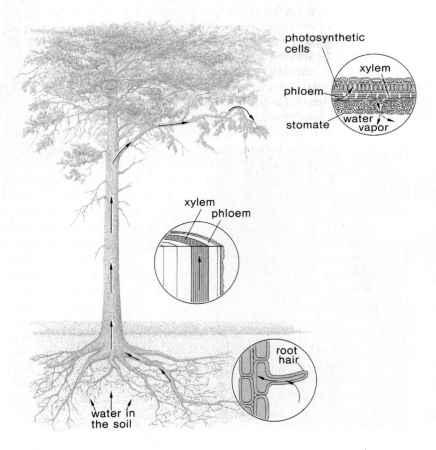

These two conditions exist within the xylem cells of plants. In addition, water molecules exhibit cohesion—they are attracted to adjoining water molecules, thus forming an unbroken chain of water molecules in each xylem tube. Tension is negative pressure that results from the pulling action of water molecules as they move out of the leaves during transpiration. The pull is transmitted through the column, and water is pulled from the roots up the plant to the leaves. Transpiration is the driving force of water transport, creating the negative pressure that keeps water moving from the soil into the roots and up through the stems into each leaf.

Concept Review

1. How are buds involved in stem growth?
2. In what way are leaves of succulents and leaves of onions alike?
3. Compare and contrast the structure and function of xylem and phloem.
4. How does the pressure-flow hypothesis explain the movement of sugars in the phloem?
5. Describe the roles of cohesion and transpiration in water movement.

INVESTIGATION 18.2   Leaves, Stems, and Roots: Structural Adaptations

The leaves, stems, and roots of flowering plants have structural adaptations that enable them to perform their specific functions efficiently. In this investigation, you will examine those plant structures and attempt to deduce the functions for which they are adapted.

Materials (per team of 2)

microscope slide
cover slip
dropping pipet
compound microscope
stereomicroscope
forceps
colored pencils (red, blue, brown, green)
prepared cross section of a leaf
prepared cross section of an herbaceous
   dicot stem

prepared cross section of a root
variety of mature plants
young bean plant
radish seedlings germinating in a petri
   dish
grass seedlings germinating on water
carrot

Procedure
PART A   Leaf Structure and Function

1. Look at the various plants and the different types of leaves. What might be the advantage of a flat, thin leaf blade to the photosynthetic capacity of a plant?
2. How are the leaves arranged on the stems of different plants? Relate this arrangement to the photosynthetic capacity of the plant.
3. Examine a prepared cross section of a leaf under the high power of your microscope. Compare your slide with Figure 18.2 on page 486 to become familiar with the various regions of the leaf. Consider its structures in relation to the functions of light absorption, water supply, and carbon dioxide absorption.
4. Locate chloroplasts in the leaf cells. What shape are they? Which leaf cells contain the chloroplasts? Among those cells, what difference in abundance of chloroplasts can you see?

5. Locate a cross section of a small vein in the center tissues of the leaf. The vein is surrounded by a sheath of cells that also are photosynthetic. Notice some empty cells with thick walls in the upper part of the sectioned vein. What might be the function of these cells?

6. Examine the covering layers of the leaf—the upper epidermis and the lower epidermis. These single layers of cells are covered by cuticle, which may have been removed when the slide was made. How might the cuticle affect the efficiency of photosynthesis?

7. Locate a stomate and its guard cells. Locate the spongy tissue just above the lower epidermis. How does the structure of the spongy tissue differ from that of the other leaf tissues?

8. Note how the stomates are located in relation to the spongy tissue. What does that relationship suggest about the function of the spongy tissue?

PART B   Stem Structure and Function

9. Observe the stem of the bean plant. Notice how the leaves are attached to the stem. The site of attachment is called the node (see Figure 18.15). What color is the stem? Judging from its color, what might be the function of an herbaceous stem? Would this be a function of all stems? Why or why not?

10. Using low power, study a cross section through the internode of a stem. Compare your slide with Figure 18.10 on page 493 to become familiar with the regions of the stem. Like a leaf, an herbaceous stem is covered by an epidermis. How many cells thick is the epidermis?

11. The most noticeable feature of the cross section is a ring of vascular bundles. Each of these vascular bundles is usually wedge-shaped, with the narrow end of the wedge directed toward the center of the stem. The center of the stem is made up of large, thin-walled cells. This is the pith. Between the outer edges of the vascular bundles and the epidermis is the cortex. Draw a circle about 15 cm in diameter. In the circle, outline the vascular bundles as they appear in your slide. Do not draw in any of the cells.

12. Using high power, look in the cortex and vascular bundles for thick-walled, empty cells. Show their location in red on your diagram. Suggest a possible function for these cells.

13. The leaves need water for photosynthesis. Water is absorbed by the roots and is conducted through the xylem of the stem to the leaves. Color the xylem in your drawing blue. In what structures of the stem did you look for xylem cells? What other functions might these cells have? What relation do these cells have to those previously noted in your stem drawing?

14. Materials manufactured in the leaves are transported to other parts of the plant through the phloem. Color the phloem in the diagram brown. Where in the stem is the phloem tissue? Was phloem also present in the leaf you observed earlier?

15. Food in the form of starch is stored in stems. Using high power, carefully examine the stem tissues to find cells that store food. Color the storage cells in the diagram green. How is the presence of stored food indicated in these tissues? What region of the stem is devoted primarily to storage?

PART C   Root Structure and Function

16. What part of a plant absorbs water?

17. Recall your study of the absorbing surfaces of animal digestive tracts (Chapter 15). What would you expect one of the characteristics of water-absorbing structures of a plant to be?

18. Some roots store more food than others. Examine the root system of the grass plant and the carrot. Which probably contains more stored food?

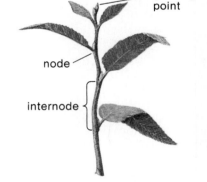

growing point

node

internode

Figure 18.15   A leafy stem.

19. Examine the radish seedlings in the petri dish. (Do not remove the cover from the dish.) Using the stereomicroscope, look through the cover at the fuzzy mass of tiny structures around the root. These structures are called root hairs. On what part of the root are the root hairs longest? What probably happens to the root hairs if a plant is pulled out of the soil?

20. With forceps, pick up a young grass seedling. Mount the seedling in a drop of water on a clean glass slide, without a cover slip. Using the low power of your compound microscope, find the root tip. (You can see that the root differs from the young shoot, which is green.) Look along the root for root hairs. Is a root hair made up of one or many cells? What relationship do root hairs have to the epidermis of the young root? (To answer this question, you may have to put a cover slip on the specimen and examine it with high power.) Are root hair cells living or dead? What important process is involved in the absorption of water by root hair cells?

    CAUTION: Forceps are sharp; handle with care.

21. Observe with low power the prepared slide of a cross section of a mature root. Locate the epidermis, cortex, xylem, and phloem. Draw a circle to represent the root cross section. Color the location of the xylem blue on your diagram. Color the location of the phloem brown. Where is the phloem in relation to the xylem?

22. Examine the cells and tissues of the root with high power, looking for stored food. Color the region of food storage on your diagram green. What region of the root contains the greatest amount of stored food? In what form is the food stored in the root?

23. Wash your hands thoroughly before leaving the laboratory.

Discussion

1. According to your observations of chloroplast location in the leaf, which are the main photosynthetic cells? How might the location of chloroplasts in these cells be advantageous in light absorption?

2. How are the leaf structures you observed adapted for leaf functions?

3. According to your observations, what are the major functions of the stem?

4. What stem structures are adapted for each of those functions?

5. How does the function of root epidermis compare with the function of stem and leaf epidermis?

6. How does the position of xylem in the root compare with its position in the stem?

7. What function of stems and roots is lacking in most leaves?

8. What function of leaves and young stems is lacking in roots?

9. What structures would a molecule of water pass through from the soil to the air just outside a leaf? List them in order.

Roots and Absorption

18.7 Roots Anchor a Plant and Absorb Materials

Roots anchor plants and help absorb water and nutrients. They also conduct materials and serve as storage organs. The vascular tissues of many roots form a cylinder in the center that is surrounded by the cortex, which may form part of the bark around an older, woody root. Sometimes food in

Guidepost: How are roots adapted for absorption of water and nutrients?

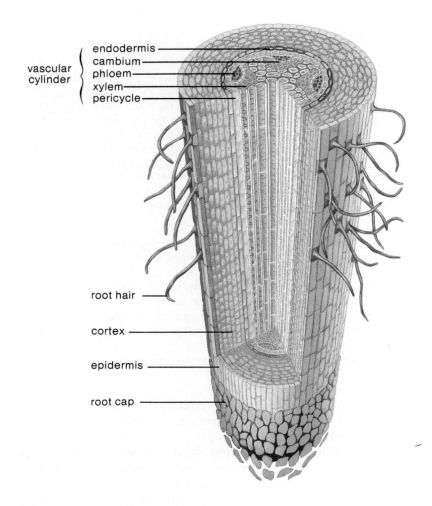

Figure 18.16   The terminal portion of a root. Colors are diagrammatic only.

the form of starch or sugars is stored in the cortex. Carrots, sugar beets, radishes, and turnips are examples of storage roots. Figure 18.16 shows the microscopic structure of the tip of a root.

In some species of plants, many branching roots come from the bottom of the plant. This is known as a fibrous root system, characteristic of such plants as corn, beans, and clover. Fibrous root systems, which are relatively close to the soil surface, hold soil in place and thus reduce erosion. In other species, the plant is anchored by a long, tapering root with slender, short, side branches. This is a taproot system, found in such plants as dandelions, mesquite, and carrots. Because their roots penetrate deep into the soil, plants with taproots can use water sources unavailable to plants with fibrous root systems. Figure 18.17 shows a variety of roots.

Most water absorption occurs through **root hairs**, which penetrate the spaces between soil particles and absorb through their huge surface area the water and nutrients required by a plant. Each root hair is a thin-walled

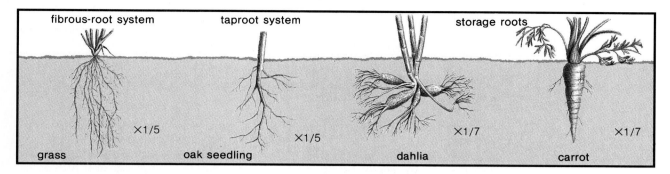

Figure 18.17 Roots are of two basic types, both of which may be adapted for storage.

extension of a single epidermal cell. In Figure 18.18 the root hairs appear as a fuzzy white zone just behind the tip of the seedling root. Because of its root hairs, a single rye plant only 60 cm tall is estimated to have a root system with a total length of 480 km and a total surface area of more than 600 m²—twice that of a tennis court.

If a plant is pulled out of the soil roughly, most of its root system may remain behind. Plants often wilt after being transplanted because most of the root hairs have been pulled off during the transplant, and without them the plant cannot absorb enough water to prevent wilting.

The cell walls of root hairs absorb water from the soil, much as paper towels soak up liquid spills. Water diffuses into the inner part of the root hair cell from the cell walls. Diffusing from cell to cell, the water eventually makes its way to a xylem cell. Here, the water begins its upward journey to the leaves of the plant, pulled by transpiration, as described in Section 18.6.

Figure 18.18 A radish seedling showing root hairs. How are root hairs adaptive?

18.8 Many Nutrients Move into a Plant by Active Transport

In addition to water, a plant requires a variety of elements for growth and maintenance. These nutrients enter the roots as ions dissolved in the soil water. Roots take in nitrogen, for example, in the form of ammonium or nitrate ions. Elements that a plant needs in relatively large amounts are called macronutrients. The most important of these macronutrients are nitrogen, phosphorus, and potassium. These three ingredients are identified by numbers such as 10-20-10 on commercial fertilizers, indicating percentage of the nutrient by weight. Elements such as zinc and copper, which are required in extremely small amounts, are called micronutrients. The essential elements have been identified by experiments in which plants were grown in solutions lacking a particular element (see Figure 18.19). These experiments have shown how the elements are absorbed and used by the plant, and what effect is produced when a particular element is missing. Table 18.2 on page 502 summarizes that information.

Nutrients usually are more concentrated in root hairs than in soil water. Absorption of these nutrients involves active transport, which requires energy. If root cells are deprived of oxygen, and thus are unable to respire,

Figure 18.19 Tomato plants showing nutrient deficiency. From left: potassium, phosphorus, and nitrogen deficient plants. The plant on the right received all essential elements.

TABLE 18.2   Elements Required by Plants

Element	Function	Deficiency
Macronutrients		
Calcium	Component of cell walls; enzyme cofactor; involved in cell membrane permeability; encourages root development.	Characterized by death of the growing points.
Magnesium	Part of the chlorophyll molecule; activator of many enzymes.	Development of pale, sickly foliage, an unhealthy condition known as chlorosis.
Nitrogen	Component of amino acids, chlorophyll, coenzymes, and proteins. An excess causes vigorous vegetative growth and suppresses food storage and fruit and seed development.	Early symptom is yellowing of leaves, followed by a stunting in the growth of all parts of the plant.
Phosphorus	Promotes root growth and hastens maturity, particularly of cereal grains.	Underdeveloped root systems; all parts of plant stunted.
Potassium	Enzyme activator; production of chlorophyll.	Pale, sickly foliage.
Sulfur	Component of some amino acids.	Chlorosis; poor root system.
Micronutrients		
Boron	Pollen germination; regulation of carbohydrate metabolism.	Darker color, abnormal growth; malformations.
Chlorine	Evolution of oxygen during photosynthesis.	Small leaves; slow growth.
Copper	Component of some enzymes.	Lowered protein synthesis.
Iron	Needed for chlorophyll production.	Chlorosis, appearing first in youngest leaves.
Manganese	Activates Krebs cycle enzymes.	Mottled chlorosis.
Molybdenum	Part of enzymes for nitrate reduction.	Severe stunting of older leaves.
Zinc	Component of some enzymes; internode elongation.	Small leaves; short internodes.

the absorption of nutrients slows. Absorbed nutrients move deeper into the root by active transport. Eventually, nutrients reach the xylem cells, through which they are conducted to the stem and leaves.

Concept Review

1. Compare and contrast the structures of fibrous root and taproot systems.
2. Explain the importance of surface area in the role played by root hairs.
3. Distinguish between macronutrients and micronutrients.
4. What is the evidence that nutrients move into and through the root by active transport?

Guidepost: What are the characteristics of growth in plants?

Plant Growth

INVESTIGATION 18.3   Seeds and Seedlings

A seed is a packaged plant. Within it is a complete set of instructions for growing a plant such as a maple tree. The seed contains everything needed to establish the plant. How does the change from seed to plant occur? What are the functions of the structures of the seed and the growing plant? What changing characteristics can you observe?

After reading the procedures, construct a hypothesis that predicts how petri dishes C and D will differ.

Materials (per team of 2)

2 pairs of safety goggles	petri dish with starch agar and boiled
2 lab aprons	corn grains—dish D
10X hand lens	Lugol's iodine solution in dropping bottle
scalpel	soaked germinating corn grains
petri dish with starch agar—dish A	soaked bean seeds
petri dish with plain agar—dish B	bean seeds germinated 1, 2, 3, and 10
petri dish with starch agar and	days
germinating corn grains—dish C	

SAFETY   **Put on your safety goggles and lab apron.**

PART A   The Seed
Procedure

1. Examine some of the external features of a bean seed. Notice that the seed is covered by a tough, leathery coat. Look along one edge of the seed and find a scar. This scar marks the place where the seed was attached to the pod.

2. Remove the seed coat and examine the two fleshy halves, the cotyledons, which are part of the embryo.

3. Using a scalpel, cut a little sliver from one of the cotyledons.

    CAUTION: Scalpels are sharp; handle with care.

    Test the sliver with a drop of Lugol's iodine solution. Record the results in your data book.

     WARNING: Lugol's iodine solution is poisonous if ingested, irritating to skin and eyes, and can stain clothing. Should a spill or splash occur, call your teacher immediately; flush the area with water 15 minutes; rinse mouth with water.

4. Separate the two cotyledons and find the little plant attached to one end of one of the cotyledons. Use a hand lens to examine the plant. You will see that this part of the embryo has two miniature leaves and a root. The small leaves and a tiny tip make up the epicotyl (*epi* = above; above the cotyledon) of the embryo. The root portion is the hypocotyl (*hypo* = below; below the cotyledon).

Discussion

1. What do you think is the function of the seed coat?
2. What would you guess is the function of a connection between a parent plant and a developing seed?
3. What would you deduce is the primary function of the cotyledons?
4. What was the original source of the substance of the cotyledons?

PART B   The Seedling
Procedure

5. Examine bean seedlings that are 1, 2, and 3 days old.
6. Compare the 10-day-old seedling with the 3-day-old seedling.

Discussion

1. What part of the plant becomes established first?
2. What part of the embryo gives rise to the root of the bean plant?
3. Where are the first leaves of the 3-day-old seedling?
4. What part of the embryo produces the first leaves?
5. What has happened to the cotyledons in the 10-day-old seedling?
6. Where is the seed coat in this plant?
7. Which part or parts of the embryo developed into the stem?
8. How are the first two tiny leaves arranged on the stem?

PART C   Corn Seeds
Procedure

7. Cut a soaked, germinated corn grain lengthwise with the scalpel.
8. Test the cut surfaces with a few drops of Lugol's iodine solution. Record the results.
9. The starch agar and plain agar in petri dishes A and B were tested with Lugol's iodine solution as a demonstration. Observe the dishes. Record the results in your data book.
10. On petri dish C, 2 or 3 corn grains have started to germinate on starch agar. Each grain was cut lengthwise, and the cut surfaces were placed on the starch agar for about 2 days. Petri dish D contains starch agar and boiled corn grains. Remove the corn grains from both dishes.
11. Cover the surface of the starch agar in petri dishes C and D with Lugol's iodine solution.
12. After a few seconds, pour off the excess.
13. Wash your hands thoroughly before leaving the laboratory.

Discussion

1. When you tested the cut surface of corn grains, what food was present?
2. What other foods might be present in the corn that were not demonstrated by the test?
3. What difference did you observe in the demonstration tests of petri dishes A and B?
4. What difference did you observe when you tested petri dishes C and D? Suggest hypotheses that might account for what you observed.
5. What kind of food would you expect to find in the areas where the germinating corn grains were?

18.9   A Seed Protects and Nourishes the Embryo Inside

A seed contains the offspring of a parent plant—the embryo—and stored food. The embryo is protected within the seed by a seed coat. Some seed coats are extremely thick, making digestion difficult for seed eaters and thus increasing the likelihood that the seed will pass through the digestive system unharmed. Other seed coats contain chemicals that inhibit germination, or sprouting, until appropriate growing conditions occur.

When most seeds germinate, the first structure to appear is the embryonic root. The root turns downward into the soil, anchoring the seedling

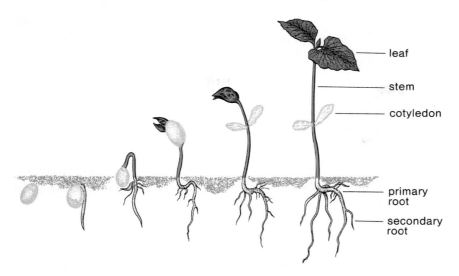

Figure 18.20   As a bean seedling develops from a seed, it grows both above and below the cotyledons.

leaf

stem

cotyledon

primary root

secondary root

and absorbing water and nutrients. The stem appears next, and in the case of many dicot seeds, it carries two green cotyledons up with it. Each cotyledon contains energy and nutrients for the seedling. As the seedling grows it uses these stored reserves. After it has grown true leaves that can produce food through photosynthesis, the cotyledons fall off. The seedling continues to grow, and its cells differentiate into all the tissues of a mature plant. Figure 18.20 shows these growth stages for a bean.

INVESTIGATION 18.4   Root Growth

In this investigation, you will observe the growth patterns of corn seedlings just a few days after they have germinated. Are all parts of the root involved in growth? Do all parts of the root grow equally? If not, which part grows more quickly? Why?

  After reading the procedures, construct a hypothesis that predicts the answers to these questions.

Materials (per team of 3)
3 petri dishes
6 pieces of paper towel to fit petri dishes
glass-marking pencil
waterproof ink
scalpel

8 cardboard tags
15-cm metric ruler
3 flat toothpicks
distilled water
12 germinated corn seeds

PART A   Setting Up
Procedure
Day 1
1. Select your seedlings. Observe the roots of one of them. Do you expect all parts of a root to grow equally? If not, which part would you expect to grow faster? Why? Label the petri dishes *1*, *2*, and *3* and add your team's symbol. Place a piece of paper towel in the bottom of each petri dish and add distilled water to dampen the paper. Pour off any excess water.
2. Select four seedlings. Using a toothpick and ink, carefully mark the shortest root with a straight, very narrow line exactly 2 mm from the tip. (You do not have to

mark the entire circumference of the root.) Be careful not to crush or damage the root. Draw as many lines as possible, 2 mm apart, behind this first mark. Repeat this procedure for the other three roots. All of the roots should have the same number of marks.

3. Measure the distance from the tip to the last mark. This is the initial root length.

4. On the moist paper towel in petri dish 1, carefully place the four marked seedlings so the markings are visible.

5. Mark the remaining eight seedlings as follows: using a toothpick and ink, place a dot 5 mm from the tip of each root. Be sure to handle the roots carefully, and do not let them dry out.

6. Using a scalpel, cut off 1 mm of the tip of two corn seedling roots; cut off 3 mm of the roots of two others; cut off 5 mm of two others. Leave the remaining two seedlings untreated.

 CAUTION: Scalpels are sharp; handle with care.

7. Label each seedling as to the amount of root cut off, using a small cardboard tag tied to each seedling. Place four of these seedlings in petri dish 2 and the other four in petri dish 3.

8. Cover the seedlings in all petri dishes with a piece of paper towel. Add water to moisten them. Press down lightly to ensure that the paper is firmly placed in the dishes and pour off any excess water. Place the dishes away from direct sunlight and heat. Do not let the paper dry out during the experiment.

9. Wash your hands thoroughly before leaving the laboratory.

Day 2

10. After 24 hours, uncover the seedlings you prepared in dish 1. Examine each seedling. Measure the distance between the tip of the root and the last mark. Measure the distances between each of the lines in order, from the tip to the base. Record your measurements.

11. Note the appearance of all the lines. Are all the lines as clear as they were yesterday?

12. Discard these seedlings, and clean or discard the petri dishes.

13. Wash your hands thoroughly before leaving the laboratory.

Discussion

1. Add together the lengths of all four roots, from tip to last mark. Divide by 4. Subtract from this length the average initial root length in Step 3. What is the average amount of growth for each seedling?

2. In your roots, did growth occur at the tip, at the base, or all along the root?

3. How much growth occurred between the tip of the root and the first mark? Between the last two marks?

4. How do you explain the smears or wider marks at or near the tip of the roots?

5. On the basis of these results, what do you predict will be the results of cutting off the root tips?

PART B   Comparing Results
Procedure
Day 3

14. After two days, examine the roots in petri dishes 2 and 3. Measure and record the distance from the original 5-mm mark to the tip of the root for each seedling.

15. Discard the seedlings and clean or discard the dishes.

16. Wash your hands thoroughly before leaving the laboratory.

Discussion

1. Add the measurements for the two seedlings that were not cut and determine the average. Has there been any growth in these two seedlings?
2. Determine the average amount of growth for the roots cut at 1 mm, 3 mm, and 5 mm.
3. Prepare a bar graph of the class data showing the amount of growth of roots cut at 1 mm, 3 mm, and 5 mm. Compare with the growth of uncut roots.
4. How important is the tip for growth of the root?
5. What information do these observations give you that the earlier observations did not?

For Further Investigation

1. Is the growth pattern of peas or beans the same as that of corn? Germinate some of these seeds and conduct the same experiment you did on corn.
2. Design and carry out an experiment to measure the growth rate of a leaf of a common houseplant.

18.10   Primary Growth Increases the Length of a Plant

Although a plant may be mature, it continues to increase in size throughout its life. A plant grows in length and in diameter, or girth. Growth in length is called primary growth because it produces primary tissues. Young plants contain only primary tissues, which always are herbaceous, or nonwoody. Growth in plant diameter is called secondary growth. Secondary tissues often are woody.

In plants, cells that have differentiated usually do not duplicate. Instead, new cells are formed by special tissues called **meristems** (MER ih stemz) that are found throughout the plant. Meristems form new cells through mitosis and cell division, which continue as long as the plant lives. Meristems are located at the tip of stems and roots, in buds, and between the xylem and phloem along the length of a stem, as shown in Figure 18.21.

The meristem responsible for growth in the length of the stem is located at the tip of the plant. Just behind the meristem is an area in which the newly formed cells grow longer. Behind this zone of elongation is a zone in which the cells differentiate into all the specialized tissues in the stem. Thus, the meristem gives rise to all stem tissues, including the cambium that produces new xylem and phloem cells.

The meristem responsible for root elongation is located just behind the root cap, a layer of protective cells covering the root tip. Part of this meristem forms the root cap cells, which are rubbed off as the root pushes through the soil. The root meristem also forms cells that differentiate into all the specialized root tissues. The meristems in the root and stem tips increase the length, or primary growth, of the plant. Thus, a plant grows from its top upward and its bottom downward. Figure 18.22 on page 508 compares the structures of a stem tip and a root tip.

In most leaves, all the cells differentiate into the leaf tissues (including xylem, phloem, and mesophyll) at an early stage of leaf formation. In late

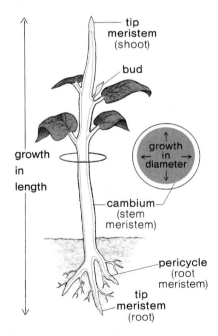

Figure 18.21 Location of plant meristems. Tip meristems bring about primary growth; cambium is responsible for secondary growth. Meristems in buds produce cells that differentiate into all the tissues of leaves, branches, and flowers.

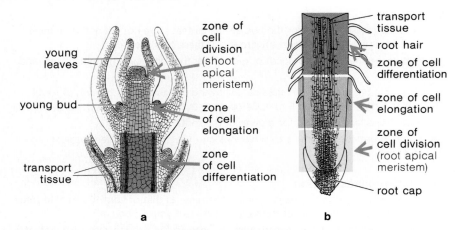

Figure 18.22 A longitudinal section of a stem tip (a) and a root tip (b). Notice the correspondence between the zones in the stem and the root.

summer, deciduous woody plants develop tiny, fully formed leaves inside their buds. In spring, the new leaves expand mostly by the enlargement of the small cells. Leaves of grasses and some other plants are an exception; they grow continuously from a meristem at the base of the leaf. These plants evolved where they were constantly grazed by large herbivores. The meristem at the leaf base is an adaptation to grazing—a leaf can grow even after most of the blade has been eaten.

18.11   Secondary Growth Increases the Diameter of a Plant

Vascular cambium is the meristem tissue that is responsible for secondary growth. Located between the xylem and phloem in the stem and root, vascular cambium produces new xylem and phloem cells throughout the life of a plant.

Ray cells, thin-walled living cells in the xylem and phloem, allow for horizontal transport in large stems. Pits in the walls of ray cells communicate with pores in both xylem and phloem. Ray cells supply water and nutrients to the phloem and sugars to the cambium cells.

Phloem cells, which are much more fragile than xylem cells, are crushed each year as the stem increases in diameter. The tough xylem cells, on the other hand, remain in the stem, forming one ring of cells each year. Figure 18.23 shows annual growth rings in a woody stem. Occurring predominantly in trees that grow in temperate zones, these annual rings can be used to determine the age of a tree as well as to reveal the growing conditions during a particular year. An extremely narrow ring could indicate bad growing conditions during the year the ring was formed. There could have been a drought or an infestation of leaf-eating caterpillars. In either case, the tree was unable to grow very much, and therefore the ring was small. The annual rings of xylem cells are the wood of the tree.

In older stems, the cambium remains as a boundary between the central core of wood (xylem) and the phloem and bark. Bark, which replaces

—ray cells

Figure 18.23   Section of a woody stem. How old was this tree? What separates bark from wood?

spring wood (abundant water)
summer wood (water not abundant)

phloem—
cambium—
—xylem

the epidermis during the first few years of a tree's life, is composed of tough, dead cells that protect the thin layers of living phloem and cambium underneath. Because the walls of cambium cells are thin and easily broken, bark usually can be peeled easily from a tree trunk.

Concept Review

1.   Describe the role of the cotyledons in a bean seed.
2.   What are the functions of cambium and other meristems?
3.   How does primary growth differ from secondary growth?
4.   What can annual rings tell about the life of a tree?

Summary

Each part of a flowering plant is adapted for a specific function. The leaf uses carbon dioxide, water, and light in photosynthesis. The mesophyll of the leaf is organized so that light and carbon dioxide can reach all the cells. Water loss through transpiration is reduced by the action of guard cells surrounding the stomates. The arrangement of the vascular tissues in stems provides support as well as transport of materials to all parts of the plant. Roots anchor the plant in the soil, and their thousands of root hairs are specialized to absorb water and nutrients. Water moves through the root cells until it reaches the xylem, where it is pulled up the root and stem by the force of transpiration. Pressure created by the sugar-water solution in phloem cells moves sugars to the nonphotosynthesizing parts of the plant. Leaves, stems, and roots may be modified for a variety of other functions.

A flowering plant begins its life as an embryo within a seed. The seed germinates, and the seedling develops characteristic plant structures. The plant grows in length and girth, forming new cells in meristems throughout its body. Growth depends on the production of sugars in photosynthesis and is regulated by the coordination of external and internal factors. These processes are discussed in the next chapter.

Applications

1. Water lily plants are rooted in mud at the bottom of ponds, but their leaves float on the water surface. How might the cellular structure of the roots, stems, and leaves of water lilies differ from that of terrestrial vascular plants?

2. In what ways do the usual environments of a plant root and a plant shoot differ?

3. What are the principal substances that pass into and out of plants through the leaves? What forces are involved with this movement?

4. During the growing season farmers spend much time cultivating their crops—loosening the soil between plants. What advantage does this have for the crop plants?

5. How would you distinguish a root, a leaf, and a stem?

6. Explain how root hairs illustrate the biological principle of the complementarity of structure and function.

7. What part of the young tree becomes a knot in the wood of an older tree? How can you tell?

8. Living plants used to be taken out of a hospital room at night for fear they would remove oxygen from the room and thus harm the patient. How would you explain this? Should people worry about this?

9. How can grasses grow back when you cut off the tips of the plants each time you mow your lawn?

10. Are root hairs more important in helping to anchor a plant or in helping to absorb water and nutrients? Explain.

11. Could wood cells be involved in active transport? Explain.

12. If you want to have *crisp*, sliced vegetables for dinner, why is it a good idea to keep the vegetables in fresh water?

13. Compare an animal's fertilized egg to a plant seed. In what ways are they similar?

14. Would you expect to find root hairs along the "zone of elongation" of a root? Explain.

15. Would you expect roots to have a cuticle? Explain your answer.

Problems

1. Investigate how growth in plants differs from growth in animals.

2. Ten years ago a farmer built a fence 1.5 m high and attached one end of it to a tree that was 7 m high. Now the tree has grown to a height of 14 m. How far above the ground is the attached end of the fence now? Explain.

3. In a middle-latitude biome, a pine and an apple tree are growing side by side. Compare the amounts of water lost by these two trees and their growth rates throughout the year.

4. The following questions concern lateral growth in woody stems:
   a. Within a given biome, how would the annual ring formed in a wet year differ from one formed in a dry year?
   b. Sometimes two rings are formed in one year. How might this happen?
   c. What is the science of dendrochronology and how is it used?

d. Would you expect to find annual rings in the bark of the tree? Why or why not?

5. Much of the grain of wood is the result of the pattern of annual rings. Examine different pieces of furniture and see if you can determine the direction in which the wood was cut. Can you clearly see annual rings?

6. Make a collection, or take photographs, or make drawings of leaves from a variety of plants, noting the type of habitat and environmental conditions in which they grow. (Make sure none of the leaves are poisonous or dangerous to touch.) What relationships can you find between (a) the size, shape, color, and coverings of leaves, and (b) where they grow?

7. Create a work of art based on the tissues of plants. Many tissues produce intricate, visually appealing patterns when seen in cross section.

8. Using any materials you like, make a working model of a stomate surrounded by a pair of guard cells. Make the guard cells control the size of the stomate.

9. Investigate the growth of roots by growing a seedling in a narrow box with two plates of glass as the sides (similar to an ant farm). Fill the box with dirt, and observe and record the growth of the root system.

10. Using a tree increment borer, take a sample of wood from a local tree. Can you determine the age of the tree, using annual rings as an indicator? What years were good growth years and what years were bad growth years for the tree? Can you link these years to any environmental factors in your area?

What do you think is the composition of the bubbles given off by this aquatic plant, called coontail (*Ceratophyllum demersum*)? What does the plant need to produce these bubbles?

The Flowering Plant: Maintenance and Coordination

Like other multicellular organisms, plants are complex systems made of millions of cells that require energy for their activities. Energy for cellular work is made available through the processes of photosynthesis and cellular respiration. During photosynthesis, green plants convert light energy into chemical energy as they form ATP. The chemical energy in ATP can be used directly for cellular activities in the plant or can be stored in sugars. Photosynthesis also produces the oxygen gas necessary for cellular respiration. During cellular respiration the energy stored in sugars is transferred to ATP. Thus, photosynthesis maintains not only the plants themselves but also most of the consumers of the biosphere.

Each plant cell contributes to the life of the entire plant. The activities of all the individual cells must be coordinated for the plant to grow, mature, flower, and form fruit. That coordination is provided by a variety of plant growth substances called hormones. This chapter describes some of the internal activity of plants, how that activity is coordinated by the diverse effects of various plant hormones, and how plants respond to environmental stimuli.

Photosynthesis

19.1 Photosynthesis Takes Place in Chloroplasts

Guidepost: How is the sun's energy changed into chemical energy in photosynthesis?

Photosynthesis is a series of reactions in which plants use the sun's energy to synthesize complex, energy-rich molecules from smaller, simpler molecules. In eukaryotic cells, all of these reactions take place in the structures known as chloroplasts. Even when they are removed from a cell in a laboratory, chloroplasts can carry on the entire process of photosynthesis by themselves.

Electron micrographs such as the one in Figure 19.1 on page 514 reveal the internal structure of chloroplasts. They show a highly organized array of internal membranes called thylakoids (THY luh koyds). Thylakoids may form stacks of flattened, disk-shaped structures called grana (GRAY nuh; singular: granum). Chlorophyll, other pigments, and enzymes are embedded in the thylakoids. Surrounding the thylakoids is the stroma (STROH muh), a colorless substance that contains other enzymes as well as DNA, RNA, and ribosomes.

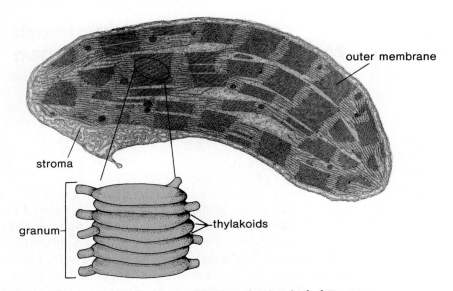

Figure 19.1   Electron micrograph of a chloroplast in a leaf of *Zea mays* (X24 000). The darker areas are stacks of thylakoids called grana; the drawing shows the structure of a granum enlarged still more.

19.2 Photosynthesis Requires Light

Most life on earth depends on the continual release of light energy from the sun. Of the sunlight that reaches our planet, only about 1 percent is involved in photosynthesis. The rest is absorbed or reflected by clouds or dust in the earth's atmosphere, or by the bare ground and bodies of water. Visible light, which is only a small fraction of the energy coming from the sun, consists of a spectrum of colors, each with a different wavelength and energy content, as shown in Figure 19.2. When light strikes an object it may be transmitted, absorbed, or reflected. In plant cells, several different pigments can absorb light energy. Each pigment is a chemical compound that

Figure 19.2   Radiations from the sun form a continuous series. The range of radiations organisms can detect with their eyes—visible light—is roughly the same range plants use. Shorter wavelengths (blue light) are more energetic than longer wavelengths (red light). Leaves are green because chlorophyll absorbs red and blue wavelengths and only a little green.

Figure 19.3 The upper curve shows the action spectrum for photosynthesis. The lower curves show absorption spectra for chlorophylls *a* and *b*. It is clear from the graph that both the chlorophylls absorb the wavelengths of light used in photosynthesis. Which wavelengths do these chlorophylls absorb most? Least?

absorbs only certain wavelengths of light and reflects or transmits all others. Green plants, for example, appear green because chlorophyll absorbs most wavelengths of visible light except green, which it reflects.

Photosynthesis depends on the green pigment chlorophyll, which occurs in several forms. Chlorophyll *a* apparently is present in all photosynthetic plants; other forms may be present in different plants in different combinations. Figure 19.3 shows an absorption spectrum—a simple graph that displays the percentage of light absorbed by a pigment at each wavelength—for chlorophylls *a* and *b*. Note that the chlorophylls absorb much of the light in the blue/violet and orange/red wavelengths and very little or none in the green/yellow wavelengths. Plants can utilize only the energy from absorbed wavelengths. The action spectrum at the top of Figure 19.3 shows this clearly. An action spectrum measures the rate of photosynthesis at certain wavelengths of light. In the wavelengths that are strongly absorbed, the rate of photosynthesis is high. In green light, the rate of photosynthesis is much lower because the pigment chlorophyll reflects those wavelengths.

In addition to chlorophylls, plants contain other pigments such as those visible in the coleus leaves shown in Figure 19.4. Some of these accessory pigments work with chlorophyll to trap and absorb additional wavelengths of light. Their absorbed light energy is transferred to chlorophyll *a* before being used in photosynthesis. Accessory pigments are responsible for fall coloration of leaves. The pigments become more visible as the chlorophyll content of the leaves declines.

Figure 19.4 These coleus leaves contain visible accessory pigments.

19.3 Photosynthesis Involves Many Interdependent Reactions

Three major events occur in photosynthesis: (1) absorption of light energy, (2) conversion of light energy into chemical energy, and (3) storage of chemical energy in sugars. The reactions by which these events occur may be considered in two distinct but interdependent groups, shown in Figure 19.5.

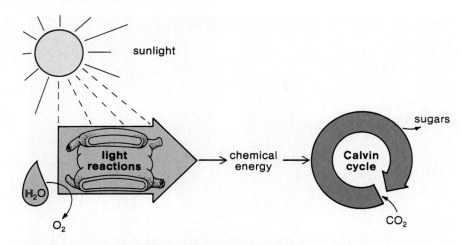

Figure 19.5   Diagrammatic summary of the reactions of photosynthesis.

In the first group, the **light reactions**, light energy is absorbed and converted into chemical energy as short-lived, energy-rich molecules are formed. These molecules then are used to make 3-carbon sugars from carbon dioxide in the second group, the series of reactions known as the **Calvin cycle**. In this cycle, chemical energy is stored in the sugars, and new carbon is incorporated into the plant for future growth.

The overall reactions of photosynthesis may be summarized as follows:

$$3\ CO_2 + 3\ H_2O \xrightarrow[\text{chlorophyll}]{\text{light energy}} C_3H_6O_3 + 3\ O_2$$

| (carbon dioxide) | (water) | | (3-carbon sugar) | (oxygen gas) |

This equation indicates only the major raw materials and some of the end products. It does not show the many chemical steps and other substances involved. The next sections explain those steps in more detail.

INVESTIGATION 19.1   Gas Exchange and Photosynthesis

Following the principles of diffusion, carbon dioxide normally would flow into a leaf during the day, and oxygen would flow out as photosynthesis proceeds. In this investigation, you will disrupt the normal gas exchange process and observe the effects.

▶ After reading the procedure for Part B, construct a hypothesis that predicts which leaf treatment will affect photosynthesis the most.

Materials Part A (per team of 2)

microscope slide	forceps
cover slip	scalpel
dropping pipet	fresh leaves, several types
compound microscope	

Materials Part B (per team of 5)

5 pairs of safety goggles	forceps
5 lab aprons	scissors
400-mL beaker	sheet of white paper
4 petri dish halves	paper towels

3 swab applicators
plastic bag labeled *waste*
500-mL beaker of water
petroleum jelly

isopropyl alcohol (99%) in screw-cap jar
Lugol's iodine solution in screw-cap jar
Histoclear™ in small corked bottle
2 potted plants

PART A   Number of Stomates
Procedure

1. While holding the lower surface of a leaf upward, tear the leaf at an angle. The tearing action should peel off a portion of the lower epidermis. It will appear as a narrow, colorless edge extending beyond the green part of the leaf.

2. Lay the epidermis on a slide and use a scalpel to cut off a small piece. Add a drop of water and a cover slip. Do not allow the epidermis to dry out.

   CAUTION: Scalpels are sharp; handle with care.

3. Using the low power of your microscope, locate some stomates. Then switch to high power. Make a sketch to show the shape of a stomate, its guard cells, and a few adjacent cells in the epidermis.

4. Count the number of stomates in the high-power field for 10 different areas of the piece of lower epidermis. Calculate the average number of stomates per mm^2 of leaf surface. (Refer to Investigation A2.2 (Appendix 2) to calculate the diameter of the high-power field. Use this figure to calculate the area of the leaf observed under the microscope.)

5. In the same manner, count the stomates on the upper epidermis of the same leaf. Examine as many other types of leaves as possible. Compare the number of stomates per mm^2 on the upper and lower surfaces of each type of leaf.

6. Wash your hands thoroughly before leaving the laboratory.

Discussion

1. How did the number of stomates per mm^2 compare in different areas of the piece of lower leaf epidermis? How did the number compare in different areas of the piece of upper leaf epidermis?

2. Did the stomates vary in the amount they were open? How can you explain your findings?

3. What would you do to assure a reliable comparison of the number of stomates per mm^2 for two species of plants?

4. What do your data suggest about the distribution of stomates in leaves of your species of plant? What assumption must you make in drawing this conclusion?

PART B   Gas Exchange and Photosynthesis

SAFETY   **Put on your safety goggles and lab apron.**

Procedure
Day 0

7. Appoint one member of your team as safety monitor to be sure the correct safety procedures are followed.

8. Three days before this investigation, one plant was placed in the dark. A second plant of the same species was placed in sunlight. Remove one leaf from each plant. Use the scissors to cut a small notch in the margin of the one placed in sunlight.

9. Using forceps, drop the leaves into a beaker of *hot* (60° C) tap water.

10. When they are limp, use forceps to transfer the leaves to a screw-cap jar about half full of isopropyl alcohol. Label the jar with your team symbol and store it overnight as directed by your teacher.

      WARNING: Isopropyl alcohol is flammable and toxic. Do not expose the liquid or its vapors to heat or flame. Do not ingest; avoid skin/eye contact. In case of spills, flood the area with water and *then* call your teacher.

11. Select four similar leaves on the plant that has been kept in the dark, but do not remove them from the plant. Using a fingertip, apply a *thin* film of petroleum jelly to the upper surface of one leaf. Check to be sure the entire surface is covered. (A layer of petroleum jelly, although transparent, is a highly effective barrier across which many gases cannot pass.) Cut one notch in the leaf's margin.

12. Apply a thin film to the lower surface of a second leaf and cut two notches in its margin.

13. Apply a thin film to both upper and lower surfaces of a third leaf and cut three notches in its margin.

14. Do not apply petroleum jelly to the fourth leaf, but cut four notches in its margin. Place the plant in sunlight.

15. Wash your hands thoroughly before leaving the laboratory.

Day 1

16. Obtain your jar of leaf-containing alcohol from Day 0. Using forceps, *carefully* remove the leaves from the alcohol and place them in a beaker of room-temperature water. (The alcohol extracts chlorophyll from the leaves but also removes most of their water, making them brittle.) Recap the jar of alcohol and return it to your teacher.

17. When the leaves have softened, place them in a screw-cap jar about half full of Lugol's iodine solution.

     WARNING: Lugol's iodine solution is poisonous if ingested, irritating to skin and eyes, and can stain clothing. Should a spill or splash occur, call your teacher immediately; flush the area with water for 15 minutes; rinse mouth with water.

18. After several minutes, use forceps to remove both leaves, rinse them in a beaker of water, and spread them out in open petri dishes of water placed on a sheet of white paper. Record the color of each leaf. Recap the jar of Lugol's iodine solution and return it to your teacher.

19. Wash your hands thoroughly before leaving the laboratory.

Day 3

20. Remove from the plant the four notched leaves prepared on Day 0 and place them on paper towels. To remove the petroleum jelly, dip a swab applicator in the Histoclear™ and *gently* rub it over the surface of the film once or twice. Then blot gently with a paper towel to remove any residue of petroleum jelly. Discard the swab applicator and the paper towel in the waste bag.

     WARNING: Histoclear is a combustible liquid. Do not expose to heat or flame. Do not ingest; avoid skin/eye contact. Should a spill or splash occur, call your teacher immediately; wash skin area with soap and water.

21. Repeat Steps 9 and 10.

22. Wash your hands thoroughly before leaving the laboratory.

Day 4

23. Repeat Steps 16–18. Compare the color reactions of the four leaves and record your observations.

24. Wash your hands thoroughly before leaving the laboratory.

Discussion

1. What was the purpose of the iodine tests on Day 1?

2. If you use these tests as an indication of photosynthetic activity, what are you assuming?

3. What is the purpose of the leaf marked with four notches?

4. In which of the leaves coated with petroleum jelly did photosynthetic activity appear to have been greatest? Least? Do your data support your hypothesis?

19.4   Oxygen Gas Is a By-Product of the Light Reactions

The light reactions of photosynthesis convert the energy in visible light into the chemical energy that powers sugar production. In these reactions, chlorophyll and other pigments in the thylakoid absorb light energy; water molecules are split into hydrogen and oxygen, releasing oxygen gas to the atmosphere; and light energy is stored as chemical energy. Figure 19.6 illustrates these reactions.

The pigments and other molecules involved in the light reactions are embedded in the thylakoid membranes and organized in a way that ensures their efficient operation. Two structurally distinct groups of molecules, called photosystem (PS) I and photosystem (PS) II, work in harmony to absorb light energy. The photosystems are connected by electron carriers that form an electron transport system. Acting together, the chlorophylls

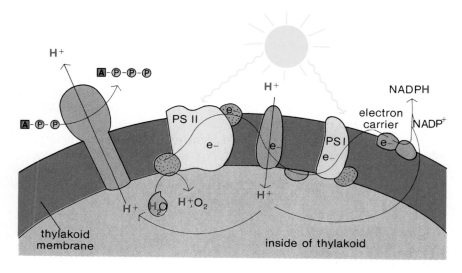

Figure 19.6   The light reactions of photosynthesis take place in the thylakoid of a chloroplast. Absorption of light energy sets up a flow of electrons from water to photosystem II to photosystem I to NADP⁺. As electrons flow along the electron transport system, protons are transported to the inside of the thylakoid. ATP forms as protons diffuse through an enzyme complex.

and accessory pigments absorb, concentrate, and transfer light energy to a special chlorophyll *a* molecule in each photosystem.

The absorption of light energy sets up a flow of electrons. In each photosystem, the special chlorophyll *a* molecule loses electrons, which are captured by electron carriers. The electrons lost by photosystem I are replaced by electrons lost from photosystem II. Electrons lost by photosystem II are replaced by electrons removed from water. As a result, water is separated into oxygen, protons, and electrons:

$$H_2O \longrightarrow 2H^+ + 2e^- + \tfrac{1}{2}O_2$$

The oxygen is given off as oxygen gas.

As electrons flow from carrier to carrier along the electron transport system, some of the energy they originally captured from the sun powers the active transport of protons across the thylakoid membrane. A relatively high concentration of protons accumulates inside the thylakoid. The protons then diffuse through an enzyme complex in the thylakoid membrane, releasing energy that powers ATP synthesis. When this occurs, light energy has been converted into chemical energy. At the end of the electron flow, the electrons, along with protons from water, combine with a hydrogen carrier called **NADP$^+$**, nicotinamide adenine dinucleotide phosphate (nik uh TEE nuh mid AD uh neen DY NOO klee oh tyd FOS fayt). NADP$^+$ is very similar to NAD$^+$, a hydrogen carrier in cellular respiration. When electrons and protons combine with NADP$^+$, NADPH is formed.

In the light reactions, the first two events of photosynthesis occur: absorption of light energy and the conversion of light energy into chemical energy. Three products are formed. One is oxygen gas, which is released to the atmosphere through the stomates of the leaf. The other two are the energy-rich molecules ATP and NADPH, both of which are used in the Calvin cycle to convert carbon dioxide into sugars.

19.5   Sugars Are Formed in the Calvin Cycle

The reactions of the Calvin cycle do not involve the absorption of light energy. They do, however, require the products of the light reactions. In addition, some of the enzymes involved in the reactions are activated in the light or deactivated in the dark. The Calvin cycle is a series of reactions in which carbon dioxide is combined with the hydrogen split from water in the light reactions. Energy for these reactions is provided by the ATP and NADPH formed in the light reactions.

Figure 19.7 is a simplified diagram of the reactions of the Calvin cycle, which take place in the stroma of the chloroplasts. As shown in Figure 19.7a, the cycle begins when a molecule of carbon dioxide combines with a 5-carbon molecule (RuBP, or ribulose bisphosphate), forming a 6-carbon molecule that immediately splits into two 3-carbon acid molecules. Each 3-carbon acid accepts hydrogen from NADPH and phosphate groups from ATP, shown in (b), forming a 3-carbon sugar phosphate. This molecule is the first sugar formed from carbon dioxide in the Calvin cycle. For each three molecules of carbon dioxide that enter the Calvin cycle, six molecules of 3-carbon sugar are formed. Of these, five molecules undergo many transformations in the Calvin cycle, thus regenerating the original 5-carbon

Figure 19.7   The Calvin cycle converts carbon dioxide into sugars. Enzymes in the stroma of the chloroplast catalyze each step of the cycle. The reactions require ATP and NADH from the light reactions.

molecules with which the cycle began, shown in (c). Without these molecules, the cycle would stop. The sixth molecule of 3-carbon sugar leaves the Calvin cycle.

Several things can happen to the 3-carbon sugar that leaves the Calvin cycle. In the stroma of the chloroplast, the sugar can be converted into starch and stored as starch grains, or it can be used to synthesize amino acids, proteins, lipids, chlorophyll, enzymes, and other compounds needed in the plant cell. Much of the 3-carbon sugar is exported to the cytosol, where it can be converted into sucrose, the major form in which sugar is transported between plant cells. The 3-carbon sugar also may enter the pathway of glycolysis and cellular respiration, providing energy for all plant cells. Thus, in the Calvin cycle, chemical energy is stored in sugars as well as in other compounds synthesized by the plant cells, as shown in Figure 19.8 on page 522.

The Calvin cycle and the Krebs cycle (see Chapter 15) are similar in involving many rearrangements of carbon chains. Both produce carbon skeletons for use in biosynthesis reactions. Carbon dioxide is used in the Calvin cycle and released in the Krebs cycle. Much ATP is used in the Calvin cycle, and a little is formed in the Krebs cycle. NADPH is used in the Calvin cycle, and NADH is produced in the Krebs cycle.

Figure 19.8 Plants use the products of photosynthesis to supply energy and carbon skeletons for biosynthesis and other energy-requiring processes.

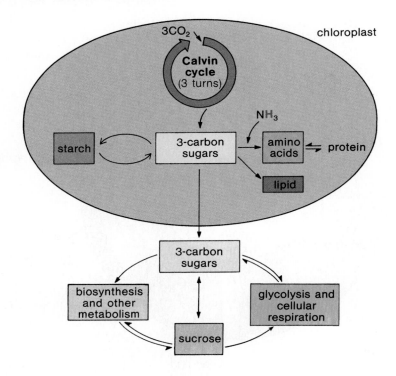

Concept Review

1. Describe the roles of the chlorophylls and accessory pigments in photosynthesis. What is the special role of chlorophyll *a*?
2. How is light used in photosynthesis?
3. Describe the roles of photosystems I and II.
4. What are the products of the light reactions, and how are they used?
5. What are the products of the Calvin cycle, and how are they used?

Guidepost: What adaptations enable plants to overcome rate-limiting environmental factors?

Photosynthesis and the Environment

INVESTIGATION 19.2 Photosynthetic Rate

There are several ways to measure the rate of photosynthesis. In this investigation, you will use elodea and pH probes with the LEAP-System™, or narrow-range pH paper. The rate of photosynthesis is determined indirectly by measuring the amount of carbon dioxide removed from water by elodea. Carbon dioxide is added by bubbling breath into the water, which absorbs the carbon dioxide. The carbon dioxide combines with water to produce carbonic acid ($H_2O + CO_2 \rightleftharpoons H_2CO_3$), a reversible reaction. As carbon dioxide is used by elodea in photosynthesis, less acid is present and the pH increases. The LEAP-System BSCS Biology Lab Pac 2 includes instructions for use of the system.

▶ After reading the procedure, construct hypotheses that predict the effects of light intensity and light color on the rate of photosynthesis.

Materials (per team of 4)
2-L beaker
100-mL beaker or small jar

250-mL flask
2 25-mm × 200-mm test tubes

2 1-mL pipets (optional)
2 wrapped drinking straws
lamp with 100-watt spotlight
Apple IIe, Apple IIGS, Macintosh, or IBM
   computer
LEAP-System™
printer
2 pH probes

1 thermistor
narrow-range pH paper (optional)
nonmercury thermometer (optional)
red, blue, and green cellophane
distilled water
tap water at 25° C and 10° C
2 15-cm sprigs of young elodea

Procedure
PART A   Equipment Assembly

1. In your data book, prepare a table similar to Table 19.1, or tape in the table your teacher provides.

2. Put 125 mL of distilled water in the flask. Blow through a straw into the water for 2 minutes. Be careful not to suck any liquid through the straw. Discard the straw after use as your teacher directs. What is the purpose of blowing into the water in the flask?

3. Place two sprigs of elodea, cut end up, into one of the test tubes (experimental tube).

4. Fill both test tubes 3/4 full with the water you blew into. What is the purpose of the second test tube?

5. Insert the pH probe from input 1 into the experimental test tube and the pH probe from input 2 into the other test tube, as shown in Figure 19.9.

6. Stand the test tubes in a 2-L beaker and add 25° C tap water to the beaker until it is about 2/3 full.

7. Insert the thermistor (or a thermometer) into the water in the 2-L beaker. Determine the temperature in the beaker and maintain this temperature throughout the experiment. (Use the small beaker to add ice water and/or to remove water from the 2-L beaker.)

8. Let the entire assembly stand for about 5 minutes to permit the temperature to become uniform throughout the system. Note the initial pH in the test tubes and record both readings in the data table. (If the LEAP-System is not available, use the

TABLE 19.1   Rate of Photosynthesis

	Experimental Condition																	
	30 cm			10 cm			50 cm			Red			Blue			Green		
Reading #	pH 1	pH 2	°C	pH 1	pH 2	°C	pH 1	pH 2	°C	pH 1	pH 2	°C	pH 1	pH 2	°C	pH 1	pH 2	°C
1																		
2																		
3																		
4																		
5																		
6																		

Figure 19.9   The LEAP-System setup.

1-mL pipets to transfer a drop of water from halfway down each test tube to a piece of narrow-range pH paper and read the pH from the comparison chart on the strip.)

PART B   Basic Photosynthetic Rate

9. Place the lamp with the 100-watt spotlight 30 cm from the beaker and illuminate the two test tubes.

10. Take pH and temperature readings for 30 minutes. In the data table, record the current readings every 5 minutes.

11. Wash your hands thoroughly before leaving the laboratory.

PART C   Effects of Light Intensity on Photosynthetic Rate

12. Repeat Steps 2–10 with the light source first 10 cm and then 50 cm away from the test-tube assembly.

PART D   Effects of Light Color on Photosynthetic Rate

13. Repeat Steps 2–10 using first red, then blue, and then green cellophane over the beaker.

14. Wash your hands thoroughly before leaving the laboratory.

Discussion

1. What chemical change occurred in the water you blew into?

2. What happens during photosynthesis that causes the pH to increase in the test tubes?

3. What two environmental factors are being controlled by the test tube without elodea?

4. How much did the pH change during the 30-minute period for each condition tested?

5. How would you use any change in pH in the test tube without elodea to correct the data for the experimental test tube?

6. Run stripcharts to produce graphs for each part of the investigation. What variables are on the horizontal and vertical axes of the graphs? (If using pH paper, prepare graphs showing pH every 5 minutes under the various conditions being tested.)

7. Use the change in pH that occurs at the three light intensities you tested to determine the effect of light intensity on the rate of photosynthesis. Do your data support your hypothesis?

8. Use the change in pH with the three colors (red, blue, and green) to determine the effect of light color on the rate of photosynthesis. Do your data support the hypothesis that you constructed?

19.6   Environmental Factors Affect the Rate of Photosynthesis

Environmental conditions profoundly affect photosynthesis and, therefore, productivity in an ecosystem or in agriculture. Four environmental factors of particular importance are light intensity, temperature, and the concentrations of carbon dioxide and oxygen. The graph in Figure 19.10a illustrates the effects of light intensity on photosynthesis. Note that before the light reaches the intensity of full sunlight, the rate of photosynthesis levels off. (Rate is the activity per unit of time.) At that point, the light reactions are saturated with light energy and are proceeding as fast as possible, so a further increase in light intensity will not result in an increased rate of photosynthesis.

Figure 19.10b illustrates the effects of temperature on the rate of photosynthesis. In general, higher temperatures speed up, and lower temperatures slow down, the rate of chemical reactions. How, then, can you explain the decrease in rate above 30° C? Recall what you have learned about the effects of temperature on enzymes and other proteins.

The concentrations of carbon dioxide and of oxygen influence the rate of photosynthesis in opposite ways. An increased level of carbon dioxide

Figure 19.10 (a) As light intensity increases, the rate of photosynthesis increases, but only up to the point that is called the light saturation point (indicated by the brown arrow). (b) As the temperature increases, the rate of photosynthesis also increases up to a point where the rate levels off and then begins to decline. What are the optimum temperatures for photosynthesis? Data are generalized to show trends.

a   b

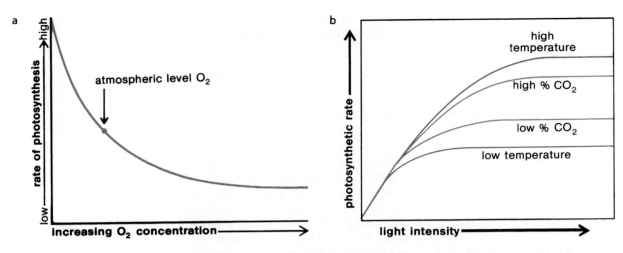

Figure 19.11   (a) Increasing concentrations of oxygen inhibit the rate of photosynthesis in most crop plants at atmospheric levels of carbon dioxide (0.035%). (b) Environmental factors have interacting effects on the rate of photosynthesis, and any one can be a limiting factor if present in less than optimum amounts.

concentration increases the rate of photosynthesis to a maximum point, after which the rate levels off. (This is similar to the effect of increased light intensity.) An increased oxygen concentration, on the other hand, slows the rate, as shown in Figure 19.11a. The present atmospheric concentrations of oxygen (about 20 percent) may slow the rate of photosynthesis by as much as 40 percent in some plants.

Because these four environmental factors interact, their combined effects must be considered. Any one of the factors, if present at less than optimal levels, may limit the rate of the reaction. For example, at optimal temperature and carbon dioxide concentration, the rate of photosynthesis may be limited by the light intensity. If temperature and light intensity are optimal, then carbon dioxide concentration may become the limiting factor. And if light intensity and carbon dioxide concentration are optimal, the temperature may control the overall rate of photosynthesis. Figure 19.11b summarizes these interactions, which explain a great deal about the distribution of plants. For example, roses do not grow at the maximum rate in cold climates even in full sunlight, because the low temperature limits their growth. Cactus plants do not grow in forests because they require large amounts of light and can survive with the small amounts of water and the extremes of temperature common to deserts.

19.7   Photorespiration Inhibits Photosynthesis

Photorespiration is a light-dependent process involving the uptake of oxygen and release of carbon dioxide that occurs along with photosynthesis in all plants. Because it prevents incorporation of carbon dioxide in the Calvin cycle and results in the loss of previously incorporated carbon dioxide, photorespiration is counterproductive to photosynthesis. Figure

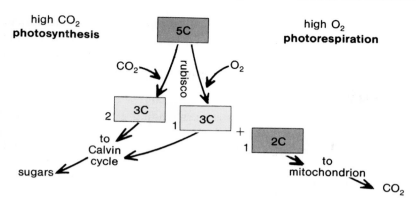

Figure 19.12   Photorespiration is the light-dependent uptake of oxygen and release of carbon dioxide in plants. It occurs simultaneously with photosynthesis and results in the loss of previously incorporated carbon dioxide. The enzyme rubisco, which can react with either carbon dioxide or oxygen, is involved in both of these processes. High levels of oxygen favor photorespiration; high levels of carbon dioxide favor photosynthesis.

19.12 compares the two processes. As described earlier in Section 19.5, each molecule of carbon dioxide that enters the Calvin cycle first combines with a 5-carbon molecule. This reaction, catalyzed by an enzyme called rubisco (roo BIS koh), forms two 3-carbon acids. Rubisco, however, can react with oxygen as well as with carbon dioxide. When rubisco reacts with oxygen, the rate of photosynthesis slows down. No new carbon dioxide is incorporated, and the 5-carbon molecule (containing previously incorporated carbon dioxide) is converted into carbon dioxide. Because there are thousands of molecules of rubisco and the 5-carbon compound in each chloroplast, photorespiration and photosynthesis can occur simultaneously in a plant. Relatively high levels of carbon dioxide favor photosynthesis. Relatively high levels of oxygen favor photorespiration.

Why would such an apparently damaging process have evolved in green plants? Evolution favors workable systems that develop from already existing systems, resulting in adequate but not optimum processes such as photosynthesis. When the Calvin cycle evolved there was little oxygen in the air, and the reaction of oxygen and rubisco was not significant. Millions of years of photosynthesis, however, have built up the atmospheric oxygen levels. Photorespiration has become a problem—one that may have a high cost because reduced plant growth reduces food production.

19.8   Special Adaptations Reduce Photorespiration

Hot, dry weather leads to high transpiration rates. In turn, high transpiration rates cause plants to lose valuable water through the stomates. In many plants, stomates are closed partially much of the time, thus avoiding excessive water loss. When stomates are closed, however, carbon dioxide levels in the leaves may drop so low that photorespiration is favored over photosynthesis.

Most plants use the Calvin cycle for the initial steps that incorporate carbon dioxide into organic material. These plants are called C-3 plants because the first stable molecule formed is a 3-carbon acid. In certain plants such as sugar cane, corn, crabgrass, and others that can tolerate hot conditions, the Calvin cycle is preceded by reactions that first incorporate carbon dioxide in a 4-carbon acid. These plants, known as C-4 plants, have two distinct types of photosynthetic cells. Surrounding each vein is a layer of tightly packed cells called bundle-sheath cells (see Figure 19.13a). Around the bundle sheath are mesophyll cells, which also are in contact with the air spaces in the leaf.

C-4 plants utilize a two-step pathway involving two separate and different enzyme systems for carbon dioxide incorporation. The first system operates in the mesophyll cells, where carbon dioxide combines with a 3-carbon acid. The resulting 4-carbon acid is rearranged and then transported to the bundle-sheath cells where the second system operates. In the bundle-sheath cells, the carbon dioxide is released from the 4-carbon acid and enters the Calvin cycle, eventually forming a 3-carbon sugar, as shown in Figure 19.13b.

The system in the mesophyll cells efficiently delivers carbon dioxide to the bundle-sheath cells. Thus, a higher concentration of carbon dioxide exists in the bundle-sheath cells than in the atmosphere—a state that favors photosynthesis and inhibits photorespiration. Even at high temperatures and at high light intensities (conditions that favor photorespiration), the high concentration of carbon dioxide around rubisco in C-4 plants overcomes photorespiration. Thus, C-4 plants can function at maximum efficiency in high temperatures and light intensities while the stomates are partially closed, avoiding excessive water loss. As a result, C-4 plants grow more rapidly than C-3 plants in warm climates and may be almost twice as efficient as C-3 plants in converting light energy to sugars under conditions of high temperature, high light intensity, and/or low water.

Figure 19.13 (a) Specialized leaf anatomy in a C-4 plant. Note the tightly packed layers of bundle-sheath and mesophyll cells surrounding the vein. (b) In C-4 plants, carbon dioxide first is incorporated into a 4-carbon acid in the outside mesophyll cells. The 4-carbon acid then is transported to the bundle-sheath cells, where carbon dioxde is released to the Calvin cycle.

Another photosynthetic specialization was discovered in desert plants such as cactus and jade plants. Called CAM, for crassulacean (kras yoo LAY see un) acid metabolism, it is an important adaptation in arid-land ecosystems or very dry conditions. CAM plants, in contrast to C-3 and C-4 plants, can open their stomates at night and incorporate carbon dioxide into organic acids in the cell vacuoles. During the day, with high temperatures and low humidity, the stomates close, preventing transpiration that could kill the plant. The organic acids release carbon dioxide that diffuses into the chloroplasts and feeds a normal Calvin cycle powered by the light reactions of photosynthesis. CAM photosynthesis, however, is not very efficient, and although desert plants that use this system can survive intense heat, they usually grow very slowly.

Concept Review

1. Describe the influence of increasing temperature on the rate of photosynthesis.
2. How do various environmental factors interact to affect the rate of photosynthesis?
3. What is photorespiration, and how does it affect photosynthesis?
4. Describe the two-step pathway of carbon dioxide incorporation in C-4 plants.
5. Construct a graph that compares the effect of oxygen on C-3 and C-4 photosynthesis.
6. What is unique about CAM photosynthesis?

Hormonal and Environmental Control of Plant Growth

19.9 Plant Hormones Interact to Regulate Plant Processes

Guidepost: How are hormones involved in plant growth and development, and what role does the environment play?

As a plant grows, it increases in size, but it also differentiates, forming a variety of cells, tissues, and organs. Although its genes determine the basic form of a plant, the environment can modify its development. Growth-regulating substances (plant hormones) that are produced by the plant play critical roles in plant responses to changing seasons and other environmental factors.

At the present time, botanists recognize five main groups of plant hormones that have some effect on plant growth. Their production is under genetic control, but environmental factors have a part in determining when the hormones are produced, how much of each hormone is produced, and which cells are sensitive to these hormones. Plant hormones are produced and are effective in extremely small concentrations, and they may interact with each other to produce their effects. The hormones may cause different effects in different parts of the plant, at different ages of the plant, or in different concentrations. Each hormone apparently has a primary effect on each affected cell type, producing a variety of responses. The response depends on poorly understood structural and physiological conditions in the affected cells. Because these conditions are not well understood, it is difficult to determine precisely how plant hormones operate.

19.10   Plant Hormones Have Diverse Effects

Auxins (AWK sins) are produced in actively growing regions such as the tip ends of shoots, from which they move by active transport to other parts of the plant. Auxins can cause cells in the growing regions of the plant to elongate, but the effects depend on a number of factors, including the concentration of the auxins. Laboratory experiments indicate that at extremely low concentrations auxins may promote elongation of roots, but at higher concentrations they may inhibit elongation (see Figure 19.14). Applied to cuttings of stems or leaves, auxins can stimulate the formation of roots; applied to certain flowers, they can promote fruit development without the need for pollination or fertilization. Such fruits lack seeds, and seedless tomatoes and cucumbers have been produced this way.

A synthetic auxin called 2,4-D is an herbicide used to kill dicot weeds. At the concentrations usually used, 2,4-D does not affect members of the grass family (monocots), so it can be used to control weeds such as dandelions in lawns and grain fields. A similar herbicide, 2,4,5-T, is a component of the infamous "Agent Orange" that was used during the Vietnam conflict to defoliate trees. Dioxin, a by-product that is produced when 2,4,5-T is made, often contaminates the 2,4,5-T and is known to be extremely toxic to humans. Dioxin is considered to be an environmental hazard all over the world.

Stem elongation in mature trees and shrubs is influenced by **gibberellins** (jib uh REL ens), which also are synthesized in the tips of stems and roots. Application of gibberellins can cause certain plants to grow faster,

Figure 19.14   The concentration of applied auxin has opposite effects on cell elongation in roots and stems.

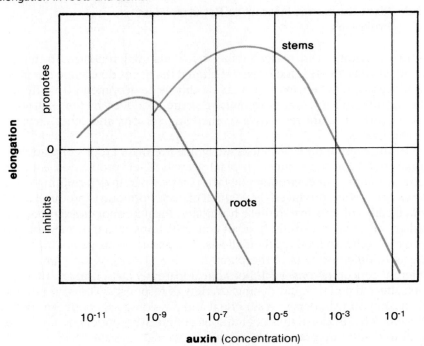

Development of New Plants

The use of plant hormones has enabled plant physiologists to devise new growing methods. In tissue culture, single cells or groups of cells are placed in sterile containers with nutrients, water, and (sometimes) hormones to stimulate growth and development. At equal concentrations of auxins and cytokinins, many cells in a culture continue to divide, forming an undifferentiated tissue called a callus. By adjusting the relative concentrations of auxins, cytokinins, and other growth substances, the callus cells can be induced to form roots or buds and, in some cases, to grow into a complete plant, as shown in the figure. Because one plant can contribute many cells, it is possible to develop many clones, or genetically identical plants. Tissue culture has been used to clone many different plants, but few plants have developed into adults. The cloning of animal cells is more difficult, and the cloning of human cells is not a present reality.

Cloning has an important advantage in the development of new plant varieties. Because there is much variation in most plant populations, there may be just one plant out of thousands that has a particular combination of desirable characteristics. It may be a new variety of rose or an unusually tasty carrot. If this one plant is crossed with another to get seeds, there is no guarantee the offspring will have the same characteristics as those of the desired parent. Can you explain why? Cloning, however, produces duplicate copies of any plant, and the cloned plants that do survive then can contribute to the culturing of many more clones. In this way an entire field of identical plants can be developed.

Tissue culture is just one of many ways scientists use to develop new and better crops. Plant and animal breeders use selective breeding, allowing only those individuals with the "best" characteristics to breed. Most of the offspring of the individuals allowed to breed are expected to have some of the characteristics of the two parents.

Another technique used to develop new and better crops is protoplast fusion, in which cells from two different plants are placed in a special liquid in the same container. Enzymes are used to digest away the cell walls, leaving intact the living contents, or protoplasts. The protoplasts may fuse, develop a cell wall, and begin to grow, forming a hybrid cell.

Usually, individuals from two different species cannot form viable offspring. For example, it has not been possible to produce a hybrid offspring between a potato and a tomato by cross-fertilization. A hybrid cell of these two plants can be formed using protoplast fusion, however. Results such as these have raised hopes that protoplast fusion could lead to the combining of extremely different plants into "superplants" or "supercrops" that might, for example, be more drought resistant or require smaller amounts of nitrogen than present varieties.

Interaction of auxins and cytokinins in plant tissue culture. The culture media were supplemented with various levels of the two hormones.

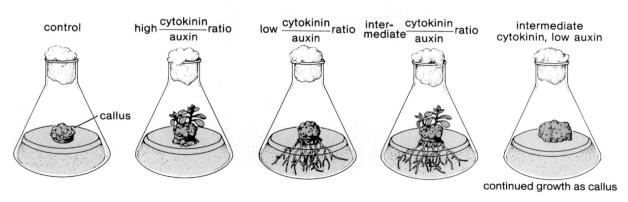

dwarf plants to grow to the height of normal varieties, and fruits to grow larger (see Figure 19.15). Gibberellins also can stimulate the production of enzymes in the endosperm of grains. The enzymes break down the stored nutrients, which are used by the embryo as it develops.

Cytokinins (syt oh KY ninz) are plant hormones that promote cell division and organ formation, at least in tissue culture. They are found mostly in plant parts with actively dividing cells, such as root tips, germinating seeds, and fruits. Cytokinins appear to be necessary for stem and root growth and to promote chloroplast development and chlorophyll synthesis. Cytokinins and auxins may interact closely to regulate the total growth pattern of the plant.

Abscisic (ab SIS ik) **acid** is thought to be synthesized primarily in mature green leaves, root caps, and fruits. Apparently it is involved in the dormancy of buds and seeds, enabling them to survive extremes of temperature and dryness. Abscisic acid also affects stomate closing and is sometimes called the stress hormone because it may help protect the plant against water loss during unfavorable environmental conditions.

A simple gas produced by ripening fruits and other plant parts, **ethylene** (ETH ih leen) promotes fruit ripening and aging of tissues. Through interaction with auxins, cytokinins, and abscisic acid, ethylene apparently plays a major role in causing leaves, flowers, and fruits to drop from a plant. Ethylene appears to counteract many effects of auxins; for example, it inhibits flowering and the elongation of stems and roots. Hundreds of years ago, Chinese farmers unknowingly used ethylene when they burned incense (which produces ethylene) to ripen fruit stored in ripening rooms. Today, many farmers pick delicate fruits such as tomatoes when they are

Figure 19.15 Grapes with and without applied gibberellic acid.

auxins and gibberellins
synthesized in young
leaves and buds may
control stem elongation

auxins may
influence
differentiation

abscisic acid can be
made in leaf in
response to water
stress; closes
stomates, reducing
water loss

cytokinins made in
young fruit; needed
for growth

ethylene accumulates
in mature fruit;
induces
ripening

cytokinins help keep
root and shoot
growth in balance

substance made in
root tip; controls
direction of root
growth

Figure 19.16 Some possible roles and interactions of hormones among the various parts of a plant.

still green and hard and thus less susceptible to damage in shipping. Then, prior to sale, the tomatoes are treated with ethylene to speed and time their ripening. Figure 19.16 summarizes the actions of the five groups of plant hormones.

INVESTIGATION 19.3 Tropisms

This investigation allows you to observe tropisms. Each team will conduct only one part of the investigation. Observe the other part so you can discuss all the results as a class.

▶ After reading the procedures, construct two hypotheses—one predicting how the germinating corn grains will be affected and the other predicting the patterns of growth of the radish seeds.

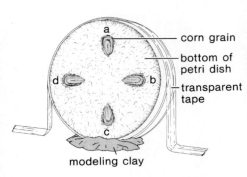

corn grain
bottom of petri dish
transparent tape
modeling clay

Figure 19.17 Petri dish setup.

Materials Part A (per team of 4)

petri dish
scissors
glass-marking pencil
nonabsorbent cotton

heavy blotting paper
transparent tape
modeling clay
4 soaked corn grains

Materials Part B

4 flowerpots, about 8 cm in diameter
4 cardboard boxes, at least 5 cm higher
   than the flowerpots
red, blue, and clear cellophane

scissors
transparent tape
40 radish seeds
soil

PART A   Orientation of Shoots and Roots in Germinating Corn
Procedure

1. Place 4 soaked corn grains in the bottom half of a petri dish. Arrange them cotyledon side down, as shown in Figure 19.17.

2. Fill the spaces between the corn grains with wads of nonabsorbent cotton to a depth slightly greater than the thickness of the grains.

3. Cut a piece of blotting paper slightly larger than the bottom of the petri dish, wet it thoroughly, and fit it snugly over the grains and the cotton.

4. Hold the dish on its edge and observe the grains through the bottom. If they do not stay in place, repack with more cotton.

5. When the grains are secure in the dish, seal the two halves of the petri dish together with cellophane tape.

6. Use modeling clay to support the dish on edge, as shown in Figure 19.17, and place it in dim light.

7. Rotate the petri dish until one of the corn grains is at the top. With the glass-marking pencil, write an *A* on the petri dish beside the topmost grain. Then, proceeding clockwise, label the other grains *B, C,* and *D.* Also label the petri dish with your team symbol.

8. When the grains begin to germinate, make sketches every day for 5 days, showing the direction in which the root and the shoot grow from each grain.

9. Wash your hands thoroughly before leaving the laboratory.

Discussion

1. From which end of the corn grains did the roots grow? From which end did the shoots grow?

2. Did the roots eventually turn in one direction? If so, what direction?

3. Did the shoots eventually turn in one direction? If so, what direction?

4. To what stimulus did the roots and shoots seem to be responding?

5. In each case, were the responses positive (toward the stimulus) or negative (away from the stimulus)?

6. Why was it important to have the seeds pointed in four different directions?

PART B   Orientation of Radish Seedlings
Procedure

10. Turn the four cardboard boxes upside down. Number them *1* to *4.* Label each box with your team symbol.

11. Cut a rectangular hole in one side of boxes 1, 2, and 3. (Use the dimensions shown in Figure 19.18.) Do not cut a hole in box 4.

12. Tape a piece of red cellophane over the hole in box 1, blue cellophane over the hole in box 2, and clear cellophane over the hole in box 3.

13. Number four flowerpots *1* to *4*. Label each with your team symbol. Fill the pots to 1 cm below the top with soil.

14. In each pot, plant 10 radish seeds about 0.5 cm deep and 2 cm apart. Press the soil down firmly over the seeds and water the soil. Place the pots in a location that receives strong light but not direct sunlight.

15. Cover each pot with the box bearing its number. Turn the boxes so the sides with holes face the light.

16. Once a day remove the boxes and water the soil. (Do not move the pots; replace the boxes in the same position.)

17. When most of the radish seedlings have been above ground for 2 or 3 days, record the direction of stem growth in each pot—upright, curved slightly, or curved greatly. If curved, record in what direction with respect to the hole in the box.

Figure 19.18   Flowerpot setup.

Discussion

1. In which pot were the radish stems most nearly upright?
2. In which pot were the radish stems most curved? In what direction were they curved? Were the stems curved in any of the other pots? Which ones and in what direction were they curved?
3. To what stimulus do you think the radish stems responded?
4. What effect, if any, did the red and blue cellophane have on the direction of the radish stem growth?

19.11   Plants Respond to Environmental Stimuli

Proper plant growth and development depend, in part, on **tropisms** (TROH piz umz)—movements of plants that occur in response to a stimulus. These movements are generally very small and take place so slowly that usually it is necessary to use time-lapse photography to see them. Most plant movements occur in response to changes in the environment. For example, when any part of the leaf of the sensitive plant, *Mimosa pudica*, is touched, the leaflets droop together suddenly, as shown in Figure 19.19. This movement results from a sudden loss of water from certain cells at the base of the

Figure 19.19   Mimosa plant before (a) and after (b) being touched.

a

b

Figure 19.20   Seedlings bending toward light.

leaves and leaflets. The significance of the movement is not completely clear, but the drooping may help prevent further water loss.

Many plant movements seem to result from differences in growth among cells of a tissue. You probably have observed that most plants can grow toward light (Figure 19.20). Charles Darwin studied this response, called **phototropism** (foh toh TROH piz um), in which cells on the lighted side of the shoot can stop growing, while those on the shaded side can continue to elongate. That differential growth seems to cause the plant to bend toward the light, but what causes the differential growth?

Experiments designed to answer that question have been carried out on oat coleoptiles (koh lee OP tilz). A coleoptile is the protective sheath around monocot seedlings. In uniform lighting or in the dark, oat seedlings grow straight upward. With one-sided lighting, bending occurs. Evidence from the experiments indicated that the tip of a growing shoot produces a chemical that causes bending. Darwin's experiments showed that if the tip was covered, the shoot would not bend. Other researchers found a substance produced in the tip when it was exposed to light. This chemical, if applied to only one side of the shoot, could cause the shoot to bend even in the dark. Eventually the chemical was identified as an auxin. It appeared to researchers that the auxin was synthesized in the tip of the shoot and then transported to the shaded side, where it stimulated elongation of the cells. That hypothesis now is being questioned, however. Improved analytical techniques fail to show any redistribution of auxin within the shoot before bending occurs. At the present time it is not clear how, or even whether, light, auxin, and perhaps other plant hormones interact to produce bending.

Gravitropism (grav ih TROH piz um) is growth movement toward or away from the earth's gravitational pull. Stems and flower stalks are negatively gravitropic, that is, they grow in a direction away from gravity. If stems are placed in a horizontal position, they bend upward (Figure 19.21), although this response can be variable. Auxins are thought to play a critical role, as are calcium ions. Roots are positively gravitropic, and the evidence indicates that sensitivity to gravity occurs in or near the root cap. There, plastids filled with starch grains migrate to the bottom of certain cells, and their movement initiates a sequence of events that leads to downward growth or curvature. Again, auxins appear to be involved, along with abscisic acid and perhaps other chemicals.

Figure 19.21   When an upright plant is turned on its side, new roots grow downward and new stem growth is upward, resulting in a bent stem. What type of gravitropism do these reactions demonstrate?

19.12 Many Plants Respond to the Length of Day and Night

The changing seasons are important environmental cues for all terrestrial organisms. Birds migrate, insects produce dormant eggs, animals hibernate, and deciduous trees lose their leaves as winter approaches. These activities are triggered by changes in day length. Organisms have mechanisms that allow them to respond to the relative amount of light and dark in a 24-hour period. This response is called **photoperiodism** (foh toh PIH ree ud iz um).

Many plants show the effects of photoperiodism in flowering. For example, spring-flowering plants, such as daffodils, can reproduce only when day length exceeds a certain number of hours. Thus, they are called long-day plants. Fall-flowering plants, such as asters, can reproduce only when day length is shorter than a certain number of hours and therefore are called short-day plants. Many plants flower whenever they become mature. These are day-neutral plants, although temperature may play an important role in their flowering. Florists use their knowledge of photoperiod to "force" short-day plants such as chrysanthemums to bloom year-round.

In fact, it is the length of the night rather than the length of the day that controls flowering and other photoperiodic responses. Experiments have demonstrated that a brief period of darkness during the daytime portion of the photoperiod has no effect on flowering. Even a few minutes of light during the dark period, however, prevents flowering, indicating that photoperiodic responses depend on a critical, uninterrupted night length. Long-day plants flower when the night is *shorter* than a critical length; short-day plants flower when the night is *longer* than a critical length (see Figure 19.22). Certain pigments that can be converted from one form to another are involved in the plant's ability to measure the length of darkness in a

Figure 19.22 An experiment illustrating the effect of different photoperiods on the flowering of Maryland mammoth tobacco.

no flowers

15½ hrs   8½ hrs

| day | night |

with flowers

excess of 9 hrs

| day | night |

photoperiod. The conversion acts as a switching mechanism that may control many events in the life of the plant, such as flowering, seed germination, and chlorophyll formation.

Concept Review

1. What characteristics of plant hormones have made them especially difficult to study?
2. Compare and contrast the effects of applied auxins and gibberellins on plant growth.
3. What distinguishes the action of cytokinins from that of auxins?
4. How do phototropism and gravitropism affect plant growth?
5. What effect does photoperiodism have on flowering in various plants?

Summary

In the light reactions of photosynthesis, light energy is absorbed by a variety of plant pigments, primarily chlorophyll. A series of reactions follows in which energy-rich molecules are formed as the light energy is converted into chemical energy. In addition to these molecules, oxygen gas is released into the atmosphere as a by-product. In the Calvin cycle, the energy-rich molecules are used to make 3-carbon sugars from carbon dioxide. During cellular respiration the energy from these sugars is released. Environmental factors such as temperature, oxygen and carbon dioxide concentration, and light intensity interact to affect the rate of photosynthesis. Photorespiration, a light-dependent process that results in the uptake of oxygen and release of carbon dioxide, occurs simultaneously with photosynthesis, resulting in loss of previously incorporated carbon dioxide and reducing photosynthetic efficiency. C-4 photosynthesis is a special adaptation that reduces photorespiration in certain warm-climate plants; CAM photosynthesis is an adaptation to hot, dry climates.

The growth and development of plants result from the interaction of plant hormones and environmental factors such as light and gravity. Auxins, gibberellins, cytokinins, abscisic acid, and ethylene are the groups of natural plant hormones. Some of these have been produced synthetically by humans and are used in research and agriculture.

Applications

1. In what ways are photosynthesis and cellular respiration opposite sets of reactions?
2. Trace the production of ATP molecules in both respiration and photosynthesis.
3. What roles do photosynthesis and respiration play in the cycling of carbon?
4. How are plant hormones used to bring about differentiation and growth of a new plant from a callus?
5. Tomatoes are fruits. Many tomatoes are picked when they are green (unripe) and packaged with plastic wrap around them. What advantage does this packaging offer?
6. How much food do plants get from the soil? Explain your answer.
7. Energy can be changed from one form to another, but it cannot be created or destroyed. Give several examples from this chapter, or any part of the text, in which energy is changed from one form to another.

8. In the overall equation for photosynthesis, why is chlorophyll written under the arrow instead of to the left or right of the arrow?

9. Trace the flow of energy in photosynthesis, beginning with the sun and ending with a sugar molecule.

10. Different species of grasses grow all over the United States. Some are C-3 species and some are C-4. Would you expect C-4 species to make up a greater percentage of all the grasses in the northern part of the United States or in the southern part? Explain.

11. Describe possible hormonal and environmental stimuli that might cause deciduous trees to lose their leaves in the fall.

12. Some plants use changes in day length to help them determine the change of seasons. Why is photoperiodism a better measure of the oncoming winter than is falling temperature?

13. Why do you think visible light is the only radiation from the sun that triggers photosynthesis? Consider the characteristics of the wavelengths that are much shorter and much longer than those of visible light.

14. Potatoes grow very well during the summer in Idaho but not very well during the summer in Texas. Why? Consider the difference in temperature and its effect on photosynthesis and photorespiration.

Problems

1. Investigate some of the practical uses of auxins and synthetic growth substances as they apply to various types of farming.

2. Many botanists are searching for plants that synthesize fuel-related carbon compounds. Part of our energy needs could be met by growing such plants commercially. What other types of useful compounds do plants produce? Report your findings to the class.

3. Experiments have shown that when photosynthesizing vascular plants are grown in an atmosphere without oxygen they take up carbon dioxide at 1.5 times the rate at which they do in natural atmosphere, which is about 20 percent oxygen. What does this indicate about the relationship between photosynthesis and photorespiration? How might this relationship affect the composition of the earth's atmosphere?

4. Plant development appears to be much more closely linked to environmental cues than does animal development. Explain how this adaptation would enhance survival of plants.

5. Write a science fiction short story in which the humans are genetically engineered to contain some of the genes of plants. What type of organisms would they be, and what advantages might they have?

6. Design and conduct an experiment that tests the effect of different colors of lights on the growth and development of plants. Make sure that the amount of light reaching each plant is the same—just different colors. What other factors must you control?

7. Investigate and report on the use of Agent Orange during the Vietnam War. What was the purpose of its use, how effective was it, and what are the long-term health consequences, if any, to the people who were exposed to it?

8. Different plant hormones may be obtained easily from your grocery store or local nursery. For example, cytokinin, along with many nutrients, may be found in coconut milk from fresh coconuts. Ethylene, a gas, is produced by ripening fruits. Auxin may be obtained in commercial root growth stimulators. Design an experiment to test the effect of one of these hormones on the growth and development of a plant.

9. Starch can be detected by bleaching the chlorophyll from a leaf and then staining it with an iodine solution. No sugar, and thus no starch, will be produced in those parts of the leaf that do not undergo photosynthesis. Designs can be made in leaves by first cutting a pattern, smaller than a leaf, out of a piece of aluminum foil and using it to cover a large green leaf of a geranium or other plant with big, sturdy blades. Leave the foil on the leaf, and leave the plant in the sun for several days. At the end of this period, remove the foil, bleach the leaf, and stain it with iodine. Can you see the design in the leaf?

10. For those of you with a solid chemistry background, look in any college biochemistry text for a complete explanation of photosynthesis, cellular respiration, and the relationship between these two processes. Make a wall chart showing a complete set of reactions relating these processes.

Section Five

Patterns in The Biosphere

Ecologists study living things as parts of interrelated patterns. Individuals are members of populations, and populations are members of communities and ecosystems. Ecosystems do not remain the same forever but change continuously. Catastrophes such as floods or fires and even the organisms themselves may change the character of entire ecosystems. The ecosystems of the earth today are the result of interacting organisms and patterns of the earth's past. The study of ecosystems—how and why they change—including how humans shape the changes, helps us predict future environments and, in a general way, the future of our world.

One of the observable patterns in the biosphere is the profound effects that organisms have on the physical environment. This coral reef in Australia illustrates their effects. Assuming that the two islands are approximately 20 km in length, estimate the magnitude of the effect these organisms have had on this environment. How could organisms as small as corals have been responsible for this effect?

Can you find the animal in this photo? What do you see that would affect its chances of being selected by a predator? How would this affect its ability to pass its genetic information to the next generation?

Behavior, Selection, and Survival 20

All living organisms are *doing* something continually. They respond to stimuli from their external environment, and they respond to stimuli within their bodies. The sum of all the activities of an organism is called behavior. An organism's behavior is influenced by other organisms of the same type, by other organisms it uses as food, and by other organisms that use *it* as food. Behavior also includes all of an organism's responses to such things as the time of year, the time of day and night, changes in the amount of light, changes in temperature, and every other characteristic of the ecosystem in which the organism lives. The behavior of an organism must be adaptive if it is to survive and reproduce.

Organisms are what they are because the genes that they share with their ancestors allowed those ancestors to survive and reproduce. Environmental selection may act in favor of individuals with certain genes, but it removes other individuals (and their genes) from the gene pool. If the surviving individuals in a population are adapted to the environment, their genes will be retained. If the environment changes, however, the individuals and their genes may be removed. This chapter discusses how behavior and different types of selection affect an organism's ability to survive and reproduce.

Behavior and Selection

Guidepost: How does an organism's behavior affect its survival?

INVESTIGATION 20.1 Conditioning

Conditioning is one type of learned response to a stimulus. In this investigation, you will have an opportunity to experience conditioning firsthand.

Materials (per team of 1)
paper and pen

Procedure
1. In your data book, prepare two tables similar to Table 20.1 and Table 20.2 on the following page, or use the tables provided by your teacher.
2. Your teacher will read a paragraph about behavior to you. Each time the word *the* is read, make a slash mark (/) on your paper. The object is to record the number of times *the* is used in this paragraph (paragraph A).

TABLE 20.1   Conditioning Tally A

	Number of Words								
	38	39	40	41	42	43	44	45	46
Paragraph A									

TABLE 20.2   Conditioning Tally B

	Number of Words								
	6	7	8	9	10	11	12	13	14
Paragraph B									

3. **Do not correct mistakes.** You will make the data inaccurate if you do. If you make a mark when the word *the* has not been read, you must leave it. You cannot skip the next *the* to make up for your mistake.
4. After your teacher has read the paragraph, count the number of marks you have made and record the total on your paper.
5. Your teacher will read a second paragraph (paragraph B). As before, make a mark on your paper each time the word *the* is read.
6. After your teacher has finished reading, total your marks for paragraph B.
7. Enter your totals for each paragraph in the tables your teacher has drawn on the chalkboard.
8. Enter the class results from the tables on the board in your data book tables. Construct a bar graph of the class results. What was the average number of marks for each paragraph?

Discussion

1. Look at the numbers your teacher has circled on the tables. They represent the actual number of times the word *the* was read for each paragraph. How do these numbers compare with the numbers you and your classmates recorded?
2. What stimulus caused you to respond when the first part of paragraph A was read?
3. What stimulus caused you to respond when the last part of paragraph A was read?
4. In the context of this investigation, what is a substitute stimulus?
5. Using what you discovered when paragraph B was read, how do you think a conditioned response can be unlearned?
6. Apply the principles of conditioned responses to describe the procedure for teaching a pet to do a trick.

20.1   Behavior Usually Falls into Two Categories

In the broadest sense, behavior is considered to be either **innate** or **learned**. Innate, or inborn, behavior is not dependent on individual experience. Sometimes it is a pattern, such as crying or sucking, that is ready to function when the organism is born. Dogs that have spent their entire lives indoors will attempt to bury a bone by scratching the carpet as if they were digging in the soil. The same dogs will turn around several times before lying down, although there is no grass to trample or insects to chase out of the bed. These behavior patterns are innate and characterize the species. Behavior is considered to be innate if it occurs in response to a particular stimulus the first time an individual is exposed to that stimulus.

Environmental influences may trigger an innate behavior, but they do not explain it. The kitten in Figure 20.1 is standing on a checkerboard surface apparently looking down a drop, or cliff. Because a plate of glass covers it, the drop is only visual, not actual. The kitten will not venture out onto the glass, however, even though it has had no experience with heights or falling. Its reaction is an innate behavior in response to the visual stimulus. Such a complex innate behavior is called an instinct.

Learned behavior, on the other hand, consists of all those responses an organism develops as a direct result of specific environmental experiences. Most animals can learn to behave in new ways. A series of experiments with the southern toad shows how experience may change an innate behavior. The toad is a natural predator of the robber fly, which looks very much like a bumblebee. A toad may capture these robber flies until it mistakenly captures a real bumblebee, and the bee stings it on the tongue. After recovering from this experience, the toad does not attempt to eat either robber flies or real bumblebees, although it continues to eat other insects. The toad has changed its innate behavior.

One intriguing type of learning is shown in Figure 20.2. Once newly hatched goslings or ducklings are strong enough to walk, they will follow their mother away from the nest, and soon they will not follow any other animal. Through the experience of seeing and following a moving object, they learn something—to follow only their mother for weeks afterward. If

Figure 20.1   This Siamese kitten has never fallen or been dropped. It has had no experience with high places. The drop you see here is only visual; a plate of glass gives the kitten firm underfooting. Yet the kitten hesitates. Such experiments are used to help determine when behavior is innate and when it is learned.

Figure 20.2   Many species of birds, such as these common mergansers, *Mergus merganser,* will follow the first moving object they see.

Camouflage and Mimicry

One evolutionary result of predator-prey relationships is the ability of many animals to "hide" in the open. Camouflage is an adaptation in form, patterning, color, or behavior that enables an organism to blend with its surroundings and thus better escape detection by predators. The canyon tree frog pictured in (a) exhibits camouflage. Camouflage, of course, is not exclusive to prey animals. Predators that rely on stealth, such as polar bears and tigers, often blend well with their background.

Another evolutionary consequence of predation is the existence of many prey species that are bad-tasting, poisonous, or able to harm their attackers in some fashion. Often, toxic prey species are colorful or have bold markings that warn potential predators to stay away. Inexperienced predators might attack a black-and-white striped skunk, a black and yellow wasp, or a poison arrow frog (Chapter 14 opening photo)—but probably only once. As a result of their experience, these predators would quickly learn to associate the colors and patterns of the skunk, wasp, or frog with pain or intestinal upsets. During the course of their evolution, some prey species without any such defenses have come to resemble distasteful ones that have bold colors or markings. If the unprotected animals are relatively few in number compared with the species they resemble, they too will be avoided by predators. The resemblance of an edible species to a relatively inedible species is one form of mimicry.

Many of the best examples of this type of mimicry occur among butterflies and moths. The larvae of the butterflies that serve as models usually feed on plants that produce strong, poisonous chemicals. The larvae eat the plants and retain the poisonous molecules in their own bodies. The mimic butterfly larvae feed on a variety of different plants, but not those protected by toxic chemicals. One well-known mimic is the viceroy butterfly, *Limenitis archippus*, shown in (b). This butterfly resembles the poisonous monarch butterfly (c). Its larvae feed on willows and cottonwoods, and neither the larvae nor the adults are distasteful to birds. Interestingly, the larvae of the viceroy are camouflaged on the leaves of their host plants, where they resemble bird droppings. The distasteful larvae of the monarch butterfly, however, are very conspicuous.

If predators are to be deceived, mimic and model not only must look alike but also must act alike. For example, the members of several families of insects that resemble wasps behave surprisingly like the wasps they mimic. Mimics also must spend most of their time in the same habitats as do their models. If they did not, predators would discover that they are quite edible.

a
b
c

Figure 20.3 A conditioned reflex experiment with a planarian. As a result of the conditioning, the planarian eventually responds to light as it would to electric shock.

the eggs are hatched away from the mother, however, the young will follow the first moving object they see, even a human. This type of learning is special because there is a sensitive period in which the learning occurs, and it requires little or no practice.

A simple type of learned behavior occurs when an animal is exposed to a stimulus over and over again. This repetitive exposure may cause the animal gradually to lose its response to the stimulus. For example, a dog responds to the sound of a car driving past by turning its head. If cars continue to drive by, the dog may learn to disregard the sound. Such a reaction can be thought of as learning how *not* to respond to unimportant stimuli.

A more complex type of learning occurs when the patterns of an innate behavior are changed. Such conditioned responses have been developed in many types of animals. One of the most interesting experiments was performed on planarians, as shown in Figure 20.3. First, the biologists shone a strong light on a planarian. Several seconds later, they administered a mild electric shock. The planarian's normal response to the light was to stretch, and its normal response to the shock was to contract or turn its head. This sequence, in which the light was followed by the shock, was repeated about 100 times. Eventually the planarian's response to the light alone was to contract or turn its head, the same as if the electric shock also had been given.

A somewhat higher level of learning occurs when an animal learns a behavior through trial and error. An animal faced with two or more possible responses eventually may learn to choose only the one for which a reward is given. Pigeons, for example, can be taught to peck only at a button of a specific shape or color if the correct choice is rewarded with food.

Learning gives an animal flexibility to adapt to a changing environment by acquiring new behavior patterns. Learning is adaptive when an animal, faced with unpredictable problems, develops new responses that prove helpful in a particular situation. Innate responses also are adaptive. For example, the kitten that avoids the cliff edge also avoids potentially fatal falls.

Most adaptive behavior has both innate and learned components. Consider the development of flight in birds. Flight is a largely innate behavior. Obviously, a young bird must be able to manage it fairly well on its first attempt, or it will crash to the ground. Young birds, however, do not fly as well as adults (see Figure 20.4 on page 548). Even with flight, then, the innate pattern can be improved by learning through practice.

Communication also can be innate or learned. The prairie dog, for example, has an innate warning call that it voices when it sees a predator. Bees have a complex set of communication signals known as the bee dance.

a

b

Figure 20.4 Although birds are born with the innate ability to fly, some maneuvers must be learned. How might this young bald eagle, *Haliaeetus leucocephalus,* benefit from practice?

Through this dance, illustrated in Figure 20.5, a bee can communicate the location of food to other bees in the hive. Humans, on the other hand, must learn their vocal communication.

Communication may confer a number of advantages on the animal, but its ultimate function is to increase the animal's reproductive success. Communication can do this directly, as when the male elk gives its special call, called a bugle, to advertise for a mate, or indirectly, as when the prairie dog warns its offspring and other group members of danger. Communication, then, helps animals to survive and reproduce. Some communication is purely visual, such as the colorful breeding plumage of many male birds.

Figure 20.5   How bees communicate.

a. A foraging honeybee returns to the hive after finding a nectar source. Other bees soon leave the hive and fly to the same source. How does the first bee inform the other bees about the food source?

b. The returning bee gives nectar to the other bees, then begins a dance on the verticle face of the hive.

c. If the dance is a round dance, other bees search for plants nearby.

d. If the dance is a figure 8 with wagging of the abdomen, bees search at a distance.

Figure 20.6 Chimp using sign language to communicate.

Other types of communication include chemical signals, such as scents, and vocal communication, such as a cat's meow.

Although humans possess the most complex brain and language, they may not be alone in the ability to learn new methods of communication or to pass on knowledge. A chimp named Washoe (Figure 20.6), who was taught some American Sign Language as an infant, has shown her own adopted infant the sign for chair and has signed *food* to him and then molded his hands with hers into the sign for food. Not all researchers agree on the meaning of these actions, but data from these and other studies have suggested that new modes of communication can be learned by some non-human animals.

No matter what method of communication is used—vocal, visual, or chemical—they all are used to communicate with other animals. The other animals may be members of the same species, predators, or other animals within the same area. Communication, then, aids the interactions animals have with one another.

20.2 Organisms Interact with One Another

Any type of interaction between two or more animals, usually of the same species, is called **social behavior**. Mating behavior is a social behavior that contributes to the reproductive success of the interacting individuals. Other social interactions are competitive rather than cooperative, usually as a result of limited resources.

Because members of a population live together and use the same resources, there is strong potential for conflict. A contest of some sort usually determines which competitor gains access to something they both need, such as food or mates. Often, the encounter involves a test of strength or, more commonly, threat displays that make individuals appear large or fierce (see Figure 20.7). Usually some type of submissive or appeasement display

Figure 20.7 This swamp crayfish, *Procambarus,* has assumed a defensive posture that makes the animal appear larger and more threatening.

Figure 20.8 Social behavior in wolves. The dominant male of the pack, called the alpha male, reasserts his dominance over his brother who, in turn, assumes a submissive or appeasing posture. The interaction is ritualized—a fight that does not result in injuries.

by the contestant who surrenders or gives up the contest follows these displays. Much of this type of threatening behavior includes **ritual**, the use of symbolic activity, with the result that little or no real harm is done to either animal.

Dogs and wolves show aggression by baring their teeth, erecting their ears, tail, and fur, standing upright, and looking directly at their opponent—all actions that make the animal look large and threatening. The eventual loser, on the other hand, sleeks its fur, tucks its tail, and looks away, as shown in Figure 20.8. This appeasement display inhibits any further aggression. Although in some cases animals injure each other, most often they end the contest as soon as the ultimate winner is clearly established. Appropriate behavior at the appropriate time is, therefore, adaptive both for the winner and the loser.

20.3 Many Animal Societies Are Hierarchical

If several chickens that are unfamiliar with one another are grouped together, they respond by pecking at each other and skirmishing. Eventually, they establish a clear "pecking order"—a more or less linear **dominance hierarchy**. The top-ranked hen drives off all others, often by mere threats rather than by actual pecking. The second-ranked hen similarly subdues all others except the top-ranked hen, and so on down the line to the lowest animal (see Figure 20.9).

In confrontations between animals, the larger, stronger, and more mature animal usually wins. Sex can make a difference: In some societies, males dominate females, whereas in others females dominate males. Animals with young tend to win out over animals without young. Animals that are ill tend to lose, and animals on winning streaks tend to continue to win. In a typical wolf pack, there is a dominance hierarchy among the females.

Figure 20.9 Dominance relationships in white leghorn hens. The hen in the upper left in her first contact has been dominated by another hen. In the foreground, the lowest hen in the flock is pecked by the alpha hen. Three other hens, intermediate in the pecking order, rushed to the food when the top hen left.

The top female controls the mating of the others. When food is abundant, she mates and allows other females to do so also; when food is scarce, she allows less mating, sometimes restricting it to herself.

Although a social hierarchy may have survival value for the society, as in the case of a pecking order that reduces energy loss by bringing the activities of the group into harmony, the hierarchy really results from individuals striving for dominance within the group. The behavior that creates such hierarchies is best understood in terms of individual reproductive success. Those individuals closest to the top of the hierarchy stand the best chance of mating and producing offspring.

20.4 Many Animals Require a Territory

Although many animals live in groups, other species live alone or in pairs. Many of these more solitary animals expend much energy to keep other members of the same species out of their territory.

Every living organism requires a certain amount of space in which to live, find food, and reproduce. The living space that an animal defends is called a territory, and the behavior of an animal defending its living area is called **territoriality**. Territories typically are used for feeding, mating, rearing young, or combinations of these activities. The size of a territory varies with the species, the use it makes of the territory, and the available resources. Song sparrows, for example, may have territories of about 3000 m², in which they live more or less permanently. Gulls and other seabirds, on the other hand, mate and nest in territories of only a few square meters or less, as Figure 20.10 illustrates.

Territories are established and defended through threat displays and challenges. An individual that has gained a territory may be difficult to dislodge. Ownership is announced continually—one function of most bird

Figure 20.10 These gannets require very little space for individual nesting sites.

songs. Other animals, such as dogs and cats, use scent marks or frequent patrols that warn potential challengers.

Although dominance hierarchies and territories are mechanisms that primarily affect the reproductive ability of the individual, they do offer some incidental advantages to the population as a whole because they tend to stabilize density. For example, if resources were distributed evenly among all members of a population, the "fair share" that each member received might not be enough to sustain any one of them, particularly if a bad year reduced food supplies. Dominance and territoriality help ensure that at least some individuals receive an adequate amount of all resources.

20.5   Sexual Selection Plays an Important Role in Many Populations

If competition and the use of threatening displays are common among animals, how do the males and females manage to get close enough to mate? In many animals, a complex courtship interaction, unique to that species, occurs before the individuals mate. Each step of the courtship must be completed correctly and in the proper sequence to assure one partner that the other is not a threat and is the appropriate age and sex. Courtship often consists of a series of small behaviors. The correct performance of one behavior on the part of one partner leads the other partner to perform an answering behavior (Figure 20.11). In many cases, the sequences probably evolved from threat displays and submissive gestures used in competitive interactions. The use of a submissive posture would allow one partner to remain close to the other.

Following the correct courtship behavior does not, however, guarantee reproductive success. A female may not pay much attention to a courting male, or she may be distracted by several males all courting her at the same time. This competition for mates and the possibility that the female will, for some reason, select one particular male from the group result in **sexual selection**.

What factors influence a female's choice of a mate? Many animals that have intricate courtship rituals also have well-developed secondary sexual characteristics, such as antlers in male deer or colorful plumage in male birds. In most vertebrates, the male is showier than the female. Sometimes, the males with the most impressive masculine features are the most attractive to females, but many of these characteristics do not appear to be adaptive in the sense of promoting survival. For example, showy feathers probably do not help male birds better cope with their environment and may even make them more noticeable to predators. If such showy feathers do not increase the individual's chance for survival, and may even decrease the chance, what function can they possibly serve? Particularly showy feathers displayed in a particularly flamboyant manner may attract the attention of the female. If such a display gives the male an advantage in gaining a mate, it improves his reproductive success.

Consider the example of the sage grouse of the central United States. During the breeding season, males gather in groups on small patches of prairie. Each male defends his territory, where he engages in a truly astonishing display. With tail feathers upright and splayed and neck pouches

Figure 20.11 The correct performance on the part of one of the blue-footed boobies, *Sula nebouxi*, signals the other to perform an answering display.

inflated, the male projects booming calls over the prairie while he prances about like a wind-up toy, as shown in Figure 20.12. Female grouse observing these males eventually select a partner from among the many competitors. Many females may choose the same male, greatly enhancing his reproductive success.

Among certain other species, males give their partners more than genes. In the case of the hang fly, the female's choice clearly benefits the female. Males of this species capture and kill a smaller fly, which they present to a female. The size of the prey may vary, but females prefer males that present them with larger prey. Moreover, the female will allow mating only as long as the food holds out.

Figure 20.12 A male sage grouse, *Centrocerus unophasianus,* in display.

20.6 Some Animals Display Cooperative Behavior

Many animals behave in ways that increase their ability to reproduce, but what about animals that give up all chances to reproduce in order to help feed another's young? In some species it is common for individuals to reduce their own reproductive success through cooperative behaviors. For example, Florida scrub jays (Figure 20.13) often flock in groups of about a dozen, but usually no more than one or two pairs mate and nest at a time. A single male and female will build a nest and produce a clutch of eggs. After the eggs hatch, the entire flock helps feed the young. Although the parents usually contribute most of the food, nonparents still contribute about 50 percent.

How can such cooperative behavior evolve if it reduces the reproductive success of the nonbreeding individuals? Consider this: Relatives share genes because they share a common ancestor. Brothers and sisters share 50 percent of their genes, the same as do parents and offspring. Which would be more effective in shaping the gene pool of a species—helping save the lives of two siblings or having one offspring? What, then, might be the composition of a flock of these cooperative scrub jays? Young jays are not easily accepted into new flocks; they tend to stay with their original flock. Therefore, a flock usually consists of closely related birds. Breeding is restricted, but nonbreeders still contribute to the survival of copies of their genes by caring for the young of the breeding pairs. The experience gained by the nonbreeders ultimately may be beneficial to the species when they have young of their own. At least among nonhuman animals, there probably is no such thing as truly unselfish behavior. Any behavior that appears to benefit another animal actually enhances the survival of the individual's own genes, directly or indirectly. If it did not, it would be selected against.

Figure 20.13 Florida scrub jays, *Aphelocoma coerulescens,* exhibit cooperative behavior.

Concept Review

1. Compare and contrast innate and learned behavior. How is each adaptive?
2. Relate dominance hierarchies and territoriality to an individual's struggle to survive and reproduce. How are they adaptive for the individual?
3. What functions do displays and rituals serve?
4. Explain how sexual selection affects reproductive success.
5. How can cooperative behavior affect an individual's reproductive success?

Guidepost: How does the
environment influence the survival of
an organism?

The Environment and Survival

INVESTIGATION 20.2   A Field Study of Animal Behavior

The study of an organism's behavior in an ecosystem is called ethology. This behavior includes the organism's responses to the abiotic environment, to other species, and to other members of its own species. You can learn a great deal about an organism by observing it under natural conditions in the field, where its behavior is likely to be typical of the species. In this investigation, you will develop a systematic field study of an organism in its natural habitat and then relate its behaviors to its basic needs. Be careful not to attribute human values and emotions to the organism.

Materials (per person)
data book
pen

Procedure

1. Select a nondomesticated animal that is available for observation. Some animals that would make good subjects are insects and spiders; birds such as robins, blue jays, pigeons, sparrows, ducks, and geese; mammals such as deer, squirrels, chipmunks, muskrats, and mice; snails, slugs, and fish.

2. Most of these organisms live in fields, ponds, parks, or forests. Even an aquarium, though not the same as a natural habitat, provides a suitable environment for observation. The best times to observe are just before sunset and just after sunrise. Many animals feed at these times. You do not need to study the same individual each time because the behaviors you observe should be typical of the species.

3. Some good behaviors on which to focus include:
   a. Orientation to external stimuli, such as wind, sun, and moisture.
   b. Communication with other members of the species or with other species.
   c. Feeding.
   d. Courtship and mating.
   e. Interaction with other members of the species.
   f. Interaction with members of other species, including protective measures for itself or its young.
   g. Reactions to the presence of other species, including humans.

4. Once you have selected an organism to study, devise a procedure and submit it to your teacher for approval. Your plan should include:
   a. The scientific and common names of your organism.
   b. Where and when you will make your observations. (If possible, the organism should be observed several different times.)
   c. The question you plan to investigate.
   d. A hypothesis related to your question.

5. Conduct your observations. Your results will reflect natural behaviors more accurately if your subject is not aware of your presence.

6. Record your observations accurately and comprehensively in your data book. You will use your data to prepare your report.

Discussion

1. Write a report relating the behaviors you observed to the organism's basic needs. Remember not to attribute human motives to the animal. A good report should include the following:
   a. Title, your name, and date.

b. The procedure approved by your teacher.

c. The actual procedure you followed if it differed from your proposal.

d. Your question and hypothesis.

e. The data you collected. Organize your notes by quantifying as much data as possible (how many, how much, how often, and so on), using tables, charts, or graphs.

f. Conclusions: interpretations or explanations of the behaviors observed in light of the basic needs of the animal.

g. Evaluation of your hypothesis: Did the data support or refute your hypothesis?

h. Recommendations for further study: What you would suggest if someone else were going to do the same study, or what you would do differently if you did the study again?

20.7   Animals Adjust to Environmental Changes in Several Ways

Environments vary according to time as well as place. The temperature of the surface of a sand dune in Death Valley, for example, may climb to over 58° C in the day and drop to below 0° C at night. Individual organisms can respond to such drastic changes in their environment through changes in their behavior, physiology, and appearance or form.

The quickest response of many animals to an unfavorable change in the environment is to move to a new location, sometimes just a few feet away. Rainbow trout, for example, can sense the heat and lower oxygen levels of the upper zone in a lake and move to deeper water that is cooler and has more oxygen. Other animals are able to modify their immediate environment by cooperative social behavior. Honeybees, for example, cool the inside of their hives on hot days by collectively beating their wings. Through such behavior, the bees can keep the hive at a set temperature, plus or minus one or two degrees.

In general, physiological responses change more slowly than behavioral reactions. If you moved from Galveston, Texas, which is at sea level, to Santa Fe, New Mexico (about 2135 m above sea level), one physiological response would be an increase in your red blood cell count, but this reaction would take several weeks to develop. Some physiological responses are faster, such as the contraction of blood vessels close to the skin surface when you go outside on a cold day. This contraction helps reduce heat loss.

Many small- and medium-sized animals in northern regions tolerate low temperature and food scarcity by entering a state of prolonged dormancy, called **hibernation** (Figure 20.14). True hibernators, such as ground squirrels, jumping mice, and marmots, prepare for hibernation by building up large amounts of body fat. Entry into hibernation is gradual. After a series of "test drops" during which body temperature decreases a few degrees and then returns to normal, the animals cool to within a degree or so of the surrounding air temperature. Metabolism decreases to a fraction of normal. Some mammals, such as bears, raccoons, badgers, and opossums, enter a state of prolonged sleep in winter with little or no decrease in body temperature. This is not true hibernation.

Figure 20.14   Golden mantle ground squirrels, *Citellus lateralis,* are true hibernators.

Usually organisms either can regulate their internal conditions or conform to the prevailing external environmental conditions. Whether an animal regulates its internal environment or conforms to the external environment largely depends on the stability of the environment in which it normally lives. For example, brine shrimp live in waters with high but variable salt concentrations and have the ability to maintain stable internal salt concentrations. Marine organisms that live where the salinity of the water is very stable have no such regulatory ability. If they are placed in water with a different salt concentration, they will gain or lose water to conform to that particular external environment.

Organisms may react to some change in the environment with growth responses that change their form or internal anatomy. One example is the arrowleaf plant, which can grow on land or in water. The leaves of submerged plants are flexible and lack a waxy cuticle. Thus, they are able to absorb nutrients from the surrounding water. Arrowleaf plants that live on land, however, have extensive root systems and more rigid leaves covered with a thick cuticle that retards water loss. In general, plants are much more capable of this type of response to environmental change than are animals.

INVESTIGATION 20.3   Environmental Tolerance

The seeds of some desert plants will not germinate (sprout) until rain washes out the chemicals in the seeds that inhibit germination. Other seeds must pass through the digestive tracts of animals before germinating. Some wheat seeds will not germinate until they have been exposed to low temperatures for a certain period. In this investigation, you will examine the tolerance of some seeds to certain environmental factors.

 Before beginning work, read through the procedure and develop hypotheses that predict which seeds will germinate best in which environment(s). Record your hypotheses in your data book.

Materials (per team of 5)

5 pairs of safety goggles	2 shoe boxes (per class)
5 lab aprons	forceps
5 pairs of gloves	glass-marking pencil
4 50-mL beakers	refrigerator
2 nonmercury thermometers (per class), −10° C to +110° C	incubator
	500 mL distilled water
5 100-mm × 15-mm petri dishes	150 mL fungicide (dilute household bleach)
5 clear plastic bags with twist ties	
20 100-mm diameter paper towel disks	50 seeds each of radish, vetch, tomato,
20 100-mm × 15-mm cardboard strips	and lettuce

SAFETY   **Put on your safety goggles, lab apron, and gloves.**

Procedure

1. In your data book, prepare a table similar to Table 20.3, or tape in the table provided by your teacher.

2. Label the beakers *Tomato*, *Radish*, *Vetch*, and *Lettuce*. In each beaker, place 40 seeds of each of the types named. Add enough fungicide to cover the seeds. Soak them for the period of time recommended by your teacher.

TABLE 20.3 Seed Germination

Dish No. _____ Environment _____

Type of Seed	Number Germinated		
	Day 1	Day 2	Day 10
Tomato			
Radish			
Vetch			
Lettuce			

CAUTION: Household bleach is an irritant and can damage clothing. Avoid skin/eye contact; do not ingest. Should a spill or splash occur, call your teacher immediately; flush the area with water for 15 minutes; rinse mouth with water.

3. Place four paper towel disks in each petri dish and moisten thoroughly with distilled water. Use the cardboard dividers to divide each dish into four sections, as shown in Figure 20.15.
4. On the bottom of each dish, label one section *Tomato*, one *Lettuce*, one *Vetch*, and one *Radish*.
5. Pour the fungicide solution from the beakers and rinse the seeds with water. (Fungicide can be poured down the drain.)
6. Using the forceps, place 10 tomato seeds in the section of each dish labeled *Tomato*. Repeat with the other three types of seeds in the corresponding three sections of each dish.
7. Label the dishes with your team symbol and number them from 1 to 5. Place each dish in a clear plastic bag and close the bag securely with a twist tie.
8. Place each dish in a different environment, as follows:
   a. Dish 1: continuous light and cold—a refrigerator with a light (10° C to 12° C).
   b. Dish 2: continuous dark and cold–a light-tight box in a refrigerator (10° C to 12° C).
   c. Dish 3: continuous light and warm—an incubator with a light (30° C to 32° C).
   d. Dish 4: continuous dark and warm—a light-tight box in an incubator (30° C to 32° C).
   e. Dish 5: variable temperature and light—on a windowsill.

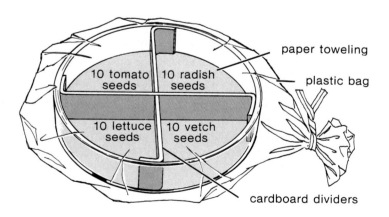

Figure 20.15 Experimental setup.

9. Wash your hands thoroughly before leaving the laboratory.
10. Each day, count the number of germinated seeds. Record the counts in your data book.
11. On the tenth day, take a final reading and combine the data of all teams.
12. Each day, wash your hands thoroughly before leaving the laboratory.

Discussion

1. Why should the data from all teams be combined?
2. In which environment did the greatest percentage of tomato seeds germinate? Radish seeds? Vetch seeds? Lettuce seeds?
3. Did any seeds of one species germinate fastest in one environment but have the greatest number of germinated seeds in another environment? If so, which species and in which environments?
4. Which species has the greatest tolerance for continuous light?
5. Which species has the greatest tolerance for low temperature?
6. Which, if any, species germinated similarly in all the experimental environments?
7. How do your results compare with your hypotheses? What do the data suggest about the ability of species to establish themselves in different environments?
8. Which factor might give a species a greater advantage—rapid germination or germination of a large percentage of seeds? Why?
9. On the basis of your experimental results, describe ecosystems in which each species you studied in this investigation might have an advantage.

For Further Investigation

From a grocery store, obtain seeds of plants that grow in a variety of climates, such as avocados, dates, grapefruits, oranges, pomegranates, lentils, and many types of beans. Test these for germination in experimental environments. In some cases, you may have to lengthen the time allowed for germination.

20.8 Organisms Vary in Their Tolerance

How do environmental factors result in the selection of certain organisms for survival? Specific environmental factors set limits for growth, which affect selection. One important environmental limiting factor is temperature. If a household geranium is left outdoors during a Minnesota winter, it will die. A blacksnake left in a shadeless cage and exposed to the August sun of Texas also will die. Household geraniums cannot tolerate long periods of freezing temperatures, and blacksnakes cannot tolerate long periods of high temperatures. If an organism cannot tolerate the highest or lowest temperature in an area, it will die or move elsewhere, if it can. Thus, **tolerance** is the ability of an individual or a species to withstand particular environmental conditions. If an individual cannot tolerate the conditions of the environment, it may not survive to reproduce.

All organisms function most efficiently under certain environmental conditions. We can determine the tolerance limits of a species for one environmental factor by varying that factor and measuring some aspect of the organism's performance. The resulting curves typically are bell-shaped, with the tails of the curve representing the limits of the organism's tolerance for

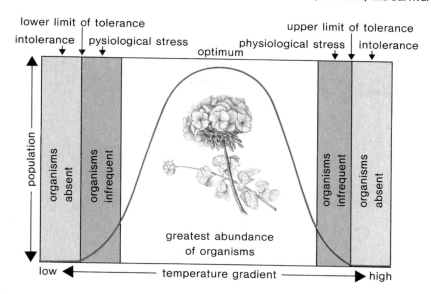

lower limit of tolerance

intolerance | pysiological stress

upper limit of tolerance

physiological stress | intolerance

optimum

population

organisms absent

organisms infrequent

organisms infrequent

organisms absent

greatest abundance
of organisms

low ◄——— temperature gradient ———► high

Figure 20.16 A typical bell-shaped curve showing the tolerance range of a given organism, in this case a geranium.

a particular environmental variable and the values in between representing the range within which the organism can live, as shown in Figure 20.16. The conditions most favorable for growth and reproduction for that species are called the optimum conditions. Tolerance limits may seem clear-cut, but complications can arise. The duration of a condition is important, especially in determining maximum and minimum limits. For example, geraniums can withstand short periods of freezing temperature but not long ones. The limiting factor of temperature selected those geraniums that can grow in moderate climates.

Within any population, individuals vary genetically for some characteristics, including tolerance. Within a species, also, populations of organisms vary in their tolerance for certain conditions. Populations that come from different parts of a species' geographic range differ in their tolerance to environmental factors. A population of jellyfish, *Aurelia aurita* (Figure 20.17a on page 560), which lives off the coast of Maine, can tolerate temperatures between 5° C and 18° C. For the *Aurelia* population off the southern coast of Florida, however, the optimum varies between 28° C and 30° C. Both the northern and the southern jellyfish have been selected for survival in their particular environments. Figure 20.17b shows the tolerance ranges for each jellyfish population. Such variation in a population arises by mutation or genetic recombination in some individuals. Not all individuals in a population are equally suited to the same environmental conditions. Those that are best suited will survive and out-reproduce the others.

The interaction of several variables also may affect an organism's tolerance. For example, several variables interact to determine the geographic range of the field sparrow. Field sparrows can survive northern winter temperatures in the United States if they have enough food for energy. Sparrows need more food in the winter because they lose more heat from their bodies than in summer. They can hunt for food only during the day, and

a

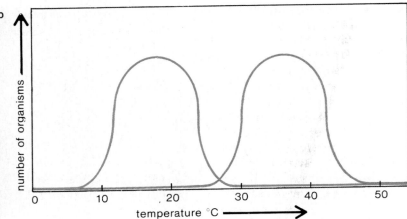

Figure 20.17 *Aurelia aurita* (a) and the tolerance ranges for two populations of *A. aurita,* one in Florida and one in Maine. Even though they are the same species, they no longer can live in water that is the same temperature (b).

winter days are short. Is it the temperature, the food supply, or the length of day that sets the northern winter boundaries for field sparrows? It could be any one or a combination of these factors.

Abiotic conditions within a species' tolerance limits may allow it to survive in a particular ecosystem, but these conditions do not guarantee that the species will survive there. The species may fail because of conditions in the biotic environment. It may have to compete with other organisms or be affected by them in other ways. For example, the geographic range of the koalas of Australia is determined not by abiotic conditions that directly affect them, such as temperature or rainfall, but by the availability of eucalyptus leaves. Only the leaves of about a dozen species are acceptable. Koalas (Figure 20.18) certainly can tolerate the abiotic conditions of other areas of Australia, but because eucalyptus trees are not found there, the koalas would not survive. Similarly, if the eucalyptus trees in the current range were destroyed to create farmland, the koalas would be unable to survive even though the abiotic conditions remained unchanged.

Figure 20.18 Koalas, *Phascolarctos cinereus,* are Australian marsupials that feed on the leaves and bark of eucalyptus trees.

20.9 Human Activity Affects the Behavior and Survival of Other Animals

The human population is growing rapidly and increasing its demand for food, housing, water, and other necessities of life. In the course of obtaining these necessities, humans affect other organisms in many, usually negative, ways. Changing the environment and destroying habitats have the greatest effect on other animals. For example, building houses on a tract of wooded land means that the resident animals either must move to another area, adapt to life among humans, or must die. Of course, most of the plants must give way to houses and lawns.

If the animals move and join another population, competition among the members of that population for food, territory, mates, and other resources probably will increase. The needs of the increased population also

Figure 20.19 These desert bighorn sheep, *Ovis canadensis,* have adapted to life near humans. What might be some of the advantages and disadvantages?

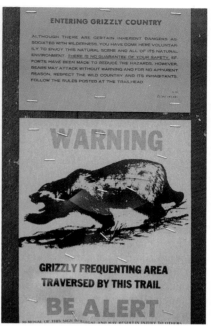

Figure 20.20 A grizzly reminder of how humans fit into the food chain. People often forget that humans are a part of the web of life. Just because humans are present in an area does not guarantee that the other animals will behave in what we consider to be an appropriate manner.

may outstrip the area's carrying capacity, leading to famine, increased disease, and higher mortality. If the new houses totally cover a unique habitat, the animals that lived there may well become extinct.

A few animals have found that living close to humans confers some benefits. The bighorn sheep in Figure 20.19 have found that the grass in lawns is more abundant, more succulent, and probably easier to obtain than wild grasses. Elk often eat the apples from trees in backyards, and raccoons are common nighttime garbage can visitors. Often, people who move into areas that were totally undeveloped a short time ago have difficulty with some of the wildlife. It is not uncommon for great horned owls (or other large birds of prey) or coyotes to dine on small dogs and cats. In fact, the human population has moved into so many different places and modified so much of the environment that we often forget our place in the web of life (see Figure 20.20).

Sometimes the animals have adapted so well to life with humans that they rarely are found away from cities and towns. Pigeons, or rock doves, such as those shown in Figure 20.21, are a good example. Other birds, such as sparrows and finches, are found in the greatest numbers near human-occupied areas. Raccoons, squirrels, skunks, pigeons, finches, and sparrows are typical representatives of urban wildlife—wildlife that has successfully adapted to life among humans.

Humans can sometimes obtain the necessities of life without totally disrupting native wildlife. In an effort to minimize harmful effects, some new housing developments hire ecologists to perform animal and plant surveys of an area. If the types of wildlife and their movements are known, provisions can be made that will encourage the animals to remain in the area. For example, greenbelts and parks can provide deer with the open areas they need to move from one place to another as well as with a rich source of food.

Figure 20.21 Rock doves, *Columba livia,* in Times Square, New York City. These birds have adapted quite completely to urban living and are rarely found away from humans.

Figure 20.22   (a) Manatees, *Trichechus manatus,* are threatened by the loss of suitable habitat and food due to dredging and other human activities. (b) Even in protected areas, encounters with motorboat propellers often prove fatal to the manatees.

Most often, however, human activities result in great damage to native animal populations. The greatest threat to most wild species is the destruction or degradation of their habitats, as in the case of the Florida manatee, shown in Figure 20.22. If a parking lot is built over a swamp, for example, nothing will help the turtles, frogs, snakes, and other animals that live there. The disruption of natural communities threatens wild species by destroying migration routes, breeding areas, and food sources. One study found that two-thirds of the original wildlife habitat in tropical Africa and in much of southeast Asia has been lost or severely degraded. The tropical rain forests in Central and South America are being destroyed at an alarming rate. In the United States, forests have been reduced by one third and wetlands by 50 percent.

Many rare and threatened species live in vulnerable, specialized habitats, such as small islands or even single trees. Human alteration of these fragile ecosystems often creates habitat "patches" that are too small to support the number of individuals needed to sustain a population. A great deal of careful planning and cooperation on a global scale is required to help wild populations survive human activity and interference.

Concept Review

1. How are physiological responses to environmental change different from changes in form or appearance?
2. What is meant by an organism's range of tolerance, and how is it important?
3. Is the range of tolerance the same for all populations of a species? Explain.
4. What factors, other than abiotic factors, may affect the survival of an organism?
5. How can humans affect natural communities and the animals that live there?

Summary

An organism's behavior is adapted to the demands of the environment in which it lives. There are two basic types of behavior—innate and learned. Both types are influenced by internal and external stimuli, but innate behavior is less dependent on environmental experience. Learned behavior consists of the responses an animal develops as a result of environmental experience. Some species exhibit cooperative behavior, which confers certain advantages on members of the group. These advantages include increased reproductive success.

Population growth usually is limited by environmental factors, which result in the selection of individuals for survival or death. Thus, limiting factors determine what the organisms' characteristics are. Through a long period, the interacting populations in an area are shaped by their biotic and abiotic environments.

Human population growth results in the destruction of habitat, which forces animals to move or to modify their behavior in order to survive. All too often the animals do not survive human interference. Worldwide, a great deal of natural habitat is being destroyed at an alarming rate. Careful planning and global cooperation are needed to help wild populations survive the impact of humans.

Applications

1. Predators actually can benefit the population of their prey as a whole, even though certain individuals will be killed. Explain the beneficial role of predators.

2. What disadvantages do you see in cooperative breeding behavior?

3. Do plants show behavior? Explain.

4. The text states that "organisms usually either can regulate their internal conditions or conform to the prevailing environmental conditions." Some people would argue that humans do neither of these things. Explain.

5. Why is it sometimes difficult to distinguish between innate and learned behaviors?

6. How would you plan an experiment to test learning in baby chicks? What useful background information might be obtained from a commercial hatchery?

7. How would you design an experiment to test whether cooperative breeding behavior in the Florida scrub jay is innate, learned, or a mixture of both?

8. The destruction of natural habitat by humans also may reduce the natural predators of a certain species. For example, mountain lions eat deer. If the habitat of mountain lions is destroyed, the cats no longer can live in the area, but the deer might remain. Explain how this might affect the deer population, the habitat, and the other animal populations in the area.

9. How is hibernation similar to sleeping? How are they different?

Problems

1. Learn how to communicate in a nonvocal system, such as signing with your hands or waving two semaphore flags.

2. Pheromones are commonly used to control insect pests. Go to your library or contact an extension agent to find information on the use of pheromones to fight certain insects.

3. Many languages are "foreign" to our ears, such as the clicking language of the !Kung bushmen in Africa and the whistling language of the Canary Islands. Listen to audiotapes of any foreign language. What do all languages have in common?

4. Investigate your community and the use of its land resources. Use the module by BSCS, *Investigating the Human Environment: Land Use* (Dubuque, IA: Kendall/Hunt, 1984), to help you study a local land use issue. Identify a local land use problem, investigate it, produce suggestions, and report these to the appropriate authorities.

5. Report on the research attempts to train chimpanzees and gorillas to use sign language to communicate with humans. How successful do you think these programs are?

6. Observe how people behave toward one another in a park, shopping center, or bank. Write a report detailing what you observed.

7. Research an extinct species in the library. Write an essay on how the behavior of that species might have influenced its ability to survive. Possible topics could be the passenger pigeon or the Carolina parakeet.

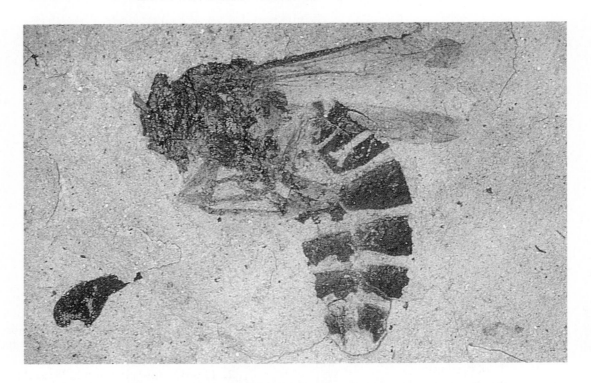

This fossil wasp is about 35 million years old. Its actual length is about 2.5 cm. What living organisms does it resemble? What would you estimate their size to be? Based on these observations, what appears to be the pattern of evolution of this group of animals?

Ecosystems of the Past

Ecologists study the interrelationships of organisms and their environments. The size, shape, and other characteristics of organisms may change quickly or may persist for hundreds of years. Selection as a result of environmental factors may promote one population's survival but extinguish another population. If the same area could be viewed once every million years, few populations would be seen to have remained the same as their ancestors of a million years before. The entire pattern of the ecosystem would have changed, or evolved. This chapter discusses how scientists study earth's history and what their studies have revealed about past ecosystems.

Reconstructing the Past

Guidepost: What evidence do researchers use to reconstruct the past?

INVESTIGATION 21.1   Evolution and Time

Life has been evolving for more than 3.5 billion years, a time span that is difficult to comprehend. Our limited experience with time involves tens and hundreds of years, not millions or billions. To understand evolution, we can represent time concretely, convert it into a linear dimension, and scale it down to a more manageable size. Time lines can be used to represent the breadth of time and the changes that have occurred during the history of earth and life. In this investigation, you will construct a scale model of major events in earth's history—a concrete time line—to gain a sense of the vast amounts of time during which evolution occurs.

Materials (per team)
Team 1
15-m tape measure
150-cm tape measure
1 each black and red broad, permanent,
    felt-tip markers
100 m of rope clothesline
Team 2
3 copies of the *Geologic Time Scale*
    (see Appendix 3)
25 sheets of construction paper
    (one color)
5 dark-colored crayons or broad felt-tip
    markers

19 clothespins
150-cm tape measure
15-cm metric ruler
pocket calculator
Team 3
3 copies of *Major Geological Events*
    (see Appendix 3)
25 sheets of construction paper
    (one color)
5 dark-colored crayons or broad felt-tip
    markers
14 clothespins

(continued)

Materials (continued)

150-cm tape measure
15-cm metric ruler
pocket calculator
Team 4
3 copies of *Major Biological Events*
(see Appendix 3)
25 sheets of construction paper
(one color)

5 dark-colored crayons or broad felt-tip
markers
20 clothespins
150-cm tape measure
15-cm metric ruler
pocket calculator

Procedure
Day 1

1. Your teacher will assign you to one of four teams for this activity. Team 1 will measure the area to be used, mark a clothesline at 1-m intervals, and help the other teams position their labels of the major events. Team 2 will prepare labels for the geologic time scale, Team 3 will prepare labels for the major geological events, and Team 4 will prepare labels for the major biological events.

2. **Team 1:** Use the 15-m tape measure to determine the length of the area, designated by your teacher, in which you will establish your scale model of geologic time. Use a 150-cm tape measure to mark the clothesline at 1-m intervals. Use the red marker for every 5 m and the black marker for the other marks.

   Teams 2, 3, and 4: Fold the construction paper in half crosswise. Each team should use a different color of construction paper. Label both sides of each sheet with one of the major events, using the underlined words on your team's table. Write large enough so the labels can be seen at a distance.

3. Select the sheets of construction paper with the following labels: *Rocky Mountains, Mississippi River, Ice Age, Life, Land Plants, Dinosaurs*, and *Modern Humans*. As a class, decide how many millions of years ago you think each of these events occurred. With the help of the measuring team, place the labels along the clothesline at the points that indicate the times you have chosen. Mark these times in your data book and compare them with the information in your text.

4. **Teams 2, 3, and 4:** The length of the measured clothesline represents the length of time since the earth was formed, and the far end represents the origin of the earth 4600 million (4.6 billion) years ago. If the length of your model is 100 m, then 1 million years is represented by a distance of 0.0217 m or 2.17 cm. Thus, an event that occurred 1 million years after the earth formed would be marked 2.17 cm from the *far* end (the past) of the time line; an event that occurred 1 million years ago would be marked 2.17 cm from the *near* end (the present) of the time line. Use the following equation to calculate the scaling factor for your model, based on information from the measuring team. (The example is based on a 100-m model.)

$$\frac{\text{length of model}}{\text{age of earth}} = \frac{\text{meters}}{\text{million years}} \ or \ \frac{100 \text{ m}}{4600 \text{ million years}} = \frac{0.0217 \text{ m (2.17 cm)}}{\text{million years}}$$

5. **Teams 2, 3, and 4:** The scaling factor allows you to convert into meters the number of years that have elapsed since any past event. For example, using the knowledge that the first mammals appeared about 225 million years ago, you can determine the proper placement for that label in the following way:

$$\text{event} \times \text{scaling factor} = \text{label placement}$$
or
$$225 \text{ million years} \times 0.0217 \text{ m/million years} = 4.88 \text{ m}$$

The label for this event should be placed 4.88 m from the near end (the present) of the time line. Events that occurred more recently may be measured in centimeters. Refer to your team's table and use the scaling factor determined in Step 4 to calculate the distance for placing labels for each geologic time period and for each geological and biological event. Enter each distance on the table and write it on the lower right corner of the appropriate label.

Day 2

7. With the help of the measuring team, use the marked clothesline, metric tape measures, and rulers to place your team's labels the correct distance from the *near* end of the time line. Use a clothespin to clip a corner of each label to the clothesline so that it can be read from either side of the line.
8. Walk through your scale model of geologic time.

Discussion

1. How close to their actual occurrence were you able to place the events on the time line? What might account for the differences?
2. Look at the end of the time line near the formation of the earth. Compare it with the opposite end, near where modern humans appear. What differences do you see?
3. When you look at the time line, do the events seem to occur one after the other, or do several things appear to happen at about the same time?
4. Some portions of the time line include very few events. What are some possible reasons for this?
5. At what point do events start to occur at an increased rate?
6. Predict the changes you think might take place in the future. How might plants and animals change? How might humans change?

21.1 Several Techniques Are Used to Date Fossils

When organisms of the distant past died, most decayed and disappeared. Here and there, however, skeletons, teeth, and other decay-resistant structures became fossils. Sometimes, parts of organisms were replaced by minerals, or petrified. At other times a mold, or cast, of an organism was formed in mud that hardened. Figure 21.1 shows a variety of fossils.

Even as fossils were forming they were covered with sand or mud that hid them from view. In some cases millions of years passed and many layers of sediment were added. The layers, or **strata** (STRAYT uh), turned to rock. Finally, some event, such as the erosion of the Grand Canyon or the building

Figure 21.1 A variety of fossils: (a) *Pecopteris,* a fossil fern; (b) *Heliobatus radions,* a stingray; and (c) a fossil bat.

a   b   c

Figure 21.2 The stratification of rock layers in this mountainside was exposed when a highway was built. Fossils of different ages are found in different layers.

Figure 21.3 A certain amount of carbon-14 exists in the atmosphere, and plants incorporate it into their tissues as carbon dioxide. Animals absorb the carbon-14 by eating plants. When an organism dies, the carbon-14 disintegrates at a known rate. It then is possible to measure the amount of carbon-14 remaining and relate that to the age of the organism.

of a highway (Figure 21.2), exposed some of the fossils so scientists could study them and learn about the past.

Fossils in layers of rock are the only remains of past ecosystems. These layers are studied by paleontologists (pay lee un TOL uh jists)—scientists who use fossils to reconstruct the history of life. These scientists have pieced together information from many parts of the world and have put most of the known rock strata in chronological sequence. The *Geologic Time Scale* in Appendix 3 is based on the ages of the rock strata. The relative ages of fossils can be determined by their position in the strata.

How old are the strata? The most reliable method for dating rocks depends on the presence of radioactive chemical elements such as potassium-40 and uranium-235. The atomic nuclei of radioactive elements are unstable and emit radiation. These radioactive elements break down to stable elements at a constant rate known as the half-life, the time it takes for half of the element to break down.

The potassium-40 dating method, which usually is used for inorganic material, indicates the age of volcanic rocks. Potassium-40 isotopes decay into argon gas (argon-40). In many types of rocks, the argon gas can accumulate as the potassium decays. In uncontaminated samples, the ratio of potassium-40 to argon-40 indicates how long it has been since the argon began to accumulate. Potassium-40 has a relatively long half-life of 1.3 billion years. Because of this long half-life, the potassium-40 technique is not accurate for dating deposits younger than 400 000 or 500 000 years.

The most important method for dating young fossils, or other organic material, measures the decay of radiocarbon, carbon-14. Figure 21.3 outlines the principles of carbon-14 decay. In the upper atmosphere, bombardment by cosmic rays transforms nitrogen into carbon-14, an unstable form of carbon. The carbon-14 filters down into the lower atmosphere where living things absorb it. When the organism dies, however, it absorbs no new carbon, and the carbon-14 slowly decays into nitrogen-14. Assuming the production of carbon-14 in the atmosphere is constant, the proportion of carbon-14 to nitrogen-14 in once-living material indicates how long it has

nitrogen atom → unstable atom → ^{14}C atom

neutron   proton

a. Nitrogen atom becomes carbon-14 atom in the atmosphere.

oxygen molecule (O_2) + ^{14}C atom = $^{14}CO_2$ molecule

b. An O_2 molecule combines with a carbon-14 atom to form carbon-14 dioxide.

c. Living organisms absorb ^{14}C.

nitrogen

^{14}C

d. When the organism dies, the ^{14}C atoms begin to disintegrate.

been since the organism died. However, carbon-14 dating has limitations. First, there is little carbon-14 in fossils and other organic material. Second, the half-life of carbon-14 is only 5730 years. If a specimen is 57 300 years old, 10 half-lives have elapsed, leaving so little of the original carbon-14 that it is hard to measure. (Currently there is no reliable radiometric technique for dating fossils between 500 000 and 50 000 years old.)

Paleontologists using radiometric dating methods have determined the age of the oldest fossils to be approximately 3.5 billion years. Despite uncertainties in radiometric dating techniques, if two or more methods yield similar data, the age usually is accepted as reliable. Such data have enabled paleontologists to construct a fairly accurate time scale.

21.2 Fossils Provide Information About Ancient Life

Inferences about ancient ecosystems are based on studies of ancient rock layers and the fossils they contain. Early paleontologists were puzzled by fossils that did not resemble any living forms. The theory of evolution by natural selection, however, made it possible to link fossils and living forms and to use fossils as clues to the history of life.

Nothing about the study of fossils is easy, however. The soft parts of organisms usually decay, and even when all the hard parts are preserved, fossils are seldom found in perfect condition. They usually must be reassembled and placed in proper relation to each other. To do this accurately, paleontologists must know the anatomy of modern organisms. Then, they must determine the placement of structures, such as muscles or leaves, that were not preserved. Finally, the paleontologists make a life-size reconstruction of the plant or animal based on its fossils. The colors used in these models are inferred from the colors of modern organisms. Different researchers may interpret the evidence differently. The farther they move from the basic evidence—the fossils themselves—the greater is the possibility of differing interpretations, as shown in Figure 21.4.

In reconstructing the evolution of species, paleontologists need to determine the relationships between modern organisms and their fossilized

Figure 21.4 An artist's reconstruction of a stegosaurus, a late Mesozoic reptile (a), and another artist's reconstruction (b).

a

b

Figure 21.5 Artist's conception of the animals along the main line of evolutionary descent from primitive bony fishes to mammals. *Eusthenopteron,* a primitive lobe-finned fish; *Ichthyostega,* a large, primitive amphibian; *Varanops,* a primitive mammal-like reptile; *Probelesodon,* a more advanced mammal-like reptile showing distinct mammalian characteristics, and *Megazostrodon,* the small, shrewlike first mammal.

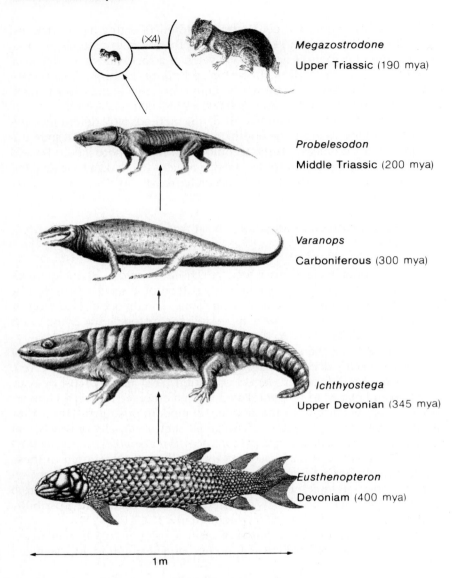

Megazostrodone
Upper Triassic (190 mya)

Probelesodon
Middle Triassic (200 mya)

Varanops
Carboniferous (300 mya)

Ichthyostega
Upper Devonian (345 mya)

Eusthenopteron
Devoniam (400 mya)

1 m

relatives. Today, relationships between fossils themselves can be studied as well. The DNA and protein of some fossils preserved in ice and mud can be compared with the DNA and protein of other fossils. Such biochemical studies, however, are a relatively recent development. Observable structural characteristics still are used widely for determining relationships and classification.

Consider the problem of classifying a fossil jaw that has both mammalian and reptilian characteristics. The fossil is from the early Mesozoic era, about 225 million years ago. Mammals were beginning to appear at that time. If mammals and reptiles shared a common ancestor, as scientists think, then there once must have been animals with a combination of mammalian and reptilian characteristics. The jaw could belong to one of these ancestral animals. Figure 21.5 shows one artist's interpretation of the evolution of mammals from bony fishes.

21.3 The Fossil Record Reveals Both Change and Stability

No one can examine a large collection of fossils without being impressed by the evidence of change, especially if the fossils of one type are arranged in chronological order. Many species have died out and left no descendants; others have changed so much through many generations that the successive forms are given different species names.

The fossil record shows that organisms in one geological period are descended from ancestral forms present in earlier periods, though not necessarily directly. For example, although mammals are thought to have evolved more recently than reptiles, the reptiles of the Mesozoic era did not give rise to mammals in the Cenozoic. Rather, the ancestors of the Mesozoic reptiles apparently also were the ancestors of the mammals.

At least 130 000 species of animals are known only from their fossils. In the Cambrian period, trilobites (Figure 21.6) were varied and numerous, but they disappeared suddenly from the geological record. There have been many large-scale extinctions. Evidence suggests the trilobites perished in one of these catastrophes. Other groups, including the dinosaurs, died out in other extinctions. What caused such extinctions? Many hypotheses have been proposed, but testing them is difficult because we are dealing with the past. Although we may never learn what caused the extinction of trilobites and dinosaurs, new evidence continues to be found.

Just as organisms have changed, whole ecosystems also have changed. Change in ecosystems depends on the changes in a particular environment. It also depends on the adaptability of the associated organisms. A characteristic that adapts an organism to one environment might be a hindrance in a different environment. Similarly, a characteristic that has no adaptive value in one situation may prove to be adaptive if conditions change.

Although the fossil record offers abundant evidence of change, it also reveals great stability. For example, twentieth-century oceans contain brachiopods (BRAY kee oh podz), animals with hinged shells, similar to those from the Paleozoic seas, as shown in Figure 21.7. Ocean environments are usually more constant than land environments. Thus, many adaptations that allowed early forms to survive apparently are still useful.

Figure 21.6 A fossil trilobite. Its actual size was about that of a paper clip.

Figure 21.7 (a) Present-day brachiopods, and (b) a fossil cast of a brachiopod.

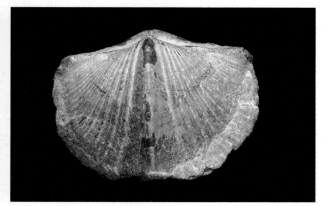

a

b

Concept Review

1. How are fossils formed, and how are they used?
2. How can the age of a fossil be determined?
3. How would a paleontologist decide whether a particular fossil represented a reptile or a mammal?
4. How does the fossil record support both the idea of change and the idea of stability in living things?
5. What problem does the extinction of large groups of organisms—such as the dinosaurs—present to paleontologists?

Guidepost: What steps led to the broad array of organisms found on earth today?

Evolution and Plate Tectonics

21.4   Evolution Has Occurred on Moving Plates

Reconstructing the history of earth and the evolution of organisms from fossils is complicated by certain geological events. Those events have been involved in cases like the ones shown in Figure 21.8. *Nothofagus* (noth oh FAY gus) may be found in New Guinea, New Zealand, Australia, and South America. Fossils of *Glossopteris* (glah SOP ter is), a genus of Paleozoic seed ferns, show a similar pattern of distribution except that they occur also in

Figure 21.8   Distribution map of extinct *Glossopteris* ferns and living *Nothofagus* trees.

distribution of *Glossopteris* flora fossils

distribution of the living plant *Nothofagus*

India and South Africa. Other Paleozoic and early Mesozoic terrestrial organisms have similar patterns of distribution. The areas in which these organisms are known to exist are separated by thousands of miles of ocean. How did the plants and their fossils get where they are now?

An answer to that and similar puzzling questions was proposed by a German scientist, Alfred Wegener, who in 1912 suggested that the present continents, including South America and Africa, had once been joined in a supercontinent he called Pangaea (pan GEE uh). He suggested further that all the continents have not always been where they are now, but rather that these big land masses might be slowly "floating" over the earth's hot liquid interior. Although Wegener's hypothesis was not well accepted by most of the geologists of his time, during the last 20 years additional data have been gathered that support his idea. We now know that about 200 million years ago all the continents were together in one huge land mass that eventually split up. Wegener's idea is now called continental drift.

The continental drift theory is now part of a broader theory of **plate tectonics** (tek TON iks), illustrated in Figure 21.9. According to the broader

Figure 21.9 Diagram illustrating the mechanism of continental drift. Magma, or molten rock from the earth's interior, forces its way through the thin oceanic crust and forms new crust. This new crust moves away from the mid-oceanic ridge as the process continues. Where the heavy oceanic crust meets the lighter continental coast it is subducted, or forced to slide beneath the continental crust, resulting in volcanoes and island arcs.

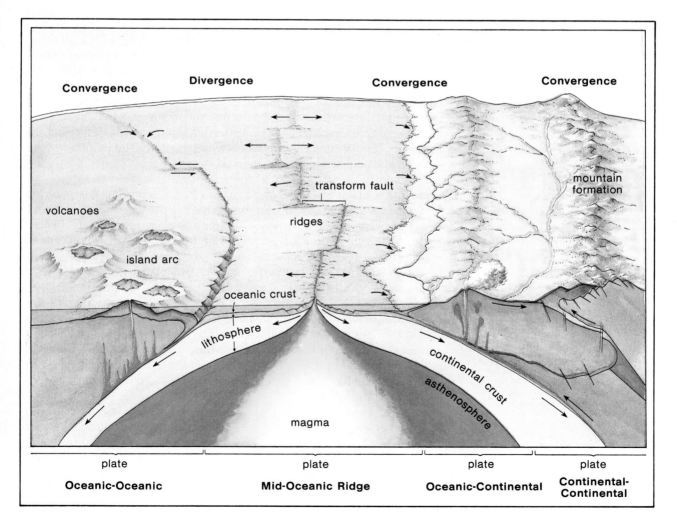

theory, land masses are the highest parts of huge floating plates that make up the earth's crust. New material from the earth's mantle is added to the plates along ridges on the ocean bottom. The mid-Atlantic ridge is one example; other ridges exist in the Pacific Ocean. The addition of new material to the earth's crust forces the plates to move slowly away from the ridges. At other locations, the plates are grinding into one another and forcing crustal materials into the mantle through deep ocean trenches. Many important geological phenomena, such as mountain building, volcanic activity, and earthquakes, happen at the boundaries of plates. The San Andreas fault in California is part of a border where two plates slide past each other. Mount St. Helens in Washington state is near another plate boundary.

21.5   Plate Tectonics Explains the Distribution of Species

The history of the earth helps to explain the current geographic distribution of species, as shown in Figure 21.10 on pages 576–77. For example, the emergence of islands, such as the Galápagos, creates new environments for founding populations, and adaptive radiation fills many of the available niches with new species. On a global scale, the drifting of continents is the major factor in the spatial distribution of life. Although these plate movements rearrange geography all the time, two such past events have been especially significant in the distribution of organisms. About 250 million years ago, plate movements brought all the land masses together to form Pangaea. Species that had been evolving in isolation now were in competition. The total amount of shoreline was reduced, and there is some evidence that the ocean basins became deeper, draining most of the shallow coastal seas. Then as now, most marine species inhabited the shallow waters, and the formation of Pangaea destroyed much of that habitat. Changing ocean currents also would have affected land and sea life.

The other major event occurred about 180 million years ago. Pangaea began to break up, and this breakup served as a giant isolating mechanism. As the continents drifted apart, each became a distinct evolutionary theater, with the organisms on each diverging from those on the other continents. The pattern of continental separation solves the puzzle of geographical distribution of such animals as marsupials and placental mammals.

Australia provides a good example of the effects of continental drift. Until about 35 million years ago, Australia was attached to a much warmer Antarctica, as was South America. Marsupials, best represented in Australia and South America today, appear to have migrated overland between these continents via Antarctica. As the two continents moved farther away from Antarctica, the land bridges that served migrating animals were lost. From then on, as Australia moved steadily northward, its peculiar plants and animals evolved in isolation.

A general principle of paleontology is, "The present is the key to the past." Conversely, as in the study of discontinuous distribution, the past often sheds light on the present. The patterns found in ecosystems today result from events that occurred in past ecosystems. The next section presents a broad overview of some of these paleoecosystems—the ecosystems of the past.

Concept Review

1. What was Pangaea?
2. How do the continental plates move?
3. How does continental drift explain some cases of discontinuous distribution of organisms?
4. What two events had the most impact on the distribution of species?
5. What principle guides the interpretation of paleontologic evidence?

A Sampling of Paleoecosystems

21.6 Cambrian Communities Were Aquatic

There is no evidence of terrestrial life during the Cambrian period, around 570 million years ago, but marine ecosystems were well developed, as shown in Figure 21.11. There were shallow- and deep-water organisms—floating, swimming, and bottom-dwelling types. The chief marine producers then, as now, probably were microscopic plankton species.

There are no chordates in the Cambrian fossil record. Otherwise, the major animal phyla known today also were represented in the Cambrian period. There were sponges and cnidarians, and the most abundant animals were marine brachiopods. Although many of the brachiopods were similar to species living today, the arthropods were so different that none of them can be placed in modern arthropod classes. Among Cambrian arthropods, the trilobites left the most abundant fossils.

21.7 Land Plants and Animals Appeared in the Devonian

After the Cambrian period, the seas receded a little but remained widespread. During this time, microscopic plants thrived, and the earliest fishes flourished. In fact, the Devonian period often is called the Age of Fishes.

Guidepost: What organisms dominated various ecosystems in the past?

EARTH

4.6 billion yrs. old

Life apeared 570 million yrs ago

Figure 21.11 An artist's reconstruction of life in a Cambrian sea. How many types of organisms can you identify?

CAMBRIAN EARTH

Four large (and two small) continents had been formed and were moving toward each other. Sea levels were high. Vermont mountains were formed.

DEVONIAN EARTH

Laurentia and Baltica had collided to form one continent. Sea levels were low. In some regions there were periods of drought.

CARBONIFEROUS EARTH

Because of high humidity, swamps and forests covered much of the land.

Figure 21.10 The formation and breakup of Pangaea correlated with events in evolution shown below.

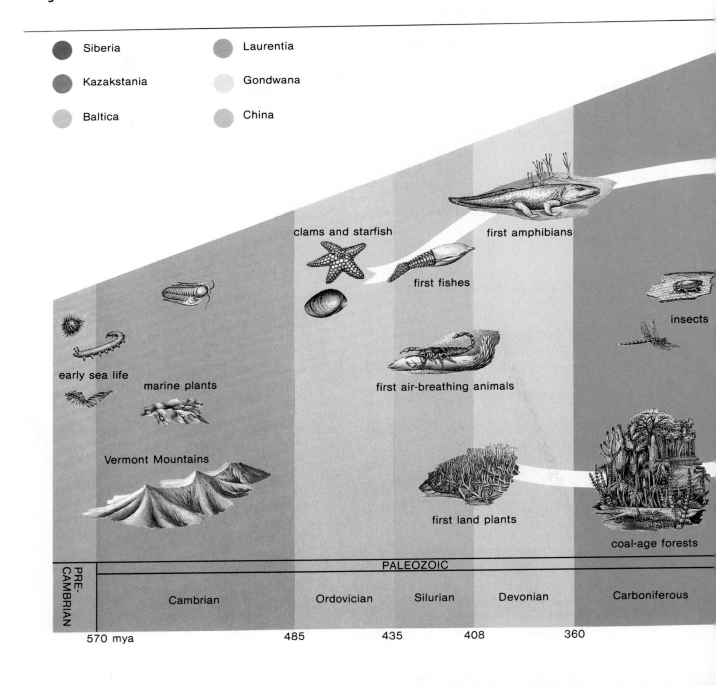

● Siberia
● Kazakstania
● Baltica
● Laurentia
○ Gondwana
● China

clams and starfish

first fishes

first amphibians

insects

early sea life

marine plants

first air-breathing animals

Vermont Mountains

first land plants

coal-age forests

PRE-CAMBRIAN | PALEOZOIC

Cambrian | Ordovician | Silurian | Devonian | Carboniferous

570 mya | 485 | 435 | 408 | 360

TRIASSIC EARTH

Pangaea had been formed by the four major continents. Sea levels were low and there were many deserts. Appalachian mountains were formed.

CRETACEOUS EARTH

North America and Europe began splitting apart and South America and Africa separated. The sea level was high. Formation of the Rocky Mountains began.

PLEISTOCENE EARTH

During each of four "ice ages" the sea was very low. Warm periods followed periods of glaciation. The continents had drifted almost to their positions of today.

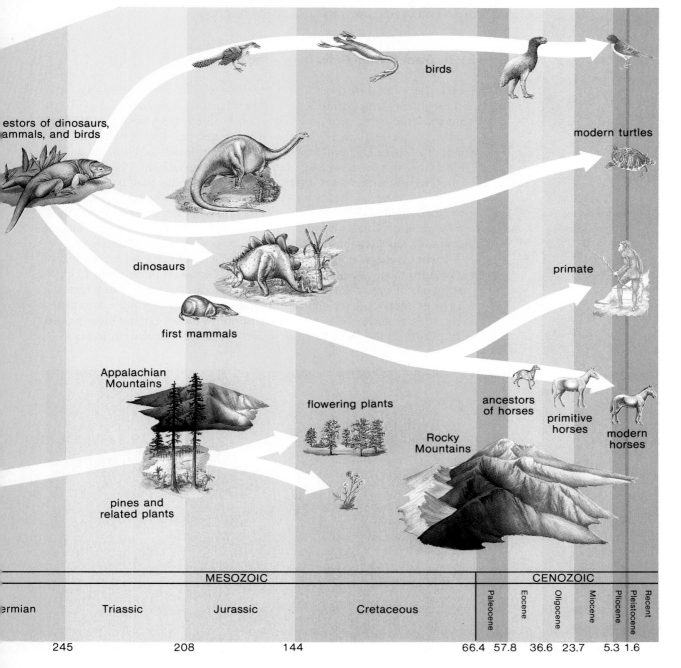

birds

ancestors of dinosaurs, mammals, and birds

modern turtles

dinosaurs

primate

first mammals

Appalachian Mountains

flowering plants

ancestors of horses

primitive horses

modern horses

Rocky Mountains

pines and related plants

	MESOZOIC			CENOZOIC						
Permian	Triassic	Jurassic	Cretaceous	Paleocene	Eocene	Oligocene	Miocene	Pliocene	Pleistocene	Recent
245	208	144		66.4	57.8	36.6	23.7	5.3	1.6	

Many fishes were covered with armorlike plates. Sharks and the ancestors of modern bony fishes evolved then also.

Land plants multiplied both in number and variety during the Devonian period. Spore-bearing ferns, seed-bearing ferns, and horsetails developed. Trees, including the ancestors of cone-bearing pines and firs, first appeared. Soon after, insects crawled around and flew above them.

Some fishes had adaptations (sometimes called preadaptations) for life on land, such as lungs, internal nostrils, and skeletons that better supported their paired fins. In some, lobe-shaped fins became modified into stumpy, leglike paddles. With the appearance of these fishes during the Devonian period, the vertebrate invasion of land had begun.

21.8   Giant Forests Grew During the Carboniferous

By the Carboniferous (coal age) period, about 360 million years ago, the earliest forests fringed the shallow seas that still covered most of North America. Unlike those in present-day forests, these trees were mostly huge relatives of present-day horsetails, club mosses, and ferns (Figure 21.12). Strata of the late Carboniferous period contain fossils of conifers and related plants, but the flowering plants of today were absent. Plant life of the Carboniferous suggests a warm and humid climate, like that in today's tropical forests. There probably was little or no seasonal change.

Invertebrates were numerous. Some predatory insects resembled modern dragonflies. One—with a wingspread of almost 75 cm—was the largest insect ever. Cockroaches abounded, some species nearly 10 cm long.

The only large animals were the true amphibians, which probably evolved from the lobe-finned fishes of the Devonian period. Most of the many types of amphibians had four legs but could not really stand. Instead, they resembled modern salamanders and snakes, and they preyed on fishes in the streams and ponds. In the seas, trilobites remained, but these would disappear from the fossil record around the end of the Paleozoic era.

Figure 21.12   Reconstruction of a forest in the Carboniferous period.

×1/23

Figure 21.13   The fossil track in the photo could have been made by a bipedal dinosaur such as the one on the right.

21.9   Dinosaurs Dominated Triassic Ecosystems

Following the Carboniferous period, Pangaea was formed. Most of the seas covering North America drained away, and by the Triassic period, about 115 million years later, large areas were dry or mountainous. By then the amphibians had some formidable competitors.

Evidence of the Triassic period (about 245 million years ago) is found in the Connecticut Valley of New England, where certain rocks contain the large, three-toed footprints seen in Figure 21.13. Characteristics of the rocks suggest that a slow, winding stream flowed through a dry valley, carrying materials from highlands on both sides to a broad, flat plain, where it deposited them. Narrow bands of coal-like rocks mark the presence of small ponds. Almost all the plant fossils are from such rocks. Aquatic insects and several types of fishes lived in the ponds. Fossils of lungfish (Figure 21.14), which could obtain oxygen from the air, suggest that many ponds sometimes dried up, just as do ponds where modern lungfish live.

Dinosaurs and other large reptiles were abundant. Lizardlike creatures scurried about. The structure of their leg bones suggests they moved quickly, and their small, sharp teeth indicate they ate insects. Several species of slender dinosaurs about 2.5 m high roamed the mud flats, leaving the three-toed tracks that first called attention to this paleoecosystem.

21.10   Large Mammals Roamed North America in the Cenozoic

The Cenozoic era, represented in Figure 21.15, often is called the Age of Mammals because mammals have become so widespread in the last 60 million years. Though the first small, ratlike mammals lived at the same time

×1/2

Figure 21.14   An Australian lungfish.

Figure 21.15 A Cenozoic landscape (Pliocene epoch, about 3.4 million years ago).

×1/25

×1/50

×1/50

Figure 21.16 Cenozoic animals. From top to bottom: *Diatryma; Smilodon,* a saber-toothed cat; and a ground sloth.

as the dinosaurs, larger mammals did not evolve until the dinosaurs died out. Whatever killed the dinosaurs spared some ancestral mammals.

In what is now downtown Los Angeles, molasseslike pools of asphalt still entrap unwary small animals, although today the La Brea tar pits are enclosed with fences that protect most animals. The huge animals that once roamed the area often bogged down in similar pools of asphalt and died. Many rabbits, rodents, shrews, and perching birds, all similar to modern species, also were trapped in the asphalt. Their similarity to modern forms suggests that the climate of that ecosystem was probably like the warm, comfortable climate of southern California today. Bishop pines and cypresses grew there, along with juniper and coast live oak trees.

Western horses and other herbivores—peccaries, tapirs, antelope, bison, camels, and deer—grazed the area. Ground sloths ate yucca and other plants. Perhaps the most impressive remains are those of mastodons and mammoths. Some of the mammoths were more than 4 m high at the shoulders. Small reptiles and amphibians also lived in the La Brea ecosystem, eating such insects as flies and crickets. There were small carnivorous mammals. The most striking carnivores, though, were saber-toothed cats and dire wolves. (Dire wolves were similar to modern timber wolves.) Figure 21.16 shows some of these animals.

Is it surprising to find no humans here, considering the relatively recent age of this ecosystem? Most evidence indicates that the first humans to reach this continent probably crossed a bridge of land or ice that connected North America with Asia. They may have been hunters who followed the mammoths and eventually brought about their extinction.

Concept Review

1. Compare the Cambrian environment with that of the Carboniferous. What types of organisms lived in each?

2. What preadaptations of some fishes may have led to the evolution of amphibians during the Devonian period?

3. Compare and contrast a forest of the Carboniferous time with a forest of today.

4. Compare and contrast the organisms of the Carboniferous period with those of the Triassic.

5. State the general trend in characteristics of animals from the Cambrian period into the Cenozoic era.

The Emergence of Humans

Guidepost: How do humans differ from other primates?

INVESTIGATION 21.2   Interpretation of Fossils

How do anthropologists learn about evolution? Fossil remains record the evolution of early humans, hominids, and other primates. Even though humans and other primates share many similar features, humans did not evolve from apes. Although primate brains have increased in both size and complexity during the course of evolution, more than an examination of braincase casts (or endocasts) is needed to help determine the relationships between modern humans and other primates and fossil hominids. Some of the most important features anthropologists use in comparing fossils include jaws and teeth; the cranium, or skull; the pelvis, or hips; and the femur, or thighbone.

This investigation simulates some of the comparisons anthropologists make when studying fossil hominid remains, including examination of the brain casts of several hominids and some skeletal measurements for a human, an early hominid, and a gorilla.

Materials (per team of 2)

protractor   graph paper
metric ruler

PART A   Comparison of Endocasts
Procedure

1. Examine the five endocasts shown in Figure 21.17 on page 582. What is the volume of each? Which species do you predict to have the largest brain volume compared with body weight?

2. Using the information in Table 21.1 and a sheet of graph paper, plot the brain volume to body mass ratio for each of the five species.

3. The area circled in color on the *Homo sapiens* endocast is Broca's area, an enlargement of the frontal lobe of the cerebrum. Broca's area sends signals to another part of the brain that controls the muscles of the face, jaw, tongue, and pharynx. If a person is injured in Broca's area, normal speech is impossible. (There are, however, two other areas of the brain that are important in language and speech.) In which species can you see Broca's area? What might the presence of Broca's area indicate?

4. Convert the ratios of brain mass to body mass shown in Table 21.2 to decimal fractions and then draw a histogram (see Investigation 9.1) of the fractions.

Discussion

1. Based on your first graph, what would you say is the relationship between evolution in primates and the ratio of brain mass to body mass?

2. What major portion of the brain has become most noticeably enlarged during primate evolution?

3. Do you think the endocast of *Australopithecus* indicates that this hominid had a Broca's area in its brain?

TABLE 21.1   Body Weight of Selected Species

Species	Average Body Mass (grams)
Tarsius (tarsier)	900
Australopithecus	22 700
Homo erectus	41 300
Pan troglodytes (chimpanzee)	45 360
Homo sapiens	63 500

TABLE 21.2   Brain-Mass-to-Body-Mass Ratios

Mammal	Brain-Mass-to-Body-Mass Ratio
Tree shrew	1 : 40
Macaque	1 : 170
Blue whale	1 : 10 000
Human	1 : 45
Squirrel monkey	1 : 12
House mouse	1 : 40
Elephant	1 : 600
Porpoise	1 : 38
Gorilla	1 : 200

Figure 21.17

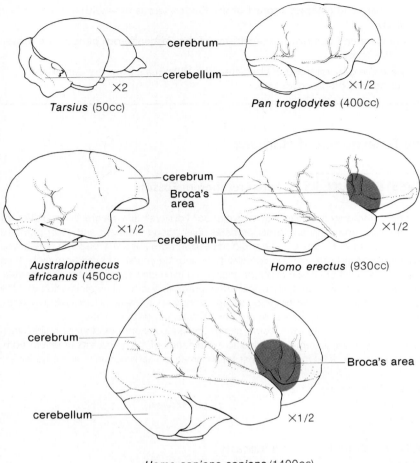

Tarsius (50cc)

Pan troglodytes (400cc)

Australopithecus africanus (450cc)

Homo erectus (930cc)

Homo sapiens sapiens (1400cc)

4. Does the presence or absence of Broca's area alone determine the language capabilities of a hominid? Explain.

5. How does your histogram affect your answer to Question 1? Are brain size and brain mass to body mass ratios reliable indicators of the course of primate evolution? Why or why not?

PART B Skeletal Comparisons
Procedure

5. Examine the three primate skulls in Figure 21.18 and the drawings of lower jaws and the pelvic area in Figure 21.19 on page 584. Imagine you are an anthropologist and these fossils have been placed before you for identification. Complete this hypothesis in your data book: "If the skulls, jaws, and pelvises are significantly different, then. . . ."

6. In your data book, prepare a table similar to Table 21.3 or tape in the table your teacher provides. Your task is to determine which skull, jaw, and pelvis belong to a human, which belong to an early hominid, and which belong to a gorilla. Record all observations and measurements in your table.

7. Brain capacity: The straight line drawn on each skull represents the brain capacity

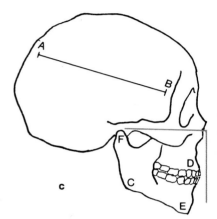

Figure 21.18

TABLE 21.3 Comparison of Primate Characteristics

Trait	Fossil A	Fossil B	Fossil C
Brain capacity			
Facial area			
Jaw angle			
Brow ridge			
Number of teeth			
Number of each type of teeth			
Relative size of teeth			
Relative size of pelvis			
Direction of flange			
Width of pelvic opening			

for each primate. Measure in centimeters the distance from point A to point B for each skull. Multiply by 175 to approximate the brain capacity in cm³.

8. Facial area: Measure from point C to point D and from point E to point F on each skull. Multiply these two measurements of each skull to determine the approximate area of the lower face. How might the size of facial area relate to evolution?

9. Facial projection: Use a protractor to determine the angle created by the colored lines. What does this measurement tell you?

Figure 21.19

10. Brow ridge: This is the bony ridge above the eye socket. Record the presence or absence of this feature for each skull. Also note the relative sizes.

11. Teeth: Study the drawings of jaws in Figure 21.19. Record the number of teeth and the number of each type of tooth. Also look at the relative sizes of the different types of teeth.

12. Pelvis: Study the drawings of the pelvis in Figure 21.19. Notice whether the flange, or lower portion, projects to the rear. Use your ruler to measure in millimeters the diameter of the pelvic opening at the widest point.

13. Read through these additional data:
    a. A large brain capacity is characteristic of humans.
    b. A smaller lower facial area is characteristic of humans.
    c. A facial projection of about 90° is characteristic of humans.
    d. Modern humans have lost most of the brow ridge.
    e. All primates have the same number of teeth and the same number of each type.
    f. Humans and other hominids have smaller canine teeth than do gorillas.
    g. A smaller pelvis, with a broad blade, a flange extending toward the rear, and a wider opening are characteristics of primates that walk on two feet.
    h. As brain size increased, the width of the pelvic opening increased to accommodate the birth of offspring with a larger head-to-body ratio.

Discussion

1. Compare the data you have assembled for each specimen. Which ones are hominid? Which traits are similar in all three primates? Which traits are similar in A and B?

2. Compare your data with the additional data. On the basis of your observations and measurements and the additional data, which fossil remains would you say are human, which are gorilla, and which are early hominid?

3. Why might an anthropologist be interested in the size of the pelvic opening?

4. Write a few paragraphs discussing the methods anthropologists use to determine human ancestry. Which of the characteristics you examined were most helpful? Which were least helpful? List any other additional observations or measurements that could be made for these specimens. Would having a more complete skeleton be helpful? Why?

21.11 Humans Are Classified in the Order Primates

Humans, monkeys, and apes are similar in the details of their anatomy. They are, therefore, grouped together in the order Primates, along with such animals as lemurs and tarsiers (Figure 21.20).

Most primates are arboreal and possess structures and behaviors that relate to the arboreal way of life. The digits of the primate, which have nails rather than claws, are well developed and give the animal a powerful grasp. In addition, the digits are sensitive and can easily tell if a surface is crumbly or slippery. The eyes of primates are directed forward instead of to the side, as in most mammals. Both eyes view the same object from slightly different angles, thus allowing the brain to perceive the object in three dimensions

Figure 21.20 Although primates share many characteristics, there is also great diversity within the group. Identify as many similarities and differences as you can.

Figure 21.21 Lucy is the most complete skeleton of an australopithecine. She stood only slightly more than a meter tall and lived about 3.2 million years ago.

and to make accurate judgments of distance. The brains of most primates are exceptionally large compared with those of all other mammals except whales, dolphins, and porpoises. Unlike many mammals, primates have a sense of color.

Bearing one offspring at a time is the rule among primates, although twins are not unusual. An active arboreal animal cannot carry many offspring. To feed these young, a female primate has only two mammary glands. Young primates, unlike young horses or jackrabbits, receive maternal care for a long time. Unlike other organisms, they do not learn all they need to know by trial and error but are taught by their parents. Primates tend to be omnivorous. They gather food in social groups and have well-established patterns of communication that range from elaborate body language to complex vocal signals and calls.

21.12 Hominids Are Primates That Walk Upright

The first humans on this continent migrated here from Asia, but the human species did not arise in Asia. Early Asians probably came from Africa, through the Mideast. Fossil evidence and early tool sites suggest that it was in Africa that **hominids** (HOM ih nidz)—primates that walked upright—developed.

The first undisputed hominids appear between 4 million and 3.75 million years ago in Tanzania and Ethiopia. Scientists uncovered the fossilized footprints of hominids that were bipedal—capable of walking on two legs. These fossils have been firmly dated to 3.75 million years ago. In addition, almost half of a complete skeleton of an individual called "Lucy" (Figure 21.21) as well as other fossil finds have been less firmly dated at between 3 and 4 million years ago. Anthropologists (scientists who study humans) generally agree that these finds represent a distinct group of hominids classified as **australopithecines** (AH strahl oh PITH uh seenz). Different forms of australopithecines have been found in East and South Africa and have been dated to different times. There is considerable doubt about the relationship of these groups to one another as well as to modern humans. Anthropologists, however, are convinced that all australopithecines were bipedal, although they may not have walked exactly as modern humans do.

Although these populations shared some traits with humans, other characteristics show how different they actually were. The length of their arms in proportion to the length of their legs and torso suggests that they were more apelike in appearance. Their cheek teeth were comparatively large and had a thick coating of enamel, possibly an adaptation to eating large quantities of fruit, seeds, tubers, roots, and pods—some of which may have been quite tough. There is no concrete evidence that they used or made tools.

At some point between 2.5 and 2 million years ago, African hominids underwent adaptive radiation. By the later date, there were at least two, possibly three, and perhaps even more species of hominids. In eastern Africa, a group of large and robust australopithecines lived at the same time as a larger-brained hominid called *Homo habilis*. Both hominids are shown in Figure 21.22. Many anthropologists believe *Homo habilis* to be the first member of our own genus *Homo*, but there is still much debate about the subject.

Figure 21.22   (a) An *Australopithecus boisei* skull, and (b) *Homo habilis* skull 1470, named for its index number at the National Museum of Kenya, East Africa. The skull is at least 1.8 million years old. Skull fragments of other individuals have been found, but skull 1470 is the most nearly complete.

21.13   The First Humans Appeared in Africa

By about 1.6 million years ago, *Homo habilis* seemingly disappeared and was replaced by an even larger-brained hominid, *Homo erectus*. *Homo erectus*, the first widely distributed hominid species, first appeared in Africa. By one million years ago, the species was present in southeastern and eastern Asia and survived in that area until about 300 000 years ago. In that span of time, the fossil record shows *Homo erectus* to have remained relatively unchanged (Figure 21.23).

Homo erectus resembled later species of *Homo* (excepting modern humans) in both body size and robustness. There is evidence that *Homo erectus* groups made large, symmetrically flaked stone tools, called hand

Figure 21.23   Two views of a reconstruction of a *Homo erectus* skull. The original fossil skull fragments were unearthed near Peking, China. Further quantities of *Homo erectus* fossils have since been discovered in China, Java, East Africa, Europe, and the Middle East.

Figure 21.24 Stone tools made by *Homo erectus*.

chopper

scrapers

points

burin
(chisel)

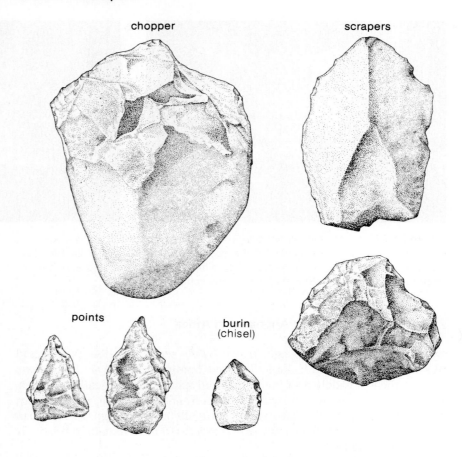

Figure 21.25 A fossil Neanderthal skull. The disappearance of these widely distributed people has never been satisfactorily explained. They survived as recently as 40 000 to 50 000 years ago.

axes, of the sort illustrated in Figure 21.24, and that some populations used fire. Toward the end of the *Homo erectus* occupation, there is fossil evidence of another group of hominids, *Homo sapiens neanderthalensis* (nee an der tal EN sis), shown in Figure 21.25. The Neanderthals, a subspecies of *Homo sapiens*, were much more robust than *Homo erectus*, or even modern humans, and the muscle attachments on their bones show that they were far stronger than modern humans. Their teeth were larger and more worn, probably from being used as tools, such as for chewing hides to soften them. The Neanderthals buried their dead and perhaps wore jewelry.

The Neanderthals spread out over much of Europe and southwestern Asia. Fossil evidence for them disappears about 30 000 years ago. Some anthropologists believe that many Neanderthals evolved into modern humans. Others insist that all Neanderthals became extinct and were replaced by modern humans who evolved from unknown genetic stock.

A case may be made, however, for the evolution of modern humans from a group of hominids that lived at the same time as the Neanderthals. These anatomically modern hominids, *Homo sapiens sapiens*, also exhibit a great deal of skeletal variation, as shown in Figure 21.26. The European fossil record suggests that these anatomically modern humans replaced the Neanderthals in a rather complicated manner and may have intermixed with the Neanderthal population. Evidence from southwestern Asia suggests that modern people had replaced the Neanderthals there by 40 000 years ago.

a

b

Figure 21.26 Two anatomically modern human skulls, possibly representing two somewhat genetically different populations. These people, like all people today, are classified as *Homo sapiens*.

Evolutionary events in Africa may have led to the emergence of these modern humans, whose subsequent intrusion into Europe may have led to the demise of the Neanderthals.

Certainly, Neanderthals and the first anatomically modern humans were quite different, and great variations also existed within each population. At least three distinct modern human cultures existed in Europe at the same time. Other cultures of humans probably inhabited every continent except Antarctica. The age of modern humans began with much of the genetic variation found in human populations today. Among present-day human populations, however, gene pools have changed significantly because of increased travel and colonization by groups from other cultures.

The study of genetics cannot provide a complete understanding of the history of human evolution, but geneticists and biochemists recently have found a new means of gathering data. They can analyze the structure of the DNA of human genes and of proteins made under the control of these genes, and they can compare these structures to DNA and protein structures of other organisms. The comparisons can yield clues about human evolution. Unfortunately, most proteins and DNA do not survive in fossils. Otherwise, biochemical tests of kinship between extinct species and the one living species of *Homo* might yield even more information about human evolution. Thus, much of our knowledge must be derived from the interpretation of fossils and other artifacts. Investigation 21.3 deals with the interpretation of evidence from two archaeological sites.

Certainly, the people of 30 000 to 35 000 years ago were *Homo sapiens*. They were cave dwellers, taller than their ancestors. Many were taller than most people today. Anthropologists also note important behavioral changes based on the archaeological record. These changes include the use of superior bone and stone tools; the use and control of fire; the use of clothing; shifts in hunting patterns, settlement patterns, population size, and ecological range; and the development of art and of ritual activity. All these characteristics point to the emergence of a species possessing modern behavioral capabilities from a *Homo erectus* line. Figure 21.27 on page 590 shows one hypothesis about the course of human evolution.

If the people who lived as recently as 30 000 to 40 000 years ago can leave behind questions that anthropologists cannot answer, it is easy to see that questions about still older populations always may exist. The origins

Biological Anthropologist

Mike Hoffman is a biological anthropologist who teaches at a liberal arts college. His major interest is in the study of the human skeleton, both prehistoric and modern.

As a child Mike was always picking up bones, and he loved to visit the dinosaur and fossil hominid displays in museums. During a family vacation, he visited Mesa Verde National Park and saw the artifacts and skeletal remains of prehistoric Southwestern Indians in the park's museum. These made a lasting impression on him.

Although he was a biology major in college, Mike took a number of anthropology courses, mostly in archaeology and biological anthropology. Through his class work, he began to see how a background in biology could be useful for understanding some of the questions asked by anthropologists. As his senior year in medical school began, he found he really was more interested in being an anthropologist than in practicing medicine. Following graduation from medical school, he started graduate studies in biological anthropology.

After completing his Ph.D. in anthropology, Mike taught for several years at universities in Arizona and California. While in California he became interested in some of the diseases found in prehistoric Native Americans of the region. Since then he has maintained a professional interest in human skeletal studies.

Today Mike teaches a variety of courses in biological anthropology, including human evolution, biological variation and adaptation, prehistory of the American Southwest, and his real love—bones. His current research on prehistoric skeletons involves material from California and from an archaeological site near Mesa Verde in Colorado. One of his major interests is in questions about the patterns of diseases in prehistoric populations, at least those diseases that are reflected in the human skeleton. Why does a population living in a particular environment at a particular time have certain diseases? Are the differences in these patterns related to the age of the individual, to male and female differences, to dietary differences, or to different activities? Diseases commonly seen in the human skeleton include arthritis, tooth cavities and abscesses, infectious diseases, and occasional tumors.

Mike's knowledge of human skeletons also allows him to be helpful to police and coroners who often call on him to help identify human skeletons or remains. He also works closely with dentists in the identification process. Using a variety of techniques, he can tell the age at death, sex, stature, ethnic group, right- versus left-handedness, general body build, and other distinguishing characteristics of individuals. This type of work is known as forensic anthropology.

In Mike's own words, "Make no bones about it, this is a very interesting profession. It also happens to be great fun."

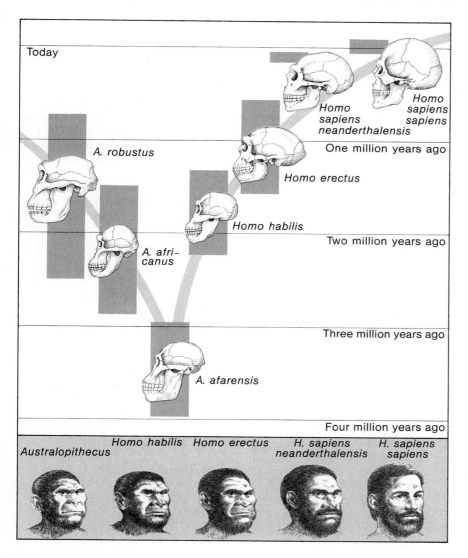

Figure 21.27 One hypothesis of human evolution. It is impossible to define the exact phylogeny of humans because numerous theories are being presented and many debates are taking place. Only recently, new discoveries have thrown doubt on previously accepted theories.

of human speech, for example, have not been mentioned because fossil evidence has not helped directly in discovering them. Yet the development of meaningful speech must have had a dramatic effect on the evolution of humans from that point onward.

Concept Review

1. What structural characteristics do primates share? How are humans different from other primates?
2. What is the best-accepted current hypothesis on the evolution of humans? How do the Neanderthals fit in?
3. In human evolution, did brain expansion or upright posture probably occur first?
4. Distinguish between living modern humans and anatomically modern humans. How are these groups different from *Homo erectus*?

INVESTIGATION 21.3 Archaeological Interpretation

Because we cannot travel back through time to see how people worked and lived thousands of years ago, we can never be sure that we understand the details of earlier cultures. Archaeologists search for clues among the remains of ancient peoples and civilizations. When they have found, dated, and studied the evidence, they formulate hypotheses to explain their findings. What emerges is a picture of life in a particular place hundreds or thousands of years ago. Although it is impossible to complete a detailed picture, many reasonable and logical conclusions can be drawn. Some findings, however, can be interpreted in a variety of ways, and archaeologists may disagree about which interpretation is correct.

This investigation requires the analysis of data found in archaeological digs. In nearly every case, several interpretations are possible. Think of as many interpretations as you can. Remember that your interpretations should account for all, not just some, of the data. In the final section of this investigation, you will predict how archaeologists far in the future might interpret today's culture and life-style.

Materials (per team of 3)
paper and pens

PART A An Indian Cemetery in Newfoundland
Procedure

1. Read the following paragraphs:

    One scientist investigated an Indian burial site located near Port aux Choix, a small village on the west coast of Newfoundland (Figure 21.28 *left*). The burial ground appeared to have been used between 4000 and 5000 years ago. The graves were strips of very fine sand, often covered with boulders or slabs of rock. One strip of sand was almost 1.6 km long, varying from about 9 to 22 m in width. The sandy strip lay about 6 m above the high water mark of the ocean shore.

    The ages of the skeletons were estimated as shown in Table 21.4. What do these data suggest about the lives of the Indians?

    The teeth of many of the adults were worn, often so worn that the pulp was exposed. Some skeletons also had teeth missing, but there was no evidence of tooth decay. Figure 21.28 (*right* a, b, c, and d) shows some of the artifacts found in the graves. In addition, a great number of woodworking tools were found.

Discussion

1. What type of an area might the Newfoundland burial site have been thousands of years ago when Indians lived there (for example, forest, mountain, beach)?
2. What might account for the sandy strip being chosen as a prime burial site?
3. Boulders or slabs of rock covered many graves. Why might that practice have been common?
4. What do data on teeth suggest about the diet and life-style of the Indians?
5. What do you think the small objects shown in Figure 21.28a were used for? Explain your reasoning. What purpose might the holes have served?
6. What might the objects in Figure 21.28b have been used for? The objects in Figure 21.28c?
7. What modern implement do the objects in Figure 21.28d resemble? What might their function have been?
8. Considering the site location, what might the Indians have built from wood?

TABLE 21.4 Port au Choix Skeletal Ages

Age	Number of Skeletons
Newborn infants	12
Under 2 years	15
2–6 years old	2
6–18	15
18–21	1
21–50	36
50+	7

Figure 21.28

PART B   Ancient People of Greece
Procedure

2. Read the following paragraphs:

Stone Age people inhabited Greece long before the dawn of classical Greek civilization. Until recently, little was known of the culture of these people who lived between 5000 and 20 000 years ago. Excavations in and around a particular cave on the east coast of Greece began in 1967. Since then, archaeologists have uncovered much evidence to show the changes that occurred in the culture and life-styles of Stone Age Greeks living there.

The oldest remains (20 000 years old) in the cave primarily are the bones of a single species of horse and some tools made from flint (rock that forms sharp edges when it breaks). What might explain the presence of these horse bones mixed with flint tools?

Newer remains (10 000 years old) include bones of red deer, bison, horses, and a species of wild goat, as well as remains of wild plants such as vetch and lentils (pealike plants), shells of land snails, marine mollusks, and some small fish bones. These newer remains were left at the time when a great ice age had come to an end. In still later finds, dated about 7250 BC, very large fish bones were found—the fish might have been 100 kg or more. At about the same time, tools made of obsidian, a type of volcanic glass, were found. The nearest source of obsidian is 150 km away from the excavation site and across a body of water. The site also yielded the oldest complete human skeleton found in Greece. A male of about 25 years of age was buried in a shallow grave covered with stones. Certain bone abnormalities show that he may have suffered from malaria.

Remains dated about 6000 BC are dominated by the bones of goats and sheep found in and around the cave. These remains were of a different type from the animal remains found away from the cave. Evidence of wheat and barley seeds also was found. Among the tools were axes and millstones. Only a few of the human graves at the site contained objects such as tools or jewelry. A 40-year-old woman, who probably died about 4500 BC, was buried with some bone tools and obsidian blades. One infant was found near a marble vessel and a broken clay pot.

Discussion

1. What happened to the diet of the cave's inhabitants through the centuries? Explain your answer.
2. What does the evidence for the period dated 7250 BC suggest about a change in the life-style of the people?
3. How do you think these ancient people obtained the obsidian?
4. What does the skeleton with the bone abnormalities suggest about the environment at that time?
5. For the remains dated 6000 BC, what do the findings suggest about the life-style of the cave's inhabitants at that time?
6. What might be the significance of finding items buried along with people?

PART C   Excavating the Present
Procedure

3. Imagine that you are an archaeologist living in the year 4000 AD. A catastrophe has destroyed the written records of the twentieth century. You have only the remains that have been preserved through 2000 years as clues to what life was like in the 1980s or 1990s. Imagine, furthermore, that you are excavating a site that was a school 2000 years ago—your own school.

Discussion

1. What things do you think you would find? Make a list and then exchange it with your partners.
2. For each object, suggest several interpretations. (Remember, you know nothing about the time when the objects were left—the only clues are the fossils.)
3. Some materials such as stone and metal can last thousands of years. Other materials, such as wood, decay rapidly. What important things would be absent in the archaeological dig that once was your school? What ideas about our society might you miss because the clues were not preserved?
4. Consider Questions 1 through 3 again. This time, imagine a different type of site, such as a mall or business park, a residential street, a farm, or an industrial area. How would your findings and interpretations differ?
5. Suppose you wanted to build a time capsule to tell people in the far distant future about our lives today. What would you put in the capsule and why?

Summary

Paleontologists interpret fossil evidence to reconstruct the organisms (and their interactions) of ancient ecosystems. The evolution from simple beginnings of the invertebrates, the flowering plants, the reptiles, the amphibians, and the mammals of today is apparent in the fossil record of those ecosystems. Evolutionary mechanisms, including the effects of natural selection of organisms occurring on continents that moved, account for the variety and distribution of today's plants and animals. Our own group, the hominids, is a subgroup of primates. Ancestors of both humans and apes radiated from early hominidlike primates, evolving into hominids and apes.

We share many characteristics with other primates but also have distinct differences. Chief among these differences are our upright posture, brain size and head shape, and speech.

Applications

1. The ocean makes up about 2/3 of the surface of our planet, yet there are only approximately 19 000 species of algae, which are mostly restricted to water. There are, however, approximately 250 000 species of flowering plants, which are mostly restricted to land. What could account for the difference in numbers of species?

2. Birds and mammals are endothermic—they use metabolic energy to maintain a constant internal body temperature. Reptiles, fishes, and amphibians are ectothermic, that is, they use environmental energy and behavioral adaptations to regulate their body temperature. Were dinosaurs endothermic or ectothermic animals, and how would this affect their lives?

3. The formation and breakup of Pangaea greatly changed the amount of shoreline in the world. How do ocean currents affect life on land? What influence do these currents have on the climate of terrestrial habitats?

4. If you plotted all of the known volcanic eruptions in the world, what would this tell you about the location of the tectonic plates?

5. Some scientists suggest that the extinction of dinosaurs was the result of an ancient "nuclear winter." What is a nuclear winter, how could it have occurred *before* humans were on earth, and how might it have killed the dinosaurs?

6. Evidence suggests our ancient ancestors walked on four limbs instead of two. Standing upright may have some advantages for our survival, but how are the following problems related to our relatively recent stand on two feet: slipped disks in the lower back, flat feet, varicose veins in the legs, and hernias?

7. Why are tropical forests poor sources of fossil evidence? Consider the conditions for fossilization and the conditions for finding fossils.

8. If you found a fossil skeleton with a cranial capacity of 550 cm³ and a foot with a high arch, would you classify it as a hominid? Explain.

9. Would a fossil skeleton with a cranial capacity of 600 cm³, holding an axelike stone, be classified as a hominid? Explain.

Problems

1. Ask a geologist to organize a field trip to an interesting site in your area. Before going on the trip, obtain information about the geology from the nearest geological survey office.

2. What type of weapons and tools did our ancient ancestors use? Try to find drawings or photographs of ancient equipment, or visit a museum to see what was used. Compare these weapons and tools with those used by primitive tribespeople alive today.

3. Go to the zoo and observe the behavior of the primates there. How easy is it to misinterpret what an animal does or to give an incorrect "reason" for a particular behavior?

4. In many museum displays, background paintings illustrate the environment and organisms of a particular time and place. Choose a prehistoric time and make a large painting showing interactions between the plants and animals living in your area at that time.

5. Using a large map of the world today, try to create the ancient continent of Pangaea. Cut out tracings of the continents and try to fit them together. Do any continents seem to fit together like pieces of a jigsaw puzzle?

6. Where and how do nonhuman primates live in the world today? What is the relationship between these wild animals and humans? Read about any of these primates and write a report.

7. Read S. J. Gould's book, *Wonderful Life* (New York: W. W. Norton, 1989), to learn about the incredible array of fossilized organisms found in Canada, many of which are extremely different from anything alive on earth today. Sketch these weird creatures and see what modern organisms they most resemble.

8. Construct a La Brea ecosystem food web.

9. Consider the causes of death in modern urban and suburban environments. Include factors that kill individuals before, during, and after the age of reproduction. Consider also factors that reduce health or impair development and factors that reduce fertility or the survival rate of offspring. Could these factors cause changes in characteristics of future human populations?

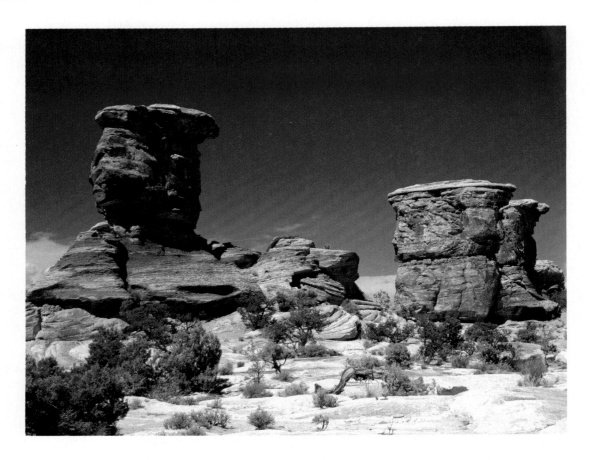

The Hoodoo rock formation in Canyonlands National Park, Utah. What types of vegetation do you see here? Based on the photo, what do you think might be some limiting factors for this biome? How might these limiting factors affect the types of plants and animals found here?

Biomes Around the World **22**

Across the face of the earth there are great differences in climate. Through billions of years, these climatic differences have resulted in the evolution of diverse ecosystems that harbor distinct groups of plants, animals, and microorganisms, even though they may be located at the same altitude or latitude. The small town of White River in Ontario, Canada, which lies about 20 km from the northeast shore of Lake Superior, is surrounded by a forest of spruce, fir, and some pine. A sign with a large thermometer, shown in Figure 22.1, proclaims that White River is the coldest town in Canada, with a low temperature of "72° below zero." A town better known than White River but at the same latitude—48° north—is Paris, France. What is the climate in Paris? Is it as cold? Do the same types of trees grow there as in White River? This chapter examines the major types of habitat, the organisms found in each, and the reasons for the differences between habitats. It also discusses some ways in which humans affect ecosystems.

Climate and Ecosystems

22.1 Abiotic Factors Influence the Character of Ecosystems

Guidepost: What abiotic factors influence the climate of an ecosystem?

The earth is a patchwork of habitats that differ in abiotic factors such as temperature and the amount of precipitation and sunlight received. The interaction of these limiting factors determines the climate of the area, which in turn affects the distribution and diversity of living organisms.

In each major type of climate, a characteristic type of vegetation develops. For example, warm, arid climates—those with little rainfall—are associated with desert vegetation. Semiarid climates usually are covered with grassland. Moist climates support forests. Each type of plant life, in turn, supports a characteristic variety of animal life. The resulting ecological community of plants and animals is called a **biome**. (Ecosystems, in contrast, also include abiotic factors such as soils and precipitation.) Biomes extend over large natural areas, as shown in Figure 22.2 on page 598.

Where does the energy that drives an ecosystem come from? Ecosystems are powered primarily by sunlight, or radiant energy, which is important for two reasons. First, radiant energy is the form of energy that most producers use to make food. Almost all organisms in an ecosystem depend

Figure 22.1 White River, Ontario.

Figure 22.2 Major biomes of the earth. Many parts of the earth have not been thoroughly studied, and even where observations are plentiful, ecologists sometimes disagree about the correct interpretation.

Figure 22.3 The distribution of solar energy on the earth's surface. At which point does North America receive the most radiant energy?

on food made by producers. Second, the temperature of an ecosystem is determined by the amount of radiant energy it receives. The energy is absorbed at the earth's surface and is radiated to the air as heat.

The position of the earth in relation to the sun affects the amount of radiant energy different parts of the earth receive. The earth is nearly spherical and is tilted with respect to its orbit around the sun. At the summer solstice, about June 21, the Northern Hemisphere is tilted 23.5° toward the sun. As a result, the Northern Hemisphere receives more solar energy than the Southern Hemisphere, so it is summer north of the equator and winter south of the equator. At the winter solstice, December 21, the Southern Hemisphere receives more solar radiation than the Northern Hemisphere, and the seasons are reversed. Figure 22.3 illustrates the position of the earth relative to the sun at different times of the year. Other factors, however, also affect the distribution of solar energy.

22.2 Air Circulation and Ocean Currents Affect Climate

The circulation of air in the atmosphere is powered by solar energy. The movement of the air helps to distribute the heat that comes to the earth as radiant energy. At the same time, circulating air currents carry water vapor from the oceans over land surfaces, where it falls as rain.

Climates occur in broad belts that encircle the earth. The boundaries of these belts are disrupted by land masses and oceans. Climates are further modified by mountains, particularly when mountain ranges block moisture-laden winds from the oceans (see Figure 22.4). When moisture-carrying wind from the ocean meets the mountains, the moisture condenses and is released as precipitation. As the air travels down the other side of the mountains, the leeward side, it warms and begins to collect moisture. As a result, the leeward sides of mountains are drier than the windward sides, and the vegetation often is very different. This phenomenon is called the rain-shadow effect.

Ocean currents also affect climate because they redistribute heat. For example, the Gulf Stream (in the North Atlantic) swings away from North America near Cape Hatteras, North Carolina, and reaches Europe near the southern British Isles. Because of the Gulf Stream, western Europe is much warmer than is eastern North America at similar latitudes. As a general rule, the western sides of continents in the temperate zones of the Northern Hemisphere are warmer and drier than the eastern sides. The opposite is true in the Southern Hemisphere. Figure 22.5 shows the major ocean currents.

It is easier to map the distribution of a particular climatic factor than it is to map a climate as a whole because climatic factors overlap and interact with one another in complex ways. To simplify climate-mapping, ecologists frequently use climatograms that summarize monthly measurements of temperature and precipitation. Appendix 3 contains several climatograms for different areas.

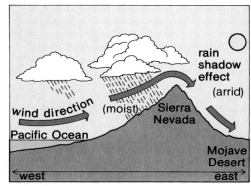

Figure 22.4 The rain-shadow effect. As air containing water vapor rises, it cools. Cool air holds less moisture than warm air; thus clouds form on the western side of the Sierra Nevada range. Rainfall tends to be heavy on this side. As the air descends on the eastern side, it picks up moisture from its surroundings rather than releasing it. This rain-shadow effect produces desert conditions on the eastern side of the Sierra Nevada range.

Figure 22.5 The major ocean currents have profound effects on climate. How does the Humboldt Current affect the western side of South America. What affects the eastern side of South America?

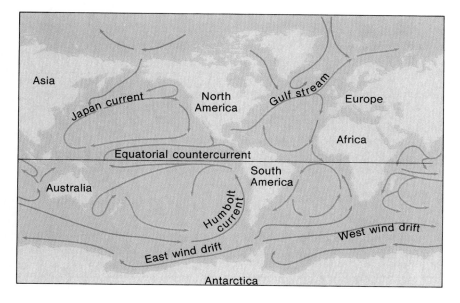

Concept Review

1. How does sunlight affect an ecosystem?
2. What causes the winter and summer solstices?
3. How do mountains and ocean currents affect precipitation and temperature?
4. What determines the characteristic plants and animals found in a biome?

INVESTIGATION 22.1   Climatograms

Climatograms, such as those in Appendix 3, show monthly variations in only two climatic factors, precipitation and temperature. Other factors also affect climate, but a climatogram gives a rough idea of the climate in a particular area. By daily observation you can associate the climate with the biome of your own locality. This investigation allows you to study the worldwide relationships between climates and biomes. Refer to the pictures and descriptions of biomes in this chapter to help you visualize the relationships between biotic and abiotic factors in some of the earth's major biomes.

Materials (per team of 1)

1–3 sheets of graph paper or
   climatogram forms

Procedure

1. Construct climatograms from the data in Table 22.1. These four, plus the six in Appendix 3, represent the major biomes on the earth.
2. Using the data in Table 22.2, construct the climatograms assigned by your teacher.
3. Obtain monthly averages of precipitation and temperature from the weather station closest to your school. These data may be expressed as inches of precipitation and degrees Fahrenheit. If so, convert the data to centimeters and degrees Celsius using the information in Figure A1.2 in Appendix 3. From these data, construct a climatogram for your area.

TABLE 22.1   Group 1

T = temperature (in degrees Celsius) P = precipitation (in centimeters)

	J	F	M	A	M	J	J	A	S	O	N	D
a. Tropical Deciduous Forest: Cuiabá, Brazil												
T	27.2	27.2	27.2	26.7	25.6	23.9	24.4	25.6	27.8	27.8	27.8	27.2
P	24.9	21.1	21.1	10.2	5.3	0.8	0.5	2.8	5.1	11.4	15.0	20.6
b. Chaparral: Santa Monica, California												
T	11.7	11.7	12.8	14.4	15.6	17.2	18.9	18.3	18.3	16.7	14.4	12.8
P	8.9	7.6	7.4	1.3	1.3	0.0	0.0	0.0	0.3	1.5	3.6	5.8
c. Savanna: Moshi, Tanzania												
T	23.2	23.2	22.2	21.2	19.8	18.4	17.9	18.4	19.8	21.4	22.0	22.4
P	3.6	6.1	9.2	40.1	30.2	5.1	5.1	2.5	2.0	3.0	8.1	6.4
d. Tropical Desert: Aden, Aden												
T	24.6	25.1	26.4	28.5	30.6	31.9	31.1	30.3	31.1	28.8	26.5	25.1
P	0.8	0.5	1.3	0.5	0.3	0.3	0.0	0.3	0.3	0.3	0.3	0.3

TABLE 22.2 Group 2

		J	F	M	A	M	J	J	A	S	O	N	D
a.	T	1.1	1.7	6.1	12.2	17.8	22.2	25.0	23.3	20.0	13.9	7.8	2.2
	P	8.1	7.6	8.9	8.4	9.2	9.9	11.2	10.2	7.9	7.9	6.4	7.9
b.	T	10.6	11.1	12.2	14.4	15.6	19.4	21.1	21.7	20.0	16.7	13.9	11.1
	P	9.1	8.9	8.6	6.6	5.1	2.0	0.5	0.5	3.6	8.4	10.9	10.4
c.	T	25.6	25.6	24.4	25.0	24.4	23.3	23.3	24.4	24.4	25.0	25.6	25.6
	P	25.8	24.9	31.0	16.5	25.4	18.8	16.8	11.7	22.1	18.3	21.3	29.2
d.	T	12.8	15.0	18.3	21.1	25.0	29.4	32.8	32.2	28.9	22.2	16.1	13.3
	P	1.0	1.3	1.0	0.3	0.0	0.0	0.3	1.3	0.5	0.5	0.8	1.0
e.	T	−3.9	−2.2	1.7	8.9	15.0	20.0	22.8	21.7	16.7	11.1	5.0	−0.6
	P	2.3	1.8	2.8	2.8	3.2	5.8	5.3	3.0	3.6	2.8	4.1	3.3
f.	T	19.4	18.9	18.3	16.1	15.0	13.3	12.8	13.3	14.4	15.0	16.7	17.8
	P	0.0	0.0	1.5	0.5	8.9	14.7	12.2	8.1	2.0	1.0	0.3	0.8
g.	T	−22.2	−22.8	−21.1	−14.4	−3.9	1.7	5.0	5.0	1.1	−3.9	−10.0	−17.2
	P	1.0	1.3	1.8	1.5	1.5	1.3	2.3	2.8	2.8	2.8	2.8	1.3
h.	T	11.7	12.8	17.2	20.6	23.9	27.2	28.3	28.3	26.1	21.1	16.1	12.2
	P	3.6	4.1	4.6	6.9	8.1	6.9	6.4	6.6	8.9	5.1	5.6	4.6
i.	T	23.3	22.2	19.4	15.6	11.7	8.3	8.3	9.4	12.2	15.1	18.9	21.7
	P	5.1	5.6	6.6	5.6	2.8	0.9	2.5	4.1	5.8	5.8	5.1	5.3
j.	T	17.2	18.9	21.1	22.8	23.3	22.2	21.1	21.1	20.6	19.4	18.9	17.2
	P	0.3	0.5	1.5	3.6	8.6	9.2	9.4	11.4	10.9	5.3	0.8	0.3
k.	T	−20.0	−18.9	−12.2	−2.2	5.6	12.2	16.1	15.0	10.6	3.9	−5.6	−15.0
	P	3.3	2.3	2.8	2.5	4.6	5.6	6.1	8.4	7.4	4.6	2.8	2.8
l.	T	−0.6	2.2	5.0	10.0	13.3	18.3	23.3	22.2	16.1	10.6	4.4	0.0
	P	1.5	1.3	1.3	1.0	1.5	0.8	0.3	0.5	0.8	1.0	0.8	1.5

Discussion

1. Compare your local climatogram with the four from Step 1 and the six in Appendix 3. Which one does it most closely resemble?

2. What similarities are there between the two?

3. What differences are there?

4. Consider the biotic characteristics of your local area. What characteristics of the local climate would be important factors in determining these biotic characteristics?

5. Does your local climatogram *exactly* match any of the original 10 climatograms?

6. Explain how any differences between your local climatogram and the one most nearly matching it might affect the biotic characteristics of your biome.

7. Compare each unidentified climatogram from Table 22.2 with the 10 identified climatograms. Label each graph with the name of the biome that you think the climatogram represents.

8. Using the climatic information shown by each climatogram constructed from Table 22.2, describe the biotic characteristics of each corresponding biome.

9. Check the location of each climatogram from the information your teacher gives you. How often were you correct in your reasoning?

Guidepost: How does the vegetation change from north to south?

Figure 22.6 Tundra in Alaska. Note the pond in the foreground and the lack of trees. What climatic factor is responsible for these features?

Biomes with Increasing Radiant Energy

22.3 Tundra Is Characterized by Low Vegetation

Beyond the northern limit of tree growth and at high altitudes is the tundra (TUN druh). Arctic tundra, shown in Figure 22.6, is determined by latitude and characterized by permafrost—continuously frozen ground that prevents plant roots from penetrating very far into the soil. Spring comes in mid-May; only plants near the sun-warmed ground can grow, flower, and produce seeds before the advent of fall in August. Most plant reproduction and growth occurs during the brief, warm summers, when daylight lasts for almost 24 hours. The tundra may receive less precipitation than many deserts, yet the combination of permafrost, low temperatures, and low evaporation leaves the soils saturated with water. These conditions further restrict the species that grow there. Dwarf perennial shrubs, sedges, grasses, mosses, and lichens are the dominant vegetation.

The change from summer to winter is rapid. Snowfall is light, only 25 to 30 cm per year, and high winds sweep open areas free of snow. There are few daylight hours, and above the Arctic Circle after the winter solstice there are three months of near darkness.

Alpine tundra (determined by altitude) is found above the tree line on high mountains. Because of the similarity in conditions, many plant and animal species found in the arctic tundra also are found in the alpine tundra. Alpine tundra can occur at any latitude if the elevation is high enough (see Figure 22.7). Day length in the alpine tundra is shorter in summer than it is in the arctic tundra, but longer in winter. Brief and intense periods of productivity occur daily in summer, but freezes may happen any night. Winter snow or gale winds result in short growing seasons.

Animals of the tundra include year-round residents such as ptarmigan, arctic foxes, and snowshoe hares. During the summer, large flocks of migratory water birds raise their young, and caribou graze on grasses and reindeer moss. Figure 22.8 shows two typical tundra animals.

Figure 22.7 Types of vegetation and animal life are affected both by altitude and latitude.

Figure 22.8 Animals of the arctic tundra: (a) snowy owl, *Nyctea scandiaca;* and (b) musk ox, *Ovibos moschatus*. How are each adapted to life in the tundra?

How do animals cope with the coming of winter? Many animals that stay in the tundra become dormant or hibernate. As winter approaches, they find shelter and their body processes slow down. The energy needed to keep them alive while they are hibernating is derived from body fat stored during the warmer months. Other animals withstand the cold by living in burrows, having a large body size, or having good insulation that retains heat. Still other species, especially birds, are migratory and leave the area during the harshest part of the winter.

22.4 Taiga Is a Coniferous Forest

At the southern edges of the arctic tundra (or at a slightly lower elevation than alpine tundra) scattered groups of dwarf trees appear in sheltered places. Eventually the tundra gives way to the great coniferous forest or taiga (TY guh), shown in Figure 22.9. This forest extends in a broad zone across Europe, Asia, and North America, but there is no similar biome in the Southern Hemisphere. The taiga receives more radiant energy, both daily and annually, than does the tundra. The summer days are shorter but warmer, and the ground thaws completely. Winters are not as long as they are in the tundra, although more snow falls.

Taiga vegetation usually is dense, and the woody plants prevent much undergrowth. Because the conifers keep out the sunlight, there are mostly mosses, lichens, and a few shrubs growing near the ground. Most food production, therefore, takes place in the upper parts of trees. Conifers, such as spruce, fir, pine, and hemlock, grow in an interesting variety of combinations with deciduous trees, such as willow, alder, and aspen. Usually only the conifers are conspicuous. The soil tends to be acidic under the coniferous trees and supports few species of decomposers. Earthworms are uncommon, but large numbers of tiny arthropods live in the soil and decompose organic matter.

Figure 22.9 The northern coniferous forests make up the taiga biome in which ponds, lakes, and bogs are abundant.

Taiga animals include deer and elk that browse the young leaves, and moose that eat aquatic vegetation (Figure 22.10). Smaller animals include squirrels, jays, hares, porcupines, and beavers. Predators of the taiga include grizzly bears, wolves, lynxes, and wolverines. What adaptations to cold might these predators have?

22.5 Midlatitude Forest Biomes Have Four Distinct Seasons

Midlatitude, or temperate, forests develop throughout the midlatitude regions where there is enough moisture to support the growth of large trees. This biome has four distinct seasons, as shown in Figure 22.11. Much of the farmland in the United States was once forest.

In summer, the daily supply of radiant energy is greater than it is in the tropics on all but a few days of the year. In winter, the earth's orientation to the sun changes, and the midlatitude forests receive less radiant energy. Thus, the total annual supply of radiant energy is much less in midlatitude forests than in the tropics. Temperature ranges from cold in the winter (−30° C) to hot in the summer (+35° C). Both rainfall and snowfall are

Figure 22.10   Moose, *Alces alces*.

Figure 22.11   Seasons in midlatitude deciduous forest. These pictures show the same area. What effects do seasonal changes have on herbivores such as deer?

Winter

Spring

Summer

Fall

abundant, with an average annual precipitation between 75 and 150 cm per year. Humidity tends to be high, growth and decomposition rates are high, and the soil is deep and rich.

The many species of deciduous trees include oaks, hickory, maple, beech, chestnut, and basswood. The tallest ones form a canopy—an upper layer of leaves that catches the full sunlight. Enough sunlight penetrates this layer to provide energy for a lower layer of trees and a layer of shrubs beneath. On the forest floor, mosses and ferns receive the remaining faint light.

The diverse vegetation supports a large number of consumers. Squirrels collect nuts and berries from trees. Deer mice climb in the shrubs and search the ground for seeds. White-tailed deer browse on shrubs and the lower branches of trees. In summer, insects are abundant in the soil and in all layers of the forest canopy. A variety of birds prey on these insects. For example, in some forests red-eyed vireos consume canopy insects, acadian flycatchers catch insects flying below the canopy, ovenbirds search out insects on the ground, and woodpeckers extract boring insects from the bark of trees (Figure 22.12). Foxes, skunks, and an occasional black bear also may be present. There are few large predators in most of this biome. Figure 22.13 on page 606 shows some animals and plants found in a midlatitude deciduous forest.

In autumn, the leaves of deciduous trees cover the ground with a thick mass of organic matter. Many mammals become fat on the abundant nuts, acorns, and berries, and some store the food. Woodchucks form thick layers of fat and then hibernate in burrows. Reptiles, much more abundant here than in the taiga, also hibernate. Many insect-eating birds migrate to the tropics. Many mammals, such as squirrels, rest during the cold spells of winter. Winter birds, which are more abundant here than in the coniferous forest, eat seeds and fruits and search out dormant insects and insect eggs from the cracks in tree bark.

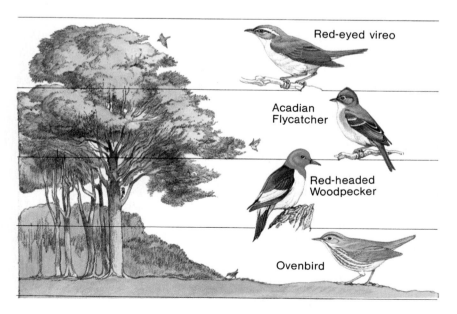

Red-eyed vireo

Acadian Flycatcher

Red-headed Woodpecker

Ovenbird

Figure 22.12 Resource partitioning in a deciduous forest. Red-eyed vireos in the canopy, flycatchers directly below, ovenbirds on the ground, and woodpeckers on the trunks. How does such partitioning affect these species?

Figure 22.13   Animals and plants of the deciduous forest: (a) fox squirrel, *Sciurus niger;* (b) white-tailed deer, *Odocoileus virginianus;* and (c) *Trillium.*

22.6   Tropical Rain Forests Have a Uniform Climate

Tropical rain forests are found on lowlands near the equator, where temperature and day length show little seasonal variation. The noon sun is almost directly overhead throughout the year, making the amount of radiant energy high and fairly constant. Rain falls almost every day, and the humidity is always high. Average rainfall is about 200 cm a year. Rainfall, however, depends on elevation and winds that bring moisture.

Vegetation is dense. Figure 22.14 shows the typical vegetation layers in a tropical rain forest. Weaving through the branches are many woody vines. Because the tall dense canopy of foliage competes for light, little light reaches the ground. Beneath the taller trees are shorter trees that can tolerate shade. Beneath these are still others that are even more shade-tolerant. Low light intensity limits the growth of vegetation on the forest floor and, hence, the animal species that live there.

Along the trunks and branches of the trees and the twisting stems of the vines are many epiphytes (EP ih fyts). These plants use the branches on which they perch for support but not for nourishment. Epiphytes have no contact with the ground, so they have special adaptations for obtaining water and nutrients. Some have roots that absorb moisture from the humid atmosphere the way blotting paper soaks up water. Many catch the daily rain in special hollow leaves. Mosquitoes, water beetles, and other aquatic insects may live in such treetop puddles.

Ripe fruits drop to the forest floor and supply food for some ground dwellers. Although the trees are always green, leaves die and fall continuously for most of the year. In the warm, moist environment huge numbers

Figure 22.14   The layering effect in a tropical rain forest. Virtually no sunlight reaches the rain forest floor, and comparatively fewer organisms are able to live on the ground than in the trees.

Figure 22.15   Arboreal rain forest animals include (a) the scarlet macaw, *Ara macao,* and (b) the spider monkey, *Ateles.*

of insects, fungi, and bacteria attack this food supply rapidly. Organic remains, therefore, do not build up on the ground. Predators and parasites are abundant at all levels of the forest. Large herbivores, such as hoofed mammals, are rare or live only near riverbanks.

What environmental factors might be related to the rarity of large, hoofed mammals? On the forest floor there is little herbaceous vegetation for herbivores to eat, and the continuously dropping debris is decomposed quickly. The shallow-rooted plants reabsorb the released nutrients through mycorrhizae. Because most of the nutrients are held in the vegetation, the soil is nutrient-poor.

All forests have some arboreal (ar BOHR ee uhl) animals—animals that live in the trees. In the tropical rain forest most animals live in the canopy (Figure 22.15). In one study of rain forests, 90 percent of the birds were found to feed mostly in the canopy. This may not be surprising in regard to birds, but in the tropical rain forest a large number of mammals, more than 50 percent of the species, also are arboreal. Because there is more food in the canopy layers than on the ground, there is a greater diversity of animals to be found in the various canopy layers. Moreover, there are many tree snakes, tree lizards and tree frogs, as well as an untold number of arboreal insects.

The seasonless rain forests with a uniform climate cover only a small part of the tropics. Most tropical regions have seasons. Instead of being warm and cold, however, the seasons are wet and dry. Many woody plants lose their leaves during the dry season. Tropical regions with uniform temperatures but wet and dry seasons produce tropical deciduous forests or tropical seasonal forests. Consult the map in Figure 22.2 to locate regions of tropical deciduous forest. The canopy is not as dense as that in the rain forest, and light filters all the way to the forest floor.

Concept Review

1. What adaptations of structure and function are found among tundra organisms?
2. What is the most noticeable difference between the tundra and taiga landscapes?
3. Compare the vegetation in a midlatitude deciduous forest and a tropical rain forest. Explain the differences.
4. Many arboreal animals live in tropical rain forests. What are the reasons for this?
5. How do epiphytes obtain water and nutrients?
6. What is the major limiting factor in the tropical rain forest? in the tropical deciduous forest?

INVESTIGATION 22.2 Studying a Piece of the Biosphere

The biosphere is composed of many parts, including the empty lots, lawns, fields, forests, and ponds near you. Wherever there is enough water, plants will grow. Wherever plants grow, there will be animals as well. Sometimes the plants and animals may not be immediately apparent, but once you begin this in-depth study of a piece of the biosphere, you will discover many different organisms and begin to learn about how they interact.

Materials (per team of 6)

2 metal coat hangers, or pieces of stiff wire, for small quadrats
4 wooden stakes and string for large quadrats
meterstick or metric tape
trowel or small-bladed garden shovel
white-enameled pan or large sheet of white paper
collection containers such as jars and plastic or paper bags
6 pairs of forceps

wire screen of approximately 1-mm mesh curved into a bowl shape
6 data books or clipboards and paper
plastic gloves
2 30-cm pieces of PVC pipe (1/2–1 inch in diameter)
4 corks to fit pipe
50-cm wooden dowel that fits inside pipe
4 sterile nutrient agar plates

Procedure
PART A Field Study Preparations
Different schools have different opportunities for outdoor studies. Your class and your teacher must work out procedures to fit the piece of biosphere that is most convenient to your school setting.

1. As a class, select a study area such as a local park, forest, or grassland. Choose an area that is convenient to get to but seldom disturbed by other people. The more natural your area of study, the greater the diversity of organisms you will find. A forest is complex and provides opportunities to collect abundant data, but it is

difficult to picture as a whole. A prairie is almost as complex and is somewhat easier to study. Cultivated areas, such as cornfields and pastures, are relatively simple to study.

Cities also contain pieces of the biosphere suitable for study. Many schools have lawns with trees and shrubs. You can study vacant lots and spaces between buildings. Even cracks in pavement, gutters, and the area around trees often contain a surprising number of organisms.

2. Make and record observations of the plants and animals in your study area. Generate a list of questions about the plants and animals. Do the same type of plants grow in all parts of your study area? Why are some plants more abundant in one part of the area than in another? Do you see any animals, and if so, what do they appear to be doing? What do you think is the relationship between the plants and the animals in your study area?

   Choose a question you would like to study. The discussion questions should give you some ideas. With your teacher, determine whether your entire class will study the same question or whether each team will study a different question in the same area. Decide what type of data to collect to answer your question and how large an area you need to study. Your teacher can direct you to other investigations where you will find techniques for studying your piece of the biosphere.

   If teams are studying different aspects of the area, one study site should be selected. For example, one team could study the plants at the site, one team the insects, and another team the other animals. Other teams might want to study the physical characteristics of the area, such as temperature, soil moisture, or light intensity. In forests, study sites for different plants should be of different sizes. In any habitat, one or several teams can study the organisms that live in the soil.

   To study organisms in the soil, take soil cores of equal length from several different parts of your sampling area. Push a length of pipe into soft soil approximately 25 cm deep. Gently twist the pipe in the ground, and then pull it up with the soil core inside. Plug up the ends of the pipe with the corks, and take all pipes back to the lab for study. In the lab, carefully push the soil core out of the pipe with a wooden dowel. Count the number of earthworms, pill bugs, or other larger organisms you see, noting at what depth in the core they are found. Sprinkle a small sample of soil from different depths of the core onto a sterile plate of nutrient agar—one sample for each plate. Cover the plates, tape closed, and incubate them at room temperature for two days. *Without opening the plates*, observe and count the number of bacterial and fungal colonies that grow.

   To study the plants, use quadrats—square or rectangular samples of the study areas. For trees, a quadrat of 10 × 10 m is common; for shrubs, 4 × 4 m is recommended. Where there are few, if any, trees or woody species, use quadrats made of stiff wire that are 0.5 m on each side, or unwind two metal coat hangers and reshape them into a quadrat. Note the size of the area inside the quadrat, and use this same size quadrat throughout the study.

3. Select a team leader to divide tasks among the team members. Decide what materials your team needs to gather the data, and how and where to obtain them. Draw up a form on which data can be quickly recorded. It may be easier to handle sheets of paper on a clipboard than to take data books into the field. Paste the sheets into your data book when you return to the laboratory. NOTE: Before you collect any specimens to take back to class, observe these important rules. First, never collect anything unless you have consulted your teacher. Second, never collect anything unless you have obtained permission from the owner of the property. Third, never collect any organism that is poisonous, harmful, or rare and endangered—make sure you know what you are touching. Finally, if you do collect organisms for the classroom, collect as few specimens as possible. These should be added to a permanent collection of plants and invertebrates that can be used as a reference by future generations of students.

4. Decide how your team will collect and identify the data. Either identify organisms only by general group names, such as trees, spiders, grasses, beetles, or turtles—or else collect specimens. Your teacher will help you decide the best method. If you collect specimens, assign each specimen a letter (A, B, C, and so on), and whenever you need to refer to that species of organism, refer to its letter. After you return to the classroom, you may be able to identify your specimen by looking it up or showing it to an expert. Your teacher can show you how to make permanent specimen displays of your plants by pressing them between pieces of paper, drying them, and mounting them on pieces of cardboard.

   Some plants you may find in your study area are:

   a. Tree—tall, woody plant with a single stem (trunk).
   b. Shrub—woody plant that branches at or near the ground and lacks a trunk.
   c. Sapling—young tree with a trunk 1 to 5 cm in diameter.
   d. Herb—nonwoody plant that dies back at least to ground level in winter.
   e. Tree seedling—very young tree with a trunk less than 1 cm in diameter.

5. If you decide to collect animal data, it should be done after the studies of plants have been completed, possibly at a later date. Work in pairs, with one person searching and the other recording. Turn over stones, logs, and other cover to find animals, being sure to return these sheltering objects to their original position. Examine plants, especially the flowers. Record notes of the types, numbers, and activities of animals you find. If possible, collect specimens of insects of unknown types for identification. In thick forests, it may be difficult to catch flying insects, but you can beat the shrubs and saplings with a stout insect net, holding the open end up. In open fields, sweep your net through the plants. Try to identify the plants on which you are making your sweeps. If you are able to identify the insect, you can release it. Otherwise, you may want to preserve it according to directions from your teacher and take it back to class for further identification.

6. During your study you may uncover toads or snakes. To study reptiles, birds, and other large animals, it may be necessary for you to cover larger areas than the ones described and observe the areas over longer periods of time. Do not collect these animals. Birds may not be present or active when your class is collecting data—around daybreak and dusk are usually the best times to observe them. Look for animal tracks and animal droppings, which also can be identified. In cities you are likely to see birds, rats, mice, dogs, and cats. Look for signs of humans, too.

PART B   Conducting the Field Study

7. If your team is studying a small area, toss the quadrat on the ground to determine a site. If not, use stakes and a metric measuring tape to define a quadrat.

8. List all the different types of plants you find within your quadrat. If you cannot name the plants, describe them well enough so that you will be able to identify them in your other quadrats, or collect samples using the system you developed in Step 4. If you collect samples, be sure to wear plastic gloves.

   CAUTION: Do not collect any plants you cannot identify as harmless.

9. For each species on your list, estimate its *cover*. This is the percentage of ground within the quadrat covered by the leaves and stems of all the plants of that type. Suppose there are three different types of plants, as in Figure 22.16. Estimate, to the nearest 10 percent, the amount of cover for each type of plant, as well as the amount of bare area. The greater the amount of cover for a plant species, the greater is its importance within the plant community.

10. To increase accuracy, at least two members of your team should estimate the cover of each plant species. Average the two estimates, and record this number. If the numbers are extremely different, reevaluate the cover. If some tall plants overshadow smaller plants within the quadrat, your total estimated cover may be greater than 100 percent when you finish.

Figure 22.16 Vegetation coverage in a quadrat.

bare

bare

bare

TABLE 22.3 Percentage of Cover Within Sample Quadrats

Species	Sample 1	Sample 2	Sample 3	Average Cover
Bluegrass	30	20	70	40.0
Dandelions	20	20	10	16.7
Ragweed	40	30	20	30.0
Bare ground	10	30	10	16.7

11. Complete at least three different quadrat samples within the study area. Calculate an average cover for each plant species you find. To do this, add up the percentages for each species and divide by the number of quadrats measured (see Table 22.3).

12. Examine a sample of the surface soil and/or organic litter. Insert a trowel or small garden shovel into the ground to a depth of approximately 10 cm, and remove a cubic sample of approximately 10 cm $\times$ 10 cm $\times$ 10 cm. Using the wire screen that has been shaped into a bowl, sift the contents onto a white-enameled pan or a large sheet of white paper. Pick through the remaining litter with forceps. Organisms present probably will include a variety of insects (especially beetles), sow or pill bugs, millipedes and centipedes, spiders or other arachnids, worms, and fungi.

13. When your field work is completed, share your team data with the rest of the class by combining data from each team in a large table drawn on the chalkboard. Answer as many of the discussion questions as possible. If different types of areas were included, make comparisons.

Discussion

1. What producers are in the area? Identify as many of the plants as you can, but answer in general terms—trees, shrubs, and so on—if you cannot name the producers.

2. The plant species with the greatest amount of cover is said to be dominant within the study area. What effect does the dominant species you found have on the rest of the study area?

3. Are the producers evenly distributed throughout the entire study area? What do you think may be responsible for the pattern of plant growth you find?

4. Do you have any evidence of seasonal changes in the types and numbers of producers?

5. Are there different groups of producers? If so, which group produces the most food in the area?

6. Are there layers of producers? If so, what relationships can you find among the layers?

7. Does the area produce all its own food, or is food carried in from beyond its boundaries? What evidence do you have for your answer?

8. What consumers are in the community? Answer in general terms or with names of the organisms you have identified.

9. Which consumers are herbivores and which are carnivores? What evidence supports your answer?

10. What relationships can you find between the numbers of a particular herbivore and the numbers of a carnivore that eats it?

11. Using the information you have, construct an energy-flow diagram for the area.

12. What is the evidence that one type of organism affects another in ways other than those involving food relationships?

13. What biome type best characterizes the area you studied? Give reasons for your choice.

14. An investigation such as this should raise more questions than it answers. In studying the data, part of your job is to look for questions that need answering. Write several questions concerning the organisms in your area. Although your data from this study are incomplete, you have taken a biologist's first step in describing a study area.

Guidepost: How does the vegetation change as the precipitation decreases?

Biomes with Decreasing Precipitation

22.7 There Are Three Distinct Types of Grassland

Along the latitudes of deciduous forest the precipitation decreases from east to west in North America. This lack of water, rather than of radiant energy, is the major factor limiting plant and animal life. Grasslands, such as the prairies of North America, typically receive 63 to 71 cm of annual precipitation, have low winter temperatures, and may be relatively high in elevation. Drought and periodic fires keep woody shrubs and trees from invading the environment, except along streams.

There are three distinct types of grasslands in North America: the tall-grass prairie in the northern Midwest, the short-grass prairie that stretches eastward from the Rocky Mountains into Kansas, and the mixed-grass prairie in the central Great Plains. Figure 22.17 shows these typical grasslands. The geographic distribution of these grasses results mostly from the nearness of the Rocky Mountains and the rain shadow the mountains produce. Tall-grass prairies receive more precipitation than do mixed-grass prairies, and mixed-grass prairies receive more than short-grass prairies. The boundaries of these grasslands are indistinct, and they share many plant and animal species.

Figure 22.17 Tall-grass prairie (a); mixed-grass prairie (b); short-grass-prairie (c). What is the most important climatic factor in determining each type of prairie?

a   b   c

Figure 22.18   Animals of the grassland include (a) meadowlarks, *Sturnella neglecta;* (b) black-tailed jackrabbits, *Lepus californicus;* and (c) badger, *Taxidea taxus.*

The most conspicuous consumers are hoofed mammals. Once there were many bison and pronghorns in North American grasslands. Now nearly all of the bison and many of the pronghorns have been replaced by cattle and sheep. Less conspicuous herbivores are jackrabbits and ground squirrels. Many types of insects also feed on the vegetation. At times grasshopper populations reach large numbers, devouring the plants down to ground level.

Wolves and coyotes were once the chief large predators. Today, wolves have been exterminated almost everywhere in the contiguous 48 states, but coyotes survive in many places. Rattlesnakes and badgers are important predators of ground squirrels and prairie dogs. Many insect-eating birds, such as meadowlarks and mountain plovers, nest on the ground under the cactus on the short-grass prairie. Figure 22.18 shows some typical grassland animals.

What adaptations might be important for animals that live in such open areas? Because of the openness of the grasslands, speed and endurance are adaptations that enable many organisms to escape predators. Small mammals seek shelter below ground, and grassland birds and reptiles escape detection because of protectively colored skins.

22.8   Midlatitude Deserts Border the Grasslands

In North America, the western edge of the grassland is bordered by desert. The climatic situation is complicated by mountains, but between the western ranges, deserts occur from eastern Washington south into Mexico. There are two types of North American deserts. The northern part of the desert biome is the cool desert of the Great Basin and has an average temperature of about 10 to 12° C in the shade. In the southwestern or hot deserts, the temperature averages around 20 to 22° C. Death Valley in the Mojave Desert has reached 57° C in the shade. Figure 22.19 shows vegetation typical of both hot and cool deserts.

Figure 22.19 Hot desert in Arizona (a). What characteristics do many of the plants have in common? Cool desert (b). How do these plants differ from those of the hot desert?

Latitude and temperature have a profound effect on the formation of deserts. The amount of precipitation that results in a desert at the equator may support a fine grassland at higher latitudes. When precipitation does occur in a desert, it is likely to be heavy but brief and is often the result of thunderstorms or cloudbursts. Much of the water runs off instead of sinking into the soil. The rate of evaporation is always high compared with precipitation. For example, although Tucson, Arizona, receives about 20 cm of rain a year, it is so hot that the evaporation rate exceeds the equivalent of about 195 cm of rain. Loss of heat from the earth's surface is greatly slowed by water vapor in the air. Because desert air is dry, the heat that builds up in the soil during the day is quickly lost at night by radiation.

Desert plants have several adaptations that enable them to survive the harsh conditions. Most desert plants resist drought by having root systems that spread out in all directions from the stems but extend only a short distance into the ground. When it rains, these widespread, shallow roots soak up the moisture rapidly. Cactus plants resist drought by storing water in the tissues of their thick stems. The seeds of some plants are covered with a chemical that prevents their germination during unfavorable conditions. The chemical is washed away during a heavy cloudburst so that germination is possible. Some plants have root systems that grow down near the water table, ensuring a continuous supply of water. Mesquite plants (Figure 22.20) can have taproots up to 25 or 30 m long and have been seen growing down through the ceilings of horizontal mine tunnels. Still other plants have small leaves with thick cuticles that reduce water loss, and some even shed their leaves when it is very dry. Some plants, such as the creosote bush, produce a substance that inhibits the growth of other plants that otherwise would compete for space and water.

Seed-eating animals, such as ants, birds, and rodents, are common in deserts. Like the desert plants, many of these animals are especially adapted to living where there are extreme temperatures and little available water. Many desert animals live in burrows and are active only at night. Some rodents wait out the hot summer days in their burrows in a state of sleep

Figure 22.20 How does mesquite escape drought?

Figure 22.21 The kangaroo rat can survive without drinking any water. What adaptations might make this possible?

called **estivation** (es tih VAY shun). Estivation is similar to hibernation, but it is a response to heat and dryness rather than to cold. Others animals are light-colored, thereby reflecting light, and have adaptations that enable them to conserve water. The kangaroo rat, shown in Figure 22.21, which lives in the American Southwest, can survive without drinking any water.

Concept Review

1. Why are there grasslands in the same latitude as forests?
2. What two abiotic factors must be considered together in describing desert climate?
3. How are plants adapted to desert conditions?
4. In what way is estivation an adaptive behavior?

Biomes with Variable Precipitation

22.9 Savannas Are Tall Grasslands in Tropical Dry Areas

The savanna (suh VAN uh) is a tropical or subtropical grassland with a few scattered trees. Three distinct seasons characterize the savanna: cool and dry; hot and dry; and warm and wet. Most savanna soils are rather poor. Savannas cover wide areas of central South America, central and South Africa, much of Australia, and portions of southern Asia.

In Africa, the savanna is home to some of the world's largest herbivores, including elephants, giraffes, and rhinos. These animals are preyed on by lions, leopards, and cheetahs. Scavengers, such as hyenas and vultures, clean up the carcasses left by the predators. Often, leopards and cheetahs will place their kill in trees so other animals cannot steal the carcass. Large herds

Guidepost: What is the effect of variable rainfall and uniform temperature on a biome?

Figure 22.22 Elephants, *Loxodonta africana,* and zebras, *Equus burchelli,* share a water hole on the savanna in Kenya.

Figure 22.23 A chaparral in California.

Guidepost: How do ecosystems change?

of zebras, wildebeest, and gazelles, as well as elephants, giraffes, and other hoofed animals, are found on the Serengeti Plains of the savanna (see Figure 22.22). At present these herds make up the largest concentration of wild animals to be found on any continent.

Consider the character of the savanna. What adaptations might savanna-dwelling animals have? Burrowing animals are common, and nest sites and shelters often are on the ground. Animals in the savanna are most active during the rainy season, and many species are nocturnal. In the winter, or dry season, when the aboveground vegetation is dry, many animals are dormant or subsist on seeds and dead plant parts.

22.10   Chaparral Covers Dry Areas with Thin Soil

Midlatitude areas along coasts where cool ocean currents circulate offshore often are characterized by mild, rainy winters and long, hot, dry summers. These areas, such as the California coast, are dominated by chaparral (shap uh RAL), or brushland, communities composed of dense, spiny shrubs with tough evergreen leaves, often coated with a waxy material. Chaparral vegetation also is found in the Mediterranean region, along the coastlines of Chile, southwestern Africa, and southwestern Australia. Plants from these regions are unrelated, but they resemble one another in form and function and show the same type of adaptations. Annual plants also are common in the chaparral during winter and early spring when rainfall is most abundant. Figure 22.23 shows a typical chaparral.

The plants of the chaparral are adapted to and maintained by periodic fires. Many of the shrubs have deep and extensive root systems that permit quick regeneration. The fires burn all the plant structures above the ground, thus releasing the nutrients in them. The nutrients then are available for use by the new plant shoots. In addition, many chaparral species produce seeds that germinate only after a hot fire. Deer, fruit-eating birds, rodents, lizards, and snakes are characteristic chaparral animals.

Concept Review

1. Describe the savanna biome and the abiotic factors that influence it. How does the savanna differ from other grasslands?
2. How does the character of the savanna influence the animals that live there?
3. What is chaparral? How are chaparral plants adapted to the abiotic factors?

Succession and Evolution in Ecosystems

INVESTIGATION 22.3   Long-Term Changes in an Ecosystem

Through many years, one biotic or abiotic factor in an ecosystem may change. This change causes a second factor to change, which, in turn, changes a third factor. Eventually, the entire ecosystem may be quite different from what it was earlier. In this investigation, you will trace the effects of a gradual climatic change on a model ecosystem. The model is much simplified from a real situation.

Materials (per team of 3)
none

Procedure

1. Read the following information, and diagram the food web in the ecosystem described.

   An oak grows next to a shallow stream in a grassy area. There are no shrubs here. Grasshoppers and blacktail jackrabbits feed on the little bluestem grass. Duckweed grows in the water, providing some of the food for wood ducks. These birds also feed on grasshoppers, acorns, and grass. The ducks are a major food source for red wolves.

   The average annual precipitation is 84 cm. The average annual temperature is 10° C.

2. In your data book, prepare a table similar to Table 22.4.

3. Working as a team, fill in the table using this additional information:

   To the west of the grassy area is a desert. There, glossy snakes eat Merriam pocket mice. These mice must burrow under shrubs or cacti to hide from the snakes. The mice, therefore, cannot live in a grassy area. In the desert, they burrow under mesquite shrubs and eat the seeds. Grasshoppers and blacktail jackrabbits eat the mesquite leaves.

   The climate is changing in the grassy area. For the next 10 000 years (50 rounds on the table), the average annual precipitation will decrease by 1.25 cm every 200 years, and the average annual temperature will increase by 0.3° C every 200 years.

   An average annual temperature of more than 20° C or an average annual precipitation of less than 34 cm will cause the stream to begin drying up and the oak trees to die.

   If the average annual temperature reaches 23° C, or the average annual precipitation is less than 24 cm, the water table will be too low for the grass.

   In filling out the table, assume that the organisms in the desert area are trying to disperse to the grassy areas at all times and that they will invade it whenever conditions allow.

4. After filling out the table, diagram the food web for the same area at the end of 10 000 years.

Discussion

1. When did the first changes in the ecosystem occur? What were they?
2. What subsequent changes occurred, and in which years?
3. What similarities or differences are there between the initial and final food webs?
4. What is a barrier to dispersal for Merriam pocket mice?
5. What limited the population of red wolves?

TABLE 22.4   Ecosystem Changes

Round	Years	Average Annual Precipitation	Average Annual Temperature	Changes in Organisms
1	200			
2	400			
etc. (to 50)				

6. What climatic factor was responsible for the other changes?

7. Why does the use of average annual values for temperature and precipitation make this model very artificial?

For Further Investigation

1. Create a computer program to determine when critical points are reached and what the outcomes are.

2. Substitute different pioneer organisms and determine the effects on the ecosystem and other organisms. For example, what would happen if creosote colonized in the new area instead of mesquite?

3. Create a new model using your local biome and nearby biomes.

22.11 Ecosystems Are Not Permanent

Do the abiotic and biotic factors that determine the character of biomes always remain the same? Landscapes, mountains, and seascapes may seem eternal, but they are not. The great sheets of ice that once covered the northern parts of North America and Eurasia have melted. In many spots, they have left bare rock and soil. In all regions of the world, erosion constantly exposes bare rock. In many areas, fire is a natural factor. Floods, blowing sand, and volcanic eruptions also may destroy the vegetation of an area and leave bare ground or create new land.

Nuclear explosions also can destroy life. Before atmospheric tests of nuclear explosions were stopped in 1958, some biologists began a five-year study of the effects of earlier nuclear explosions at the Nevada test site. The explosions had removed the topsoil at ground zero and destroyed all the desert vegetation within a 1 km radius, as shown in Figure 22.24b. Farther out from ground zero, only a few hardy species (such as the creosote bush

Figure 22.24 The aftereffects of atomic testing: (a) before testing at ground zero, showing typical vegetation; (b) after testing. The exposed rocks indicate extensive soil removal. Russian thistle (c) was the only species that grew within 0.5 km of ground zero in the first year. Its presence indicates the start of recovery.

a   b   c

and the ground thorn) survived. A year after the explosion, some spring annuals (stickleaf, stork's bill, and desert pincushion) appeared. These grew from seeds carried in by wind or birds. Small plants began covering the area, as shown in Figure 22.24c.

As the plants died, their remains were added to the sandy soil. The plants changed the abiotic environment in other ways, also. Their roots held more moisture in the soil, which made the temperature suitable for the germination of seeds of other species. By 1961, there was a community of plants, including foxtail chess, wild buckwheat, and others. This was not the pre-explosion community, however. It was similar to the community that eventually appears in a grassland area previously swept by fire.

22.12 Communities Change Through Relatively Long Periods of Time

The process by which one type of community replaces another is called **succession**. The traditional idea of succession starts with some bare lifeless substrate, such as rocks. Lichens may grow first, forming small pockets of soil and breaking down the stone. Mosses then may colonize, eventually followed by ferns and the seedlings of flowering plants. Through many thousands of years, or even longer periods, the rocks may be completely broken down, and the vegetation in an area where there once was a rock outcrop may be just like that of the surrounding forest or grassland. Figure 22.25 on page 620 illustrates succession in a woodland. You can observe succession in your own region in a fairly short time by comparing pieces of land that have been abandoned for varying numbers of years.

The community that ends a succession, at least for a long period of time, is called the climax community. Unlike the other communities in the succession process, the climax community is not replaced by any other community unless a disturbance or a change in the climate occurs. If a single spruce dies of old age, the space that it occupied is too shaded by neighboring spruces for aspen to grow. A young spruce, tolerant of shade, probably will take its place. Thus, once established, the climax community is relatively permanent. Although other factors (such as soil) may play important roles, the climax community is determined mainly by the climate of the region. In some areas, for example, the climate is unsuitable for trees, and the climax community is dominated by grasses. With the increasing realization that the climate keeps changing, that the process of succession often is very slow, and that the nature of a region's vegetation is greatly determined by human activities, many ecologists consider the concept of climax communities to be less useful than it once was.

Whatever causes succession, early stages are likely to be dominated by plants that are dispersed readily, grow rapidly, and have adaptations that favor colonization. Most colonizers are small annuals that grow close to the ground. They use most of their energy to produce seeds, rather than to develop large root, stem, and leaf systems. Weeds are common colonizers that can tolerate a wide range of adverse environmental conditions such as poor soils, high temperatures, and low precipitation. Later stages of succession are likely to be dominated by longer-living species that are able to endure increased competition and increased shade.

bare rock → lichen → meadow → aspens → spruce and fir ← aspens ← meadow ← marsh ← pond

succession from rock   climax   climax   succession from water

burned ground

succession from fire

annuals

Figure 22.25   Successional stages and animals in a coniferous forest. Typical animals of the coniferous forest include (clockwise from the left) voles, elk, squirrels, porcupines, whitetail deer, crossbills, myrtle warblers, otters, beavers, woodpeckers, and moose.

In any community, the plants and the abiotic environment determine which animals can live there. When the former Nevada nuclear test site was invaded by plants again, animals soon followed. No forest or tropical animals appeared—only pocket mice, kangaroo rats, and other species that feed on desert plants. Through time, the climate in an area may change. If it grows colder or drier, some animals may die or move elsewhere. Other animals may find the area more hospitable than it was earlier. The selective effect on the animals depends on the adaptations they possess. The adaptations may be to climatic factors themselves or to certain plants that are, in turn, adapted to a certain climate.

INVESTIGATION 22.4   Effects of Fire on Biomes

Fire is an important ecological factor in terrestrial ecosystems. Some fires start from natural causes, but many are caused by humans—deliberately or accidentally. No matter how they begin, fires have many effects on the organisms in their paths. The most easily observed effects are on vegetation.

    This investigation considers three different North American biomes and the changes that fire might cause in each.

Materials (per team of 3)

Procedure

PART A   Fire in a Midlatitude Grassland

1. Figure 22.26 depicts a series of events in the southern part of the North American grassland. Two types of populations are involved: grasses of various species and mesquite shrubs. Study Figure 22.26a and 22.26b. What is happening to the sizes of the populations?

2. Roots of mesquite have been found in mine shafts many meters below the surface of the soil. What competitive advantage might this type of root growth give mesquite over grasses?

3. If occasional droughts strike the area (as actually happens), what type of community do you think might result?

4. Refer to Figure 22.26c and 22.26d. In these figures, plant parts shown in light color represent unharmed tissue. Do both types of plants survive fires?

5. In which type of plant has more growing tissue been killed?

6. Grasses usually reach maturity and produce seeds in 1 to 2 years; mesquite usually requires 4 to 10 years. Which type of plant has lost more growing time?

7. In Figure 22.26d, which type of plant occupies the most land?

Figure 22.26

a. grasses and mesquite bush

b. same area 10 years later

c. fire

d. 2 years after fire

8. What might you expect this area to look like 4 or 5 years after a fire?
9. Make a generalization about the effect of fire on this community. Describe the probable landscape if fires did not occur at all.
10. How would the landscape look if fires occurred every few years?
11. What seems to be necessary for maintaining grassland in this region?

PART B Fire in a Forest of the Great Lakes Region

Around the Great Lakes of North America is a region of transition between the coniferous and midlatitude deciduous forests. In many places much of the forest consists of pines (Figure 22.27a). Early in the settlement of the region, Europeans brought about a great change in the landscape (Figure 22.27b).

12. What was this change?
13. Fires apparently had been rare in this region, but following the change in landscape, they became more frequent. What might have brought about the increase in the number of fires?
14. If a fire does not occur, what might the area shown in Figure 22.27b look like in later years?
15. Study Figures 22.27b, 22.27c, and 22.27d, which depict jack pine. What characteristic of jack pine gives that species a competitive advantage when there is a fire? (Hint: Compare the cones in Figure 22.27b and 22.27c.)

Figure 22.27

a. before settlement

b. after settlement

c. fire

d. 3 years after fire

16. Describe the probable appearance of the area shown in Figure 22.27d 5 or 6 years later.

17. Jack pines produce cones in 8 to 10 years but do not live to a very great age. Their seedlings do not thrive in shade. Suppose no fires occur for 200 years. What changes in appearance might take place in this area during that period?

18. Suppose fires occur about once every 20 years. What might the area look like at the end of 200 years?

PART C   Fire in a Forest of the Southeastern United States

In the southeastern United States are great forests in which longleaf pine is almost the only large tree. Occasionally there may be seedlings and saplings of deciduous trees. Between 3 and 7 years of age the longleaf pines look somewhat like a clump of grass (Figure 22.28a). While in this stage, the young trees develop deep roots in which food is stored. (See the cutaway, lower corner of Figure 22.28a.)

19. What is the effect of fire on young longleaf pines?

20. What is the effect on the deciduous shrubs and saplings?

21. Which plants have a competitive advantage after a fire?

22. After the grass stage, longleaf pines grow rapidly in height and develop a thick, fire-resistant bark. What is the effect of ground fires at this stage in the development of the pines (Figure 22.28c)?

Figure 22.28

a. longleaf pines, deciduous saplings and shrubs

b. fire in same area

c. fire in mature pine forest

d. pine forest - no fire for 5 years

23. Which plants have a competitive advantage when fires do not occur (Figure 22.28d)?

24. What factor seems to maintain a forest of longleaf pines within the deciduous forest biome?

Discussion

1. If you were interested in raising cattle in a midlatitude grassland, would occasional fires be an advantage or a disadvantage? Why?

2. Jack pine is not as valuable a lumber tree as other trees of the Great Lakes region. If you were a landowner in that region, would fire be an advantage or a disadvantage? Why?

3. If you were interested in maintaining a longleaf pine forest to obtain turpentine, would ground fires be an advantage or a disadvantage? Why?

4. Suppose you wanted bobwhites (game birds that nest on the ground) in your turpentine forest. What effect might this have on your management of the forest?

5. What things must ecologists know before deciding whether to recommend fire as a method of management to a landowner?

22.13 Succession May Not Always Be Orderly

The traditional idea of succession holds that the diversity of species in an ecosystem increases as succession progresses. As a community matures, species that are better competitors usually replace the initial colonizers. Interactions between animals, such as predation, symbiosis, and competition, become more complex and extensive during successions, thus opening new niches and increasing the diversity of animals that can live in the community. According to this model, succession reaches climax when the web of interaction becomes so intricate that no additional species can fit into the community unless niches should become available through localized extinction.

Recently, many ecologists have challenged this traditional model of succession in which predictable and orderly changes result in a climax community. Most communities are in a continual state of flux, or change, and the types and number of species change throughout all the stages of succession, even at the so-called climax stage. The course of succession may vary depending on which species happen to colonize the area in the early stages and on the nature of the disturbance that causes further change. Disturbances such as fire, flood, hurricanes, and even unseasonable temperatures can prevent a community from ever reaching a state of equilibrium. When disturbances are severe and frequent, community membership may be restricted to those organisms that are good colonizers. If disturbances are mild and rare, then the species that are most competitive will make up the community. Species diversity, then, would be greatest when disturbances are intermediate in frequency and severity, because organisms typical of different successional stages are likely to be present.

The recovery of the Mount St. Helens area after the volcanic eruption in 1980 is an interesting study (see Figure 22.29). Scientists were divided on their predictions of how the recovery would progress. One group of

Figure 22.29 An area of Mount St. Helens colonized by *Penstemon*.

Biodiversity in Danger

All organisms on earth, including humans, are dependent on each other. The interplay between biotic and abiotic factors makes possible the diversity of organisms that exists today.

Natural ecosystems have a vast amount of genetic diversity, yet researchers have strong evidence that ecosystems are in the opening stages of an extinction episode. There is a pronounced and sudden decline in the worldwide abundance and diversity of ecologically different groups of organisms, particularly those in tropical rain forests. Such a rate of extinction is the most extreme in the past 65 million years. Some researchers estimate that at the current rate more than one million species may become extinct by the end of this century.

The primary cause of this decline in diversity is not direct human exploitation but the habitat destruction that results from the expansion of human populations and activities. As the human population grows, it places greater demands on the ecosystem for space and goods. As a result, land is cleared for cultivation and living space, destroying the habitat of many organisms.

Organisms require appropriate habitats if they are to survive. Just as humans cannot live in an atmosphere with too little oxygen, trout cannot breed in water that is too warm or too acidic. Red-eyed vireos, scarlet tanagers, and other birds must have mature tropical forest in which to spend the winters. Black-footed ferrets (see figure) require prairie that still supports prairie dogs, their major source of food.

Although many organisms provide humans with crops, domesticated animals, industrial products, and medicines, the roles that microorganisms, plants, and animals play within the natural ecosystem are more important. All these organisms exchange gases with their environments and thus help to maintain the mix of gases in the atmosphere. Changes in that mix, such as increases in carbon dioxide, nitrogen oxides, and methane, may lead to rapid climatic change and agricultural disaster.

The extent to which humans have devastated the diversity of organisms and the effect this devastation will have on the ecosystems of the world are as yet undetermined. Currently, however, procedures to alleviate some of the problems are more widely practiced than ever before. Some of these procedures require a change in life-style. People are learning to recycle aluminum, plastic, paper, and glass, and to refuse to purchase items intended to be used only once and then thrown away. Other procedures involve new techniques to perpetuate diversity, such as embryo transplants between an endangered species and a domestic species (see Chapter 6, Biology Today, page 149). In the months and years to come, researchers will continue to develop new techniques and products to help alleviate the human pressure on natural resources and to create a "sustainable-earth" economy.

A black-footed ferret. Attempts are being made to breed these endangered animals in captivity so they can be reintroduced in the wild.

scientists expected an orderly progression beginning with pioneer plants and eventually ending in a mature forest. Others thought recovery would begin as a matter of chance—a seed would arrive, perhaps blown by the wind, take root, and then play a role in influencing what other plants would grow nearby. To the surprise of both groups, recovery is happening in both ways. Seeds of alpine lupine, a pioneer plant, began to grow in areas that were nearly sterile. After they died, their stalks and dead leaves collected sand and dust particles, which allowed middle-stage plants to root. This supports the orderly succession idea. In contrast, individual trees, mostly alder, have become established out in the middle of very desolate areas without benefit of any pioneer plants.

Concept Review

1. What is succession?
2. What types of conditions start natural ecological succession?
3. What is a climax community? Why is the idea of a climax community less useful than it once was?
4. How does geography help determine the members of climax plant communities?
5. How does the traditional model of succession differ from the newer model?

Guidepost: How are human activities changing biomes?

Human Influence on Biomes

22.14   The Need for Land Has Changed Some Biomes

A flight from Cleveland, Ohio, to Nashville, Tennessee, crosses the mid-latitude deciduous forest, yet only a few traces of forest can be seen. A flight from Chicago, Illinois, to Lincoln, Nebraska, crosses the eastern portion of North American grassland, but here, too, only traces of the original biome remain. In fact, in these two flights, the landscapes below appear remarkably similar. In both cases the climate remains an important factor, but the present landscape has been shaped by humans, as shown in Figure 22.30. The changes have been caused by the need for food and homes. In Ohio, for example, only about 2 percent of the original deciduous forest remains, and the land now supports farms and cities instead. This is the inevitable consequence of a growing population.

Biomes modified for human use are simplified. For example, when grasslands are plowed, thousands of interrelated plant and animal species in that ecosystem may be replaced with a greatly simplified single-crop ecosystem, or with structures such as highways and parking lots. Modern agriculture is based on the practice of deliberately keeping ecosystems in the early stages of succession, which are highly productive but vulnerable because of the diminished diversity of species. For example, destruction of some insects by insecticides can lead to the failure of crops that depend on insect pollination. Extermination of the enemies of insect pests, such as some birds and beneficial insects, can lead to severe pest outbreaks. The extermination of soil organisms by fungicides and herbicides can destroy

Figure 22.30 The rich, flat plains of the Midwest became the corn and wheat belt of the United States.

the fertility of the soil. Because of such problems, there is an increased interest in "organic" or natural farming techniques that maintain productivity but are much less destructive in their effects on the soil and the ecosystem.

22.15 Tropical Forests May Disappear

The demands of a growing population have placed the tropical deciduous forests (Figure 22.31) and tropical rain forests of the world in jeopardy. These tropical rain forests, which account for two-thirds of all tropical forests, are home to more than 50 percent of all species on earth. Already, more than half the original area of tropical rain forest has been cleared for timber, cattle grazing, fuelwood, and farming. If clearing and degradation continue at the current rate, almost all tropical forests will be gone or severely damaged in about 30 years.

The present destruction and degradation of tropical forests is considered one of the world's most serious environmental and resource problems. Why? Two-thirds of the plant and animal species and as much as 80 percent of the plant nutrients in a tropical rain forest are in the canopy. Hence, removing all or most of the trees in a forest destroys the habitat and food supply of most of its plant and animal life, as shown in Figure 2.16, page 47. Tropical deforestation also exposes the nutrient-poor soil to drying by the hot sun and to erosion by torrential rains. Ecologists warn that the loss of these diverse biomes could cause the premature extinction of as many as one million plant and animal species by the beginning of the next century.

From 50 to 80 percent of the moisture in the air over tropical forests comes from trees through transpiration. If large areas of these forests are cleared, average annual precipitation would decrease and the climate would

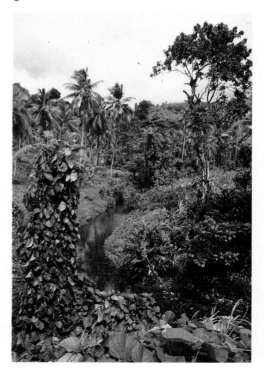

Figure 22.31 Tropical deciduous forest. Many food products and medicines are made from plants that grow in such forests.

become hotter and drier. Any rain that fell would run off the bare soil very rapidly instead of being absorbed and slowly released by vegetation. Eventually these changes could convert a tropical forest into a sparse grassland or even a desert. The carbon dioxide emitted by deforestation and burning also would affect global climate, food production, and sea levels by warming the earth.

Deforestation also will have a serious economic impact. Tropical forests supply a variety of food products including coffee, cocoa, spices, nuts, and fruits. They also provide the raw materials for latexes, gums, resins, dyes, waxes, and essential oils. These substances are used in the manufacture of ice cream, toothpaste, shampoo, deodorant, perfume, varnish, phonograph records, tires, tennis rackets, jogging shoes, and many other products. In addition, the raw materials of one-fourth of the prescription drugs used, such as vincristine (see Figure 3.20), come from plants that grow in tropical rain forests. The loss of tropical forest would severely impact the manufacture of these products and medicines. In addition, plants that could be vitally important in the development of new drugs and medicines would be gone.

22.16 Desertification Is Spreading Rapidly

Presently, about 35 percent of the earth's surface is classified as arid or semi-arid desert. In the drier parts of the world, desert areas are increasing at an alarming rate from a combination of natural processes and human activities. The conversion of rangeland, rain-fed cropland, or irrigated cropland to desertlike land is called **desertification** (Figure 22.32). A grassland that once would support a large herd of livestock becomes unable to support any at all. Natural desertification occurs near the edges of existing deserts and is caused by the dehydration of the top soil layers during periods of prolonged drought and evaporation. Drought is the absence of sufficient precipitation to make the plants grow. The few plants that remain soon are eaten by the livestock. With the ground cover gone, the winds begin moving the soil of the grazing lands into dunes, and in some places the sands of the deserts encroach on the margins of the grasslands. This is happening in the sub-Saharan region called the Sahel and has brought great misery and death to those nomads who live there (see Chapter 2, Biology Today). It is estimated that hundreds of millions of hectares of land that border deserts are at risk of becoming deserts by the year 2000. Desertification also is occurring in the Midwest, the region that currently produces the bulk of the United States' grains. Experts predict that unless soil erosion is controlled, another dust bowl may result.

Natural desertification, however, is accelerated by such short-sighted practices as overgrazing. Overgrazing is a result of keeping too many head of livestock on too small an area of land. Poor soil and water resource management and deforestation also can speed up desertification. It is estimated that, worldwide, an area 12 times the size of Texas has become desert during the past 50 years. Despite attempts to halt or sharply reduce the spread of desertification, each year an estimated 6 million hectares (15 million acres)—an area the size of West Virginia—of new desert are formed, mostly in Africa.

Figure 22.32 Crops being covered by sand near Portales, New Mexico.

22.17 Acid Deposition Has Many Harmful Effects

When electric power plants and industrial plants burn coal and oil, they emit large amounts of sulfur dioxide, particulate matter, and nitrogen oxides. As these emissions are transported by winds, they form secondary pollutants such as nitrogen dioxide, nitric acid vapor, and droplets containing solutions of sulfuric acid, sulfate, and nitrate salts. These chemicals fall to the earth's surface as acid rain or snow, or as gases, fog, dew, and solid particles. Gases in this mixture can be absorbed directly by leaves. The combination of dry and wet deposition and absorption of acids and acid-forming compounds is called acid deposition, or acid rain. Normal rain has a pH of 5.6. The pH of acid rain ranges from 5.5 to as low as 2.4. Precipitation over much of the world now is as much as thirty times more acid than normal.

What happens to forests that are watered with acid rain? The soils supporting the forests no longer can buffer or neutralize the acid precipitation, which then leaches nutrients out of the soil. This deprives the trees of the elements they require for growth, and they begin to weaken and die. The waxy cuticle on leaf surfaces can be destroyed directly by the acid rain. Needles and leaves then turn yellow and drop prematurely, and the trees eventually die (Figure 22.33). Whole forest ecosystems are threatened by acid rain; the forests of the eastern United States, Canada, and Europe show the greatest destruction so far.

Acid deposition has other harmful effects. It damages statues, buildings, and car finishes. It can kill fish, aquatic plants, and microorganisms in lakes and streams, at the same time increasing populations of giardia, a parasite associated with a severe gastrointestinal disease. It also can affect the ability of trout and salmon to reproduce. The acid stunts the growth of

Figure 22.33 Trees exposed to acid rain eventually die. How does their death affect the atmosphere?

crops, such as tomatoes, soybeans, spinach, carrots, and cotton, and it leaches toxic metals from water pipes into the drinking water itself.

Concept Review

1. What has happened to the deciduous forests of the Midwest?
2. How does the cutting of a tropical rain forest affect the soil under the trees?
3. What factors, human and natural, increase the size of deserts?
4. What causes acid deposition, and how is it harmful?

Summary

In each type of climate, a characteristic type of vegetation develops. That vegetation supports a characteristic variety of animal life. The tundra biome of the Northern Hemisphere slowly blends with the taiga in a southerly direction. In the middle latitudes are the deciduous forest biome in the east, the grasslands to the west, and the desert biome in the southwest.

The tropical rain forest biome, located just north and south of the equator, receives abundant rainfall and has warm temperatures all year long. The tropics also support a deciduous forest, but unlike the rain forest, this biome has wet and dry seasons. Still within the tropics is the savanna biome, a grassland that supports a few trees and the world's largest concentration of hoofed animals. Southern California and other scattered areas support evergreen shrubs that make up the chaparral biome. Because of changes in altitude, the mountains have a number of biomes that form bands of vegetation ranging from grasslands at the base to alpine tundra on the top of the tallest peaks.

No ecosystem is permanent. Natural and human-made catastrophes can change the character of an ecosystem. As a result, a succession of communities may occur, ending, at least temporarily, in a relatively stable climax community. All ecosystems around the world are undergoing changes as a consequence of human activities. The cutting of the tropical rain forests, the loss of grasslands to desertification, and the acid rain problem in North America and Europe are just a few examples.

Applications

1. Explain why the four seasons are more pronounced in the midlatitudes than in the tropics.
2. What biomes would you encounter if you were to travel from Washington, DC, to San Francisco?
3. Give several examples of changes in ecosystems caused by human activities.
4. Describe what you think would happen if all humans left the following places: a farm in Nebraska; a sidewalk in New York; a swimming pool in Seattle, Washington; a landscaped park in Las Vegas, Nevada.
5. Make a list of terrestrial organisms that humans have brought into your locality. Divide the list into two parts; organisms that survive because of human activities, and organisms that would survive without human help. Give reasons for your decision on each organism.
6. Why do you think biomes are characterized by the types of plants that grow there rather than the types of animals that live there?

7. During the winter in the United States, the earth is closer to the sun than at any other time of the year. How do you explain that this is the coldest season of the year?

8. What do you think would happen if all of the major ocean currents reversed direction? Speculate as to the changes in climate we would see in the United States.

9. In which biomes would you expect to find the most productive farms? Why?

10. In which area would you expect the food webs to be more complex—the tundra or the tropical rain forest? Explain.

11. Would you expect the tree line to be lower or higher in the mountains of Mexico than in the mountains of Montana? Explain.

12. Why might a large animal have an easier time keeping warm than a small animal?

13. Is there such a thing as a "good" forest fire?

14. Compare and contrast hibernation and estivation. How are they adaptive?

Problems

1. Investigate your area and determine the biome in which you live.

2. The prairie dog once was abundant in the prairie ecosystem. What is the role of the prairie dog in that ecosystem? How is it being exterminated? How will other species be affected by the extermination of the prairie dog?

3. The search for oil in the Arctic has ecologists concerned about damage to the tundra. What happens to the tundra when heavy equipment is driven over it? What might be the effects on migrating herbivores such as the caribou?

4. Find out what different countries are doing to protect their particular ecosystems. Some things that could be explored are: national parks, forest preserves, wildlife preserves, and city parks.

5. Investigate the ecological recovery from the 1988 Yellowstone fires.

6. Before Europeans settled the grasslands of Australia, kangaroos were the ecological equivalent of the bison in North American grasslands. These animals had similar niches in the community structure of their regions—they were the largest grazing herbivores. What are the ecological equivalents of these animals in most of the Australian and North American grasslands today? What were the ecological equivalents of these animals in the steppes of Asia, the pampas of Argentina, and the veldt of South Africa before these regions were highly modified by humans? What are the ecological equivalents of these animals in the tundra?

In the desert of South Africa, what are the ecological equivalents of the cacti of North American deserts? In the tropical forests of the Old World, what are the ecological equivalents of the hummingbirds of the New World tropical forests?

7. Contact an environmental group and ask about their programs to help preserve parts of biomes around the world.

8. Start a collection of photographs and slides illustrating different biomes around the world. Display groups of these photographs around your classroom, or create a slide show about biomes for your class.

9. Arguably, the tropical rain forest is the most important biome in the world. Estimates are that 20.24 hectares/minute are being destroyed. What can you, your school, and your community do to help save this endangered biome?

10. Choose any biome and compare the plants and animals in that biome in different parts of the world. Are the species the same all around the earth?

11. Movies and television shows that are filmed "on location" use the natural vegetation as a backdrop. The next time you view a movie or TV program, check to see in what type of biome the filming was done. Is the vegetation consistent with the location identified in the script?

12. How has your community's ecosystem changed through time? Visit a local museum or library to find out what types of plants grew in your community 1000 years ago, 100 years ago, or even 10 years ago. What types of plants replaced the native plants?

13. Use BSCS, *Land Use* (Dubuque, IA: Kendall/Hunt, 1984) to study a land use issue in your community. What is the problem, what are alternative solutions, and how would you manage the problem?

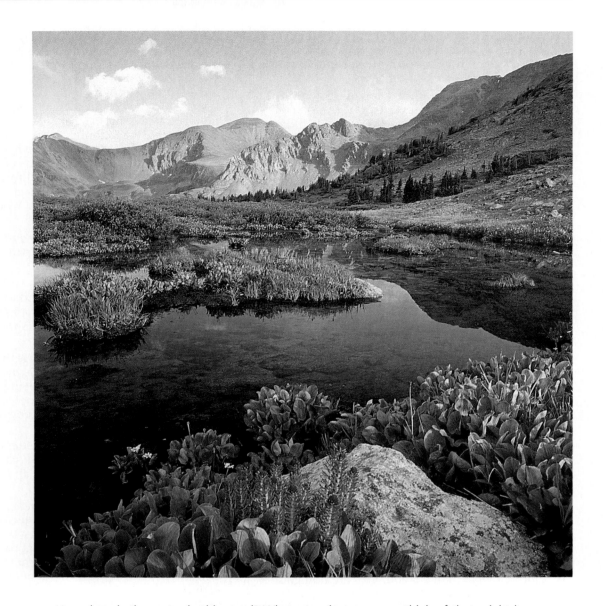

How deep is the water in this pond? What organisms can you think of that might be able to live here? Where do you think this pond is found? What are some limiting factors in this particular biome?

Aquatic Ecosystems

The earth, when viewed from outer space, is blue with water. All this water makes up the hydrosphere (HY droh sfir), just as all the air makes up the atmosphere. The hydrosphere is a vast heat reservoir that absorbs, stores, and circulates the heat that results when sunlight strikes the earth. The hydrosphere also is a reservoir of chemical elements and compounds that are continuously dissolved in water that eventually drains into the oceans. The oceans contain 97 percent of all water found on earth and form the great interconnecting water system that surrounds the continents, covering 70 percent of the face of the earth.

Many inland waters are found on the surface of the land. Inland-water ecosystems are classified either as standing waters or flowing waters, although the boundary between these classifications often is not distinct. Standing waters range in size from roadside puddles to the Caspian Sea, which occupies more than 350 000 km². These waters also range from very shallow to very deep, from clear to muddy, and from fresh to salty. A river is an example of flowing water, yet in some places a river may have such a slow current it is difficult to see any movement at the surface.

Although humans are terrestrial organisms, they always have lived close to a source of fresh water. Throughout the world, villages, towns, and cities have developed close to bodies of water, which have supplied food, transportation, and a waste removal system. This chapter examines the major aquatic ecosystems and the impact human activities have on them.

Standing Fresh Waters

Guidepost: What are the characteristics of ponds and lakes?

23.1   Ponds Are Shallow Enough for Rooted Plants

A pond is an example of standing water, yet many ponds are fed by springs or brooks, and many have outlets. Ponds usually are defined as small, completely enclosed bodies of water that are shallow enough so that light penetrates to the bottom. Rooted plants can grow throughout a pond, and temperatures from top to bottom are relatively uniform during the warm months of the year. Whether ponds are natural or artificial, their characteristics are much the same. Figure 23.1 shows a cross section of a typical pond in the northeastern United States.

oaks and maples

willows and alders

muskrat

burreeds and arrowheads

FOREST   SHRUBS   EMERGEN

Figure 23.1   Cross section through the edge of a natural pond in the northeastern United States. Where is the boundary of the ecosystems?

Figure 23.2   Plankton.

In studying any ecosystem, an ecologist first looks for its source of energy. In a pond, the most important producers are not the most obvious ones. The large rooted plants, such as cattails and rushes, are conspicuous, but they produce little of the ecosystem's food. Indeed, in some ponds—especially artificial ones—rooted plants may be scarce or absent. Many plants live within the water itself—some floating on the surface, some submerged. These plants may become so numerous that they hamper a swimmer, yet even these are not the most important pond producers.

Light shining into a pond reveals moving specks that look like dust in a beam of sunlight. These specks, shown in Figure 23.2, are living organisms called **plankton** (PLANK ton), which may belong to any of the five kingdoms. Few of these organisms actually can swim, but they are carried about by the currents in the water. Plankton make up for their small size by their incredible numbers; research shows that most food production in all parts of the hydrosphere—not just in ponds—is the result of photosynthesis by plankton.

The plankton producers collectively are called phytoplankton (FYT oh PLANK ton). Different types of phytoplankton vary in their abundance from one body of water to another. Diatoms usually are the most numerous, although not the most conspicuous. Green algae may become so abundant in late summer that the entire surface of a pond appears green.

Microscopic herbivores, called zooplankton (ZOH oh PLANK ton), include a variety of protists and also tiny animals such as rotifers and small

ils. and bulrushes

egret   frog

water snake   water lilies

pondweeds

pickerel

ANTS | FLOATING PLANTS | SUBMERGED PLANTS

crustaceans. Zooplankton are not the only pond herbivores, however. Many young fishes, such as minnows, and bottom dwellers, such as mussels, also are herbivores. Other pond fishes are carnivores; the larger fishes eat the smaller ones. Fishes of all sizes eat aquatic insects, many of which are carnivores themselves. Thus, pond food chains have numerous links. Most of the macroscopic carnivores stay near the edge of a pond, where floating plants provide a hiding place. Larger predatory fishes swim throughout the entire pond.

Layers of organic matter—mostly dead organisms and waste products—build up on pond bottoms. Decomposers such as tubifex worms (annelids) reduce the dead organic matter to smaller bits and pieces, which are further broken down by other decomposers such as fungi and bacteria.

It often is difficult to determine the boundaries of an ecosystem. In a pond ecosystem, the surface of the water is not a distinct boundary. Insects usually are flying above and around ponds. Many of them hatch from eggs laid in the water and spend most of their lives there. A mayfly's entire life, for example, is spent in the water except for the last few hours, when the adults mate in the air. The females then lay eggs in the water, and the immature mayflies grow there for a year or more. Figure 23.3 shows both the adult and immature forms of a mayfly. Frogs and many fish live primarily on these insects. Water snakes eat fish and frogs. Many species of birds and some mammals feed on fish, crayfish, and mussels. These consumers, although they spend a great deal of their time on the land, actually are a part of pond ecosystems. Their supply of energy can be traced back to the phytoplankton.

Figure 23.3 The mayfly lives its entire life in and near the water. Nymph (top) and adult (bottom).

INVESTIGATION 23.1   Water Quality Assessment

How clean is the water in a lake, pond, or stream near your school? Technical tests can determine exact types and amounts of water pollution, but a simple survey of algae and cyanobacteria can be a reliable index of water quality. Clean water usually has many different species of algae and cyanobacteria, and no single species predominates. Polluted water has fewer species and high numbers of each. This investigation consists of three parts: learning to identify key genera of algae and cyanobacteria; testing known samples of clean, intermediate, and polluted water; and testing a local water source.

Materials (per team of 2)

3 microscope slides
3 22-mm × 22-mm cover slips
dropping pipet
1 L beaker
plankton net
compound microscope
protist/cyanobacteria key
A Catalog of Living Things
2-cm² piece of graph paper with 1-mm squares

4 flat toothpicks
pocket calculator (optional)
Detain™ in squeeze bottle
water sample with algae and
    cyanobacteria
water samples—clean, intermediate, and
    polluted

Procedure
PART A   Field of View Calculations

1. Place the 2 cm² piece of graph paper on a microscope slide and add a cover slip. Examine the slide with the low power (100X) of your microscope. Align the squares so one edge of a square just touches the bottom edge of the field of view. Count the number of squares from the bottom to the top of the field (see Figure 23.4a). The number of squares equals the diameter of the field of view in millimeters. In your data book, record the results and the low-power magnification.

2. Repeat Step 1 with the high power (400X). (Do not use the oil immersion lens.)

3. You will count individual organisms on a slide made from your water samples. To do so, you will move the slide across the microscope stage, the width of one field of view at a time, as illustrated in Figure 23.4b. The area of the strip of fields of view equals the diameter of the field of view multiplied by the width of the cover slip (22 mm). Practice moving the slide one field of view at a time, going from one side of the cover slip to the other.

PART B   Algae/Cyanobacteria Identification

4. Place a drop of Detain™ (protist slowing agent) on a clean microscope slide. Using a dropping pipet, add one drop (0.1 mL) of a sample, and use a toothpick to mix the two drops. Carefully add a cover slip.

Figure 23.4   (a) Microscopic view of 1 mm² graph paper at X100, field diameter = 2 mm. (b) The area of a strip = the length of the cover slip (22 mm) × diameter of the microscope field.

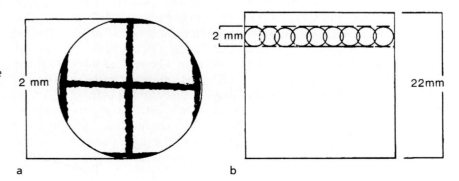

a   b

5. Practice locating and identifying different organisms using the protist/cyanobacteria key and *A Catalog of Living Things*. Locate organisms under low power of the microscope and then use high power to make a positive identification.

6. Move the slide, one field of view at a time, going from one side of the cover slip to the other. In each strip, count the numbers of individuals for each genus you identify. Ask your teacher for help if you cannot identify some of the organisms.

7. Wash your hands thoroughly before leaving the laboratory.

PART C   Identification of Water Samples

8. Table 23.1 lists the pollution index of each genus used to determine water quality with the Palmer method. You will use this information to determine the pollution index of several samples of water.

9. For each of the three water samples, you will count the number of organisms in each genus in three strips of fields of view. Then you will compute the pollution index for each sample in order to decide which sample is clean, which is intermediate, and which is polluted. In your data book, prepare four worksheets with the headings in Table 23.2 or tape in the worksheets your teacher provides. Use the worksheets for your calculations.

10. Put one drop of Detain™ and one drop (0.1 mL) of Sample 1 on a slide. Mix with a clean toothpick. Add cover slip. In your worksheet, note the sample number.

11. Count and record the number of individuals from each genus present in a strip of fields of view as you move the slide from one side to the other. Count *units* (colonies or filaments are counted as one unit). Large filaments and colonies only partly lying in the strip should be counted as fractions. For example, if a filament of *Stigeoclonium* is only half in your field of view, record 0.5 instead of 1.

12. Repeat Steps 10 and 11 two more times for a total of three strips.

13. The calculations for the Palmer score assume the sample from which you extracted the organisms was 1 L (1000 mL) of water that was filtered and concentrated in 100 mL of water. Study the example in Table 23.2. The first calculation (column F) determines the number of individuals from each genus in your 0.1-mL concentrated sample. The second calculation (column I) calculates the number of units per mL in the original 1-L sample. *You must obtain at least 50 units per mL for the 1-L sample or you cannot enter a Palmer score for that genus*. If you are having difficulty with the calculations, ask your teacher for help.

14. After obtaining the final count for each genus with a frequency of at least 50 units per mL, look up the pollution index for that genus and enter it in your worksheet. (Only genera included in the Palmer list are scored as indicator organisms.) A total score (the sum of the Palmer scores for all species in a sample) of 20 or more for a sample is evidence for high organic pollution. Scores between 15 and 19 indicate moderate pollution, and scores less than 15 indicate absence of pollution.

TABLE 23.1 Algae/Cyanobacteria Genera Pollution Index (Palmer 1969)

Genera	Pollution Index
Anacystis	1
Ankistrodesmus	2
Chlamydomonas	4
Chlorella	3
Closterium	1
Cyclotella	1
Euglena	5
Gomphonema	1
Lepocinclis	1
Melosira	1
Micractinium	1
Navicula	3
Nitzschia	3
Oscillatoria	5
Pandorina	1
Phacus	2
Phormidium	1
Scenedesmus	4
Stigeoclonium	2
Synedra	2

TABLE 23.2   Worksheet for Determining Water Quality

Genus	Number of Cells or Units Strip 1	Strip 2	Strip 3	A Total	B Area of Sample	C Strip Area	D Number of Strips	E Volume of Sample	F $\frac{A \times B}{C \times D \times E}$	G Sample Volume	H Water Filtered	I $\frac{F \times G}{H}$	Palmer Score
(Example) Oscill	卌 卌 卌	卌 卌 0.2	卌 卌 III	38.2	484 mm²	44 mm²	3	0.1 mL	1400/mL	100 mL	1000 mL	140/mL	5

15. Repeat Steps 10 through 13 for Samples 2 and 3.
16. Check with your teacher to determine whether your total Palmer scores correctly identify the water quality of the three samples.

PART D   Investigation of a Local Water Source

17. Using water from a local pond or lake, filter 1 L through a plankton net to trap the organisms. Rinse the organisms from the net into the 1-L beaker and add clean water to total 100 mL. Repeat Steps 10 through 13 to determine the pollution index.
18. Wash your hands thoroughly before leaving the laboratory.

Discussion

1. Which genera can tolerate the highest levels of organic pollution?
2. If a genus has a Palmer score of 1, what does that indicate?
3. If your local water source has moderate or high pollution, what might be the source of pollution?
4. How can algae and cyanobacteria be helpful in polluted water?
5. How can algae and cyanobacteria be harmful in polluted water?

23.2   Lakes Are Larger and Deeper than Ponds

Lakes, such as the one illustrated in Figure 23.5, are larger than ponds and have some deep areas that remain dark. Most of the world's lakes were formed by the action of glaciers and, through the process of succession, eventually will become terrestrial ecosystems, as shown in Figure 23.6. Phytoplankton are the major producers in lakes, as they are in ponds. The phytoplankton of a lake, however, can exist only near the water's surface, where there is enough light for photosynthesis. In deep lakes, therefore, all the food is produced in the upper part of the water.

Figure 23.5 A mountain lake. What are the environmental conditions here, and what are their effects?

Figure 23.6   An example of lake succession. At what stage is the lake apt to be nutrient-rich?

   The existence of aquatic consumers is limited not by light but by the amount of oxygen in the water. Water at the surface of a lake constantly receives dissolved oxygen from photosynthetic organisms and from the air. This oxygen diffuses slowly downward from the surface, but most of it is used up by zooplankton before it diffuses very far. The spread of oxygen into deep water thus depends on other factors.

   During the summer, the sun heats the lake surface and the water expands, becoming lighter or less dense than water at the lake bottom. This expansion forms a warm water layer that floats on the colder water below.

a

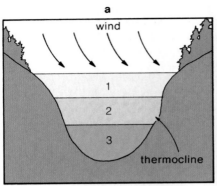

layers formed during midsummer

b

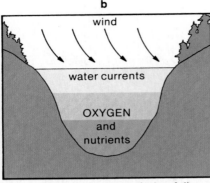

mixing of water currents during fall and early spring in a northern lake

c

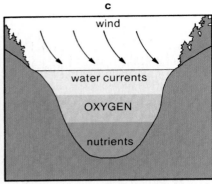

a northern lake in late spring and early summer.

Figure 23.7   The formation of a thermocline in a lake.

Figure 23.8   Mono Lake in California is an example of the salinization of a lake. Salt accumulation is responsible for the large tufa formations on the right. The water is so saline that only brine shrimp live in it.

Separating these layers is a zone, called the thermocline (THER moh klyn), across which both temperature and oxygen concentration drop sharply. The thermocline prevents vertical mixing of the upper and lower layers of water, resulting in important differences between these layers. The concentration of dissolved oxygen remains high in the upper layer because of photosynthesis but decreases in the lower layer as oxygen is used up by consumers, thus restricting the activities of aerobic decomposers. As a result, the dead bodies of organisms that fall from above are only partly decomposed, and the nutrients produced through decomposition remain on the bottom of the lake.

In the fall, as the air temperature drops, the surface water becomes colder and more dense. The heavier surface water sinks, forcing deep water to the top. Wind striking the surface helps mix the water and distribute oxygen throughout the lake. During winter, the surface of the lake may freeze, and a warmer layer remains at the bottom. The oxygen concentration declines, but anaerobic decomposers continue to release nutrients from dead organic matter. In the spring, rising temperatures again produce movement of the waters, which mixes oxygen and nutrients throughout the lake. Figure 23.7 illustrates this mixing of water layers at different times of the year.

If a lake has no outlet, nutrients washed in from the surrounding land become concentrated in the water because only water molecules can evaporate from the surface. The water may become saturated with nutrients, the most abundant of which is usually sodium chloride. The buildup of salt and other compounds in a lake, called salinization (sal ih nih ZAY shun), takes place faster in arid regions where evaporation is more rapid, resulting in an environment unfavorable to most organisms, such as that shown in Figure 23.8.

Some of the water that falls on land runs directly into lakes, ponds, and streams; some evaporates. Most of the water, however, soaks into the ground. This groundwater reappears in springs, from which the course to the ocean may be short or long. This is the way water cycles in the biosphere, as shown in Figure 23.9.

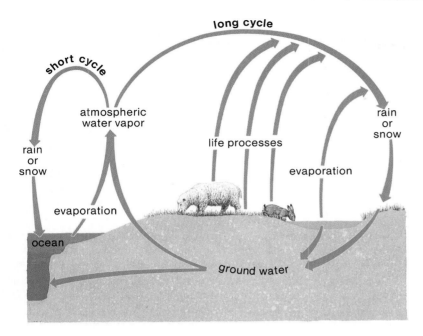

Figure 23.9   The water cycle. Water is necessary to life in many ways. Land plants absorb water from the soil, and land animals drink water or obtain it from their foods. Water constantly bathes organisms that live in ponds, lakes, rivers, and the oceans. Water also evaporates from the surface of all bodies of water and from land. Plants and animals also lose water, but through different processes. All the water comes from the nonliving environment. What is the difference between the long and the short cycles?

Concept Review

1.   In what ways are the inland waters different from one another?
2.   How do phytoplankton affect a pond ecosystem?
3.   How is a pond ecosystem linked to surrounding terrestrial ecosystems?
4.   What is a thermocline, and how does it affect the distribution of oxygen and nutrients in lakes?

Flowing Waters

Guidepost: How do flowing water ecosystems differ from those of standing water?

INVESTIGATION 23.2   A Field Study of a Stream

The surface of bodies of water usually hides the diversity and activity of life below. Animals may live in the stream or just come to the stream to use its resources. Dead plants and animals that fall into the running water are used as food by stream dwellers. Underwater, an entire community of plants, animals, and other organisms interact with one another and their environment. This investigation examines the types of organisms that inhabit a local stream, and the physical environment in which they are found.

Materials (per team of 5)
2 10X hand lenses
nonmercury thermometer
3 pairs of forceps
stopwatch or wrist watch with second hand
6 30-cm wooden or metal stakes

100-m measuring tape
2 laminated black-and-white cards
5 pairs of thick cotton gloves
5 pairs of waders, hip boots, or old sneakers
1-m × 0.75-m screened frame

(continued)

Materials (continued)

1-m × 0.75-m screenless frame   keys for aquatic organisms
plankton net (optional)   data books or clipboards
2 large trays filled with stream water   lemon

PART A   Stream Flow and Physical Characteristics
Procedure

1. Mark a 50-m section of a stream with the six stakes, 10 m apart. At each of the six stakes, measure the width of the stream and its depth midway between the two banks. Determine the average depth and width.

    CAUTION: Do not walk into the stream with bare feet or sandals. Wear waders, hip boots, or an old pair of sneakers.

2. Determine the area of the stream by multiplying the average depth by the average width.

3. To estimate the velocity of the water, toss the lemon into the water upstream from the first stake. Measure the time it takes for the lemon to travel from the first stake to the last. Repeat this step twice and average your results. Remove the stakes.

4. Determine the rate of flow of water through this section of the stream by multiplying the average velocity by the average area.

5. Determine the water temperature at several locations and record these data.

6. Make note of other physical characteristics of the stream, such as whether the bottom is rocky or sandy.

Discussion

1. What is the average width of the stream? What is its average depth?

2. What is the area of the stream?

3. What is the average time it takes for the lemon to travel 50 m?

4. What is the volume of the water that passes by any stake per minute? (Biologists record these data in cubic feet.) What is the rate of flow?

PART B   Stream Organisms
Procedure

7. Put on your gloves. Place the screenless frame on a flat part of the stream bottom so the entire frame rests on the streambed. Within the frame, estimate the area that is covered by algae, leaf litter, submerged plants, emergent plants, and the area that is bare. Refer to Investigation 22.2 for information about estimating ground cover.

8. Place the screened frame upright immediately downstream from the open frame so the two frames touch and the water will pass through the screen, as shown in Figure 23.10. If a plankton net is available, hold it a little downstream from the screened frame to collect smaller organisms.

9. Disturb the substrate within the screenless frame, turning over rocks and stirring up gravel. Dislodge any organisms found on the rocks or underneath them. Allow these organisms to flow downstream to be caught on the screen. One team member should hold the screenless frame on the streambed, one should hold the screened frame upright, one should hold the plankton net, and one should turn over rocks.

10. The fifth team member should examine rocks carefully for any invertebrates clinging to them. Place rocks and gravel into one tray on the bank for further examination.

11. Carefully lift the screened frame with its collected material. Try not to let any material be washed away before observing it. Using forceps, transfer any organisms on the screen into the trays of fresh stream water on the bank. Also transfer any organisms in the tray containing the gravel and rocks.

water flow

streambed

Figure 23.10 Screen setup.

 CAUTION: Do not touch any of the organisms with your bare hands. Be careful not to damage the organisms by squeezing them too tightly with the forceps. Forceps are sharp; handle with care.

12. Identify the organisms and divide them into groups based on what they eat. Your teacher may have you develop your own classification system for the organisms you collect. One suggested grouping is the following:
    a. Collectors—animals that live on decomposing organic matter.
    b. Grazers—herbivores that scrape algae off rocks.
    c. Predators—animals that eat part or all of other animals.
    d. Shredders—herbivores that chew on plant material, or mine or bore through leaves.

13. When you have finished viewing and counting organisms, return them to the stream.

14. Place a drop or two of the water from the jar at the end of the plankton net on the laminated card and examine this sample using the hand lens. Identify the organisms present.

Discussion
1. What types of organisms did you collect?
2. Compare your results with those of other teams in your class.
3. Determine the number of organisms collected per m².

PART C Hypothesis Forming and Testing
Procedure
15. After making observations of the stream and its surroundings, make a list of factors that might affect life in the stream. Does the rate of water flowing downstream make a difference to stream life? For example, do some organisms prefer fast-moving water? Are there many plants growing along the banks or in the water, and what type are they? How important to life in the stream is the amount of rock, gravel, or sand on the bottom? How do physical conditions above the water affect conditions in the water? What abiotic and biotic factors may affect life in the stream? Which of these factors might be related to the types of organisms you collected from the stream?

Discussion

1. What were the most common organisms you found? Are they found in the same abundance throughout the stream? If not, why not? What factor do you think is important in controlling the population size of the most common organism? Develop a hypothesis about the types of organisms you found and one factor that might affect their lives and their abundance within the stream. How would you design a study to test your hypothesis?

2. Each team should develop a different hypothesis to study. Return to the stream to conduct your investigation and report your findings to the entire class.

For Further Investigation

Colonization studies can yield additional information about microorganisms in the stream ecosystem. Place a commercial sponge in the streambed. After about two weeks, remove the sponge, place it in a jar, and return it to the lab for examination. Suspend glass microscope slides in the stream, remove them two weeks later, and return them to the lab for staining and examination.

23.3   Stream Size and Flow Affect Aquatic Organisms

Figure 23.11 A mountain brook. What evidence can you see that this aquatic ecosystem has few producers?

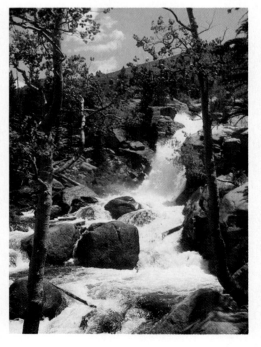

Most springs give rise to small brooks. Such flowing water usually is cool, although some hot springs do exist. Tumbling through rapids and falls, the water traps many air bubbles from which oxygen easily dissolves. Because cool water can hold relatively large amounts of gases in solution, the water in brooks usually contains much oxygen.

Plankton generally are absent in the swift-flowing water of brooks (Figure 23.11) because they are swept away. Producers, such as cyanobacteria, diatoms, and water mosses, grow on the stones at the edge or on the bottom of the brook. These producers provide some food and much shelter for aquatic insects. The insects, in turn, are food for small fish. Most of the food supply in a brook ecosystem comes from the land around it. Small insects fall in the brook and are eaten by stream inhabitants. Every rain washes in dead organic matter. Anything not used immediately is washed downstream, and very little food is left for decomposers.

Brooks meet and form larger streams with wider beds. The water in streams usually moves more slowly, and solid substances carried along by the swift brooks are deposited as sediments. Bits of organic matter that accumulate among the sediments provide food for decomposers. As the stream widens, the relative amount of water surface shaded by trees and shrubs along its banks decreases, and direct sunlight reaches most of the water surface. Some phytoplankton organisms live in this slower-moving water, but many still are carried downstream. Rooted plants grow in the sediments of stream bottoms, although they may be uprooted and washed away during floods.

Streams support a larger number of consumers than do brooks because they contain more producers. Mussels, snails, crayfish, and many immature insects inhabit the bottom. Larger consumers, such as turtles and fish, are dependent on these bottom dwellers. Figure 23.12 illustrates some typical stream consumers.

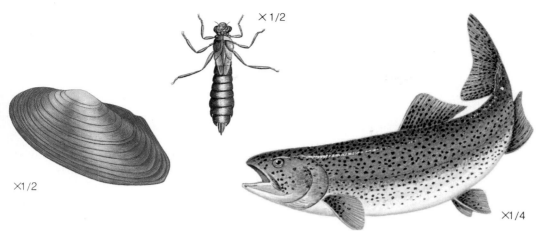

×1/2

×1/2

×1/4

Figure 23.12 Stream consumers (left to right): freshwater mussel, dragonfly nymph, and rainbow trout.

Many streams come together to form a river. As a river approaches the sea, it usually moves more slowly and deposits larger amounts of sediments. Thus, instead of eroding the land, a river often builds up land near its mouth, as does the Mississippi River, shown in Figure 23.13. Riverbanks actually may become higher than the land behind them. During floods a river often breaks through these natural levees. The water left behind is slow to drain away and forms a swamp where many rooted and floating plants grow. Subsequent floods may sweep fruits, seeds, and other plant parts back into the river and thus contribute to the food supply of the river ecosystem. Phytoplankton, which grow well in the unshaded, slow-moving swamp water, supply much of the food for consumers in large rivers.

Figure 23.13 Part of the Mississippi delta. What major types of ecosystems can you see here?

Consumers in rivers are varied and numerous. Zooplankton are food for larger consumers. Mollusks, crustaceans, and fish often grow very large. Crocodiles are common in tropical rivers. Many birds and mammals obtain their food from rivers. Humans also have long taken advantage of the abundant food in rivers.

23.4   Flowing Waters May Serve as a Laboratory

Not all springs give rise to brooks. Some springs produce so much water that large streams flow from them. Many of the short rivers of northwest Florida, for example, originate in this manner. These short Florida rivers are excellent outdoor laboratories. Because these rivers have areas in which the water volume, current, chemical composition, and temperature are stable, ecologists have used several of them in studies of ecological productivity.

The productivity of any community depends on the photosynthetic rate of its producers. To determine productivity indirectly, biologists first measure the amount of oxygen produced during a given time. Because photosynthetic organisms give off oxygen in proportion to the amount of organic substance they produce, biologists can use this figure to calculate the amount of organic substance produced by photosynthesis during the same period.

Any increase in the amount of oxygen in the spring water as it flows downstream comes largely from photosynthesis. Figure 23.14 shows the oxygen content measured at given places along the river flowing from Rainbow Springs. The biologists also measured the amount of water that flowed past each point every day. Based on this information, they calculated the total productivity of the water between its source in the springs and the point of measurement. This productivity is expressed as grams of organic substance produced per square meter of stream surface per day. The investigators found an average productivity of 17.5 g/m²/day.

In further studies the investigators estimated the **biomass**—the total mass of all individuals of a species in a given area—for each major species in the community. Figure 23.15 shows the result of biomass measurements

Figure 23.14   Diagram of Rainbow Springs, Florida, and the river that flows from them. Ecologists measured the amounts of oxygen in the springs and in the river water at several places downstream. Such measurements are expressed as parts per million (ppm). One ppm = 0.0001 percent.

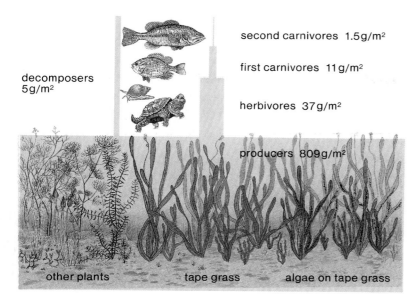

decomposers
5g/m²

second carnivores 1.5g/m²

first carnivores 11g/m²

herbivores 37g/m²

producers 809g/m²

other plants   tape grass   algae on tape grass

Figure 23.15 Average biomass measurements from Silver River, Florida. Largemouth bass are top carnivores. Compare this figure with Figure 3.12.

in the Silver River, another short Florida river. Such measurements help ecologists understand the interrelationships in an ecosystem.

Concept Review

1. How do abiotic conditions in a brook differ from those in a river?
2. What is the source of energy for organisms in a brook?
3. Why is plankton more abundant in a river than in a brook?
4. How do biologists measure productivity?
5. What usually happens to biomass between one link in a food chain and the next?

Ocean Ecosystems

Guidepost: What factors affect ocean ecosystems?

INVESTIGATION 23.3   Effects of Salinity on Aquatic Organisms

If you were to move freshwater organisms into the ocean or ocean organisms into fresh water, they probably would die very quickly. Although salmon migrate from the open ocean into freshwater rivers in order to reproduce, they first spend several days in water with decreasing salinity. In this way, their bodies gradually adjust to the lower salt concentrations.

Small freshwater ponds become salty as the summer sun evaporates the water. Small aquatic organisms that cannot tolerate the change either die or go into a dormant stage.

This investigation examines the tolerances of some freshwater organisms to various concentrations of salt solutions.

Review the procedure and construct hypotheses predicting the reaction of your assigned organisms to changes in salinity.

Materials (per team of 2)

2 microscope slides
2 cover slips
dropping pipet
compound microscope
paper towel
5% NaCl solution in dropping bottle

aged tap water
living specimens of small aquatic
    organisms in aged tap water and in
    1%, 3%, and 5% NaCl solutions
elodea or filamentous algae

Procedure

PART A   Plants and Salinity

1. Remove an elodea leaf from near the tip of the plant and place it upside down on a microscope slide. Add a drop of aged tap water and a cover slip.

2. Use the low power of your microscope to find a cell you can observe clearly. Describe the cell contents. Does the cell seem to be completely filled?

3. Add a few drops of 5% NaCl solution to one side of the cover slip. Draw the solution under the cover slip by applying a small piece of paper towel to the opposite side of the cover slip. Describe what you observe.

4. Rinse the dropping pipet. Determine if the process is reversible by drawing aged tap water under the cover slip. Record any changes you observe.

PART B   Animals and Salinity

5. On a clean slide, place a drop of the freshwater sample assigned by your teacher. Add a cover slip.

6. Use the low power (100X) of your microscope to observe the organisms for a few minutes to determine their normal appearance and actions.

7. Prepare a slide of the organisms in 1% NaCl solution. Observe the organisms and record any difference in movements and shape from the freshwater sample.

8. Repeat Step 7 with the organisms in 3% and 5% NaCl solutions. Which of these solutions is most like ocean water?

9. Replace the 5% NaCl solution with water as you did in Step 4. The organisms may be drawn up by the paper towel. If any organisms remain on the slide, record any changes in movement or shape you observe.

Discussion

1. Did all individuals of the species react in the same way? If not, what differences were noted?

2. Did the type of reaction differ with different NaCl solutions? Try to explain any differences.

3. Compare observations of teams that worked on different types of organisms. Which type was most tolerant to changes in NaCl concentration? Which was least tolerant?

4. What type of aquatic habitat do you think each type of organism normally inhabits?

5. Which of your observations support your hypothesis?

6. Do any of your observations weaken your hypothesis?

7. Taking all your observations into account, draw a conclusion that relates your hypothesis to the experiment.

23.5   Many Abiotic Factors Influence the Ocean

Ocean environments, such as the one shown in Figure 23.16, differ in many ways from inland-water environments, but the principal difference lies in the chemical composition of the water itself. Seawater is about 3.5 percent dissolved substances, most of which are salts. Because sodium chloride accounts for more than 75 percent of the substances present, the content of dissolved substances in seawater is referred to as salinity (suh LIN ih tee). Salinity varies somewhat at different depths and in different places on the surface. Table 23.3 shows the average content of various elements in seawater.

Evidence from rocks and fossils shows that oceans have had a high salinity for hundreds of millions of years. Dissolved substances constantly washed into oceans from the land appear to have had only a slight effect on the composition of the water. Ocean water represents a steady state. Substances are added to it continually, but substances also are removed continually by marine organisms and remain in the accumulated skeletons of these organisms for long periods of geologic time. Ocean water is the environment of marine organisms, but it is also a product of their activities. The hydrosphere, like the atmosphere, undoubtedly would have a much different composition if life were absent.

Rapid changes in the atmosphere result in great differences in the climates of terrestrial ecosystems. Such rapid changes do not occur in the hydrosphere. Temperature changes between day and night and between the seasons are very small at the ocean surface. Although the temperature of the surface water at the equator is very different from that of antarctic and arctic waters, those differences diminish with increasing depth. At great depths the oceans have uniform temperature.

TABLE 23.3   Average Content of Different Elements in Seawater*

Element	Seawater (Parts per thousand)
Chlorine	18.98
Sodium	10.56
Magnesium	1.27
Sulfur	.88
Calcium	.40
Potassium	.38
Bromine	.065
Carbon	.028
Strontium	.013
Silicon	.003
Fluorine	.001
Aluminum	.0005
Phosphorus	.0001
Iodine	.00005

*These elements occur as parts of compounds.

Figure 23.16   Drake Passage between Tierra del Fuego and the Antarctic pennisula.

Figure 23.17   Penetration of seawater by sunlight. Why does the water appear blue?

The amount of radiant energy available to photosynthetic organisms is greatest at the water's surface. The depth to which light penetrates into the water depends on several factors, including the angle at which light strikes the water's surface and whether the surface is smooth or broken by waves. Another factor is the rate at which the water absorbs the light. As light passes through the water, the longer wavelengths are absorbed most rapidly. Red and yellow disappear first; blue and violet penetrate farthest (Figure 23.17). The rate at which water absorbs light also depends on the clarity of the water. The cloudier the water, the faster the rate of light absorption.

Ocean currents distribute chemicals that are useful to organisms. Currents also affect water temperatures and salinities at any given place in the ocean, and they affect climate (see Chapter 22). Currents, in turn, are affected by the world pattern of winds and by the earth's rotation.

23.6   Productivity Is Limited in the Open Ocean

The chief producers of the open ocean are diatoms, other microscopic algae, and certain flagellates. The zooplankton depend directly on these phytoplankton. Many food chains are based on these producers and include large consumers such as tuna, sharks, whales, and oceanic birds such as albatrosses. The oceans vary greatly in productivity. One study of phytoplankton population density, conducted at a laboratory in Plymouth, England, found that the ocean water there contains, at the very least, 4.5 million phytoplankton organisms in each liter of water. Biologists do not have enough data to give accurate density averages for the ocean as a whole.

A limiting factor in the ocean, as on land, is the availability of chemical elements, especially phosphorus and nitrogen. Phytoplankton constantly

Figure 23.18   Diagram of an ocean in cross section. What is the most productive zone?

exhaust the supply of these elements, and their continued growth depends on constant resupply. Elements may be added from sediments washed from the land or from water welling up from below. Upwelling may result from seasonal changes or from offshore winds moving surface water and deep water rising to take its place. Because of upwelling, ocean areas at higher latitudes and close to the land—such as the North Sea and the Grand Banks of Newfoundland—are very productive. These areas of upwelling are small, however, compared with the entire ocean. Fifty percent of all fish and shellfish harvested from the ocean are taken from 0.1 percent of its surface. In contrast, ocean regions far from land, especially those in the tropics, have very low productivity.

Regardless of the availability of nutrients, phytoplankton need light. All of the photosynthesis in the open ocean occurs in the upper layer of water, where light penetrates beneath the surface. This layer is very thin compared with the total depth of ocean water, as shown in Figure 23.18. Despite the deep penetration of blue light (550 m near Bermuda), phytoplankton are largely limited to the upper 100 m of water.

23.7   Ocean Depths Support Some Forms of Life

For a long time biologists thought life could not exist in the dark, cold ocean depths because of the tremendous pressure of the water. The first clear evidence that this idea was wrong came in 1858, when one of the telegraph cables lying on the bottom of the Mediterranean Sea broke and was hauled up for repair. It was encrusted with bottom-dwelling animals, mostly sponges, some of which had grown at depths as great as 2000 m. Water pressure has no ill effect on organisms unless they contain spaces filled with air or other gases. Because deep-sea organisms lack such spaces, pressure is exerted equally from both inside and outside the organism, thus preventing damage.

The ocean depths are special ecosystems that require unusual adaptations in the organisms that survive there. The depths are cold, dark, and quiet, and food is scarce. Life depends on organic substances that settle from the water above. The adaptations of organisms to the pressures of the deep make their ascent to upper levels fatal. Thus, the consumers in the ocean depths form one of the most isolated communities of the biosphere.

In the eternal darkness of the ocean depths, most animals are either black or dark red and have very sensitive eyes. Many are bioluminescent (by oh loo mih NES ent)—they produce their own light. One species of fish dangles a special luminescent organ in front of its mouth that apparently lures unwary prey close enough to be caught and eaten. Deep-sea shrimp escape some predators by emitting clouds of a luminescent secretion they give off when disturbed. Patterns of light on an animal's body also may help organisms recognize one another in the dark. Figure 23.19 shows some fishes that live in the deep ocean.

Figure 23.19 Some fishes of the deep ocean and the depths at which they have been caught. All are small. They are shown here close to actual size.

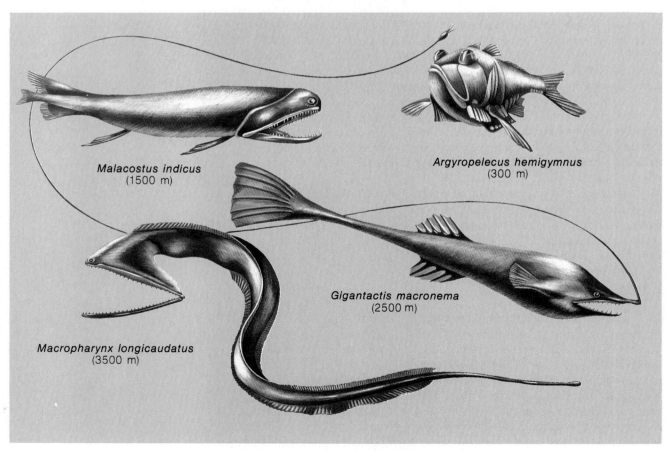

Malacostus indicus
(1500 m)

Argyropelecus hemigymnus
(300 m)

Gigantactis macronema
(2500 m)

Macropharynx longicaudatus
(3500 m)

Nearly all life on the earth depends on radiant energy that is converted to chemical energy by photosynthetic organisms. A few organisms, however, are not dependent on photosynthesis or photosynthetic organisms for their chemical energy. A type of ecosystem based on chemosynthesis was discovered by researchers aboard the submersible *Alvin* in 1977 at the Galápagos Rift, a boundary between tectonic plates. Seawater seeps into the vents created by volcanic rocks erupting along this boundary, becomes heated, and rises again. As a result, oases of warmth are created in the normally near-freezing waters. More important, the water reacts chemically, under extreme heat and pressure, with the rocks deep in the earth's crust. The crucial reaction appears to be the conversion of sulfate (SO_4^{2-}) in the seawater to sulfite (S^{2-}). Heat supplies the energy necessary for the reaction. Chemosynthetic bacteria obtain energy by converting the sulfite back to sulfate. These bacteria use the energy to extract carbon from carbon dioxide in the seawater and incorporate it in organic compounds, thereby becoming the primary producers in the ecosystem.

About a dozen such vent ecosystems have been identified, each with a somewhat different animal community. Among the most spectacular inhabitants are clams measuring some 20 cm in diameter. Other permanent inhabitants include crabs and mussels. The clams and mussels feed on the chemosynthetic bacteria and are, in turn, fed on by the crustaceans. One oasis, called the Garden of Eden, is dominated by huge tube worms, shown in Figure 23.20. Apparently these worms, which lack mouths and digestive tracts, utilize carbon compounds synthesized by chemosynthetic bacteria living within them.

Figure 23.20   Among the organisms that live in vent ecosystems are giant tube worms, *Riftia pachyptila,* many more than a meter long.

23.8   Coastal Waters and Coral Reefs Support a Diversity of Life

With few exceptions, oceans are relatively shallow near the continents. These bands of shallow water, which average less than 200 m in depth, are widest at the mouths of large rivers and along areas of broad lowlands. Along mountainous coasts, as in California, shallow waters may be almost absent.

In shallow water some light reaches the bottom. Where there is plenty of light, a luxuriant growth of algae may be found. In middle and higher latitudes, the brown algae are most common and conspicuous. Among the brown algae are kelps (Figure 23.21), which may reach lengths of 35 m or more.

The physical characteristics of the ocean bottom—sand, mud, or rock—determine the types of organisms that live there. Plants usually are not abundant on unstable sandy bottoms, but many types of animals burrow into the sand, especially crustaceans, mollusks, and annelid worms. Muddy bottoms have even more burrowers, and most of the species are unlike those adapted to sand. Sea cucumbers, clams, and some types of crabs are common in the mud.

Because they are shallow and close to land, coastal waters offer more opportunity for human use than does the open ocean. Much of the marine food supply of humans comes from the coastal waters. Underwater farming of kelp, fish, crustaceans, and mollusks is a future possibility. It is unlikely, however, that the ocean can supply all the food humans will need in the

Figure 23.21   A diver in a kelp bed.

future. Except in coastal waters (including coral reefs), the ocean is very limited in some of its dissolved nutrients. Rich fertile soils may contain 0.5 percent nitrogen, an important nutrient for growth, whereas the richest ocean water, in comparison, contains only about 0.00005 percent nitrogen.

Among the most productive and diverse ecosystems are coral reefs (Figure 22.22), found only in coastal tropical waters where there is adequate light and oxygen. The reefs are large formations of calcium carbonate laid down by individual coral polyps, which harbor symbiotic green algae in their tissues. These algae carry out most of the photosynthesis of the reef community, and they also assist in the cycling of chemicals such as nitrogen. Because of the diversity of life forms that coral reefs support, they have been called the tropical rain forests of the sea.

Figure 23.22   A Caribbean coral reef. What organisms can you identify?

23.9   The Intertidal Zone Is a Difficult Place to Live

Everywhere along the margins of the oceans you can see the effects of waves and tides. In the Bay of Fundy (Nova Scotia) the maximum vertical change between high and low tides is 15.4 m. At the other extreme, the average tidal difference in the Mediterranean is only 0.35 m. Wave action, too, varies depending on the place and time. Some shores, such as those of Maine, are pounded by heavy surf, as shown in Figure 23.23. Others, especially in small, protected bays, may be no more exposed to wave action than the shores of small lakes. High and low tides each occur twice a day. The zone between high and low tides—the intertidal zone—is a difficult environment for life. Twice a day intertidal organisms are submerged in salt water and then exposed to air and the hot sun or freezing wind. Between these times the intertidal zone is pounded by advancing or retreating surf.

On rocky coasts, life in the intertidal region is surprisingly abundant, as shown in Figure 23.24. In cold water, different species of brown algae,

Figure 23.23   The ocean shore. What is the effect of wave action?

Figure 23.24   A tide pool community including sea stars, *Piaster ocuraceus*, rockweed, and barnacles. What conditions must these organisms be able to tolerate?

Figure 23.25   A ghost crab, *Ocypode puadrata,* showing defensive posture.

protected from the drying sun by a jelly-like coating, cling to the rocks. These tangled algae provide protection and support for other algae, protists, and many animals. In addition, barnacles, chitons, and snails cling firmly to rocks or seaweeds, closing up tightly during the periods when they are exposed to air.

On sandy coasts, life in the intertidal zone is limited to organisms that can burrow in the sand or skitter over it. There are no attached producers, but phytoplankton may be present when the tide is in. The burrowers, such as small crustaceans, eat food particles brought by the high tides. Shorebirds, which prey on the sand burrowers or forage in the debris left behind by the retreating tide, are a link with land ecosystems. Land crabs, such as those in Figure 23.25, release their young in the water. Sea turtles crawl out on the beach and bury their eggs. Terrestrial and marine ecosystems merge in the intertidal zone.

Wherever ocean meets land there is constant change. In only a few years sediments deposited at the mouths of rivers stretch the land into the ocean. Elsewhere you can observe the ocean pushing back the land as it carves out sandy beaches with its wave action. Such changes have occurred throughout most of the earth's history.

Concept Review

1. Why do some parts of the oceans have a higher productivity than others?
2. Describe the environment of the ocean depths and the organisms that live there.
3. How do coastal ecosystems differ from those of the open ocean?
4. How do marine organisms affect the amount of dissolved substances in ocean water?
5. Compare variations of temperature in the oceans with those in the atmosphere.
6. What factors determine the amount and color of light found beneath the water surface?

Guidepost: How do human activities affect aquatic ecosystems?

Human Influences on Aquatic Ecosystems

23.10 Wetland Ecosystems Are Threatened

Wetlands, including marshes, swamps, shallow ponds, and coastal areas near estuaries, are flooded with fresh or salt water all or part of the year. Among the most productive of all ecosystems, wetlands form the base of many aquatic and terrestrial food chains.

Inland wetlands provide freshwater habitats for a diversity of fish, waterfowl, and other organisms, including a number of threatened and endangered species such as the whooping crane. Wetlands near rivers help control erosion by storing water during periods of high runoff and by regulating stream flow. Because wetlands trap stream sediments and absorb, dilute, and degrade many toxic pollutants, they improve water quality. If overburdened, however, they can accumulate too many toxins and be destroyed. Wetlands also play a significant role in the cycling of carbon, nitrogen, and sulfur. Important crops, such as cranberries, blueberries, and rice, are grown in wetlands. Such areas also are used for recreation—especially hunting. Figure 23.26 shows a typical inland wetland.

Coastal wetlands, such as the one in Figure 23.27, are found near estuaries where rivers join the oceans. These areas supply food and serve as spawning and nursery grounds for many species of marine fish and shellfish, including salmon, shrimp, oysters, clams, and haddock. Coastal wetlands also are important breeding grounds and habitat for waterfowl and other wildlife, and they help protect coastal areas by absorbing the impact of damaging waves caused by storms and hurricanes. Like inland wetlands, coastal areas dilute and filter out large amounts of water-borne pollutants. Some researchers estimate that one acre of coastal wetland is equivalent to a $75 000 waste treatment plant.

Because the importance of wetlands may go unnoticed, these areas often are dredged or filled in and used as cropland, garbage dumps, and sites for

Figure 23.26 Swamps and marshes such as this cypress swamp in North Carolina, are important inland wetlands.

Figure 23.27 Low tide in a coastal estuary. What conditions must organisms tolerate?

Plant Ecologist

Joy Zedler is a plant ecologist and the director of the Pacific Estuarine Research Laboratory (PERL) in San Diego, California. PERL operates two field research facilities that conduct a wide range of applied ecological studies involving wetlands, including research projects related to wastewater management. Two such projects currently are under way in Southern California. One project is aimed at determining at what level and for how long sensitive estuarine organisms, such as halibut, arrow gobies, and other native fishes and invertebrates, can tolerate reduced salinity of their water. Historically, these species have experienced population reductions when freshwater inflows have occurred following sewage spills. The second project uses wetlands to improve the quality of wastewater discharged to coastal areas. The wastewater is discharged in pulses, thus reducing the amount of time that the inflow of fresh water has to affect coastal waters.

Joy's personal research interests include salt marsh ecology, the structure and functioning of coastal wetland ecosystems, the effects of rare events on estuarine ecosystems, the interactions of native and exotic species in coastal wetlands, and the use of scientific information in the management of coastal habitats. Currently, Joy and her associates are researching the restoration of salt marsh ecosystems. In addition to studying how well-restored salt marshes function compared with natural salt marshes, they also are developing new methods to accelerate the restoration process. By understanding what factors act to slow ecosystem development in human-made wetlands, they are hoping to correct similar problems in new restoration sites.

Joy grew up on a farm where she always felt a close relationship with the land—a factor that greatly influenced her decision to become an ecologist. She attributes her interest in figuring out how to restore wetlands to the "fix-it" ethic practiced on the farm by her parents. As a high-school student in rural South Dakota, she was expected to get a secretarial job, marry, and raise children. Joy rebelled against that stereotype and pursued her education in botany and plant ecology. Along the way, however, her secretarial skills helped her earn money for college. She did get married, and her children now work on wetland experiments for their high-school science fair projects.

Joy's interest in biology, and eventually in wetlands, is the consequence of various opportunities and many teachers who encouraged her. In Joy's words, "Very little of my career was planned, but I have always followed leads and tried to find a match between what I can do and what someone needs to have done. You have to find the link between skills and available funds, and then be willing to grow with the opportunities that develop. I doubt if it's very different in any other field."

urban and industrial development. So far, about 56 percent of coastal and inland wetland acreage in the contiguous 48 states—an area about four times the size of Ohio—has been destroyed. Most of this loss is due to drainage and clearing for agriculture, and the remainder has been lost to real estate development and highways. Although concerned groups are making attempts to slow the rate of wetland loss, only a small fraction (8 percent) of inland wetlands is under federal protection.

Although the drainage of standing waters continues, the surface area of inland waters has actually increased during the last 50 years as a result of the construction of dams across running waters. Dams create waterfalls that turn electric generators, and most dams, especially those in the western states, also form reservoirs that store floodwaters for irrigation and for supplying water to cities. Major changes in the environment, such as reservoirs and controlled stream flow, bring about major changes in communities within that environment. A large dam may block the passage of fish, such as salmon, that swim up rivers to lay their eggs in headwater streams. On the other hand, dams may greatly increase the amount of habitat favorable for other species, such as pike.

INVESTIGATION 23.4   Eutrophication

Many ponds and lakes suddenly turn green at various times of the year, fish die, and the water smells bad. This process, known as eutrophication, occurs frequently around areas where fertilizers and/or detergents enter streams that drain into the ponds or lakes. Are the fertilizers and/or detergents responsible for this "greening" effect? Using the LEAP-System™ with a turbidity probe, or density indicator strips if a LEAP-System is not available, you can test the effects of various substances on the growth of euglena, a common pondwater inhabitant. The LEAP-System BSCS Biology Lab Pac 2 includes instructions for use of the system.

 After reading the procedure, construct hypotheses that predict the effects of the test substances on the growth of euglena.

Materials (per team of 3)

3 pairs of safety goggles	3 density indicator strips (optional)
3 lab aprons	3 stoppers to fit test tubes
3 18-mm × 150-mm test tubes	test-tube rack
dropping pipet	glass-marking pencil
Apple IIe, Apple IIGS, Macintosh, or IBM computer	aged tap water
	aged fertilizer water
LEAP-System™	aged detergent water
turbidity probe	euglena culture

SAFETY   **Put on your safety goggles and lab apron.**

Procedure

1. Label the three test tubes *T*, *D*, and *F*. Fill each of the test tubes about 3/4 full of the appropriate aged water—tap water in the T tube, detergent water in the D tube, and fertilizer water in the F tube. Place all three test tubes in the test-tube rack. (If you are using density indicator strips, be sure to affix a strip in each test tube before filling with water.)

CAUTION: The fertilizer water may contain chemicals that are irritating to skin. Flush spills or splashes with water.

2. Use the turbidity probe to take readings of each test tube for about 10 seconds. Record the readings in your data book. (Or, record initial readings from the density indicator strips.) What is the purpose of this reading?

3. Thoroughly mix the euglena culture and transfer one full dropping pipet of culture into each of the three labeled test tubes.

4. Securely cork each test tube and invert several times to distribute the euglena evenly. Remove the cork and immediately take turbidity readings as in Step 2. Record the readings in your data book. What is the purpose of this reading?

5. Label the test tubes with your team symbol and store them, uncorked, in the location specified by your teacher.

6. Repeat Step 4 each school day for one or two weeks, or for the time specified by your teacher.

7. Wash your hands thoroughly before leaving the laboratory.

Discussion

1. Account for any differences in turbidity readings between test tubes T, D, and F.

2. Describe any visual differences between the three test tubes.

3. What is the purpose of the test tube labeled *T*?

4. Prepare a graph of the readings you recorded for each test tube. Label the *Y* axis *percent turbidity* and the *X* axis *time*. Summarize the effects of detergent and fertilizer on the growth of euglena.

5. Describe what happens in eutrophication.

6. Why would the ponds on golf courses be especially susceptible to the process of eutrophication?

7. Would farm ponds be susceptible to eutrophication? Explain.

23.11 Sewage and Industrial Wastes Affect Aquatic Ecosystems

Although fresh water is a renewable resource, it can become so contaminated by human activities that it is no longer safe for many purposes and is harmful to living organisms. Water pollution is a local, regional, and global problem connected with how we use the land. Some of the fertilizers and pesticides humans apply to croplands run off into nearby surface waters or leach into groundwaters. Poor land use accelerates erosion, and wastes produced on land often are dumped into the oceans. In large, rapidly flowing rivers, relatively small amounts of pollutants are diluted quickly to low concentrations, and the supply of dissolved oxygen needed for aquatic life and the decomposition of organic wastes is renewed quickly. One common pollutant, sewage, provides nutrients for the growth and reproduction of aquatic producers, as shown in Figure 23.28 on pages 660–61. If the amount of sewage is large relative to the volume of river water, the growth of phytoplankton is great. Phytoplankton have short life spans, and the organisms that decompose the phytoplankton use up the available oxygen. Anaerobic conditions result, a situation that is deadly for all aerobic organisms. As the water flows farther downriver, the sewage is diluted by tributary streams, and eventually decomposers use up the sewage materials.

Figure 23.28   Diagram of sewage pollution in a stream. Organisms are not drawn to scale and distances are decreased greatly. At the left, sewage is attacked by decomposers and the oxygen supply decreases. Aerobic organisms die. Further downstream the amount of sewage decreases and the oxygen supply increases. The phytoplankton and aerobic consumers increase in number.

Every day, rivers also receive enormous amounts of natural sediment runoff and surface runoff from urban and agricultural uses of land. Factories and mills contribute alkalis, chromium, lead, and mercury ions—all of which are poisonous to some living things, particularly to the organisms that decompose organic wastes. Industrial processes, especially the generation of electricity, often result in the discharge of hot water into rivers. Such thermal pollution abruptly changes the abiotic environment and makes it intolerable for many aquatic organisms. The warmer water also may bring about a change in the types of species that can inhabit the area. All of these pollutants also affect the health of the human populations that depend on the water.

In contrast to rivers, lakes and reservoirs act as natural traps that collect nutrients, suspended solids, and toxic chemicals—such as those that result from acid rain—in their bottom sediments. Many pollutants, such as pesticides, fertilizers, and sewage, contain high levels of plant nutrients, specifically nitrates and phosphates. The accumulation of these nutrients in lakes and other standing bodies of water encourages the growth of photosynthetic organisms, including algae. These organisms produce great quantities of oxygen during the day, but at night cellular respiration greatly reduces the amount of oxygen. As the organisms die and organic material accumulates, the metabolism of the decomposers uses up the available oxygen. A body of water, such as that shown in Figure 23.29, that is nutrient-rich yet oxygen-poor is **eutrophic** (yoo TROH fik; from the Greek *eutrophos*, for well nourished). Although eutrophication is a natural process, it is accelerated by the intrusion of pollutants into the water. Eutrophic conditions may make it impossible for some organisms to survive. For example, eutrophication of Lake Erie resulting from pollution completely wiped out commercially important fish species such as blue pike, lake trout, and whitefish in the 1960s. Tighter regulations on the dumping of wastes into the lake have enabled many fish populations to rebound, although some fish and invertebrate populations have not recovered.

Figure 23.29 A highly eutrophic pond. How does the fast growth of the algae affect other aquatic organisms?

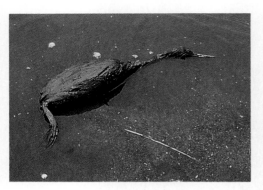

Figure 23.30 An oil-soaked grebe in west San Francisco Bay.

23.12 The Oceans Receive the Major Part of Human Waste

Oceans are the ultimate dump for much of the wastes produced by humans. In addition to natural runoff, oceans receive agricultural and urban runoff, atmospheric fallout, garbage and untreated sewage from ships, and accidental oil spills from tankers and offshore oil drilling platforms. Barges and ships also dump industrial wastes, sludge from sewage plants, and other wastes.

Oceans can dilute, disperse, and degrade large amounts of some organic pollutants, especially in deep-water areas. Other pollutants, however, do not readily degrade. What happens to these pollutants when they reach the ocean? Some, such as pesticides and mercury, enter the food chains and can eventually end up in the seafood we eat. The people of Minamata, Japan, learned firsthand what mercury pollution can do. They ate fish with high concentrations of methyl mercury in their tissues, and before long there were several thousand people, many of them children, who were left crippled as a result of damage to the nervous system.

Oil spills are a constant threat to ocean ecosystems. After an oil spill, hydrocarbons such as benzene and toluene are the primary killers of fish and shellfish. Other chemicals remain on the surface and form floating tar-like globs that adhere to marine birds, sea otters, seals, sand, rocks, and almost all objects they encounter (Figure 23.30). This oily coating destroys the animal's natural insulation and buoyancy, and many animals drown or die from exposure due to loss of body heat. These tarlike globs gradually are degraded by naturally occurring bacteria. Heavy oil components that sink to the ocean floor probably have the greatest impact on marine ecosystems. These components can kill bottom-dwelling organisms, such as crabs, oysters, mussels, and clams, or else can make them unsuitable for human consumption.

Everything that is dumped into a river or stream eventually reaches the ocean, and many of those pollutants will end up in the food chain. Most pollutants, however, stay in coastal waters, the most productive parts of the ocean. They could cause the destruction of important species of fish that many people depend on for food.

23.13 Some Practices Can Reduce Water Pollution

Several measures can reduce water pollution, both directly and indirectly. The indirect pollution of water by fertilizer runoff and leaching can be controlled in several ways. Farmers should avoid using excessive amounts of fertilizer and should use no fertilizers on land with steep slopes. Contour plowing, such as that shown in Figure 23.31, can control erosion and thus the amount of sediment- and nutrient-rich runoff that reaches lakes and streams and that leaches into groundwater. Periodically planting crops that restore the nitrogen content of the soil can reduce the need for fertilizers. The need for pesticides can be reduced by employing biological pest control methods, such as using spiders to reduce the numbers of insect pests. Already more than 300 biological pest control projects have been successful worldwide.

Figure 23.31 How does contour plowing reduce water pollution?

Directly controlling water pollution involves improving the current methods of waste treatment and disposal. When sewage reaches a treatment plant, it can undergo up to three levels of purification, depending on the type of plant and degree of purity desired. Once the desired level of purity is achieved, the wastewater is disinfected before being released into the environment. Primary treatment of wastewater screens out debris, such as sticks and leaves, and the suspended solids are allowed to settle out. Secondary treatment uses aerobic bacteria to remove degradable organic wastes. The wastewater then goes into a tank where most suspended solids and microorganisms settle out. The resulting sludge is removed and disposed of by burning, dumping in the ocean or landfill, or using it as fertilizer. The third level of treatment uses specialized chemical and physical processes that lower the quantity of specific pollutants remaining after primary and secondary treatment. For example, wastewater from secondary treatment plants can be purified further by allowing it to flow slowly through long canals filled with water hyacinths or bulrushes. These plants remove toxic organic chemicals and metal compounds not removed by conventional treatment. Ensuring that all wastewater receives secondary treatment and is disposed of properly would greatly reduce water pollution.

As individuals, we can reduce water pollution by safely disposing of products that contain harmful chemicals, by using less water, and by using low-phosphate or nonphosphate detergents and cleansers. Inorganic fertilizers, pesticides, detergents, bleaches, and other chemicals should be used only when necessary and then in the smallest effective amounts.

Concept Review

1. What benefits do humans get from draining inland water? From damming flowing waters?
2. How are wetlands important ecosystems?
3. How may draining and damming waters be harmful to human interests?

4. Compare the effects of sewage dumped in a river with the effects of sewage dumped in a lake.
5. How did the people of Minamata, Japan, get mercury poisoning?
6. How do oil spills affect ocean ecosystems?

Summary

Humans always have depended on various sources of water to meet their needs. Inland or freshwater ecosystems are classified as standing or flowing waters. Ponds are shallow and smaller than lakes, but phytoplankton are the primary producers in both. Plankton are usually absent in brooks and streams but may be present in slow-moving rivers. Rivers carry sediments and pollutants into the sea. Drainage, dam building, and pollution may have adverse effects on the quantity and quality of water and on the organisms that use it.

The oceans cover about 70 percent of the earth's surface. The concentration of salt and other substances in the oceans remains in a steady state through the action of marine organisms. Temperature, the amount of light available to photosynthetic organisms, and currents also affect ocean waters. The four major zones in the ocean blend together, but each one is different in terms of physical factors and biota. All of the zones are affected by human activities, making ocean pollution one of our most demanding challenges.

Applications

1. Why does the concentration of oxygen in a small pond differ from night to day?
2. How would a cloudy, windy day affect the productivity of phytoplankton in a lake?
3. Is a layer of ice on the surface of a lake harmful to fish? Explain.
4. When ponds or lakes have no inlet or outlet, how do plants and animals find their way into these bodies of water?
5. Compare the sources of oxygen in a brook, a slow-moving river, a pond, and a lake.
6. Many rivers overflow their banks almost every year. The Amazon River is one such ecosystem in which fish, during the flood season, are able to swim among the trees growing near the banks of the river. What effect might these fish have on the trees, and what effect might the trees have on the fish?
7. Are decomposers likely to be more active on the bottom of a pond or of a lake? Explain.
8. Why is the deep sea blue? What makes the Red Sea red? What other descriptive names are appropriate, or inappropriate, for bodies of water with which you are familiar?
9. Some people have called the open ocean (the part away from the coastline) a desert. Why?
10. Why is a poisonous substance in a pond—such as a mercury compound—likely to be more concentrated in the flesh of a bass than in the flesh of a ciliate?
11. A program designed to improve the fishing was introduced in a midwestern pond. First, a fish poison was used to kill all the many small fish that were in the pond. Then the pond was restocked with game fish. Instead of large game fish, the new population contained many stunted individuals. Explain.
12. If you collect 167 snails of a given species from five plots totaling 5 m² and find their total mass is 534 g, what would be the biomass of the snail population?
13. What do you think happens to the household cleansers that you and your family wash down the sink? Where do they go, and how much do you throw away per year? Also, what happens to the unused household cleansers and other chemicals that you throw in your garbage?

14. A common saying about the problem of pollution used to be, "Dilution is the solution to pollution." The idea was that people could get rid of wastes simply by dumping them into the ocean or flushing them down the sink with a lot of water. What do you think is the problem with this philosophy?

Problems

1. Current is one of the most important limiting factors in a stream ecosystem. Stream organisms must be adapted for maintaining a constant position. Cite at least five specific organisms showing five different adaptations to stream currents.

2. Differences in the physical characteristics of water and air are important in understanding the contrast between aquatic and terrestrial environments. Consider such questions as these:
   a. What land organisms have the most streamlined bodies? With what form of locomotion do you associate this streamlining?
   b. What water organisms have the most streamlined bodies? What niches in aquatic ecosystems do these organisms occupy?
   c. Why do most plankton organisms have little or no streamlining?
   d. How does locomotion by walking on the bottom of the ocean differ from locomotion by walking on land?.

3. The St. Lawrence River in North America is about 1197 km long. Find out what changes have been made in the river in the last 25 years and what these changes have done to the river's ecosystem.

4. Investigate the effects of the El Niño current.

5. Estuaries, such as Chesapeake Bay and San Francisco Bay, represent a special type of aquatic environment. Find out what characteristics distinguish estuarine environments from marine and inland-water environments and how these affect aquatic life.

6. Investigate the various types of adaptations that allow organisms to survive in the intertidal zone.

7. Though the seas are immense, marine waters can become polluted. Investigate the types of oceanic pollution and their effects on marine ecosystems.

8. What does the damming of a river do to the terrestrial habitats upstream from the dam? What does the damming do to the aquatic and wetland habitats downstream? Study a dam that has been built in your state, or a projected dam to be built, and investigate what impact it had (or may have) on other ecosystems related to the river.

9. What are organizations such as "Ducks Unlimited" doing to help preserve wetlands in the United States? Ask a spokesperson from the group to inform your class.

10. What were the environmental effects of the oil spilled along the coast of Alaska in 1989 and into the Persian Gulf in 1991? What were the predicted effects, and how many of these predictions were realized?

11. What are the most important game fish in the streams and lakes of your area? What do they eat, and where do they live? Create a food web relating the game fish of your area to their food supply, or draw a profile of a lake showing the types of habitats in which you might expect to find different fish species.

12. One of the great freshwater ecosystems in the world is the Everglades, found at the tip of Florida. Unfortunately, this community is in great danger of being completely destroyed. What types of plants and animals live in the Everglades, and what makes them unusual? What threatens the existence of the Everglades?

13. Create an intertidal ecosystem in an aquarium. A real challenge is to maintain such organisms as sea anemones, sea cucumbers, and sea urchins in your classroom. Before any organisms are ordered, make certain you have all of the necessary equipment and materials to maintain a healthy environment for them.

What has been planted and what is happening to it? This area was once part of a prairie. As humans modified this ecosystem, many changes took place that require about twice the energy needed for a natural ecosystem. Where does this added energy come from? What might be the benefits of the extra expenditure of energy? What are some of the other costs associated with maintaining this artificial ecosystem? For how many years in the future do you think humans can afford to sustain such an ecosystem? What are some alternatives for the future?

Managing Human-Affected Ecosystems

For thousands of years, humans have lived apart from their natural environment. We have dominated the land and other organisms by building houses and roads, cutting trees and clearing natural vegetation for cropland and cities, damming rivers, and polluting the air and water. The consequences of our actions now are all too clear.

Although the outlook is not as bright as we might hope, there still is time to change the course of events. By passing and enforcing tougher laws, developing more energy-efficient practices, and changing our attitudes and personal habits, we can help ensure that our species, and countless others, will continue for generations to come.

Previous chapters have examined the principles of biology and ecology and have shown how human activities disrupt natural ecosystems. This chapter summarizes the relationships between human culture, technology, and resources, how this interaction affects ecosystems, and what we can do to lessen our effect on the earth and help manage it in a responsible fashion.

Human Culture and Technology

Guidepost: How has the relationship of humans to our environment changed?

24.1 Human Culture Has Affected Our Role in the Ecosystem

Early hominids lived in a comparatively simple fashion. If they ate meat, it probably was scavenged from animals killed by other predators. Their main sources of food most likely were berries, fruits, nuts, and roots, which they gathered. Until about 25 000 years ago, humans and their ancestors evolved as a stable part of the equilibrium that existed among plants, animals, and other organisms. The size of human groups probably fluctuated according to the amount of resources in the immediate environment—in good years the populations grew, in bad years the populations declined. Figure 24.1 on page 668 illustrates such a self-regulating population control model.

Through time, the cranial capacity of hominids increased from about 400 cm³ some 5 million years ago to between 1000 and 2000 cm³ today. This relatively rapid increase in brain size resulted in a tremendously complex brain, which allowed hominids to respond to environmental stresses in a totally new fashion. Previously, if a group was biologically suited to its environment, it survived; if it was not well suited, the group either adapted

Figure 24.1 A simplified diagram of population control. How is this system self-regulating?

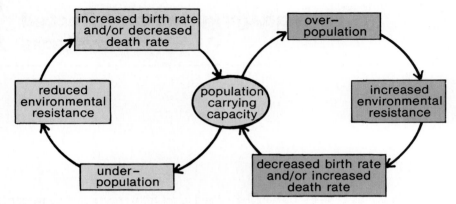

to meet the conditions of its environment, moved somewhere else more suitable, or became extinct. A complex brain, however, allowed our ancestors to devise other means of coping with environmental stressors, such as using furs to keep warm in colder climates. These responses are cultural, not biological, adaptations. Human culture consists of those ideas and behaviors perpetuated through teaching and learning. Probably the most important development in human evolution was the use of symbolic language, which provides a means of remembering and transmitting culture.

New ideas for coping with problems and changes required new tools. The earliest technological developments probably came slowly. First using naturally occurring stone flakes as knives, our ancestors learned to make their own simple stone tools. Through millions of years, they improved tools to serve different purposes. The rate of technological change eventually quickened, and cultural developments came more rapidly.

TABLE 24.1 Change in Atmospheric CO_2 in Hawaii

Year	Average Annual CO_2 Level (ppm)
1958	315
1960	317
1962	318
1964	319
1966	321
1968	322
1970	324
1972	326
1974	331
1976	333
1978	335
1980	337
1982	340
1984	343
1986	346
1988	350
1990	353

INVESTIGATION 24.1 Views of the Earth from Afar

Aerial photos give the "big picture" of events in an area. Infrared film and computer imaging disclose events invisible to the naked eye. Infrared film is sensitive to heat. Consequently, objects giving off heat appear dark in black-and-white photos or show up as a different color in color photos. Some color prints show warmer areas as green; others show them as red. In either case, the contrast with cooler areas is striking.

This investigation uses infrared photos and other modern technology to examine a current environmental problem.

Materials (per team of 1)
none

Procedure

1. Examine the data in Table 24.1, which were obtained from studies of "old air" in tightly sealed containers. These data were collected in Hawaii, and similar data have been collected in other areas. Between 1958 and 1990, by what percentage did the proportion of carbon dioxide in Hawaii's air change and in which direction did it change? If this trend is typical of the industrial United States, by what percentage has the atmospheric level of carbon dioxide increased in the last 100 years?

2. Study the maps shown in Figure 24.2, which were produced by a computer from satellite data. What color has been used for your area for January? For July? What do these colors indicate about temperatures?

Figure 24.2

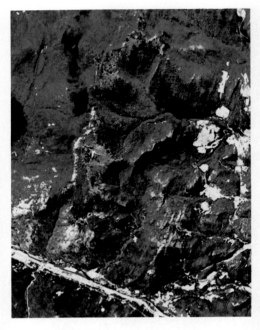

Figure 24.3

3. Study the aerial photo of the Vermont forest (Figure 24.3). In this photo, healthy trees appear green and damaged trees appear red. In areas west of this forest, industries are producing acid rain and other forms of air pollution.

Discussion

1. What is the general trend in atmospheric concentration of carbon dioxide from 1958 to 1990?
2. How might this trend be related to acid rain? To air pollution in general?
3. During the last 100 years, measurements taken around the world show that the mean global temperature has risen 0.4° C. What is the relationship between the level of atmospheric carbon dioxide and temperature?
4. If the trend represented in Table 24.1 continues, what colors might be used in a map of your area 100 years from now? How might the food web in your area change as a result of a change in temperature?
5. What cause-and-effect relationship, if any, is demonstrated here?

24.2   Agriculture Had Wide-Reaching Effects

Eventually humans learned to change their environment in a fundamental way. One of the most significant changes in human history began about 10 000 years ago when groups of people began to switch from hunting wild animals and gathering other foods to domesticating—taming, herding, and breeding—wild game. They also began domesticating selected wild food plants, consciously planting and harvesting them. Many crop species first were domesticated in areas called centers of origin, including Southeast Asia, the Middle East, Northeast Africa, and northern South America. Most of the plant foods we eat originally came from these centers of origin.

Some plant cultivation may have begun in tropical forest areas. To prepare for planting, early agriculturists cleared small patches of forest by cutting down trees and other vegetation. They allowed the cut vegetation to dry and then burned it. The resulting ashes added plant nutrients to the normally nutrient-poor soils of the tropical forest. This type of agriculture, shown in Figure 24.4, is called slash-and-burn agriculture. After a plot had been cultivated for several years, few if any crops could be grown because of the depleted condition of the soil. When crop yields dropped, growers shifted to a new area of forest and cleared a new plot for planting.

These early growers practiced subsistence agriculture, growing only enough food to feed their families. Because their dependence on human labor and crude stone tools allowed them to cultivate only small plots, they had relatively little impact on their environment.

Large-scale agriculture began about 7000 years ago with the invention of the plow, pulled by domesticated animals and steered by the farmer, as illustrated in Figure 24.5. Animal-drawn plows allowed farmers to break up fertile grassland soils, which could not be cultivated before because of the thick, extensive root systems of grasses. In some arid regions, farmers further increased crop yields by diverting water into hand-dug ditches and canals to irrigate crops. With animal- and irrigation-based agriculture, families usually grew enough crops to survive, and sometimes they had enough left over for sale or for storage.

Figure 24.4   Slash-and-burn agriculture in the Brazilian rain forest.

Figure 24.5 Animal-powered agriculture. Preparing rice fields with water buffalo in Java, Indonesia. Compare its effects on the biosphere with the effects of the types of agriculture shown in Figure 24.6.

The gradual shift from hunting and gathering to farming had several major effects. The human population began to increase because of the larger, more stable food supply. Humans cleared increasingly larger areas of land and began to control and shape the surface of the earth to suit their needs. The formation of cities—urbanization—began because a small number of farmers could produce enough food to feed their families as well as a surplus that could be traded to other people. Specialized occupations and long-distance trade developed as people not involved in farming learned new crafts, such as weaving, toolmaking, and pottery. These goods could be exchanged for food. Small villages grew into towns and cities, which served as centers for trade, government, and religion.

About 5500 years ago, the interdependence of farmers and city-dwellers led to the gradual development of a number of agriculture-based urban societies near early agricultural settlements. The growing populations of these emerging civilizations needed more food and more wood for building and for fuel. To meet these needs, more forests were cleared and grasslands were plowed. Such massive land clearing destroyed and degraded many habitats. Poor management of the cleared areas led to increased deforestation, soil erosion, and overgrazing. The gradual degradation of the soil, water, forests, grazing land, and wildlife probably was a factor in the downfall of many great civilizations in the Middle East, North Africa, and the Mediterranean.

24.3 The Industrial Revolution Increased the Demand for Resources

The next major cultural change, the Industrial Revolution, spread to the United States from England in the 1800s. Goods formerly produced on a small scale by hand now were being produced on a much larger scale by

Figure 24.6 Industrialized agriculture.

machines. Eventually, locomotives, cars, trucks, tractors, and ships powered by fossil fuels replaced horse-drawn wagons, plows, grain harvesters, and wind-driven ships.

Within a few decades, agriculture-based urban societies became early industrial-based urban societies. The growth of industries increased the flow of raw materials into the industrial centers. The increased demand for resources caused the environment of the nonurban areas supplying these resources to be further degraded. Industrialization also produced greater amounts of smoke, ash, garbage, and other wastes in urban areas.

Farm machines powered by fossil fuels, the use of commercial fertilizers, and new plant-breeding techniques greatly increased the amount of food that could be grown on each acre of land. Fewer people were needed to raise food and, thus, the number of people migrating from rural to urban areas increased. The development of more efficient machines and mass production techniques led to the advanced industrialized societies of today, such as Japan, Germany, and the United States.

The combination of industrialized agriculture (Figure 24.6), increased mining, and urbanization has accelerated the degradation of potentially renewable topsoil, forests and grasslands, and wildlife populations—the same problems that contributed to the downfall of earlier civilizations.

24.4 Human Culture and Technology Have Created Many Problems

Today, the world's population exceeds 5 billion and another billion probably will be added before 1997. Most of these people will live in urban areas, such as that shown in Figure 24.7, which are ill-equipped to supply them with the necessities of life. Meeting the demands of the growing population has increased economic activity worldwide but also has increased

Figure 24.7   An urban slum. Every day, hundreds of people leave rural areas for cities. What effect does this immigration have on the city?

pollution levels, depleted many of the world's resources, and degraded a large portion of the world's farmlands, rangelands, forests, and other non-urban areas. Industrialized countries have assumed a throwaway life-style in which they consume the bulk of the world's energy and goods and create the bulk of the world's waste and pollution.

Burning fossil fuels increases the emission of carbon dioxide, sulfur, nitrogen oxide, and other harmful gases. Acid rain, damage to the ozone layer, and other problems created by this pollution now are damaging public health and harming forests, fish, and crops. Increased carbon dioxide levels could cause the climate to warm, thus disrupting agriculture and fisheries and affecting weather patterns. Fewer plant and animal species can survive as favorable habitats are either degraded or destroyed.

Toxic metals and synthetic chemicals from industrial wastes, fertilizers, and pesticides pollute many rivers and lakes and leach into groundwater supplies. Oil spills threaten ocean and coastal habitats. Excessive amounts of sediment from erosion destroy fish-spawning areas. Nutrients entering water supplies with runoff can cause eutrophication, thus threatening fish and other aquatic organisms. Housing developments and industrial developments, along with environmental pollution, degrade and destroy important wetland ecosystems.

The human species still is evolving, and invisible and immeasurable changes slowly are taking place. The agents of selection may no longer be climate and disease, but factors such as atmospheric nitrogen oxide and emotional stressors. Humans as a species still depend on their culture to help them adapt and survive, a process called cultural adaptation. This reliance, however, creates a problem: Each new cultural adaptation alters the environment to which we are adapting. New adaptations increase the population, and greater population densities require further adaptations. New adaptations deplete world resources, and resource depletion requires more

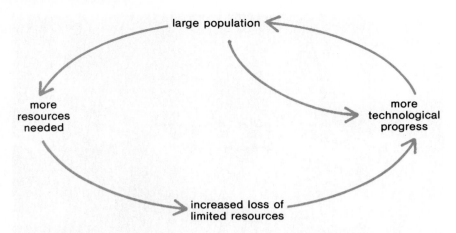

Figure 24.8   The positive feedback cycle of technology, population growth, and resource use. How can the cycle be broken?

adaptations. Figure 24.8 shows this feedback cycle. What is vital now is a concerted effort to break out of the cycle and bring about a new stability in population size, in technology, and in the consumption of resources.

24.5   All Environmental Issues Are Connected to Population Growth

Environmental degradation occurs slowly, and the early stages are not always obvious. In fact, individuals who stand to profit from exploiting the environment often deny that their projects will degrade it. As the population continues to increase, there is more pressure to take more resources from the limited biotic and abiotic environment. To what extent are we justified in satisfying our needs at the expense of future generations?

Between 1958 and 1962 an estimated 30 million people died from famine in China. China has since made tremendous efforts to feed its people and bring its population growth under control. Between 1972 and 1985, China managed to lower its birthrate from 5.7 children per woman to 2.1 children, and today many experts believe that China's population actually will decline after peaking early in the twenty-first century. To accomplish the drop, China established an extensive and strict population control program, which has since been modified. The program encourages couples to postpone marriage and provides easy and free access for married couples to sterilization, contraception, and abortion. If a couple already has two children, one of the parents is encouraged to be sterilized. Couples who sign and adhere to an agreement to have no more than one child receive economic rewards such as salary bonuses, extra food, larger pensions, better housing, and free medical care and schooling for their child. Although this population control program greatly restricts individual freedom, the unchecked growth of the population would have far worse consequences for the entire population.

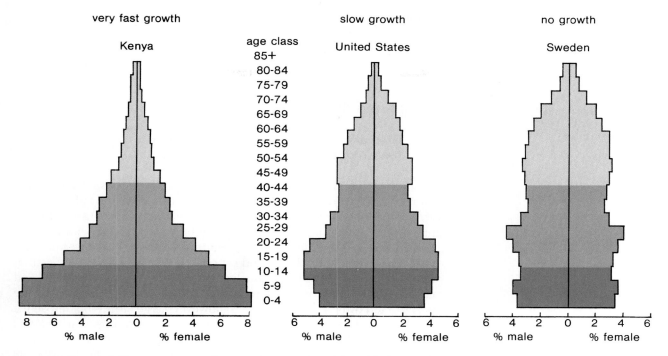

Figure 24.9 Population age structure diagrams for countries with rapid, slow, and zero population growth rates. What does each shaded portion represent?

Simply reducing the number of children each woman has, however, does not immediately slow population growth. Any country with a large number of people below age 15 has a powerful built-in momentum toward increasing population size unless death rates rise sharply (see Figure 24.9). Because the number of women who can have children increases greatly as the large number of young people reach their reproductive years, the number of births rises even if women have only one or two children. The effects of family planning practices such as having only one or two children are not felt for an entire generation—until the children themselves reach reproductive age.

Most other countries cannot or do not want to employ the major features of China's population program. The best lesson other countries, including the United States, can learn from China's experience is to curb population growth now through family planning and economic incentives and not wait until the choice is between mass starvation and strict coercive measures.

Concept Review

1. Explain the relationship between culture and technology.
2. What effects did the invention of agriculture have on the human population? How has it affected the environment?
3. What is the relationship between growing population numbers, technology, and resource depletion?

4.   How has technology led to increased levels of pollution?
5.   Describe the feedback cycle involving population size, resources, and technology.

Guidepost: What measures are needed to alleviate the current environmental crisis?

Change and the Future

24.6   International Cooperation Is Necessary to Solve Modern Problems

One nation alone cannot curb worldwide pollution or ensure the preservation of endangered species. Just as ecological problems fail to stop at state boundaries, so they fail to stop at national boundaries. Air pollution drifts into other countries, and water pollution flows into the oceans.

In April 1986, a nuclear power plant in Chernobyl, in the western part of the Soviet Union, exploded and caught fire (Figure 24.10). The carbon rods used to control the rate of nuclear fission burned for several days, and many deaths and injuries were reported. The effects of the accident were not restricted by national boundaries, however. Large clouds of radioactive particles, released by the fire in the reactor, were carried by wind to Scandinavia and countries in eastern Europe—over 2000 km away.

Soviet and Western medical experts estimate that 5000 to 100 000 people in the Soviet Union and the rest of Europe will die prematurely during the next 70 years from cancer caused by exposure to the radiation released at Chernobyl. Thousands of others will be afflicted with tumors, cataracts, and sterility. The death toll would have been much higher if the

Figure 24.10   The Chernobyl nuclear power plant. This photo was taken May 9, 1986.

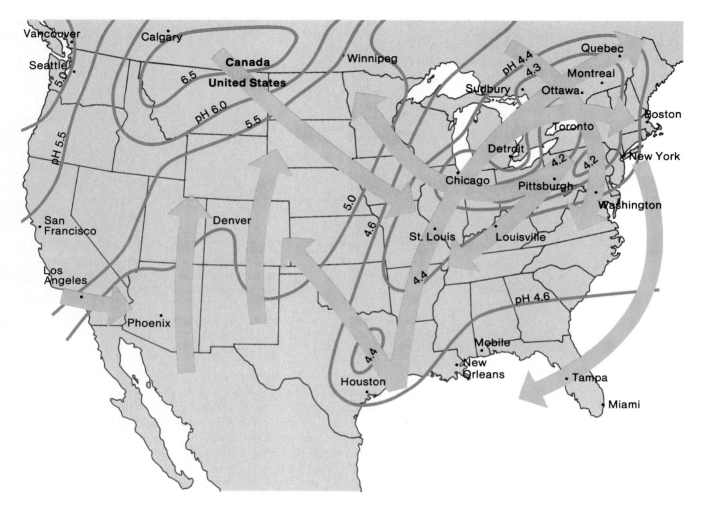

Figure 24.11 Direction traveled by pollutants responsible for acid rain, and the average acidity of rainfall. Recall that low pH figures indicate high acidity. What other factors should be taken into account when considering damage done by acid rain?

accident had occurred during the day, when people were outdoors, and if the wind had been blowing toward Kiev and its 2.4 million people.

Another problem that has global environmental implications is acid rain, which is carried by winds across national boundaries, as shown in Figure 24.11. Mexican copper smelters produce acid rain that falls on the United States, and U.S. industries produce acid rain that falls on Canada. All three countries must cooperate to solve this problem.

24.7 Technology Can Help Solve Some Problems

As humans we are an integral part of nature, yet we have attempted to "conquer" nature. Rather than trying to conquer nature, we must learn to coexist with it. We have an obligation to our successors and to the offspring of all

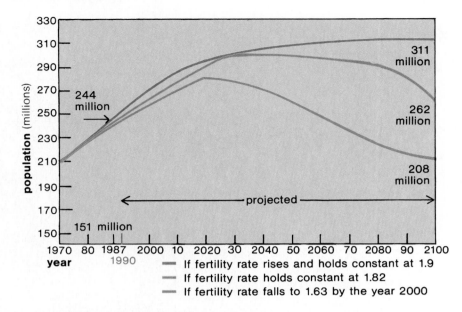

Figure 24.12 Projections of changes in the population size of the United States for different fertility rates.

species to preserve the environment in which we live. The time for correcting our past mistakes is running out. In many parts of the world, biomes have been changed permanently to deserts or urban wastelands, but there still is some wilderness left, there still is farmland that can be restored, and there still are countless species that share this planet with us. The technology that helped create the current situation can be used to develop solutions to some, if not all, of the problems we now face. It is not too late to take steps to reverse the environmental downward spiral.

Population growth can be controlled by decreasing the fertility rate of women through intensive family planning and increased access to inexpensive or free contraceptives. New birth control devices can prevent pregnancy for up to five years, and other methods, such as male contraceptives, currently are under development. Figure 24.12 shows three population predictions for different fertility rates in the United States.

Air pollution can be controlled by regulating population growth, reducing the unnecessary waste of metals, paper, and other resources, increasing energy efficiency, reducing energy use, switching from coal to natural gas, and switching from fossil fuels to energy from the sun, wind, and flowing water or nuclear power.

In the past 20 years, the United States has come a long way toward increased energy efficiency. The typical house built today is 35 percent more energy efficient than one built before 1973, and improvements still can be made. Cost-effective conservation measures currently available would reduce the amount of energy consumed in buildings by 30 to 50 percent. New car mileage now averages about 27 miles per gallon of gasoline—nearly double that of new cars in 1973. New automobile designs with cleaner-burning engines, improved lubricants, electronic emissions controls, improved tires, and improved aerodynamics would result in still more

Bioremediation

Bioremediation is a process in which certain eubacteria are used to degrade hazardous substances found in the environment. Recently, bioremediation has been used to treat oil spills. Research on this process has been conducted on several oil spills, including the *Exxon Valdez* disaster in Prince William Sound, Alaska. How are eubacteria used in treating oil spills?

Biodegradation is a natural process in which certain eubacteria break down, or consume, petroleum and release carbon dioxide. These eubacteria are sometimes called hydrocarbon-oxidizing eubacteria because they destroy petroleum molecules by adding oxygen to them. The newly oxygenated molecules then are further broken down by the eubacteria until only end products remain.

Hydrocarbon-degrading eubacteria occur naturally, although their numbers vary from one environment to another. In Prince William Sound and nearby areas, the community of such degraders contains hundreds of different organisms that act together to degrade the petroleum and its products.

Eubacteria that degrade hydrocarbon require three components to grow: hydrocarbons, oxygen, and nutrients. Oxygen is plentiful on the surface layers of the beaches in Prince William Sound because of their contact with the air. Normally, the population of hydrocarbon-degrading eubacteria is limited by the availability of hydrocarbons. Before the *Exxon Valdez* spill, these eubacteria accounted for less than 0.1 percent of the naturally occurring eubacteria found on the beaches. After the oil spill, the availability of hydrocarbons was unlimited, and the hydrocarbon-degrading eubacteria accounted for more than 10 percent of all eubacteria on the beaches. Any further growth in the population of these eubacteria is limited only by the availability of nutrients such as nitrogen and phosphorus.

One way to increase the population of these naturally occurring eubacteria is to add the nutrients as fertilizers. The application of fertilizers to beaches is difficult, however, because tidal and wave action quickly wash away water-soluble nutrients. In addition, introducing nitrogen and phosphorus into sheltered bays and estuaries might cause algal blooms and eutrophication. To minimize these problems, researchers have developed

A worker spraying a section of beach with fertilizers to promote the growth of oil-degrading bacteria.

several special methods of applying fertilizers. One method uses fertilizers that stick to or dissolve in the oil, another uses granules that release the fertilizers over several days or weeks, and a third uses an automatic spray system to apply fertilizers at low tide.

In May 1989, the Environmental Protection Agency and Exxon agreed to test the different fertilizer application strategies on an island in Prince William Sound. After application of the fertilizer developed to stick to the oil, the beaches showed a perceptible improvement within 10 days (see Chapter 11 opening photo). The beaches were monitored during and after the test applications to determine any adverse effects. To date, none have been recorded. Apparently, the numbers of eubacteria return to normal levels after the fertilizer has been used or washed away. Researchers continue to be concerned, however, about the effects of introducing high levels of fertilizers into marine ecosystems.

679

energy-efficient cars. Increasing automobile efficiency, however, will help reduce energy use only so far. Alternative modes of transportation also are necessary.

The Clean Air Act of 1990 sets standards for the maximum allowable levels of several key air pollutants. As a result of earlier air pollution control laws, levels of most major air pollutants in the United States either have decreased or have not risen significantly. Much more needs to be done, however, to reduce levels of ozone-destroying chemicals and greenhouse gases. The 1990 Clean Air Act limits certain emissions that cause acid rain, calls for a reduction of toxic air pollutants and tailpipe emissions, and addresses the phasing out of ozone-destroying chemicals by the year 2000. Some experts believe restrictions should be even tighter and the use of some chemicals should be banned altogether.

Water pollution can be decreased by decreasing air pollution, by improving erosion controls, farming techniques, and human waste treatment and disposal techniques, and by strictly controlling the disposal of industrial wastes. Runoff from feedlots and barnyards can be collected in detention ponds. The nutrient-rich water from these ponds can be applied to croplands or forests as fertilizer.

Biotechnology will continue to aid us in the future, although bioethical questions surely will follow. Investigation 24.2 examines some of the bioethical issues involved in environmental decisions.

24.8 Sustainable Agriculture Is Important

One way to reduce world hunger and the harmful environmental effects of industrialized agriculture is to develop a variety of sustainable agricultural systems that produce adequate amounts of high-quality food but conserve resources. Such systems combine appropriate aspects of existing agricultural systems with new agricultural techniques to take advantage of local climates, soils, and resources (Figure 24.13). Sustainable agricultural systems do not require large inputs of fossil fuels, and they conserve topsoil, build new topsoil with organic fertilizers, conserve irrigation water, and control pests with little, if any, use of pesticides.

The primary focus of sustainable agricultural systems is on the long-term condition of the soil, rather than on short-term agricultural productivity. Instead of producing a single type of crop, sustainable systems produce a diverse mix of fruit, vegetable, and fuelwood crops and use organic fertilizers made from animal and crop wastes. Special fast-growing trees planted along with the food crops supply fuelwood and add nitrogen to the soil. Whenever possible, locally available alternatives to fossils fuels—such as wind energy or solar power—are used.

A central component of sustainable agricultural systems is crop rotation—the planned succession of various crops on one field. Rotation provides better weed and insect control, improves nutrient cycling, and thus improves crop yields. A typical seven-season rotation might call for three seasons of planting alfalfa and plowing it back into the soil, followed by four seasons of harvested crops—one of wheat, then one of soybeans, another of wheat, and finally one of oats. The cycle then would start over. The wheat grown the first season would remove some of the nitrogen produced

Figure 24.13 Crops that are plowed under instead of being harvested for sale, as shown here, enrich the soil and improve future crop productivity. Many of these crops contribute nitrogen to soil, thus reducing the need for synthetic nitrogen fertilizers.

by the alfalfa (a legume). Soybeans, which are nitrogen-fixing legumes also, restore some nitrogen to the soil. Oats are grown at the end of the cycle because they demand fewer nutrients than does wheat. Figure 24.14 illustrates a crop rotation system.

A second important component of sustainable agricultural systems is the regular addition of crop wastes, manures, and other organic materials

Figure 24.14 Growing combinations of crops can make farms more sustainable by maintaining the productivity of the soil and reducing the farm's reliance on a single crop. On the farm shown here, the parallel strips of land have been planted on the contour of the terrain with oats (yellow), corn, and alfalfa (both green). Within each strip, the crops are rotated on a four-year cycle. Corn is grown for one year and then replaced by an oat crop, which also is grown for one year. The oat crop is then replaced by alfalfa, which is grown for two years.

to the soil. Organic matter improves soil structure, increases its ability to store water, and enhances fertility. Soil in good condition is easier to till, and it allows seedlings to emerge and root more easily. Water readily seeps into rich soil, thus reducing surface runoff and erosion. Organic materials also provide food for earthworms and other soil organisms, which in turn improve the condition of the soil.

In industrialized countries such as the United States, a shift from industrialized to sustainable agriculture is difficult to accomplish. Already, however, 50 000 to 100 000 of the 650 000 full-time farmers in the United States have shifted partly or completely away from conventional industrialized agriculture to forms of sustainable agriculture. Some experts believe that with governmental support of research, subsidies and tax breaks, and training programs for farmers, the shift from industrialized agriculture to sustainable agriculture could be accomplished in 10 to 20 years.

INVESTIGATION 24.2 Decisions About Land Use

Land is a finite resource necessary to support human life. Because many interests compete for the limited available land, the problems involved in deciding how land will be used are complex. Land use decisions involve not only biology and technology but also political, economic, legal, and attitudinal problems. The solutions often require trade-offs—compromises of values or life-styles in exchange for a portion of the available land.

Before a community land use decision is made, the public should be given the opportunity to express opinions and offer information or insights. This investigation simulates a real-life situation and asks you to evaluate different land use proposals in a fictitious case.

Materials (per team of 10)

set of role cards
map

Procedure

1. Read the following background information for a situation involving proposed land uses:

   Winterville is a growing city, located somewhere in the United States, that has just annexed 2 square miles of land. Although Winterville owns the land, it may be sold or used for any purpose the city wishes.

   Ten different groups are interested in what happens to this land. Speaking for some of these interest groups are a farmer, a college student, a single parent, a retired citizen, and an industrialist. Other groups are represented by a laborer, an environmentalist, a construction worker, a developer, and a city manager.

   Figure 24.15 is a topographic map of the annexed land. The lake on the annexed land is unpolluted and is one source of irrigation and drinking water for the city. The present sewage treatment plant and sewer lines of Winterville are working at their top capacity. The garbage dump is only large enough to handle the garbage produced by the current population.

   The population of Winterville is 100 000, and the growth rate during the last 10 years was 10 percent. Projections for the future show a similar growth rate for the next 10 years. The birthrate is 20 births per 1000 people per year. Table 24.2 shows the present age structure of the community.

   Average per capita income per household for the city is $25 000, and there are approximately 40 000 jobs in the city. The unemployment rate is 6

TABLE 24.2 Age Structure

Age	Number
Under 5	14 000
5–9	12 000
10–14	10 000
15–19	11 000
20–24	10 000
25–34	13 000
35–44	8000
45–65	15 000
Over 65	7000

Figure 24.15

percent; that is, 6000 people of working age need jobs. Table 24.3 shows the income distribution of the city.

There are 18 elementary schools, each of which can accommodate up to 700 students. Each of the 6 junior high schools can serve 1000 students. The 3 high schools have room for 3000 students each.

2. In your assigned team, select a role card at random and read about the individual described on your card. Take particular notice of the proposed land use this individual wants. From this person's point of view, think of arguments and consequences that will persuade the others in your team to support your cause. The total amount of land needed for each use is shown on the role card, but the total area of all the proposed land uses is much greater than the 1280 acres available for development (see Table 24.4). This means some of the proposed land uses cannot be accommodated or will have to be reduced greatly in size.

TABLE 24.3   Family Income Distribution

Annual Income	Percent of Population
Less than $10 000	10
$10 000 to 15 000	11
$15 000 to 20 000	15
$25 000 to 30 000	27
$30 000 to 35 000	12
More than $35 000	6

TABLE 24.4   Proposed Land Uses

Proposed Land Use	Amount of Land Needed (acres)
Shoe manufacturing plant	40
Low-cost housing	40
Dump and sewage treatment plant	10
Luxury homes	640
Park	320
Shopping center	40
Open space	1280
School	20
Houses and condominiums	240
Farm	240
Total	2870
Total land available	1280

TABLE 24.5   Land Use Considerations

Will the Proposed Land Use Have an Effect on Individuals?

1. Will it change individual life-styles? Will it cause people to live in different ways? Will it allow them to do something that they want to do? Will it force people to do something that they do not want to do, or to stop doing something they currently do?

2. Will it change individual financial situations? How will the proposed land use be financed? Will it affect the cost of goods and services? Will it affect taxes? Will the action affect people's jobs—create new jobs or eliminate current jobs?

3. Will it change the health of individuals? Will the land use have a foreseeable positive or negative effect on the physical or mental health of individuals in the area? For example, will it disrupt the social structure of a neighborhood?

What Are the Consequences for the Whole Community?

1. Will the action conflict with the land's economic (agricultural, mining, industrial) use or recreation use? Will the action conflict with the land's aesthetic value?

2. Is the land especially well-suited for one use and less appropriate for other uses?

3. Can the local government provide necessary services: water, sewers, electrical and gas connections, waste disposal, transportation, police and fire protection, and schools?

4. What about conflicting interests? Will the action satisfy the wants or needs of only a small part of the community's population?

How Will Various Parts of the Natural Environment Be Affected?

1. Does the area have unique geological, biological, archeological, or historical features that should be preserved?

2. What effect will the action have on vegetation, wildlife, and wildlife habitats?

3. What effect will the action have on temperature, climate, or air quality?

4. What effect will the action have on water quality?

5. Are there natural hazards in the area that should be considered? These may include floods, types of soil, erosion, landslides or snow avalanches, sinking or settling of the earth, earthquakes, and high winds or other dangers due to extreme weather.

3. List the five most important positive consequences of the land use you have proposed. Some relevant ideas you may want to consider are listed in Table 24.5. Remember, you have assumed the role of the person on your card and are speaking from that position.

4. Discuss the concerns of each member of your team, one at a time. Record in your data book each person's occupation, the use of the land he or she proposes, and the five positive consequences he or she claims will result.

5. After each person has presented his or her proposal, evaluate each of the five arguments, using the values analysis scale shown in Figure 24.16. Assign only positive scores. Do not evaluate your own arguments.

6. Total the five positive scores for each proposal and compare your evaluations with those of the other team members.

Figure 24.16

−10 −9 −8 −7 −6 −5 −4 −3 −2 −1	0	+1 +2 +3 +4 +5 +6 +7 +8 +9 +10
very ... negative ... slightly	don't	slightly ... positive ... very
negative .... negative	know	positive .... positive

7. Form a small group with other members of your team who share similar opinions and values concerning the use of the annexed land. If you feel you have nothing in common with the other citizens, continue to act as an individual.

8. In your small groups, discuss the negative consequences of all the land proposals your group is against. List the five most serious negative consequences for each proposal your group is against.

9. Record the negative consequences and evaluate them as you did in Step 6, but assign only negative scores. Total the negative scores for each land use.

10. Add your negative evaluation score to your positive evaluation score for each land use. Compare your results with those of the others in your team.

11. As a team, decide which of the proposals to accept. You may eliminate any of the land uses you wish, or you may keep all of them. Remember, however, that the total area of land available for use is limited and that your final decision must be made as a team.

12. Using a large copy of Figure 24.15, identify where each land use will be located. Keep the following information in mind as you locate each item: (a) the area required for each land use (the area given includes roads that must be built for access); (b) the location of each use in relation to the physical aspects of the land; and (c) the location of each use in relation to the other uses.

Discussion

1. How does your team's decision on land use compare with the other teams'?
2. Compare your completed map with those of the other teams. How do they differ? Do you agree with the placement of land uses on other maps?
3. Which land use map do you think is best? Why?
4. What consequences were different? Did you overlook any consequences that others included? Did you include consequences that others did not think of?
5. As a class, develop a master plan for the land based on input from all the teams.

For Further Investigation

Invite someone from the city planning department to talk to your class about how land use decisions are made in your community.

24.9 Many Measures Require Ethical Decisions

Each environmental issue has unique, local aspects, in addition to biotic, abiotic, and social aspects. Decisions about specific problems cannot be made from a distance or on a broad basis only. They must, in part, be made by those who are close to the situation and who are themselves affected.

Some people do not believe we should send food to countries suffering from famine unless they agree to control their growing populations, thus helping to prevent future famines. In contrast, some people oppose abortion and feel we should not give any financial aid to countries that promote the use of abortion for population control. Do we have the right to interfere with other countries' ethical decisions when those decisions affect the global environment?

There are similar difficult bioethical decisions to be made in other areas. Does a childless couple have the right to use fertility drugs that may produce multiple births? Do we have the right to use medical technology to

Figure 24.17   The power of humans to shape our environment may lead to continuing usefulness, or this power may be directed toward waste, depletion, and ruin. Consider the choices shown here and in the world around you.

correct potentially fatal genetic disorders? Do we have the right to use other primates as sources of organs for transplants or as subjects for research that may be painful? Do industrialized countries have the right to maintain high-consumption life-styles at the expense of other, less-developed countries? Do we have the right to destroy the habitats of other animals for our own purposes (Figure 24.17)? These are but a few of the issues we need to address if we are to be successful in turning around the current course of events.

24.10   Managing the Earth Effectively Requires Personal Commitment

Most large-scale measures to combat the global population and pollution crisis require legislation and worldwide cooperation. There are, however, choices that individuals can make now that also will make a difference. To move from causing these environmental problems to correcting them, we must each ask ourselves what type of world we want, what type of world we can get, and what changes we are willing to make to get it.

Technology cannot cure *all* the ailments humans have created, but if we become ecologically informed, choose a simpler life-style by reducing our consumption of resources and our production of wastes, and limit the number of children we have, there is hope for the future.

For example, you can improve the energy efficiency of your home in numerous ways. Simple things include replacing incandescent bulbs with fluorescent bulbs, replacing the weather stripping around the doors, and turning down the thermostat on the furnace and water heater. Lessen household water use: Use low-flow shower heads and replace the washers on all leaky faucets. Do not let the water run the entire time you brush your teeth or shave. Turn off lights when you leave the room, and do not leave on outdoor lights all night.

Further measures include riding a bicycle, walking, or using public transportation instead of driving. If you have to drive, use a car that gets good gas mileage and carpool whenever possible. In the store, do not buy overpackaged and overprocessed products. Eat lower on the food chain by consuming more fruit, vegetables, and whole grains, and less meat and dairy products. Recycle paper, aluminum, steel, glass, and plastic. Try to distinguish between needs and wants before you purchase any item.

Each of these small acts constitutes a personal economic and political decision that, when combined with those of many others, can lead to large-scale changes with profound effects on the earth. Remember that all organisms on the earth, including ourselves, depend on many types of energy flows and chemical cycles that are affected by human activity. These life-sustaining processes and cycles are a beautiful and diverse web of interrelationships—the web of life.

Concept Review

1. Explain why countries must cooperate to solve environmental problems. Give examples.
2. How can technology help solve some of the environmental problems humans have created?
3. How does sustainable agriculture differ from present agricultural practices?
4. What measures can be taken to control population growth?
5. How can ethics affect environmental decisions? Give examples.
6. What steps can individuals take to help correct environmental problems?

Summary

The earliest humans changed the biosphere little, living within it much like other animals. With the development of culture and technology, however, our ancestors began adapting environments to make them more hospitable to themselves. The advent of agriculture and industry greatly increased the size of the human population and the impact of humans on their surroundings.

Today, the exploding population and the subsequent demand for resources have created severe environmental problems that threaten life. Changes in governmental policies, industrial practices, and life-styles must be implemented soon to slow the environmental decline. Shifting agricultural practices away from industrialized systems to sustainable systems will provide food without degrading the soil and depleting resources. Using our knowledge of ecology and the environment, we can help manage the earth in a responsible manner. Humans are a small but powerful part of the web of life. We have the chance to help determine the future. By learning about ecology and caring about the earth and the living things on it we can help bring about the rebirth of our planet.

Applications

1. What are the major disadvantages of the life of a hunter-gatherer?

2. Many hunter-gatherers were semi-nomadic people—they settled in one place for a while and then moved to another. Why do you think hunter-gatherers moved from one location to another within a broad territory?

3. Do you think agriculture developed in one part of the world and then spread to other parts, or do you think it developed independently in many parts of the world? What evidence might support your position?

4. Your text states that one of the most important developments in human evolution was the use of symbolic language. Why is this considered to be so important?

5. Very few plants and animals have been domesticated by humans compared with the great number of species in the wild. What do you think are important characteristics of plants and animals that have been domesticated?

6. Does the United States have a population problem? Support your answer.

7. In what ways have automobiles contributed to the human impact on the biosphere?

8. A population map of the United States would show large areas that are barely inhabited. If someone argued that the map showed there is no population problem here, what arguments might you present for another conclusion?

9. How does annual crop rotation—the alternation of crops such as corn with leguminous plants such as soybeans—aid the soil?

10. Do you think the United States should send food to starving children in other countries? Support your answer.

11. Certain Native American people believe that no human has a right to "own" land and to do with it what he or she wants. Rather, everyone has a right to use the land, and people merely are caretakers or stewards who have the responsibility of preserving the land for use by future generations. How does this philosophy differ from that of most Americans? Is this an important or realistic philosophy for America? Why or why not?

12. If Americans were to eat more corn muffins and less corn-fed beef, what would be the effect of such a change on corn consumption and production?

Problems

1. Suppose you are among the first colonizers on the moon. You must build a habitat, large enough for 20 people, in which you can live for several months. What environmental factors should you take into consideration in your building design, and what would your habitat look like?

2. Make a list of the plants you eat during a week. Find out from where the ancestors of these food plants originally came. How many of them are from what is now the United States?

3. Demographers are concerned not only with the total numbers of persons but also with the *structure* of populations—the relative numbers of individuals of various types. For example, two populations of the same size may have different proportions of males and females. Or, two populations of the same size may have different proportions of children and adults. Such data often provide useful information about a population.

a. The total populations of the United States and Sri Lanka are quite different. More important, the United States' population has a smaller proportion of children than that of Sri Lanka. What hypotheses about the two countries can you suggest on the strength of this information?

b. We can divide the population of the United States into three age groups: (1) persons under 20, most of whom are not self-supporting; (2) persons 20 to 65, most of whom are working; (3) persons over 65, most of whom are retired. In recent years the first and third groups have been increasing more rapidly than the second group. What hypotheses can you suggest to explain this? Can you see a future economic problem in this situation?

c. In human females, reproduction occurs mostly between the ages of 15 and 45. Suppose this age group increases more slowly than the age group over 45 but the number of children per female remains the same. What will happen to the birthrate in the population, expressed as births per 1000 of population?

d. The average age at which a female has her first child is higher in nation A than in nation B. The average age at death is about the same in both nations. From this information, make a guess about the rate of population growth in the two countries. What additional information would make your guess more reliable?

4. If enough nuclear power plants were built, they could provide all of the energy the United States needs. The U.S. public, however, has been reluctant to support the building of such plants. Is there a place for nuclear energy in the United States of the future? What are the advantages and disadvantages of nuclear power plants?

5. Suppose the United States was not able to import any natural resources from other countries. What effect would this have on our economy? What products would not be made, and what activities would you have to stop?

6. Research what changes have taken place during the last 50 years in the farming industry, and report to the class.

7. What can you do to help reduce the amount of resources you consume and the amount of wastes you produce each year? What can you, your family, your school, your city, and your state do? Make a list of suggestions, and carry out as many as possible.

8. The term *ecology* has become a household word, but it is often used as if it were a synonym for *environment*. Sometimes it is used merely as a vague indication of something good. How would you explain the scientific meaning of the word *ecology* to a person who has never studied biology?

9. Identify a local environmental problem in your community. Research the biotic, abiotic, and social aspects of the problem.

General Procedures
for the Laboratory

The laboratory is a scientist's workshop—the place where ideas are tested. In the laboratory portion of this course, you will see evidence that supports major biological concepts. To pursue your investigations effectively, you need to learn certain basic techniques, including safe laboratory practices, record keeping, report writing, and measurement. The information on the following pages will help you acquire these techniques.

Laboratory Safety

The laboratory has the potential to be either a safe place or a dangerous place. The difference depends on your knowledge of and adherence to safe laboratory practices. It is important that you read the information here and learn how to recognize and avoid potentially hazardous situations. Basic rules for working safely in the laboratory include the following:

1. Be prepared. Study the assigned investigation before you come to class. Resolve any questions about the procedures before you begin to work.

2. Be organized. Arrange the materials needed for the investigation in an orderly fashion.

3. Maintain a clean, open work area, free of everything except those materials necessary for the assigned investigation. Store books, backpacks, and purses out of the way. Keep laboratory materials away from the edge of the work surface.

4. Tie back long hair and remove dangling jewelry. Roll up long sleeves and tuck long neckties into your shirt. Do not wear loose-fitting sleeves or open-toed shoes in the laboratory.

5. Wear safety goggles and a lab apron whenever working with chemicals, hot liquids, lab burners, hot plates, or apparatus that could break or shatter. Wear protective gloves when working with preserved specimens, toxic and corrosive chemicals, or when otherwise directed to do so.

6. Never wear contact lenses while conducting any experiment involving chemicals. If you must wear them (by a physician's order), inform your teacher *prior* to conducting any experiment involving chemicals.

7. Never use direct or reflected sunlight to illuminate your microscope or any other optical device. Direct or reflected sunlight can cause serious damage to your retina.

8. Keep your hands away from the sharp or pointed ends of equipment such as scalpels, dissecting needles, or scissors.

safety goggles

Safety goggles are for eye protection. Wear goggles whenever you see this symbol. If you wear glasses, be sure the goggles fit comfortably over them. In case of splashes into the eye, flush the eye (including under the lid) at an eyewash station for 15 to 20 minutes. If you wear contact lenses, remove them *immediately* and flush the eye as directed. Call your teacher.

lab apron

A lab apron is intended to protect your clothing. Whenever you see this symbol, put on your apron and tie it securely behind you. If you spill any substance on your clothing, call your teacher.

gloves

Wear gloves when you see this symbol or whenever your teacher directs you to do so. Wear them when using *any* chemical or reagent solution. Do not wear your gloves for an extended period of time.

sharp object

Sharp objects can cause injury, either a cut or a puncture. Handle all sharp objects with caution and use them only as your teacher instructs you. *Do not* use them for any purpose other than the intended one. If you do get a cut or puncture, call your teacher and get first aid.

irritant

An irritant is any substance that, on contact, can cause reddening of living tissue. Wear safety goggles, lab apron, and protective gloves when handling any irritating chemical. In case of contact, flush the affected area with soap and water for at least 15 minutes and call your teacher. Remove contaminated clothing.

reactive

These chemicals are capable of reacting with any other substance, including water, and can cause a violent reaction. *Do not* mix a reactive chemical with any other substance, including water, unless directed to do so by your teacher. Wear your safety goggles, lab apron, and protective gloves.

corrosive

A corrosive substance injures or destroys body tissue on contact by direct chemical action. When handling any corrosive substance, wear safety goggles, lab apron, and protective gloves. In case of contact with a corrosive material, *immediately* flush the affected area with water and call your teacher.

flammable

A flammable substance is any material capable of igniting under certain conditions. Do not bring flammable materials into contact with open flames or near heat sources unless instructed to do so by your teacher. Remember that flammable liquids give off vapors that can be ignited by a nearby heat source. Should a fire occur, *do not* attempt to extinguish it yourself. Call your teacher. Wear safety goggles, lab apron, and protective gloves whenever handling a flammable substance.

poison

Poisons can cause injury by direct action within a body system through direct contact (skin), inhalation, ingestion, or penetration. *Always* wear safety goggles, lab apron, and protective gloves when handling any material with this label. Before handling any poison, inform your teacher if you have any preexisting injuries to your skin. In case of contact, call your teacher *immediately.*

biohazard

Any biological substance that can cause infection through exposure is a biohazard. Before handling any material so labeled, review your teacher's specific instructions. *Do not* handle in any manner other than as instructed. Wear safety goggles, lab apron, and protective gloves. Any contact with a biohazard should be reported to your teacher *immediately.*

NO FOOD OR DRINKS SHOULD BE PRESENT IN THE LAB AT ANY TIME

Figure A1.1

9. Observe all cautions in the procedural steps of the investigation. **CAUTION, WARNING,** and **DANGER** are signal words used in the text and on labeled chemicals or reagents that tell you about the potential for harm and/or injury. They remind you to observe specific safety practices. *Always read and follow these statements.* They are meant to help keep you and your fellow students safe.

    CAUTION statements advise you that the material or procedure has *some potential risk* of harm or injury if directions are not followed.

    WARNING statements advise you that the material or procedure has a *moderate risk* of harm or injury if directions are not followed.

    DANGER statements advise you that the material or procedure has a *high risk* of harm or injury if directions are not followed.

10. Become familiar with the caution symbols identified in Figure A1.1.

11. Never put anything in your mouth and never touch or taste substances in the laboratory unless specifically instructed to do so by your teacher.

12. Never smell substances in the laboratory without specific instructions. Even then, do not inhale fumes directly; wave the air above the substance toward your nose and sniff carefully.

13. Never eat, drink, chew gum, or apply cosmetics in the laboratory. Do not store food or beverages in the lab area.

14. Know the location of all safety equipment and learn how to use each piece of equipment.

15. If you witness an unsafe incident, an accident, or a chemical spill, report it to your teacher immediately.

16. Use materials only from containers labeled with the name of the chemical and the precautions to be used. Become familiar with the safety precautions for each chemical by reading the label before use.

17. When diluting acid with water, *always add acid to water.*

18. Never return unused chemicals to the stock bottles. Do not put any object into a chemical bottle, except the dropper with which it may be equipped.

19. Clean up thoroughly. Dispose of chemicals and wash used glassware and instruments according to the teacher's instructions. Clean tables and sinks. Put away all equipment and supplies. Make sure all water, gas jets, burners, and electrical appliances are turned off. Return all laboratory materials and equipment to their proper places.

20. Wash your hands thoroughly after handling any living organisms or hazardous materials and before leaving the laboratory.

21. Never perform unauthorized experiments. Do only those experiments assigned by your teacher.

22. Never work alone in the laboratory, and never work without the teacher's supervision.

23. Approach laboratory work with maturity. Never run, push, or engage in horseplay or practical jokes of any type in the laboratory. Use laboratory materials and equipment only as directed.

In addition to observing these general safety precautions, you need to know about some specific categories of safety. Before you do any laboratory work, familiarize yourself with the following precautions.

Heat

1. Use only the source of heat specified in the investigation.
2. Never allow flammable materials, such as alcohol, near a flame or any other source of ignition.
3. When heating a substance in a test tube, point the mouth of the tube away from other students and yourself.
4. Never leave a lighted lab burner, hot plate, or any other hot object unattended.
5. Never reach over an exposed flame or other heat source.
6. Use tongs, test-tube clamps, insulated gloves, or potholders to handle hot equipment.

Glassware

1. Never use cracked or chipped glassware.
2. Use caution and proper equipment when handling hot glassware; remember that hot glass looks the same as cool glass.
3. Make sure glassware is clean before you use it and clean when you store it.
4. When putting glass tubing into a rubber stopper, moisten the tubing and the stopper. When putting glass tubing into or removing it from a rubber stopper, protect your hands with heavy cloth. Never force or twist the tubing.
5. Broken glassware should be swept up immediately (never picked up with your fingers) and discarded in a special, labeled container for broken glass.

Electrical Equipment and Other Apparatus

1. Before you begin any work, learn how to use each piece of apparatus safely and correctly in order to obtain accurate scientific information.
2. Never use equipment with frayed insulation or with loose or broken wires.
3. Make sure the area under and around electrical equipment is dry and free of flammable materials. Never touch electrical equipment with wet hands.
4. Turn off all power switches before plugging an appliance into an outlet. Never jerk wires from outlets or pull appliance plugs out by the wire.

Living and Preserved Specimens

1. Specimens for dissection should be properly mounted and supported. Do not cut a specimen while holding it in your hand.
2. Wash down your work surface with a disinfectant solution both before and after using live microorganisms.
3. Always wash your hands with soap and water after working with live or preserved specimens.
4. Animals must be cared for humanely. General rules are as follows:
   a. Always follow carefully the teacher's instructions concerning the care of laboratory animals.

b. Provide a suitable escape-proof container in a location where the animal will not be constantly disturbed.

c. Keep the container clean. Cages of small birds and mammals should be cleaned daily. Provide proper ventilation, light, and temperature.

d. Provide water at all times.

e. Feed regularly, depending on the animal's needs.

f. Treat laboratory animals gently and with kindness in all situations.

g. If you are responsible for the regular care of any animals, be sure to make arrangements for weekends, holidays, and vacations.

h. When animals must be disposed of or released, your teacher will provide a suitable method.

5. Many plants or plant parts are poisonous. Work only with the plants specified by your teacher. Never put any plant or plant parts in your mouth.

6. Handle plants carefully and gently. Most plants must have light, soil, and water, although requirements differ.

Accident Procedures

1. Report *all* incidents, accidents, and injuries, and all breakage and spills, no matter how minor, to the teacher.

2. If a chemical spills on your skin or clothing, wash it off immediately with plenty of water and notify the teacher.

3. If a chemical gets into your eyes or on your face, wash immediately at the eyewash fountain with plenty of water. Wash for at least 15 minutes, flushing the eyes, including under each eyelid. Have a classmate notify the teacher.

4. If a chemical spills on the floor or work surface, do not clean it up yourself. Notify the teacher immediately.

5. If a thermometer breaks, do not touch the broken pieces with your bare hands. Notify the teacher immediately.

6. Smother small fires with a wet towel. Use a blanket or the safety shower to extinguish clothing fires. Always notify the teacher.

7. All cuts and abrasions received in the laboratory, no matter how small, should be reported to your teacher.

Record Keeping

Science deals with verifiable observations. No one—not even the original observer—can trust the accuracy of a confusing, indefinite, or incomplete observation. Scientific record keeping requires clear and accurate records made *at the time of observation*.

The best method of keeping records is to jot them down in a data book. It should be a hardcover book, permanently bound (not loose-leaf), and preferably with square-grid (graph-paper type) pages.

Keep records in diary form, recording the date first. Keep observations of two or more investigations separate. Data recorded in words should be brief but to the point. Complete sentences are not necessary, but single words are seldom descriptive enough to represent accurately what you have observed.

You may choose to sketch your observations. A drawing often records an observation more easily, completely, and accurately than words can. Your sketches need not be works of art. Their success depends on your ability to observe, not on your artistic talent. Keep the drawings simple, use a hard pencil, and include clearly written labels.

Data recorded numerically as counts or measurements should include the units in which the measurements are made. Often numerical data are most easily recorded in the form of a table.

Do not record your data on other papers to be copied into the data book later. Doing so might increase neatness, but it will decrease accuracy. You data book is *your* record, regardless of the blots and stains that are a normal circumstance of field and laboratory work.

You will do much of your laboratory work as a member of a team. Your data book, therefore, will sometimes contain data contributed by other team members. Keep track of the source of observations by circling (or recording in a different color) the data reported by others.

Writing Laboratory Reports

Discoveries become a part of science only when they are known to others. Communication, therefore, is important. In writing, scientists must express themselves so clearly that a person can repeat their procedures exactly. The reader must know what material was used (in biology this includes the species of organism) and must be able to understand every detail of the work. Scientific reports frequently are written in a standard form as follows:

1. Title
2. Introduction: a statement of how the problem arose, often including a summary of past work
3. Materials and equipment: usually a list of all equipment, chemicals, specimens, and other materials
4. Procedure: a complete and exact account of the methods used in gathering data
5. Results: data obtained from the procedure, often in the form of tables and graphs
6. Discussion: a section that demonstrates the relationship between the data and the purpose of the work
7. Conclusion: a summary of the meaning of the results, often suggesting further work that might be done
8. References: published scientific reports and papers that you have mentioned specifically

Your teacher will tell you what form your laboratory reports should take for this course. Part of your work may include written answers to the discussion questions at the end of each investigation. In any event, the material in your data book will be the basis for your reports.

Measurement

Measurement in science is made using the *Système Internationale d'Unités* (International System of Units), more commonly referred to as "SI." A modification of the older metric system, SI was used first in France and is now the common system of measurement throughout the world.

Among the basic units of SI measurement are the meter (length), the kilogram (mass), the kelvin (temperature), and the second (time). All other SI units are derived from these four. Units of temperature that you will use in this course are degrees Celsius, which are equal to kelvins.

SI units for volume are based on meters cubed. In addition, you will use units based on the liter for measuring volumes of liquids. Although not officially part of SI, liter measure is used commonly. Like meters cubed, the liter is metric (1 L = 0.001m³).

Some of the SI units derived from the basic units for length, mass, volume, and temperature follow. Note especially those units described as most common in your laboratory work.

Length

1 kilometer (km) = 1000 meters
1 hectometer (hm) = 100 meters
1 dekameter (dkm) = 10 meters
1 meter (m)
1 decimeter (dm) = 0.1 meter
1 centimeter (cm) = 0.01 meter
1 millimeter (mm) = 0.001 meter
1 micrometer (μm) = 0.000001 meter
1 nanometer (nm) = 0.000000001 meter

Measurements under microscopes often are made in micrometers. Still smaller measurements, as for wavelengths of light used by plants in photosynthesis, are made in nanometers. The units of length you will use most frequently in the laboratory are centimeters (cm).

Units of area are derived from units of length by multiplication. One hectometer squared is one measure that is often used for ecological studies; it commonly is called a hectare and equals 10 000 m². Measurements of area made in the laboratory most frequently will be in centimeters squared (cm²).

Units of volume also are derived from units of length by multiplication. One meter cubed (m³) is the standard, but it is too large for practical use in the laboratory. Centimeters cubed (cm³) are the more common units you will see. For the conditions you will encounter in these laboratory investigations, 1 cm³ measures a volume equal to 1 mL.

Mass

1 kilogram (kg) = 1000 grams
1 hectogram (hg) = 100 grams
1 dekagram (dkg) = 10 grams
1 gram (g)
1 decigram (dg) = 0.1 gram
1 centigram (cg) = 0.01 gram
1 milligram (mg) = 0.001 gram
1 microgram (μg) = 0.000001 gram
1 nanogram (ng) = 0.000000001 gram

Measurements of mass in your biology laboratory usually will be made in kilograms, grams, centigrams, and milligrams.

°F | °C

210 — 100
200 —
190 — 90
180 —
170 — 80
160 —
150 — 70
140 — 60
130 —
120 — 50
110 —
100 — 40
90 —
80 — 30
70 — 20
60 —
50 — 10
40 —
30 — 0
20 —
10 — -10
0 —
-10 — -20

°F | °C

Figure A1.2 A comparison of Fahrenheit and Celsius temperature scales.

Volume

1 kiloliter (kL) = 1000 liters
1 hectoliter (hL) = 100 liters
1 dekaliter (dkL) = 10 liters
1 liter (L)
1 deciliter (dL) = 0.1 liter
1 centiliter (cL) = 0.01 liter
1 milliliter (mL) = 0.001 liter

Your volume measurements in the laboratory usually will be made in glassware marked for milliliters and liters.

Temperature

On the Celsius scale, 0° C is the freezing point of water, and 100° C is the boiling point of water. (Atmospheric pressure affects both of these temperatures.) Figure A1.2 illustrates the Celsius scale alongside the Fahrenheit scale, which still is used in the United States. On the Fahrenheit scale, 32° F is the freezing point of water, and 212° F is the boiling point of water. Figure A1.2 is useful in converting from one scale to the other.

Another type of measurement you will encounter in this course is molarity (labeled with the letter M). Molarity measures the concentration of a dissolved substance in a solution. A high molarity indicates a high concentration. The solutions you will use in the investigations frequently will be identified by their molarity.

If you wish to learn more about SI measure, write to the U.S. Department of Commerce, National Institute of Standards and Technology (NIST), Washington, DC 20234.

APPENDIX TWO
Supplementary Investigations

INVESTIGATION A2.1 Chemical Safety

All chemicals are hazardous in some way. A hazardous chemical is defined as a substance that is likely to cause injury. Chemicals can be placed in four hazard categories: flammable/combustible, toxic, corrosive, and reactive. Table A2.1 summarizes their characteristics. In the investigations for this course, every effort is made to minimize the use of dangerous materials. Many "less hazardous" chemicals can cause injury, however, if not handled properly. This investigation will help you become aware of the four types of chemical hazards and of how you can reduce the risk of injury when using chemicals. Be sure to review the safety rules in Appendix One before you work with any chemical.

Materials (per team of 2)
2 pairs of safety goggles   2 pairs of plastic gloves
2 lab aprons

PART A
none

PART B
3 microscope slides   concentrated biological stain in dropping
3 cover slips   bottle
dropping pipet   dilute biological stain in dropping bottle
3 flat toothpicks   Detain™ in squeeze bottle
compound microscope   paramecium culture

PART C
2 glass dropping pipets   100 mL red food dye
2 100-mL beakers   scissors
white paper

PART D
100-mL glass beaker   baking soda (sodium bicarbonate)
spatula   vinegar (acetic acid)

PART A Flammable/Combustible Substances
Procedure
Flammable/combustible substances are solids, liquids, or gases that will burn or that will sustain burning. The process of burning involves three interrelated components: fuel (any substance capable of burning), oxidizer (often air or a specific chemical), and ignition source (a spark, flame, or heat). The three components are represented in the diagram of a fire triangle in Figure A2.1. For burning to occur, all three components (sides) of

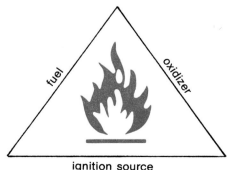

Figure A2.1 The fire triangle.

699

TABLE A2.1   Categories of Hazardous Materials

Type of Material	Categories
Flammable/Combustible Substance Common Lab Hazard 	**Flammable Liquid**—Liquid having a flashpoint* less than 100° F (37.8° C) **Combustible Liquid**—Flashpoint* equal to or greater than 100° F (37.8° C) but less than 200° F. 　*Flashpoint—the lowest temperature at which a liquid gives off vapor that forms an ignitable mixture with air near the surface. **Flammable Solid**—Causes fire through friction, absorption of moisture, or spontaneous chemical change. Ignites readily; burns vigorously. **Flammable Gas**—Forms a flammable mixture with air at ambient temperature and pressure.
Reactive Chemical Violent reaction under certain ambient or induced conditions. Spontaneous generation of great heat, light, and flammable and nonflammable gases or toxicants. 	**Acid-Sensitive**—Reacts with acids or acid fumes to generate heat, flammable or explosive gases, or toxicants. **Water-Sensitive**—Reacts with moisture to generate heat and/or flammable or explosive gases. **Oxidizer**—Promotes combustion in other materials through release of oxygen or other gases.
Corrosive Chemical Injures body tissue and corrodes metal by direct chemical action. 	**Corrosive Liquid, Solid, Gas**—Causes visible destruction or irreversible alterations in living tissue. **Irritant**—Causes reversible inflammation in living tissue. **Sensitizer**—Causes allergic reaction in normal tissue of a substantial number of individuals after repeated exposure.
Toxic Chemical Injures by direct action with body systems when tolerable limits are exceeded. Exposure routes: direct contact, inhalation, ingestion, and penetration. 	**Inhalation**—Toxic gases may pass rapidly into capillary beds of lungs and be carried to all parts of the body via circulatory system. **Ingestion**—Toxics may damage tissues of mouth, throat, and gastrointestinal tract; produce systemic poisoning if absorbed through these tissues. **Skin/Eye Contact**—Hair follicles, sweat glands, as well as cuts and abrasions are the main portals of entry. Eyes are acutely sensitive to chemical irritants as well as corrosives. **Injection/Penetration**—Exposure to toxics by injection seldom occurs; but wounds by broken glass or metal are frequent entry points for injected chemicals.

Information courtesy of Ward's Natural Science Establishment.

the fire triangle must be present. To control fire hazard, one must remove, or otherwise make inaccessible, at least one side of the fire triangle. Flammable chemicals should not be used in the presence of ignition sources such as lab burners, hot plates, and sparks from electrical equipment or static electricity. Containers of flammables should be closed when not in use. Sufficient ventilation in the laboratory will help to keep the concentration of flammable vapors to a minimum. Wearing safety goggles, lab aprons, and protective gloves are important precautionary measures when using flammable/combustible materials.

1. Use your knowledge to draw fire triangles with the appropriate side removed to represent the following control measures.

   a. Isolate flammable materials from reactive chemicals.
   b. Do not store more than 600 mL of flammable material in glass in the laboratory.
   c. Transfer flammable liquids in a working fume hood whenever possible.
   d. Eliminate sources of ignition.
   e. Store flammable substances in a cool area, at a maximum temperature of 27° C (80° F).
   f. Use special absorbent materials to reduce vapor pressure when wiping up spills.
   g. Transport flammable liquids in metal or other protective containers.

2. Use your knowledge to circle the side of the fire triangle that is most vulnerable to these common safety problems.

   a. Improperly stored glass containers of flammable solvents.
   b. A lack of adequate grounding to prevent sparking (by static electricity) generated by flowing liquids.
   c. Flammable liquids mixed in a waste disposal can with other chemicals.
   d. Open flames on a laboratory bench near flammable liquids.
   e. Transporting glass containers containing flammable liquids on top of a cart.
   f. Flammable substances stored on a shelf with acids and other reactive chemicals.

Discussion

1. What fire control measures are used at gas stations when fuel is pumped?
2. Why do firefighters apply foam when a flammable material is spilled?

PART B   Toxic Substances

SAFETY   **Put on your safety goggles, lab apron, and plastic gloves.**

Procedure

Most of the chemicals in a laboratory are toxic, or poisonous to life. The degree of toxicity depends on the properties of the substance, its concentration, the type of exposure, and other variables. The effects of a toxic substance can range from minor discomfort to serious illness or death. Exposure to toxic substances can occur through ingestion, skin contact, or through inhaling toxic vapors. Wearing safety goggles, lab apron, and protective gloves are important precautions when using toxic chemicals. A clean work area, prompt spill cleanup, and good ventilation also are important.

3. Prepare a wet mount with one drop each of paramecium culture, concentrated biological stain, and Detain™ (protist slowing agent). Mix with a toothpick and label the slide *A*.
4. Repeat Step 3, but this time use dilute biological stain. Label this slide *B*.
5. Repeat Step 3, but do not add any stain. Label this slide *C*.
6. Observe each of the three slides for 15 to 20 minutes under various magnifications of a compound microscope.
7. Wash your hands thoroughly before leaving the laboratory.

Discussion

1. What is the purpose of the third slide?
2. Which concentration of stain is toxic to paramecia?
3. What observations would yield criteria to measure cell toxicity of the stain?

PART C   Corrosive Substances

SAFETY   Put on your safety goggles, lab apron, and plastic gloves.

Procedure

Corrosive chemicals are solids, liquids, or gases that by direct chemical action either destroy living tissue or cause permanent change in the tissue. Corrosive substances can destroy eye and respiratory-tract tissues, causing impaired sight or permanent blindness, severe disfigurement, permanent severe breathing difficulties, and even death. Safety goggles, lab apron, and protective gloves should be worn when handling corrosive chemicals to prevent contact with the skin or eyes. Splashes on the skin or in the eyes should be washed off immediately while a classmate notifies the teacher.

I   Corrosive solids

Corrosive solids may appear relatively harmless because they can be removed more easily than liquids and because they may not cause an immediate destructive effect. The effect depends largely on their solubility in skin moisture and even more rapid solubility in the moisture present in the respiratory and intestinal tracts.

Caustic alkalies (sodium or potassium hydroxides) are perhaps the greatest potential hazard because of their wide use in general science laboratories. The hazards of corrosive solids include the following:

   a. Solutions of corrosive solids are readily absorbed through the skin.
   b. Caustic alkalies and other corrosive solids may produce delayed reactions rather than immediate, painful reactions.
   c. Many corrosive solids dust easily, thus increasing the hazard through other exposure routes (inhalation).
   d. Molten corrosive solids greatly increase the threat of exposure by acting as liquid corrosives.

The most serious hazard associated with corrosives is from materials in the gaseous state. In this state, corrosives are rapidly absorbed into the body by dissolving skin moisture and by inhalation. Many corrosives give off dangerous vapors either by themselves or during spills or chemical reactions.

8. Your teacher has set up a demonstration that shows the action of corrosive solids. Agar is used to represent body tissue. Observe the effects of each substance on the agar during the next 30 to 45 minutes.

II   Inorganic acids

Everyone knows that acids can cause burns, but few people realize the extent to which they can damage body tissues. For example, sulfuric acid can destroy skin and eye tissues and dissolve teeth, and even the vapor can destroy lung and other respiratory tissues.

Every student has heard the admonishment, "Do as you oughta . . . add acid to water." The following investigation demonstrates why this practice is critical to chemical safety.

9. For illustration purposes, the red food coloring will represent hydrochloric acid. Cut and position 7-cm wide white paper strips around the top edge of two beakers.

Fill one beaker with red dye, the other with tap water. Fill the beakers to the same height, about 2 cm from the lip.

10. Use a dropping pipet to add 5 to 10 drops of water to the beaker containing the "acid" (red dye). Hold the dropping pipet 15 to 20 cm above the beaker—normal dispensing height.

11. Use a dropping pipet to add 5 to 10 drops of "acid" (red dye) to the beaker containing the water. Hold the dropping pipet at normal dispensing height when adding the red dye.

12. Remove the white paper from around each beaker and check carefully for signs of splashing.

Discussion

1. Which solid(s) had the greatest corrosive effect? Cite evidence for your observation.

2. Which solid(s) had the least corrosive effect?

3. What factor present in the agar is responsible for initiating the corrosive effect?

4. Are some substances noncorrosive? On what observed characteristics is your answer based?

5. Based on your observations, what can you conclude about those solids that are corrosive?

6. What can you conclude from the evidence (the color of the splash marks) about acid dilution?

PART D   Reactive Substances

SAFETY   **Put on your safety goggles, lab apron, and plastic gloves.**

Procedure

Reactive chemicals (see Table A2.1) promote violent reactions under certain conditions. A chemical may explode spontaneously or when mechanically disturbed. Reactive chemicals also include those that react rapidly when mixed with another chemical, releasing a large amount of energy. Keep chemicals separate from each other unless they are being combined according to specific instructions in an investigation. Heed any other cautions your teacher may give you. Always wear safety goggles, lab apron, and protective gloves when handling reactive chemicals.

13. Pour a small amount of vinegar in the glass beaker.

 CAUTION: Vinegar is an irritant. Avoid skin/eye contact; do not ingest. In case of spills or splashes, call your teacher immediately; flush with water 15 minutes; rinse mouth with water.

14. Use the spatula to scoop up a pinch of baking soda and add it to the vinegar. Record what happens.

15. Wash your hands thoroughly before leaving the laboratory.

Discussion

1. What reaction category did this chemical reaction demonstrate?

2. Would you expect this reaction to proceed faster if sulfuric acid were used? Explain.

eyepiece

arm

low-power
objective

stage
clip

coarse
adjust-
ment

fine
adjust-
ment

base

revolving
nosepiece

high-power
objective

scanning
power

stage

diaphragm

substage lamp

lamp switch

electric
cord

10×   4X 40X

Figure A2.2   Parts of a compound
microscope.

INVESTIGATION A2.2   The Compound Microscope

The human eye cannot distinguish objects much smaller than 0.1 mm in size. The microscope is a biological tool that extends vision and allows observation of much smaller objects. The most commonly used compound microscope (Figure A2.2) is monocular (one eyepiece). Light reaches the eye after being passed through the objects to be examined. In this investigation, you will learn how to use and care for a microscope.

Materials (per person or team of 2)

3 cover slips	scissors
3 microscope slides	transparent metric ruler
dropping pipet	lens paper
compound microscope	newspaper

Procedure
PART A   Care of the Microscope

1. The microscope is a precision instrument that requires proper care. Always carry the microscope with both hands, one hand under its base, the other on its arm.

2. When setting the microscope on a table, keep it away from the edge. If a lamp is attached to the microscope, keep its wire out of the way. Keep everything not needed for microscope studies off your lab table.

3. Avoid tilting the microscope when using temporary slides made with water.

4. The lenses of the microscope cost almost as much as all the other parts put together. Never clean lenses with anything other than the lens paper designed for this task.

5. Before putting the microscope away, *always* return it to the low-power setting. The high-power objective reaches too near the stage to be left in place safely.

PART B   Setting Up the Microscope

6. Rotate the low-power objective into place if it is not already there. When you change from one objective to another you will hear the objective click into position.

7. Move the mirror so that even illumination is obtained through the opening in the stage, or turn on the substage lamp. Most microscopes are equipped with a diaphragm for regulating light. Some materials are best viewed in dim light, others in bright light.

    DANGER: Never use a microscope mirror to capture direct sunlight when illuminating objects viewed under a microscope. The mirror concentrates light rays, which can permanently damage the retina of the eye. Always use indirect light.

8. Make sure the lenses are dry and free of fingerprints and debris. Wipe lenses with lens paper only.

PART C   Using the Microscope

9. In your data book, prepare a data table similar to Table A2.2.

10. Cut a lowercase letter o from a piece of newspaper. Place it right side up on a clean slide. With a dropping pipet, place one drop of water on the letter. This type of slide is called a wet mount.

11. Wait until the paper is soaked before adding a cover slip. Hold the cover slip at about a 45° angle to the slide and then slowly lower it. Figure A2.3 shows these first steps.

12. Place the slide on the microscope stage and clamp it down. Move the slide so the letter is in the middle of the hole in the stage. Use the coarse-adjustment knob to lower the low-power objective to the lowest position.

TABLE A2.2   Microscopic Observations

Object Being Viewed	Observations and Comments
Letter *o*	
Letter *c*	
Etc.	
mm ruler	

a   b   c

Figure A2.3   Preparing a wet mount with a microscope slide and cover slip.

13. Look through the eyepiece and use the coarse-adjustment knob to raise the objective slowly, until the letter *o* is in view. Use the fine-adjustment knob to sharpen the focus. Position the diaphragm for the best light. Compare the way the letter looks through the microscope with the way it looks to the naked eye.

14. To determine how greatly magnified the view is, multiply the number inscribed on the eyepiece by the number on the objective being used. For example, eyepiece (10X) × objective (10X) = total (100X).

15. Follow the same procedure with a lowercase *c*. In your data book, describe how the letter appears when viewed through a microscope.

16. Make a wet mount of the letter *e* or the letter *r*. Describe how the letter appears when viewed through the microscope. What new information (not revealed by the letter *c*) is revealed by the *e* or *r*?

17. Look through the eyepiece at the letter as you use your thumbs and forefingers to move the slide slowly *away* from you. Which way does your view of the letter move? Move the slide to the right. Which way does the image move?

18. Make a pencil sketch of the letter as you see it under the microscope. Label the changes in image and movement that occur under the microscope.

19. Make a wet mount of two different-colored hairs, one light and one dark. Cross one hair over the other. Position the slide so the hairs cross in the center of the field. Sketch the hairs under low power; then go to Part D.

PART D   Using High Power

20. With the crossed hairs centered under low power, adjust the diaphragm for the best light.

21. Turn the high-power objective into viewing position. Do *not* change the focus.

22. Sharpen the focus with the *fine-adjustment knob only. Do not focus under high power with the coarse-adjustment knob.*

23. Readjust the diaphragm to get the best light. If you are not successful in finding the object under high power the first time, return to Step 20 and repeat the whole procedure carefully.

24. Using the fine-adjustment knob, focus on the hairs at the point where they cross. Can you see both hairs sharply at the same focus level? How can you use the fine adjustment knob to determine which hair is crossed over the other? Sketch the hairs under high power.

PART E   Measuring with a Microscope

25. Because objects examined with a microscope are usually small, biologists use units of length smaller than centimeters or millimeters for microscopic measurement.

One such unit is the micrometer, which is one thousandth of a millimeter. The symbol for micrometer is μm, the Greek letter μ (called mu) followed by m.

26. You can estimate the size of a microscopic object by comparing it with the size of the circular field of view. To determine the size of the field, place a transparent metric ruler on the stage. Use the low-power objective to obtain a clear image of the divisions on the ruler. Carefully move the ruler until its marked edge passes through the exact center of the field of view. Now, count the number of divisions that you can see in the field of view. The marks on the ruler will appear quite wide; 1 mm is the distance from the center of one mark to the center of the next. Record the diameter, in millimeters, of the low-power field of your microscope.

27. Remove the ruler and replace it with the wet mount of the letter e. (If the mount has dried, lift the cover slip and add water.) Using low power, compare the height of the letter with the diameter of the field of view. Estimate as accurately as possible the actual height of the letter in millimeters.

Discussion

1. Summarize the differences between an image viewed through a microscope and the same image viewed with the naked eye.
2. When viewing an object through the high-power objective, not all of the object may be in focus. Explain.
3. What is the relationship between magnification and the diameter of the field of view?
4. What is the diameter in micrometers of the low-power field of view of your microscope?
5. Calculate the diameter in micrometers of the high-power field. Use the following equations:

$$\frac{\text{magnification number of high-power objective}}{\text{magnification number of low-power objective}} = A$$

$$\frac{\text{diameter of low-power field of view}}{A} = \text{diameter of high-power field of view}$$

For example, if the magnification of your low-power objective is 12X and that of your high-power is 48X, A = 4. If the diameter of the low-power field of view is 1600 μm, the diameter of the high-power field of view is $\frac{1600}{4}$, or 400 μm.

INVESTIGATION A2.3   The Compound Microscope:
Biological Material

This investigation provides practice in using a compound microscope.

Materials (per team of 2)

2 pairs of safety goggles	compound microscope
2 lab aprons	lens paper
microscope slide	paper towel
cover slip	Lugol's iodine solution in dropping bottle
dropping pipet	yeast culture
100-mL beaker containing water	small piece of white potato

SAFETY   **Put on your safety goggles and lab apron.**

Procedure

For setting up the microscope and preparing wet mounts, follow the directions in Investigation A2.2.

1. Place a *small* piece of potato in the center of a clean slide. Place the slide on your laboratory table and carefully press the potato with a finger until some juice is forced out. Distribute the juice evenly over the center of the slide by moving the piece of potato in a circle. Discard the potato. Add a drop of water and a cover slip. Avoid getting air bubbles in the wet mount.

2. Examine the wet mount under low power. To increase contrast between the starch grains and the water surrounding them, decrease the size of the opening in the substage diaphragm. Move the slide on the stage until you locate a field containing well-separated grains. Center several grains in the field and switch to high power. Describe the shape of an individual starch grain. Describe any internal structure you can see in these grains.

3. Turn again to low power. Stain the grains by placing a small drop of Lugol's iodine solution on the slide at one side of the cover slip, as shown in Figure A2.4. Tear off a small piece of paper towel. Place the torn edge in contact with the water at the opposite edge of the cover slip. As water is absorbed by the paper towel at one edge, the Lugol's iodine solution will be drawn under the cover slip at the opposite edge. Continue until the Lugol's solution covers half the space under the cover slip. The Lugol's solution will continue to spread slowly throughout the wet mount.

    WARNING: Lugol's iodine solution is poisonous if ingested, irritating to skin and eyes, and can stain clothing. In case of spills or splashes, call your teacher immediately; flush area with water 15 minutes; rinse mouth with water.

4. Examine various regions of the wet mount to observe the effects of different concentrations of Lugol's iodine solution on starch grains. Examine under low power and then high power. What changes occur in the starch grains exposed to relatively high concentrations of Lugol's iodine solution? What differences do you see between these grains and others exposed to lower concentrations of Lugol's solution? Describe any internal structures you can see in the stained grains. Using the method described in Investigation A2.2, estimate the size (in micrometers) of the larger starch grains.

5. Remove the slide, lift off the cover slip, and wash and dry them. Carefully wipe off any liquid from the microscope stage.

Figure A2.4   Putting liquid under a cover slip.

6. Place one drop of yeast culture on the clean slide and add a cover slip. Examine first under low power and then under high power. Describe the shape of the yeast organisms.

7. Study the arrangement of small groups of these organisms. From your observations, what can you conclude about how new yeast organisms develop? Sketch any internal structures you see.

8. Using Lugol's iodine solution, stain the yeast as you stained the starch grains. Compare the effects of the Lugol's solution on the yeast organisms with its effects on starch grains. Describe any structures that were not visible in the unstained yeast organisms.

9. Using the method from Investigation A2.2, estimate the size (in micrometers) of an average yeast organism.

10. Wash your hands thoroughly before leaving the laboratory.

Discussion

1. Based on your observations, what do you think is the purpose of adding stain to specimens to be examined with a compound microscope?

2. What problems may arise as a result of adding stain to a specimen?

Tables

Table of Relative Humidities

Dry-bulb Temperature	Differences Between Dry-bulb and Wet-bulb Temperatures																																		
(°C)	0.5	1.0	1.5	2.0	2.5	3.0	3.5	4.0	4.5	5.0	5.5	6.0	6.5	7.0	7.5	8.0	8.5	9.0	9.5	10.0	10.5	11.0	11.5	12.0	12.5	13.0	13.5	14.0	14.5	15.0	16.0	17.0	18.0	19.0	20.0
10	94	88	82	77	71	66	60	55	50	44	39	34	29	24	20	15	10	6																	
11	94	89	83	78	72	67	61	56	51	46	41	36	32	27	22	18	13	9	5																
12	94	89	83	78	73	68	63	58	53	48	43	39	34	29	25	21	16	12	8																
13	95	89	83	79	74	69	64	59	54	50	45	41	36	32	28	23	19	15	11	7															
14	95	90	84	79	75	70	65	60	56	51	47	42	38	34	30	26	22	18	14	10	6														
15	95	90	85	80	75	71	66	61	57	53	48	44	40	36	32	27	24	20	16	13	9	6													
16	95	90	85	81	76	71	67	63	58	54	50	46	42	38	34	30	26	23	19	15	12	8	5												
17	95	90	86	81	76	72	68	64	60	55	51	47	43	40	36	32	28	25	21	18	14	11	8												
18	95	91	86	82	77	73	69	65	61	57	53	49	45	41	38	34	30	27	23	20	17	14	10	7											
19	95	91	87	82	78	74	70	65	62	58	54	50	46	43	39	36	32	29	26	22	19	16	13	10	7										
20	96	91	87	83	78	74	70	66	63	59	55	51	48	44	41	37	34	31	28	24	21	18	15	12	9	6									
21	96	91	87	83	79	75	71	67	64	60	56	53	49	46	42	39	36	32	29	26	23	20	17	14	11	8	6								
22	96	92	88	83	79	76	72	68	65	61	57	54	50	47	44	40	37	34	31	28	25	22	19	16	13	11	8	6							
23	96	92	88	84	80	76	72	69	65	62	58	55	52	48	45	42	39	36	33	30	27	24	21	19	16	13	11	8	6						
24	96	92	88	84	80	77	73	69	66	62	59	56	53	49	46	43	40	37	34	31	29	26	23	20	18	15	13	10	8	5					
25	96	92	88	84	81	77	74	70	67	63	60	57	54	50	47	44	41	39	36	33	30	28	25	22	20	17	15	12	10	8	5				
26	96	92	89	85	81	78	74	71	68	64	61	58	55	51	49	46	43	40	37	34	32	29	26	24	21	19	17	14	12	10	7	5			
27	96	92	89	85	82	78	75	71	68	65	62	58	56	52	50	47	44	41	38	35	33	30	28	25	23	21	19	16	14	12	9	7			
28	96	93	89	86	82	78	75	72	69	65	62	59	56	53	50	47	45	42	39	36	34	31	29	26	24	22	20	18	16	14	11	9			
29	96	93	89	86	82	79	76	72	69	66	63	60	57	54	51	49	46	43	41	38	36	33	31	28	26	24	22	19	17	15	12	10	8		
30	96	93	90	86	83	79	76	73	70	67	64	61	58	55	52	50	47	44	42	39	37	35	32	30	28	25	23	21	19	17	14	12	10		
31	96	93	90	86	83	80	77	74	71	68	65	62	59	56	54	51	48	46	43	41	38	36	33	31	29	27	25	23	21	19	16	14	11		
32	96	93	90	87	83	80	77	74	71	68	66	63	60	57	55	52	50	47	45	42	40	38	35	33	31	29	27	25	23	21	18	16	13		
33	97	93	90	87	84	81	78	75	72	69	66	64	61	58	56	53	51	48	46	44	41	39	37	35	32	30	28	26	24	22	19	17	14	12	
34	97	93	90	87	84	81	78	75	72	70	67	64	61	59	56	54	51	49	47	44	42	40	38	36	33	31	29	27	25	23	20	17	15	12	
35	97	94	90	87	84	81	79	76	73	70	68	65	62	60	57	55	53	51	48	46	44	42	39	37	35	33	31	29	26	24	21	19	16	13	
36	97	94	90	87	84	81	79	76	73	71	68	66	63	61	58	56	54	51	49	47	45	43	40	38	36	34	32	30	28	25	23	20	17	15	12
37	97	94	91	87	84	81	79	76	73	71	68	66	63	61	59	56	54	52	50	48	46	43	41	39	37	35	33	31	29	27	24	20	18	16	13
38	97	94	91	88	84	81	79	76	74	71	69	67	64	62	60	58	55	53	51	49	47	45	43	41	39	37	35	33	31	28	25	22	19	17	14
39	97	94	91	88	85	82	79	77	74	72	69	67	64	62	60	58	56	54	51	49	47	45	43	41	39	36	34	32	30	28	25	22	20	18	15
40	97	94	91	88	85	82	80	77	74	72	70	67	65	62	60	58	56	53	51	49	47	45	43	41	39	36	34	32	30	28	25	22	19	16	14

Periodic Table of the Elements

(based on $^{12}_{6}C = 12.0000$)

Key:
- 14 — Atomic number
- Si — Symbol
- Silicon — Name
- 28.0855 — Atomic mass

TRANSITION METALS

INNER TRANSITION METALS

Lanthanide series

Actinide series

Group 1	2	3	4	5	6	7	8	9	10	11	12	13	14	15	16	17	18
1 H Hydrogen 1.007																	2 He Helium 4.0026
3 Li Lithium 6.941	4 Be Beryllium 9.012											5 B Boron 10.81	6 C Carbon 12.0111	7 N Nitrogen 14.0067	8 O Oxygen 15.9994	9 F Fluorine 18.998	10 Ne Neon 20.179
11 Na Sodium 22.98977	12 Mg Magnesium 24.305											13 Al Aluminum 26.9815	14 Si Silicon 28.0855	15 P Phosphorus 30.973	16 S Sulfur 32.06	17 Cl Chlorine 35.453	18 Ar Argon 39.948
19 K Potassium 39.098	20 Ca Calcium 40.08	21 Sc Scandium 44.955	22 Ti Titanium 47.88	23 V Vanadium 50.9415	24 Cr Chromium 51.996	25 Mn Manganese 54.938	26 Fe Iron 55.847	27 Co Cobalt 58.933	28 Ni Nickel 58.69	29 Cu Copper 63.546	30 Zn Zinc 65.39	31 Ga Gallium 69.72	32 Ge Germanium 72.59	33 As Arsenic 74.92	34 Se Selenium 78.96	35 Br Bromine 79.904	36 Kr Krypton 83.80
37 Rb Rubidium 85.467	38 Sr Strontium 87.62	39 Y Yttrium 88.905	40 Zr Zirconium 91.224	41 Nb Niobium 92.906	42 Mo Molybdenum 95.94	43 Tc Technetium (98)	44 Ru Ruthenium 101.07	45 Rh Rhodium 102.906	46 Pd Palladium 106.42	47 Ag Silver 107.868	48 Cd Cadmium 112.41	49 In Indium 114.82	50 Sn Tin 118.71	51 Sb Antimony 121.75	52 Te Tellurium 127.60	53 I Iodine 126.905	54 Xe Xenon 131.29
55 Cs Cesium 132.905	56 Ba Barium 137.3	57 La Lanthanum 138.906	72 Hf Hafnium 178.49	73 Ta Tantalum 180.948	74 W Tungsten 183.85	75 Re Rhenium 186.207	76 Os Osmium 190.2	77 Ir Iridium 192.22	78 Pt Platinum 195.08	79 Au Gold 196.967	80 Hg Mercury 200.59	81 Tl Thallium 204.383	82 Pb Lead 207.2	83 Bi Bismuth 208.980	84 Po Polonium (209)	85 At Astatine (210)	86 Rn Radon (222)
87 Fr Francium (223)	88 Ra Radium (226.0)	89 Ac Actinium 227.028	104 (261)	105 (262)	106 (263)	107 (262)	108	109 (266)									

Lanthanide series:

58 Ce Cerium 140.12	59 Pr Praseodymium 140.908	60 Nd Neodymium 144.24	61 Pm Promethium (145)	62 Sm Samarium 150.36	63 Eu Europium 151.96	64 Gd Gadolinium 157.25	65 Tb Terbium 158.925	66 Dy Dysprosium 162.50	67 Ho Holmium 164.930	68 Er Erbium 167.26	69 Tm Thulium 168.934	70 Yb Ytterbium 173.04	71 Lu Lutetium 174.96

Actinide series:

90 Th Thorium 232.038	91 Pa Protactinium 231.036	92 U Uranium 238.029	93 Np Neptunium (244)	94 Pu Plutonium (244)	95 Am Americium (243)	96 Cm Curium (247)	97 Bk Berkelium (247)	98 Cf Californium (251)	99 Es Einsteinium (252)	100 Fm Fermium (257)	101 Md Mendelevium (258)	102 No Nobelium (259)	103 Lr Lawrencium (260)

Protist/Cyanobacteria Key

Identification may be made either by comparing the observed cell to the line drawings or by using the key. Start at number 1, comparing the observed cell to each of the characteristics listed (1a and 1b). Advance to the number corresponding to the characteristic you have chosen (2 or 17). Proceed according to the key until it terminates in the name of an organism.

1a
White or colorless
.... 2

1b
Colored
.... 17

2a
Slow-creeping or floats without apparent motion
.... 3

2b
Exhibits other motion
.... 8

3a
Slow-creeping
.... 5

3b
Appears to float without apparent motion
.... 4

4a
Cell large, spherical in shape; with central core and radiating spines (axopods)
Actinosphaerium

(200–1000 μm)

4b
Cell small, spherical in shape with radiating spines; with no central core
Actinophrys

(40–50 μm)

5a
Cell possesses a test or shell sometimes composed of small grains of sand
.... 6

5b
Cell naked, has no shell
.... 7

6a
Shell flattened; without attached or embedded material; pale brown in color; with colorless pseudopods ("false feet")
Arecella

(45–100 μm, shell diameter)

6b
Shell dome-shaped; with attached particles, usually of sand; with colorless pseudopods
Difflugia

(60–580 μm x 40–240 μm shell)

7a
Small cell; creeps along using pseudopods; changes shape constantly, usually by extending a single pseudopod. Cell with a single, large nucleus
Amoeba

(to 600 μm)

7b
Large cell; creeps using pseudopods; changes shape constantly, usually by extending many pseudopods. Cell with many (1000+) nuclei
Chaos

(150–450 μm)

8a
Cell has many hairlike structures (cilia) used for movement
.... 9

8b
Cell moves by one or more whiplike flagella
.... 15

9a
Cell body uniformly covered with cilia
.... 10

9b
Cell body not uniformly covered with cilia; instead has either bands or tufts of cilia or distinct pointlike projections (fused groups of cilia called cirri)
.... 16

10a
Cell body elongate, either cylindrical or flattened
.... 11

10b
Cell body other shape
.... 12

11a
Large cell, visible to naked eye, with elongate, flattened body with blunt ends; swims slowly. Cell can contract to 1/4 body size when stimulated.
Spirostomum

(1–3 mm)

11b
Small cell with cylindrical (cigar-shaped) body having rounded ends; swims rapidly in corkscrew fashion
Paramecium

(200–300 μm)

12a
Cell body oval in shape
.... 13

12b
Body other shape
.... 14

13a
Large oval cell, visible to naked eye, with large round mouth at anterior end
Bursaria

(0.5–1 mm)

13b
Small oval cell, mouth on side; rapid swimmer
Colpidium

(50–70 μm)

14a
Large cell with an elongate body having a pointed tip and a pronounced long neck; mouth at base of neck
Dileptus

(250–500 μm)

14b
Small cell with pear-shaped body; mouth at top of cell
Tetrahymena

(40–60 μm)

15a
Small, elongate colorless cell with single long flagellum. Posterior end of cell rounded. Cell highly plastic when

Protist/Cyanobacteria Key   (continued)

stationary; often appears to vibrate when in motion

Parenema

(27–70 μm)

15b
Small, elongate colorless cell; with two visible flagella. Posterior end of cell narrowed

Chilomonas

(20–40 μm)

16a
Cell on stalk
.... 42

16b
Cell not on stalk
.... 44

17a
Cell green, golden brown, or blue-green in color
.... 18

17b
Color not green
.... 47

18a
Cells contain chloroplasts
.... 19

18b
Cells without chloroplasts; blue-green pigment seemingly distributed and diffused throughout the cell or nearly so (Cyanobacteria)
.... 33

19a
Cells arranged in filaments (threads) or branching filaments
.... 30

19b
Cells not arranged in filaments
.... 20

20a
Single cell or a group of cells that show no apparent motion; with distinct glasslike walls etched with grooves or holes forming delicate patterns
.... 21

20b
Single cell or group of cells with or without motion; without distinct glasslike walls or delicate etchings
.... 25

21a
Single cell
.... 22

21b
Cells in groups or clusters
.... 23

22a
Cell cigar-shaped, or cigar-shaped with end knobs.

Navicula

(200 μm)

22b
Cell bean pod-shaped

Hantzschia

(900 μm)

23a
Cell rectangular in shape
.... 24

23b
Cells barrel-shaped, arranged in a filament with clear center plates

Melosira

(14–25 μm, cell)

24a
Rectangular-shaped cells arranged in a zig-zag pattern

Tabellaria

(50 μm, cell)

24b
Rectangular-shaped cells arranged in ribbons

Fragilaria

(100 μm, cell)

25a
Single cell (sometimes may occur in clusters)
.... 26

25b
Groups of cells arranged in a definite colony
.... 39

26a
Cell with flagella
.... 27

26b
Cell usually single; no flagella (sometimes may occur in clusters)
.... 37

27a
Cell colored golden-brown; differentiated into two distinct cones, each composed of plates. Cell has two small flagella in an area between cones

Peridinium

(50–65 μm)

27b
Cell green-colored without cones made of plates
.... 28

28a
Cell with single flagellum
.... 29

28b
Small, pear-shaped cell with two flagella and a horseshoe-shaped green chloroplast

Chlamydomonas

(20 μm)

29a
Green-colored cell flattened and oval in shape. Cell has conspicuous eyespot with a long, conspicuous, single flagellum. Cell sometimes is twisted

Phacus

(40–170 μm)

29b
Small green-colored cell, cylindrical in shape with a pointed end; with a long, single flagellum

Euglena

(35–55 μm)

30a
Cells arranged in a single filament (thread); cells mostly longer than wide
.... 31

30b
Cells arranged in a branching filament
.... 32

31a
Individual cells in filament each with two star-shaped chloroplasts

Zygnema

(40 μm, diameter)

31b
Individual cells in filament each with many chloroplasts that are arranged in a definite spiral

Spirogyra

(50–60 μm, diameter)

Protist/Cyanobacteria Key   (continued)

32a
Visible to naked eye. Branching filaments appear as tufts. Each filament branch with a thornlike tip. Each cell with a single chloroplast usually at the side
Stigeoclonium

(12–22 μm, diameter)

32b
Visible to naked eye. Branching filaments, each with a rounded tip. Each cell contains numerous netlike chloroplasts
Cladophora

(75–100 μm, diameter)

33a
Cells arranged in a filament (thread)
.... 49

33b
Cells not arranged in a filament
.... 34

34a
Oval-shaped cells arranged in pairs as a flat, platelike colony
Merismopedia

(10 μm, diameter)

34b
Cells arranged in a round, globular colony
.... 35

35a
Group of oval-shaped cells arranged within concentric rings of mucilage
Gloeocapsa

(1–2 μm, cell diameter)

35b
Cells not arranged in concentric rings of mucilage
.... 36

36a
Numerous tiny round cells crowded together within a colony inside a mucilage matrix. Cells appear brownish black or purplish
Microcystis

(2–4 μm, cell diameter)

36b
Spherical-shaped cells arranged in pairs and evenly distributed throughout a mucilage matrix
Anacystis

(2–4 μm, cell diameter)

37a
Cells green-colored and spherical in shape. Sometimes may occur in clusters.
Chlorella

(7–10 μm, diameter)

37b
Cells curved or needlelike
.... 38

38a
Cells curved or crescent-shaped; green in color
Closterium

(30–70 μm)

38b
Cells needlelike, slightly curved
Ankistrodesmus

(25–100+ μm)

39a
Colony flat and disc-shaped; usually containing 16 cells
Gonium

(5–15 μm, cell diameter)

39b
Colony spherical in shape
.... 40

40a
Colony contains 32 cells or less
.... 41

40b
Colony visible to naked eye; contains more than 32 cells (hundreds)
Volvox

(2.5–6 μm, cell; 300–500 μm, colony)

41a
Colony contains 32 cells
Eudorina

(10–20 μm, cell; to 200 μm, colony)

41b
Colony contains 16 cells
Pandorina

(10–20 μm, cell; to 220 μm, colony)

42a
Stalk not branched; cells contract (stalk appears to contract like a spring)
Vorticella

(50–150 μm)

42b
Stalk branched; cells may or may not contract
.... 43

43a
Cells contract independently (stalk appears to contract like a spring)
Carchesium

(100–250 μm, cells; 3 mm, colony)

43b
Cells do not contract (lack myoneme)
Epistylis

(50–150 μm, cells; 3 mm, colony)

44a
Cell with tufts or bands of cilia
.... 46

44b
Cell with distinct pointlike projections (cirri)
.... 45

45a
Cell with oval body, flattened on ventral side, convex on dorsal side; with 4 posterior cirri. Travels by walking on cirri
Euplotes

(80–200 μm)

45b
Cell with oval body, flattened on ventral side, convex on dorsal side; with 3 posterior cirri. Travels by walking on cirri
Stylonychia

(100–300 μm)

46a
Cell body barrel-shaped with dark-colored plates. Tuft of cilia around the mouth; with 8 posterior spines (four at and four near the rear)
Coleps

(80–110 μm)

46b
Cell oval-shaped with two distinct ciliary bands—one anterior and one in the middle of the body. Swims with a spiral motion.
Didinium

(80–200 μm)

Protist/Cyanobacteria Key (continued)

47a
Single cell with cilia
.... 48

47b
Red-colored tufted mass visible to naked eye. Colony composed of branched filaments with each filament having numerous bead-shaped cells
Batrachospermum

(5–10 cm, plant mass)

48a
Pink or rose-colored ciliate
Blepharisma

(130–200 μm)

48b
Dark bluish-green ciliate; trumpet-shaped
Stentor

(1–2 mm extended)

49a
Flat, rectangular-shaped cells arranged in a filament or thread
.... 51

49b
Round or barrel-shaped cells arranged in a filament or chain
.... 50

50a
Single straight chain of beadlike cells; looks like a necklace
Anabaena

(3–11 μm, diameter)

50b
Numerous coiled filaments or chains of beadlike cells which may comprise a visible globular (ball-shaped) mass looking like a hollow egg or bead
Nostoc

(5–7 μm, cell; 5–10 mm, colony)

51a
Filaments inside a clear covering or sheath; cells at ends of each filament rounded
Phormidium

(4–6 μm, diameter)

51b
Filament not inside a clear sheath; cells at ends of filament tapered or rounded
Oscillatoria

(5–6 μm, diameter)

Elements Important in Humans

Name	Food Sources	Function	Deficiency	Excess
Major Elements				
Calcium (Ca)	Dairy products, green vegetables, eggs, nuts, dried legumes, broccoli, canned salmon, hard water	Development of bones and teeth, muscle contraction, blood clotting, nerve impulse transmission, and enzyme activation	Osteoporosis, stunted growth, poor quality bones and teeth, rickets, convulsions	Rare—excess blood calcium, loss of appetite, muscle weakness, fever
Chlorine (Cl)	Table salt, most water supplies, soy sauce	Acid-base balance, hydrochloric acid in stomach, bone and connective tissue growth	Rare—metabolic alkalosis, constipation, failure to gain weight (in infants)	Vomiting; elemental chlorine is a poison used for chemical warfare
Magnesium (Mg)	Whole grains, liver, kidneys, milk, nuts, green leafy vegetables	Component of chlorophyll; bones and teeth, coenzyme in carbohydrate and protein metabolism	Infertility, menstrual disorders	Loss of reflexes, drowsiness, coma, death
Phosphorus (P)	Soybeans, dairy foods, egg yolk, meat, whole grains, shrimp, spinach, peas, beef, green vegetables	Development of bones and teeth, energy metabolism, pH balance, cold adaptation	Rare—bone fractures, disorders of red blood cells, metabolic problems, irritability, weakness	As phosphorus increases, calcium decreases; muscle spasm, jaw erosion
Potassium (K)	Whole grains, meat, bananas, vegetables, cantaloupe, milk, peanut butter	Body water and pH balance, nerve and muscle activity, insulin release, glycogen and protein synthesis	Muscle and nerve weakness, poor digestion	Rare—abnormalities in heartbeat or stoppage, muscle weakness, mental confusion, cold and pale skin
Sodium (Na)	Table salt, dairy foods, eggs, baking soda and powder, meat, vegetables	Body water and pH balance, nerve and muscle activity, glucose absorptive	Weakness, muscle cramps, diarrhea, dehydration, nausea	High blood pressure, edema, kidney disease
Sulphur (S)	Dairy products, nuts, legumes, garlic, onion, egg yolk	Component of some amino acids, enzyme activator, blood clotting	Related to intake and deficiency of sulfur-containing amino acids	Excess sulfur-containing amino acid intake leads to poor growth
Trace Elements				
Cobalt (Co)	Common in foods, meat, milk, fish	Component of Vitamin B_{12}, essential for red blood cell formation	Rare	Dermatitis, excessive production of red corpuscles
Copper (Cu)	Liver, meat, seafood, whole grains, legumes, nuts	Production of hemoglobin, bone formation, component of electron carriers, melanin synthesis	Anemia, bone and connective tissue disorders, scurvylike conditions, an early death	Toxic concentrations in liver and eyes of persons with genetic inability to metabolize

Elements Important in Humans   (continued)

Name	Food Sources	Function	Deficiency	Excess
Fluoride (F)	Most water supplies, seafood	Prevents bacterial tooth decay, aids in hardening and stability of bones and teeth	Osteoporosis, tooth decay, bone weakness	Mottling and brown spots on teeth, deformed teeth and bones
Iodine (I)	Seafood, iodized salt, dairy products	Component of thyroid hormone, which controls cellular respiration	Inadequate synthesis of thyroid hormone, goiter (enlarged thyroid), cretinism, fatigue	Antithyroid compounds
Iron (Fe)	Liver, meat, eggs, spinach, enriched bread and cereals, asparagus, raisins	Component of hemoglobin (oxygen and electron transport system); immune system	Iron-deficiency anemia, chronic fatigue, weakness, poor digestion	Accidental poisoning of children, cirrhosis of liver
Manganese (Mn)	Liver, kidneys, legumes, cereals, tea, coffee, green vegetables	Ions necessary in protein and carbohydrate metabolism; Krebs cycle	Infertility, menstrual problems	Brain and nervous system disorders
Molybdenum (Mo)	Organ meats, milk, whole grains, leafy vegetables, legumes	Enzyme component, uric acid excretion	Edema, lethargy, disorientation, and coma	Weight loss, growth retardation, and changes in connective tissue
Zinc (Zn)	Liver, seafoods, common foods, nuts, chicken, carrots, peas	Essential enzyme component, necessary for normal senses of taste and smell, disease resistance	Slow sexual development, loss of appetite, retarded growth, loss of taste and smell, lower disease resistance	Nausea, bloating, and cramps; depresses copper absorption

Kcal Tally

Meat
140	Beef pot roast, lean, 2 thin slices
85	Ham, boiled, 1 slice
245	Hamburger patty, plain
170	Hot dog, plain
115	Meat loaf, 1 slice
150	Pork chop, 1 lean

Poultry
215	Chicken breast, fried, 1/2
185	Chicken, broiled, 1/4
150	Turkey, light, 1 slice
175	Turkey, dark, 1 slice

Fish
50	Fish stick, breaded, 1
230	Halibut, broiled, 1 steak
195	Ocean perch, breaded, fried, 1 piece
215	Tuna, in oil, 1/2 cup

Dairy
100	Butter or margarine, 1 tbs
59	Cheese, American, 1 slice
60	Cottage cheese, 1/4 cup
150	Milk, whole, 1 cup

Main Dishes
345	Bean tostada with cheese
460	Beef potpie, 1 individual
180	Beef stew, 1 cup
95	Chicken chow mein, 1 cup
365	Chicken with noodles, 1 cup
545	Chicken potpie, individual
170	Chili con carne with beans, canned, 1/2 cup
365	Enchiladas, 2
380	Lasagna, 8 ounces
240	Macaroni and cheese, 1/2 cup
153	Pizza, cheese, 1/8 of a pie
157	Pizza, sausage, 1/8 of a pie
260	Spaghetti w/meat sauce, 3/4 cup

Salads
285	Chef's salad (lettuce, ham, cheese, dressing)
50	Coleslaw, 1/2 cup
245	Potato salad, 1/2 cup
180	Tuna salad, 1/2 cup
185	Waldorf salad, 1/2 cup

Sandwiches
280	Bacon, lettuce, and tomato
560	Big Mac™
360	Bologna
330	Cheese
545	Cheeseburger
410	Fish fillet, fried
445	Hamburger
280	Hot dog, 1 bun
310	Roast beef
360	Salami

Soups
100	Beef, 1 cup
100	Chicken noodle, 1 cup
140	Split pea, 1 cup
90	Tomato, 1 cup
80	Vegetable, 1 cup

Vegetables
165	Beans, baked, with pork, 1 cup
15	Beans, green, snap, wax, or yellow, 1/2 cup
50	Broccoli, 1 cup
35	Beets, diced, 1/2 cup
30	Carrots, 1 medium, raw
15	Cauliflower, 1/2 cup
70	Corn, 1 medium ear
210	Corn, creamed, 1 cup
5	Lettuce, 4 small leaves
38	Onion, raw, 1/2 cup
60	Peas, green, 1/2 cup
15	Pepper, green, 1 medium
145	Potato, baked
115	Potato chips, 10 medium
155	Potatoes, French-fried, 10
135	Potatoes, mashed, 1/2 cup

Desserts
330	Apple pie, 1 slice
103	Brownie, frosted, with nuts
420	Chocolate cake, fudge icing, 2" wedge
120	Devil's-food cupcake with chocolate icing
80	Gelatin, plain, 1/2 cup
135	Ice cream, vanilla 1/2 cup
300	Lemon meringue pie, 1 slice
265	Pumpkin pie, 1 slice

Breakfast Foods
95	Bacon, 2 strips
145	Bran flakes, with raisins, 1 cup
60	Bread, whole wheat, 1 slice
76	Bread, white, 1 slice
85	Cornflakes, 1 cup
135	Doughnut, cake
85	Egg, fried in butter
80	Egg, poached or hardcooked
95	Egg, scrambled (milk added)
110	Muffin, blueberry
110	Oatmeal, 3/4 cup
60	Pancake, 4" cake
170	Pork sausage, 2" patty
95	Pork sausage, 1 link
200	Toaster pastry
210	Waffle

Fruits
80	Apple, raw, 1 medium
380	Avocado, raw, 1 whole
100	Banana
5	Cucumber, 6 thin slices
75	Grapefruit, 1/2
70	Orange
40	Peach
100	Pear
30	Strawberries, 1/2 cup
30	Tomato, 1 medium

Drinks
117	Apple juice, 1 cup
175	Cocoa (all milk), 3/4 cup
0	Coffee or tea, unsweetened
145	Cola, 12 ounces
1	Cola, diet, 12 ounces
340	Eggnog, 1 cup
75	Grape juice, 1/2 cup
100	Grapefruit juice, unsweetened, 1 cup
210	Milk, whole, chocolate, 1 cup
135	Orange juice, 1 cup
350	Shake, chocolate, 8 ounces

Candies/Cookies/Snacks
145	Candy bar, chocolate, 1 ounce
200	Chocolate chip cookies, 4
100	Chocolate kisses, 5
200	Vanilla cookie, sandwich type, 4
90	Marshmallow
840	Peanuts, roasted, 1 cup
23	Popcorn, plain, 1 cup
41	Popcorn, oil and salt, 1 cup
50	Pretzels, 3-ring, 10
810	Sunflower seeds, dry, hulled, 1 cup

Vitamins Important in Humans

Name	Food Sources	Function	Deficiency Symptoms
A, retinol	Liver, green and yellow vegetables, fruits, egg yolks, butter	Formation of visual pigments; cell growth, especially of epithelial cells	Night blindness, flaky skin, lowered resistance to infection, growth stunting, faulty reproduction
D, calciferol	Fish oils, liver, action of sunlight on lipids in skin, fortified milk, butter, eggs	Increases calcium absorption from gut; important in bone and tooth formation	Rickets—defective bone formation
E, tocopherol	Oils, whole grains, liver, mayonnaise, margarine	Protects red blood cells, plasma membranes, and vitamin A from destruction; important in muscle maintenance; DNA and RNA synthesis	Fragility of red blood cells, muscle wasting, sterility
K, menadione	Synthesis by intestinal bacteria; green and yellow vegetables	Synthesis of clotting factors by liver	Internal hemorrhaging (deficiency can be caused by oral antibiotics, which kill intestinal bacteria)
B_1, thiamine	Whole grains, legumes, nuts, liver, heart, kidney, pork, macaroni, wheat germ	Carbohydrate metabolism, nerve transmission, RNA synthesis	Beriberi, loss of appetite, indigestion, fatigue, nerve irritability, heart failure, depression, poor coordination
B_2, riboflavin	Liver, kidney, heart, yeast, milk, eggs, whole grains, broccoli, almonds, cottage cheese, yogurt, macaroni	Forms part of electron carrier in electron transport system, production of FAD, activates B_6 and folic acid	Sore mouth and tongue, cracks at corners of mouth, eye irritation, scaly skin, growth retardation
Pantothenic acid	Yeast, liver, eggs, wheat germ, bran, peanuts, peas, fish, whole grain cereals	Part of Coenzyme A; essential for energy release and biosynthesis; stimulates antibodies and intestinal absorption	Fatigue, headache, sleep disturbances, nausea, muscle cramps, loss of antibody production, irritability, vomiting
B_3, niacin	Yeast, liver, kidney, heart, meat, fish, poultry, legumes, nuts, whole grains, eggs, milk	Coenzyme in energy metabolism; part of NAD^+ and NADP	Pellagra, skin lesions, digestive problems, nerve disorders, diarrhea, headaches, fatigue
B_6, pyridoxine	Whole grains, potatoes, fish, poultry, red meats, legumes, seeds	Coenzyme for amino acid and fatty acid metabolism; synthesis of brain chemicals, antibodies, red blood cells, DNA; glucose tolerance	Skin disorders, sore mouth and tongue, nerve disorders, anemia, weight loss, impaired antibody response, convulsive seizures
Biotin (vitamin B)	Cauliflower, liver, kidney, yeast, egg yolks, whole grains, fish, legumes, nuts, meats, dairy products; synthesis by intestinal bacteria	Fatty acid, amino acid, and protein synthesis; energy release from glucose, insulin activity	Skin disorders, appetite loss, depression, sleeplessness, muscle pain; elevated blood cholesterol and glucose
Folacin (folic acid)	Liver, yeast, leafy vegetables, asparagus, salmon	Nucleic acid synthesis, amino acid metabolism, new cell growth	Failure of red blood cells to mature, anemia, intestinal disturbances and diarrhea
B_{12} (cyanocobalamin)	Liver and organ meats, muscle meats, fish, eggs, shellfish, milk; synthesis by intestinal bacteria	Nucleic acid synthesis; maintenance of nervous tissue; glucose metabolism	Pernicious anemia, fatigue, irritability, loss of appetite, headaches
C, ascorbic acid	Citrus fruits, tomatoes, green leafy vegetables, sweet peppers, broccoli, cauliflower	Essential to formation of collagen and intercellular substance; protects against infection; maintains strength of blood vessels; increases iron absorption from gut; important in muscle maintenance; stress tolerance	Scurvy, failure to form connective tissue, bleeding, anemia, slow wound healing, joint pain, irritability, premature wrinkling and aging

Sources, Uses, and Effects of Some Psychoactive Drugs

Psychoactive drugs affect body and nervous functions and may change behavior. This table lists some of the drugs most commonly abused in the United States today. Effects vary a great deal from person to person; only a few are described.

Name	Source and Use	Effects
Depressants		
Alcohol (in wine, beer, ale, spirits, liquors, mixed drinks, etc.)	Synthesized or produced naturally by fermentation of fruits, vegetables, or grains.	Affects brain in proportion to amount in bloodstream. Small dose produces euphoria, drowsiness, dizziness, flushing, and release of inhibitions and tensions. Slurred speech, staggering, and double vision may result from large dose. Impairs driving in any dose and exaggerates effects of other drugs. Liver, heart, and brain damage, and vitamin deficiencies can occur with excessive use. Use during pregnancy may result in underweight babies with physical, mental, and behavioral abnormalities. May produce tolerance and addiction. Most common of all forms of drug abuse today.
Tranquilizers—Valium, Librium, Miltown	Synthetic chemicals. Used to treat anxiety, nervousness, tension, and muscle spasm.	Produce mildly impaired coordination and balance, reduced alertness and emotional reactions, loss of inhibition. May produce sleep disturbances or depression. Use during pregnancy may cause congenital malformations. Use with alcohol or while driving is dangerous. Tolerance-inducing and addictive.
Opiates—opium, morphine, codeine, heroin; also called dope, horse, smack, skag	Seed capsules of the opium poppy (*Papaver somniferum*) produce a milky juice. From this juice, many drugs can be prepared, including morphine and codeine. Heroin is formed by chemical synthesis from morphine. Taken by smoking, by mouth, or by injection. Used in anti-diarrhea and anti-cough medications and as pain killers.	Injection produces surge of pleasure, state of gratification, and stupor. Sensations of pain, hunger, and sexual desire are lost. Physical effects include nausea, vomiting, insensitivity to pain, constipation, sweating, slowed breathing. Long-term effects include severe constipation, moodiness, menstrual irregularities, and liver and brain damage. Use with alcohol very dangerous. Tolerance-inducing and highly addictive.
Barbiturates—pentobarbital, secobarbital; also called goof-balls, downers, yellow jackets, red devils, etc.	Many compounds, developed to suppress pain, treat sleeplessness, anxiety, tension, high blood pressure, convulsions, and to induce anesthesia. Taken by mouth or injected into muscles.	Effects similar to alcohol: release from tension and inhibition, followed by interference with mental function, depression, and hostility. Overdoses cause unconsciousness, coma, and death. Extremely dangerous in combination with alcohol. Long-term effects include liver damage and chronic intoxication. Addictive.
Stimulants Caffeine	White, bitter, crystalline substance, found in coffee beans, tea leaves, coca leaves, and kola nuts. Ingredient in many pain relievers, cold remedies, and beverages.	Increases metabolic rate, blood pressure, urination, and body temperature; shortens sleep, decreases appetite, and impairs coordination of movement. Long-term or heavy use can cause insomnia, anxiety, depression, or birth problems. Regular consumption of 4 cups of coffee a day can produce a form of physical dependence.

Sources, Uses, and Effects of Some Psychoactive Drugs (continued)

Name	Source and Use	Effects
Tobacco	Shredded, dried leaves of tobacco plant. Contains tar and nicotine and cancer-causing substances. Smoke contains carbon monoxide. Generates 2000 different chemical compounds. Tobacco is smoked or chewed.	Increases heart and breathing rates and blood pressure; decreases appetite. Nicotine enters bloodstream; tars accumulate in lungs. Carbon monoxide interferes with oxygen transport. Long-term effects include blockage of blood vessels, respiratory disease, cancers of lung, mouth, or throat, stomach ulcers, and reduced immunity. Chewing tobacco increases the risk of oral cancer by 40 times. Smoking during pregnancy increases risk of miscarriage, low birth weight, and developmental problems. Causes psychological and physical dependence.
Amphetamines—benzadrine, dexedrine, methedrine, also called speed, splash, uppers, bennies, dexies, crystal	Synthetic drugs taken by mouth or injected into veins.	Increases alertness and energy, feeling of well-being, heart and breathing rates, and blood pressure. Restlessness and hostility common. Very large doses can produce irregular heartbeat, hallucinations, or heart failure. Long-term use can produce mental illness, kidney damage, or lung problems. Produce powerful psychological and physical dependence.
Cocaine—also called coke, snow, nose candy, crack	Comes from the leaves of the coca plant. Sniffed, ingested, injected, or smoked as freebase.	Effects similar to amphetamines but of shorter duration. Toxic; causes heart irregularities that can lead to death, lung damage, hallucinations, muscle spasms, and convulsions. Use produces restlessness, insomnia, delusions, and weight loss. Produces powerful psychological and physical dependence.

Hallucinogens

Name	Source and Use	Effects
Cannabis—hashish and marijuana; also called pot, grass, weed, reefer, and joint	Flowers, leaves, oils, or resin of *Cannabis sativa*. Smoked in pipe or hand-rolled cigarette, or ingested by mouth.	Impairs concentration, short-term memory, coordination, and motor skills. Enhances sensory perception and feelings of relaxation. Distorts space-time sense and induces withdrawal, fearfulness, anxiety, depression, or hallucinations. Long-term effects include loss of motivation, interest, memory, and concentration. The smoke can cause chronic lung disease and lung cancer. Psychological dependence common.
LSD—lysergic acid diethylamide or acid; mescaline and psilocybin, also called mind-expanding drugs	LSD is an artificial chemical compound developed in the 1940s. Mescaline is a natural product of peyote (a cactus) long used by American Indians in religious practices. Psilocybin is formed by a mushroom. Usually taken by mouth.	Vivid psychic effects, including hallucinations. Sensations intensified and distorted. Extreme mood swings, including joy, inspiration, depression, anxiety, terror, and aggression. Long-term effects of LSD include decreased motivation and interest, prolonged depression, and anxiety. Fetal abnormalities may result from use during pregnancy.

Geologic Time Scale

Eon	Era	Period	Epoch	Beginning Time (Millions of Years Ago)	Distance From Near End
			Holocene	10 000 years ago to present	
		Quaternary	Pleistocene	1.6	
			Pliocene	5.3	
			Miocene	23.7	
			Oligocene	36.6	
			Eocene	57.8	
	Cenozoic	Tertiary	Paleocene	66.4	
		Cretaceous		144	
		Jurassic		208	
	Mesozoic	Triassic		245	
		Permian		286	
		Carboniferous		360	
		Devonian		408	
		Silurian		438	
		Ordovician		505	
Phanerozoic	Paleozoic	Cambrian		570	
Proterozoic				2500	
Archaean				3800	
Hadean				4600	

Adapted from the Geological Society of America, "A Decade of North American Geology," compiled in 1983.

Major Geological Events

First evidence of . . .	Millions of Years Ago	Distance From Near End
Pleistocene ice age begins	1.6	
land bridge between North and South America	5.7	
Antarctic ice cap	24	
Mississippi River	35	
separation of Antarctica and Australia	50	
collision of India with Asia and formation of the Himalayan Mountains	55	
formation of the present Rocky Mountains	70	
breakup of Pangaea and formation of the Atlantic Ocean	165	
formation of the supercontinent Pangaea	225	
extensive coal deposits	350	
free oxygen (O_2) nears present levels (1–20 percent)	600	
free oxygen (O_2) building up in the atmosphere	2500	
an atmosphere and hydrosphere	3800	
the earth as a solid planet	4600	

Major Biological Events

First evidence of . . .	Millions of Years Ago	Distance From Near End
modern humans (*Homo sapiens sapiens*)	.04	
"Lucy" (*Australopithecus afarensis*)	3.2	
members of the human family (hominids)	4.0	
monkeys	35	
primates	65	
flowers	140	
birds	150	
mammals	225	
dinosaurs	235	
reptiles	300	
seeds	350	
amphibians	360	
land animals	400	
land plants	430	
vertebrates (jawless fishes)	520	
animals with hard shells (marine invertebrates)	590	
animals (soft-bodied marine invertebrates)	680	
multicellular organisms (algae)	1000	
cells with a nucleus (eukaryotes)	1400	
life (bacteria)	3500	

Climatograms

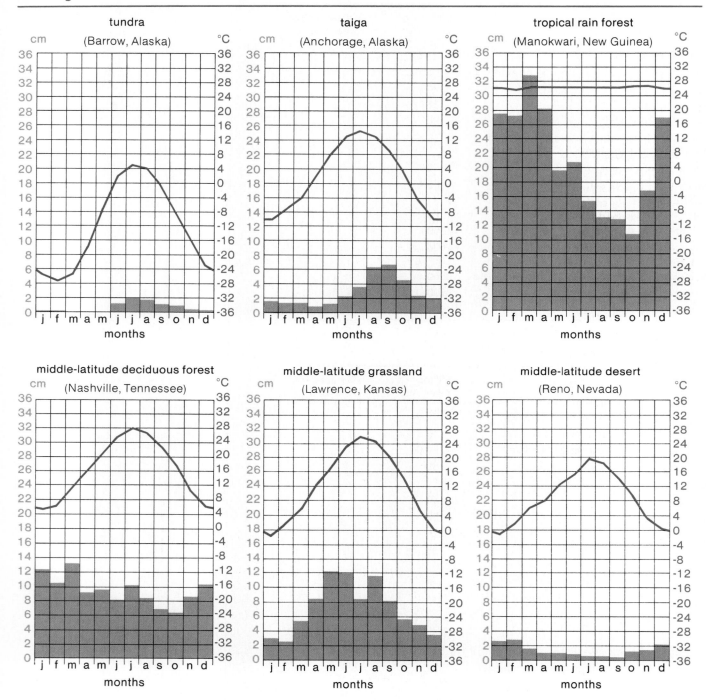

Industrialist

Position: wants city to grow

Land use: shoe factory

Area: 40 acres

Cost to city: none

Income to city: $10,000/acre; property taxes, 750 jobs

Concerns: low property and income taxes; availability of work force; source of water for cooling purposes; good transportation (railroad, highway, or river) for bringing in raw materials and distributing manufacturing goods; price and availability of raw materials

Retired Citizen

Position: wants a nice, affordable home

Land use: low-rent housing

Area: 40 acres (400 units, 1200 people)

Cost to city: 20 percent of building costs (80% from federal government)

Income to city: increased utility use

Concerns: economical housing with low maintenance costs; shopping nearby; safety of neighborhood; availability of social programs; medical facilities nearby; quiet neighborhood; mass transportation; libraries nearby; restrictions of a fixed income

Developer

Position: wants city to grow

Land use: shopping center

Area: 40 acres

Cost to city: extension of services

Income to city: property taxes and sales tax from sale or leasing of land

Concerns: property and business taxes; availability of retail leasing sites; promotion of business and residential development; expansion of commercial zoning; good roads to area; public transportation to area; population size

College Student

Position: interested in environment and recreation

Land use: park with recreational facilities

Area: 320 acres

Cost to city: landscaping, construction, maintenance

Income to city: minor, such as entrance fee to swimming pool

Concerns: environmental concerns; using park for jogging and swimming; strong controls and restrictions on land development; politics; availability of part-time jobs

Single Parent

Position: wants good education for child

Land use: elementary school

Area: 20 acres

Cost to city: cost of buildings and maintenance

Income to city: none

Concerns: does not want crowded schools; good transportation to school; safety of neighborhood; jobs for teachers and other personnel; no increase in taxes

Construction Worker

Position: wants job security

Land use: middle income homes and condominiums

Area: 240 acres (720 homes, 2800 people)

Cost to city: extension of services to these homes

Income to city: $1000/acre paid by developer; property taxes

Concerns: home to live in; property taxes; promotion of community growth and expansion; construction (building) codes

City Manager

Position: wants growth of city carefully managed

Land use: landfill site and sewage treatment plant

Area: 10 acres total (5 acres each)

Cost to city: building and maintenance

Income to city: dumping fees, increased rate for sewage service

Concerns: all aspects of development; services for all people (water, sewers, electricity, street maintenance); safety and health of population; police and fire protection; increased income to city from property and sales taxes

Environmentalist

Position: wants to preserve land in its natural condition

Land use: open space

Area: 1280 acres

Cost to city: maintenance

Income to city: none

Concerns: city needs open space; organisms living on the land now are worth saving; keep pollution under control; strong controls and restrictions on land development; political activist

Farmer

Position: wants to grow cash crops (fruits and vegetables) for market

Land use: farm

Area: 240 acres

Cost to city: none

Income to city: $1000/acre; annual property tax

Concerns: topography of land: length of growing season and climate; type of soil; amount of water available for irrigation; market for crops; not very interested in politics; wants to limit expansion of city; needs good roads into city to transport produce

Laborer

Position: wants a job

Land use: low-middle income houses and apartments

Area: 640 acres (100 homes, 400 people; 5 apartment complexes, 800 people)

Cost to city: extension of services to these homes

Income to city: $20 000/acre paid by developer; property taxes

Concerns: apolitical; interested in meeting immediate needs of feeding and housing family; not interested in environmental issues.

A Catalog of Living Things

This appendix shows one way taxonomists arrange the major groups of living organisms. It does not take into account the many extinct groups known only from fossils.

In general the classification is not carried below class level. In some cases, however, examples are given at the family level. In two groups—insects and mammals—a more detailed classification at the order level is given. To show how complicated classification can become at lower levels, the primate order of mammals is carried to the family level.

In examples, common names of groups are used wherever appropriate, as, for example, phylum Chlorophyta: green algae. The illustrated organisms are identified by common names if they have them, or by names of genera.

A Catalog of Living Prokaryotes
Prokaryotic; lack membrane-enclosed organelles; circular DNA without protein; plasmids usually present. Autotrophic (photosynthetic or chemosynthetic) or heterotrophic; some parasitic; important decomposers; essential to many chemical cycles. Prevalent in all environments. Minute (usually 1 to 5 μm); single cells, chains, colonies, or slender branched filaments; most have rigid cell walls that lack cellulose. Locomotion, if present, by unique flagella or gliding. Great diversity in metabolic pathways; many anaerobic; photosynthetic species may use H_2, H_2S, or other substances. Reproduction by fission; exchange of genetic material occurs in some species. More than 5000 species described, two major groups; classification to phyla difficult.

Archaebacteria
Biochemically unique; composition of cell walls and plasma membranes differs from that of eubacteria; transfer RNA and ribosomal RNA also unique. May represent the first organisms; adapted to extreme conditions. Three major groups: methanogens, thermoacidophiles, extreme halophiles.

X5833

X7500

methanogens

X4000

halophiles

729

Eubacteria

Three major groups based on characteristics of cell wall: wall-less, gram-positive, and gram-negative. Mycoplasmas lack a cell wall; have extremely simple cell structure. Many gram-positive eubacteria produce lactic acid; others are source of antibiotics. Gram-negative eubacteria include photosynthetic and chemosynthetic types; many are anaerobic.

X3830

Streptococcus
(gram-positive)

X680

Streptomyces
(gram-positive)

X3170

Micrococcus
(gram-positive)

X6071

Desulfovibrio
(gram-negative, sulfur-
excreting eubacteria)

X1500

Rhodomicrobium vannielli
(gram-negative, purple eubacteria)

X10 000

Escherichia coli
(gram-negative)

X7692

Pseudomonas
(gram-negative)

X8000

Mycoplasma

X790

Spirillum
(gram-negative)

X1980

Chloroflexus
(green nonsulfur eubacteria)

cyanobacteria

X485

Rivularia

X1500

Gloeocapsa

X500

Merismopedia

X760

Nostoc

X300

Tolypothrix

A Catalog of Living Protists

Eukaryotic, with membrane-enclosed nuclei and organelles; DNA organized into chromosomes complexed with protein. Autotrophic or heterotrophic; many can switch from one form to the other; many parasitic. Aquatic; marine, freshwater, or in watery tissues of other organisms. Size ranges from microscopic to 100 m; unicellular, colonial, or multicellular. Great variability in cell organization, chromosome structure, mitosis, meiosis, and life cycles. Reproduction asexual, by fission; or sexual, by conjugation or fertilization; often involving complex life cycles. Algae, protozoa, slime molds, slime nets, water molds, and others; 27 phyla; more than 100 000 species.

Green Algae   (Phylum Chlorophyta) [Greek *chloros*, green; *phyton*, plant]
Autotrophic; chlorophyll seldom masked by other pigments; food usually stored as starch. Microscopic to macroscopic; unicellular, filamentous, ribbons, sheets, tubes, or irregular masses. Probably ancestral to plants. Reproduction asexual and sexual. About 7000 species.

Diatoms   (Phylum Bacillariophyta) [Latin *bacillus*, little stick; Greek *phyton*, plant]
Autotrophic; chlorophyll usually masked by yellow pigments; food stored as oil. Mostly microscopic; unicellular, colonial, or multicellular; shells (tests) made of two parts impregnated with silica. Reproduction asexual and sexual. As many as 10 000 species.

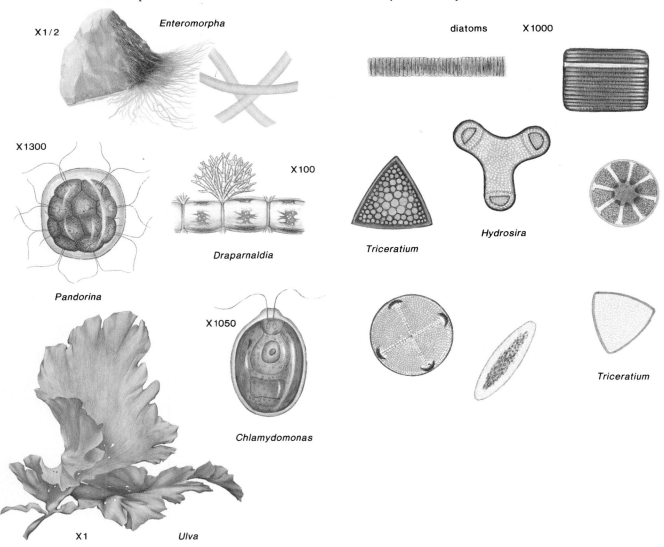

X1/2   *Enteromorpha*

X1300

X100

Draparnaldia

Pandorina

X1050

Chlamydomonas

X1   *Ulva*

diatoms   X1000

Triceratium

Hydrosira

Triceratium

Brown Algae   (Phylum Phaeophyta) [Greek *phaios*, brown; *phyton*, plant]
Autotrophic; chlorophyll usually masked by brownish pigments; food stored as carbohydrates, but not as starch. Almost all marine and macroscopic (up to 100 m); many with specialized structures. Reproduction mostly sexual. About 1500 species.

Red Algae   (Phylum Rhodophyta) [Greek: *rhodos*, red; *phyton*, plant]
Autotrophic; chlorophyll usually masked by red pigments; food stored as carbohydrates but not as starch. Almost all marine and macroscopic. Reproduction sexual, with complex life cycles; reproductive cells nonmotile. About 4000 species.

X1/8

Laminaria

X1/100

Nereocystis

X1/4

X1/4

Acsophyllum

X1

Agardhiella

X40

X2

Polysiphonia

Postelia

X1/3

X12

X1/2

Corallina

Porphyra

Sargassum

X1/2

Chondrus

Flagellates   (several phyla) [Latin *flagellatus,* whipped]
Heterotrophic, autotrophic, or both; many form symbiotic
relationships; many parasitic species. Locomotion by
whiplike flagella; microscopic or almost so; unicellular or
colonial. Reproduction asexual. About 2000 species.

Sarcodines   (several phyla) [Greek *sarx,* flesh; *eidos,* form]
Heterotrophic; some parasitic species. Locomotion by
pseudopods; microscopic or almost so; unicellular; many
with intricate skeletal structures. Reproduction asexual.
About 8000 species.

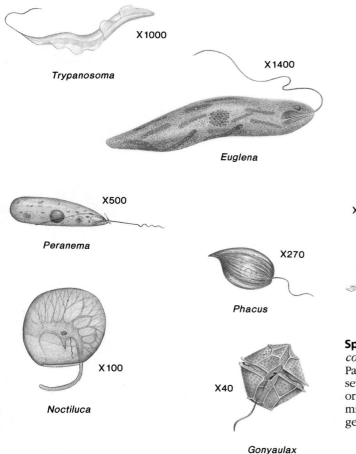

X1000

Trypanosoma

X1400

Euglena

X500

Peranema

X270

Phacus

X100

Noctiluca

X40

Gonyaulax

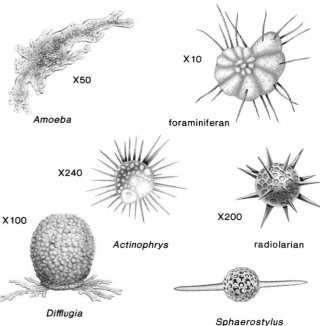

X50

Amoeba

X10

foraminiferan

X240

X100

Actinophrys

X200

radiolarian

Difflugia

Sphaerostylus

Sporozoans   (Phylum Apicomplexa) [Latin *apex,* summit;
complexus, an embrace]
Parasites of animals, with complex life cycles often involving
several species of hosts. Usually nonmotile, but pseudopods
or flagella may occur in certain stages of some species;
microscopic. Reproduction sexual, with alternation of
generations. About 2000 species.

X1500

Plasmodium vivax

X800

Gregarina

X860

Eimeria

Ciliates   (Phylum Ciliophora) [Latin *cilium*, eyelash; Greek *phorein*, to bear]
Heterotrophic. Locomotion by cilia; microscopic or almost so; unicellular; cells with macro- and micronuclei. Reproduction asexual, by fission or budding, or sexual by conjugation. About 5000 species.

Slime Molds   (Phylum Myxomycota) [Greek *myxa*, mucus; *mykes*, fungus]
Heterotrophic (saprotrophic). Amoeboid, multinucleate stage engulfs decaying vegetation in damp habitats. Sexual reproduction by spores formed in sporangia. About 450 species.

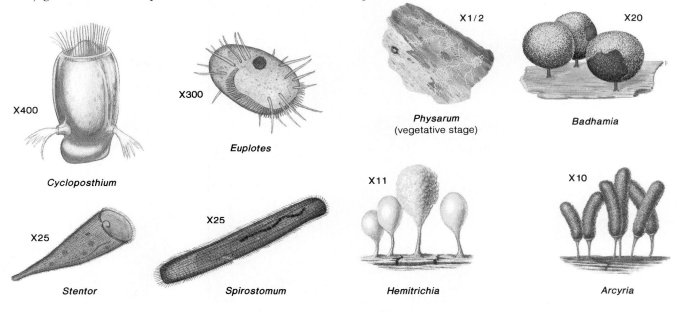

X400
Cycloposthium

X300
Euplotes

X1/2
Physarum
(vegetative stage)

X20
Badhamia

X25
Stentor

X25
Spirostomum

X11
Hemitrichia

X10
Arcyria

A Catalog of Living Fungi
Eukaryotic; heterotrophic (saprotrophic), many parasitic; important decomposers; many are sources of antibiotics. Mostly terrestrial; a few marine species. Nonmotile; mostly multicellular; structure primarily a system of threadlike cells or hyphae. Reproduction by asexual and sexual spores that develop directly into hyphae. Five phyla; more than 100 000 species.

Conjugating Fungi   (Phylum Zygomycota) [Greek *zygon*, pair; *mykes*, fungus]
Hyphae usually not divided by cross walls. Reproduction asexual, by spores, and sexual, by conjugation and formation of thick-walled zygospores. About 600 species.

Sac Fungi   (Phylum Ascomycota) [Greek *askos*, bladder; *mykes*, fungus]
Hyphae divided by cross walls; a few unicellular species. Reproduction asexual, by conidiospores, and sexual, by conjugation and formation of a saclike structure, the ascus, in which sexual spores (ascospores) form by meiosis. Often form lichen partnerships with green algae or cyanobacteria. About 25 000 species.

Rhizopus

X3200

X1/6
morel

yeast

X60

apple scab

Club Fungi (Phylum Basidiomycota) [Greek *basidion,* small base; *mykes,* fungus]
Hyphae divided by cross walls. Reproduction sexual, by spores produced on the surface of a clublike structure, the basidium. Many form mycorrhizae. About 25 000 species.

Imperfect Fungi (Phylum Deuteromycota) [Greek *deuteros,* second; *mykes,* fungus]
Hyphae divided by cross walls. Reproduction asexual, by conidiospores; sexual reproduction stages unknown. Taxonomic grouping of convenience rather than of relationship. Includes members of the genus *Penicillium.* About 25 000 species.

X1/2

X400

Coprinus

Puccinia

X300

X260

Fusarium

Alternaria

X260

Cordana

X1/2

bracket fungus

Lichens (Phylum Mycophycophyta) [Greek *mykes,* fungus; *phykos,* alga; *phyton,* plant]
Symbiotic relationship between a fungus and an alga or cyanobacterium; extremely slow-growing. Reproduction sexual, by formation of an ascus or basidium, depending on the fungal partner. About 25 000 species.

X1

lichens

A Catalog of Living Plants

Eukaryotic, autotrophic; nonmotile; multicellular. All have cellulose-containing cell walls and chloroplasts containing chlorophylls *a* and *b* plus other pigments. Mostly terrestrial; some aquatic. Reproduction sexual, with alternation of generations; in some cases asexual, by vegetative means; develop from a multicellular embryo that does not have a blastula stage. Nine divisions; more than 500 000 species.

Bryophytes   (Division Bryophyta) [Greek *bryon*, moss; *phyton*, plant]

Small (less than 40 cm tall); mostly terrestrial but require moist habitats; lack vascular (conducting) tissue. Well-developed alternation of generations; haploid gametophyte predominant; sporophyte more or less dependent on it. About 24 000 species.

Class Hepaticae   (liverworts) [Greek *hepatikos*, liverlike (from the shape of the leaves)]

Gametophytes flat, often simple, branching masses of green tissue, sometimes with leaflike structures. About 8500 species.

X1

Marchantia

Class Anthocerotae   (Hornworts) [Greek *anthos*, flower; *keras*, horn]

Gametophytes similar to those of liverworts; sporophytes live longer and are capable of continuous growth. About 50 species.

X1

Anthoceros

Class Musci   (true mosses) [Latin *muscus*, moss]

Plants usually erect (not flat), with leaflike structures arranged in radial symmetry around a stalk. Gametophytes develop from algalike masses of green threads. About 15 000 species.

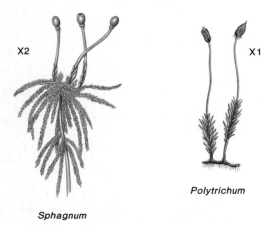

X2

X1

Polytrichum

Sphagnum

Whisk Ferns   (Division Psilophyta) [Greek *psilo*, bare; *phyton*, plant]

Vascular; forking stems produce spore cases at the tips of short branches; no roots. Oldest fossil land plants. Alternation of generations with conspicuous sporophyte. Three species.

X1/2

X1

Psilotum

Tmesipteris

Club Mosses   (Division Lycophyta) [Greek *lycos,* wolf; *phyton,* plant (so named because the roots of a lycopod were thought to resemble a wolf's claw)]
Low-growing evergreen vascular plants with roots, horizontal stems, and small leaves. Spore cases borne in various ways, usually on modified leaves grouped to form structures something like cones. About 1000 species.

X2/3

Lycopodium

Horsetails   (Division Sphenophyta) [Greek *sphen,* wedge (from the shape of the leaves); *phyton,* plant]
Vascular plants with roots and jointed hollow stems containing silica; small scalelike leaves arranged in a circle around each stem joint. Spores borne in conelike structures. About 40 species.

X1/3

Equisetum

Ferns   (Division Pterophyta) [Greek *pteron,* feather; *phyton,* plant]
Vascular plants with roots, stems, and leaves; compound leaves grow from horizontal underground stems. Sporophyte generation dominant, but gametophytes independent; free-swimming sperm cells. About 12 000 species.

X1/7

water fern

X1/10

cinnamon fern

X1/6

adder's tongue

X1/3

grape fern

X1/4

climbing fern

Cyads　(Division Cycadophyta) [Greek *kyos,* a palm; *phyton,* plant]
Tropical and subtropical vascular plants with palmlike or fernlike compound leaves; unbranched stems. Naked seeds borne in cones. About 100 species.

X 1/10　　X 1/30

cycad

Ginkgoes　(Division Ginkgophyta) [Japanese *ginkyo,* silver apricot; Greek *phyton,* plant]
Ancient division of vascular woody plants with fan-shaped leaves; cultivated worldwide. Male trees produce pollen that is carried by wind to female trees. Pollen that lands on a fleshy ovule is pulled through its opening; a sperm is released and fertilizes the egg nucleus. The embryo develops within a naked seed (not enclosed in an ovary wall). One species.

X 1/2

Ginkgo

Gnetophytes　(Division Gnetophyta) [Latin *gnetum* from Moluccan Malay *ganemu,* a gnetophyte species found on the island of Ternate; Greek *phyton,* plant]
Vascular plants with naked seeds borne in cones; some characteristics like those of flowering plants. About 70 species.

X 1/50

Welwitschia

Conifers　(Division Coniferophyta) [Latin *conus,* cone; *ferre,* to bear; Greek *phyton,* plant]
Vascular plants; most have needle-shaped leaves; many are evergreen. Gametophytes microscopic, within tissues of sporophytes; naked seeds borne on scales of a cone. About 700 species.

X 1/4　　X 1/900　　X 1/2

blue spruce

hemlock

X 1/4　　X 1/2

X 1/1000

bald cypress

X 1/950

larch

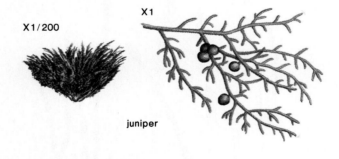

X 1/200　　X 1

juniper

X 1/900　　X 1/4

white pine

Flowering Plants (Division Anthophyta) [Greek *anthos,* flower; *phyton,* plant]
Vascular plants with seeds enclosed in a fruit; sperm cells in pollen tubes. Gametophytes microscopic, within tissues of sporophytes. About 230 000 species. (There are more than 300 families in the division Anthophyta. A few of the common families are listed for the two classes. Orders are omitted.)

Monocots (Class Monocotyledoneae) [Greek *monos,* one; *cotyledon,* a cavity]
Seed with one cotyledon; leaves with parallel veins; flower parts usually in threes or multiples of three; mostly herbaceous. About 34 000 species.

Family Alismataceae (Water Plantain Family) Aquatic or marsh plants; flowers radially symmetrical; 3 sepals, 3 petals, 6 to many stamens, 6 to many carpels, ovary superior (located *above* the attachment of sepals, petals, and stamens).

X1/4

arrowhead

Family Amaryllidaceae (Amaryllis Family) Flowers radially symmetrical, 3 sepals, 3 petals, 6 stamens, one carpel, ovary inferior (located *below* the attachment of sepals, petals, and stamens).

1/3

narcissus

1/100

century plant

Family Commelinaceae (Spiderwort Family) Flowers radially or somewhat bilaterally symmetrical; 3 sepals, 3 petals, 3 or 6 stamens, one carpel, ovary superior.

X1/4

spiderwort

Family Cyperaceae (Sedge Family) Stems usually solid; leaves sheath the stem. Flowers radially symmetrical; no sepals or petals (but scalelike structures present), 1 to 3 stamens, one carpel, ovary superior.

X1

X1/10

sedge

umbrella plant

Family Iridaceae (Iris Family) Flowers radially or somewhat bilaterally symmetrical; 3 sepals, 3 petals, 3 stamens, one carpel, ovary inferior.

X1/5

X1/5

gladiolus

iris

Family Liliaceae (Lily Family)   Flowers radially symmetrical; 3 sepals, 3 petals (sepals and petals often colored alike, thus appearing to be 6 petals), 3 or 6 stamens, one carpel, ovary superior.

X1/8

tulip

X1/9

tiger lily

Family Orchidaceae (Orchid Family)   Flowers bilaterally symmetrical; 3 sepals, 3 petals (united), 1 or 2 stamens, one carpel, ovary inferior.

X1/2

orchid

Family Poaceae (Grass Family)   Stems usually hollow; leaves sheath the stem. Flowers radially symmetrical; no sepals or petals (but scalelike structures present), 1 to 6 stamens, one carpel, ovary superior.

X1/10

bamboo

X1/3

bluegrass

X1/20

corn

Dicots   (Class Dicotyledoneae) [Greek *dis,* two; *cotyledon,* a cavity]
Seed with two cotyledons; leaf veins form a network; flower parts usually in fours or fives or multiples of these numbers. About 166 000 species.

Family Apiaceae (Parsley Family)   Herbaceous; small, usually numerous, flowers in an umbel (umbrella-shape); flowers radially symmetrical; 5 small sepals, 5 petals, 5 stamens, one carpel, ovary inferior.

X1/2

Queen Anne's Lace

Family Asteraceae (Sunflower Family)   Herbaceous; small flowers in dense groups forming a head that appears to be a single, large flower. Individual flowers radially or bilaterally symmetrical; sepals reduced to bristles or scales, 5 petals (united), 5 stamens, one carpel, ovary inferior.

X1

X1/8

dandelion

sunflower

Family Brassicaceae (Mustard Family)   Herbaceous; flowers radially symmetrical; 4 sepals, 4 petals arranged in the form of a cross, 2 sets of stamens (4 long and 2 short), one carpel, ovary superior. Often having a turniplike or cabbagelike odor.

X1/2

field mustard

Family Caprifoliaceae (Honeysuckle Family) Woody shrubs or vines; flowers radially or bilaterally symmetrical; 4 or 5 sepals, 4 or 5 petals (united), 4 or 5 stamens, one carpel, ovary inferior.

twinflower

snowberry

Family Fabaceae (Pea Family) Woody or herbaceous; flowers bilaterally symmetrical; 5 sepals, 5 petals, 10 stamens, one carpel, ovary superior. Legumes.

sweet pea

Family Fagaceae (Oak Family) Trees and shrubs; flowers radially symmetrical, 4 to 7 sepals, no petals, few to many stamens, one carpel, ovary inferior; carpels and stamens in separate flowers.

white oak

Family Lamiaceae (Mint Family) Herbaceous; stems usually square in cross section; flowers bilaterally symmetrical; 5 sepals (united), 5 petals (united), 2 or 4 stamens, one carpel, ovary superior.

mint

coleus

Family Polemoniaceae (Phlox Family) Herbaceous; flowers radially symmetrical; 5 sepals (united), 5 petals (united), 5 stamens, one carpel, ovary superior.

Jacob's ladder

phlox

Family Ranunculaceae (Buttercup Family) Herbaceous; flowers radially symmetrical; few to many sepals and petals, many stamens and carpels, ovary superior.

larkspur

buttercup

Family Rosaceae (Rose Family)   Herbaceous or woody; flowers radially symmetrical; 5 sepals, 5 petals, numerous stamens, one to many carpels, ovary either superior or less inferior.

X1/2

wild rose

Family Scrophulariaceae (Figwort Family)   Herbaceous; flowers bilaterally symmetrical; 5 sepals, 5 petals, (2 forming an upper lip and 3 forming a lower lip), 4 stamens (in 2 unlike pairs), one carpel, ovary superior.

X2/3

X2/3

snapdragon   butter-and-eggs

A Catalog of Living Animals

Eukaryotic, heterotrophic; most are motile at some stage in the life cycle; multicellular. Reproduction usually sexual, with organisms developing from an embryo that has a blastula stage. 33 phyla; more than one million species.

Sponges   (Phylum Porifera) [Latin *porus*, pore; *ferre*, to bear]
Free-living; mostly marine. Sessile—adults attached to solid object. Asymmetrical; two cell layers; body with pores connected to an internal canal system. Reproduction asexual and sexual. About 10 000 species.

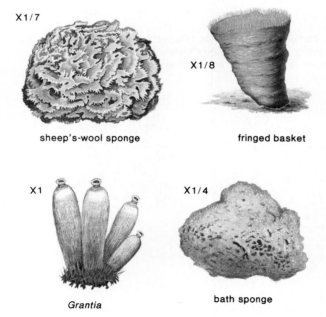

X1/7

X1/8

sheep's-wool sponge   fringed basket

X1

X1/4

Grantia   bath sponge

Cnidarians   (Phylum Cnidaria) [Greek *knide*, nettle]
Free-living; mostly marine. Motile stages. Radially symmetrical; two cell layers with jellylike material between; digestive sac; mouth surrounded by stinging tentacles; nerve network. Reproduction asexual and sexual. About 9500 species.

Class Hydrozoa   (hydras) [Greek *hydra*, water serpent; *zoon*, animal]
Single individuals or colonies; digestive sac undivided; simple sense organs. About 3100 species.

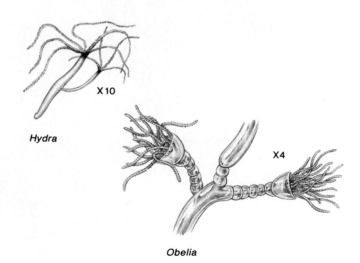

X10

Hydra

X4

Obelia

Class Scyphozoa (jellyfish) [Greek *skyphos,* cup; *zoon,* animal]

Single individuals that float or swim; a few species have attached stages in the life cycle. Digestive sac divided; rather complex sense organs. About 200 species.

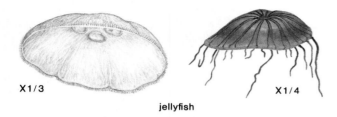

jellyfish

Class Anthozoa (corals and sea anemones) [Greek *anthos,* flower; *zoon,* animal]

Single individuals or massive colonies; no floating or swimming stages. Often produce limy skeletons; digestive sac divided. About 6200 species.

sea anemones

brain coral

coral

Comb Jellies (Phylum Ctenophora) [Greek *kteis,* comb; *pherein,* to bear]

Free-living; marine. Floating or free-swimming by means of eight bands of paddlelike comb plates; somewhat resemble jellyfish without stinging cells. Reproduction sexual. About 90 species.

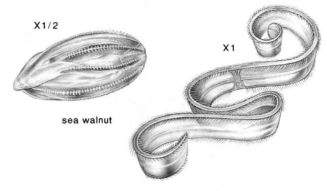

sea walnut

Venus's-girdle

Mesozoans (Phylum Mesozoa) [Greek *mesos,* middle; *zoon,* animal]

Parasitic in flatworms, mollusks, and annelids. Small, wormlike, bilaterally symmetrical; two cell layers; lack organs and systems except for ovaries and testes. About 45 species.

Pseudicyema

Flatworms (Phylum Platyhelminthes) [Greek *platys,* flat; *helmins,* worm]
Free-living or parasitic; motile. Bilaterally symmetrical; flat; three cell layers; branched or unbranched digestive sac or no digestive cavity; excretion by flame cells. Reproduction asexual and sexual. About 15 000 species.

Class Turbellaria (planarians) [Latin *turba,* disturbance (so named because the cilia cause tiny currents in the water)]
Free-living; usually with external cilia. Mostly marine, but some freshwater or terrestrial species.

X8

planarian

Class Trematoda (flukes) [Greek *trematodes,* with holes]
Parasitic; no external cilia. Usually possess hooks and suckers; digestive system present.

X4

liver fluke
(stained)

Class Cestoda (tape worms) [Greek *kestos,* girdle]
Internal parasites of animals; complex life cycles involving alternate hosts; no digestive system.

X2

tapeworm

Ribbon Worms (Phylum Nemertina) [Greek *Nemertes,* a sea nymph]
Free-living; mostly marine. Bilaterally symmetrical; flat; unsegmented; digestive tube with two openings; circulatory system. About 900 species.

X1/2

Cerebratulus

Gastrotrichs (Phylum Gastrotricha) [Greek *gaster,* stomach; *thrix,* hair]
Free-living; freshwater or marine. Bilaterally symmetrical; flat; unsegmented; spiny, bristly, or scaly. Reproduction sexual. About 400 species.

X22

Chaetonotus

Rotifers (Phylum Rotifera) [Latin *rota,* wheel; *ferre,* to bear]
Mostly free-living; some symbiotic or parasitic; freshwater or marine. Mostly motile; microscopic. Bilaterally symmetrical; body cavity; digestive tube with two openings; cilia form a wheel around mouth; no circulatory system. Reproduction asexual and sexual. About 2000 species.

X30

Asplanchna X32

Philodina

Kinorynchs (Phylum Kinorhyncha) [Greek *kinein,* to move; *rhynchos,* beak]
Free-living; marine; minute. Spiny, retractable head; pull themselves along by rings of recurved spines. About 150 species.

X23

Echinoderella

Spiny-headed Worms (Phylum Acanthocephala) [Greek *akantha,* spine; *kephale,* head]
Young parasitic in arthropods; adults parasitic in intestine of vertebrates; no digestive system. Reproduction sexual. More than 600 species.

X2

Oncicola

Roundworms (Phylum Nematoda) [Greek *nema,* thread]
Free-living or parasitic, especially on roots of plants; about 30 species parasitic in humans; many species are decomposers; motile. Bilaterally symmetrical; cylindrical; digestive tube with two openings; no circulatory system. Reproduction sexual. More than 80 000 species.

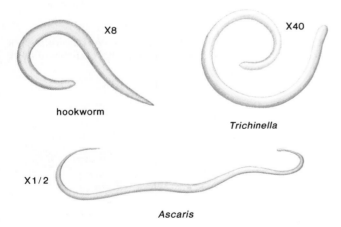

X8

hookworm

X40

Trichinella

X1/2

Ascaris

Horsehair Worms (Phylum Nematomorpha) [Greek *nema,* thread; *morphe,* form]
Young are parasitic in arthropods; adults free-living, with much reduced digestive tubes. About 240 species.

X1/2

Gordius

Bryozoans (Phylum Ectoprocta) [Greek *ektos,* outside; *proktos,* anus]
Free-living; mostly marine, forming attached colonies. U-shaped digestive tube; mouth encircled by a crown of tentacles. Reproduction sexual. About 5000 species.

X10

Electra

Phoronids (Phylum Phoronida) [Greek *pherein,* to bear; Latin *nidus,* nest]
Free-living; marine. Live in tubes in mud; U-shaped digestive tube; ring of tentacles forms a food-gathering organ; closed circulatory system with hemoglobin-containing red blood cells. Reproduction sexual. About 15 species.

X1/2

Phoronis

Brachiopods (Phylum Brachiopoda) [Latin *brachion,* arm; Greek *pous,* foot]
Free-living; attached or burrowing; marine. Bilaterally symmetrical; dorsal-ventral shells enclosing tentacle-bearing arms; open circulatory system. Reproduction sexual; sexes separate. About 335 living species.

X1

Lingula

Mollusks (Phylum Mollusca) [Latin *molluscus,* soft]
Free-living; marine, freshwater, or terrestrial; sessile or motile. Bilaterally symmetrical or asymmetrical; no segmentation; true body cavity; well-developed digestive, circulatory, and nervous systems. Soft-bodied, with external or internal shell lined with a fold of tissue, the mantle, which secretes the hard, limy shell. Reproduction sexual. About 110 000 species.

Class Monoplacophora [Greek *monos,* one; *plax,* tablet, flat plate; *phora,* bearing]
Marine; found in deep ocean trenches. Single shell with a curved apex; broad, flattened foot. Ten species.

X2

Neopilina

Class Polyplacophora (chitons) [Greek *polys,* many; *plax,* plate; *phora,* bearing]
Marine; shell composed of eight overlapping plates (exposed or hidden); no distinct head.

chiton

X1/2

Class Scaphoda (tusk shells) [Greek *skaphe,* boat; *pous,* foot]
Marine; shells form a tapering tube; food-catching tentacles on head.

X1/2

tuskshell

Class Gastropoda (slugs and snails) [Greek *gaster,* stomach; *pous,* foot]
Marine, freshwater, or terrestrial; shell (if present) coiled; head usually distinct.

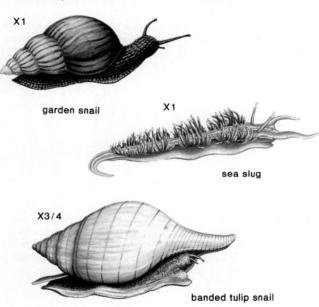

X1

garden snail

X1

sea slug

X3/4

banded tulip snail

Class Bivalvia (bivalves) [Latin *bi,* two; *valva,* folding door]
Marine or freshwater; some attached; others burrow in mud or sand. Shells in two parts, hinged.

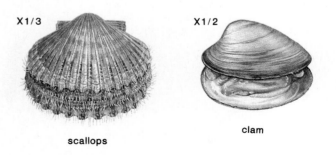

X1/3

X1/2

scallops

clam

Class Cephalopoda (octopus and squid) [Greek *kephale,* head; *pous,* foot]

Marine; small, internal shell. In a few cases shell is external, coiled, and internally divided. Several tentacles on head; locomotion by jet of water.

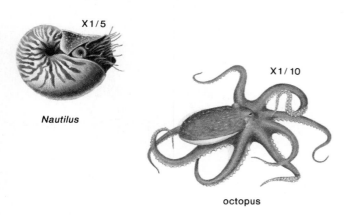

X1/5

Nautilus

X1/10

octopus

Priapulids (Phylum Priapulida) [Latin *priapos,* phallus]

Free-living; marine; burrowing; mouth with curved spines that grasp prey; body covered with rings of cuticle. Reproduction sexual. About 5 species.

X1/2

Priapulus

Segmented Worms (Phylum Annelida) [Latin *annellus,* little ring]

Free-living or parasitic; marine, freshwater, or terrestrial; motile. Bilaterally symmetrical; body internally and externally segmented; appendages either not jointed or lacking; main nerve cord ventral. Reproduction asexual and sexual. About 8800 species.

Class Oligochaeta (earthworms) [Greek *oligos,* few; *chaite,* hair]

Mostly freshwater or terrestrial. Appendages small or lacking.

X1

earthworm

Class Hirudinea (leeches) [Latin *hirudo,* leech]

Parasitic, with suction disks at each end; flat; appendages lacking.

X1

leech

Class Polychaeta [Greek *polys,* many; *chaite,* hair]

Mostly marine; burrowers or tube-builders; usually with paddlelike appendages on each body segment.

X1/2

clam worm

Velvet Worms (Phylum Onychophora) [Greek *onyx,* claw; *pherein,* to bear]

Free-living; terrestrial; tropical; require high humidity. Wormlike, unsegmented; paired legs; one pair short antennae. Combine many annelid and arthropod characteristics. About 80 species.

X1/2

Peripatus

Arthropods (Phylum Arthropoda) [Greek *arthron*, joint; *pous*, foot]
Free-living; important pollinators; many transmit parasites; marine, freshwater, or terrestrial; motile. Bilaterally symmetrical; segmented body fused to two or three parts; jointed appendages; jointed exoskeleton; ventral nerve cord. Reproduction sexual, metamorphosis common in life cycle. About 800 000 described species.

Subphylum Chelicerata [Greek *chele*, claw]
First pair of appendages modified to grasp food; four pairs of legs; no antennae; two body parts (head-thorax and abdomen); no wings.

Class Merostomata (horseshoe crabs) [Greek *meros*, thigh; *stoma*, mouth]
Aquatic; compound lateral eyes; gas exchange through gills.

X1/4

horseshoe crab

Class Arachnida (spiders, ticks, and mites) [Greek *arachne*, spider]
Terrestrial; simple eyes; gas exchange through trachea, lungs, or gill books.

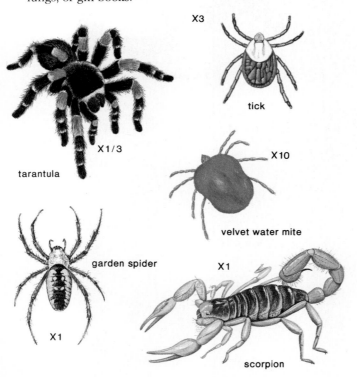

X3

tick

X1/3

tarantula

X10

velvet water mite

garden spider

X1

X1

scorpion

Subphylum Crustacea [Latin *crusta*, shell]
Mostly aquatic; compound eyes; gas exchange through gills; two pairs of antennae; jaws; three distinct body parts.

Class Branchiopoda (fairy shrimp, brine shrimp, water fleas) [Greek *branchia*, gills; *pous*, foot]
Reduced first antennae; gas exchange through gills on flattened, leaflike legs.

X15

X15

Daphnia

brine shrimp

Class Copepoda (cyclops) [Greek *kope*, oar; *pous*, foot]
Free-living and parasitic. Major component of plankton; important food item for whales, sharks, many fish.

X13

Cyclops

Class Cirripedia (barnacles) [Latin *cirrus*, curl of hair; *pes*, foot]
Free-living filter feeders or parasites; marine; sessile as adults; frequently foul ship bottoms in such large numbers that ship speed is reduced 30–40 percent.

X1

barnacle

X1/2

rock barnacle

Class Malacostraca (sow bugs, crabs, lobsters) [Greek *malakos,* soft; *ostrakon,* shell]
Mostly marine; sow bugs (isopods) terrestrial in moist soil litter under stones; decapods (lobsters, crabs, true shrimp) exist in great variety of forms; important food source.

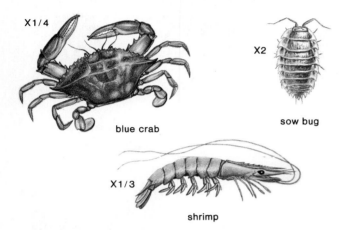

X1/4

blue crab

X2

sow bug

X1/3

shrimp

Subphylum Uniramia [Latin *unus,* one; *ramus,* branch]
Three distinct body parts; one pair of antennae.

Class Diplopoda (millipedes) [Greek *diploos,* double; *pous,* foot]
Important forest decomposers. Many-segmented, cylindrical body; one pair of short antennae; two pairs of legs per segment; gas exchange through trachea.

X1/2

millipede

Class Chilopoda (centipedes) [Greek *cheilos,* lip; *pous,* foot]
Predators, mostly of insects. Many-segmented, flattened body; one pair of long antennae; one pair of legs per segment; gas exchange through trachea.

X1

centipede

Class Insecta (insects) [Latin *insectus,* cut into (from the segmented bodies)]
Three body parts (head, thorax, and abdomen); three pairs of legs on thorax; generally one or two pairs of wings; gas exchange through trachea. Perhaps 10 million species. (The following are the more common orders in the class Insecta.)

Order Thysanura [Greek *thysanos,* tassel; *oura,* tail]
Small; wingless; soft scales on the body; three long bristles at posterior end.

X4

silverfish

Order Ephemeroptera [Greek *ephemeros,* temporary (literally, lasting but a day); *pteron,* wing]
Two pairs of transparent wings, hind wings smaller; two or three long tails; immature forms aquatic.

X2

mayfly

Order Odonata [Greek *odous,* tooth; *ata,* characterized by]
Two similar pairs of long wings; antennae short; abdomen long and slender; immature forms aquatic.

X1

damselfly

Order Orthoptera [Greek *orthos,* straight; *pteron,* wing]
Front wings leathery; hind wings folded, fanlike; chewing mouth parts; terrestrial.

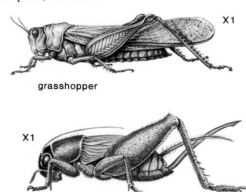

X1

grasshopper

X1

cricket

Order Isoptera [Greek *isos,* equal, *pteron,* wing]
Four wings alike in size, with many fine veins, or wings
lacking; chewing mouth parts; social.

termites
X2

(winged)   (wingless)

Order Anoplura [Greek *anoplos,* unarmed; *oura,* tail]
Wingless, flat body; sucking or piercing mouth parts;
parasitic on mammals.

X15

louse

Order Hemiptera [Greek *hemi,* half; *pteron,* wing]
Front wings thick at the base, thin at the tips; hind wings
thin; jointed beak on front of head.

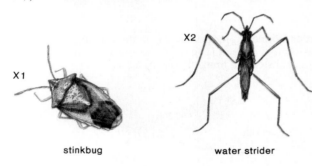

X2

X1

stinkbug   water strider

Order Homoptera [Greek *homos,* same; *pteron,* wing]
Two pairs of wings arched above the body, or wings lacking;
jointed sucking beak at base of head

X6   X5

leafhopper   aphid

Order Coleoptera [Greek *koleos,* sheath; *pteron,* wing]
Front wings form a hard sheath; hind wings folded beneath
them; chewing mouth parts; young usually wormlike
(grubs).

X2   X1   X2

hister beetle   scarab beetle   cucumber beetle

Order Lepidoptera [Greek *lepidos,* scale; *pteron,* wing]
Two pairs of wings covered with soft scales; coiled, sucking
mouth parts in adults; young wormlike (caterpillars).

X1/2

X2

moth   butterfly

Order Diptera [Greek *dis,* two; *pteron,* wing]
One pair of wings; hind wings reduced to small rods;
antennae short; sucking mouth parts; young wormlike and
either terrestrial (maggots) or aquatic.

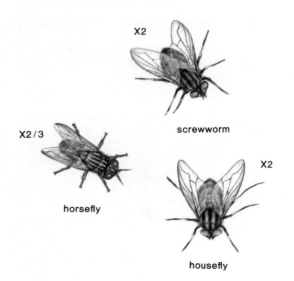

X2

screwworm

X2/3

horsefly

X2

housefly

Order Hymenoptera [Greek *hymen,* membrane; *pteron,* wing]
Front wings much larger than hind wings, with the two pairs hooked together, or wings lacking; chewing or sucking mouth parts; young wormlike; many social species.

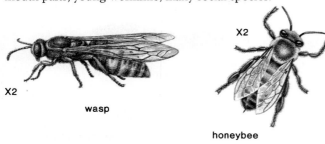

X2

wasp

X2

honeybee

Echinoderms (Phylum Echinodermata) [Greek *echinos,* sea urchin; *derma,* skin]
Free-living; marine; feeding, locomotion, and gas exchange by tube feet. Adults radially symmetrical; unsegmented; no head or brain. Internal, limy skeleton, usually with many projecting spines. Reproduction asexual and sexual; larvae bilaterally symmetrical; chordatelike. About 6000 species.

Class Crinoidea (crinoids) [Greek *krinon,* lily; *eidos,* appearance]
Many highly branched arms; attached (at least when young).

X1/4

sea lily

X1/6

Antedon dentata

Class Asteroidea (sea stars) [Greek *aster,* star; *eidos,* form]
Usually five arms, jointed to the body at broad bases.

X1/2

sea star

Class Ophiuroidea (brittle stars) [Greek *phis,* snake; *oura,* tail; *eidos,* form]
Usually five long, slim arms (sometimes branched), clearly distinguished from the body; pharyngeal gill slits, at least in embryo. Reproduction sexual; sexes separate. About 45 000 species.

X1/2

brittle star

Class Echinoidea (sand dollars) [Greek *echinos,* sea urchin; *eidos,* form]
No arms; spherical or disk-shaped; long spines or short, hairlike projections from body; skeleton of interlocking plates.

X1/4

sand dollar

X1/4

sea urchin

Class Holothuroidea (sea cucumbers) [Greek *holothourion,* sea cucumber; *eidos,* form]
No arms; somewhat cylindrical; tentacles around mouth; no spines; skeleton consists of particles embedded in the leathery skin.

X1/4

sea cucumber

Arrowworms   (Phylum Chaetognatha) [Greek *chaite,* hair; *gnathos,* jaw]
Predators; grasp prey with moveable hooks; marine; free-swimming or floating. Bilaterally symmetrical; straight digestive tube. Reproduction sexual; hermaphroditic. More than 100 species.

X1

Sagitta

Acorn Worms   (Phylum Hemichordata) [Greek *hemi,* half; Latin *chorda,* cord]
Predators; marine. Bilaterally symmetrical; unsegmented; wormlike; conspicuous proboscis used for burrowing in mud and sand; dorsal nerve cord and pharyngeal slits; notochord doubtfully present. Reproduction asexual, by budding, or sexual; sexes separate. About 90 species.

X1/2

acorn worm

Chordates   (Phylum Chordata) [Latin *chorda,* cord]
Marine, freshwater, or terrestrial. Bilaterally symmetrical; usually with paired appendages; some segmentation, especially in arrangement of muscles and nerves. Hollow dorsal nerve tube; cartilaginous notochord that may be lost or replaced by vertebral column during development.

Subphylum Urochordata   (tunicates) [Greek *oura,* tail; Latin, *chorda,* cord]
Marine; larvae free-swimming; adults usually attached. Notochord disappears during development; no brain.

X1/2

tunicate

Subphylum Cephalochordata   (lancelets) [Greek *kephale,* head; Latin *chorda,* cord]
Marine; free-swimming; translucent; well-developed hollow dorsal nerve cord, notochord, and pharyngeal slits in adults.

X3/4

lancelet

Subphylum Vertebrata   (vertebrates) [Latin *vertebra,* joint]
Notochord replaced by a backbone of vertebrae during development; brain protected by cartilage or bone; usually with paired appendages.

Class Agnatha   (jawless fishes) [Greek *a,* without; *gnathos,* jaw]
Cartilaginous skeleton; no true jaws or paired appendages; two-chambered heart; gas exchange through gills.

X1/7

X1/10

hagfish

lamprey

Class Chondrichthyes   (cartilaginous fishes) [Greek *chondros,* cartilage; *ichthyes,* fish]
Cartilaginous skeleton; gill slits externally visible; jaws; paired fins; two-chambered heart; gas exchange through gills.

X1/25

X1/30

stingray

shark

Class Osteichthyes (bony fishes) [Greek *osteon,* bone; *ichthyes,* fish]
Bony skeleton; gill slits covered; jaws; paired fins; scales; two-chambered heart; gas exchange through gills.

Class Reptilia (reptiles) [Latin *repere,* to creep]
Bony skeleton; dry, scaly skin; most with two pairs of limbs; leathery-shelled eggs; incompletely divided four-chambered heart; gas exchange through lungs. About 5000 species.

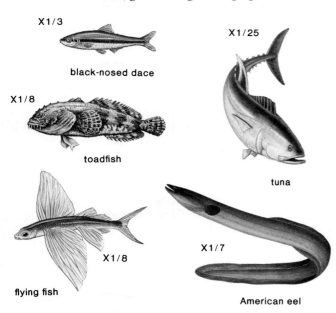

X1/3
black-nosed dace

X1/8
toadfish

X1/25
tuna

X1/8
flying fish

X1/7
American eel

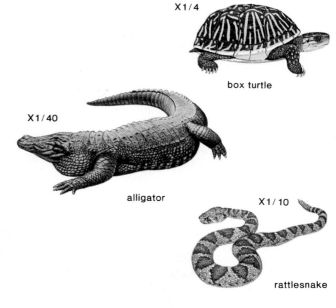

X1/4
box turtle

X1/40
alligator

X1/10
rattlesnake

X1/5
Gila monster

X1/3
chameleon

Class Amphibia (amphibians) [Greek *amphis,* both; *bios,* life]
Bony skeleton; moist glandular skin; most with two pairs of limbs; three-chambered heart; gas exchange through skin or lungs in adult; through gills in aquatic larvae. About 2000 species.

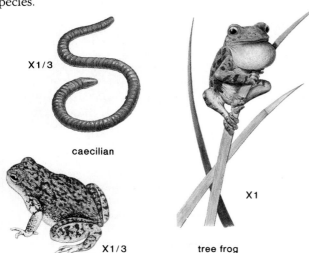

X1/3
caecilian

X1
tree frog

X1/3
green toad

X1/2
six-lined race runner

X1/2
tiger salamander

Class Aves   (birds) [Latin *avis,* bird]
Bony skeleton; scales modified as feathers; no teeth; forelimbs modified as wings; hard-shelled eggs; four-chambered heart; endothermic; gas exchange through lungs. Nearly 9000 species.

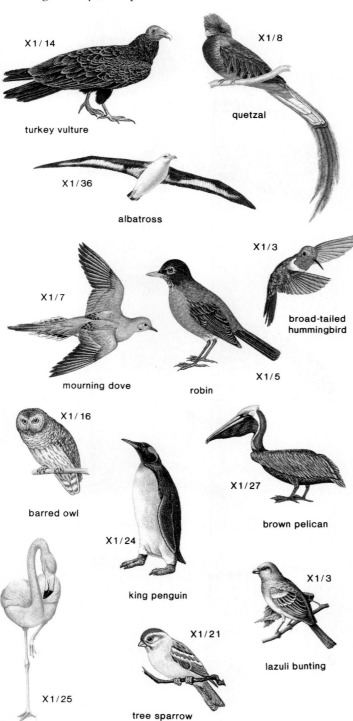

X1/14

turkey vulture

X1/8

quetzal

X1/36

albatross

X1/7

mourning dove

robin

X1/3

broad-tailed hummingbird

X1/5

X1/16

barred owl

X1/24

king penguin

X1/27

brown pelican

X1/3

lazuli bunting

X1/21

tree sparrow

X1/25

flamingo

Class Mammalia   (mammals) [Latin *mamma,* breast]
Bony skeleton; scales modified as hairs; mammary glands in females secrete milk; four-chambered heart; endothermic; gas exchange through lungs; teeth usually of four well-defined types (incisors, canines, premolars, molars). About 4500 species.

Order Monotremata   [Greek *monos,* single; *trema,* hole]
Egg-laying; mammary glands without nipples. 5 species.

X1/8

platypus

Order Marsupialia   [Greek *marsypos,* pouch]
Young born in undeveloped state and transferred to a pouch, where they remain tightly attached to the nipples. About 250 species.

X1/16

opossum

X1/43

X1/16

great red kangaroo

koala

Order Insectivora   [Latin *insectum,* insect; *vorare,* to eat]
Numerous teeth of all four mammalian types, none highly specialized. About 400 species.

X1/4

mole

X1

shrew

Order Chiroptera [Greek *cheir*, hand; *pteron*, wing]
Web of skin between fingers and between front limbs and
hind limbs allows flight. About 900 species.

X1/4

bat

Order Carnivora [Latin *carnis*, flesh; *vorare*, to eat]
Incisors small; canines large; premolars adapted for
shearing; claws usually sharp. About 280 species.

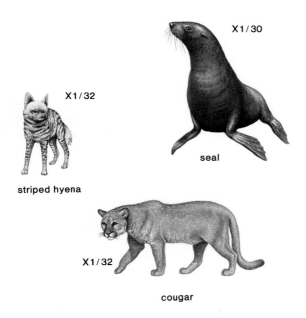

X1/30

seal

X1/32

striped hyena

X1/32

cougar

Order Tubulidentata [Latin *tubulus*, tube; *dens*, tooth]
Teeth few in adults but numerous in embryos; toes ending
in nails that are intermediate between claws and hoofs. One
species.

X1/27

aardvark

Order Rodentia [Latin *rodere*, to gnaw]
Chisel-like incisors that grow continually from the roots; no
canines; broad molars. About 1700 species.

X1/10

X1/10

woodchuck

squirrel

Order Pholidota [Greek *pholis*, horny scale]
No teeth; body encased in scales formed from modified
hairs. 8 species.

X1/12

pangolin

Order Lagomorpha [Greek *lagos*, hare; *morphe*, form]
Harelike mammals; teeth similar to those of rodents, but
with four upper incisors instead of two; tail very short. About
60 species.

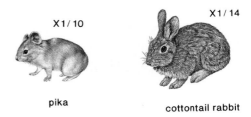

X1/10

X1/14

pika

cottontail rabbit

X1/20

porcupine

Order Edentata   [Latin *edentatus,* toothless]
No front teeth; molars in some species. About 30 species.

X1/13

armadillo

X1/24

sloth

Order Cetacea   [Latin *cetos,* whale]
Marine; front limbs modified as flippers; hind limbs absent; no hair on adults; eyes small; head very large. About 80 species.

X1/450

blue whale

Order Proboscidea   [Greek *pro,* before; *boskein,* to feed]
Herbivorous; upper incisors modified as tusks; molars produced two to four at a time as older ones wear out; nose and upper lip modified as a trunk. 2 species.

X1/80

elephant

Order Sirenia   [Greek *seiren,* sea nymph]
Herbivorous; aquatic; no hind limbs; broad, flat tail expanded as a fin; few hairs. 5 species.

X1/80

manatee

X1/64

dugong

Order Perissodactyla   [Greek *perissos,* odd; *daktylos,* toe]
Herbivorous; odd-toed, hoofed mammals; well-developed molars. About 15 species.

X1/57

rhinoceros

X1/30

American tapir

Order Artiodactyla   [Greek *artios,* even; *daktylos,* toe]
Herbivorous, most with complex stomachs; even-toed, hoofed mammals; often have horns or antlers. About 170 species.

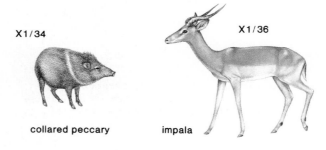

X1/34

collared peccary

X1/36

impala

X1/55

hippopotamus

Order Primates [Latin *prima,* first]

Eyes usually directed forward; nails usually present instead of claws; teeth much like those of insectivores. About 200 species.

To show how complex classification can become, a complete classification of the order Primates follows.

X1/30

chimpanzee

X1/40

gorilla

X1/24

baboon

X1/10

red howler

X1/8

marmoset

X1/32

human

Suborder Prosimii
Infraorder Lemuriformes
Superfamily Tupaioidea
Family Tupaiidea: tree shrews
Superfamily Lemuroidea
Family Lemuridae: lemurs
Family Indriidae: indris
Superfamily Daubentonioidea
Family Daubentoniidae: aye-ayes
Infraorder Lorisiformes
Family Lorisidae: lorises, pottos, galagos
Infraorder Tarsiiformes
Family Tarsiidae: tarsiers
Suborder Anthropoidea
Superfamily Ceboidea
Family Cebidae: New World monkeys
Family Callithricidae: marmosets
Superfamily Cercopithecoidea
Family Cercopithecidae: Old World monkeys, baboons
Superfamily Hominoidea
Family Pongidae: apes
Family Hominidae: humans

Glossary

A

abiotic (AY by OT ik) **factor:** a physical or nonliving component of an ecosystem. p. 33

abscisic (ab SIS ik) **acid:** a plant hormone that protects a plant in an unfavorable environment by promoting dormancy in buds and seeds and closing of stomates. p. 529

absorption: the movement of water and of substances dissolved in water into a cell, tissue, or organism. p. 395

absorption spectrum: the characteristic array of wavelengths (colors) of light that a particular substance absorbs. p. 515

acid deposition: The falling of acids and acid-forming compounds from the atmosphere on the earth's surface; commonly known as acid rain. p. 629

acidic: having a pH of less than 7, reflecting more dissolved hydrogen ions than hydroxide ions. p. 80

acquired characteristics: characteristics acquired during the life of an organism; once thought to be heritable. p. 222

actin: a protein in a muscle fiber that, together with myosin, is responsible for muscle contraction and relaxation. p. 455

action spectrum: a representation of the rate of an activity, especially photosynthesis, under different wavelengths of light at a given light intensity. p. 515

active site: the specific portion of an enzyme that attaches to the substrate through weak chemical bonds. p. 96

active transport: the movement of a substance across a biological membrane against its concentration gradient with the help of energy input and specific transport proteins. p. 122

adaptation: in natural selection, a hereditary characteristic of some organisms in a population that improves their chances for survival and reproduction in their environment compared with the chances of other organisms in the population. p. 222

adaptive radiation: the emergence of numerous species from a common ancestor introduced to an environment presenting a diversity of conditions. p. 234

adenosine diphosphate: see ADP.

adenosine triphosphate: see ATP.

ADP: adenosine diphosphate (uh DEN oh seen dy FOS fayt); the compound that remains when a phosphate group is transferred from ATP to a cell reaction site requiring energy input. p. 87

adrenal cortex: the outer portion of the adrenal gland. p. 476

adrenal medulla: the central portion of the adrenal gland that secretes epinephrine. p. 476

aerobic (eh ROH bik): occurring or living in the presence of free or dissolved oxygen. p. 273

agglutinate (uh GLOO tin ayt): to unite in a mass; clump. p. 439

aggression: hostile behavior. p. 548

algae: unicellular or multicellular photosynthetic organisms lacking multicellular sex organs. p. 306

alkaline: basic; having a pH greater than 7, reflecting more dissolved hydroxide ions than hydrogen ions. p. 80

allele (al LEEL): one of two or more possible forms of a gene, each affecting the hereditary trait somewhat differently. p. 181

alpine tundra: a biome roughly similar to arctic tundra that occurs above the timberline on mountains. p. 602

alternation of generations: a reproductive cycle in which a haploid (n) phase, the gametophyte, gives rise to gametes that, after fusion to form a zygote, produce a diploid ($2n$) phase, the sporophyte; spores from the sporophyte give rise to new gametophytes. p. 340

alveoli (al VEE oh ly): air sacs in a lung. p. 442

amino (uh MEEN oh) **acid:** an organic compound composed of a central carbon atom to which are bonded a hydrogen atom, an amino group (—NH_2), an acid group (—COOH), and one of a variety of other atoms or groups of atoms; amino acids are the building blocks of polypeptides and proteins. p. 92

ammonia: (1) a toxic nitrogenous waste excreted primarily by aquatic organisms, p. 267; (2) a gas in the earth's early atmosphere. p. 384

amniocentesis (AM nee oh sen TEE sus): a technique used to detect genetic abnormalities in a fetus through the presence of certain chemicals or defects in fetal cells in the amniotic fluid; the fluid is obtained through a needle inserted into the amniotic sac. p. 167

amnion: a sac or membrane, filled with fluid, that encloses the embryo of a reptile, bird, or mammal. p. 167

amniotic fluid: the liquid that bathes an embryo or fetus. p. 167

amphibian: any of the various ectothermic, smooth-skinned vertebrate organisms, belonging to the class Amphibia, that characteristically hatch as aquatic larvae with gills and that develop into adult forms that use lungs for gas exchange. p. 374

anaerobic (an eh ROH bik): occurring or living in conditions without free oxygen. p. 272

anaphase: the stage in mitosis in which chromosomes on the spindle separate and are pulled toward opposite ends of the cell. p. 128

Animalia (an ih MAYL yuh): the animal kingdom. p. 262

Annelida (an NEL ih duh): the phylum that includes segmented worms, such as earthworms and leeches. p. 370

annelids (AN ih lidz): worms belonging to the phylum Annelida. p. 370

anorexia nervosa (an oh REX ee uh ner VOH suh): a condition characterized by abnormal loss of appetite and induced self-starvation to reduce body weight. p. 419

anterior: situated toward the front; having a head end. p. 364

anther: the enlarged end of a stamen, inside which pollen grains with male gametes form in a flower. p. 340.

Anthophyta (an THOF ih tuh): the division containing the flowering plants. p. 350

anthropologist (an thruh PAH loh jist): a scientist who studies humans—human evolution, variability, and both past and present cultures and behavior. p. 386

antibody: a blood protein produced in response to an antigen, with which it combines specifically. p. 289

antigen: a macromolecule that is recognized as foreign and elicits an immune response. p. 431

anus (AY nus): the outlet of the digestive tube. p. 397

Apicomplexa (ap ih kum PLEX suh): the phylum containing the sporozoans. p. 314

appendage (uh PEN dij): a structure attached to the main part of a body; in animals, a tentacle, a leg, a flipper, a wing, etc. p. 365

aquatic (uh KWAH tik): describing a water environment or organisms that live in water. p. 364

arboreal: adapted for living in or around trees, as are monkeys and some apes. p. 586

archaebacteria (AR kee bak TIR ee uh): the more ancient of the two major lineages of prokaryotes represented today by a few groups of bacteria inhabiting extreme environments. p. 287

archaeology: the systematic study of the human past; locating and interpreting the cultural products of prehistoric humans. p. 592

arctic tundra: biome at the northernmost limits of plant growth, where plant forms are limited to low, shrubby, or matlike vegetation. p. 602

artery: a vessel that transports blood away from the heart. p. 424

arthropod: any of the numerous invertebrate organisms of the phylum Arthropoda. p. 371

Arthropoda (ar THRAP uh duh): the phylum that includes insects, crustaceans, arachnids, and others. p. 371

artificial selection: the selective breeding of domesticated plants and animals to encourage the occurrence of desirable traits. p. 221

Ascomycota (AS koh my KOH tuh): the phylum containing the sac fungi. p. 321

ascus (AS kus): a reproductive structure formed by sac fungi when two hyphae conjugate. p. 155

asexual reproduction: any method of reproduction that requires only one parent or one parent cell. p. 137

atherosclerosis (ATH uh roh skleh roh sis): a chronic cardiovascular disease in which plaques develop on inner walls of arteries, restricting the flow of blood. p. 415

atom: the smallest particle of an element; in turn, an atom is made of smaller particles that do not separately have the properties of the element. p. 14

ATP: adenosine triphosphate (uh DEN oh seen try FOS fayt); a compound that has three phosphate groups and is used by cells to store energy. p. 86

atrium (AY tree um; plural *atria*): a chamber of the heart that receives blood from the veins. p. 425

australopithecine (AH strahl oh PITH uh seen): any of the earliest known species of hominids that walked erect and had humanlike teeth but whose skull, jaw, and brain capacity were more apelike; may include several species. p. 586

autoimmune: a response in which antibodies are produced against some of the body's own cells. p. 437

autonomic nervous system: a division of the nervous system that controls involuntary activities of the body such as blood pressure, body temperature, and other functions necessary to the maintenance of homeostasis. p. 468

autosome: a chromosome that is not directly involved in determining sex. p. 200

autotroph (AWT oh trohf): an organism able to make and store food, using sunlight or another nonliving energy source. p. 260

auxin (AWK sin): a plant hormone produced in an actively growing region of a plant and transported to another part of the plant, where it produces a growth effect. p. 528

axon: a structure that extends out from a neuron and conducts impulses away from the cell body. p. 463

B

Bacillariophyta (buh SIL er ee OFF ih tuh): the phylum containing the diatoms. p. 308

basic: alkaline; having a pH greater than 7, reflecting more dissolved hydroxide ions than hydrogen ions. p. 80

basidia (buh SID ee uh): specialized reproductive cells of basidiomycetes, often club-shaped, in which nuclear fusion and meiosis occur. p. 322

Basidiomycota (buh SID ee oh my KOH tuh): the phylum containing the club fungi. p. 322

B cell: a type of lymphocyte that develops in the bone marrow and later produces antibodies. p. 431

bilateral symmetry (by LAT er ul SIM eh tree): a body having two corresponding or complementary halves. p. 364

bile: a secretion of the liver, stored in the gall bladder, and released through a duct to the small intestine; breaks large fat droplets into smaller ones that enzymes can act upon more efficiently. p. 396

binomial nomenclature (by NOH mee ul NOH men klay chur): the two-word naming system used in systematics or taxonomy. p. 257

biocide: a poisonous substance produced and used to kill forms of life considered to be pests to humans or that spread diseases. p. 67

biodiversity: the diversity of different species and the genetic variability among individuals within each species. p. 69

biology: the study of living organisms and life processes. p. 8

biomass: the dry weight of organic matter composing a group of organisms in a particular habitat. p. 646

biome: the distinctive plant cover and the rest of the community of organisms associated with a particular physical environment; often the biome is named for its plant cover. p. 597

biosphere (BY oh sfir): the outer portion of the earth—air, water, and soil—where life is found. p. 17

biosynthesis (by oh SIN thuh sis): the process of putting together or building up the large molecules characteristic of a particular type of cell or tissue. p. 406

biotic (by OT ik) **factor:** an organism or its remains in an ecosystem. p. 33

bipedal: capable of walking erect on the hind limbs, freeing the hands for other uses. p. 586

blade: the broad, expanded part of a leaf. p. 485

blastocyst: the mammalian embryonic stage that corresponds to the blastula of other animals. p. 166

blastula (BLAS choo luh): an animal embryo after the cleavage stage, when a pattern of cell movements toward the outside of the ball of cells results in a fluid-filled cavity inside. p. 159

brachiopods (BRAY kee oh podz): marine invertebrates with dorsal and ventral shells and a pair of tentacled structures on either side of the mouth. p. 571

Bryophyta (bry OFF ih tuh): the division containing the bryophytes. p. 343

bryophytes (BRY oh fyts): the mosses, liverworts, and hornworts; a group of nonvascular plants that inhabit land but lack many of the terrestrial adaptations of vascular plants. p. 336

bulimia (buh LEE mee uh): an abnormal craving for food beyond the body's needs; frequently expressed as gorging followed by forced vomiting. p. 419

bundle sheath cells: tightly packed cells surrounding veins in the leaves of C-4 plants that receive a rearranged 4-carbon acid in the first stage of C-4 photosynthesis; within the bundle sheath cells, carbon dioxide is released and reincorporated by an enzyme into PGA. p. 527

C

callus: a mass of dividing, undifferentiated cells at the cut end of a shoot or in a tissue culture, from which adventitious roots develop. p. 530

calorie: the amount of heat required to raise the temperature of one gram of water $1°$ C. p. 46

Calvin cycle: the cycle (named for its discoverer) by which carbon dioxide is incorporated in sugars during photosynthesis; uses chemical energy previously converted from light energy. p. 516

CAM: crassulacean (kras yoo LAY see un) acid metabolism; an adaptation for photosynthesis in arid conditions, in which carbon dioxide entering open stomates at night is converted into organic acids, which release carbon dioxide during the day when the stomates are less open. p. 527

cambium (KAM bee um): a layer of meristem tissue in the stems and roots of plants that produces all growth in diameter, including new xylem and phloem cells. p. 494

cancer: malignancy arising from cells that are characterized by profound abnormalities in the plasma membrane and in the cytosol, and by abnormal growth and division. p. 170

capillary (KAP ih layr ee): a microscopic blood vessel that penetrates the tissues and that has walls consisting of a single layer of cells that allows exchange between the blood and tissue fluids. p. 424

carbohydrate (kar boh HY drayt): an organic compound made of carbon, hydrogen, and oxygen, with the hydrogen and oxygen atoms in a 2 : 1 ratio; examples are sugars, starches, glycogen, and cellulose. p. 89

carbon cycle: the biogeochemical cycle in which carbon compounds made by some organisms are digested and decomposed by others, releasing the carbon in small inorganic molecules that can be used again by more organisms to synthesize carbon compounds. p. 101

carbon-14 dating: a technique to date fossils that uses the known disintegration rate of radioactive carbon; the amount of carbon-14 remaining in fossils indicates their age. p. 586

cardiac (KARD ee ak) **muscle:** a specialized type of muscle tissue found only in the heart. p. 457

cardiovascular disease: disease of the heart and/or blood vessels, such as atherosclerosis. p. 415

cardiovascular fitness: the relative state of health or fitness of the lungs, heart, and blood vessels, especially as regards their ability to deliver oxygen to the cells. p. 460

carnivore (KAR nih vor): any organism that eats animals; a meat-eater, as opposed to a plant-eater, or herbivore. p. 55

carpel (KAR puhl): the female reproductive organ of a flower, consisting of the stigma, style, and ovary. p. 340

carrying capacity: the maximum population size that can be supported by the available resources of a given area. p. 35

cartilage: a tough, elastic connective tissue that makes up the skeleton of cartilaginous fishes, but that in other vertebrates is replaced mostly by bone as the animal matures. p. 373

catalyst (KAT uh list): a chemical that promotes a reaction between other chemicals and may take part in the reaction but emerges in its original form. p. 80

cell cycle: an ordered sequence of events in the life of a dividing cell, composed of the M, G_1, S, and G_2 phases. p. 125

cell theory: the theory that organisms are composed of cells and their products and that these cells all are derived from preexisting cells. p. 107

cellular respiration (SEL yoo ler res pih RAY shun): the series of chemical reactions by which a living cell breaks down food molecules and obtains energy from them. p. 85

cellulose (SEL yoo lohs): a carbohydrate found in cells walls. p. 90

cell wall: a nonliving covering around the plasma membrane of certain cells, as in plants, many algae, and some prokaryotes; in plants the cell wall is constructed of cellulose and other materials. p. 118

central nervous system: the brain and spinal cord in vertebrates. p. 464

centriole (SEN tree ohl): one of two structures in animal cells, composed of cylinders of nine triplet microtubules in a ring; centrioles help organize microtubule assembly during cell division. p. 117

centromere (SEN troh meer): the specialized region of a chromosome that holds two replicated chromosomal strands together and that attaches to the spindle in mitosis. p. 140

cerebellum (ser eh BEL um): the part of the brain in vertebrates that is associated with regulating muscular coordination, balance, and similar functions. p. 466

cerebrum (seh REE brum): the largest portion of the brain in humans and many other animals; controls the higher mental functions such as learning. p. 465

chancre (SHAN ker): an ulcer located at the initial point of entry of a pathogen; a dull red, hard lesion that is the first manifestation of syphilis. p. 296

chaparral (shap uh RAL): a scrubland biome of dense, spiny evergreen shrubs found at midlatitudes along the coast where cold ocean currents circulate offshore; characterized by mild, rainy winters and long, hot, dry summers. p. 616

chelicerates (chih LIS er ayts): arthropods possessing paired appendages near the mouth that are modified for grasping. p. 371

chemical bond: the attraction between two atoms resulting from the sharing or transfer of outer electrons from one atom to another. p. 79

chemical energy: energy stored in the structure of molecules, particularly organic molecules. p. 12

chemosynthesis (KEE moh SIN thih sis): a pathway that uses energy from the oxidation of inorganic substances to drive the formation of organic molecules. p. 286

chitin (KYT in): a hard organic substance secreted by insects and certain other invertebrates as the supporting material in their exoskeletons. p. 320

chlorophyll (KLOR uh fil): the green pigments of plants and many microorganisms; converts light energy (via changes involving electrons) to chemical energy that is used in biological reactions. p. 84

Chlorophyta (kloh ROF it uh): the phylum containing the green algae. p. 302

chloroplast (KLOR oh plast): an organelle found only in plants and photosynthetic protists; contains chlorophyll that absorbs light energy used to drive photosynthesis. p. 117

cholesterol (koh LES ter ol): a lipid that is associated particularly with animal plasma membranes and is linked to deposits in blood vessels and to corresponding disorders of the heart. p. 91

Chordata (kohr DAH tuh): the phylum containing the chordates. p. 373

chordates: a diverse phylum of animals that possess a notocord, a dorsal hollow nerve cord, a tail, and pharyngeal gill slits at some stage of the life cycle. p. 373

chorion (KOR ee on): an embryonic membrane that surrounds all the other embryonic membranes in reptiles, birds, and mammals. p. 166

chorionic villi sampling (CVS): a procedure by which a small piece of membrane surrounding a fetus is removed and analyzed to detect genetic defects. p. 169

chromatid (KROH muh tid): one of two strands of a replicated chromosome before their separation during mitosis or meiosis. p. 140

chromosome (KROH moh sohm): a long, threadlike group of genes found in the nucleus of all eukaryotic cells and most visible during mitosis and meiosis; chromosomes consist of DNA and protein. p. 117

chromosome theory of heredity: the ideas that both eggs and sperm make equal genetic contributions to a new organism, that genes are located in the nucleus of both gametes, and that chromosomes contain the genes. p. 184

cilia (SIL ee uh): short, hairlike cell appendages specialized for locomotion and formed from a core of nine outer doublet microtubules and two inner single microtubules. p. 315

ciliates: unicellular, heterotrophic organisms that move by means of cilia. p. 315

Ciliophora (sil ee OFF er uh): the phylum containing the ciliates. p. 315

class: the third largest grouping, after kingdom and phylum or division, in the biological classification system. p. 251

cleavage (KLEE vaj): the process of cell division in animal cells, characterized by rapid cell divisions without growth that occur during early embryonic development and that convert the zygote into a ball of cells. p. 159

climatogram: a graph of monthly measurements of temperature and precipitation for a given area during a year. p. 599

clone: a lineage of genetically identical individuals. p. 531

closed circulatory system: a type of internal transport in which blood is confined to vessels. p. 380

closed population: a population in which there is no immigration or emigration; a completely isolated population. p. 27

club fungi: fungi belonging to the phylum Basidiomycota and possessing basidia, on which spores are produced. p. 322

Cnidaria: the phylum containing cnidarians. p. 367

cnidarians (nih DAR ee unz): radially symmetrical, stinging-tentacled animals that possess nerve networks and digestive sacs. p. 367

CoA: an enzyme present in all cells and necessary for cellular respiration and fatty acid metabolism. p. 403

coacervate (koh AS er vayt): a cluster of proteinlike substances held together in a small droplet within a surrounding liquid; used as a model for a precell to investigate the formation of the first life on the earth. p. 269

codominance: a condition in which both alleles in a heterozygous organism are expressed. p. 198

codon: the basic unit of the genetic code; a sequence of three adjacent nucleotides in DNA or mRNA. p. 190

coevolution: the evolution of two different species interacting with each other and reciprocally influencing each other's adaptations. p. 238

cohesion-tension hypothesis: the hypothesis that water is raised upward in a column in the xylem of plants by the force holding the water molecules together and by the continuing evaporation of water from the top of the column, in leaves. p. 496

communication: the exchange of thoughts, messages, or information by vocalizations, signals, or writing. p. 547

community: all the organisms that inhabit a particular area. p. 9

companion cell: a specialized elongated cell in the phloem of flowering plants that is associated with a sieve-tube cell. p. 494

competition: interaction between members of the same population or of two or more populations to obtain a mutually required resource in limited supply. p. 60

complement system: a complex group of serum proteins that can destroy antigens. p. 433

complete protein: a protein that contains all the essential amino acids in the proportions required by humans. p. 419

compound: a substance formed by chemical bonds between atoms of two or more different elements. p. 14

concentration gradient: a difference in the concentration of certain molecules over a distance. p. 120

conditioning: training that modifies a response so that it becomes associated with a stimulus different from the stimulus that originally caused it. p. 543

Coniferophyta: the division containing conifers. p. 347

conjugation: the process by which sexual exchange and reproduction take place in many microscopic organisms, principally protists and fungi. p. 320

consumer: a heterotroph; an organism that feeds on other organisms or on their organic wastes. p. 9

contact inhibition: the phenomenon observed in normal animal cells that causes them to stop dividing when they come in contact with one another. p. 171

continental drift: change in distance between continents and in their relative positions, caused by the spreading of new crustal material in sea-floor and by land rifts. p. 573

control: the norm (unchanged subject or group) in an experiment. p. 22

convergent evolution: the independent development of similarity between species as a result of their having similar ecological roles and selection pressures. p. 235

cooperative behavior: behavior that reduces the reproductive fitness of the performing individual and increases the fitness of the recipient. p. 553

cornea (KOR nee uh): the transparent outer layer of the vertebrate eye. p. 164

corpus luteum (KOR pus LOOT ee um): the structure that forms from the tissues of a ruptured ovarian follicle and secretes female hormones. p. 150

cortex: (1) the outer layers of the adrenal gland, p. 476; (2) a layer of cells under the epidermis or bark of some plant stems. p. 492

cotyledons (kot ih LEE dunz): the single (monocot) or double (dicot) seed leaves of a flowering plant embryo. p. 355

C-4 plants: plants with a photosynthetic pathway that incorporates carbon dioxide into 4-carbon compounds before beginning the Calvin cycle. p. 528

cranium: the skull (without the jaw). p. 581

creatine phosphate (KREE uh tin FOS fayt): an energy-storage compound used by muscle cells of vertebrates to replenish ATP supplies; replenished by the breakdown of glycogen. p. 457

crossing-over: during prophase I of meiosis, the breakage and exchange of corresponding segments of chromosome pairs at one or more sites along their length, resulting in genetic recombination. p. 141

cultural adaption: an adaptation to new pressures or situations resulting from cultural innovation. p. 668

culture: a system of learned behaviors, symbols, customs, beliefs, institutions, artifacts, and technology characteristic of a group and transmitted by its members to their offspring. p. 668

cuticle: in plants, a noncellular, waxy outer layer covering certain leaves and fruits. p. 336

cyanobacteria (SY an oh bak TIR ee uh): the blue-green bacteria, which carry on oxygen-producing photosynthesis much like plants, but without membrane-enclosed chloroplasts isolating their chlorophyll. p. 285

cytokinesis (syt oh kih NEE sus):   the division of the cytoplasm of a cell after nuclear division. p. 126

cytokinin (syt oh KY nin):   a plant hormone that promotes cell division, stem and root growth, chlorophyll synthesis, and chloroplast development. p. 532

cytoplasm (SYT oh plaz um):   the entire contents of the cell, except the nucleus, bounded by the plasma membrane. p. 116

cytoskeleton: a network of microtubules, microfilaments, and intermediate filaments that run throughout the cytoplasm and serve a variety of mechanical and transport functions. p. 118

cytosol (SYT oh sol):   the gelatinlike portion of the cytoplasm that bathes the organelles of the cell. p. 118

D

data (singular, *datum*):   observations and experimental evidence bearing on a biological question or problem. p. 19

decomposer: an organism that lives on decaying organic material, from which it obtains energy and its own raw materials for life. p. 9

decomposition (de kom poh ZISH un):   the process of taking molecules apart; heat and chemicals are the chief agents. p. 79

deforestation: the removal of trees from a forested area without adequate replanting. p. 627

dendrite: a structure that extends out from a neuron and transmits impulses toward the cell body. p. 463

denitrifying (DEE ny trih fy ing) **bacteria**:   bacteria that break down nitrogen compounds in the soil and release nitrogen gas to the air. p. 288

density: the number of individuals in a population in proportion to the size of their environment or living space. p. 34

dental caries: cavities; tooth decay caused, in part, by lactic acid bacteria. p. 290

dental plaque: a film composed of eubacteria in a sugar matrix on the surface of teeth. p. 290

deoxyribose (dee OK sih RY bohs):   a sugar used in the structure of DNA; it contains one less oxygen atom than does ribose. p. 98

depressant: a drug that slows the functioning of the central nervous system. p. 481

desert: a biome characterized by lack of precipitation and by extreme temperature variation; there are both hot and cold deserts. p. 613

desertification: the conversion of rangeland or cropland to desertlike land with a drop in agricultural productivity; usually caused by a combination of overgrazing, soil erosion, prolonged drought, and climate change. p. 628

development: (1) cell division, growth, and differentiation of cells from embryonic layers into all the tissues and organs of the body, p. 159; (2) later changes with age, including reproductive maturity with its effects on appearance and body function. p. 170

diabetes mellitus (MEL luh tus):   a disease resulting from insufficient insulin secretion by the pancreas or impairment of insulin receptors on body cells; characterized by abnormal absorption and use of glucose. p. 471

diaphragm (DY uh fram):   (1) the sheet of muscle that separates the chest and abdominal cavity in mammals and, along with rib muscles, is important in breathing, p. 442; (2) a caplike rubber device inserted in the vagina and used as a contraceptive. p. 149

diatom (DY uh tom):   any of a large group of algae with intricate, patterned, silica-containing shells made in two halves; unusually shaped pores create the patterns. p. 308

dicot (DY kot):   flowering plant that has two embryonic seed leaves, or cotyledons. p. 355

differentially permeable: open to passage or penetration by some substances, but not all, depending on molecule size, charge, and other factors. p. 122

differentiation: specialization, as when developing cells become ordered into certain tissues and organs. p. 159

diffusion (dih FYOO zhun):   the movement of a substance down its concentration gradient from a more concentrated area to a less concentrated area. p. 120

digestion: the process by which larger food molecules are broken down into smaller molecules that can be absorbed. p. 378

dinoflagellate: any of the numerous, chiefly marine, protozoans characteristically possessing two flagella; one of the major types of plankton. p. 313

dipeptide (DY PEP tyd):   two amino acid molecules bonded to one another; the dipeptide may be the start of a chain for a protein or a product of digestion of a protein and a polypeptide. p. 92

diploid (DIP loyd):   a cell containing both members of every chromosome pair characteristic of a species (*2n*). p. 140

dispersal: (1) the spreading of organisms from a place of concentration, p. 39; (2) the scattering of spores and seeds that promotes spreading of nonmotile organisms. p. 354

divergent evolution: evolutionary change away from the ancestral type, with selection favoring newly arising adaptations. p. 235

division: the second largest grouping, after kingdom, in the biological classification system for plants. p. 251

DNA: deoxyribonucleic (dee OK sih ry boh noo KLEE ik) acid; the hereditary material of most organisms; DNA makes up the genes; these nucleic acids contain deoxyribose, a phosphate group, and one of four bases. p. 97

dominance hierarchy: a linear "pecking order" of animals in which position dictates characteristic social behaviors. p. 550

dominant: a trait that is visible in a heterozygous organism. p. 180

dormant: inactive but alive or viable. p. 326

dorsal: in animals, situated toward the top or back side. p. 364

E

ecology: the study of living and nonliving components of the environment and of the interactions that affect biological species. p. 8

ecosystem (EE koh sis tum): a biological community in its abiotic environment. p. 53

ectoderm (EK toh derm): the outer layer of cells in the gastrula stage of an animal embryo. p. 160

effector: a muscle or gland activated by nerve impulses or hormones. p. 464

electron (ee LEK trahn): a negatively charged particle that occurs in varying numbers in clouds surrounding the nuclei of atoms. p. 77

electron transport system: the process in which electrons are transferred from one carrier molecule to another in photosynthesis and in cellular respiration; results in storage of some of the energy in ATP molecules. p. 401

element: a substance composed of atoms that are chemically identical—alike in their proton and electron numbers. p. 14

embryo: an organism in its earliest stages of development. p. 146

emigration (em uh GRAY shun): departure by migration; out-migration. p. 32

emphysema (em fuh SEE muh): a lung condition in which the air sacs dilate and the walls atrophy, resulting in labored breathing and susceptibility to infection. p. 442

endocast: a plaster or plastic cast of the inside surface, or the contents, of a covering or enclosing structure; examples include endocasts of a skull's braincase, a plant's spore case, or a marine microorganism's glasslike shell. p. 581

endocrine (EN doh krin) **gland:** a ductless gland that secretes one or more hormones into the bloodstream. p. 469

endocrine system: the system of glands that secrete their products from their cells directly into the blood. p. 469

endoderm (EN doh derm): an inner layer of cells, as in an embryo. p. 160

endoplasmic reticulum (en doh PLAZ mik reh TIK yoo lum): an extensive membranous network in eukaryotic cells composed of ribosome-studded (rough) and ribosome-free (smooth) regions. p. 116

endoskeleton (EN doh SKEL eh tun): a hard skeleton buried in the soft tissues of an animal, such as the spicules of sponges and bony skeletons of vertebrates. p. 386

endosperm (EN doh sperm): a nutrient-rich structure, formed by the union of a sperm cell with two polar nuclei during double fertilization, that provides nourishment to the developing embryo in seeds of flowering plants. p. 341

endospore (EN doh spor): a thick-walled spore of a particular type, like that produced by the anaerobic bacterium that causes botulism. p. 280

energy: the active component of all atoms; the ability to do work or to cause change. p. 82

energy pyramid: a graphic representation of the energy available for use by producers and consumers as levels of a pyramid: at the bottom, with the greatest amount of available energy, are the producers, followed by primary consumers, secondary consumers, and top-level consumers; only about 10 percent of the energy available at the preceding level is available for use at the next higher level. p. 65

environment (en VY run ment): everything living and nonliving in an organism's surroundings, including light, temperature, air, soil, water, and other organisms. p. 33

enzyme (EN zym): a protein or part-protein molecule made by an organism and used as a catalyst in a specific biochemical reaction. p. 80

enzyme-substrate complex: an enzyme molecule together with the molecules on which it acts, correctly arranged at the active site of the enzyme. p. 96

epinephrine (ep ih NEF rin): an adrenal hormone, also called adrenaline, that speeds up heart rate and raises blood sugar level and blood pressure; the "fight-or-flight" hormone that is secreted during a sudden fright or emergency. p. 476

epiphyte (EP ih fyt): a plant that takes its moisture and nutrients from the air and from rainfall and that usually grows on a branch of another plant; not a parasite. p. 606

esophagus: the tubular portion of the digestive tract that leads from the pharynx to the stomach. p. 393

essential nutrient: a nutrient that an organism cannot synthesize, or cannot synthesize in the quantities it requires; plants obtain these nutrients from the soil, animals from food they ingest. p. 414

estivation (es tih VAY shun): a physiological state, characterized by decreased metabolism and inactivity, that permits survival during long periods of elevated temperature and diminished water supplies. p. 615

estrogen (ES troh jen): a hormone that stimulates the development of female secondary sexual characteristics. p. 150

ethylene (ETH ih leen): a gaseous plant hormone that promotes fruit ripening while inhibiting further plant growth in roots and stems. p. 532

eubacteria (YOO bak TIR ee uh): the bacterial group including the cyanobacteria but not the archaebacteria; sometimes called the "true bacteria"; they differ from archaebacteria in their ribosomal RNA, tRNA, and in other ways. p. 281

euglenoids: a small group of freshwater flagellates similar to the organisms of the genus *Euglena;* many contain chloroplasts; others absorb organic substances or ingest prey; all lack rigid cell walls. p. 313

eukaryote (yoo KAIR ee oht): an organism whose cells have a membrane-enclosed nucleus and organelles; a protist, fungus, plant, or animal. p. 110

eutrophic (yoo TROH fik): characterizing a body of water that is nutrient-rich, leading to population explosions of photosynthetic organisms and then of decomposers, which deplete the dissolved oxygen and cause fish and other aquatic organisms to die. p. 661

excretion (ek SKREE shun): the elimination of wastes, especially by-products of body metabolism, by organisms. p. 382

exon (EKS on): a segment of DNA that is transcribed into RNA and translated into protein, specifying the amino acid sequence of a polypeptide; characteristic of eukaryotes and some prokaryotes. p. 206

exoskeleton (EK soh SKEL eh tun): a hard encasement deposited on the surface of an animal, such as the shell of a mollusk, that provides protection and points of attachment for muscles. p. 386

extensor: a muscle that extends a limb or skeletal part. p. 460

extinct: no longer surviving as a species. p. 69

F

F_1: the first filial, or hybrid, offspring in a genetic cross-fertilization. p. 179

F_2: offspring resulting from interbreeding of the hybrid F_1 generation. p. 179

family: the fifth largest grouping after kingdom, phylum or division, class, and order, in the biological classification system; a group of related genera. p. 250

famine (FAM in): severe shortage of food, causing widespread hunger and starvation within a population. p. 45

feces (FEE seez): the waste material expelled from the digestive tract. p. 397

feedback system: a relationship in which one activity of an organism affects another, which in turn affects the first, yielding a regulatory balance. p. 147

femur: the thighbone. p. 581

fermentation (fer men TAY shun): the incomplete breakdown of food molecules, especially sugars, in the absence of oxygen. p. 409

fertilization: the union of an egg nucleus and a sperm nucleus. p. 140

fetus: a vertebrate embryo in later stages of development when it has attained the recognizable structural plan and features of its type. p. 167

fibrin (FY brin): an insoluble, fibrous protein that forms a network of fibers around which a blood clot develops. p. 428

fibrinogen (fy BRIN oh jen): a soluble blood protein that is changed into its insoluble form as fibrin during the blood-clotting process. p. 428

fight-or-flight response: a response of the neuroendocrine system to stressors in which the body prepares to fight or flee; includes the secretion of epinephrine that speeds up heart rate and elevates blood pressure. p. 480

flagella (fluh JEL uh): singular, flagellum; the long cellular appendages specialized for locomotion and formed from a core of nine outer doublet microtubules and two single inner microtubules; many protists and certain animal cells have flagella. p. 259

flame cell: in planarians, a cell with cilia in a network connected by tubules; absorbs fluid and wastes and moves fluid through the tubules to eliminate excess water. p. 383

flatworm: any member of the phylum Platyhelminthes. p. 368

flexor: a muscle that bends a joint; its action is opposite to that of an extensor. p. 460

follicle: in mammals, an ovarian sac from which an egg is released. p. 150

follicle-stimulating hormone (FSH): a substance secreted by the anterior lobe of the pituitary that stimulates the development of an ovarian follicle in a female or the production of sperm cells in a male. p. 150

food: a substance containing energy-rich organic compounds made by organisms and used as a source of energy and matter for life. p. 15

food chain: the transfer of food from one feeding level to another, beginning with producers. p. 10

food web: food chains in an ecosystem taken collectively, showing partial overlapping and competition for many food organisms. p. 10

foraminifera (FOR am ih NIF er uh): any of the unicellular microorganisms of the order Foraminifera, characteristically possessing a calcareous shell with perforations through which numerous pseudopodia protrude. p. 314

fossil: a cast of an organism preserved in rock that formed where it died; the organism itself in ice, volcanic glass, or amber; or the skeleton preserved in deposits. p. 265

Fungi (FUN jy): a kingdom of heterotrophic organisms that develop from spores; many are decomposers; some are parasites of other organisms. p. 262

G

G_1 (Growth 1): the first growth phase of the cell cycle, starting just after offspring cells form. p. 125

G_2 (Growth 2): the second growth phase of the cell cycle, beginning after DNA synthesis. p. 126

gamete (GAM eet): a sex cell, either an egg cell or a sperm, formed by meiosis. p. 137

gametophyte (guh MEET oh fyt): a plant of the gamete-producing generation in a plant species that undergoes alternation of generations. p. 336

gastric juice: mixed secretions of the glands in the stomach wall; in humans, principally mucus, hydrochloric acid, and protein-fragmenting enzymes. p. 394

gastrula (GAS truh luh): an early embryo at the stage when infolding of cells from the outside occurs. p. 160

gastrulation: the formation of the gastrula stage of an embryo; the stage at which the embryo acquires three layers. p. 160

gene: the fundamental physical unit of heredity, which transmits a set of specifications from one generation to the next; a segment of DNA that codes for a specific product. p. 117

gene flow: the loss or gain of alleles in a population due to the emigration or immigration of fertile individuals. p. 229

gene pool: the total aggregate of genes in a population at any one time. p. 228

genetic drift: changes in the gene pool of a small population due to chance. p. 229

genome: the total genetic content or complement of a haploid cell from any given species. p. 206

genotype (JEE noh typ): the genetic makeup of an organism. p. 181

genus (JEE nus): the next largest grouping after kingdom, phylum or division, class, order, and family in the biological classification system; a group of related species. p. 249

gibberellin (jib uh REL en): a plant hormone that stimulates elongation of the stems, triggers the germination of seeds, and with auxin, stimulates fruit development. p. 530

gills: the respiratory organs chiefly of aquatic organisms such as fish; the chief excretory organs in many ocean fish. p. 380

gizzard: a muscular sac in the digestive system of birds, earthworms, and other animals, that mechanically changes food by grinding it against sand particles. p. 378

glomerulus (glah MER yoo lus): a ball of capillaries surrounded by a capsule in the nephron and serving as the site of filtration in the kidneys. p. 445

glucagon (GLOO kuh gahn): a pancreatic hormone that acts to raise the blood glucose level. p. 470

glycerol: a 3-carbon alcohol molecule that combines with fatty acids to form fats and oils. p. 91

glycogen: the chief carbohydrate used by animals for energy storage. p. 91

glycolysis (gly KAWL uh sis): the initial breakdown of a carbohydrate, usually glucose, into smaller molecules at the beginning of cellular respiration. p. 401

Golgi (GOHL jee) **apparatus:** an organelle in eukaryotic cells consisting of stacked membranes that modifies and packages materials for export from the cell. p. 116

grana (GRAY nuh): stacks of thylakoids within a chloroplast. p. 513

gravitropism (grav ih TROH piz um): a positive or negative response of a plant or animal to the acceleration force of gravity. p. 536

guard cells: a pair of cells that surround a stomate in a leaf's epidermis; turgor pressure in the guard cells regulates the opening and closing of the stomate. p. 490

gut: the alimentary canal, or portions thereof, especially the stomach and intestines. p. 394

H

habitat (HAB ih tat): place where an organism lives; even in the same ecosystem, different organisms differ in their habitats. p. 57

half-life: in radioisotopes, the time required for half of a specified quantity to decay. p. 568

hallucinogen: a drug that causes the user to see, feel, or hear things that are not there. p. 482

halophile (HAL uh fyl): an organism that requires a salty environment; usually refers to a type of archaebacteria. p. 283

haploid (HAP loyd): a cell containing only one member

(*n*) of each chromosome pair characteristic of a species. p. 427

hemoglobin (HEE moh gloh bin): the pigment in red blood cells responsible for the transport of oxygen. p. 427

herbaceous (her BAY shus): herblike; without woody tissues. p. 355

herbivore (HER bih vor): a plant-eating consumer; one of the class of consumers most closely associated with producers. p. 55

heterotroph (HET eh roh trohf): an organism that obtains carbon and all metabolic energy from organic molecules previously assembled by autotrophs; a consumer. p. 260

heterotroph hypothesis: the hypothesis that the first life forms used the supply of naturally occurring organic compounds for food. p. 270

heterozygous (HET er oh ZY gus): having two different alleles for a given trait. p. 181

hibernation: a physiological state that permits survival during long periods of cold and diminished food, in which metabolism decreases, the heart and respiratory system slow down, and body temperature is maintained at a lower level than normal. p. 555

histamine: a substance released by injured cells that causes blood vessels to dilate during an inflammatory response. p. 430

homeostasis (hoh mee oh STAY sis): the tendency for an organism, or a population of organisms, to remain relatively stable under the range of conditions to which it is subjected. p. 35

hominid (HOM ih nid): a primate of the family Hominidae, which includes modern humans, earlier subspecies, and australopithecines. p. 586

Homo erectus: an extinct tool-using human species that lived from 1.6 million to 300 000 years ago; the first undisputed human species. p. 587

Homo habilis: a fossil hominid, larger than an australopithecine, thought to be between 1.6 and 2 million years old. p. 586

homologies (hoh MOL uh jeez): likeness in form, as a result of evolution from the same ancestors. p. 244

Homo sapiens neanderthalensis (nee an der tal EN sis): a subspecies of modern humans that lived in Europe, Africa, and Asia from about 100 000 to 35 000 years ago; their relationship to modern humans is debated. p. 588

Homo sapiens sapiens: a subspecies of anatomically modern humans composed of present-day human groups. p. 588

homozygous (HOH moh ZY gus): having two identical alleles for a given trait. p. 181

hormone: a substance, secreted by cells or glands, that has a regulatory effect on cells and organs elsewhere in the body; a chemical messenger. p. 147

host: an organism that serves as a habitat or living food source, or both, for another organism. p. 289

hybrid: having different alleles for a given trait, one inherited from each parent; heterozygous. p. 181

hyphae (HY fee): threadlike growth of a fungus; in an irregular mass they compose the mycelium of many fungi; in an orderly and tightly packed arrangement they compose the body of a mushroom or bracket fungus. p. 320

hypoglycemia (hy poh gly SEE mee uh): low blood sugar level. p. 470

hypothalamus: a specialized part of the base of the brain that in humans combines neuron and hormone activity; it links that autonomic nervous system with the endocrine system in regulating many body functions. p. 150

hypothesis (hy POTH uh sis): a statement suggesting an explanation for an observation or an answer to a scientific problem. p. 20

I

immigration (im uh GRAY shun): arrival by migration; inmigration. p. 32

immunity: disease-resistance, usually specific for one disease or pathogen. p. 289

imperfect fungi: any of the fungi not belonging to other phyla, in which the method of reproduction, if known, is asexual; a taxonomic grouping of convenience. p. 323

incubation period: the time from the entry of an infection or its initiation within an organism up to the time of the first appearance of signs or symptoms. p. 296

industrialized agriculture: using large inputs of energy primarily derived from fossil fuels to produce large quantities of crops and livestock. p. 672

infectious disease: a disease caused by viruses or microorganisms that can be transmitted directly from an affected individual to a healthy individual. p. 289

ingestion: the process of taking a substance from the environment, usually food, into the body. p. 376

innate: behavior that is genetically determined, as in the organization of an ant society; also called instinctive behavior. p. 544

instinct: the capacity of an animal to complete a fairly complex, stereotyped response to a key stimulus without having prior experience. p. 545

insulin (IN suh lin): a pancreatic hormone that promotes cell absorption and use of glucose; impairment of its secretion or its action results in diabetes. p. 470

interneuron: associative neuron; a neuron located between a sensory neuron and a motor neuron. p. 464

interphase (IN ter fayz): a normal interval between successive cell divisions when the only evidence of future divisions is that chromosomes begin to be replicated; a cell at work, rather than a cell dividing. p. 125

intertidal zone: the shallow zone of the ocean where land meets water; also called the littoral zone. p. 654

intestinal juice: secretions of the glands in the small intestinal wall containing enzymes that act on dipeptides and double sugars. p. 396

intron (IN tron): a segment of DNA that is transcribed into precursor mRNA but then removed before the mRNA leaves the nucleus. p. 206

ion (EYE on): an atom or molecule that has either gained or lost one or more electrons, giving it a positive or negative charge. p. 79

ionization: the conversion of a nonionic substance, such as water, into ions. p. 79

J

joint: a point of movement, or of fixed calcium deposits preventing movement, marking where two bones meet in the skeleton. p. 458

K

karyotype: a method of organizing the chromosomes of a cell in relation to number, size, and type. p. 200

kcal: kilocalorie; a measure of food energy equal to 1000 calories. p. 46

kingdom: the largest grouping in the biological classification system. p. 251

Krebs cycle: the cycle in cellular respiration that completes the breakdown of intermediate products of glycolysis, releasing energy; also a source of carbon skeletons for use in biosynthesis reactions. p. 401

L

larva (LAR vuh): an immature stage of development in offspring of many types of animals. p. 373

latent: present but not evident or active. p. 297

learned behavior: behavior developed as a result of experience. p. 545

lens: in the eye, a transparent tissue layer behind the iris that focuses light rays entering through the pupil to form an image on the retina. p. 164

lenticel (LENT ih sel): an opening in the surface of a plant stem through which air can diffuse. p. 491

life cycle: all the events that occur between the beginning of one generation and the beginning of the next generation. p. 135

ligaments: a cord or sheet of connective tissue by which two or more bones are bound together at a joint. p. 459

light reactions: the energy-capturing reactions in photosynthesis. p. 516

limiting factor: an environmental condition such as food, temperature, water, or sunlight that restricts the types of organisms and population numbers that an environment can support. p. 33

lipid (LIP id): a fat, oil, or fatlike compound that usually has fatty acids in its molecular structure; an important component of the plasma membrane. p. 89

luteinizing hormone (LH): a hormone secreted by the anterior lobe of the pituitary gland that controls the formation of the corpus luteum in females and the secretion of testosterone in males. p. 150

Lycophyta (ly KOF ih tuh):   the division containing the club mosses. p. 344

lymph (LIMF):   the fluid transported by the lymphatic vessels. p. 396

lymphatic (lim FAT ik) **system:**   a system of vessels through which body lymph flows, eventually entering the bloodstream where the largest lymph duct joins a vein. p. 396

lymph node: a tiny, twisted portion of a lymph vessel in which white blood cells attack any pathogenic organisms in the lymph and engulf any foreign particles. p. 429

lymphocyte (LIM foh syt):   a type of small white blood cell important in the immune response. p. 431

lymphokine (LIM foh kyn):   any of a class of proteins by which the cells of the vertebrate immune system communicate with one another. p. 432

lymph vessel: a vessel in which lymph from the body tissues, or from villi in the intestine, flows until it enters the largest lymph duct, which empties into a vein. p. 424

lysosome (LY soh zohm):   a cell vesicle that contains digestive enzymes. p. 117

M

macronucleus: the larger of two types of nuclei in ciliates; one or more of the macronuclei may be present in each organism. p. 315

macronutrient: a nutrient required in large amounts by a plant or other organism. p. 501

macrophage (MAK roh fayj):   a large white blood cell that ingests pathogens and dead cells. p. 431

mammary glands: the milk-producing organs in female mammals, consisting of clusters of alveoli with small ducts terminating in a nipple or teat. p. 376

marsupials: a group of mammals, such as koalas, kangaroos, and opossums, whose young complete their embryonic development inside a maternal pouch called a marsupium. p. 376

mating behavior: the behaviors, including courtship rituals and displays, that lead to mating; such behavior usually is specific for each species. p. 549

medulla (meh DUL lah):   (1) the inner portion of an organ, such as the kidney, p. 445; (2) the most posterior region of the vertebrate brain, the hindbrain. p. 466

meiosis (my OH sis):   two successive nuclear divisions (with corresponding cell divisions) that produce gametes (in animals) or sexual spores (in plants) having one-half of the genetic material of the original cell. p. 140

memory cell: B- or T-lymphocyte, produced in response to a primary immune response, that remains in the circulation and can respond rapidly if the same antigen is encountered in the future. p. 434

menopause (MEN oh pahz):   in human females, the period of cessation of menstruation, usually occurring between the ages of 45 and 50. p. 151

menstrual cycle: the female reproductive cycle that is characterized by regularly recurring changes in the uterine lining. p. 148

menstruation: periodic sloughing of the blood-enriched lining of the uterus when pregnancy does not occur. p. 150

meristem (MER ih stem):   plant cells at the growing tips of roots and stems and in buds and cambium that divide and produce new cells that can differentiate into various plant tissues. p. 507

mesoderm (MEZ oh derm):   in most animal embryos, a tissue layer between the ectoderm and endoderm that gives rise to muscle, to organs of circulation, reproduction, and excretion, to most of the internal skeleton (if present), and to connective tissue layers of gut and body covering. p. 160

mesophyll (MEZ oh fil):   the green leaf cells between the upper and lower epidermis of a leaf; the primary site of photosynthesis in leaves. p. 486

mesosome (MEZ oh sohm):   an infolding of the plasma membrane of a prokaryotic cell. p. 279

metabolism (meh TAB oh liz um):   the sum of all the chemical changes taking place in an organism. p. 119

metamorphosis (met uh MOR phuh sis):   in the life cycles of many animals, marked changes in body form and functions from the newly hatched young to the adults. p. 373

metaphase: the stage in mitosis in which replicated chromosomes move to the center of the spindle and become attached to it. p. 128

metastasize: to spread, as in the spread of cancer cells. p. 171

methanogen (meh THAN uh jen):   a methane-producing archaebacterium. p. 283

micronucleus: the smaller of two types of nuclei in ciliates; one or more micronuclei may be present in each organism. p. 315

micronutrient: a nutrient required in only small amounts by a plant or other organism. p. 501

microorganism: an organism too small to be seen with the unaided human eye. p. 9

microsphere: a cooling droplet from a hot water solution of polypeptides; the droplet forms its own double-layered boundary as it cools; used as a model for precells to study the formation of the first life on the earth. p. 269

microtubule: a hollow rod of protein found in the cytoplasm of all eukaryotic cells, making up part of the cytoskeleton and involved in cell contraction. p. 118

midlatitude forest: a biome located throughout midlatitude regions where there is sufficient moisture to support the growth of large trees, mostly of the broad-leaved deciduous type; also called temperate forest. p. 604

mitochondria (my toh KON dree uh):   the cell organelles in eukaryotic cells that carry on cellular respiration, releasing energy from food molecules and storing it in ATP. p. 117

mitosis (my TOH sis):   the replication of the chromosomes and the production of two nuclei in one cell; usually followed by cytokinesis. p. 126

mixed-grass prairie: the biome composing the central Great Plains that contains both short and tall grasses and is characterized by low precipitation, low winter temperatures, and relatively high elevation. p. 612

molecule (MOL uh kyool): a particle consisting of two or more atoms of the same or different elements chemically bonded together. p. 14

Mollusca (muh LUS kuh): the phylum containing mollusks. p. 370

mollusk: any invertebrate of the phylum Mollusca, typically having a hard shell that wholly or partly encloses a soft, unsegmented body; includes chitons, snails, bivalves, squid, octopuses, and similar animals. p. 370

monocot (MON oh kot): a subdivision of flowering plants whose members have one embryonic seed leaf, or cotyledon. p. 355

monotremes: a group of egg-laying mammals, represented by the platypus and echidnas. p. 376

mortality (mor TAL ih tee): death rate, measured as the proportion of deaths to total population over a given period; often expressed as number of deaths per 1000 or 10 000 individuals. p. 32

motile (MOH tuhl): capable of movement from place to place, characteristic of most animals. p. 363

motor neuron: a specialized neuron that receives impulses from the central nervous system and transmits them to a muscle or gland. p. 464

multicellular: composed of many cells, as contrasted with a single-celled organism. p. 107

multifactorial inheritance: a pattern of inheritance in which characteristics are determined by at least several genes with a large number of environmental variables. p. 199

multiple alleles: the existence of several known alleles for a gene. p. 198

mutation (myoo TAY shun): a chemical change in a gene, resulting in a new allele; or, a change in the portion of a chromosome that regulates the gene; in either case the change is hereditary. p. 182

mycelium (my SEE lee um): the densely branched network of hyphae in a fungus. p. 320

mycoplasma (my koh PLAZ muh): eubacteria that lack cell walls; probably the smallest organisms capable of independent growth. p. 284

mycorrhizae (my koh RY zee): symbiotic associations of plant roots and fungi. p. 326

myosin: a protein that, together with actin, is responsible for muscular contraction and relaxation. p. 455

Myxomycota (mik soh my KOH tuh): the phylum containing the slime molds. p. 315

N

NAD+: nicotinamide adenine dinucleotide (nik uh TEE nuh myd AD uh neen DY NOO klee oh tyd); an electron and hydrogen carrier in cellular respiration. p. 403

NADP+: nicotinamide adenine dinucleotide phosphate; a hydrogen-carrier molecule that forms NADPH in the light reactions of photosynthesis. p. 520

natality (nay TAL ih tee): birthrate, as measured by the proportion of new individuals to total population over a given period; often expressed as new individuals per 1000 or 10 000 in the population. p. 32

natural selection: a mechanism of evolution whereby members of a population with the most successful adaptations to their environment are more likely to survive and reproduce than members with less successful adaptations. p. 222

nectar: a secretion, mainly a sugar solution, produced by small glands in the petals of many flowers. p. 351

Nematoda (nee muh TOH duh): the phylum containing the unsegmented worms. p. 369

nephron (NEF rahn): the functional unit of a kidney, consisting of a long, coiled tubule, one end of which forms a cup that encloses a mass of capillaries and the other end of which opens into a duct that collects urine; the entire nephron is surrounded by a network of capillaries. p. 444

nerve impulse: a wave of chemical and electrical changes that passes along a nerve fiber in response to a stimulus. p. 463

nerve: a bundle of nerve fibers; the cell bodies from which the fibers extend usually are located together at one end of the fiber. p. 463

nervous system: a coordinating mechanism in all multicellular animals, except sponges, that regulates internal body functions and responses to external stimuli; in vertebrates, it consists of the brain, spinal cord, nerves, ganglia, and parts of receptor and effector organs. p. 463

neural tube: the structure formed in the embryo that eventually gives rise to the brain and spinal cord. p. 160

neuron (NYOO rahn): a nerve cell; a name usually reserved for nerve cells in animals that have a complex brain and specialized associative, motor, and sensory nerves. p. 463

neurotransmitter (NYOOR oh trans MIT er): a chemical messenger, often similar to or identical with a hormone, that diffuses across the synapse and transmits a nerve impulse from one neuron to another. p. 463

neutral: a solution in which the concentrations of hydrogen and hydroxide ions are equal. p. 80

neutron (NOO trahn): a particle carrying no electrical charge; found in the nuclei of all atoms except those of hydrogen. p. 77

niche (NITCH): the total of an organism's utilization of the biotic and abiotic resources of its environment. p. 57

nitrifying (NY trih fy ing) **bacteria:** bacteria that use ammonium ions to produce nitrite and nitrate ions. p. 288

nitrogen-fixing bacteria: bacteria that take in free nitrogen from the atmosphere and use it to produce ammonia; subsequent reactions by other bacteria produce ammonium ions and nitrates, from which plants obtain their nitrogen. p. 286

nodule (NOD yool): a rounded growth of tissue that usually contains microorganisms or some other agent associated with the growth; in certain plants, nitrogen-fixing bacteria live in nodules on the plant roots. p. 287

norepinephrine: a hormone secreted by the adrenal medulla in response to nerve signals from the sympathetic division of the autonomic nervous system. p. 476

notochord (NOH toh kord): in chordates, a flexible, dorsal, rodlike structure that extends the length of the body; in vertebrates it is replaced in later stages of development by the vertebrae. p. 373

nuclear envelope: the membrane in eukaryotes that encloses the genetic material, separating it from the cytoplasm. p. 117

nucleic (noo KLEE ik) **acid:** DNA or RNA; an organic compound composed of nucleotides and important in coding instructions for cell processes. p. 89

nucleoid (NOO klee oyd): a region in a prokaryotic cell consisting of a concentrated mass of DNA. p. 280

nucleotide (NOO klee oh tyd): a subunit or building block of DNA or RNA, chemically constructed of a 5-carbon sugar, a nitrogen base, and a phosphate group. p. 97

nucleus (NOO klee us): in atoms, the central core, containing positively charged protons and (in all but hydrogen) electrically neutral neutrons, p. 77; in eukaryotic cells, the membranous organelle that houses the chromosomal DNA, p. 111; in the central nervous system, a cluster of nerve cell bodies. p. 463

nutrient: (1) a food substance in a form, usually digested or not requiring digestion, that is interchangeable among organisms, p. 46; (2) certain chemicals used by plants. p. 501

O

omnivore (OM nih vor): a consumer organism that feeds partly as an herbivore and partly as a carnivore, eating plants, animals, and fungi. p. 55

oncogene: a gene found in viruses or as part of the normal genome that is crucial for triggering cancerous characteristics. p. 172

open circulatory system: a system in which the blood does not travel through the body completely enclosed in vessels; the return to the heart is usually through open spaces, or sinuses, as in the grasshopper. p. 380

open population: a population that gains members by immigration or loses them by emigration, or both. p. 36

opportunistic infections: infections that occur in organisms with weakened immune responses; normally these infections would be repelled by healthy immune systems. p. 297

organelle (or guh NEL): an organized structure within a cell, with a specific function; a chloroplast and a mitochondrion are examples. p. 111

organic compounds: compounds built of carbon combined with other elements. p. 77

osmosis (os MOH sis): the movement of water across a selectively permeable membrane. p. 121

ovaries (singular, *ovary*): the primary reproductive organs of a female; egg-cell-producing organs. p. 146

oviduct: a tube that leads from an ovary to the uterus. p. 148

ovulation: in vertebrates, the release of one or more eggs from an ovary. p. 147

ovule (OHV yool): a structure that develops in the plant ovary and contains the female egg. p. 340

ovum: a female gamete, or egg. p. 139

P

P$_1$: the parental organisms in a genetic cross-fertilization. p. 179

paleontologist (pay lee un TOL uh jist): a specialist who studies extinct organisms through fossils. p. 568

palisade layer: the layer of tightly packed and column-shaped mesophyll cells containing many chloroplasts. p. 486

pancreatic (pan kree AT ik) **juice:** fluid secreted by the pancreas. p. 396

Pangaea (pan GEE uh): in the Triassic period, the supercontinent formed by the four major continents; the breakup of Pangaea affected the distribution and evolution of organisms. p. 573

parallel evolution: the type of evolution that occurs when organisms from a single species become so different that they no longer can interbreed, yet both groups continue to evolve in a similar direction; the ostrich and the rhea are the result of parallel evolution. p. 235

parasitism (PAIR uh sih tiz um): an ecological niche in which one organism is the habitat and the food for another, which lives and feeds on the host organism, usually without killing it. p. 61

parasympathetic division: in vertebrates, one of the two divisions of the autonomic nervous system; stimulates resting activities, such as digestion, and restores the body to normal after emergencies. p. 468

passive transport: the diffusion of a substance across a biological membrane by means of a transport protein. p. 122

pathogen: a disease-causing organism. p. 289

penicillin: any of several antibiotic compounds obtained from penicillium mold and used to prevent or treat a wide variety of diseases and infections. p. 226

penis: in vertebrates, the male organ through which sperm are passed to the female and through which nitrogenous wastes from the kidneys—in the form of urine—are discharged outside the body. p. 146

peptidoglycan (PEP tid oh GLY kan): a substance found only in eubacteria cell walls that consists of modified sugars cross-linked by short polypeptides. p. 281

peripheral nervous system: the sensory and motor neurons that connect the central nervous system to the rest of the body. p. 466

peristalsis (per ih STAWL sis): rhythmic waves of contraction of smooth muscle which push food along the digestive tract. p. 393

permafrost: the sublayers of soil that remain frozen through the summer thaw in the tundra of northern latitudes. p. 602

petal: one of the leaflike, often brightly colored, structures within the ring of green sepals in a developing flower; in the mature flower the petals may form their own ring or fuse to form a cuplike or tubular structure. p. 340

petiole (PET ee ohl): the slender structure at the base of a leaf that attaches it to a plant stem. p. 485

Phaeophyta (fay OFF it uh): the phylum containing the brown algae. p. 308

phenotype (FEE noh typ): the expression of a genotype in the appearance or function of an organism; the observed trait. p. 181

pheromone (FAYR uh mohn): a small chemical signal functioning in communication between animals and acting much like hormones to influence physiology and behavior. p. 155

phloem (FLOH em): a portion of the vascular system in plants, consisting of living cells arranged into elongated tubes that transport sugar and other organic nutrients throughout the plant. p. 494

photoperiodism (foh toh PIH ree ud iz um): a physiological response to day length, such as in flowering plants. p. 537

photorespiration: a metabolic pathway that uses oxygen, produces carbon dioxide, generates no ATP, and slows photosynthetic output; generally occurs on hot, dry, bright days, when stomates are nearly closed and the oxygen concentration in the leaf exceeds that of carbon dioxide. p. 526

photosynthesis: the process by which living cells that contain chlorophyll use light energy to make organic compounds from inorganic materials. p. 13

photosystems (PS) I and II: the light-harvesting units in photosynthesis, located in the thylakoid membranes of the chloroplast. p. 519

phototropism (foh toh TROH piz um): movement or growth curvature toward light. p. 536

pH scale: a scale from 0 to 14 reflecting the concentration of hydrogen ions; the lower numbers denote acidic conditions, 7 is neutral, and the upper numbers denote basic or alkaline conditions. p. 80

phylum (FY lum): the second largest grouping, after kingdom, in the biological classification system, for all organisms except plants, which are classified in divisions. p. 251

phytoplankton (FYT oh PLANK ton): very small aquatic organisms, many microscopic, that carry on photosynthesis. p. 634

pigment: any coloring matter or substance. p. 514

pith: cells at the center of the young stems of many plants; in some plants the pith disappears as the stems age. p. 493

pituitary (pih TOO ih ter ee): endocrine gland attached to the base of the brain, consisting of an anterior and a posterior lobe. p. 150

placenta (pluh SEN tuh): a structure in the pregnant uterus for nourishing a fetus with the mother's blood supply, formed from the uterine lining and embryonic membranes. p. 166

plankton (PLANK ton): very small aquatic organisms, many microscopic, that usually float or feebly swim near the surface. p. 634

Plantae (PLAN tee): the plant kingdom, containing multicellular, autotrophic eukaryotes; all members have cellulose-containing cell walls and chloroplasts; reproduction may be sexual or asexual. p. 260

plaque: a growth that develops on inner arterial walls when lipids, such as cholesterol, infiltrate a matrix of smooth muscle; plaques are a product of atherosclerosis. p. 415

plasma (PLAZ muh): the liquid portion of the blood in which the cells are suspended. p. 427

plasma cell: an antibody-producing cell that is formed as a result of the proliferation of sensitized B-lymphocytes. p. 432

plasma membrane: the membrane at the boundary of every cell that serves as a selective barrier to the passage of ions and molecules. p. 116

plasmid: in prokaryotes, a small ring of DNA that carries accessory genes separate from those of the bacterial chromosome. p. 207

plasmodium (plaz MOHD ee um): the motile sheetlike stage of life formed by the fusion of many amoebalike slime molds. p. 315

platelet (PLAYT let): a small plate-shaped blood factor that contributes to blood clotting at the site of a wound; platelets release substances that begin formation of a network in which the platelets are caught, forming a clot. p. 428

plate tectonics: the theory and study of the great plates in the earth's crust and their movements, which produce earthquakes, sea-floor spreading, continental drift, and mountain building. p. 573

Platyhelminthes (plah tih hel MIN theez): the phylum containing bilaterally symmetrical, somewhat flattened worms. p. 368

pollen grain: a haploid spore produced by a flowering plant; it gives rise to sperm nuclei. p. 340

pollen tube: a tube that develops from a germinating pollen grain and penetrates the carpel until it reaches the ovary; sperm nuclei pass through the tube. p. 341

pollination: the placement of pollen onto the stigma of a carpel by wind or animal vectors, a prerequisite to fertilization. p. 341

polypeptide (POL ee PEP tyd): a long chain of chemically bonded amino acids. p. 92

population: a group of organisms of one species that live in the same place at the same time. p. 31.

Porifera (poh RIH fer uh): the phylum containing sponges and their relatives. p. 366

posterior: situated toward the rear, or coming last; the tail end in most animals. p. 364

potassium-40 dating: a radiometric dating technique that involves measuring the relative amounts of potassium-40 to argon-40 in a volcanic rock sample; the ratio is compared to the known half-life of potassium-40, thus yielding an estimate of the age of the rocks. p. 568

predator-prey relationship: the relationship between organisms in which one, the predator, feeds on the other, the prey. p. 60

pressure-flow: the hypothesis that food is transported through the phloem as a result of differences in pressure. p. 494

primary growth: growth in the length of plant roots and stems. p. 507

principle of independent assortment: the inheritance of alleles for one trait does not affect the inheritance of alleles for another trait. p. 197

principle of segregation: during meiosis, chromosome pairs that separate into different gametes such that each of the two alleles for a given trait appears in a different gamete. p. 180

probability: the chance that any given event will occur. p. 176

producer: an autotroph; any organism that produces its own food using matter and energy from the nonliving world. p. 9

productivity: the amount of available solar energy converted to chemical energy by producers during any given period. p. 646

progesterone (proh JES tuh rohn): a female hormone secreted by the corpus luteum of the ovary and by the placenta that acts to prepare and maintain the uterus for pregnancy and to prepare the breasts for lactation. p. 150

Prokaryotae (proh kayr ee OH tee): the kingdom of bacteria; their cells are prokaryotic and lack membrane-enclosed organelles such as nuclei and mitochondria. p. 260

prokaryote (pro KAIR ee oht): an organism whose cells do not have membrane-enclosed organelles, such as nuclei, mitochondria, and chloroplasts; a bacterium. p. 110

prophase: the stage in mitosis during which replicated strands of chromosomes condense to become thicker and shorter, the nuclear envelope begins to disappear, and a spindle forms. p. 127

protein (PROH teen): an organic compound composed of one or more polypeptide chains of amino acids; most structural materials and enzymes in a cell are proteins. p. 89

prothrombin (proh THROM bin): a plasma protein that functions in the formation of blood clots. p. 428

prothrombin activator: in the clotting process, a substance that catalyzes the conversion of prothrombin to thrombin. p. 428

Protista (pro TIST uh): a kingdom of mostly aquatic organisms whose cells are eukaryotic, but which are mostly microscopic; exceptions in size include the larger "seaweeds" such as kelp. p. 263

proton (PRO tahn): a particle bearing a positive electrical charge, found in the nuclei of all atoms. p. 77

pseudopod (SOO doh pod): literally, a false foot; a term used to describe an amoebalike extension of a cell or unicellular organism; pseudopods are used in locomotion and obtaining food, or, as in white blood cells, in engulfing bacteria and foreign particles. p. 314

psychoactive (sy koh AK tiv) **drug:** a drug that affects the mind or mental processes. p. 481

Pterophyta (ter OFF ih tuh): the division containing the ferns. p. 346

puberty: the stage of development in which the reproductive organs become functional. p. 147

punctuated equilibria: a theory of evolution positing spurts of relatively rapid change followed by long periods of stasis. p. 235

pure breeding: genetically pure; homozygous. p. 178

pyloric valve: the valve located between the stomach and the small intestine that regulates the passage of food from the stomach. p. 395

R

radial (RAYD ee ul) **symmetry:** correspondence in size, shape, and position of parts as though they all radiated equally from a center point, or from a center line or axis. p. 364

rate: change per unit of time; the amount of change measured over a period of time, divided by the length of time. p. 32

ray cell: any of the cells in the stems of woody plants that form a channel allowing the passage of material laterally back and forth between the xylem and the phloem. p. 508

receptor: a specialized sensory cell, as in the eye or the skin, that is sensitive to a particular type of stimulus. p. 464

recessive: a term used to describe an allele or trait that is masked by a dominant allele or trait. p. 180

recombinant DNA: DNA that incorporates parts of different parent DNA molecules, as formed by natural recombination mechanisms or by recombinant DNA technology. p. 206

reduction division: the first meiotic division, in which the chromosome number is reduced from diploid ($2n$) to haploid (n). p. 141

reflex: an involuntary reaction or response to a stimulus. p. 465

replication: the process of making a copy of the chromosome in a cell nucleus, and other genes in certain organelles outside the nucleus—particularly chloroplasts and mitochondria; the process is unlike duplication in that each gene and each chromosome in the double set is partly new but also includes part of the old gene or chromosome. p. 126

reproductive isolation: the state of a population or species in which successful mating outside the group is impossible because of anatomical, geographical, or behavioral differences. p. 217

reptile: any of the ectothermic, usually egg-laying vertebrates that belong to the class Reptilia; includes snakes, lizards, crocodiles, turtles. p. 216

resistance: relative immunity; the ability of a host organism to destroy a pathogen or prevent the disease symptoms it causes. p. 289

resource: in ecology, an environmental supply of one or more of an organism's requirements (light energy, food energy, water, oxygen or carbon dioxide, living space, protective cover, and so on); in a human society, a resource may be anything useful. p. 34

retina (REH tin uh): the photosensitive layer of the vertebrate eye that contains several layers of neurons and light receptors (rods and cones); the retina receives the image formed by the lens and transmits it to the brain via the optic nerve. p. 164

retrovirus: certain RNA viruses that must make copies of DNA from their RNA before they can reproduce. p. 293

reverse transcriptase: an enzyme that transcribes RNA into DNA, found only in association with retroviruses. p. 293

rhizoid (RY zoid): the threadlike structure that in nonvascular plants absorbs water and nutrients from the soil and helps hold the plants in place. p. 343

Rhodophyta (roh DOF it uh): the phylum containing the red algae. p. 309

ribose (RY bohs): a 5-carbon sugar found in RNA molecules. p. 98

ribosome (RY boh sohm): a cell organelle constructed in the nucleus, consisting of two subunits and functioning as the site of protein synthesis in the cytoplasm. p. 117

ritual: a detailed method of procedure that is faithfully or regularly followed. p. 505

RNA: ribonucleic (ry boh noo KLEE ik) acid; the hereditary material of certain viruses, and the material coded by the DNA of other cells to carry out specific genetic functions, for example, messenger RNA and transfer RNA. p. 97

root cap: a layer of protective cells that covers the growing tip of a plant root. p. 507

root hair: a hairlike extension of an epidermal cell of a plant root that absorbs water and nutrients. p. 510

roundworm: an unsegmented, cylindrical worm having a digestive tube with two openings; a nematode. p. 369

rubisco (roo BIS koh): ribulose bisphosphate carboxylase (RY byoo lose BIS fos fayt kar BOX uh lays); an enzyme that catalyzes the initial incorporation of carbon dioxide in the Calvin cycle. p. 527

rumen (ROO men): an enlargement of the digestive tract of many herbivorous mammals, in which microorganisms that can digest cellulose live. p. 283

S

sac fungi: fungi belonging to the phylum Ascomycota, in which sexual spores are formed in a saclike structure, the ascus. p. 321

sarcodines: microscopic, or almost microscopic, protists that move by using pseudopods; many produce intricate shells or skeletal structures; includes amoebas, foraminiferans, and radiolarians. p. 314

saturated fat: a fat containing fatty acids with no double-carbon bonds; usually solid at room temperature. p. 415

savanna (suh VAN uh): a tropical grassland biome with scattered individual trees and large herbivores; water is the major limiting factor. p. 615

scavenger (SKAV en jer): a consumer organism that feeds on the dead carcasses of other consumer organisms it did not kill. p. 62

scrotum: a pouch of skin that encloses the testes. p. 146

scurvy: a disease caused by a deficiency of vitamin C, characterized by spongy, bleeding gums, bleeding under the skin, and extreme weakness. p. 289

secondary growth: growth in girth or diameter of plant and root. p. 507

secondary sexual characteristics: differences in appearance or behavior associated with male or female hormones; in many animal species, males are larger or more colorful. p. 552

seed: an embryo plant, along with food storage tissue (endosperm), both enclosed within protective coatings formed of tissues from an ovule in the parent plant. p. 340

seed coat: the tough, protective outer covering of a seed. p. 504

semen (SEE men): the thick fluid in which sperm are transported in mammalian males. p. 152

sensory neuron: a neuron that receives impulses from a sensory organ or receptor and transmits them toward the central nervous system. p. 464

sepal (SEE pul): one of the leaflike structures that enclose and protect a flower bud; in the mature flower they are on the underside, next to the stem; they often are green. p. 340

sessile (SES il): not free to move about in the environment; sessile animals usually are attached by the base to an object in the environment. p. 364

sex chromosome: one of a pair of chromosomes that differentiates between and is partially responsible for determining the sexes. p. 200

sexual reproduction: reproduction involving the contribution of genetic material from two parents. p. 136

sexual selection: selection based on variation in secondary sexual characteristics, leading to the enhancement of individual reproductive fitness. p. 229

short-grass prairie: a grassland biome that stretches from the central Great Plains to the Rocky Mountains; grasses usually are less than 0.5 m tall. p. 612

sieve cell: a type of phloem cell in plants that forms a column. p. 494

skeletal muscle: a type of muscle tissue found in muscles attached to skeletal parts and responsible for voluntary muscle movement. p. 455

slash-and-burn cultivation: cutting down trees and other vegetation in a patch of forest, leaving the cut material to dry, and then burning it; the ashes add plant nutrients to the nutrient-poor soils of most tropical forest areas. p. 670

slime molds: members of the phylum Myxomycota; slime molds form amoebalike colonies, are heterotrophic, and reproduce by spores. p. 315

small intestine: a digestive organ of vertebrates and some invertebrates; in vertebrates it is located between the stomach and the large intestine and is the organ in which the digestive processes are completed. p. 395

smooth muscle: the type of muscle tissue found in the walls of hollow internal organs. p. 455

social behavior: animal behavior that shows evidence of differing individual roles in the organization of a group and of cooperation or division of labor in tasks. p. 549

somatic nervous system: those nerves leading from the central nervous system to skeletal muscles. p. 468

species (SPEE sheez): all individuals and populations of a particular type of organism, maintained by biological mechanisms that result in their breeding mostly with their type. p. 69

sperm cell: a male gamete, usually motile in swimming movements; its motility increases its chance of encountering and fertilizing an egg. p. 139

S phase (Synthesis): the synthesis phase of the cell cycle during which DNA is replicated. p. 126

Sphenophyta (sfen OFF ih tuh): the division containing horsetails. p. 345

sphincter (SFINK ter): a circular muscle that functions to close an opening of a tubular structure. p. 394

spina bifida (SPY nuh BIH fih duh): a birth defect resulting from the failure of the spinal cord and brain to close properly. p. 164

spinal cord: a complex band of neurons that runs through the spinal column of vertebrates to the brain. p. 464

spongy layer: a layer of round, chloroplast-containing mesophyll cells surrounded by air spaces. p. 486

sporangia (spoh RAN jee uh): spore-producing structures formed in one stage of the life cycle of many organisms, but not animals. p. 315

spores: one-celled reproductive bodies that are usually resistant to harsh environmental conditions and may remain dormant, in a dry covering, for long periods; in some organisms, spores are asexual and may initiate the growth of a new organism under favorable conditions; in other organisms, spores are sexual and must unite with those of the other sex before producing a new organism. p. 155, p. 315

sporophyte (SPOR oh fyt): a plant of the spore-producing generation in a plant species that undergoes alternation of generations; in some species, the sporophyte is reduced to a dependent structure that grows from the gametophyte plant. p. 336

sporozoans: usually nonmotile, heterotrophic, parasitic protists of the phylum Apicomplexa, with complex life cycles. p. 314

stamen (STAY men): the pollen-producing male reproductive organ of a flower, consisting of an anther and filament. p. 340

sterile: not capable of reproducing. p. 217

stigma: the tip of the carpel in a flower; it secretes a sticky substance that traps pollen. p. 340

stimulant: a drug that increases the activity of the central nervous system. p. 481

stimulus (STIM yoo lus): a change or signal in the internal or external environment that causes an adjustment or reaction by an organism. p. 384

stomach: the digestive organ located between the esophagus and the small intestine. p. 394

stomate (STOH mayt): the opening between two guard cells in the epidermis of a plant leaf; gases are exchanged with the air through stomates. p. 336

strata (STRAYT uh; singular, *stratum*): layers, usually of deposited earth sediments carried by erosion; many strata become mineralized into rock layers. p. 567

stress: a physiological response to factors causing disruptive changes in the body's internal environment. p. 478

stressor: a factor capable of stimulating a stress response. p. 478

stroma (STROH muh): the colorless substance in a chloroplast surrounding the thylakoids; the enzymes of the Calvin cycle also are in the stroma. p. 513

stromatolite (stroh MAT uh lyt): a rock made of banded domes of sediment in which are found the most ancient forms of life, fossil prokaryotes dating back as far as 3.5 billion years. p. 266

subsistence agriculture: producing only enough food to feed oneself and family members; in good years, enough may be left over to sell or store. p. 670

substrate (SUB strayt): a molecule on which enzymes act. p. 96

succession: the replacement of one community by another in a progression to a climax community. p. 619

sustainable agriculture: a method of growing crops and raising livestock based on organic fertilizers, soil conservation, water conservation, biological pest control, and minimal use of nonrenewable energy sources. p. 680

symbiosis (sim by OH sis): an ecological relationship between organisms of two different species that live together in direct contact. p. 61

symmetry: correspondence in form and arrangement of parts on opposite sides of a boundary. p. 364

sympathetic division: a division of the autonomic nervous system of vertebrates that functions in an alarm response; increases heart rate and dilates blood vessels while placing the body's everyday functions on hold; mobilizes the body for response to stressors, danger, or excitement. p. 468

synapse (SIN aps): an open junction between neurons, across which an impulse is transmitted by a chemical messenger, a neurotransmitter. p. 463

synthesis (SIN thih sis): the process of putting together or building up; applicable to ideas, chemical compounds, and so on. p. 79

T

taiga (TY guh): the coniferous or boreal forest biome, characterized by much snow, harsh winters, short summers, and evergreen trees. p. 603

tall-grass prairie: a grassland biome found at the western edge of the deciduous forest in the United States; grasses may grow 1.5 to 2 m tall. p. 612

target organ: a specific organ on which a hormone acts. p. 469

taxonomy (tak SAHN uh mee): the study of species and their classification by genus, family, order, class, phylum, and kingdom. p. 243

T cell: a lymphocyte that in the thymus stimulated by the presence of a particular antigen; it differentiates and divides, producing offspring cells (killer cells) that attack and kill the cells bearing the antigen. p. 431

telophase: the final stage in mitosis; two new cell nuclei are completed as nuclear envelopes form around the two clusters of chromosomes at opposite ends of the cell, and the cell itself divides. p. 128

tendon: a cordlike mass of white fibrous connective tissue that connects muscle to bone. p. 460

terrestrial (ter ES tree uhl): living on land. p. 365

territoriality: the behavior pattern in animals consisting of the occupation and defense of a territory. p. 551

testes: the primary reproductive organs of a male; sperm-cell-producing organs. p. 146

thermoacidophiles (THER moh a SID uh fylz): archaebacteria requiring high temperatures and/or acidic conditions for life. p. 282

thermocline (THER moh klyn): a layer in a thermally stratified body of water that separates upper, oxygen-rich and nutrient-poor warm water from lower, oxygen-poor and nutrient-rich cold water. p. 640

threat display: a behavior in which an animal attempts to make itself appear larger or fiercer than it actually is in an attempt to discourage intruders. p. 549

thrombin: a blood protein that is important in the clotting process. p. 428

thylakoid (THY luh koyd): a flattened sac in a chloroplast; many of the thylakoids are arranged in stacks known as grana; the pigments and enzymes for the light reactions of photosynthesis are embedded in the sac membrane. p. 513

thyroid gland: an endocrine gland in the neck region of most vertebrates that controls the rate of cell metabolism in the body through one of its hormones, thyroxine. p. 476

thyroxine (thy ROK sin): a principal hormone of the thyroid gland; regulates the rate of cell metabolism. p. 476

tolerance: the ability to withstand or survive a particular environmental condition. p. 558

toxin: a substance produced by one organism that is poisonous to another. p. 290

trachea (TRAY kee uh): the windpipe of an air-breathing vertebrate, connecting the air passage in the throat with the lungs. p. 442

tracheid (TRAY kee id): a water-conducting and supportive element of xylem composed of long, thin cells with tapered ends and hardened walls. p. 494

transpiration (trans pih RAY shun): the loss of water to the atmosphere by a plant through the stomates in its leaves. p. 489

trimester: a period or term of three months; the gestation period of humans usually is divided into trimesters. p. 170

tropical deciduous forest: a tropical forest biome with wet and dry seasons and constant temperatures. p. 608

tropical rain forest: the most complex of all communities, located near the equator where rainfall is abundant and harboring more species of plants and animals than any other biome in the world; light is the major limiting factor. p. 606

tropism (TROH piz um): a change in the orientation of a plant, or part of a plant, in response to light, gravity, or other environmental factors. p. 535

tumor: a mass that forms within otherwise normal tissue, caused by the uncontrolled growth of a transformed cell. p. 171

turgor (TER ger) **pressure:** pressure exerted by plant cells against their cell walls whenever the plant is adequately supplied with water. p. 489

U

ultrasound: high-frequency sound waves; a method used to determine some fetal abnormalities. p. 169

ultraviolet light: the range of radiation wavelengths just beyond violet in the visible spectrum, on the border of the X-ray region. p. 162

umbilical (um BIL ih kul) **cord:** in placental mammals, a tube connecting the embryo with the placenta. p. 166

unicellular: one-celled. p. 107

Uniramia (yoo nih RAY mee uh): the subphylum containing animals with three distinct body parts and one pair of antennae; includes millipedes, centipedes, and insects. p. 371

unsaturated fat: a fat containing fatty acids with one or more double-bonded carbon atoms; each double bond in the carbon chains reduces by one the number of hydrogen atoms that can be bonded to the carbons; unsaturated fats usually are liquid at room temperature. p. 415

uranium-235 dating: a radiometric dating technique in which the decay of uranium into lead isotopes is measured; used to date rocks 2 billion years old. p. 568

urban wildlife: wildlife living in urban and suburban areas, including raccoons, skunks, pigeons, songbirds, raptors, mice, rats, and similar animals. p. 561

urea (yoo REE uh): a nonprotein nitrogenous substance produced as a result of protein metabolism. p. 384

ureter (YOOR ee ter): a muscular tube that carries urine from the kidney to the urinary bladder. p. 445

urethra (yoo REE thruh): the tube through which urine is carried from the bladder to the outside of the body in vertebrates. p. 445

uric (YOOR ik) **acid:** the insoluble precipitate of nitrogenous waste excreted by land snails, insects, birds, and some reptiles. p. 384

uterus: a hollow muscular organ, located in the female pelvis, in which a fetus develops. p. 148

V

vaccine: a substance that contains antigens and is used to stimulate the production of antibodies. p. 289

vacuole (VAK yoo ohl): a membrane-enclosed structure in the cytoplasm of a cell or a unicellular organism; different types of vacuoles serve different functions. p. 117

vagina: a tubular organ that leads from the uterus to opening of the female reproductive tract. p. 148

valve: a membrane or similar structure in an organ or passage, such as an artery or vein, that retards or prevents the return flow of a bodily fluid. p. 426

variable: a condition that varies from one organism to another (size, shape, color) or is subject to change for an individual organism (humidity, temperature, light intensity, fatigue). p. 21

variation: small differences among individuals within a population or species that provide the raw material for evolution. p. 218

vascular tissue: plant tissues specialized for the transport of water and nutrients; also for support. p. 336

vegetative reproduction: asexual reproduction by plants that also may reproduce sexually; examples include potato plants from "eyes" and grass plants from runners. p. 137

vein: a vessel that carries blood toward the heart. p. 424

ventral: in animals, situated toward the lower or belly side. p. 364

ventricle (VEN trih kul): one of two chambers of the heart that pump blood out of the heart. p. 426

vertebra (VER teh bruh): an articulated bone; the vertebrae make up the backbone, or spinal column, of vertebrates. p. 373

vertebrate: a chordate animal with a backbone; mammals, birds, reptiles, amphibians, and various classes of fishes are examples. p. 215

vesicle: a small, intracellular membrane-enclosed sac in which various substances are transported or stored. p. 117

vessel: in plants, a type of water-conducting xylem cell. p. 494

villi (VIL eye): fingerlike projections of the small intestine that increase surface area. p. 166

virulence (VIR yuh lents): the relative ability of a pathogen to overcome body defenses and cause disease. p. 289

virus: a submicroscopic pathogen composed of a core of nucleic acid surrounded by a protein coat that can reproduce only inside a living cell. p. 292

W

wetland: land that stays flooded all or part of the year with fresh or salt water; includes coastal and inland wetlands. p. 656

X

X-linked trait: a trait determined by a gene carried on the X chromosome. p. 201

xylem (ZY lem): conducting tissue that transports water and dissolved nutrients in vascular plants. p. 493

Z

zooplankton (ZOH oh PLANK ton): very small, feebly swimming aquatic organisms that are herbivorous or carnivorous or both. p. 634

Zygomycota (ZY goh my KOH tuh): the phylum containing conjugation fungi. p. 320

zygospore (ZY goh spor): a zygote that forms a spore; produced in some fungi and many plants following the union of sexual cells or nuclei. p. 321

zygote (ZY goht): the diploid product of the union of haploid gametes in conception; a fertilized egg. p. 140

Index

Illustrations and tables are referenced in italics.

Acknowledgements continued from page iv

Inv. 3.2 Adapted with permission from the National Association of Biology Teachers; **Inv. 3.3** Adapted with permission from the National Association of Biology Teachers; **Inv. 10.1** Adapted from Cleveland P. Hickman, Jr., Larry S. Roberts, and Frances M. Hickman, *Integrated Principles of Zoology,* 8th ed. (St. Louis: Times Mirror/ Mosby College Publishing, 1988): 118; **Inv. 17.1** Reprinted by permission from Peter DeDecker, *The Responsible Use of Animals in Biology Classrooms,* (Hastings High School, Reston, Virginia: National Association of Biology Teachers.).

Illustration Acknowledgements

All illustrations by BSCS except the following: **3.4** Adapted with permission from the National Association of Biology Teachers; **3.14** From *Living in the Environment,* 6th ed. by G. Tyler Miller, Jr., Copyright © 1990 by Wadworth, Inc. Adapted with permission of the publisher; **4.BT** From *Living in the Environment,* 6th ed. by G. Tyler Miller, Jr., Copyright © 1990 by Wadsworth, Inc. Adapted with permisson of the publisher; **10.3** Dr. Walter M. Fitch; **11.7** C. R. Woese; **11.32** Adapted from *Science,* 22 December 1989, Copyright © by AAAS; **18.8c** From *Biology,* by Leland G. Johnson. Dubuque, IA: Wm. C. Brown, Copyright © 1983. Reprinted by permission. **21.24** "Peking Man," by Wu Rukang and Lin Shenglong. Copyright © 1983 by Scientific American, Inc. All rights reserved; **21.27** "Peking Man," by Wu Rukang and Lin Shenglong. Copyright © 1983 by Scientific American, Inc. All rights reserved; **22.14** From "The Canopy of the Tropical Rainforest," by Donald R. Perry. Copyright © 1984 by Scientific American, Inc. All rights reserved; **24.1** From *Living in the Environment,* 6th ed. by G. Tyler Miller, Jr., Copyright © 1990 by Wadsworth, Inc. Adapted with permission of the publisher; **24.12** From *Living in the Environment,* 6th ed. by G. Tyler Miller, Jr., Copyright © 1990 by Wadsworth, Inc. Adapted with permission of the publisher.

Photograph Acknowledgements

Section One: Copyright © 1989 by Scientific American/George V. Kelvin. All rights reserved. **Chapter 1 Opener:** Jeff Foott; **1.1**(*a*) C. Allan Morgan; **1.1**(*b*) Milton Rand/Tom Stack & Associates; **1.2** BSCS by Carlye Calvin; **1.3** USDA-APHIS; **1.6** Gordon E Uno; **1.9**(*a*) BSCS by Carlye Calvin; **1.9**(*b*) Ronald Hathaway; **1.14** Copyright © 1990 by Gary Braasch; **1.15**(*a*) Staley Studios; **1.15**(*b*) BSCS; **1.15**(*c*) BSCS; **1.15**(*d*) Staley Studios; **1.BT** Ariel E. Lugo, USDA Forest Service. **Chapter 2 Opener:** C. Allan Morgan (detail, *vii*); **2.1** BSCS by Richard Tolman; **2.3** C. Allan Morgan; **2.4** BSCS by Doug Sokell; **2.5** BSCS by Jacqueline Ott-Rogers; **2.11**(*a*) Paul Hewlett; **2.11**(*b*) James McClintock, Department of Biology, University of Alabama, Birmingham; **2.12** © Dwight R. Kuhn; **2.15**(*a*) John D. Cunningham/Visuals Unlimited; **2.15**(*b*) Gary Milburn/Tom Stack & Associates; **2.16** Copyright © 1990 by Gary Braasch; **2.18**(*a*) Tim McCabe/Soil Conservation Service; **2.18**(*b*) Tim McCabe/Soil Conservation Service; **2.BT** NASA. **Chapter 3 Opener:** Dwight Kuhn; **3.1** BSCS by Jacqueline Ott-Rogers; **3.7** Jeff Foott; **3.8** Brian Parker/Tom Stack & Associates; **3.9** BSCS by Jacqueline Ott-Rogers; **3.10** Florida Department of Natural Resources; **3.11** Katherine A. Winternitz; **3.13** USDA; **3.16** John D. Cunningham/Visuals Unlimited; **3.17** National Park Service/Richard Frear; **3.18** C. Allan Morgan; **3.19** Kerry T. Givens/Tom Stack & Associates; **3.20** G. Prance/Visuals Unlimited; **3.21** Doug Sokell; **3.BT** El Paso County Park Department's Bear Creek Nature Center. **Chapter 4 Opener:** Doug Sokell; **4.8** Brian Parker/Tom Stack & Associates; **4.21** BSCS by Doug Sokell. **Section Two:** Jeff Foott. **Chapter 5 Opener:** Ed Reschke; **5.1**(*a*) Ed Reschke (detail, *viii*); **5.1**(*b*) Ripon Microslides, Inc.; **5.1**(*c*) Dwight Kuhn; **5.1**(*d*) Ed Reschke; **5.2**(*a*) Kevin and Betty Collins/Visuals Unlimited; **5.2**(*b*) Ed Reschke; **5.3**(*a*) Ripon Microslides, Inc.;

5.3(*b*) Carolyn Noble; **5.4**(*a*) BPS/Tom Stack & Associates; **5.4**(*b*) BPS/Tom Stack & Associates; **5.5** SPL/Photo Researchers; **5.8** BPT/ Tom Stack & Associates; **5.9** David M. Phillips/Visuals Unlimited; **5.10** K. G. Murti/Visuals Unlimited; **5.12** BSCS by Doug Sokell; **5.18** Dr. Andrew S. Bajer, Department of Biology, University of Oregon; **5.19** BSCS by Werner Heim; **5.21** David M. Phillips/Visuals Unlimited; **5.22** David M. Phillips/Visuals Unlimited; **5.BT**(*a*) Linda Hall Library. **Chapter 6 Opener:** David M. Phillips/Visuals Unlimited; **6.2** Ed Reschke; **6.3**(*a*) BSCS by Jacqueline Ott-Rogers; **6.3**(*b*) Brian Parker/ Tom Stack & Associates (detail, *ix, tp*); **6.5** David M. Phillips/Visuals Unlimited; **6.6** C. Allan Morgan; **6.11** BSCS by Jacqueline Ott-Rogers; **6.12** Ed Reschke. **Chapter 7 Opener:** Dwight Kuhn; **7.3** Carolina Biological Supply Company; **7.6** Ward's Natural Science Establishment; **7.11** Carolina Biological Supply Company; **7.13** Carnegie Institute of Washington, Department of Embryology, Davis Division; **7.15**(*a*) Carnegie Institute of Washington, Department of Embryology, Davis Division; **7.15**(*b*) Lennart Nilsson, *A Child Is Born,* Dell Publishing Company; **7.15**(*c*) Roberts Rugh; **7.15**(*d*) Roberts Rugh; **7.15**(*e*) Roberts Rugh; **7.15**(*f*) Lennart Nilsson, *A Child Is Born,* Dell Publishing Company; **7.18** Dr. James W. Hanson, M.D.; **7.19**(*a*) Leukemia Society of America; **7.19**(*b*) Leukemia Society of America. **Chapter 8 Opener:** Jeff Foott; **8.1**(*a*) Cystic Fibrosis Foundation; **8.1**(*b*) Cystic Fibrosis Foundation; **8.2** BSCS by Doug Sokell; **8.3** M. Hofer; **8.8** Peter J. Gryant/BPS/Tom Stack & Associates; **8.11** BSCS by Jacqueline Ott-Rogers; **8.22** David C. Peakman, Cytogenetics Laboratory, Reproductive Genetics Center, Denver, Colorado; **8.23** David C. Peakman, Cytogenetics Laboratory, Reproductive Genetics Center, Denver, Colorado; **8.26**(*b*) Douglas P. Haug; **8.28** Courtesy of C. Trunca, Ph.D.; **8.34** Cystic Fibrosis Foundation; **8.BT** Cystic Fibrosis Foundation. **Chapter 9 Opener:** (*a*) Paläontologisches Museum, Museum für Naturkund der Humboldt-Universität, Berlin, Germany (detail, *ix, btm*), (*b*) John Shaw/Tom Stack & Associates; **9.2** Ripon Microslides, Inc.; **9.5** Kaylene Silvester, Walnut Creek, CA; **9.6**(*a–c*) W. D. Bransford/National Wildflower Research Center; **9.8** C. Allan Morgan; **9.9** Darwin Museum Down House, Courtesy of the Royal College of Surgeons of England; **9.12**(*a*) BSCS; **9.12**(*b*) Barbara Fleischaker; **9.13** J. A. Bishop & L. M. Cook; **9.21**(*a*) Jeff Foott/Tom Stack & Associates; **9.21**(*b*) W. Perry Conway/Tom Stack & Associates; **9.24**(*a*) Walt Anderson/Visuals Unlimited; **9.24**(*b*) C. Allan Morgan; **9.25**(*tp,l*) John D. Cunningham/Visuals Unlimited; **9.25**(*btm,l*) Gordon E. Uno; **9.25**(*btm,r*) Gordon E. Uno; **9.26**(*a*) BSCS; **9.26**(*b*) C. Allan Morgan; **9.BT**(*a*) Wildlife Safari, Winston, Oregon; **9.BT**(*b*) Wildlife Safari, Winston, Oregon. **Section Three:** Dwight Kuhn. **Chapter 10 Opener:** George D. Dodge/Tom Stack & Associates; **10.2** Robert C. Simpson/Tom Stack & Associates; **10.4** Martin G. Miller/Visuals Unlimited; **10.5** BSCS by Doug Sokell; **10.13** BSCS by Jacqueline Ott-Rogers; **10.17**(*a*) B. Ormerod/Visuals Unlimited; **10.17**(*b*) Ripon Microslides, Inc.; **10.17**(*c*) J. G. Zeikus; **10.18**(*a*) Doug Sokell; **10.18**(*b*) BSCS; **10.18**(*c*) Rich Buzzelli; **10.19**(*a*) Stan Elems/Visuals Unlimited; **10.19**(*b*) Dr. Werner W. Schulz; **10.19**(*c*) BSCS; **10.20**(*a*) C. Allan Morgan (detail, *x, tp*); **10.20**(*b*) Athalie E. Barzee; **10.20**(*c*) Kenneth D. Whitney/Visuals Unlimited; **10.21**(*a*) J. Robert Waaland/BPS/Tom Stack & Associates; **10.21**(*b*) © T. E. Adams/Visuals Unlimited; **10.21**(*c*) John Shaw/Tom Stack & Associates; **10.24** S. M. Awramik/ BPS/University of California/Tom Stack & Associates; **10.25** S. M. Awramik/BPS/University of California/Tom Stack & Associates; **10.28** Dr. Glen M. Mason, Indiana University Southeast; **10.30** Dr. Sidney Fox, Southern Illinois University at Carbondale; **10.33** Virus Laboratory; **10.BT** Cincinnati Zoo Center for Reproduction of Endangered Wildlife.